ar Bases
Activities of the 21st Century

W. MENDELL, EDITOR

Lunar Bases
and Space Activities of the 21st Century

W. W. MENDELL, EDITOR

Lunar and Planetary Institute
Houston

Library of Congress Cataloging-in-Publication Data
Main entry under title:

Lunar bases and space activities of the 21st century.

   Papers from a NASA-sponsored, public symposium
hosted by the National Academy of Sciences in
Washington, D.C., Oct. 29-31, 1984.
   1. Lunar bases--Congresses. 2. Mars (Planet)--
Exploration--Congresses. I. Mendell, W. W. (Wendell W.),
1941-   . II. Lunar and Planetary Institute.
III. United States. National Aeronautics and Space
Administration.
TL799.M6L83 1985   919.9'104   86-50
ISBN 0-942862-02-3

Published by the Lunar and Planetary Institute, 3303 NASA Road One, Houston, TX 77058-4399. Printed in the U.S.A. Library of Congress CIP data available from the Library of Congress, CIP Division, or from the publisher.

Cover illustration: Two inhabitants of the Moon overlook an advanced lunar installation from a museum construction site. The original, primitive lunar base lies to the left of a large electromagnetic launch facility, which dominates the vista. An array of solar dynamic generators on the horizon supplement the power from a nuclear reactor to operate greenhouses, industrial processing plants, scientific research laboratories, and a spaceport. Artist: Pat Rawlings, Eagle Engineering Co., Houston, Texas.

Associate Editors

Michael B. Duke
NASA/Johnson Space Center

Harold P. Klein
University of Santa Clara

Chris W. Knudsen
Carbotek, Inc.

John M. Logsdon
George Washington University

Wendell W. Mendell
NASA/Johnson Space Center

Barney Roberts
NASA/Johnson Space Center

Richard Tangum
University of Texas

Richard Williams
NASA/Johnson Space Center

David Vaniman
Los Alamos National Laboratory

CONTENTS

Prologue

Lunar Bases and Space Activities of the 21st Century *is a collection of short papers dealing with various aspects of a manned lunar base and the concomitant expansion of humanity into near-Earth space. Most of these papers were delivered at a symposium on the subject, sponsored by NASA and hosted by the National Academy of Sciences in Washington, DC, October 29–31, 1984. The program of the symposium reflected the structure of the Report of the Lunar Base Working Group, the output of a workshop sponsored by NASA and hosted by the Institute of Geophysics and Planetary Physics of the University of California. The Lunar Base Working Group, consisting of approximately 50 scientists, engineers, industrialists, and scholars, met during the week of April 23–27, 1984, at the Los Alamos National Laboratory to discuss the scientific, technological, and social issues associated with a permanently crewed facility on the lunar surface.*

Although these meetings have been conducted and these papers are being published under NASA auspices, the work presented here is not a formal part of an advanced planning activity within the agency. Official NASA studies of lunar bases ceased in 1972. Since the early 70's, the program has focused on operations in low-Earth orbit (LEO) with the space shuttle and, more recently, with the space station. The past decade of fiscal, technical, and political pressure has discouraged all but very near-term planning and has given birth to pervasive conservatism in the body bureaucratic. This book, the symposium, and the precursor workshop are part of an ongoing effort to expand the discussion of future activities in the United States space program to include the era following the LEO space station.

In late 1981, Jeff Warner, Mike Duke, and I realized that the space transportation technology of the year 2000 would be capable of routinely carrying payloads to the Moon. At first our interest centered on the scientific groundwork for future lunar activities, but we soon became aware that a permanent lunar base has far-reaching implications for national policy, international relations, and American technology. In a conservative scenario, a lunar program would become the major space policy issue upon initiation of routine operations at the LEO space station, i.e., in the mid 1990's. However, the capabilities—and limitations—of the Space Transportation System are being established today. Thus, we argued, a low level research program ought to be instituted now to develop rationale, investigate strategies, and explore technology associated with lunar development.

Over the past four years, such arguments have made small headway within NASA, in an institutional sense, for a variety of reasons. A lunar base project does not fall within the clear purview of any of the Associate Administrators. Therefore, sympathetic program managers find it difficult to provide steady support in the context of currently identified budget categories. Some involved in NASA planning believe that all of the future options should be characterized equally at a low level of detail. Finding commonality among all possible requirements is seen as the guiding principle for designers of the Space Transportation System. Stressing any one program could embarrass the agency, or worse, if future policy directions pointed elsewhere. Finally, initiation of discussion of manned lunar activities is frequently haunted by the ghost of Project Apollo. A common, and erroneous, assumption is that a lunar base requires a national commitment of the scale invoked by President Kennedy in 1961 during a time of perceived national crisis.

If NASA has been timid in advocating an ambitious civilian space program, perhaps the reason lies in the annual travail of the federal budget process. Any long-term line item raises concern in the Office of Management and Budget that another "entitlement" has been added to the uncontrollable Federal budget. NASA also has been described as an agency without a constituency. When civilian space expenditures are cut, no special interest group rises to protest. Yet, the public image of the space program is very positive. NASA is one of the most visible and, arguably, most successful Federal departments.

The fundamental question to be addressed squarely is what role civilian space activities will play in the future of the nation and, indeed, the world. If we plan to continue small, expensive, experimental programs in low-Earth orbit with its limited volume and dearth of resources, then a myopic policy will suffice, ignoring the next program until the present one is done. However, the scope of discussion in this book demonstrates that we stand on the brink of an explosion in space development potentially an order of magnitude greater than the current official vision. The increase in scale will be enabled by the utilization of space resources, e.g., from the Moon, to sustain and capitalize the space infrastructure. Inclusion of the Moon in the system will create new ways of doing business in the same sense that the space shuttle and the space station create new opportunities.

The promise of space development can be lost or seriously delayed by absence of planning. New manned landings on the Moon probably will result from politically inspired decisions, but human presence can be permanent only if lunar enterprises have economic value. An efficient space transportation infrastructure for shipping supplies and products is required. Extensive private sector involvement and *exploitation of lunar resources is implied. Present day space law and space policy are not structured to deal with future realities.*

In this book many questions are raised, and a few answers are suggested. A collection of papers such as this one cannot be complete in coverage of the subject matter. Contributions were required to be short to keep the level of detail approximately equivalent, even when some aspects of lunar development are quite well studied and others have only been recently recognized. In reality, each section could be expanded into a book; many authors found the constraints on length burdensome. I have written short introductions to the sections to aid the reader in picking out common threads among manuscripts grouped together.

It is my hope that the discussions in this book will seed more interest and thought. The technological strength of this country is based on its breadth and depth of expertise, and space cannot be utilized effectively without attracting specialists to the relevant problems. The scope of the technical presentations here may increase the awareness of a broader community to the challenge of space development. Students who ask what fields of study lead to careers in the space program will find a wealth of topics. Policy makers may glimpse the gateway to a complex future with new solutions to old problems as well as new problems. The citizen who participates in the political process can find information to aid evaluation of various options offered to his representatives. In the long run, opportunities for participation abound not only for all walks of society in the United States but, indeed, throughout the world.

The successful effort to initiate and sustain public dialogue on the issues of lunar development has been the product of the time and interest of many capable, busy people. I particularly wish to acknowledge the work of Michael Duke and Barney Roberts of the NASA Johnson Space Center, Paul Keaton of the Los Alamos National Laboratory, and Larry Haskin of Washington University. I also extend my thanks for interest and encouragement from others too numerous to list. I anticipate sharing with all of you the reality of this vision.

Wendell Mendell
November, 1985

1 / THE SYMPOSIUM: KEYNOTE SPEECHES

F ROM OCTOBER 29 TO 31, 1984 a NASA-sponsored, public symposium entitled "Lunar Bases and Space Activities of the 21st Century" was hosted by the National Academy of Sciences in Washington, DC. Approximately 300 attendees registered to hear 135 papers on a variety of topics relevant to space program goals in the era following establishment of the LEO space station. Since very little research on these issues is currently being funded, the many participants who traveled to the meeting tended to have a very personal, as well as professional, interest in the theme. As indicated by the title of the conference, the bulk of the discussion centered on lunar occupation, its implementation, and its implications. However, other future scenarios were not excluded; several contributed papers suggested manned exploration of Mars as an alternate or complementary long range goal.

To those unfamiliar with the state of space technology, discussion of colonies on the Moon and Mars have the ring of science fiction. Persuasive arguments are made within the pages of this book and elsewhere that permanent human presence in the space environment can be established and maintained within the bounds of contemporary technology. While technological capability is a necessary condition for a lunar outpost, it is not a sufficient one. Space transportation is still expensive and therefore achievable only by institutions with significant financial resources.

In all likelihood, private capital will not be invested in lunar development until near-term profitability is more than speculation. Similarly, public funds will not be spent until a clear case can be made in terms of the national interest. The nature of the national interest is complex, ranging from national security to stimulus of the private sector of the economy. For that reason, several individuals with experience in the formulation of national policy were invited to address the Symposium in the opening plenary session. Their views are particularly relevant to that part of the feasibility issue that addresses public investment in the future of society.

Mr. James Beggs is Administrator of NASA. Dr. George Keyworth is scientific advisor to President Reagan and directs the Office of Science and Technology Policy. Mr. Walter Hickel of the Yukon Pacific Corporation is former Governor of Alaska and Secretary of the Interior. Dr. Arthur Kantrowitz of Dartmouth College, former chairman of Avco Corporation, has in recent years been involved in studies of formulation of public policy with respect to technology. Dr. Edward Teller is a prominent nuclear physicist serving on the President's Science Advisory Committee.

Two keynote speakers chose to develop their remarks into more formal papers. Dr. Philip Smith, Executive Officer of the National Academy of Sciences, discusses the issue of international relations in the section Societal Issues. Dr. Harrison Schmitt, Apollo 17 astronaut and former Senator, looks at the implications of manned missions to Mars for national policy in the section Mars.

Remarks on the Lunar Base

James M. Beggs

Administrator, NASA/Headquarters, Washington, DC 20546

Over the years, the National Academy and its Space Science Board have played a pivotal role in NASA's advanced science planning. Our joint efforts have been extremely valuable in helping to chart a course not only for our future efforts in astronomy, astrophysics, and planetary exploration, but in the earth sciences as well.

Little by little, our efforts to explore space are becoming broader and more encompassing. Step by step, we are learning about the physical universe. By no means do we know all there is to know about the chain that connects the first appearance of the universe some 15 billion years ago with its evolution, the formation of matter, the galaxies, the stars, the solar system, the planets, and ourselves, but we are learning fast.

Our solar system exploration has been exceptionally successful and productive. We have observed the Sun from above the Earth's atmosphere; we have explored the Moon; by the end of the decade, we will have sent spacecraft to all of the planets except distant Pluto.

Ironically, we probably know more about Mars and Jupiter as respective planetary systems than we do about our own planet. We are just beginning to understand Earth as a system and the complex interaction of its various subsystems—atmosphere, oceans, natural resources—and the forces that built them and that determine their evolution and destiny.

Our goal on Earth, as in space, is to push scientific frontiers forward, to the cutting edge of understanding. In this effort, we are fortunate, indeed, that the Academy's Space Science Board recently agreed to provide additional guidance for NASA's long-term space science. During the next 2 years the Board will be undertaking a broad study of science priorities for the next 25 or 30 years and identifying the technology advances necessary to meet those priorities.

That effort is only one of several under way to help us define what we should be doing and where we should be going in space during the next generation. Another ongoing project is our internal NASA long-range planning study on post-space station options.

This symposium is still another timely effort to define where we should be going in space in the early years of the 21st Century. It follows closely the Lunar Base Working Group meeting, held last April at Los Alamos, which debated the pros and cons of establishing a permanently manned base on the Moon's surface. As you know, the

Working Group concluded that such a base should be adopted by NASA as a long-term goal for the 21st Century.

This goal has long been our vision, and it still is. Even before Apollo, our studies concluded that such a base could serve as a facility for scientific research, economic exploitation of the Moon's resources, and colonization of the Moon.

Today, more than 15 years after we first set our footprints on the Moon, we have learned much about it. Twelve Apollo astronauts walked on the lunar surface. They returned more than 2,000 samples of lunar rocks and soil from six locations. Soviet unmanned spacecraft have provided us with samples from three other sites. Spacecraft have photographed the entire Moon from lunar orbit and performed chemical analysis of more than a fourth of its surface.

Our lunar exploration revealed no water, no organic matter, and no living organisms, but the Moon rocks turned out to be lunar time capsules. They contain the secrets of more than four and a half billion years of lunar history, a history we are beginning to understand using radioactive measurement techniques. The lunar soil, more akin to fine, broken rubble, is teaching us much about our Sun because it has absorbed billions of years of its radiations, unprotected by any atmosphere. We know now that the Moon not only has plentiful oxygen in its rocks but also silicon and possibly valuable metals such as iron and titanium.

The Moon, our nearest neighbor in the solar system, is becoming a familiar place. For more than four billion years, it lay there, quiet and still. Today we are exploring the possibility of going back there to live and work, a truly extraordinary development in the short history of the space age.

I believe it is highly likely that before the first decade of the next century is out, we will, indeed, return to the Moon. We will do so not only to mine its oxygen-rich rocks and other resources but to establish an outpost for further exploration and expansion of human activities in the solar system, in particular, on Mars and the near-Earth asteroids.

One does not have to be an historian of the space age to recognize that a return to the Moon would be a rational extension of our program to expand human activities in space. From Mercury, and on through Gemini, Apollo, Skylab, and the Space Shuttle, we have moved steadily into larger and more ambitious programs. Each has been a logical extension of what came before, and each has been built on past experience.

Now that the space shuttle is proving to be the reliable and versatile machine its designers intended, we will use it to help meet our next major challenge: to develop a permanently manned space station in low-Earth orbit within a decade, as the President directed us to do.

We expect that by the year 2000, the space station will be equipped with a supporting infrastructure that will enable us to operate routinely at both low-Earth and geostationary orbits and between them and, eventually, at distances as far as the Moon and the inner planets. Two key elements of this infrastructure will be reusable and might be compared to a local taxi and an intercontinental airline.

The former is called the Orbital Maneuvering Vehicle. It will be used to service satellites close to the space station and for other tasks. The latter, known as the Orbital Transfer Vehicle, will ferry payloads to and from geosynchronous orbit or launch spacecraft to the Moon and other points in the solar system.

Wernher Von Braun used to say that if people are ever to open vast uncharted regions to detailed exploration and permanent human habitation, they need an enabling technology. In the case of Antarctica, as Wernher rightly pointed out, it was the airplane. In the case of space, it will be the space station and its infrastructure.

This enabling technology will permit us to engage in a variety of manned and unmanned activities in space. It will spur exploration and the commercial use of space. It will invigorate Earth applications and stimulate sustained research and development on innovative systems and techniques. It could also trigger extensive initiatives to benefit life on Earth, such as satellite power systems and nuclear waste disposal systems in space. And it will be the key to future, more ambitious missions, such as a manned mission to Mars, the capture of an asteroid, or large, automated, deep space and planetary probes.

One of those missions, as I have said, could very well be the establishment of a permanently manned lunar base. That's why this conference is so important. It will help us move toward a national dialogue in the scientific and technical community on the uses, feasibility, and significance of a lunar base. It is essential to start such a dialogue now if we are to lay the groundwork for NASA's consideration of a long-range program plan to include a permanent human presence on the Moon.

These proceedings undoubtedly will be both stimulating and productive. And, as you continue your discussions over the next few days, you will be covering many crucial considerations—technical, scientific, political, economic, and social—that will guide future public policy decisions on a permanently manned lunar base. I'd like to single out just three, because they symbolize both the philosophical and practical realities of this new era of opportunity in space.

First, a whole generation of people is coming of age, not only in the United States, but around the world, who are barely able to remember that it was once thought impossible to go to the Moon. We now know that we can get there. The question is, what should we be doing if we establish permanent roots there to make our presence most productive and beneficial for mankind?

Next, we know that any enterprise of the magnitude and scope of a permanently manned lunar base would be an enormous challenge. We would have to develop not only new technologies, but the managment techniques and economic analyses that would make it both feasible and profitable. This implies even greater international cooperation and international sharing of risks and benefits in the future.

In this connection, we expect our friends and allies to accept President Reagan's invitation to join with us in developing the space station. Such cooperation could lay the groundwork for even greater international collaboration in space for the future. Indeed, an internationally developed lunar base might prove an irresistible lure to the Soviets. And, if they were to join with us, I believe that it would certainly enhance the prospects for peace in the world and in space.

A third consideration is technological. If we were to mine the Moon, how would we go about it? Many methods have been proposed, but none have been proven.

At present, we know how to extract valuable materials from ore deposits on Earth. But Earth's ore deposits are unusual, in that their valuable elements are highly concentrated and relatively easy to extract. Moon rocks and meteorites are different. Their key elements are not concentrated and are hard to extract. We have no Earth-based technology at present that could do the job.

Clearly, such a technology will have to be developed if we are ever to mine the Moon. That's why we should begin soon, on a small scale and in a preliminary way, to study how to extract useful minerals from lunar rock and soil. In space, as on Earth, there are rich dividends and enormous benefits for those who are able to muster the resources, know-how, and vision to follow where their dreams may lead.

On September 16, 1969, Astronaut Michael Collins closed an extraordinary chapter in his life and in the history of the world—the Apollo 11 mission—with these words before a joint session of Congress:

> We have taken to the Moon the wealth of this nation, the vision of its political leaders, the intelligence of its scientists, the dedication of its engineers, the careful craftmanship of its workers, and the enthusiastic support of its people.
> We have brought back rocks, and I think it is a fair trade . . .
> Man has always gone where he has been able to go. It's that simple. He will continue pushing back his frontier, no matter how far it may carry him from his homeland.

Mike's words sum up as eloquently as any the opportunities and challenges that lie ahead. If we are to be true to our own past and our own future, we will continue to pursue them.

The Challenges and Opportunities of a New Era in Space: How Will We Respond?

G. A. Keyworth, II

Science Advisor to the President and Director, Office of Science and Technology Policy
Executive Office of the President, Washington, DC 20500

N ow, more than at any time since the dawn of the space age a quarter-century ago, we're poised at the edge of great advances in our understanding and use of space. Twenty-five years ago we rose magnificently to the challenge presented to us from Sputnik. In barely more than a decade, we created a wholly new technology—the technology of space travel and space exploration.

In the 1970s, the decade that followed, we consolidated our gains and refined the technologies. We started demonstrating that we could develop increasingly reliable technologies—both manned and unmanned—to operate in that new environment. What had been solely a government venture began to become attractive to the private sector, and soon that attractiveness turned to profitability, particularly with the proliferation of communications satellites. I would characterize the development, and now almost routine operation, of the shuttle as the beginning of the mature phase of our early space technologies. We've now achieved a degree of confidence and predictability in space operations that gives us a calculable basis for considering investment in space enterprises.

The real question now is where we go from here. Are our grand exploring days just beginning? Or will we devote our efforts from now on to consolidating our gains and expanding the commercial applications of space technologies? Or will we do both?

Just four or five years ago we might well have answered that question differently than we would today. I doubt, for instance, that this meeting could have taken place in the atmosphere of those days. Five years ago, the country as a whole was in a trough. Our industries were encountering competition of the sort they had never seen before—competition from foreign firms in the very high technology areas we had always dominated and taken for granted. The Japanese had discovered that Japan was a great place to make cars and televisions and that America was a great place to make money. American companies had good reason to wonder what the future would bring them. At the same time, government had let our leadership in science and technology begin to erode by failing to invest in the basic research that underlies so much of our progress in today's world. That hardly encouraged industry to count on glowing prospects for technological innovation in the long term.

Perhaps worst of all, we were hearing an incessant drumbeat that the world was running out of room and resources, that we faced a world of limits, a zero-sum game in which our children would have to settle for less in life than we had. Many young people who picked up that message—that success would be elusive—concluded that study and hard work might not be the kind of blue-chip investments they were a generation earlier. The space program itself was viewed by many as a too-expensive series of stunts that would be unaffordable in the bad times ahead. It was certainly not seen in terms of an investment in breaking out of those limits and into new and better times.

That was five years ago. You may have noticed some changes in attitude since then. American business is just as respectful of the foreign challenge as ever, but it's no longer paralyzed by irrational fear, and it's planning for the future with the confidence of people who are determined to meet any challenge. The federal government, during the five years of the Reagan Administration, has turned that decline in support for basic research into the greatest rate of growth since the early post-Sputnik years, an increase of 55% in four years. Our universities are well on their way to regaining the health that had been jeopardized, and there's a strong renewed commitment to excellence in education, especially in science and math, in our public schools. No one is talking any more about limits to growth.

So what better time could there be for a conference like this? I said earlier that there are two paths our space program could take at this point—the practical and the visionary—and I asked which we would choose. My firm opinion is that we have to follow both paths aggressively. That's why we're now seeing federal agencies other than NASA assuming responsibilities for commercializing space activities. That's the best evidence that you'll see that we intend to reserve NASA for what it does best: research and development. I'll add my firm belief that only by continuing to push at the boundaries of that vast space frontier will we be able to assure our world leadership in the relatively more mundane practical space technologies. In a very real sense, the economic future of space is tied to how much we stretch our vision and creativity to respond to the grand challenges.

Some of you may have read President Reagan's response in this month's Omni *magazine when he was asked the question, "What interests you about space?" His answer sets the tone for this conference better than anything I could say:*

> What interests me about space is quite simple; space is a part of the future, of a future that captures the imagination of young and old alike. Think of the remarkable achievements of the American Space Program—the manned Moon landings, Viking's landing on Mars, the spectacular Voyager missions to Jupiter and Saturn, the continual rediscovery of our own planet, and the space shuttle, with its many important scientific achievements.

In what other areas have American science and technology succeeded so well in literally moving us from one age to the next? Our space successes have proved that although all but a handful of us are physically bound to the Earth, our spirits and our national pride can soar along with astronauts and spacecraft. To me, that's the ultimate attraction of the space program—the elevation of the human spirit as we demonstrate our unmatched capability to reach out to new worlds.

It's that context, one of a whole people reaching out to new worlds, that should underlie these discussions of a manned lunar base. Today's new era of competition, of commitment to excellence in science and technology, and of development of the new talents we need for the future, demands and will support *new space initiatives if they match the boldness of our times.*

Today we stand on the threshold of that boldness. We need only create a door that opens out on this new world of challenges and opportunities that space offers us, and, last January, President Reagan committed the nation to building that doorway in the form of the manned space station.

Well, what possibilities wait for us when we step through that doorway into the next space age? I think we can begin to see that future at this conference, because the lunar base is one of the more obvious of the bold, exciting goals we can reach through the space station doorway.

In looking at the program that Mike Duke and the others have put together, I was excited by the diversity of the topics and the expertise of the participants. Either the planetary science community is farther along in its thinking about a lunar base than more people give it credit for, or this conference itself is providing a potent stimulus to thought. In either case, this conference is probably going to be remembered as a landmark in the evolution of thinking on the subject.

But as you listen and talk over the next three days, or afterwards, I'd like for you to ask yourselves two questions that I think will help to develop the concept of a lunar base as a possible national initiative. The first question may seem presumptuous, but I think we have to ask, right at the outset, where we go from the lunar base? What steps should we be taking in parallel with the lunar base, and what comes after it? Do we go to Mars, and if so, why? Do we try to visit an asteroid? Remember that much of the momentum of our space program was lost after Apollo because we treated the Moon landing as an end in itself. This time we should know enough to define and update our goals in space in broad terms related to our future, not in terms of individual projects. And we should cast as wide a net as possible in creating this vision of our future, involving the American public and being driven by their enthusiasm as well as our own.

Now, if my first question seemed presumptuous, then my second one may seem backward, but try it on anyway. How would a base on the Moon affect life on Earth? Again, let's remember Apollo and how it changed our concept of stretching our horizons. I think those of us in science often fail to understand the impact on society as a whole of well-defined and clearly articulated goals. Scientists, of course, don't need much encouragement to beat a path to a frontier. After all, little more than a hint of a new sub-atomic particle will send my fellow physicists off on years of day-and-night dedication at some microscopic frontier, just as geologists will devote lifetimes working at the macroscopic frontier of piecing together the patterns of the movement of continents. But mere scientific curiosity can't hold a candle to the kind of inspiration that space has provided to our nation. It's hard to think of any peacetime program that ever inspired a nation's imagination and enthusiasm the way the Apollo program did—and that's the spirit that we should be trying to generate in tomorrow's space program.

Let me add that both those broad inspirational benefits and the attraction of space as a scientific frontier have a very tangible counterpart. Space can be the kind of intellectual and technological endeavor that attracts the best young minds in our country and challenges their creativity. The response to those challenges, in the form of innovation and new technologies, will help keep the United States at the forefront in science and technology—and will strengthen our ability to compete aggressively in the world marketplace.

In spite of our best efforts to detail the spinoffs and benefits from the space program, I don't think the chief benefit of Apollo can ever be quantified. That benefit was a glorious elevation of the human spirit and of national pride, a reminder that our society, giving free rein to human creativity and enterprise, can achieve herculean goals if it's challenged to do so.

That kind of inspiration doesn't come along very often, so it's the kind of fire to keep burning. In an age of fierce and growing competition in the world commercial marketplace, there's no inspiration I'd rather be able to pass on to the youth of our nation. In fact, I can't think of a more eloquent statement of that idea than the President offered earlier this year when he spoke to the graduates of the Air Force Academy. He said, "Our willingness to accept the challenge of space will reflect whether America's men and women today have the same bold vision, the same courage and indomitable spirit that made us a great nation . . .The only limits we have are those of our own courage and imagination. And our freedom and well-being will be tied to new achievements and pushing back new frontiers."

To me that's the real theme of any conference on space—the fact that it challenges our vision today and will challenge it for as long as we can imagine.

In Space: One World United

Walter J. Hickel

Chairman, Yukon Pacific, P.O. 101700, Anchorage, AK 99510

In the next few days you're going to hear a group of learned individuals address the prospects of permanent habitation of space. The experts will consider what tools we might take with us, what resources we might find out there, and what we might bring back to the Earth that is useful. Yet before we get into those details, I want to talk with you, as a layman, about the nature of frontiers and big projects in general.

Frontiers summon the creativity, imagination, and inventiveness of the human mind to conquer them . . .and in this age, the big project is the means to do so. A big project is what you get when you ask for the Moon. We called the first project Apollo. Today we're no longer just asking for the Moon; we've been there. We're looking further—to begin living outside Earth for good, wherever opportunity, curiosity, and need will take us. Our goal is more than another big project. It's time to open up the final frontier, permanently, to the point of no turning back. Our journey is really infinite, and a big project is what civilization needs to begin this journey and perfect the commitment to go beyond.

Big projects don't start with a lot of money. They start in the mind. Obviously, then, our journey begins here on Earth. We need to remember we're not here on this Earth just to be a user. We were made in the image of a Creator, and we're here to create. When we create, we explore the frontiers of the mind and the soul, for we must first acknowledge and open our inner frontiers before we can approach the outer frontiers of space. Charles DeGaulle once said, "We may well go to the Moon, but that's not very far. The greatest distance we have to cover still lies within us."

Thus, the immediate challenge before us now as Americans is to allow ourselves frontiers again. The parallel challenge is to train ourselves to bring about the big projects necessary to explore those frontiers. Candidly, we've been lagging on both fronts.

Frontiers have nothing to do with a coon-skin cap and a rifle. They have to do with limits . . .things we set for ourselves. You either believe in limits or you don't. Lately, we've only begun to emerge from a time in our history when many people believed economic growth on Earth has natural, predictable limits. Just a few years ago, we believed we were running out of everything. Many even ruled out hope. Frontiers were forgotten. About the same time in our history—ten years or so ago—America lost enthusiasm for its space program. The country began looking inward and became uncertain of itself.

I remember giving a series of talks in 1976 at the University of California. One of them talked about "the inexhaustible Earth." I believed then, as I do now, that the Earth is inexhaustible because God made man's mind inexhaustible. As long as we don't run out of imagination, we're not going to run out of anything. As I was saying that, a man in the audience scoffed. "Don't you know we're even running out of cement?" he asked. Imagine that. He seriously thought there was a shortage of cement. Cement—sand and gravel, ash and limestone!

Frontiers have always existed, and they always will. We approach a limited world with fear, but an infinite universe with confidence. A frontier is where you live with more questions than answers. You make decisions with your gut and your heart because your head can't help, but the important thing is to make those decisions: to go and to stay. Those left behind won't understand why we go or why we stay. But when a leap of faith is coupled with confidence in the future, progress results and all mankind benefits.

In the next few days we will probably hear more questions than answers. But let's respond with a decision. Let the questions point the way, not stop us from going.

In most businesses, it's the inventors, the scientists, the most knowledgeable people associated with a project who advocate moving forward the quickest. Politicians and external business forces usually act to slow things down. The space program, since the days of Sputnik, has seen just the opposite. Here, scientists are cautious in their advocacy. It takes politicians, and sometimes a national crisis, to speed things up.

Today we don't have a crisis. We just have an opportunity, and if the believers who are knowledgeable about space aren't advocates, how can we expect the non-believers to be? No one in this room should wonder if we're going too far, too fast in discussions about settling the universe. How can you go too fast when where you're going is infinite? I submit you can't go fast enough! And while I'm concerned that America is lagging in its space pioneering thinking these days, I believe we also need to examine our ability to undertake the big projects necessary to approach that frontier.

Look at the megabillion dollar projects on our plate right now. As an engineering society, we're performing miserably. Hesitation and indecision are the roots of the problem. America's nuclear power program is substandard to France, Japan, and Korea, in part, I believe, because we've refused to standardize. We're left with billion-dollar flower pots, many just close to being completed, that will never produce nuclear power. It seems that we can control the atom, but we can't control our costs. I entered a business recently to see that a $20 billion pipeline project is built across Alaska. The last group who tried spent $600 million before deferring construction indefinitely. They spent $600 million on paperwork . . .and there's still not a piece of pipe in the ground. There's a dam on the

*drawing board in Alaska right now that has more than a hundred million dollars in it—
and yet the state has yet to make a decision to build it. As a newcomer to the details of the
space program, I was appalled to find out that we've dismantled the tools, and more
importantly the collection of minds, that can produce a Saturn rocket.*

*America, we can't do business this way. Civilization needs big projects, the kind that
ignite the mind and inspire the soul. Remember, only eight years elapsed between Alan
Shephard's first space flight and Neil Armstrong's first step on the Moon. We got there
because John F. Kennedy made a decision. Big projects need decisions, not dollars, to get
started. They need continuity and commitment. Civilization could never afford to spend
billions of dollars dabbling in this or that . . .leading everywhere and going nowhere. It's
jump or don't jump. Do it or don't.*

*Consider wars for a moment. Wars unite. They forge alliances among allies. They force
an urgency of focus. They forge a common purpose, and they mobilize a will to achieve
that purpose. But the fruit of war is destruction. Nevertheless, a war is a decision. We can
go for years looking at ways to clean up our cities, but drop a bomb—or have an
earthquake—one day, and suddenly we're cleaning up and building the next.*

*Decisions telescope time. I saw one space scenario recently, a planning document that
moved us forward from a space station in 1991 to a lunar base in the year 2000 to a
landing on Mars in 2030, forty-five years from now. That plan was made without a
decision. I'm not an expert, but many of you are. Let me ask a question. How much further
could you move in just ten years if we gave you a mandate?*

*For those of you who make your livings on the prospects of space exploration and
development, I join you today as a kindred spirit. Forty-four years ago, as a Kansas farm
boy, I went west, looking for a country. Going north, I found it, in Alaska, with enough
opportunities to last many lifetimes. As I arrived, there were people leaving. Many of them
felt that everything that was going to happen had happened already. Their frontier was
over, because they never had one. But we believed in Alaska. We really believed.*

*We put together an economic base from the oil discoveries that were made by
pioneering companies. Next came a megaproject that rivaled the costs and the
organizational challenge of Apollo. We built a nine-billion-dollar oil pipeline eight hundred
miles across virgin territory. It required new technology, new ways of doing business. But it
worked, because a decision was made to do it.*

*Still, we've really only started. Flowing out of the ground with the oil is natural gas,
almost as much in energy value. Our challenge today is not only to build another
pipeline—the world's largest private project—but also to bring four nations together in this
enterprise. In the maturing of Alaska, we depend not just on America, but on Japan, Korea,*

and Taiwan, who must trust enough in the future to buy the gas and join the project. So, as we embark on the frontier in space, we might benefit from the experiences of one of the great frontiers on Earth.

The arctic regions and the Moon have much in common. They're both remote. Not everyone understands living there. They're both rich in potential, potential that requires bold decisions to realize . . .but life is exciting and beautiful nevertheless. Leaps forward are made on the infrequent occasions when the decisionmaker has vision and isn't protecting the status quo. Decisions start with believers who really believe.

In starting most things anew, long term development thinking is absolutely necessary. Think of a child. If I had to pay someone to do what my wife did to bring up our children, waiting for a return on our investment with compound interest, no child would be economically feasible. During the next few years there will undoubtedly be continuous debate over whether or not living in space makes economic sense. I contend it won't make sense unless civilization commits to build a railroad.

A laska's history provides an excellent example. We had a delegate to Congress in 1914 named Wickersham. He knew the territory needed a railroad. The richest private interests in the world had tried and failed to make it work. However, Wickersham was a believer. He convinced Congress not to seek a return on the railroad right away. The nation would get its returns from everything else the railroad made possible. The bill that passed authorized the President to build up to a thousand miles of rail, "its primary purpose to open up the country." Today, our civilization needs to make a similar commitment to build a railroad into space to begin to open up the universe. We need similar champions to tell us where we are.

Buzz Aldrin called the Moon "magnificent desolation." Its long term value may not be what it is but where it is.

I've talked about infinite thinking that establishes frontiers. I've discussed essential decisions that bring big projects about. There's one final thought I'd like to leave with you. In these discussions, the question of international cooperation has always surfaced. In my mind, it's no issue at all: leaving the Earth can unite us.

The President recently talked about sharing "Star Wars" technology with Russia to help free civilization from the threat of nuclear holocaust. We should ask the question, couldn't civilization also benefit by our united effort in the exploration of space? Should we ask the Russians to join us in going to Mars, and beyond?

Leaving the Earth can unite our purposes and peoples . . .our energies, talents, and technologies . . .and especially our minds. The exploration of space may well be civilization's last chance to join together in a great undertaking bigger than any of us— technologies for a common purpose, a great undertaking that can heal the divisions between men—as together we seek to go vast distances, and yet, to cover those inner distances that yearn for explanation.

Like a small boy asking, "Mother, what's out there?" let's join together and find out. What difference does it make which political party is in power or whose ideology is in vogue when we finally find God and the universe among us?

Good luck and Godspeed.

An Opportunity for Openness

Arthur Kantrowitz

Dartmouth College, Hanover, New Hampshire, 03755

W e are gathered to discuss a momentous undertaking. A lunar base would certainly begin a new chapter in world history. With a lunar base, humanity would for the first time be in a position to utilize extraterrestrial matter (ETM). Many important civilian uses for ETM will be dealt with in this symposium. In view of today's level of international tension, the application of large quantities of ETM in the hardening of space-based military assets inevitably comes to mind. What could this new capability imply for the debate now raging around the stability of Mutual Assured Destruction and the President's Strategic Defense Initiative (SDI)?

Several prominent scientists have come forward with negative theorems about SDI, easily produced and loudly announced, and at least partially based on the fragility of today's spacecraft without including the possibility of ETM. Perhaps on this occasion we should point out again that they really should have announced only that they did not see how an effective strategic defense could be built.

The threat of nuclear Armageddon has darkened the world for decades. To escape the darkness, arms control is usually offered as the world's only hope. It provides an opportunity for dialogue, and talk is safer than the alternatives. But verifiable arms control has a problem that must be faced. Starting with the tens of thousands of existing nuclear weapons, it may well be true that it is easier to verify adequate compliance with a freeze or a reduction agreement than to hide evasion. But hiding evasion gets progressively easier and verification gets progressively harder as the numbers are reduced, and the numbers must be reduced by a factor of thousands to bring us back to the destructive capabilities of WWII. While the dialogue on arms control must be pursued, we must not let it blind us to other possibilities, perhaps now only barely perceptible, for escaping the darkness.

Arms control is the control of hardware. I would like to discuss a software possibility. In an instructive recital of the dismal history of arms control (New York Times, April 18, 1982), Barbara Tuchman quoted Salvador de Madariaga (Chairman of the League of Nations Disarmament Commission) as concluding that ". . .Nations don't distrust each other because they are armed; they are armed because they distrust each other."

I find Madariaga's proposition exciting because it provides a new light in which to view a striking exchange on SDI in the recent presidential election debate. In response to a question concerning sharing ". . .the best of America's technology with our principal adversary," the President began his answer with "Why not?" Mondale declared it was ". . .in my opinion a total non-starter."

F ollowing Madariaga's emphasis on the primacy of trust, a good way to judge a new military technology or policy in the nuclear age would be: Can it help nations to trust each other?

Surveillance satellites pass this difficult test, and indeed they constitute an important stabilizing influence in the present tension. Competitive deployment of a BMD system on the other hand is violently destabilizing. It is just this frightening destabilization that impelled Reagan to offer a new openness. This new openness could indeed help nations to trust each other. It could be the first light of dawn.

Technological change drives social change. Remember the trauma of the early industrial revolution and William Blake crying out against ". . .dark satanic mills." In just those countries and in just those times, the most open societies the world has ever known were growing. Openly published science soon left magic and its secrets far behind. Industrial competitiveness forced a transition from arts passed secretly from generation to generation to the mass education needed for a technological society.

In contrast, let me tell you briefly a story of rejection of a new opportunity for openness that occurred in early Ming Dynasty China. In the massive scholarly work Science and Civilization in China (Cambridge, 1971, Vol. 4, Part III, pp. 392–535), Joseph Needham recounts the spectacular rise and fall of Chinese naval power. Up to the early 15th Century A.D., Chinese technology led the world. In his utopian New Atlantis, Francis Bacon exhibited three great inventions—printing, gunpowder, and the compass—as evidence that humanity could advance beyond the heights reached in antiquity. These were the foundation of his wonderfully fruitful proposal for the organization of applied research. Needham points out that these remarkable inventions were all well-known in China centuries before they appeared in Europe.

In the early Ming Dynasty, this flowering technology culminated in a remarkable series of naval expeditions made by fleets of 1500-ton "treasure ships," each able to carry about 500 men. In the early years of the 15th Century, these fleets sailed around Southeast Asia to Bengal, Ceylon (from which they brought back a defiant king as prisoner) and finally down the east coast of Africa in 1420. About fifty years later, in a similar adventure, the Portugese sailed down the west coast of Africa in a series of voyages that opened up a new chapter in world history.

China made its last great voyage in 1431–33. According to Jung-pang Lo, in 1436, when the Cheng-t'ung Emperor came to the throne, an edict was issued that not only forbade the building of ships for overseas voyages but also cut down the construction of *warships and armaments ("Decline of the Early Ming Navy,"* Oriens Extremus, *Hamburg, 1958, Vol. 5).*

What changes took place with such repressive effect that China recoiled from its seaward expansion and career as a naval power? "This was, indeed," says Needham, "but one aspect of a general decline which reflected itself severely in many branches of science and technology. The navy simply fell to pieces."

After that happened, China lost its leadership in technology and isolated itself as long as it could from the explosive growth of Europe. This isolation became impossible when the military strength of the "barbarians" forced itself upon the Chinese in the Opium Wars four centuries later.

This was only one episode in a great administrative battle waged for thousands of years between the Confucian scholars who led the bureaucracy and the Grand Eunuchs of the Imperial Court. The eunuch Admiral Cheng Ho had led the last great voyages, and clearly their spectacular successes constituted a great threat to the power of the scholarly bureaucracy. But the scholars had a trump card—they had exclusive control of the education of the next emperor. Thus when he came to the throne, one of his first acts was to destroy the navy, and with it he destroyed China's leadership in technology. The scholars had argued that China had no need for contact with the rest of the world. The bureaucrats, by stopping technological advance, maintained their feudal power.

T*echnological change forces social change. Nuclear weapons, access to space, information technology, and genetic technology constitute great and threatening technological changes. What social changes are they forcing?*

I will hazard a conjecture. Technological advance will continue to force us to a more open world. The SDI hardware, even if completely successful, would only protect us from ballistic missiles, but the increased trust, the new openness, that could come from cooperating with the Soviet Union and the world in a massive technological undertaking would have much more far-reaching effects. It could move us toward a world in which any secret enclave that could hide weaponry will be seen and dealt with as an aggressive act. It could help to create a world open enough to prevent all varieties of nuclear terrorism.

Thoughts on a Lunar Base

Edward Teller

Lawrence Livermore Laboratories, University of California, Livermore, CA 94550

I would like to start with a statement that I expect, and even hope, may be controversial. I believe there is a very great difference between the space station now being planned and any activity on the Moon now under discussion. I believe that in the space station we should do as much as possible with robots for two simple reasons. There is nothing in space—practically nothing—except what we put there. Therefore, we can foresee the conditions under which we are going to work, and, in general, I think robots are less trouble than people.

The other reason is that, apart from experiments and special missions that we have in space, we do not want to proceed to change anything in space, whereas on the Moon we will want to change things. Likewise on the Moon, we will find many things that we do not expect. Adapting robots to all the various tasks that may come up, and that we do not even foresee, is not possible.

The space station is obviously extremely interesting for many reasons. However, that is not what I want to talk about except to state that, of course, the space station is apt to develop into a transfer station to the Moon. Therefore, its establishment is not independent of what we are discussing here.

I would like to look forward to an early lunar colony. I do not want to spend time in making estimates but simply want to say that it would be nice to have a dozen people on the Moon as soon as possible. I think we could have it in ten years or so. When I say 12 people, I do not mean 12 people to stay there but to have 12 people at all times, to serve as long as it seems reasonable. To me, 3 months is the kind of period from which you could expect a good payoff for having made the trip. Longer rotations than that might be a little hard, and efficiency might come down. But all this is, of course, a wild estimate on my part.

What kind of people should be there? It will be necessary to have all of them highly capable in a technical manner, and I believe that they should perform all kinds of work. Probably at least half of them, after coming back to Earth, should get the Nobel Prize. The result will be that we will soon run out of Nobel Prizes because I believe there will be very considerable discoveries.

Also, if you have 12 people you probably ought to have a Governor. I have already picked out the Governor to be, of course, Jack Schmitt. Furthermore, I would like to tell you that when I first testified about space, and was asked whether there should be women astronauts, I proposed that all astronauts should be women. The packaging of intelligence in women is more effective in terms of intelligence per unit weight. However, in view of the strong sentiment for ERA, I think I might compromise with an equal number of women and men. That arrangement has all kinds of advantages.

I believe that the discussion here has had plenty of emphasis on what I know will be the main practical result of a lunar base—use as a refueling station. It will supply both portable energy in a concentrated form and portable fuel for refueling rockets, primarily in the form of oxygen extracted out of lunar rocks. The only question is how to do it. My first idea was, of course, we should do it with nuclear reactors. Perhaps the environmental movement, the Sierra Club, may not have an arm that extends beyond one light second. On the other hand, we will have some problems, problems of cooling. However, most of the energy might be needed to squeeze oxygen out of iron oxide, and that simply means a high temperature. You may not need a lot of machinery, and some of the energy can be, in this way, usefully absorbed right inside the reactor. What remains probably should be converted to electricity.

The other possibility is solar energy. I am strongly inclined to believe that solar energy will be quite useful for two reasons. First, great advances have been made in solar cells, particularly with regard to Ovshinsky's idea of utilizing amorphous semiconductors. The point is that they are not very good conductors of electricity and therefore must be thin, but, on the other hand, amorphous materials are very good absorbers of light and therefore can be thin. Methods to fabricate them have indicated that you can, with practical certainty, get down to one dollar per peak watt.

There is, however, another advantage to solar power and that is if you do not want power, but just want high temperature for driving oxygen out of oxides, you may not need mirrors that have to be moved. It might be sufficient to have the right kind of surface that absorbs and emits ultraviolet but is highly reflective in the visible and the infrared. In equilibrium with solar radiation, this will give high temperatures; the farther you go in the ultraviolet the more you can approach the maximum temperature obtainable, the surface temperature of the Sun. If you try to approach this limit, then the energy content—the power—will be small because it utilizes a smaller portion of the solar spectrum. But the temperature you can get is high. What the optimum is where you will want to compromise, I do not know.

Let me extend this idea one step further. I would not only like to get very high temperatures; I also want to get very low temperatures as cheaply as possible. You can achieve the latter during the 14-day lunar night. If you isolate yourself from the surface of the Moon, put your apparatus on legs and put some space in between—all very cheap arrangements—you can approach temperatures in the neighborhood of 2.7 degrees Absolute. In this way you can get low temperature regions of large volume and high temperature regions of large volume.

Now, I would like to talk about one practical point that may not have been discussed, namely, the question of where on the Moon the colony should be. I would like to go to one of the poles because I would like to have the choice between sunlight and shade with little movement. Furthermore, it would be a real advantage to establish the colony in and around a crater where you might have even permanent shade in some places and where moving away from the rim on one side or the other you can vary conditions quite fast. Of course, it is of importance not only to position yourself in regard to the Sun but also in regard to the Earth. For many purposes, you want to see the Earth in order to observe it. For other purposes, for instance astronomy, you want to be shielded from the Earth, not to be disturbed by all the terrestrial radio emission. All these conditions will be best satisfied in a crater near a pole.

I have a little difficulty in reading the lunar maps. There seem to be three good craters in the immediate vicinity of the south pole but no good craters near the north pole, or vice versa. I am not quite sure. At any rate I want to go to the pole that has the craters.

The purpose of all this is obviously what I have said to begin with and what you all realize—refueling and energy. Oxygen is the main point, but it would be nice also to have hydrogen. Hydrogen we could get from the Earth much more cheaply than the oxygen, but still it is one-ninth the cost of oxygen plus the considerable weight of the tank. Hydrogen has been deposited in the lunar dust by the solar wind over geologic time, and the mass of hydrogen in that lunar dust, as far as I know, is not much less than one part in ten thousand. Without having made a decent analysis, my hunch is that it is easier to move the lunar dust a few miles on the Moon than to come all the way from the Earth even though you have to move ten thousand times the mass. If you can distill oxygen out of iron oxide, you certainly can distill hydrogen out of the lunar dust. Furthermore, Jack Schmitt tells me that there is a possibility of finding hydrogen, perhaps even hydrogen that is four and one-half billion years old, in other parts of the Moon in greater abundance than what we see in the average lunar dust.

All of this is, of course, of great importance and perhaps serves as a little illustration of what kind of constructions we are discussing. Obviously, we will have to try to make these constructions with tools as light as can be transported from the Earth. In planning the lunar colony, special tools and special apparatus have to be fabricated on the Earth, specifically adapted to the tasks already described as well as others.

I would like to make a special proposal. I believe that surveillance of the Earth—permanent, continuous surveillance that is hard to interfere with—is an extremely *important question, important to us, important for the international community, important for peace keeping. There have been proposals, and I am for them, to guarantee present observation of facilities by treaties. On the other hand, treaties not only can be broken; treaties have been broken. It is in everyone's best interest to have observation stations that are not easy to interfere with.*

I would like to take the biggest chunks that I could get off the Moon and put them into a lunar orbit, perhaps 120 and 240 degrees away from the Moon. Of course, they will be very small compared to the Moon but maybe quite big compared to other objects that we put into space. If the Moon and these two additional satellites are available for observation, then we can have a continuous watch on all of the Earth with somewhat lesser information around the pole. The latter also can be obtained with additional expenditure, but to have 95 percent of the most interesting part of the Earth covered continuously would be already a great advantage. I would be very happy if, on these observation stations, we would do what we should have done with our satellites and are still not doing, namely, make the information of just the photographs obtained from the satellites universally available. I believe that would be a great step forward in international cooperation, international relations, and peace keeping.

*T**raveling to these artificial satellites from the Moon is a much smaller job than reaching them from the Earth. Since you stay on the same orbit, you just have to have a very small additional velocity after leaving the Moon, wait until you are in the right position, and then use a retrorocket. The total energy for that is small, and if you produce the rocket fuel on the Moon, then I think you have optimal conditions.*

I also would like to have a satellite with a special property. It should have as big a mass as possible, built up from a small mass in the course of time. But, furthermore, I want it to rotate in such a manner that instead of turning the same face all the time to the Earth it should turn the same face all the time to the Sun. If you can do that, then half of the surface will be in permanent night, half in permanent illumination, and whatever we can do on the Moon, for instance setting up a permanent low-temperature establishment, you can do that very much better on these satellites.

Now, I would like to finish up by making a very few remarks on purely scientific work that will become possible. In the vacuum of the Moon we can work with clean surfaces. It is obvious that surface chemistry could make big strides. This can be done equally well in the space station, and, in this respect, the Moon does not have an obvious advantage.

Where you do get an obvious advantage is in astronomical observations where you want the possibility to collimate in a really effective manner. When you want to look at x-rays or gamma rays from certain directions, all you need to do is to drill a deep hole that acts as a collimator and have the detectors at its bottom. You would have to have a considerable number of these holes, but I believe that it will be much cheaper than to have a considerable number of observation apparatus shot out from the Earth, particularly because the mass for collimation will be not available in space stations except at a considerable cost. The same holes may be used for high energy cosmic rays.

Another obvious application is in high energy physics. As the size of accelerators kept going up, many years ago our very good friend Enrico Fermi at a Physical Society meeting, as far as I know, made the proposal in completely serious Italian style that sooner or later we will make an accelerator around the equator of the Earth. Well, we are approaching that—at least we are planning an accelerator that takes in a good part of Texas. I am not quite sure that we should do that. Let us wait until we get to the Moon. (That might happen almost as soon as a giant accelerator can be constructed.) We actually could have an accelerator around the equator of the Moon. Taking advantage of the vacuum available, you only need the deflecting magnets and the accelerating stations, and these can be put point for point rather than continuously.

I have been interested for many years in the remarkable discovery of Klebesadel at Los Alamos of gamma ray bursts that last for longer than 15 milliseconds and less than 100 seconds, have their main energy emission between 100 and 200 kilovolts, but seem to have components far above a million volts, too. I believe everybody is in agreement that these come from something hitting neutron stars and converting the energy into gamma rays. But most people believe that they come from nearby regions of our galaxy and are, therefore, isotropic. Actually the number of observations depends on the intensity in such way as though from more distant places we do not get as many as expected. The usual explanation is that we get these from farther places and we get them only from the galactic disc rather than a sphere. Unfortunately these bursts are so weak that the directional determinations cannot be made. On the Moon you could deploy acres of gamma ray detectors of various kinds and leave them exposed to the gamma rays or cover them up with one gram per square centimeter, five grams per square centimeter, or ten grams per square centimeter so that with some spectral discrimination you will get a greater intensity from perpendicular incidence than from oblique incidence. As this apparatus will look into the plane of the galaxy, into the main extension of the galaxy, or toward the galactic pole, you should see a difference, a deviation from spherical distribution, for these weakest bursts, essentially bursts of 10^{-5} to 10^{-7} ergs/cm^2/s.

A very good friend, Montgomery Johnson (who unfortunately died a few months ago) and I had made an assumption that these radiations really do not come from the galaxy but from outer space, from regions where the stars are dense and where collisions between neutron stars and dense stars like the white dwarfs may occur. Good candidates are the globular clusters, but there may be other dense regions in the universe as well. If *this hypothesis turns out to be correct, then the reason you find fewer events at great distances are cosmological reasons—curvature of space, a greater red-shift, lesser numbers of neutron stars and white dwarfs in the distant past, which was closer to the beginning of the universe. Actually, if this hypothesis is correct, then the gamma-ray bursts would, in the end, give us information about early stages of the universe. No matter which way it goes, the gamma-ray bursts are interesting phenomena, and the Moon is one of the places where they could be investigated with real success.*

I am sure that in these ways and many others an early lunar colony would be of great advantage.

2 / LUNAR BASE CONCEPTS

T HE TERM "LUNAR BASE" can refer to a spectrum of concepts ranging from a mannable "line shack" to a multifunctional, self-sufficient, populous colony. In general, the authors contributing to this book discuss the earliest stages of a permanently manned facility with the capability for scientific investigations and some ability to support its own operation with local materials. The exact form of the "final" configuration usually is not critical to the discussion until cost is included. Costs of a lunar base can be similar to the space station program or can be at the level of the Apollo project. Since cost is such a sensitive topic in the advocacy phase, it becomes very important to understand not only the total cost but also the spending rate and the basic assumptions about what is charged to the project. The costs derived by Hoffman and Niehoff in their study presented in this section differ from costs referenced by Sellers and Keaton in a later section. The final configurations in the two studies differ considerably, but in both cases the spending rates over the duration of the project are well within the rate of expenditure of the current space program and are substantially less than rates associated with Project Apollo.

Because lower cost is a major strategy goal, design concepts generally adopt hardware from prior programs. For example, the studies conducted by NASA in the 1960's and described by Lowman and by Johnson and Leonard depict habitats inspired by the Apollo transportation system. Contemporary drawings show space station modules emplaced on the lunar surface. Maximizing design inheritance to decrease uncertainties in technology development builds confidence in estimates of feasibility and affordability of a lunar base or any other program. Conversely, awareness of a long range lunar goal during design of the space station can increase the "inheritability" of the technology. Duke *et al.* propose a model for long range development based on three distinct choices for programmatic objectives.

The selection of the location for the first base on the Moon will be heavily influenced by programmatic priorities. Some argue that a return to one of the Apollo landing sites will suffice. The geology and the environment of a landing site are well known, obviating the need for any expense or delay associated with precursor survey missions. If scientific investigations have the highest priorities, then the major questions in lunar science would drive the selection process. Since radio astronomy from the farside of the Moon has long been a prime candidate for a surface investigation, a good location might be somewhere on the limb, where communication with the Earth can be maintained while the radio telescope is still nearby. On the other hand, the long-term strategy for building the surface infrastructure might require the early exploitation of local resources. An unmanned polar orbiting satellite would make sense as a precursor resource survey mission.

Some scientists have advocated a base at a lunar pole. The nearly perpendicular orientation of the lunar rotation axis to its orbital plane results in a continual twilight at the poles and, consequently, constant access to solar energy. A polar base would reside on the limb and would be continuously accessible from a station in lunar polar orbit. Unfortunately, the polar regions are the least known either in terms of geology or resources. Jim Burke reviews the difficulties and advantages of polar living in more detail and presents some concepts for exploiting that unique environment.

Lunar bases at any other latitude will suffer through the diurnal cycle of two weeks of daylight followed by two weeks of night. A power system based entirely on solar energy will require massive energy storage facilities for night-time usage and must be oversized to generate the stored energy during the daytime operation. Principally for this reason, nuclear energy appears to be the best solution for early stage lunar bases. Buden and Angelo discuss the evolution of the power plant with growing needs at the base, while French reviews some practical considerations of siting nuclear reactors on the Moon.

LUNAR BASES: A POST-APOLLO EVALUATION

Paul D. Lowman, Jr.

Geophysics Branch, Code 622, Goddard Space Flight Center, Greenbelt, MD 20771

A lunar base would be an extremely productive choice for future American space efforts. Further exploration of the Moon is scientifically important; the Moon offers a stable and radio-quiet platform for astronomy and space physics, material resources (chiefly Si, Al, Fe, O, Mg, and Ti) are available for use in near-Earth space or on the Moon itself, and Earth-Moon operations offer the technological stimulus of interplanetary missions at lower cost and with less risk. It is recommended that the Lunar Geochemical Orbiter be given high priority, that space station modules be designed for use on the Moon as well as in space, that design studies of a manned orbital transfer vehicle be started, and that continued analysis of lunar samples and meteorites be strongly supported.

INTRODUCTION

The Apollo Program, whose six lunar landing missions began at Tranquillity Base, could have led to the establishment of a permanent base on the Moon. It did not, for reasons that are well documented, and there have been no American lunar missions of any sort since 1972. However, with the revival of the American space program, marked by the first flight of the space shuttle in 1981, has come a revival of interest in lunar and planetary missions in general. The Solar System Exploration Committee (1983) has recommended an ambitious but fiscally conservative set of missions that is now being acted upon, the first two new starts being the Venus Radar Mapper and the Mars Observer. A parallel development has been renewed interest in lunar bases (Duke *et al.*, 1984), demonstrated by the 1984 Lunar Base Symposium held in Washington and its preparatory workshop held in Los Alamos.

This paper was presented at the 1984 symposium in abbreviated form. Its objective is to reevaluate the desirability of an American lunar base in light of the many scientific, technological, and political developments since the last Apollo mission in 1972. The term "lunar base" will be used here to cover a wide range of possible programs, from small facilities for short-term occupations by a few people up to large complexes at several locations occupied semi-permanently by large staffs. It will not include large autonomous colonies on the Moon, since one of the objectives of a lunar base program would be to explore the technical and economic feasibility of such colonies.

BACKGROUND

Technically sound and essentially modern descriptions of possible lunar bases were published as early as 1946 (Harper, 1947; Clarke, 1951; von Braun *et al.*, 1953; Burgess, 1957). Detailed planning for such bases began in the United States shortly after the Apollo

Program was started in 1961, and dozen of studies were carried out by NASA, its contractors, the U.S. Air Force, and other organizations. A biliography of these studies has been compiled by Lowman (1984), but only a few main concepts can be summarized in this paper.

The great majority of lunar base concepts proposed in the 1960s were predicated on use of the Saturn V system or direct derivatives thereof. Two of these, the Apollo Logistics Support System (ALSS) and Lunar Exploration Systems for Apollo (LESA), are illustrated in Figs. 1 and 2. The LESA was the most ambitious base proposal of the 1960s,

Figure 1. Artists' concepts of ALSS and LESA with comparative statistics. From Anonymous (1964).

Figure 2. Two versions of LESA modules emplaced on the Moon. From Boeing (1963).

being planned for expansion by landing of separate 25,000-lb. modules. The LESA could have formed the nucleus for a large permanent colony had it been carried out. A complete set of supporting parametric studies was concluded by various contractors, covering all aspects of base establishment and operation, including logistics, life support, and scientific missions. Some aspects of the LESA studies are by now quite outdated, and the Saturn V system is no longer available. The missions proposed for LESA have now been carried out to some extent by the Apollo Program or, for Moon-based astronomy, by instruments in Earth orbit. Nevertheless, the surface environmental model assumed was fundamentally correct and the life-support parameters reasonably accurate. It seems safe to say that the LESA studies are still of value for baseline planning of future lunar programs.

Probably the most important lunar base study in terms of possible future work is the Lunar Base Synthesis Study, completed in 1971 by North American Rockwell (1971a). This study is of interest in several respects. First, it was done late enough to take into account results from early Apollo missions, to say nothing of experience gained from the many Earth-orbital missions flown by then. Second, it was the only major lunar base study done to assume use of transportation to low Earth orbit by the space shuttle rather than the Saturn V. The study produced a conceptual design for crew modules (Fig. 3) derived from a related study of an orbiting lunar station (North American Rockwell, 1971b) that could be used either for a modular space station or a lunar surface base. It is clear that if a lunar base program should be started by the U.S. within the next 10 years, something like the 1971 studies must be the starting point, given the continued use

- 4 CREWMEN STATEROOM
- HYGIENE FACILITY
- INITIAL GALLEY & CREW CARE

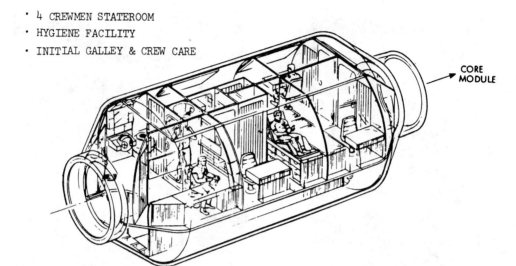

CORE MODULE

Figure 3.   Module for a lunar surface base (LSB) derived from a modular space station (MSS) module designed for shuttle launch. From North American Rockwell (1971a).

of the space shuttle and the presumed establishment of a permanent space station in low Earth orbit.

A large number of studies was carried out in the 1960s on the nature and use of lunar resources, many of which are still of value. For example, lunar soil has proven easily workable, so parametric studies involving its bulk use for shielding are still applicable. One important exception, however, relates to the use of indigenous lunar water. Many studies, some published as late as the month before the Apollo 11 mission (Lowman, 1969), assumed that lunar rocks would contain at least a small amount of combined water that could be extracted for life support or, optimistically, for use as rocket fuel. Unfortunately, it was discovered immediately on return of actual lunar samples that at least the surficial rocks and soils are completely anhydrous. Even more discouraging is the fact that all lunar igneous rocks, such as basalts, are highly reduced in a chemical sense, implying extremely dry magmas and magma source areas. In this respect the early lunar base studies appear to have been optimistic, although the possibility of deep-seated lunar water or hydrogen, or even polar ice (Arnold, 1979; O'Keefe, 1985), cannot be ruled out completely.

It is clear that a substantial fraction of the work on possible lunar bases done before 1972 has permanent value and should be taken account of in any planning for new lunar programs. Let us now consider factors that have changed greatly since 1972.

POST-APOLLO DEVELOPMENTS RELEVANT TO LUNAR BASES

Several developments and trends since 1972 have direct importance for the lunar base concept and must be taken into account in a reevaluation such as future lunar operations and missions.

Resumption of U.S. Planetary Exploration

After a period of several years in the 1970s when no new starts were approved, the United States has resumed a modest program of planetary missions, starting with the Venus Radar Mapper. The VRM is the first element of a long-term planetary program outlined by the Solar System Exploration Committee (SSEC, 1983) and utilizes relatively low-cost common hardware. The program includes a Lunar Geoscience Orbiter as one priority item, reaffirming the scientific value of renewed lunar exploration.

Space Transportation System

The fundamental inefficiency of the "ammunition philosophy," *i.e.*, expendable boosters, was always obvious despite the success of the Saturn V-launched Apollo missions. With the successful development of the space shuttle, the U.S. is now back on what should eventually be a more cost-effective course of manned space exploration. The shuttle now provides routine and frequent access to low Earth orbit (LEO). Furthermore, the last lunar base study performed before 1972 (North American Rockwell, 1971a) showed that the shuttle could carry space station modules adaptable to use on the lunar surface.

Space Station

The permanent manned space station, a traditional space goal, has now become an official U.S. program, and Phase B studies have begun. The basic feasibility of such a station is unquestioned, although support for it has been less than unanimous. Given its relatively low cost and operational utility for satellite servicing and microgravity manufacturing, the space station in some form should be in orbit by the 1990s. Once deployed, it can serve as a transportation node, one function being to launch lunar and planetary missions. The space station will then be another link in a space transportation system that could establish and support a lunar base. Furthermore, two related studies on a lunar base and an orbiting lunar station (North American Rockwell, 1971a,b) called attention to the possibility of designing space station modules for use with minor modifications on the Moon as well. It thus appears that the only major items necessary for a lunar base not already in use or being planned are a reusable orbital transfer vehicle (OTV) and a comparable lunar landing vehicle (LLV).

Assimilation of Apollo Scientific Results

Although study of the 385 kg of returned lunar samples can profitably continue for many years with increasingly refined or completely new analytical techniques, it is safe to say that the scientific results of the Apollo Program have now been largely assimilated (French, 1977; Taylor, 1982). With respect to a lunar base, our knowledge of the Moon, although incomplete in many respects, is now comprehensive and firm enough to say that the environmental feasibility of such a base is established. The surface environment, the dominant topography, and the gross composition of the main crustal rocks are all known well enough to permit detailed planning for at least a temporary lunar base supporting perhaps a few dozen people. The feasibility of a permanent autonomous colony, of course, is at this time far from established.

General Technical Progress

It is hardly necessary to point out that there has been great progress since 1972, in space technology as well as in several other fields, all having some bearing on lunar base feasibility. Among the most obvious are remote sensing, rocket propulsion, telemetry, materials, communications, and computer technology (Lowman, 1979). The field of microelectronics in particular has advanced several generations since 1972, its most spectacular application being in the small but powerful computers of the mid 1980s. Even a cursory review of such technological progress would be out of place here; it is mentioned simply to point out that lunar base concepts that would have been feasible with the technology of 1970 will obviously be far more so with the technology of the 1990s. It is worthwhile citing here an historic analogy: the 1927 New York-to-Paris flight of Charles Lindbergh. What had been accomplished marginally and with great difficulty in 1919 by Alcock and Brown, who barely managed to fly from Newfoundland to Ireland, was technically fairly easy for Lindbergh eight years later because of the rapid improvement in aircraft and especially aircraft engines.

Reorientation of the Earth Sciences

From about 1965 to the early 1980s terrestrial geology and geophysics were dominated by plate tectonics. Plate tectonic theory is fundamentally concerned with oceanic crust and its margins (past and present), and its deserved acceptance led to a certain eclipsing of interest in studies of the continental crust. This imbalance is starting to correct itself, and there has been a strong revival of research in purely continental geology and geophysics. Particularly relevant in the context of this paper is the growing number of studies of early crustal genesis, focused on Precambrian shields and the lower continental crust as revealed by reflection profiling. The study of lunar geology has already had influence on concepts of continental crust formation (Lowman, 1976; Frey, 1980; Solomon, 1980) by demonstrating early global differentiation in even a small body. Many questions remain, however, about the early evolution of the Earth; and detailed exploration of the Moon, particularly the highlands, will certainly contribute to our understanding of the Earth.

Soviet Space Operations

Although the United States was the first to land men on the Moon, the Soviet Union has continued its long tradition of ambitious and persevering efforts in astronautics that began in the 1930s and is reportedly developing new launch vehicles far more capable than the American shuttle. Several Salyut space stations have given the Soviet Union unique experience with long-duration manned spaceflight. There has been considerable speculation about the next direction of Russian space activities; one informed view (Schmitt and Silver, 1984) is that the U.S.S.R. will attempt a manned Mars mission (though not necessarily a landing) in the early 1990s. However, a Russian lunar program seems equally possible, one indication being the recent announcement of a lunar polar orbiter to be launched late in this decade. The late Charles Sheldon, Library of Congress authority on Soviet space programs, wrote (1967) that colonization of the entire habitable solar system was the probable ultimate goal of Russian space ventures. Viewed in this context, the

Moon can be considered the most easily accessible and "habitable" extraterrestrial body. The justification for a lunar base will be as valid for the Soviet Union as for the United States, which is a factor that must be considered in planning post-space station American programs.

THE NEED FOR A LUNAR BASE: 1985

The desirability of a lunar base was repeatedly asserted during the Apollo Program. But the very success of that program and of other space activities raises the possibility that formerly proposed lunar base functions and objectives have already been accomplished or can be better accomplished in other ways. The question can be conveniently discussed under the following headings.

Unanswered Scientific Questions

The Apollo missions and several years' analysis of data and samples therefrom have obviously settled many pre-Apollo controversies about lunar geology (Taylor, 1982). However, as had been expected (Lowman, 1966a), an even greater number of new questions have been raised. Most general is the question of how the Moon formed; even 15 years after Apollo 11 there appear to be insuperable difficulties with all traditional theories. The composition and structure of the highland crust are only approximately known, and there is consequently strong disagreement on whether it formed by differentiation in a magma ocean or by partial melting and subsequent igneous activity. The mare basins are now agreed to be large basalt-flooded impact craters, but it is not known whether they were formed by a discrete late heavy bombardment or as part of a continuously declining flux. Furthermore, the source and nature of the basin-forming bodies, a key problem in solar system evolution, are quite unknown. The question of whether the Moon is still internally active is as yet unanswered. If it is, and if volatiles such as water or hydrogen are being emitted, the answer has extremely important implications for the ultimate habitability of the Moon.

Moon-based Astronomy

The spectacular success of Earth-orbiting astronomical satellites and the hoped-for success of the forthcoming Hubble Space Telescope raise the obvious question of whether the traditional goal of an observatory on the Moon is still worth pursuing. The answer appears to be an unequivocal "yes" for two main reasons. First, the existing and planned Earth-orbiting telescopes are heavily over-subscribed; there are far more important observations to be made than these instruments can accommodate. Second, the Moon has unique advantages over any instrument in Earth orbit, as summarized by Walker (1984). The farside offers shielding from terrestrial radiation, and in particular is probably the most radio-silent location in the accessible solar system. The Moon offers a slow rotation rate for continuous observations up to 14 days long, as well as a stable platform for interferometry. Finally, it may also be a good place for observation of neutrinos and gravity waves; an attempt was actually made during the Apollo 17 mission to detect gravity waves (Giganti *et al.*, 1972). In summary, it appears that Moon-based astronomy can still provide much of the scientific justification for a lunar base.

Stimulus to Space Technology

The Apollo Program proved to be, as had been hoped, an immense stimulus to technology in fields far removed from the obvious ones, such as rocket propulsion and inertial guidance (Lowman, 1975). However, it also gave the U.S. a commanding, though perhaps temporary, lead in space technology. Rocket engines using cryogenic hydrogen and oxygen, for example, are in routine use in the space shuttle, whereas the Soviet Union has not at this writing successfully used such stages at all for its launch vehicles.

A renewed lunar program focusing on a base would have unique value as a continuing technological stimulus. The reasoning is as follows. Lunar missions are in all qualitative essentials *short interplanetary missions* involving escape trajectories, deep space tracking, landings on another body, and return to Earth. They can thus provide most of the technological challenge of planetary missions but are quicker, cheaper, and safer. (Consider the outcome of the Apollo 13 Service Module explosion had it occurred halfway to Mars.) A long-term lunar base program would provide an operational matrix into which new developments in space technology could profitably be fitted as they became available: reusable orbital transfer vehicles (OTV), closed ecology life support systems, nuclear reactors for both stationary and propulsion use, advanced space suits, and the like. Furthermore, lunar capabilities can rather easily be reoriented for use in Earth orbit if necessary, just as the Apollo system was modified to produce Skylab. Systems designed solely for low Earth orbit, however, do not necessarily have the same two-way versatility. Deep-space capability implies Earth-orbit capability; the reverse is not true.

There is increasing concern that the United States may be losing its position as the preeminent space-faring nation. The Soviet challenge in space is still the most pressing one. A lunar base would have little military application, and, in fact, could provide a natural area for Soviet-American cooperation, but the technological stimulus it would provide should help prevent another "Sputnik surprise," which is never to be forgotten by anyone who lived through it.

Evaluation and Use of Lunar Resources

The only natural resource available in interplanetary space is energy, but all solid bodies offer materials as well. The Moon is the most convenient source of extraterrestrial materials, either for use on the Moon or in near-Earth space, with the possible exception of Earth-crossing asteroids. The use of lunar hydrogen and oxygen for rocket fuel was studied extensively during the 1960s (Lowman, 1966b), but the absence of indigenous water in all lunar rocks and soils so far sampled has discouraged this prospect. On the other hand, we now know that there are essentially inexhaustible amounts of aluminum, iron, silicon, titanium, magnesium, and oxygen (Phinney *et al.*, 1977), and the possibility of lunar hydrogen (as water or otherwise) has by no means been ruled out (Arnold, 1984). Furthermore, lunar materials, electromagnetically launched from the surface, are more competitive than they might appear because the launch energy per unit mass (*i.e.*, the kinetic energy) depends on the *square* of the escape velocity. To place a given mass of material in near-Earth space from the surface of the Earth would thus take very roughly 22 times as much energy as it would from the Moon. For this reason, there has been a strong revival of interest in lunar raw materials for construction or shielding of large structures in Earth orbit. A satellite solar power system, for example, will probably only be economically feasible with the use of lunar materials (Glaser, 1974; O'Neill, 1975). The most valuable lunar resource would be hydrogen, which would greatly facilitate the establishment of an autonomous permanent colony, and perhaps a lunar refueling facility for chemical or nuclear rockets.

A lunar base will be an essential step toward establishing the nature and value of the Moon's resources by permitting long-range surface traverses, extensive sampling, and *in situ* analysis of data and samples. There is now no doubt that the Moon has useful resources; the questions are exactly what and where they are and how they can be utilized.

Establishment of a U.S. Presence

The exploration of the Moon has been frequently compared with that of Antarctica, a valid comparison with political implications. Both areas are nominally protected by treaties against national appropriation. However, it is generally recognized that a major reason for maintaining Antarctic bases is to lend substance to the Antarctic Treaty by presenting a national presence on the continent, preventing territorial claims from being established by default. Joyner and Schmitt (1984) have proposed a governing body similar to INTELSAT for lunar development, but pending such an arrangement, a lunar base would be valuable to maintain an American or international presence on the Moon.

Long-range Uses of the Moon

Two completely independent findings in recent years point to the probability that the physical universe is more hostile to intelligent life than had previously been believed. First is the very strong evidence that some and perhaps many of the great biological extinctions on Earth have been caused by catastrophic impacts (Alvarez *et al.*, 1980). The interval between such impacts may be only a few tens of millions of years, and if they are non-periodic, most life on Earth could be destroyed at any time, as has happened many times in geological history. The second finding is a negative one: radio astronomers to date have detected no intelligent signals from elsewhere in the universe, and there is no believable evidence that extraterrestrials have ever visited the Earth. It has been argued by Tipler (1980) that because interstellar colonization by von Neumann machines or otherwise is possible, and would be rapid in terms of cosmic time, the absence of extraterrestrials (or their machines) in the solar system means that there are none anywhere. Tipler suggests that we may be the only intelligences in the universe.

An obvious way of explaining the apparent absence of other intelligent life (assuming there to be some) is that the communicative lifetime may be reduced, perhaps to zero, by natural and uncontrollable catastrophes such as asteroidal impacts, supernovae, externally initiated glaciations, or the like. In other words, intelligent life in any given system may not last very long, in terms of geologic time, if it even arises at all.

The point of this discussion has been made many times before, as by Bernal (1939), for example: "If human society, or whatever emerges from it, is to escape complete destruction by inevitable geological or cosmological cataclysms, some means of escape from the Earth must be found. The development of space navigation, however fanciful it may seem at present, is a necessary one for human survival. . ." The Moon furnishes a uniquely convenient first step for the "escape" called for, to say nothing of providing material for a completely space-borne civilization of the kind visualized by O'Neill (1975). In the most basic terms, then, a lunar base could be the beginning of a long-range program to ensure the survival of the species.

RECOMMENDED MEASURES FOR A REVIVED LUNAR PROGRAM

There are several immediate technical measures that could be undertaken without major funding commitments to lay the groundwork for a lunar base program. The most obvious of these, described in the SSEC (1983) report, is the Lunar Geoscience Orbiter. This spacecraft would carry out a comprehensive program of remote sensing from lunar polar orbit intended to produce global composition maps as well as indirect data on the Moon's structure. Although not strictly necessary for establishment of a minimal lunar base, the LGO would be an important measure to rebuild American scientific capability for renewed lunar exploration. A concurrent measure with the same general objective would be continued or increased funding for study of Apollo samples and meteorites and for Earth-based astronomical studies of the Moon.

Another step that could be taken almost immediately would be to design modules for the proposed space station for eventual adaptation to use on the surface of the Moon, as mentioned earlier. Since most space station proponents agree that one of its functions should be to launch lunar and planetary missions, the proposal for dual use models should be reasonably welcome. Furthermore, it is consistent with the SSEC report's emphasis on "spacecraft inheritance and commonality of systems" as means of reducing costs.

A much more challenging step, but one with long-range implications, would be development of a reusable manned orbital transfer vehicle, possibly nuclear-powered. Solid-core fission rockets of the sort static-fired in the 1960s, using hydrogen as propellant, can produce specific impulses in the 800 second range (Glasstone, 1965), roughly twice the performance of hydrogen-oxygen engines. Gas-core fission rockets can in principle produce much higher I_{sp} figures (Hunter, 1965). However, the apparent scarcity of hydrogen on the Moon tends to lessen the advantage of nuclear rockets for Earth-Moon transportation, and it is possible that advanced chemical rockets will be more practical. At this point, one can say only that a permanently manned and periodically resupplied lunar base would provide ample justification for some sort of reusable orbital transfer vehicle. Such an OTV could, in principle, be adapted for planetary missions when desirable, widening the range of options for American space programs of the 21st Century.

SUMMARY AND CONCLUSIONS

Most of the reasons advanced for establishing a lunar base as the successor to the Apollo missions are still valid 12 years after the last of those missions. The Moon is still a scientifically interesting object in itself; it is a valuable site for observing the universe, and it represents usable territory with potentially valuable raw materials. Establishment and operation of a lunar base or complex of bases would be a unique and continuing stimulus to space technology, because the Moon is actually a conveniently close planet that can be reached in less time than it took the first scheduled airliner to cross the Pacific Ocean (a 5½ day trip by China Clipper in 1935). Lunar missions are thus in all essentials short interplanetary missions, unlike low Earth orbit operations, and development of an Earth-Moon transportation system would to a large degree produce a planetary capability as well.

When these predictable benefits are added to the feasibility of an initial lunar base, demonstrated by the six missions that began at Tranquillity Base, it is clear that the Moon is a leading contender for the focus of American space activities of the near future.

Acknowledgments. *I thank M. B. Duke and W. W. Mendell for their encouragement in writing this paper, and J. P. Loftus, M. G. Goodhart, B. M. Christy, R. W. Johnson, and M. B. Duke for helpful reviews. I also wish to acknowledge my debt to my former colleagues on the Working Group for Extraterrestrial Resources, whose work in the late 1960s provided a foundation for much planning since then.*

REFERENCES

Alvarez L. W., Alvarez A., Asaro F., and Michel H. V. (1980) Extraterrestrial cause for the Cretaceous–Tertiary extinction. *Science, 208,* 1095–1108.

Anonymous (1964) *The Hearings Before the Subcommittee on Manned Spaceflight of the Committee on Science and Astronautics,* U. S. House of Representatives, 88th Congress, 2nd Session on H. R. 9641 (1965 FY NASA Authorization, Feb. 20, 1964). U. S. Government Printing Office, Washington, D. C.

Arnold J. R. (1979) Ice in the lunar polar regions. *J. Geophys. Res., 84,* 5659–5668.

Bernal J. D. (1939) *The Social Function of Science.* Routledge and Kegan Paul, Ltd., London. 482 pp.

Boeing (1963) *Initial Concept of Lunar-Exploration Systems for Apollo,* Vol. 1. The Boeing Company, Seattle. 168 pp.

Burgess E. (1957) *Satellites and Spaceflight.* Macmillan, New York. 159 pp.

Clarke A. C. (1951) *The Exploration of Space.* Harper, New York. 199 pp.

Duke M. B., Mendell W. W., and Roberts B. B. (1984) Toward a lunar base. *Aerosp. Amer., 22,* 70–73.

French B. M. (1977) *The Moon Book.* Penguin Books, New York. 287 pp.

Frey H. (1980) Crustal evolution of the early Earth: The role of major impacts. *Precambrian Res., 10,* 195–216.

Giganti J. J., Larson J. V., Richard J. P., and Weber J. (1972) Lunar surface gravimeter experiment. In *Apollo 17 Preliminary Science Report,* NASA SP-330, NASA, Washington, D.C. 4 pp.

Glaser P. E., Maynard O. E., Mackovcink J. Jr., and Ralph E. L. (1974) *Feasibility Study of a Satellite Solar Power Station.* NASA CR-2357. NASA, Washington, D.C. 199 pp.

Glasstone S. (1965) *Sourcebook on the Space Sciences.* Van Nostrand, New York. 937 pp.

Harper H. (1947) *Dawn of the Space Age.* Sampson Low, Marston and Co., London. 142 pp.

Hunter M. W. II (1965) *Thrust into Space*. Holt, Rinehart, and Winston, New York. 224 pp.

Joyner C. C. and Schmitt H. H. (1984) Lunar bases and extraterrestrial law: General legal principles and a particular regime proposal (abstract). In *Papers Presented to the Symposium on Lunar Bases and Space Activities of the 21st Century*, p. 40. NASA/Johnson Space Center, Houston.

Lowman P. D. Jr. (1966a) The scientific value of manned lunar exploration. *Ann. N.Y. Acad. Sci., 140*, 628–635.

Lowman P. D. Jr. (1966b) Lunar resources: their value in lunar and planetary exploration. *Astronaut. Acta, 12*, 284–393.

Lowman P. D. Jr. (1969) The Moon's resources. *Sci. J., 6*, 90–95.

Lowman P. D. Jr. (1975) The Apollo Program: Was it worth it? *World Resources: The Forensic Quarterly, 49*, 291–302.

Lowman P. D. Jr. (1976) Crustal evolution in silicate planets: Implications for the origin of continents. *J. Geol., 84*, 1–26.

Lowman P. D. Jr. (1979) Impact of technology on planetology. In *Impact of Technology on Geophysics*, pp. 110–121. National Academy of Sciences, Washington, D.C.

Lowman P. D. Jr. (1984) Lunar bases and post-Apollo lunar exploration: An annotated bibliography of federally funded American studies, 1960–82. Preprint, NASA/Goddard Space Flight Center, Greenbelt. 41 pp.

North American Rockwell (now Rockwell International) (1971a) *Lunar Base Synthesis Study*. SC 71-477, North American Rockwell, Downey. 103 pp.

North American Rockwell (now Rockwell International) (1971b) *Orbiting Lunar Station, Vol. 4*. SD 71-207. North American Rockwell, Downey. 634 pp.

O'Keefe J. A. (1985) The coming revolution in planetology. *EOS, Trans. AGU, 66*, 89–90.

O'Neill G. K. (1975) Space colonies and energy supply to the Earth. *Science, 190*, 943–947.

Phinney W. C., Criswell D., Drexler E., and Garmirian J. (1977) Lunar resources and their utilization. In *Space-based Manufacturing from Nonterrestrial Materials* (G. K. O'Neill, ed.) pp. 97–123. *Prog. Aeronaut. Astronaut., 57*. American Institute of Aeronautics and Astronautics, New York.

Schmitt H. H. and Silver L. T. (1984) Man and the planets (abstract). In *Papers Presented to the Symposium on Lunar Bases and Space Activities of the 21st Century*, p. 67. NASA/Johnson Space Centere, Houston.

Sheldon C. (1967) Review of the Soviet space program. *Report of the Committee on Science and Astronautics, U.S. House of Representatives, Ninetieth Congress*. U.S. Government Printing Office, Washington, D.C.

SSEC (Solar System Exploration Committee) (1983) *Planetary Exploration Through Year 2000*. U.S. Government Printing Office, Washington, D.C. 167 pp.

Solomon S. C. (1980) Differentiation of crusts and core of the terrestrial planets: Lessons for the early Earth? *Precambrian Res., 10*, 177–194.

Taylor S. R. (1982) *Planetary Science: A Lunar Perspective*. Lunar and Planetary Institute, Houston. 481 pp.

Tipler F. J. (1980) Extraterrestrial intelligent beings do not exist. *Q. J. Roy. Astron. Soc., 21*, 267–281.

von Braun W., Whipple F. L., and Ley W. (1953) *Conquest of the Moon*, Viking, New York. 126 pp.

Walker A. B. C. (1984) Astronomical observatories on the Moon (abstract). In *Papers Presented to the Symposium on Lunar Bases and Space Activities of the 21st Century*, p. 72. NASA/Johnson Space Center, Houston.

EVOLUTION OF CONCEPTS FOR LUNAR BASES

Stewart W. Johnson

The BDM Corporation, 1801 Randolph Road, S.E., Albuquerque, NM 87106

Ray S. Leonard

The Zia Company, P.O. Box 1539, Los Alamos, NM 87544

Man's aspiration to colonize the Moon has a very long history. In the second century, Lucian wrote of a trip to the Moon. In 1638, Wilkins mentioned lunar colonies after Kepler's early telescopic observations of the lunar surface. The literature continues almost uninterrupted with the imaginations of writers such as Cyrano de Bergerac and Jules Verne. Verne's fictional launch from Florida was a strange parallel to the Apollo program. Lunar base concepts include those of Rodney Johnson based on 1960s technology (from a May, 1969 *Science Journal* special issue featuring man on the Moon) and the associated Lunar Exploration Systems for Apollo (LESA) and Mission Modes and Systems Analysis (MIMOSA) studies. Might the strong drive of mankind to push outward and explore help mankind set a common goal? Could mankind unite in the future to bring about construction of a lunar base and exploration of the space beyond?

INTRODUCTION

This paper traces the evolution of lunar base concepts. The time span is at least 2000 years. The paper is divided into the following parts: early history to the 1870s (when Mars replaced the Moon as the foremost destination of cosmic voyagers), early 1900s through World War II, and the pre-Apollo and post-Apollo period.

Many old tales about trips to the Moon appear in the literature of most space-using nations. They provide a common bond that the creation of a lunar base would help focus. Effort involved in building such a base could be a beginning of the establishment of a shared culture. An innovative and properly planned initiative toward a lunar base could be a significant element in reducing international tensions (Johnson and Leonard, 1984). Through constructive interaction in the establishment of a lunar colony, nations may learn to work together in ways helpful in promoting world peace.

EARLY HISTORY

The earliest, but legendary, casualty of space was Icarus, the astronaut son of Daedalus, who flew too close to the sun. His wings of wax and feathers disintegrated, and he fell back into the sea.

In the second century A.D., a Syrian resident of Athens named Lucian gave an account of a lunar trip by fifty men aboard a ship in a book called *True History*. They were caught in an Atlantic storm that carried them toward the Moon (a large, circular illuminated island) where, after seven days of travel, they landed. They became involved in a war between creatures of the Moon and those from the Sun. After a truce in the war, Lucian sailed back to Earth for more adventures (Clarke, 1968).

One night in 1609 Galileo turned a telescope toward the Moon to make observations that showed the Moon not to be a smooth disk but rather a world of mountains and valleys. The impact of that discovery was soon noted in lunar base literature.

In the 17th Century, Johannes Kepler wrote fiction to disseminate his explanations of planetary motion. In *Somnium*, or *Astronomy of the Moon*, he avoided hypothesizing a transportation system. His travelers were carried to the Moon during sleep by spirits who crossed on a bridge of shadow during an eclipse. They found on the Moon creatures who spent their days in caves to avoid the sun's heat. Kepler emphasized the extremes of the lunar climate and the observed nature of its topography (Clarke, 1968).

Cyrano de Bergerac wrote *Voyages to the Moon and Sun* a few years after Kepler's *Somnium*. His hero was lifted with the morning dew in one form of launch. Another approach was to coat the body with beef marrow, which the waning Moon attracted. He also used solar energy to power jet propulsion, and in one instance tossed a magnetized ball upward so that his metal ship was pulled along. Francis Godwin, in his 1638 book, had as a hero an individual who traveled by goose-power. This Godwin character was encountered on the Moon in Cyrano de Bergerac's book.

The first discussion in print about a lunar colony is attributed to Bishop John Wilkins. In his 1638 book, *A Discourse Concerning a New World and Another Planet*, he voiced the opinions that man would one day learn to fly and would plant a colony on the Moon.

In 1783 two Frenchmen flew over Paris in a balloon and caused space flight to appear attainable. However, hopes were dashed as men encountered the difficulties of high altitude balloon flight; several would-be astronauts died from lack of oxygen. Man had begun to learn that the road to the Moon must be through the vacuum of space.

It was not until 1827 that a story stressed that the ship "must cross an airless void in bitter cold." *A Voyage to the Moon*, by Joseph Atterley (George Tucker) is also notable in that it is the first American entry, that its spacecraft was the first to be coated with an anti-gravity material, and that it includes scientific data for its own sake (Barron, 1981).

Jules Verne wrote *From the Earth to the Moon* in 1865. Verne combined science, speculation, and imagination in a tale of a launch from Florida with 400,000 pounds of guncotton in a 900-foot cannon. The escape velocity attained was 25,000 miles per hour, and air friction caused the vehicle to heat in the atmosphere; subsequently it was steered by rockets, and weightlessness was encountered in space.

THROUGH WORLD WAR II

Soon after 1900, H.G. Wells wrote *Men in the Moon*. His book contains elements of science fiction (*e.g.*, negative gravity and lunar inhabitants called Selenites). In Russia, Konstantin Tsiolkovsky, a serious space travel exponent, had dreams of spaceflight that he developed in many technical articles and in the novel *Outside the Earth*.

Literature relating to lunar exploration, as it is adjusted to increased technical knowledge, is found in the *Journal of the British Interplanetary Society*, which was first founded in 1933 and then revived after World War II ended. In the United States, Willy Ley conducted a space flight symposium at New York's Hayden Planetarium on October 12, 1951. This event led to a famous set of *Collier's* magazine articles by Werner von Braun that popularized the idea of a future in space.

After World War II, the lunar base literature began to focus on the technical requirements, the critical issues to be resolved, and the associated technology. Rocketry had come of age at Peenemünde. It became clear that even larger rockets with the capability of providing transportation to the Moon could be developed. The trend was away from the fanciful to the more technologically reasonable (but sometimes extreme) solutions to the challenge of establishing a lunar base.

A book first published in 1949 (Bonestell and Ley, 1958) described a trip to the Moon by a winged single-stage rocket ship with an atomic rocket motor. A single layer of shielding separated the crew's cabin in the nose from the rest of the ship. The ship on the ground was surrounded by concrete shielding open at the top. Takeoff was from a desolate location on a mountain top near the equator, above the densest layers of the Earth's atmosphere, with acceleration at 4 g for about 500 seconds. The crew encountered a difficult 4-day trip in weightlessness with very little to do. The pilot was relegated to monitoring instruments for the lunar landing, while preset controls did the work. Bonestell painted a concept of the beginnings of a lunar base adjacent to the winged rocket that provided weekly transport to Earth.

APPROACHES TO BUILDING LUNAR SHELTERS

Szilard (1959) noted that the building of a lunar base might occur in five to ten years. This article emphasized that the structural design of the base would be influenced by its purpose, the lunar conditions, and the construction materials used. The recommended lunar base structure was a completely prefabricated sphere supported by four or more adjustable legs attached to a circular edge beam. A crew of five could be accommodated in a closed biological circuit including plants. Beryllium alloys and annealed aluminosilicate glasses provided structural integrity and transparency. Shielding was provided by coolants circulated between inner and outer walls of the sphere.

Holbrook (1958) likened the establishment of a lunar base to a combination Normandy Invasion and Mt. Everest Expedition. Many ideas have been offered for the design of shelters for a lunar base. One suggestion was inflated shells rigidized by a plastic hardening compound (Helvey, 1960), with lunar soil radiation shielding and micrometeoroid bumpers installed over them. To accommodate the deep dust anticipated by some scientists, Rinehart (1959) proposed a cylinder that would lie on its side floating half-submerged in dust.

DiLeonardo (1962) investigated transporting structural tension members from Earth because they are relatively light, and mixing alkali transported from Earth with pumiceous lunar dust to make glass to be used for compression members. An alternative was to quarry lunar rock for use in shelter construction. The Army Lunar Construction and Mapping

Program (1960) proposed cylinders for shelter to be laid end-to-end in a trench under several feet of lunar soil. Blasting might be used to assist excavation, but placement and backfill would be accomplished by a multi-purpose construction vehicle.

DeNike and Zahn (1962) suggested tunneling and then lining the tunnel to retain pressurization. DiLeonardo (1962) proposed impacting a projectile on the Moon that would penetrate to a predetermined depth where an explosive charge would detonate to form a cavity. The cavity, when excavated and lined, would provide shelter. Reedy (1961) used existing crevasses or caverns for protection from the lunar environment. Johnson (1964) investigated criteria for the design of structures to be erected at a permanent lunar base with particular attention to effects of the lower (1/6 terrestrial) gravitational field.

PRE-APOLLO AND POST-APOLLO UNTIL THE PRESENT

The early 1960s was a time of intense study of lunar base concepts. The Apollo program was being pursued with enthusiasm and vigor. Planners sought to find a way to transition with Apollo technology from short-term manned visits to the Moon, to extended visits associated with semi-permanent lunar bases and colonies. Figure 1 shows one

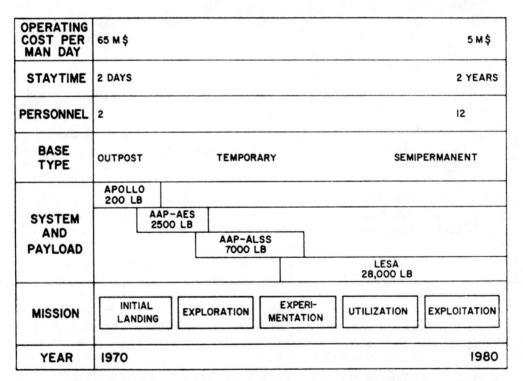

Figure 1. Projected evolutionary trends in lunar base development. Based on an orderly process of upgrading facilities, obtaining longer stay-times, and reducing operating costs. See Table 1 for acronyms (Johnson, 1966, page 368).

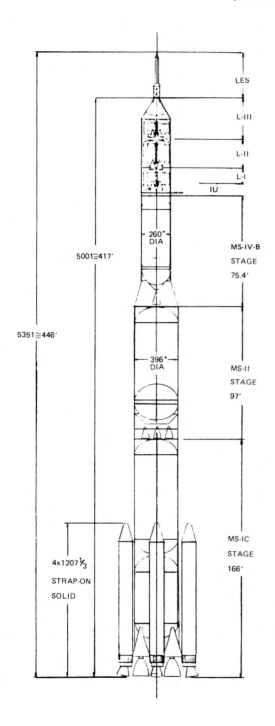

Figure 2. Extended Launch Vehicle (ELV) for six-man direct manned lunar landing. First stage five F-1 engines with 1.522 million pounds thrust each at liftoff plus four 120-inch 7-segment rocket motors. Overall system envisioned as having translunar payload capability of about 85,000 kg (unpublished NASA Report of 1966).

6 MEN — 6 MONTHS

Figure 3.   LESA Base Model 2 for accommodating six men for six months on the Moon. Involved is about 46,000 pounds of payload on the Moon, including a 10-kilowatt nuclear power unit, a 3765-lb. roving vehicle with an extended mobility module, and equipment to place lunar soil shielding (Boeing, 1963b).

example of the development flow that was considered feasible based on a systematic upgrading of the Saturn V transportation system to 111 percent and then 188 percent of basic. Figure 2 shows a view of an advanced Saturn V with strap-on solid rockets and a multi-stage stack extending to a height of approximately 446 feet. Some acronyms of the lunar basing investigations of the 1960s are presented in Table 1.

Figure 3 from the LESA Initial Concept (Boeing, 1963b) shows an artists' rendition of Base Model 2 on the surface of the Moon. This model envisioned six men on the Moon for six months. The LRV would have the capability for 3000 miles travel. LESA contemplated a building block approach based on several different modules. Two feet of soil for shielding is shown over the top and within expandable caissons on the sides of the shelter. A later study (Lockheed, 1965) portrayed the soil emplacement sequence (Fig. 4). Base Model 2 was considered a favorable approach following Apollo landings (Johnson, 1966). It would have required eight Saturn V launches per year to furnish 72 man-months per year on the lunar surface. Anticipated were one reconnaissance trip,

Table 1. Acronyms for Proposed Lunar Basing Options for Apollo

Program	Full Title
AAP–AES	Apollo Applications Program – Apollo Extension Systems
AAP–ALSS	Apollo Applications Program – Apollo Logistics Support System
LESA	Lunar Exploration System for Apollo
MIMOSA	Study of Mission Modes and System Analysis for Lunar Exploration
MOBEV	Lunar Surface Mobility Systems Comparison and Evolution Study
MOLAB	A concept for a long–range mobile laboratory that was associated with ALSS

one manning trip (assuming six men per vehicle), two rotation trips, and four logistics payloads. In contrast, the LESA Base Model 4 would have required 18 Saturn V launches per year and yielded 216 man–months per year.

Prime power for Base 2 was to come from a nuclear thermoelectric powerplant concept with a rating of 10 kilowatts. Standby power was envisioned to be furnished

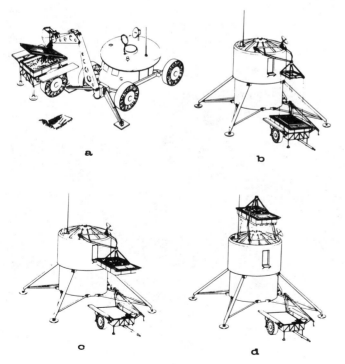

Figure 4. Soil operations sequence for shielding placement. (a) Soil excavation by backhoe. Soil box pulled by LRV trailer for soil transportation. (b) and (c) A-frame with hoist to lift box. (d) Release of soil (Lockheed, 1965).

by a fuel cell with a 15-kilowatt average load power rating. Larger bases (LESA 3 and 4) were to be served by a 100-kilowatt nuclear (Rankine cycle) SNAP 8 reactor. Nuclear reactor power plants were to be remotely placed and shielded with lunar material (Boeing, 1963a).

In 1969 the lunar colony concept (Fig. 5) was developed to encompass a lunar base buried under lunar soil (Johnson, 1969). The sequence of events thought to be possible was a landing in 1969, resources development in 1973–75, a scientific station in 1975, and the lunar colony by 1978.

Lockheed (1967) outlined three exploration programs, each emphasizing science accomplishments. Program III was to establish four temporary lunar bases, including two at Grimaldi Crater, one at the center of the farside, and another at the south pole. At Grimaldi, a base of two years duration was to perform astronomy, biology, and applied

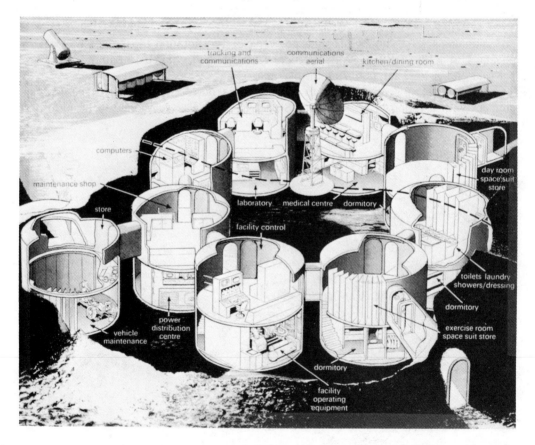

Figure 5.   Projected lunar colony including nuclear power plant at 760-m safety radius, future housing, hydroponic algae farm, sewage treatment, photosynthesis, housing and command structure, telescope, cryogenic storage, launch pads, guidance antennae, and communications (Johnson, 1969).

science experiments. The program would have required 63 Saturn V launches between 1971 and 1988.

Recent studies by Duke *et al.* (1984), Staehle (1983), and Carroll *et al.* (1983) emphasize the exploitation of materials found on the Moon. Usage of lunar materials such as oxygen is sought to offset high Earth–to–orbit transportation costs.

CONCLUSION

Science fiction was not only the domain of the scientist turned writer but also of the poet, as evidenced by the poet Alfred, Lord Tennyson, who expressed his view of the future world in "Locksley Hall:"

> For I dipt into the future, far as human eye could see,
> Saw the Vision of the world and all the wonder that would be;
> Saw the heavens fill with commerce, argosies of magic sails,
> Pilots of the purple twilight, dropping down with costly bales.

It is possible that in the future we will see spacecraft returning to space stations in Earth orbit with "costly bales" from bases on the Moon. Man has developed the capability of colonizing the Moon. Whether he will do so, and for what reasons and when, remain unanswered questions.

Today we live in a tremendously fragmented world. As Margaret Mead observed, mankind needs a mutually shared body of materials, events and efforts that gives everyone, whether from a technologically advanced society or a primitive one, a basis for communication and understanding (Mead, 1965). To date the unifying force is weak. The conquest of space, of which a lunar colony or observatory is but a small first step, could be the beginning of the development of important elements of a shared culture. Much of mythology deals with humanity's battle with nature. Science fiction deals with man's use of technology in this effort. Creation of a lunar base would be but one more step by man to bring to reality positive elements of mythology and science fiction. Perhaps the establishment of a lunar base could be an element in the dawning of a bright new age for mankind—the age of space exploration.

Acknowledgments. Stewart Johnson and Ray Leonard acknowledge the support of The BDM Corporation and The Zia Company in the preparation of this paper.

REFERENCES

Barron N. (editor) (1981) *Anatomy of Wonder, A Critical Guide to Science Fiction*, 2nd Edition. R. R. Bowker, New York. 724 pp.

Army Lunar Construction and Mapping Program (1960) 86th Congress, 2nd Session, House Report No. 1931, Washington DC, U.S. Government Printing Office.

Bonestell C. and Ley W. (1958) *The Conquest of Space*. Viking, New York. 160 pp.

Boeing (1963a) *Initial Concept of Lunar Exploration Systems for Apollo, Vol. II, Systems Considerations*. The Boeing Company, NASA CR-55764, Seattle. 322 pp.

Boeing (1963b) *Summary Digest, Initial Concept of Lunar Exploration Systems for Apollo*. The Boeing Company, NASA CR-55763, Seattle. 168 pp.

Carroll W. F., Steurer W. H., Frisbee R. H., and Jones R. M. (1983) Should we make products on the Moon? *Astronaut. Aeronaut., 21*, 80–85.

Clarke A. C. (1968) *Man and Space*. Time-Life Books, New York. 200 pp.

DeNike J. and Zahn S. (1962) Lunar Basing. *Aerosp. Eng., 21*, 8–14.

DiLeonardo G. (1962) Lunar Constructions. *ARS J, 32*, 973–975.

Duke M. B., Mendell W. W., and Roberts B. B. (1984) Toward a lunar base. *Aerosp. Am., 22*, 70–73.

Helvey T. C. (1960) *Moon Base: Technical and Psychological Aspects*. J. F. Rider, New York. 72 pp.

Holbrook R. D. (1958) *Lunar Base Planning Considerations*, Rand Paper P-1436. The Rand Corporation, Santa Monica, CA.

Johnson R. W. (1969) The lunar colony. *Sci. J., 5*, 82–88.

Johnson R. W. (1966) Planning and development of lunar bases. *Astronaut. Acta, 12*, 359–369.

Johnson S. W. (1964) Criteria for the Design of Structures for a Permanent Lunar Base. Ph.D. Dissertation, University of Illinois, Urbana. 177 pp.

Johnson S. W. and Leonard R. S. (1984) Lunar-based platforms for an astronomical observatory, in *Proceedings of the National Symposium and Workshop on Optical Platforms*, vol. 493 (C. L. Wyman, ed.), pp. 147–158. Bellingham, Washington.

Lockheed (1965) *Deployment Procedures, Lunar Exploration Systems for Apollo, Vol. 1, Summary*. Lockheed Missiles and Space Company, LMSC-665606-I, Sunnyvale, CA. 131 pp.

Lockheed (1967) *MIMOSA, Study of Mission Modes and System Analysis for Lunar Exploration, Final Report, Recommended Lunar Exploration Plan*, Vol. III. Lockheed Missiles and Space Company, LMSC-A847942, Sunnyvale, CA.

Mead M. (1965) The future as the basis for establishing a shared culture. In *Science and Culture: A Study of Cohesive and Disjunctive Forces* (G. Holton, ed.), pp. 163–183. Houghton Mifflin, Boston.

Reedy C. M. (1961) Engineering problems of lunar exploration. *Mil. Eng., 53*, 107–109.

Rinehart J. S. (1959) Basic criteria for Moon building. *J. Br. Interplanet. Soc., 17*, 126–129.

Staehle R. L. (1983) Finding paydirt on the Moon and asteroids. *Astronaut. Aeronaut., 21*, 44–49.

Szilard R. (1959) Structures for the Moon. *Civ. Eng., 29*, 46–49.

STRATEGIES FOR A PERMANENT LUNAR BASE

Michael B. Duke, Wendell W. Mendell, and Barney B. Roberts

NASA/Johnson Space Center, Houston, TX 77058

Planned activities at a manned lunar base can be categorized as supporting one or more of three possible objectives: scientific research, exploitation of lunar resources for use in building a space infrastructure, or attainment of self-sufficiency in the lunar environment as a first step in planetary habitation. Scenarios constructed around each of the three goals have many common elements, particularly in the early phases. The cost and the complexity of the base, as well as the structure of the Space Transportation System, are functions of the chosen long-term strategy. A real lunar base will manifest some combination of characteristics from these idealized end members.

A MOON IN AMERICA'S FUTURE

The Earth is unique in the solar system, not only for harboring life, but also for its relatively massive satellite. It is speculative that the two attributes are somehow related, but certainly the Earth's companion has left cultural and biological imprints on humanity. As cumulative application of the scientific method has increased our understanding and awareness of the physical universe, fascination with the habitability of the Moon has blossomed. As late as the last century, newspaper stories reported telescopic observations of the daily lives of lunar creatures. The manned lunar landings of the last decade have dispelled such romanticism forever but in turn have provided the technology and the information necessary to fulfill a greater dream—the transport of civilization beyond the confines of the Earth.

Cultural expansion is a recurring theme in human affairs. Motivations for exploration or conquest vary from resource limitations (Mongol invasions) to religion (Turkish probings of medieval Europe) to commerce (global circumnavigations of the Sixteenth and Seventeenth Centuries). American history especially is permeated by the doctrine of manifest destiny. The concept of the frontier has come to symbolize for Americans the exercise of individual freedom, which in collective expression leads to social renewal. Contemporary popular writings cater to this mythos by describing for an overpopulated and confused world the "high frontier" of space. So far, the promise of space has been a reality for a few and only a vicarious experience for most. However, humanity, and the United States particularly, stands today at the threshold of a truly new world—the Moon.

The promise of the Moon is not immediately evident from examination of the current American space program. However, the space shuttle and the proposed space station can be viewed as building blocks in a general purpose space transportation infrastructure (Fig. 1). To service geosynchronous orbit, an upper stage is needed in addition to the shuttle. If that upper stage is provided in the form of a reusable orbit-to-orbit transfer

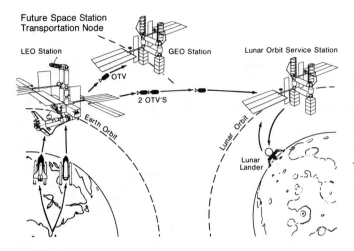

Future Space Station
Transportation Node

LEO Station

GEO Station

Lunar Orbit Service Station

OTV

2 OTV'S

Earth Orbit

Lunar Orbit

Lunar
Lander

Figure 1. The Space Transportation System of the future may service a station in geosynchronous orbit as well as a lunar base via a station in lunar space. The lift capacity of the Shuttle fleet may be augmented by an unmanned heavy lift vehicle, designed to ship fuel and consumables to space.

vehicle docked at the space station, the transportation system can be multipurpose. In particular, a rudimentary lunar transportation system then will exist because the propulsion requirements for attaining geosynchronous orbit and lunar orbit are essentially identical. A lunar landing vehicle is required to place payloads on the lunar surface, but its design can be a straightforward adaptation of the orbital transfer vehicle (OTV). The space station and the reusable OTV constitute a natural evolutionary path that, when achieved, will make accessible all near-Earth space including the Moon. This "enabling technology" is a NASA target for the mid 1990's.

When the requisite technology exists, the American political process inevitably will include lunar surface activities as a major space objective. In fact, some sort of declaration may well precede the actual establishment of the space station. It is therefore prudent to consider the nature of a permanent manned presence on the Moon and its potential impact on the evolution of the Space Transportation System (STS).

Although the lunar base program is one in which the United States can assert its leadership in space, it is inherently international in scope and should involve as much participation as possible from other countries. Opportunities for international cooperation exist in the planning stages, in the science and technology development, and in operations at the lunar base. A legal framework will be needed to guarantee that potentially profit-making ventures adequately consider the concerns of the international community.

USES OF THE MOON

A manned lunar base can be discussed in terms of three distinct functions. The first involves the scientific investigation of the Moon and its environment and the application of special properties of the Moon to research problems. The second produces the capability to utilize the materials of the Moon for beneficial purposes throughout the Earth-Moon system. The last, and perhaps the most intriguing, is to conduct research and development leading to a self-sufficient and self-supporting lunar base, the first extraterrestrial human colony. Although these activities take place on the Moon, the developed technology and the established capability will benefit society on Earth as well as the growing industrialization of near-Earth space.

Scientific Research

A lunar base will create new opportunities for investigating the Moon and its environment and for using the Moon as a platform for scientific investigations. Analogous to the function of McMurdo Base in Antarctica, the lunar base will provide logistical and supporting laboratory capability to rapidly expand knowledge of lunar geology, geophysics, environmental science, and resource potential through wide-ranging field investigations, sampling, and placement of instrumentation. Access to large, free vacuum volumes may enable new experimental facilities such as macroparticle accelerators. The firm, fixed platform will enable new astronomical interferometric measurements to be obtained (Fig. 2). The challenge of long-term, self-sufficient operations on the Moon can spur scientific and technological advances in materials science, bioprocessing, physics and chemistry based on lunar materials, and reprocessing systems. These concepts are explored by other papers in this volume.

Exploitation of Lunar Resources

It has been argued that major industrialization of space cannot occur without access to the resources of the Moon. Studies of immense projects such as solar power satellites

Figure 2. A radio telescope located on the farside of the Moon would be shielded from background noise generated by terrestrial sources. Although depicted here as a parabolic dish in a convenient crater, an initial lunar instrument may well be a phased array of dipole antennas.

have demonstrated that at a sufficiently large scale, it is reasonable to develop the resource potential of the Moon to offset the high Earth-to-orbit transportation costs (Hearth, 1976). The lower gravitational field of the Moon and the absence of an atmosphere that retards objects accelerated from the surface provides a potential 20- to 30-fold advantage for launching from the Moon instead of Earth. For example, at liftoff, about 1.5% of the space shuttle's mass is payload. Most of the mass is propellant. From the Moon, approximately 50% of the mass can be payload.

The commodity currently envisioned to be most in demand in Earth-Moon space over the next three decades is liquid oxygen, which makes up 6/7 of the mass of propellant utilized by cryogenic (hydrogen-oxygen) rockets, such as the Centaur or postulated OTV's. Although it would appear unlikely that an atmosphereless body is a source for oxygen, it is actually an abundant element on the Moon (Arnold and Duke, 1978). It must be extracted, however, from silicate and oxide minerals into its liquid form for use as a propellant. Several processes have been suggested (Criswell, 1980) for accomplishing this, including reduction of raw soil by fluorine (which is recovered) or reduction of iron-titanium oxide (ilmenite) by hydrogen (also recovered). Preliminary laboratory studies have verified the concepts behind some of these processes.

Systems studies (*e.g.*, Carroll *et al.*, 1983) show that oxygen production on the Moon could benefit STS in the early years of the next century, even if the hydrogen component of the propellant needed to be brought from Earth (Fig. 3–5). Finding concentrations of

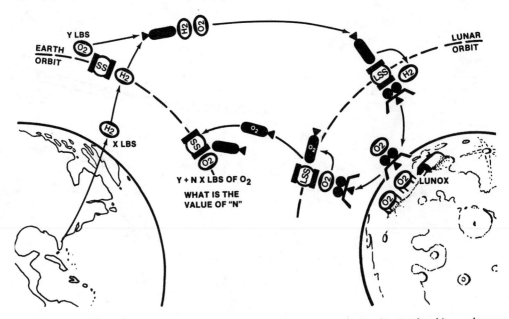

Figure 3. Liquid oxygen fuel (LOX), manufactured on the Moon and delivered to low-Earth orbit may become a profitable export for a lunar base. A critical parameter in analyses of the system is the mass payback ratio, defined as the ratio of the excess lunar LOX in LEO to the liquid hydrogen fuel delivered from Earth to LEO.

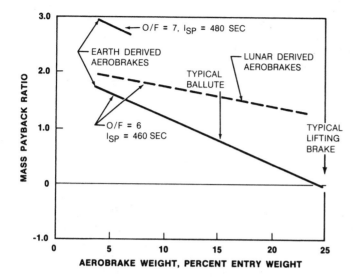

Figure 4. The mass payback ratio for lunar LOX delivered to LEO is sensitive to the design characteristics of the OTV used as a lunar freighter. The fractional mass of the OTV aerobrake and the oxidizer to fuel ratio are key parameters. Manufacture of aerobrakes on the Moon would enhance system performance.

water at the lunar poles (Arnold, 1979) or extracting the dispersed solar wind–derived hydrogen in the lunar regolith would greatly improve the economics of the transportation system.

Other commodities also could be produced. Metals, such as iron or titanium, can be extracted from the lunar soil or from specific rocks or minerals with differing degrees of difficulty. For example, small quantities of metal (primarily iron) from meteorites can be concentrated with a magnetic device from large amounts of lunar soil, or, with much larger energy inputs, titanium can be obtained from ilmenite. These products could find applications in large space structures. Lunar titania or alumina might be used to produce aerobrakes (heat shields) used in OTV's. In the long term, at relatively high levels of development, production of components for solar electric power generation in space (*e.g.,* solar power satellites) could be made feasible (Bock, 1979).

Lunar Autarky

A self-sufficient lunar base is a possible long-term objective that creates new challenges in planning and development. In the near term, emplacement of a controlled environment capsule on the Moon involves known technology. The initial concept for a lunar habitat module is simply an extension of the design experience from Apollo, Skylab, the space shuttle, and space station (Fig. 6). A different perspective is required to plan systems that can utilize the Moon's native materials and energy sources to produce a self-sufficient capability.

Most of the generic technologies for an advanced system are similar to those employed in general space operations (life support, power, thermal control, communications, logistics, and transportation, *etc.*), but they must be modified to utilize lunar materials for growth and extension. Ultimately, the desire to minimize or to eliminate the resupply link from

COST BENEFIT ANALYSIS:
LUNAR LOX FOR TRANSPORTATION SYSTEM

LLOX PLANT AMORTIZED OVER 10 YEARS
OPERATIONS COST = OC = 0, 100 $/LB   (PC, OC)
PLANT COST = PC = 3, 5, 10$B

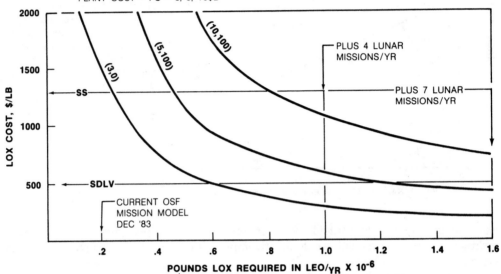

Figure 5. *A simple cost-benefit analysis assumes that a lunar oxygen production facility has its capital costs amortized solely by "profits" on delivery of LOX to LEO. While lunar oxygen is competitive with shuttle delivery in all cases, introduction of a cost-efficient heavy lift vehicle reduces the advantage under more conservative cost estimates for the lunar operation. If costs of lunar LOX are shared with other activities, the advantage is restored.*

Earth requires a host of applications, new to the space program, carried to new levels of system reliability. Exploration of technologies such as lunar metallurgy, ceramics, manufacturing processes, power systems, and others, will reveal whether autarky is a realistic objective and can prepare the way for achieving it at an operational base. Perhaps this is the most compelling rationale for a lunar base program, as it promises eventual self-sufficiency elsewhere in the solar system.

PHASED EVOLUTION OF A LUNAR BASE

We loosely define three scenarios, each based on one of the long-term rationales described above: scientific research, production, and self-sufficiency (Tables 1–3). Each scenario passes through several phases, some of which are common to the other scenarios. The distinction among the three views lies with the culminating phase of each.

Precursor Exploration. Because the scientific data base is incomplete, particularly in the polar regions, the first step in Phase I is global mapping of the Moon, both with relatively high resolution imagery and with remote-sensing measurements to determine

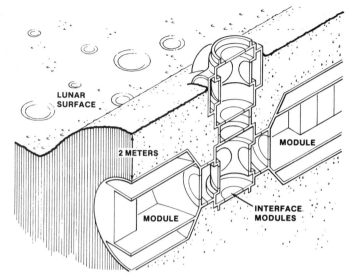

Figure 6. The first lunar base habitats and laboratories could be space station modules, buried in the lunar regolith for protection from solar flare radiation. Interface modules not only interconnect the buried structures but also can be stacked to create exits to the surface.

the chemical variability. This task can be accomplished with an unmanned satellite, a Lunar Geochemical Orbiter, or LGO (Minear *et al.*, 1977), which is a proposed mission in NASA's planetary program and could be flown in the 1990–1992 time frame. The LGO is in the Planetary Observer mission class, a low-cost approach to planetary exploration recommended by the report of the Solar System Exploration Committee (1983). Secondly, Phase I should include research on technologies necessary to exploit lunar resources. Technology development in resource problems on Earth is typically a long lead time process. At the conclusion of Phase I, the initial site for a base will have been defined and planned activities understood in some detail. Concurrently with this preliminary phase in the lunar program, development of a space station and an OTV capable of supporting a lunar base would be carried out in NASA's STS program.

Research Outpost. At Phase II, an initial surface facility would establish limited research capability for science, materials processing, or lunar surface operations. Depending on the long-term objectives of the lunar base program, the detailed studies and the experimental plans start to diverge at this phase for the different scenarios. A focus on lunar science and astronomy would result in local geological exploration, the establishment of a small astronomical observatory, and emplacement of automated instruments. If production were to be the focus, a pilot plant for lunar oxygen extraction could be set up instead, and study of the fabrication of aerobrakes from lunar material could be initiated. If the program goal pointed to achieving self-sufficiency, the emphasis at this stage could be on agricultural experiments utilizing lunar soil as substrate and recycling water, oxygen, and carbon dioxide.

To accomplish Phase II in any of the scenarios, the STS must have the capability of landing and taking off from the Moon, transporting manned capsules (about 10,000 kg) to and from the lunar surface, and delivering payloads of about 20,000 kg to the

Table 1.   Lunar Base Growth Phases: Science Base Scenario

A growing capability to do lunar science and to use the Moon as a research base for other disciplines, using lunar resources to a limited extent to support operations.

Phase I: Preparatory exploration
- Lunar orbiter explorer and mapper
- Instrument and experiment definition
- Site selection
- Automated site preparation

Phase II: Research outpost
- Minimum base, temporarily occupied, totally resupplied from Earth
- Small telescope/Geoscience module
- Short range science sorties
- Instrument package emplacement

Phase III: Operational base
- Permanently occupied facility
- Consumable production/Recycling pilot plant
- Longer range science sorties
- Geoscience/Biomedical laboratory
- Experimental lunar radiotelescope
- Extended surface science experiment packages

Phase IV: Advanced base
- Advanced consumable production
- Satellite outposts
- Advanced geoscience laboratory
- Plant research laboratory
- Advanced astronomical observatory
- Long-range surface exploration

---

lunar surface. This involves delivering approximately 40,000 kg into lunar orbit using OTV's. The requirement for storage of the return vehicle on the Moon for extended periods (14 days to 3 months) may require new high-performance, storable propellant systems at this phase of development.

Permanent Occupancy. At Phase III, permanent occupancy is the objective. The surface infrastructure would include greater access to power, better mobility in and away from the base, and more diversified research capability. Still, depending on the long-term objectives, the nature of the base can vary. A science base might emphasize long-range traverses for planetological studies or extension of observational capability with larger telescopes. A production base will incorporate highly automated systems to produce and transfer liquid oxygen for use in the transportation system. Advanced research for a self-sufficient base would be making the first extensions of the base utilizing indigenous materials. The production and the self-sufficiency scenarios require a small cousin to

Table 2. Lunar Base Growth Phases Production Base Scenario.

A lunar base that is intended to develop one or more products for commercial use. Manned activity may be continuous, but a high degree of automation is expected.

Phase I: Preparatory exploration
  - Lunar orbiter explorer and mapper
  - Lunar pilot plant definition
  - Site selection
  - Automated site preparation

Phase II: Research outpost
  - Minimum base, temporarily occupied, totally resupplied from Earth
  - Surface mining pilot operation
  - Lunar oxygen pilot plant
  - Lunar materials utilization research module

Phase III: Operational base
  - Permanently occupied facility
  - Expanded mining facility
  - Consumables supplied locally
  - Oxygen production plant
  - Lunar materials processing pilot plant(s)

Phase IV: Advanced base
  - Large scale oxygen production
  - Ceramics/Metals production facility
  - Locally derived consumables for industrial use
  - Industrial research facility

the Earth-orbit space station in lunar space (lunar orbit or an Earth-Moon libration point) to provide for transfer, refueling, and maintenance of the lunar lander and the OTV's.

Advanced Base. The advanced base, Phase IV, is even more specialized. Depending on the long-term plan, it produces more types or a greater range of scientific investigations, adds products to the growing lunar industrial base, or enters a phase of significant expansion of capabilities using lunar materials as the majority of the feedstock. This is the terminal phase for the science and production scenarios. Future growth may occur by enlarging the number of experiments or products produced on the Moon, but a self-sustaining capability is not included. The production base might even develop toward a highly automated state where permanent occupancy was unnecessary. For the production and independence scenarios, the base should begin paying its own operational costs. In the self-sufficiency scenario, research and development of pilot plants aimed at a broad range of indigenous lunar technologies would be pursued. The final phase of the self-sufficient scenario is a truly autarkic settlement, a lunar colony, in which the link to Earth can be discretionary.

Table 3. Lunar Growth Phases: Lunar Self-sufficiency Research Base Scenario

A lunar base that grows in its capacity to support itself and expand its capabilities utilizing the indigenous resources of the Moon, with the ultimate objective of becoming independent of Earth.

Phase I: Preparatory exploration
- Lunar orbiter explorer and mapper
- Process definition
- Site selection
- Automated site preparation

Phase II: Research outpost
- Minimum base, temporarily occupied, totally resupplied from Earth
- Surface mining pilot operation
- Lunar oxygen production pilot plant
- Closed systems research module

Phase III: Operational base
- Permanently occupied facility
- Expanded mining facility
- Lunar agriculture research laboratory
- Lunar materials processing pilot plant(s)

Phase IV: Advanced base
- Lunar ecology research laboratory
- Lunar power station–90% lunar materials–derived
- Agricultural production pilot plant
- Lunar manufacturing facility
- Oxygen production plant
- Lunar volatile extraction pilot plant

Phase V: Self-sufficient colony
- Full-scale production of exportable oxygen
- Volatile production for agriculture, Moon–orbit transportation
- Closed ecological life support system
- Lunar manufacturing facility: tools, containment systems, fabricated assemblies, *etc.*
- Lunar power station–100% lunar materials–derived
- Expanding population base

EVOLUTION OF THE PROGRAM

Figure 7 ties the possible development of a lunar base to the growth of lunar resource support of the transportation system. Initially, the base is totally dependent on terrestrial supply where 7 kg in low–Earth orbit is required to place 1 kg on the lunar surface. With the introduction of lunar oxygen first into near-Moon operations and then into the return leg of the transportation system, the slope of the curve changes from 7:1 to 3.5:1.

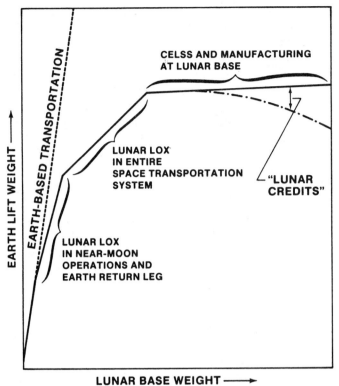

Figure 7. Initially, almost 7 kg must be lifted into LEO for every kg landed on the Moon. As lunar oxygen is introduced into the transportation system, the ratio improves as a unit mass goes from Earth to Moon with only little overhead in the system. In a Phase IV advanced base, the growth of lunar surface infrastructure becomes only weakly dependent on imports from Earth. A favorable balance of trade is ultimately conceivable.

As the lunar manufacturing capability increases to the point where aerobrakes can be manufactured, the slope decreases to something slightly greater than 1:1. Further growth of lunar capability allows expansion of base mass to be more or less independent of the quantity of imported terrestrial mass. At the point of self-sufficiency, only trace minerals and crew changeout are chargeable weights to lunar operations; the slope of the curve in Fig. 7 is essentially flat.

Another consideration in the growth of lunar activities is the economic "balance of trade" between Earth orbit and the lunar surface. The value of lunar products may support lunar operations before a true mass balance is achieved. It is difficult to calculate the economic value of lunar oxygen and other products in low-Earth orbit. However these "lunar credits" are shown qualitatively in Fig. 7 at the point where a closed ecological life support system (CELSS) and a significant manufacturing capability are available. The slope of the "credits" line will be a function of many things, such as the amount of oxygen required to support non-lunar activities, the value and quantity of lunar resources required in low-Earth orbit, and the more intangible value of science and research enabled by the lunar base. Finally, the dashed line of constant slope indicates the continued total dependency that would exist if these technologies are not pursued on the Moon, that is, if a self-sufficiency element is not included in the lunar base program.

The real lunar base will evolve as some combination of the above scenarios. Determination of the right mix requires research, development, and debate. Even if a program is started now, several years should be devoted to study of the detailed lunar base scenario. The time is available because the development of the space transportation infrastructure and the completion of the orbital science survey will take 7–10 years. Proper preparation will make it possible to decide on a specific lunar base design in the early 1990's. That time frame is consistent with the development of the infrastructure that will enable the lunar base program to be carried out to its full potential. The first manned landings could occur early in the first decade of the next century; permanent occupancy could be achieved by the year 2007, the fiftieth anniversary of the Space Age.

There are potential technological problems that may slow the development of the lunar base, and at each phase there will be serious questions as to whether to proceed and how and when to proceed. A commitment need not be made now to the whole plan. Nevertheless, the long-term objective is one of immense significance in human history and should not be casually discarded. It is inevitable that humankind will settle the Moon and other bodies in the Solar System. We live in a generation that has already taken very significant steps along that path. With careful planning, we can nurture the capability to move from the planet, to provide benefits to the Earth, and to satisfy humanity's spirit of adventure.

REFERENCES

Arnold James R. (1979) Ice in the lunar polar regions. *J. Geophys. Res., 84,* 5659–5668.

Arnold James R. and Duke Michael B. (1978) *Summer Workshop on Near-Earth Resources.* NASA CP-2031, NASA, Washington, DC. 107 pp.

Bock Edward H. (1979) *Lunar Resources Utilization for Space Construction.* Final Report for Contract NAS9-15560, General Dynamics Convair Div., Advanced Space Programs, San Diego, CA.

Carroll W. F., Steurer W. H., Frisbee R. H., and Jones R. M. (1983) Extraterrestrial materials—Their role in future space operations. *Astronaut. Aeronaut., 21,* 80–85.

Criswell David R. (1980) *Extraterrestrial materials processing and construction.* Final Report Contr. NSR 09-051-001. Lunar and Planetary Institute, Houston, TX.

Hearth Donald P. (1976) *Outlook for Space.* NASA SP-386, NASA, Washington, DC. 237 pp.

Minear J. W., Hubbard N., Johnson T. V., and Clarke V. C. Jr. (1977) *Mission Summary for Lunar Polar Orbiter.* JPL Document 660-41, Rev. A, Jet Propulsion Laboratory, California Institute of Technology, Pasadena, CA. 36 pp.

Solar System Exploration Committee (1983) *Planetary Exploration Through Year 2000: A Core Program.* U.S. Government Printing Office, Washington, DC. 167 pp.

PRELIMINARY DESIGN OF A PERMANENTLY MANNED LUNAR SURFACE RESEARCH BASE

Stephen J. Hoffman and John C. Niehoff

Science Applications International Corporation, 1701 E. Woodfield Road, Suite 819, Schaumburg, IL 60195

A brief study has been performed to assess the advantages and/or disadvantages of a lunar surface space base for civilian research and development. The suitability of undertaking scientific investigations in the diverse fields of astronomy, high-energy physics, selenology, planetary exploration, Earth sciences, and life sciences was considered. A lunar base was conceived to conduct the identified science, along with transportation requirements to establish and support continued operations at the base. A rough order of magnitude (ROM) estimate of the cost to deploy and operate the lunar base for a period of three years was made. Starting with the space station will assure performance of important low-Earth-orbit science and would also set in place certain elements of the transportation infrastructure found necessary to deploy and sustain a lunar base at a reasonable cost level. It is suggested that a lunar base be given serious consideration as a longer term goal of space policy, capable of providing important direction to the space station initiative.

INTRODUCTION

The purpose of this study is to define a concept for a permanently manned research base on the lunar surface and a manned reconnaissance mission that would precede base construction. A key study assumption limits the technology used for these two missions to that which is currently available, such as the space shuttle and spacelab, or to technology that will be available in the near term, such as space station and aerobraking. The remainder of this paper highlights the details of the two missions along with the science experiments to be carried out during each phase. The transportation network needed to accomplish these missions is also presented. A more complete discussion of these topics can be found in the references cited at the end of this paper.

TRANSPORTATION SYSTEM

Three major components of a transportation network were assumed to be in existence before the reconnaissance mission began. These elements included the space shuttle, a low-Earth-orbit staging point (presumably the space station), and a high performance space-based OTV. Only the OTV element required further definition for the purposes of this study. The OTVs used here are configurations proposed by NASA/JSC (Lineberry, personal communication, 1983) for use in Earth-orbital applications and for high-energy interplanetary missions. Each OTV has a maximum 27,216 kg (60,000 lb) of usable propellant and an I_{sp} of 460 seconds (LH_2-LO_2). A thrust level of 147,000 N (33,000 lbf) was assumed, which is representative of two RL-10 engines. The gravity losses corresponding to the

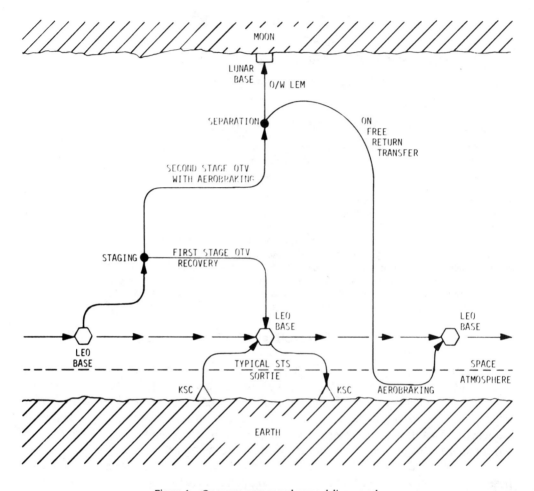

Figure 1. One way, unmanned cargo delivery mode.

resultant burn times are approximately 3%. As shown in Figs. 1 and 2, the first stage returns to LEO propulsively while the second stage returns using an aero-assisted maneuver. These two figures also show the two methods that would be used to deliver cargo and personnel to the Moon. For unmanned cargo sorties, a mission-unique, expendable lander is placed on an intercept course for the Moon and lands on the surface using its own propulsion system. After separation from the lander, the second stage of the OTV is retargeted for a free return to near-Earth space. For manned missions, the second stage will rendezvous in low lunar orbit with a prepositioned lunar excursion module (LEM) where the crew and LEM propellant will be transferred for descent to the surface. Crew retrieval will be accomplished by reversing this procedure.

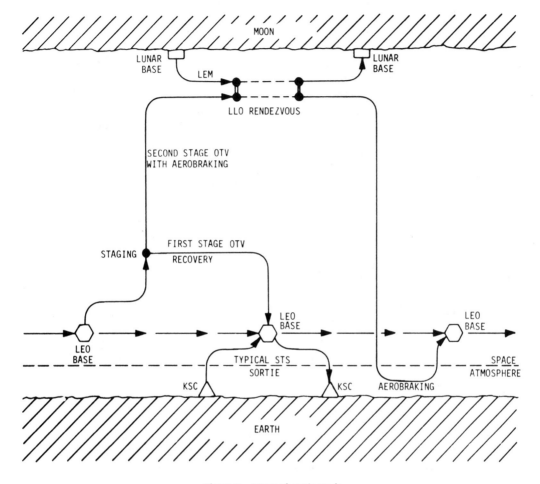

Figure 2. Manned sortie mode.

SITE RECONNAISSANCE AND SELECTION

An exploration team consisting of four individuals would spend up to 30 days exploring a region 50 km in radius that has been previously selected from remotely obtained data. Two surface vehicles would be used with two crew members per vehicle to carry out the exploration (Science Applications International Corp., 1984a). For safety reasons these vehicles would operate in tandem rather than individually. The two surface exploration vehicles would each consist of a rover and a trailer, the latter containing crew quarters and experiment facilities. The rovers would have the capability to move moderate amounts of lunar soil in order to expose subsurface strata. With the exception of the science experiments, both rovers and trailers would be identical and capable of supporting the entire crew under emergency conditions. A mass budget of 2400 kg has been assumed for the instruments needed to ascertain which site is best suited for the base. These instruments would focus on the local composition, seismic characteristics, and stratigraphic make-up of each candidate site. Preliminary data analysis would be conducted on board the two trailers with more detailed analyses to be carried out upon return to Earth. These analyses would support the final site selection for the permanent base.

This segment of the base deployment mission is anticipated to require 60–90 days from first shuttle launch to crew recovery and would require a total of 12 shuttle launches. The shuttle launches would lift the two rover/trailer combinations, their lander, the LEM, and all necessary propellant into low-Earth orbit. Four sorties by the two-stage OTV would then be needed to complete the reconnaissance. The first two sorties would deploy the rovers and trailers to the surface and place the unfueled LEM in low orbit. The remaining two sorties would be used to deliver and subsequently recover the surface team. The LEM and all surface equipment would remain for use by the research base personnel.

OPERATIONAL BASE

Figure 3 shows the proposed configuration for the initial operational base (Science Applications International Corp., 1984b). Each of the three main modules would be buried

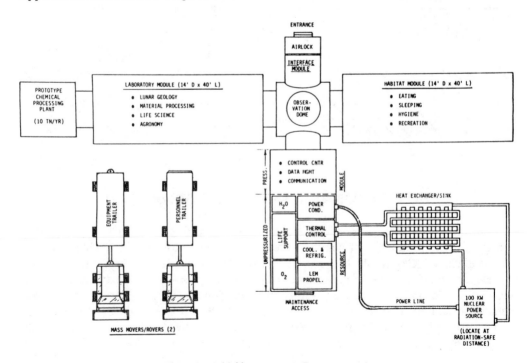

Figure 3.   Initial base concept (7-person crew).

to provide both thermal and radiation protection for the crew. The rovers left by the reconnaissance crew would be used to position these modules and cover them with soil. The three main modules are connected by the airlock/interface module and are supplied with power from a 100-kW nuclear power source. This power system has been oversized for the base configuration shown here but provides for future growth of the facility. Table 1 shows a possible strategy for the deployment and initial operation of this base.

A seven-member crew, consisting of six scientist/technicians and a LEM pilot, would operate the base, each serving a four-month tour of duty. Half of the crew would be replaced every two months to maintain a core of experienced crewmembers at the station at all times. In addition, two unmanned logistical resupply missions would be flown each year to replace base consumables. This translates into an annual requirement of 18 shuttle flights and eight OTV sorties.

A diverse range of experimental investigations would be carried out at this base. As can be seen in Fig. 3, a chemical processing plant has been included in the initial configuration. This facility will be used to determine the extent to which usable resources can be extracted from lunar soils. Extensive selenology experiments can be carried out using the rovers and trailers from the reconnaissance mission. The trailer facilities can be enhanced using equipment brought from Earth and excursions in these units can be used to place automated sensing packages at sites far removed from the base. Radio astronomy and VLBI experiments in particular can be carried out from a base of this scale. Finally, life science experiments in health maintenance and food production could be conducted. As operational experience is gained with the base, each of the experiments cited above can be expanded and enhanced. Experiments in high-energy physics, gravity

Table 1. Suggested Strategy for Deployment and Initial Operation of a Lunar Science Base

No.	Mission Description	Personnel		LEM Status*		No. of People On the Moon*
		Going	Returning	In LLO	On Surface	
1	Deploy interface module and power plant	0	0	1	0	0
2	Deploy laboratory module	0	0	1	0	0
3	Deploy habitat module and processing plant	0	0	1	0	0
4	Deploy resources module	0	0	1	0	0
5	Deploy second LEM	0	0	2	0	0
6	Send first construction team	4	0	1	1	4
7	Send second construction team	3	0	1	1	7
8	Switch 1st construction team and 1st station team	4	4	1	1	7
9	Switch 2nd construction team and 2nd station team	3	3	1	1	7

*At completion of mission

waves, and space plasmas can also be added in such a way as to take advantage of the unique conditions found on the lunar surface. These experiments are complementary to those already being conducted at the space station (Science Applications International Corp., 1984b).

SUMMARY

This study has highlighted two missions designed to establish a permanent research facility on the lunar surface. A manned reconnaissance mission was believed to be necessary to conduct final siting of the base prior to its construction. This first mission is entirely complementary to the later operational base since all equipment developed for reconnaissance would be used at the permanent facility. Table 2 shows a cost breakdown

Table 2.   Manned Lunar Surface Base Cost (Present Year $B)

	Reconnaissance	Surface Base	Total
Surface modules			10.2
Shelter	0.1		
Trailer (2)	1.5		
Rover (2)	1.4		
Permanent modules (4)		5.8	
Chemical processing plant		0.9	
Nuclear power plant		0.5	
Propulsion stages			16.4
Lunar excursion module	2.7	1.4	
Lunar logistics lander	2.7	3.6	
OTVs	0.8	3.0	
OTV crew module	1.6	0.6	
On-orbit assembly and test	1.0		1.0
STS			10.0
Reconnaissance (12 launches)	1.3		
Base deployment (25 launches)		2.7	
Base operations (18 launches per year)		6.0	
Operations*			14.5
Mission control center	0.5	2.1	
Training/operations development/ management	2.0	5.0	
Mission (orbital and flight operations)	0.7	3.1	
Logistics	0.2	0.9	
Totals	16.5	35.6*	52.1

*Includes 3^y ops at surface base

for both of these missions, assuming the use of existing or near-term technology. It should be noted that the cost of the surface base includes three years of operations. The base could and probably would function for a much longer time than this. The total cost of approximately $52 billion would only be slightly less without the initial reconnaissance mission. For comparison, the cost of the Apollo Program in equivalent dollars is $75 billion. Both the concept and the cost suggest that this facility is programmatically feasible and would make a worthwhile national or international goal in the post space station era.

REFERENCES

Science Applications International Corp. (1984a) *Manned Lunar, Asteroid, and Mars Missions; Visions of Space Flight: Circa 2001.* Report No. SAIC/84-1448. SAIC, Schaumburg, IL. 82 pp.

Science Applications International Corp. (1984b) *A Manned Lunar Base: An Alternative to Space Station Science?* Report No. SAIC/84-1502. SAIC, Schaumburg, IL. 31 pp.

MERITS OF A LUNAR POLAR BASE LOCATION

James D. Burke

Jet Propulsion Laboratory, California Institute of Technology, Pasadena, California 91109

Because the Moon's spin axis is inclined only 1½ degrees off normal to the plane of the ecliptic, there are no seasons; there are regions near the poles in permanent shadow and, possibly, regions where the Sun never fully sets. The permanent-shadow regions theoretically should be very cold and, with continuous sunlight nearby, are inviting sites for thermodynamic systems. If located near a pole, a lunar base can have solar electric power and piped-in solar illumination continuously available except during occasional solar eclipses. Habitat and farm conditions in underground facilities are easily kept constant. Access from lunar orbit is good because a polar orbiter would pass overhead about every two hours. Waste heat rejection should be much easier than in the widely varying thermal environment of lower latitudes. Polar cold-trapped volatiles may be available. Even if useful volatiles are not naturally present, the cold regions provide convenient storage sites for volatile products of material processing in the base—important when transport logistics are considered. The polar sites offer excellent astronomical opportunities: half of the sky is continuously visible from each pole, and cryogenic instruments can readily be operated there. A geochemical and topographical survey from polar orbit is the next logical step in determining the real merits of a polar base site on the Moon.

INTRODUCTION

The Moon's polar regions offer some unique advantages for human living. Because the Moon's polar axis is inclined only 1½ degrees off normal to the plane of the ecliptic (Fig. 1), there are no seasons on the Moon. Near both poles, portions of the surface such as crater bottoms are permanently dark, and there may be places—this is not a certainty—where some part of the solar disk is always above the horizon. A solar power plant built upon such a "mountain of perpetual light" would supply continuous power except during brief solar eclipses when the Earth would shut off the sunlight.

This circumstance, unique to polar sites, is sufficient justification for considering all aspects, both positive and negative, of locating bases at the poles. Previous studies of this and related subjects are reported in Watson *et al.* (1961), Gary *et al.* (1965), Arnold (1979), Culbertson (1961), Dalton and Hohmann (1972), Green (1978), and Burke (1977, 1978).

WHAT IS KNOWN AND UNKNOWN ABOUT THE LUNAR POLES

Figures 2 and 3 are the best available overhead photos of the lunar polar regions, obtained by Lunar Orbiter IV in 1967. The ground resolution in these images is of the order of 100 m and, of course, nothing is seen in the large shadowed areas. Though the low sun angle exaggerates the roughness of the surface, it is true that the topography at both poles is fairly rugged for the Moon; the surface morphology is that of the ancient, heavily-cratered highlands. Geologic maps of the polar regions have been published based

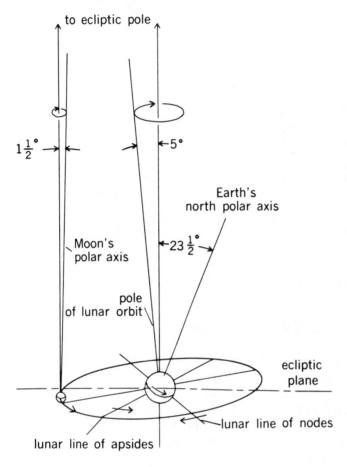

to ecliptic pole

$1\frac{1}{2}°$

←5°

Earth's
north polar axis

Moon's
polar axis

←$23\frac{1}{2}°$→

pole
of lunar orbit

ecliptic
plane

lunar line of nodes

lunar line of apsides

Figure 1. An illustration of the motions of Earth and Moon with reference to the pole of the ecliptic. While Earth's polar axis is inclined 23½ degrees and precesses with a period of about 25,000 years, giving us seasons and the progression of signs of the Zodiac, the Moon's polar axis is inclined only 1½ degrees. Thus, despite the five-degree inclination of the lunar orbit plane and the eighteen-year precession of the lunar polar axis and orbit plane (as discovered in the 18th century by Cassini), sunlight is always nearly horizontal at the lunar poles.

on the lunar orbiter photos (Lucchitta, 1978; Wilhelms *et al.*, 1979). Geochemical mapping, however, awaits the flight of a remote–sensing polar orbiter.

By analogy with the data obtained at lower latitudes by Apollo and other missions, we have reason to believe that lunar highland rocks and soils, rather than mare types, will predominate in the polar regions, with the south pole having the more strongly highland character. Thus, the industrial resources peculiar to the maria may not be abundant near the poles. Detailed surface properties will, however, remain unknown pending orbital and surface exploration. At lower latitudes there is a twilight haze, detected by both U.S. and Soviet spacecraft, that is believed to be due to small particles moving in electrostatic suspension within a few meters of the surface (De and Criswell, 1977). This particle haze and also some of the gas clouds detected by ALSEP instruments are associated with terminator passage (sunrise and sunset). Since the terminator is always present in the polar regions, the local environment due to these particle and gas effects may be different

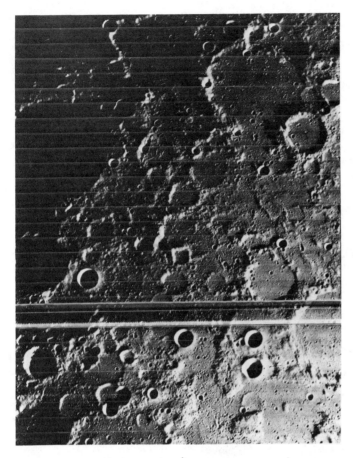

Figure 2.   North polar region of the Moon. Craters Peary and Byrd, at top center and upper right, are about 80 km across. Pole is at upper left.

and may, for example, influence the choice of sites for astronomical instruments where minimizing scattered light is a criterion.

An important unknown about the polar regions is the presence or absence of surface and subsurface ices (Watson *et al.*, 1969; Arnold, 1979; Lanzerotti *et al.*, 1981). Very low temperatures must prevail in the permanently-shadowed regions (perhaps as low as 40 K), raising the prospect that trapped water and other ices could survive there over geologic time. However, there is no way, other than spacecraft exploration, to ascertain whether or not useful quantities of such ices are present. If they are, they will provide an overwhelming reason for locating at least some part of a base complex near a pole.

ADVANTAGES, OTHER THAN POSSIBLE NATIVE VOLATILES, OF A POLAR BASE LOCATION

The dominant advantage of a polar site, from the standpoint of habitat design, is the constant thermal and illumination environment. Anywhere else on the Moon, the

Figure 3.   South polar region of the Moon. Crater Amundsen, near center, is about 100 km across. Pole is about halfway from Amundsen to bottom of frame.

base design must cope with two-week days and two-week nights. Engineering solutions to this problem, including thermal insulation and control measures and energy management, are available in principle; some of them have even been used on the Moon. For example, the Soviet Lunokhod rovers each had a hinged solar panel that served also as a thermal cover at night when closed, and they had radio-isotope heaters to maintain internal temperatures through the night. However, in a human habitat—especially one that includes agriculture, even on an early experimental scale—continuous sunlight and a stable thermal environment would permit much simpler support systems and would remove several possible sources of failure.

Figure 4 is an artist's concept for an underground habitat powered, warmed, and illuminated by the nearly-horizontal sunlight at a lunar pole. A heliostat mirror directs the solar beam into a periscope-like tunnel whose shape provides shielding against cosmic and solar ionizing radiation. Within the base, other mirrors direct the sunlight as desired,

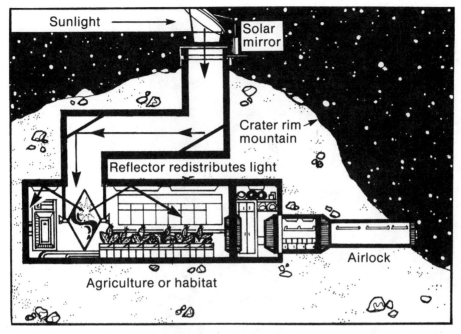

Figure 4. Underground polar habitat with sunlight piped in from heliostat. Drawing by L. Ortiz, provided courtesy of NASA.

avoiding some of the energy conversions that would otherwise be needed. In Fig. 5, a solar power tower is shown. On a common base, rotating one-half degree per hour to point continuously toward the Sun, are installed a heliostat mirror for a solar furnace, a cylindrical collector for lower-temperature heat, and a solar photovoltaic panel. With systems of this sort for energy supply and with a small reserve power plant to handle solar eclipses (whose duration is typically a couple of hours) a base should be able to operate with a nearly constant energy flow. When one considers this energy throughput another advantage of the polar site appears: heat rejection at the bottom end of the thermodynamic cycle may well be done through a surface radiating to space, insulated on its bottom side, and located in a cold, permanently-shadowed crater bottom, thus simplifying and reducing the size of this large and important system element.

Apart from these thermal and illumination advantages, other benefits may be found in the polar regions. Even if no natural ices are present, cold products can be kept in the dark crater bottoms—an important opportunity if the base's functions include producing and storing volatile life-support or propellant materials. At any warmer location, heavy pressure tanks would be needed for such storage. Because the natural equilibrium temperatures in the polar cold traps are unknown now, it is too early to tell how important this storage prospect may be. Storage of water as ice is likely to be practical; storage of liquid or solid oxygen may or may not be; storage of hydrogen will probably require containers.

Figure 5. Solar power tower in permanent sunlight near a lunar pole. Painting by Maralyn Vicary, provided courtesy of NASA.

The equilibrium temperature of such a passive storage unit would be set by the balance among heat generated locally, natural heat flow from the lunar interior upward through foundation insulation, scattered sunlight from any lunar surfaces or other nearby illuminated objects not shielded from the radiator's view, and energy from stars and other cosmic sources falling on the radiator. Temperatures below 100 K should be readily achieved [as suggested by radio brightness temperature measurements such as those of Gary *et al.* (1965)], but how low it is ultimately feasible to go, with practical lunar engineering design solutions, is at present unknown.

Another inviting aspect of the polar sites is their potential with respect to astronomy. Cryogenic telescopes located in continuous darkness could view celestial objects for as long as desired—observing, of course, only a bit less than half of the sky from each pole. If a choice of poles had to be made, the south pole would probably be preferred because the southern sky is less explored and includes unique objects such as the galactic center.

For radio astronomy, a polar location seems to offer no particular advantages over a lunar farside site, shielded from the radio noise of Earth and located at low latitude for viewing the entire sky. However, if both poles are occupied it may be more convenient to locate radio telescopes there, and the rough topography probably provides sites adequately occulted from Earth.

A solar tower telescope is another astronomical instrument that would benefit from a polar location, from whence continuous viewing of the Sun (as is now being done during the austral summer from Antarctica) would be possible. However, the advantages, if any, of such installations relative to observatories in heliocentric orbit are debatable and depend on assumptions about the supporting infra-structure. For example, a fixed lunar site offers refurbishment and maintenance advantages if, and only if, the base is capable of supporting those functions. This raises the familiar question of how to allocate and compare costs for orbital versus lunar observatories.

Another advantage, available at the poles and at the equator but nowhere else on the Moon, is quick access to and from lunar orbit. A lunar-orbiting space station in polar orbit would pass over a polar base every two hours, facilitating the schedules of both routine and emergency transport from the lunar surface to lunar and Earth orbits. Also, a polar orbit for the lunar space station is advantageous because it gives overhead coverage and access, from time to time, for expeditions anywhere on the Moon.

DISADVANTAGES OF LIVING AT THE POLES

The lunar environment, unfamiliar at best, will be even stranger for the first pioneers who settle near the poles. As the glaring sun creeps endlessly around on the horizon, most surfaces will be dark, unless illuminated by lights or mirrors for local work. Earth will hover in one direction moving from side to side and up and down a few degrees on the sky but remaining below the horizon from many nearby regions. Communications to and from Earth will, therefore, have to involve orbital or surface relays. While these are quite practical, they should be carefully designed to preserve the radio silence of the lunar farside, which, being shielded from the radio noise of Earth, is a prime site for radio astronomy and searches for radio evidence of extraterrestrial intelligence (SETI). Whether or not the polar environment presents any real hazard to human activity remains to be seen; the question can probably be answered in part by simulation experiments on Earth.

Accommodations to the natural disadvantages of a polar base site seem rather straightforward and unlikely to outweigh the many advantages of living there. We should explore the Moon's polar regions both from orbit and on the surface, and we should seriously consider one or both poles as base locations. Because the regions of primary interest are likely to be small, with dimensions of only tens or hundreds of kilometers on the lunar surface, it will probably be important to control access to them and to protect their unique natural environments under some accepted, international, legal regime. What a happy outcome it would be if international crews occupied both lunar poles, exploiting the experimental advantages of each and building toward the time when humanity can fully realize the benefits of living comfortably and productively in these unique environments of the Moon.

REFERENCES

Arnold J. R. (1979) Ice in the lunar polar regions. *J. Geophys. Res., 84*, 5659–5668.

Burke J. D. (1977) Where do we locate the Moon base? *Spaceflight, 19,* 363–366.

Burke J. D. (1978) Energy conversion at a lunar polar site. In *Radiation Energy Conversion in Space* (K. W. Billman, ed.), p. 95–103. Prog. Astronaut. Aeronaut., 61. AIAA. Princeton, New Jersey.

Culbertson P. E. (1961) Lunar base concepts and operational modes. *Astronaut. Acta, 12,* 345–351.

Dalton C. and Hohmann E. (editors) (1972) Conceptual design of a lunar colony. 1972 NASA/ASEE Systems Design Institute, University of Houston/MSC/Rice University, Houston.

De B. R. and Criswell D. R. (1977) Intense localized photoelectric charging in the lunar sunset terminator region. *J. Geophys. Res., 82,* 999–1004.

Gary B. L., Stacey J. M., and Drake F. D. (1965) Radiometric mapping of the moon at 3 millimeters wavelength. *Ap. J. Suppl., 12,* 239.

Green J. (1978) The polar lunar base. In *The Future United States Space Program, Proceedings of the 25th Anniversary Conference*, pp. 385–425. Amer. Astronaut. Soc. AAS Paper 78-191. Univelt, San Diego.

Lanzerotti L. J., Brown W. L., and Johnson R. E. (1981) Ice in the polar regions of the Moon. *J. Geophys. Res., 86,* 3949–3950.

Lucchitta B. (1978) *Geologic Map of the North Side of the Moon.* U.S.G.S. Map #I-1062. United States Geological Survey, Reston.

Watson K., Murray B. C., and Brown H. (1961) The behavior of volatiles on the lunar surface. *J. Geophys. Res., 66,* 3033.

Wilhelms D., Howard K., and Wilshire H. G. (1979) *Geologic Map of the South Side of the Moon.* U.S.G.S. Map #I-1162. United States Geological Survey, Reston.

NUCLEAR ENERGY—KEY TO LUNAR DEVELOPMENT

David Buden

Science Applications International Corporation, 505 Marquette NW, Suite 1200, Albuquerque, NM 87102

Joseph A. Angelo, Jr.

Space Technology Program, Florida Institute of Technology, 150 W. University Blvd., Melbourne, FL 32901

The Moon will play a very central role in man's exploitation of cislunar space. Energy, especially nuclear energy in the form of advanced radioisotope and fission reactor power systems, will play an equally major role in any lunar development program. This paper explores the relationship between man's successful return to the Moon as a permanent inhabitant and the use of nuclear energy. It is done within the context of a five-stage lunar development scenario. The technical discussion extends from the use of radioisotope-powered vehicles for mineral resource exploration and automated site preparation, to the reactor-powered early manned bases in which scientific investigations and prototype manufacturing projects are undertaken, to the rise of a fully autonomous lunar civilization, nourished by its own nuclear fuel cycle. If the use of nuclear energy is properly integrated into lunar development strategies, it will not only greatly facilitate the industrial development of the Moon, but may also represent a major lunar industry in itself. It is distinctly possible that very large nuclear-powered communication platforms located throughout cislunar space will be designed, constructed, and fueled by future lunar inhabitants. The same may be said for the advanced multimegawatt class reactors that will power electric propulsion vehicles, carrying human explorers to Mars and sophisticated robot explorers to the outer reaches of our solar system and beyond. The Moon is humanity's gateway to the Universe— and nuclear energy is the technical key to that gate!

INTRODUCTION

Energy is key to the exploration and development of lunar resources. High space transportation costs from Earth make it necessary to utilize energy systems that minimize mass; the clever use of *in situ* lunar materials is another way to minimize the amount of equipment that must be transported from the Earth to the Moon. An initial use of Moon materials will be used for building shields or barriers around nuclear power plants. Eventually, as lunar bases grow, their mining and manufacturing capabilities will take further advantage of native lunar resources to meet expanding energy requirements.

The development of man's permanent civilization on the Moon can be partitioned into five distinct stages: (1) automated surface exploration/site preparation, (2) the initial lunar base, (3) early lunar settlements, (4) mature lunar settlements, and (5) the autonomous lunar civilization.

STAGES OF LUNAR DEVELOPMENT AND POWER NEEDS

The stages of lunar development are given in Table 1 (Angelo and Buden, 1983), and scientific objectives are given in more detail in Table 2.

Automated Surface Exploration/Site Preparation

The first objective is to complete the mapping of the Moon's surface and to establish the chemical, mineralogical, and petrological characteristics of the Moon in order to locate optimal locations for permanent manned bases. The lunar polar regions have not been explored, so the presence or lack of surface water (ice and other frozen volatiles) must still be verified. A Lunar Polar Orbiter could be used to complete this mapping (Duke *et al.*, 1984). Power requirements are quite low (hundreds of watts); they can probably be met using existing solar energy photovoltaic technology.

Surface and subsurface exploration can be achieved with both manned and unmanned rovers. The early rovers might be solar powered, instead of battery powered as in the Apollo program, to provide longer operating times. This choice, however, limits exploration to the lunar day (14 Earth-days long) or requires mass intensive energy storage systems. Drill/manipulator robotic systems could be used to drill holes to determine subsurface

Table 1.   Stages of Lunar Development

Stages	Population
Automated Surface Exploration/Site Preparation	Robotic
Detailed surface exploration	2–5 people (semi–permanent)
Subsurface exploration	
Site preparations for initial lunar base	
Initial Lunar Base	6–12 people (permanent)
Initial scientific base	
Expanded resource exploration	
Extraterrestrial materials experiments	
Early Lunar Settlements	100–1000 people
Expanded research activities	
Prototype lunar materials processing	
Start of lunar agriculture	
Materials source for space station industrialization	
Mature Lunar Settlement	1000–10,000 people
Large scale mining and materials processing	
Manufacture products	
Cislunar trading	
Limited food production	
Autonomous Lunar Civilization	10,000–100,000 people
Self-sufficiency in raw materials and manufactured goods	
Self-sufficiency in food production	
Lunar fuel cycle	
Radioisotope generators and reactor fuel	
Lunar waste repository	

Table 2. Prioritized Scientific Objectives for Continued Exploration of the Moon

Objectives
1. Assessment of global resources, including a search for global volatiles at the poles
2. Intensive study of local areas to establish the chemical, mineralogical, and petrological character of the lunar surface
3. Measurement of global figure and surface topography
4. Exploration of the nature and dynamics of the Moon's interior
5. Establishment of the Moon's gravitational field

mineral content. Either conventional drilling techniques or hot–dry rock techniques as developed in the subterrene program at Los Alamos National Laboratory might be used during this surface exploration to explore more effectively the subsurface resource potential of the Moon (Hanold *et al.*, 1977).

This stage would culminate in the establishment of a site for a permanent lunar base (see Fig. 1). Site preparation will probably need robotic regolith-moving equipment. Power levels, based on construction equipment built for space transportation, are expected to be a few kilowatts. Part of the site preparation may include a barrier composed of regolith material that will function as a radiation shield for a nuclear reactor. Power levels on the order of a hundred kilowatts will be needed once an initial lunar base is established.

Figure 1.   Automated site preparation.

Initial Lunar Base

The initial lunar base will serve as a science center to better understand the Moon and our solar system. This base will also support investigations concerning processes for mining lunar materials, materials beneficiation, and manufacturing (see Fig. 2). Oxygen is frequently mentioned as a leading candidate for lunar manufacturing, both because it is abundant in lunar rocks and because it has a high value as a chemical propellant in space transportation systems. A major goal of this stage is to learn how to live and work on the lunar surface. The initial lunar base is expected to require power levels similar to the early space station—on the order of 100 kWe.

Early Lunar Settlements

As the initial lunar bases expand their activities, selected, small-scale mining and beneficiation of lunar materials will take place (see Fig. 3). Selected Moon materials will start being exported to support overall space industrialization activities (see Fig. 4 for one scenario). In addition, certain products will be manufactured on the Moon. The lunar population, or "selenians," will number from 100 to 1000 permanent inhabitants. This

Figure 2.   Early lunar base.

Figure 3. Early lunar settlements.

time period will witness the initiation of extraterrestrial agriculture, with food being produced in special lunar greenhouses to accomodate the expanding population. Power consumption during this period of lunar development is expected to be in the megawatts, based on the number of inhabitants present and on-going activities.

Mature Lunar Settlements

Nourished by native resources, the lunar population will eventually swell to 10,000 or more permanent inhabitants (see Fig. 5). A semi-autonomous status will be achieved as much of the manufactured goods and significant quantities of food will be produced on the Moon for both domestic consumption and "export." Power demand will reach tens of megawatts and parts of a nuclear power industry will be in place. The power level estimates are based on the size of population and power demands of a similar population on Earth. Lunar settlements will be more dependent on electric power because of the greater limitations of alternate energy sources.

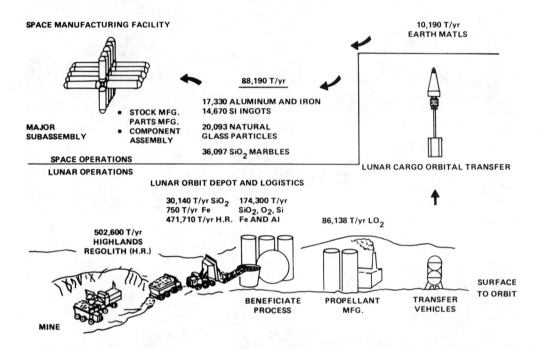

SPACE MANUFACTURING FACILITY

10,190 T/yr
EARTH MATLS

88,190 T/yr

• STOCK MFG.
PARTS MFG.
• COMPONENT
ASSEMBLY

17,330 ALUMINUM AND IRON
14,670 SI INGOTS

20,093 NATURAL
GLASS PARTICLES

36,097 SiO_2 MARBLES

MAJOR
SUBASSEMBLY

SPACE OPERATIONS

LUNAR OPERATIONS

LUNAR CARGO ORBITAL TRANSFER

LUNAR ORBIT DEPOT AND LOGISTICS

30,140 T/yr SiO_2 174,300 T/yr
750 T/yr Fe SiO_2, O_2, Si
471,710 T/yr H.R.   Fe AND Al

86,138 T/yr LO_2

502,600 T/yr
HIGHLANDS
REGOLITH (H.R.)

SURFACE
TO ORBIT

BENEFICIATE
PROCESS

PROPELLANT
MFG.

TRANSFER
VEHICLES

MINE

Figure 4.   General processing and space-based manufacturing showing flow of lunar materials.

Autonomous Lunar Civilization

The lunar civilization will reach maturity (see Fig. 6). Its population will no longer be dependent on Earth for manufactured and agricultural goods. With an energy-rich, dynamic lunar civilization now feeding further expansion into heliocentric space, the selenians will eventually become the space–faring portion of the human species. Hundreds of magawatts of power will be needed in this phase to provide the necessary energy for such a large population.

In the full scale exploitation of cislunar space and the Moon, nuclear electric propulsion systems (NEPS) will serve a critical enabling role in the efficient transport of massive, non–priority cargoes (Buden and Garrison, 1985). NEPS will serve not only as the propulsive means of placing a massive payload in an appropriate operating orbit, but once the operational location is reached the nuclear reactor would then serve as the prime power supply for many years of continuous, profit–making operation of the payload. Nuclear electric propulsion systems could also be used as reusable orbital transfer vehicles (OTVs) or "space tugs." These propulsive workhorses of tomorrow would gently move massive cargoes, supplies and materials, large and fragile space systems that had been assembled in lunar orbit, or even entire (unoccupied) habitats, and ferry these cargoes to their final destinations throughout cislunar space. Habitat fabrication may be a major lunar industry by the 22nd Century.

NUCLEAR POWER OPTIONS

The various stages of lunar development and associated power requirements are given in Table 3.

Because the Moon experiences long diurnal cycles (a 14 Earth-day "day" and a 14 Earth-day "night") solar energy becomes an awkward energy source to rely on in a continuously inhabited and operated lunar settlement. This is mainly because massive energy storage devices would be needed for power in the night cycle. Nuclear energy offers a relatively compact power source that is not affected by the diurnal day/night cycle, and the technology should be available if current development plans proceed as now scheduled.

Radioisotope generators have been used where long life, high reliability, solar independence, and operation in severe environments are critical. These use the spontaneous

Figure 5. Semi-autonomous lunar settlement.

Figure 6. Autonomous lunar civilization.

decay of plutonium-238 as a heat source. The energy has traditionally been converted to electricity by means of thermoelectrics placed next to the heat source. Radioisotope generators have been launched in 21 spacecraft, beginning with the successful flight of a SNAP-3B power source in 1961. Their reliability and long life is demonstrated by the Pioneer satellite, which after 11 years of operation left our solar system still in a functioning state. The recent magnificent pictures of Saturn taken from the Voyager spacecraft, powered by radioisotope generators, are also testimonials to the longevity and reliability of this type of power supply. Improved versions of the generators with thermoelectric electrical conversion devices will increase their performance. However, radioisotope thermoelectric generators will probably be restricted to under 500 W. Higher power levels of maybe 1–10 kW are possible by using dynamic electric converters for power conversions. A 1.3 kW version has been tested for several thousand hours prior to program termination (Bennett *et al.*, 1981).

For initial lunar bases and early lunar settlements, the SP-100 Program technology is applicable. If we use native lunar materials for radiation shield, all other nuclear cycle components can be transported from Earth with contemporary space transportation system

Table 3.   Stages of Lunar Development and Power Requirements

Stage	Activity	Power Levels	Probable Nuclear Power Supply
1	Automated Surface Exploration/ Site Preparation	kWe	Radioisotope generators
2	Initial Lunar Base (6–12 persons)	~100 kWe	Nuclear Reactor (SP-100)
3	Early Lunar Settlements (100–1000 persons)	~1 MWe	Expanded SP-100 (Advanced Design)
4	Mature Lunar Settlement (~10,000 persons)	~100 MWe	Nuclear Reactor (Advanced Design)
5	Autonomous Lunar Civilization (Self-sufficient Lunar Economy: >100,000 persons)	hundreds of MWe	Nuclear Reactors (Advanced Design, Complete Lunar Nuclear Fuel Cycle)

vehicles, *i.e.*, Shuttle plus advanced orbital transfer vehicles. The SP-100 Program is a joint program of the Departments of Defense and Energy and NASA to develop space nuclear power systems technology. Following screening of over a hundred potential space nuclear power system concepts by the SP-100 Program, the field has now been narrowed to three candidate systems (Ambrus *et al.*, 1984).

One concept uses a fast spectrum, lithium–cooled pin–type fuel element reactor coupled to thermoelectrics for power conversion (see Fig. 7). The reactor, which is a right circular cylinder, approximately one meter in diameter and one meter high, is at the apex of the conical structure. It is controlled by twelve rotatable drums, each with a section of absorbing material and a section of reflective material, to control the criticality level. Control of the reactor is maintained by properly positioning the drums. The shield is mounted directly behind the reactor and consists of both a gamma and a neutron shield. Thermal transport is accomplished by a lithium working fluid that is pumped by a thermoelectrically driven electromagnetic pump. The reactor thermal interface with the heat distribution system is through a set of heat exchangers. Thermoelectric elements are coupled to the internal surfaces of the heat rejection panels and accept heat from the source heat pipe assembly. The heat rejection surfaces are beryllium sheets with titanium potassium heat pipes brazed to the surface to distribute and carry the heat to the deployable panels that are required to provide additional heat rejection surfaces.

A second approach is an in–core thermionic system with a pumped sodium–potassium eutectic coolant. The general arrangement of the in–core thermionic space power system design is shown in Fig. 8. The design forms a conical frustum that is 5.8 m long with minor and major diameters of 0.7 m and 3.6 m, respectively. The reactor contains the thermionic fuel element (TFE) converters within a cylindrical vessel, which is completely surrounded by control drums. Electrical power is generated in the space between the tungsten emitter and niobium collector, and the electrical current output is conducted from one cell to the next through the tungsten stem of the emitter and the tantalum transition piece. The NaK primary coolant routing to and from the reactor vessel is arranged so that the hot NaK leaves the reactor at the aft end and the cold NaK is returned

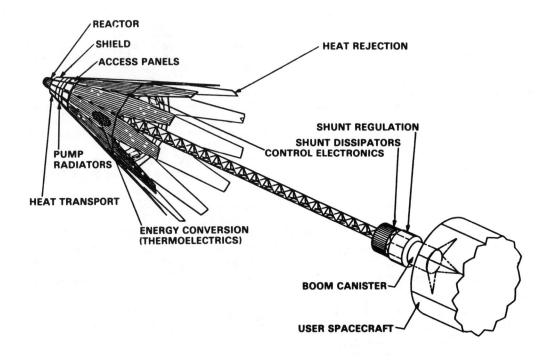

Figure 7. High-temperature reactor with thermoelectric power conversion concept.

to the forward end. The TFE consists of six cells connected in series with end reflectors of BeO. The reactor is also surrounded by an array of long, thin cylindrical reservoirs, which collect and retain the fission gases generated in the reactor core during the operating life of the system. Waste heat is removed from the primary reactor loop through the heat exchanger. The energy is transferred through the heat sink heat exchanger to heat pipes that form the radiating surfaces for rejection of heat to space.

The third approach uses a Stirling engine to convert heat from a lower-temperature (~1000 K), fuel-pin reactor design to electricity (see Fig. 9). This design emphasizes the use of state-of-the-art fuel pins of stainless steel and UO_2 with sodium or sodium potassium eutectic as the working fluid. Such fuel pins have been developed for the breeder reactor program with 1059 days of operation and 8.5% burn-up demonstrated. If the use of stainless steel fuel pins is not possible, a refractory alloy such as Nb-1Zr could be substituted. The reactor can be similar in design to the high-temperature reactor but utilizes lower temperature materials. In Fig. 9, the reactor is constructed as a separate module from the conversion subsystem. Four or five Stirling engines, each rated to deliver 25–33 kWe, are included in the design concept. This provides some redundancy in case of a unit failure. Normally the engines operate at partial power to produce a 100 kWe output. Each engine contains a pair of opposed motion pistons that operate 180° out of phase.

This arrangement eliminates unbalanced linear momentum. Each engine receives heat from a pumped loop connected to the reactor vessel. The heat is supplied to heater heads integral with the engine. Waste heat is removed from the cooler heads and delivered to a liquid–to–heat pipe heat exchanger. The heat pipes, in turn, deliver the waste heat to the radiator, where it is ejected to space.

The advanced stages of lunar settlements will require new reactor designs to satisfy demands for megawatts of power. Power plants will probably need to be refuelable, have a 30–year lifetime, and provide multimegawatts of power. Several technology approaches are possible including solid core, fluid cores, particle bed, and gaseous core reactors. For space, solid–core reactors were most extensively developed as part of the nuclear rocket program. The Rover design featured a graphite-moderated hydrogen-cooled core (Buden and Angelo, 1983). The 93.15% ^{235}U fuel was in the form of UC_2 particles, coated with a pyrolytic graphite. The fuel was arranged in hexagonal-shaped fuel elements, coated with ZrC; each element had 19 coolant channels. Electric power on the order of 100 megawatts could be generated by replacing the rocket thrust nozzle with power conversion equipment. This is a limited-life system, however.

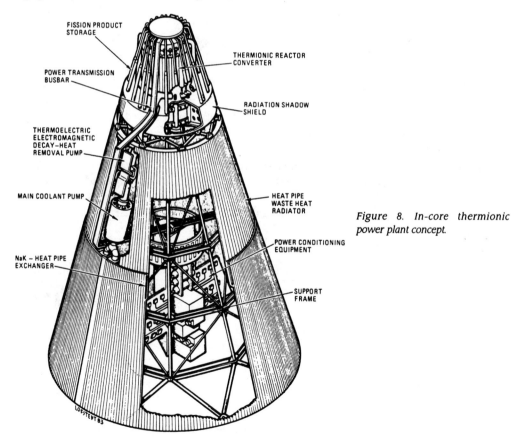

Figure 8. In-core thermionic power plant concept.

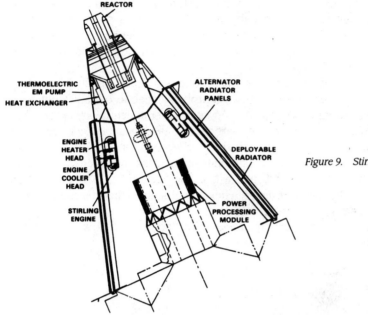

Figure 9. Stirling engine conversion concept.

High-power requirements might also be met by fluidized bed reactors, in either the rotating or fixed-bed forms. The former was investigated as a rocket propulsion concept with modest research effort in fluidized bed reactors carried out from 1960 to 1973.

Another candidate for megawatt-power reactors is a gaseous core reactor system (Thom and Schwenk, 1977). The central component of such a gaseous core reactor is a cavity where the nuclear fuel is in the gaseous state. One reactor concept is an externally moderated cavity assembly that contains the uranium fuel in the gaseous phase. For temperature requirements less than a few thousand degrees Kelvin, the appropriate nuclear fuel would be uranium hexaflouride, UF_6. It is desirable to keep the gaseous fuel separate

Table 4.   Average Heavy Nuclide Content of Lunar Regolith

Lunar Surface Material	Lunar Mission	Thorium (ppm)	Uranium (ppm)
Mare	Apollo 11	2.24	1.37
"	Apollo 17	0.82	0.26
"	Apollo 12	6.63	1.61
"	Apollo 15	1.76	0.483
"	Luna 16	1.07	0.300
Highland	Apollo 16	1.87	0.52
"	Luna 20	1.44	0.45
Basin Ejecta	Apollo 14	13.5	3.48
" "	Apollo 15	4.15	0.99
" "	Apollo 17	3.01	0.90

from the cavity walls. This is accomplished through fluid dynamics by using a higher velocity buffer gas along the wall. Power is extracted by convection or radiation heat transfer. Gaseous core reactors offer simple core structures and certain safety and maintainability advantages. The basic research development was completed prior to program termination, including the demonstration of fluid mechanical vortex confinement UF_6 at densities sufficient to sustain nuclear criticality.

New developments in the next several decades will probably have a strong influence on the design approaches for advanced lunar settlements. These designs may more nearly resemble advanced terrestrial reactor central power plant designs. On-line refueling and robotic maintainability features are envisioned with a 30-year or more useful lifetime. Another characteristic of this new generation of lunar reactors would be "inherent safety"— that is, if a malfunction should occur in any part of the power plant, it is designed so that no human operator action or even mechanical automatic control mechanism is needed to achieve a safe condition.

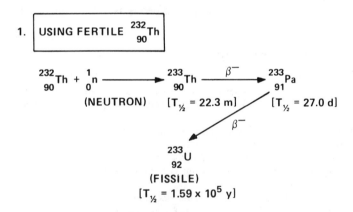

Figure 10. Classic nuclear fuel breeding reactions.

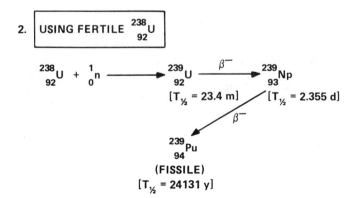

Finally, as the lunar settlement expands and grows, a point will eventually be reached when the lunar civilization, for all practical purposes, becomes fully self-sufficient. As part of this autonomy in Stage 5, a lunar nuclear fuel cycle will evolve, taking advantage of native thorium and uranium minerals [see Table 4 from References (Phinney *et al.*, 1977) and various NASA publications] and the classic nuclear fuel breeding reactions for the "fertile" nuclides, $^{232}_{90}\text{Th}$ and $^{238}_{92}\text{U}$ (see Fig. 10).

CONCLUSIONS

Power will be the key to lunar development. Without adequate power, the rate and size of lunar development will be severely limited. Nuclear power is a prime technology capable of satisfying these power requirements. Radioisotope generators need further development with an improved electrical conversion system. The SP-100 Program should provide an acceptable power plant for the initial lunar base and early lunar settlements, especially if it satisfies station growth needs. An advanced space power systems now under consideration—including solid core, particle bed, and gaseous core reactor concepts—should be capable of satisfying multimegawatt power needs as lunar civilization evolves.

REFERENCES

Ambrus J. H., Bunch D. F., and Wright W. E. (1984) The tri-agency space nuclear reactor system technology program. *Proceedings of the 19th Intersociety Energy Conversion Engineering Conference*, pp. 561–569. Amer. Nucl. Soc., Washington.

Angelo J., Jr. and Buden D. (1983) Lunar base requirements (abstract). In *Lunar and Planetary Science XIV* (Special Session Abstracts), p. 3. Lunar and Planetary Institute, Houston.

Bennett G. L., Lombardo J. J., and Rock B. J. (1981) Nuclear electric power for space systems: Technology background and flight systems program. *Proceedings of the 16th Intersociety Energy Conversion Engineering Conference*, pp. 362–368. Amer. Soc. Mech. Engrs., New York.

Buden D. and Angelo J., Jr. (1983) Space reactors—Past, present and future. *Proceedings of the 18th Intersociety Energy Conversion Engineering Conference*, pp. 61–68. Amer. Inst. Chem. Engrg., New York.

Buden D. and Garrison P. W. (1985) Design of a nuclear electric propulsion orbital transfer vehicle. *J. Propul. Power, 1*, 70–76.

Duke M. B., Mendell W. W., and Roberts B. B. (1984) Toward a lunar base. *Aerospace America, Oct., '84, pp.* 70–73.

Hanold R. I., Alfseimer, J. H., Armstrong P. E., Fisher H., and Krupka, M. C. (1977) Rapid excavation by rock melting—LASL subterrene program, September 1973–June 1976. *Los Alamos National Laboratory Report LA-5979-SR*, 83 pp., LANL, Los Alamos.

Phinney W. C., Criswell, D., Drexler E., and Garmirian J. (1977) Lunar resources and their utilization. *Spacebased Manufacturing from Nonterrestrial Materials., 57,* (G.K. O'Neill and B. Leary, eds.) pp. 97–131, Progress in Astronautics and Aeronautics. AIAA, New York.

Thom K. and Schwenk F. C. (1977) Gaseous fuel reactor systems for aerospace applications. *AIAA Conference on the Future of Aerospace Power Systems, St. Louis, Missouri*, Paper No. 77-513. AIAA, New York.

NUCLEAR POWERPLANTS FOR LUNAR BASES

J. R. French

Jet Propulsion Laboratory, 233–307, 4800 Oak Grove Drive, Pasadena CA 91109

INTRODUCTION

After a hiatus of some ten years, the United States is again involved in development of a nuclear reactor-based electric power plant for space. The SP-100 Program, having been in existence for some two years, has as its goal the development of system concepts and technology for power plants capable of generating 100–1000 kWe for two years initially, with potential growth to seven years. The current phase of the program is devoted to evaluating the various possible design concepts and assessing the status of critical technology. Limited technology development is being carried out as well.

The goal of this phase is to select the most promising system concept and to develop the plan for the system and technology development program that will form the next phase of the project. This phase will begin in FY 1986 with a goal of demonstrating critical technologies, subsystems, and system interfaces for the chosen concept by the early 1990s.

The SP-100 Program is being jointly conducted by DARPA (Defense Advanced Research Projects Agency), NASA, and the Department of Energy with technical leadership provided by the Jet Propulsion Laboratory, NASA/Lewis Research Center, and Los Alamos National Laboratory.

SP-100 CHARACTERISTICS

The SP-100 system is defined as being a nuclear reactor-based electrical power plant designed for space use. Design performance requirements specified for the 100 kWe system under study are listed in Table 1. As noted in the Table, the initial lifetime

Table 1. SP-100 Design Performance Requirements

Power	100 kWe
Mass	< 3000 kg
Launch dimensions	STS bay diameter $\times$ 1/3 bay length
Radiation to payload	500 K rad
	7 y at full power
	$N-10^{13}$ nvt

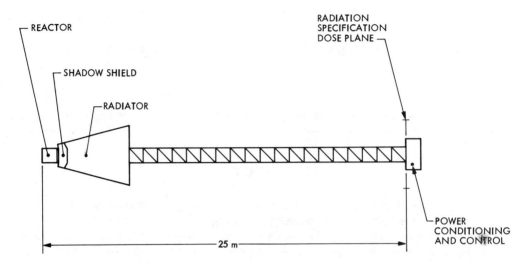

Figure 1. The generic SP-100 design is depicted showing the reactor, shield, conical radiator, and the power conditioning and control unit mounted on a deployable boom at the dose specification plane.

goal is 2 years; however, the design should incorporate no known characteristics that would preclude seven years of full power operation. Launch volume is restricted to the full diameter of the shuttle cargo bay and 1/3 of the length; however, there is no restriction upon deployed size, and variable geometry concepts are viable candidates.

Radiation toward the payload is attenuated by a shadow shield that subtends a solid angle, usually in the 12°–17° half angle range. The radiation levels are specified over a 4.5 m diameter plane 25 m from the opposite end of the reactor (see Fig. 1). Note that this does not mean that the payload must be at 25 m; that is simply the dimension chosen to specify the radiation. The payload may be nearer or more distant as requirements and radiation resistance dictate. The system designers working on the concepts have all chosen to place the power conditioning and control subsystem at the 25 m point. This spacing would be obtained by deployment of an erectable boom structure following deployment from the shuttle but before reactor start-up.

While the values presented in Table 1 are being used as a focus for present design, the SP-100 Project is maintaining awareness of possible need for growth. Power output levels of 100–1000 kWe are possible by allowing mass and volume to exceed the Table 1 limits. Some concepts lend themselves to large growth factors better than others, and selection of growth concepts would depend upon trade-off of a variety of requirements.

SYSTEM CONCEPTS

Three (or possibly four) concepts are presently being evaluated for possible selection. These are listed in Table 2. The first three concepts, usually distinguished by reference to their means of conversion (thermoelectric, thermionic, and Stirling) are presently under

Table 2. System Concepts

Reactor	Heat Conversion	Heat Transfer	Rejection
Fast, compact	thermoelectric	pumped lithium	deployable heat pipe radiator
Fast, thermionic	in-core thermionic	pumped Nak	fixed heat pipe radiator
Fast, compact	Brayton, alternator	pumped Nak	heat pipe or gas, probably deployable

active study by system contractor teams: General Electric, G.A. Technologies/Martin Marietta, and Rockwell, respectively. The fourth concept, the Brayton or gas turbine cycle, was not selected as a prime candidate because of specific technical concerns that appeared to be life- or reliability-limiting in space applications except at the cost of considerable mass. It still appears as a "dark-horse" candidate because of its high efficiency and high level of technical development. In general external appearance, the systems are much the same configuration, differing mainly in radiator size.

The thermoelectric concept, a derivative of the technology used with great success in the Radio-isotope Thermoelectric Generators (RTG) for the Pioneer, Viking, and Voyager spacecraft, offers the advantage of being a static system requiring no moving parts for thermal-to-electric conversion. Materials constraints limit the inlet and outlet temperatures. This, together with the relatively low efficiency of conversion, leads to a large deployable waste heat radiator. The potentially very high reliability and the years of RTG experience are major strong points of this system. The only "moving part" in the entire system is the lithium coolant that is pumped electromagnetically. This carries thermal energy through intermediate heat pipes to the hot side of the thermopiles. The cold side is cooled by the radiator.

The thermionic concept also contains no moving parts, unless the NaK (sodium-potassium eutectic) coolant is so considered. Thermal-to-electric conversion takes place directly in the reactor with the nuclear fuel heating the thermionic emitters and the collectors cooled by NaK. The waste heat from the collectors goes directly to a heat exchanger and then to the radiator. Radiator temperature as well as conversion efficiency is higher than for thermoelectric systems, allowing a smaller radiator. Concern exists as to possible life-limiting mechanisms in the thermionic converters. Inclusion of the conversion capability in the core tends to make the reactor larger and heavier. This may be less significant at higher power levels.

The Stirling system concept involves a free-piston Stirling engine coupled to a linear alternator. This system offers the highest conversion efficiency of any of the concepts, perhaps five times that of the thermoelectric system. The technology is less well-developed than that of the static concepts; however, the high performance and possibility of operating at much lower reactor temperature makes the system of great interest. Areas of concern center about the dynamic nature of the system and the resulting potential for wear and vibration. Efficient means of heat input and withdrawal to and from the multiple engines is an area requiring attention.

The Brayton system, not currently being studied by a contractor, offers efficiency close to that of the Stirling as well as the potential for growth. Concern about life expectancy resulting from attack of refractory metals by oxygen impurities in the noble gas working fluid as well as the vibration and creep problems inherent in dynamic systems must be dealt with. The system mass may be a problem if waste heat withdrawal must use heat exchange from the gas to a heat pipe radiator. The lighter gas radiator has lifetime/ reliability problems due to micrometeoroids. This system could be a viable candidate for some applications.

APPLICATION OF SP-100 TO A LUNAR BASE

For a variety of reasons, it would be difficult to apply the presently developing SP-100 configuration designs directly to a lunar base. The technology and components developed, however, could certainly be applied to a lunar base power plant.

Problems that might be involved in a lunar surface application are generally due to the presence of the lunar surface. The system is being designed for space use, and proper performance is based upon there being no material in the vicinity except for the system itself and its user spacecraft. Because radiation can be scattered back from surrounding materials, there could be very small changes in criticality for given control settings because of the presence of the lunar surface. This effect should be well within the reactor control capability of SP-100. Radiation will tend to scatter around the shadow shield as well.

The proximity of the lunar surface will detrimentally affect the functioning of the waste heat radiator because of the diminished view factor to space. This will be especially severe during the lunar day when the surface will be itself a heat source in the infra-red range. While all candidate systems will reject heat at temperatures substantially above those of the lunar surface, the effect will generally be such as to dictate larger radiator area. This problem will be less severe if the base is situated at high latitudes.

The lunar environment can be most useful, however, in another role. The shadow shield in the current design only protects a very small area. This would be unsuitable for a manned lunar surface application unless the power plant were to be placed at

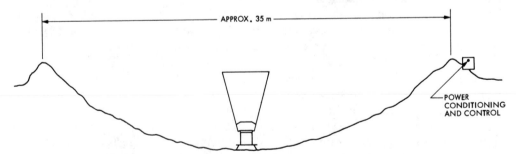

Figure 2. A generic space-type SP-100 mounted in a crater for shielding as often suggested is depicted. Based upon geometry of the Barringer crater in Arizona, a 6.1 m long SP-100 would require an approximate 35 m diameter crater. (Note: Because of scattering, even the radiator must be shielded from manned presence. Thus, unless a large quarantine area is specified, the crater must encompass the complete system.) A longer SP-100, i.e., deployable radiator, would require a still larger crater. Since most heat is rejected from the outside of the cone, the radiator view factor is poor.

a substantial distance from the base (with attendant transmission losses) and approach to it prohibited. Use of lunar materials, as has often been suggested, can solve this problem. While lunar soil or rock is not an especially efficient shielding material, the fact that it is available and "free" (ignoring the effort of collecting it) renders this application very attractive compared with hauling 20,000–30,000 kg of more efficient material from Earth. Whether shielding might be accomplished by placing the reactor at the bottom of a convenient crater or by building up a structure around it needs to be investigated.

The former approach (Fig. 2), if feasible, would certainly be the simpler of the two but may have some inherent disadvantages. For example, in order to provide adequate depth the crater would have to be fairly large. To minimize plumbing lengths for liquid metal, the entire system would have to be within the crater. Once the reactor goes critical, the interior of the crater and, thus, the entire system are off limits. After extensive operations the accumulation of radioactive fission products would preclude close approach even with the reactor off. While it could be argued that this is not important in a system designed for 2–7 years of unattended operations, most individuals with practical experience in operating systems will admit the importance of access for manned inspection and maintenance. An additional problem is reflected radiation from the surface increasing the dose to electronics. The power conditioning and control would have to be outside the crater or specially shielded.

Figure 3. Possible reconfigurations of the SP-100 for use in a crater-shielded installation are shown. In all cases, the radiator is reconfigured as a flat disc rejecting heat mostly from the top side. The thermoelectric concept asssumes that the thermoelectric converter remains a cone shape and that the converter must be shielded from direct reactor radiation and at least part of the ground scatter, hence the large shield. This requires a fairly large crater (say 25 m). If it should not be necessary to shield the converters, they could be configured in a disc as well, resulting in a shorter stack and reducing shield mass. The other concepts are more compact. In the thermionic concept, a shield is assumed only for control electronics, actuators, etc. Similarly, the shield for the Stirling system as shown does not fully shield the Stirling engines and alternators. If this is unacceptable, the shield

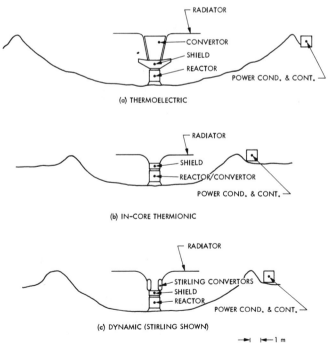

(a) THERMOELECTRIC

(b) IN-CORE THERMIONIC

(c) DYNAMIC (STIRLING SHOWN)

would have to increase in diameter, but the stack would remain relatively short. (Note: These sketches represent concepts not designs and are presented to show possible options.)

Further inspection of Fig. 2 indicates that this shielding configuration worsens the heat rejection problem by further degrading the view of space. The three parts of Fig. 3 depict conceptually how an SP-100 system might be reconfigured to function up to its full potential in a crater environment. Note that the thermoelectric concept may require a shadow shield to protect the thermoelectric converters.

A custom built shield in combination with a reconfigured SP-100 system is sketched in Fig. 4, showing slightly different configurations for the various conversion concepts. At the cost of slightly longer liquid metal plumbing runs, the conversion systems can be shielded from the reactor allowing for the possibility of maintenance or replacement. Similar comments apply to the radiator.

As previously observed, it is not clear which of the two approaches would be preferred. Certainly the second requires more preparation. They may in fact represent two stages

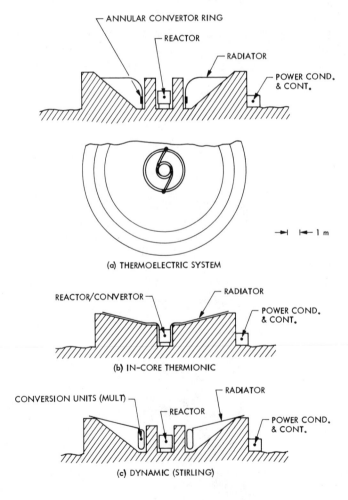

(a) THERMOELECTRIC SYSTEM

(b) IN-CORE THERMIONIC

(c) DYNAMIC (STIRLING)

Figure 4. The concepts of Fig. 3 are depicted as reconfigured to operate in a custom shield of lunar material. Such shields might be lunar brick or concrete fabricated by various techniques under study. Most probably, they would be brick or concrete retaining walls filled in with compacted regolith. Since there is relatively little external equipment involved in the thermionic concept, the shield is shown as a solid mass. The other concepts could be similarly configured; however, any attempt to work on the converters would expose personnel to the shut-down but still radioactive reactor. By contrast, the concepts shown in (a) and (c) would allow personnel to enter the cavity for maintenance with the reactor shut-down because of the protection afforded by the inner shield. No shield is shown over the top because it is assumed that control electronics and actuators could be located "around the corner" of the shield. If this is not practical, a small top shield as shown in Fig. 3 (b and c) would be needed.

of evolution, since it seems probable that more than one power plant will ultimately be required for the base.

It is important to remember that the changes to adapt the space version of the SP-100 to lunar surface use are either simply reconfiguration of the major components or relatively minor changes in the components. A complete new development in parallel to SP-100 should not be necessary in order to provide a lunar base power plant.

It seems reasonable to suppose that a 100 kWe nuclear power plant for use on the Moon should not exceed the 3000 kg proposed for the space version. In fact, it might well be considerably lighter since typically the shadow shield mass is 500–700 kg, and this component will probably be unnecessary for some options. The relatively low mass and compact size offer substantial potential advantage compared to solar/storage systems in terms of transportation. Since payload mass and volume will be at a premium, especially during the early years, this consideration should be of substantial importance.

OPERATIONAL CONSIDERATIONS

The obvious advantage of a nuclear power source for a lunar base is the fact that it operates night or day. For a solar-based system, the two-week day/night cycle means either a massive storage system or else shutdown of all but the most vital systems during the night. Even a powered-down lunar base would require substantial power. It has been proposed that solar arrays on high elevations at the lunar poles would always be in the sun. However, even if one wishes to postulate such a construction project, the location of the base will be limited to sites near the poles, which may or may not fit other needs. Nuclear power makes site selection independent of power requirements.

Another advantage is the "waste" heat left over from the conversion process. In the deep space version, this heat is indeed waste, being radiated away to space. For the lunar application in its simplest form, the heat could be radiated to space as shown in the Figs. 3 and 4. This is not necessarily the case, however. The waste (or excess) heat energy might be applied directly to base heating, materials processing, or other functions, thus reducing the need for electrical energy for these applications.

GROWTH POTENTIAL

The excess heat energy might also be used to generate additional electrical energy. It would be possible to couple additional conversion units to accept the excess heat from the primary unit. The in-core thermionic primary system may be best adapted to this purpose, since it rejects heat at the highest temperature of the candidate concepts. A possibility would be coupling one or several Stirling or Brayton units to the system in a "bottoming cycle." These highly efficient units operate at inlet temperatures quite compatible with the heat rejection temperature of a thermionic primary system: about 1000 K. As an example, a thermionic primary system generating 100 kWe would be rejecting an excess of 1000 kWe of thermal energy. If this could be efficiently directed into a Stirling or Brayton system, another 250–350 kWe could be produced. Similar arrangements are possible with a thermoelectric system; however, the rejection temperature is lower. Substantial waste heat will still be available from the bottoming cycle converters but at substantially lower temperature, say 500–700 K. This will result in much larger radiators. A major advantage of the bottoming cycle approach is that, with proper design, these converters can be added later to an already functioning power system, thus incorporating substitute/power growth potential without requiring a new reactor installation. Coupling of a bottoming system to the primary system might be done by installing some type of coupling plates over the radiators (essentially converting them to heat exchangers). While this approach might not be highly efficient from the heat transfer standpoint, it would avoid the necessity of breaking into the existing plumbing. Alternatively, a heat exchanger and valving could be built into the original system to allow for later integration of the bottoming cycle.

SUMMARY

It appears that SP-100 concepts and technology under development for a deep space system could be readily adapted to provide a lunar base power plant. A nuclear power plant offers substantial advantage because of its compact size and night and day operation. Heat rejected by the primary conversion system may be used for industrial purposes or to generate additional power via a bottoming cycle.

Acknowledgments. *This document is based upon work performed by the Jet Propulsion Laboratory, Los Alamos National Laboratory, NASA/Lewis Research Center, and their industrial contractors. Conclusions regarding lunar adaptions are the author's own. The author wishes to thank Mr. Monte Parker of Los Alamos National Laboratory for consultation concerning nuclear and shielding considerations.*

3 / TRANSPORTATION ISSUES

T HE SINGLE EXISTING ELEMENT of the Space Transportation System (STS) is the Earth-to-orbit vehicle, the Space Shuttle. Passengers can be carried to low-Earth orbit (LEO) with equipment for activities of limited duration. Satellites can be placed into LEO from the payload bay or can be launched beyond LEO using upper stages.

The next element of the STS will be the LEO space station, which will increase the allowed duration of scientific and commercial activities in orbit. Just as the shuttle introduced the concept of reusable space vehicles, the space station will establish permanent human presence as an element of space operations. By the turn of the century, reusable orbit-to-orbit vehicles will be added to the space station. The Orbital Transfer Vehicles (OTV) will be designed to lower the cost of shipment of payloads to deep space through reuse of the upper stage. The space station then will incorporate the function of a freight dock, where cargo is readied for transshipment to destinations beyond.

At the same time, the shuttle will be nearing the end of its operational life; and a second generation of Earth-to-orbit transportation will be readied. Will this second generation consist of an updated shuttle, or will the functions of transporting freight and carrying passengers be explicitly separated to alleviate the crippling expense of launching to LEO? For example, an unmanned launch vehicle based on shuttle propulsion technology could carry many times more cargo for less launch cost than does the shuttle today. However, the cost of developing such a vehicle can be justified only if the projected traffic into LEO is large.

The buildup of a lunar base places demands on the STS that must be reflected in the design of the elements. The first three papers describe the impact of a lunar base and other deep space activities. Woodcock delineates the fundamentals of the orbital mechanics involved in the simple shipment of payloads to the lunar surface and suggests some modes of operation for accomplishing the emplacement of lunar surface facilities. Babb *et al*, examine the role of the LEO space station and define resources that must be made available at that transportation node to support freight operations. Keaton takes a fresh look at the whole infrastructure of the STS from the point of view that a variety of destinations will be desirable in the next century. His elegant analyses of orbital mechanics in the Sun–Earth–Moon system are supported experimentally by the ingenious targeting maneuvers derived by Bob Farquhar for the interception of Comet Giacobini–Zinner by the ICE spacecraft.

The final three papers in this section describe innovative uses of the lunar environment to reduce the operational burden on the terrestrially based STS. Heppenheimer reviews elements of his own work over the past decade on mechanics of launching raw materials and other products from the Moon with large, fixed electromagnetic "guns." Rosenberg discusses the production of silane from lunar materials as a propellant to combine with lunar-produced oxygen in an engine design uniquely suited to the space environment. Finally, Anderson et al., describe a simple lunar surface launching technique for putting small payloads into lunar orbit with solid fuel rockets for scientific investigations and other purposes.

MISSION AND OPERATIONS MODES FOR LUNAR BASING

Gordon R. Woodcock

Boeing Aerospace Company, 4910 University Square, Suite 3, Huntsville, AL 35805

Future lunar operations may be directed to permanent or long-term presence on the Moon and supported by a space station as an intermodal transportation complex. Flight mechanics constraints on station-supported lunar operations are described and analyzed, and the implications to space station are presented. Mission modes supportable by the NSTS and its derivatives, and by the space station, are described and compared. Sensitivities of the modes and their uses for lunar operations are described.

INTRODUCTION

A U.S. civil space station is planned for low-Earth orbit by the early to mid 1990's. By the late 1990s it will serve as an intermodal transportation complex, *i.e.*, a spaceport that could, among other things, support advanced lunar operations. The space shuttle and its future derivaries—support by the space station, new technologies for reusable orbit transfer vehicles (ROTVs), and the potential of obtaining propellant oxygen from the Moon itself—all combine to dramatically lower the calculated cost of future lunar operations compared to the "brute force" of the Apollo missions. Where a permanent human presence on the Moon was once viewed as an option only for a heavily funded civil space program, it now appears (as also observed by Paul Keaton and others) as an achievable option within the funding scope of present civil space activities.

Effective use of new mission and operational schemes is not without challenge. The Apollo missions, especially after the requirement for free-return trajectories was removed, were relatively unconstrained. Launch opportunities were frequent, launch windows up to hours in length, and any return trajectory that terminated in a rather large recovery zone in the Pacific Ocean was acceptable.

INFLUENCES OF FLIGHT MECHANICS ON OPERATIONS DESIGN

With an orbiting spaceport, departure to the Moon is constrained to a particular Earth orbit. Further, if oxygen derived from lunar surface materials is to be used by the transportation system, it is important to have an oxygen cache in the lunar vicinity. This necessitates a particular lunar staging point. Although a range of choices (including the L1 and L2 Lagrangian points as well as lunar orbits) is possible, a polar lunar orbit has been selected by this as well as earlier studies, because it permits access to all of the lunar surface and does not constrain base location. Thus lunar base support operations are likely to be constrained to operations from a particular Earth orbit and a particular lunar orbit. If this is to be economical and effective, these constraints must not introduce large performance penalties.

Motion of the Moon

The Moon revolves about the Earth in a nearly circular orbit of approximately 384,000 km average radius, with a period of 27.3 days. Its orbit is inclined 5 degrees to the ecliptic plane. The angular momentum of the Moon in its orbit is torqued by the attraction of the Sun, causing a slow precession of about 18.5 years' period. Therefore, the inclination of the Moon's orbit to Earth's equator varies as shown in Fig. 1., from 18° to 28°.

Synchronized Orbital Motions

The NASA space station orbit is planned as 28.5° inclination. Since this exceeds the Moon's maximum inclination, in-plane transfers from the space station orbit to the Moon will always be possible, *i.e.*, whenever a vector from the Earth to the Moon lies in the space station orbit plane. The frequency of occurrence depends not only on the motion of the Moon, but also on the motion of the space station orbital plane.

Inclined Earth orbits precess in a motion called regression of the line of nodes. This motion is illustrated in Fig. 2. The rate varies with inclination and altitude according to the equation given in the figure and is between six and seven degrees per day (retrograde) for typical space station orbits. As the Moon revolves posigrade at about 13° per day, it passes through the space station orbit plane about every 9 days on the average, permitting

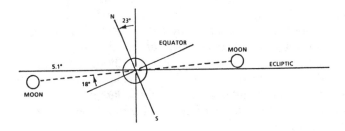

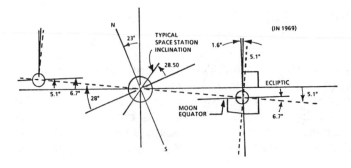

Figure 1. Variation in the Moon's orbital inclination.

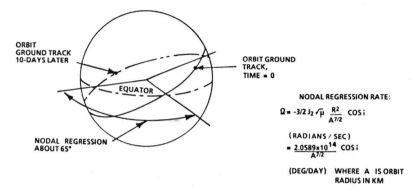

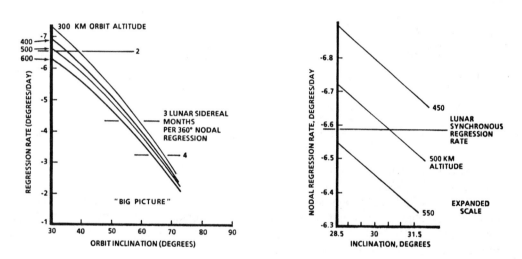

Figure 2.   Earth orbit nodal regression.

an in-plane departure from the space station to the Moon. An additional constraint is that the lunar polar orbit needs to be aligned with the incoming and departing transfer vectors so that lunar orbit insertion and departures are at least approximately in-plane. Because of the Moon's motion, the alignment involves a vector sum as illustrated in Fig. 3. The combined motions of the Moon and the space station orbit must be synchronized with the lunar period (lunar month) so that alignments for Earth departure and lunar arrival recur regularly. Synchronism is possible at orbit altitudes and inclinations very close to the NASA space station baseline of 500 km and 28.5°, as also shown in Fig. 2.

For in-plane (minimum ΔV) departure from Earth orbit, the lunar vehicle is constrained to travel to the Moon in the plane representing the space station orbit at the instant

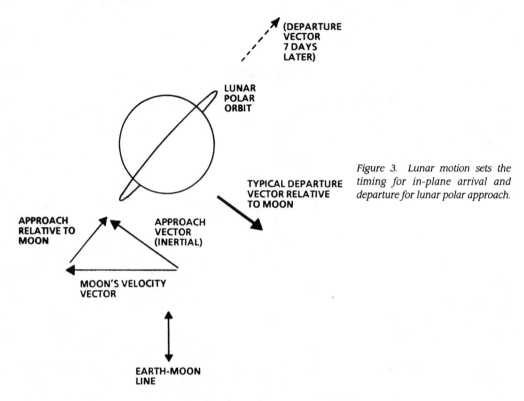

Figure 3. *Lunar motion sets the timing for in-plane arrival and departure for lunar polar approach.*

of departure (translunar injection, TLI). The plane of the translunar path is nearly "frozen" in inertial space (because of the energy of the transfer orbit), while the space station orbit continues its nodal regression. Earth departure and lunar arrival are timed so that the Moon, traveling in its orbital plane, arrives at the intersection of the two orbital planes at the same time the vehicle does. A similar timing constraint applies to the return trip; the launch to Earth (transEarth injection, TEI) occurs when the Moon crosses the intersection of its orbit plane and the orbit plane that the space station *will be in* at the time of Earth arrival. A typical synchronized round trip is depicted in Fig. 4.

Mission timing may be analyzed by depicting motions in the lunar orbit plane—the Moon moving posigrade at 13.1764° per day, and the intersection line of the lunar and space station orbits traveling retrograde, at half this rate (the slow movement of the lunar orbit plane is neglected). Because the space station orbit is regressing about the Earth's equator and not about a normal to the Moon's orbit, the rate of motion of the intersection line varies, and the dihedral angle between the orbits varies from the sum to the difference of their inclinations. Synchronism occurs whenever the combined angular motions, *i.e.*, displacements, add up to a complete circle, such as $N\pi$ where N is an even integer.

A practical synchronism occurs with N = 4 as illustrated in Fig. 5. In this example, the mission begins when the dihedral angle between the orbit planes is at minimum

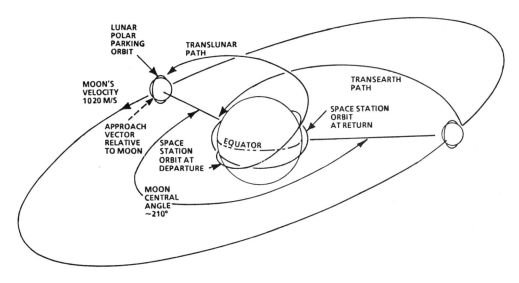

Figure 4. Pictorial of typical synchronized Earth-Moon round trip.

and the rate of motion of the intersection line is therefore maximum. The lunar vehicle paths traverse angles slightly less than π because typical trip times are less than the 5 days needed for a 180° transfer. This example represents the minimum practical mission duration under the constraints that we are dealing with, about three weeks.

For lunar polar orbits, the staytime in lunar orbit must be at least 13.66 days if out-of-plane descents and ascents are to be avoided. The Moon's rotation is locked to its orbital motion; the Moon rotates under the orbit at one revolution per 27.3 days; any particular landing site passes through the orbit plane twice per revolution. If further constraint is applied, a particular surface site, or to have the surface stay take place in lunar daylight (landing at dawn and liftoff at dusk), a longer orbital stay will usually be needed in waiting for orbital alignment with the lunar terminator. In such a case, a total mission duration on the order of 35 days may be needed (the mission time increases by increments of 13.66 days because a landing site passes under a polar lunar orbit at that interval). In general, one cannot have at the same time a particular lunar orbit, a particular surface site, and daylight surface stay.

Method of Analysis

The key issue for mission mode analysis is whether all these constraints can be accommodated without large performance penalties for propulsive plane changes to achieve the necessary orbital alignments. The problem can be visualized as a large nomograph representing the constraints and their interrelationships. Most of these can be represented in graphical form.

The complete analysis includes 36 interrelationships and is too lengthy to display in this short paper. As shown in Fig. 6, the analysis divides into four sub-problems.

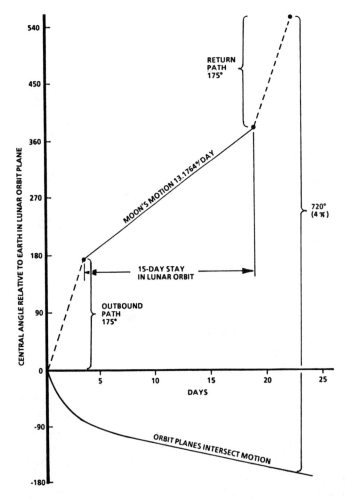

Figure 5. *Synchronizing orbital plane and path alignments.*

Determination of the outbound paths and ΔVs is also illustrated in Fig. 6. The out-of-plane angle between the arrival path and the Moon's orbit is just the dihedral angle between the space station and lunar orbit planes and is determined by the timing of mission start. TLI ΔV and the geometry and velocity of lunar approach are determined by the outbound transit time and the above out-of-plane angle. The lunar orbit insertion (LOI) ΔV depends on approach velocity relative to the Moon, which is in turn dependent on transit time (the Moon's orbital velocity of about 1020 m/s must also be taken into account), the altitude of the lunar orbit, and the plane change accomplished entering lunar orbit. For this study, lunar orbit altitude was fixed at 100 km. Figure 7 graphs ΔV entering lunar orbit at this altitude.

For this study, plane changes departing from or returning to Earth orbit were ruled out. They are expensive and don't help much to establish the needed synchronisms. (They

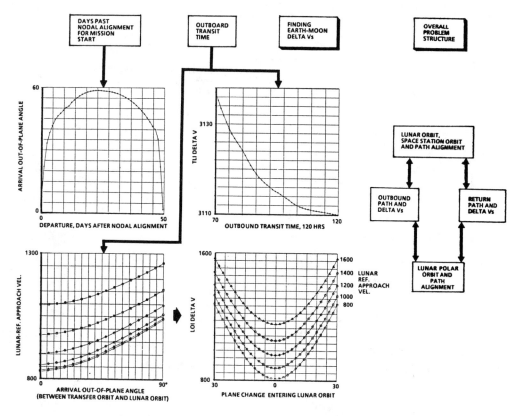

Figure 6. Problem structure with a partial example of detailed structure.

may help by relaxing timing constraints.) Satisfaction of lunar orbit, space station orbit, and path alignments sets conditions for satisfying other constraints. Fixing outbound transit time and mission duration permits iteration on lunar orbit staytime and return transit time to satisfy the constraints. The transit times and staytime set conditions for satifying lunar polar orbit and path alignments. Fixing the plane change entering lunar orbit leaves the plane change departing lunar orbit as a free parameter adjusted to satisfy polar orbit/ path constraints. Outbound transit time, mission duration, and plane change entering lunar orbit are thus *optimization parameters* that can be adjusted to minimize mission ΔV.

The analysis was conducted using a computer code analogous to a modern "spread sheet" code, but having capability to accept graphical (table lookup) relationship descriptions, to internally iterate to resolve interdependencies, and to scan optimization parameters to optimize an objective function, in this case total mission ΔV. Representative results are shown in Fig. 8. (Values shown include 75 m/s margin above the ideal values for finite-burn losses, *etc.*) It was found that adjusting the optimization parameters permitted

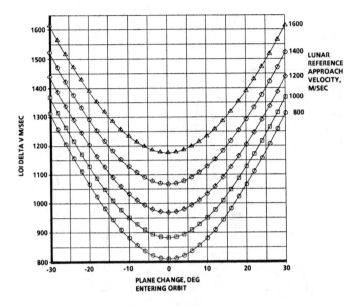

Figure 7. LOI ΔV parametrics.

lunar arrival and departures with very small plane changes, yielding total mission ΔV no greater than typical for an unconstrained mission. The added constraints of the mission can be met by timing and do not impose performance penalties. The timing constraints are, however, stringent; near-optimal missions have windows on the order of a few hours.

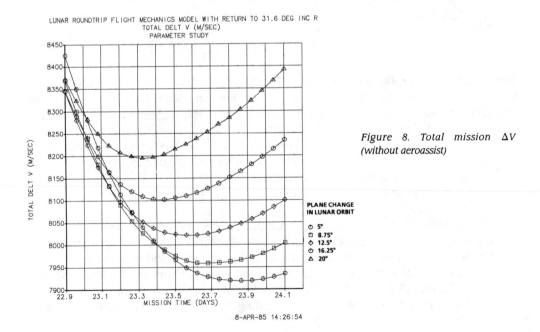

Figure 8. Total mission ΔV (without aeroassist)

Table 1.   Delta V Budgets

	Rough	Calculated
TLI	3200 m/s	3139 (includes 75 m/s g loss)
LOI	800 m/s	867
Landing	2100 m/s	2100 (from Apollo)
Ascent	2000 m/s	2000 (from Apollo)
TEI	800 m/s	906
EOI	3200 m/s	3061 (propulsive)
	or 3200 m/s	200 (aeroassisted)
Total	12100 m/s	12073 (propulsive)
	or 9100 m/s	9212 (aeroassisted)

A figure of 915 m/s for LOI/TEI is frequently given in the literature. This value was used in Apollo ΔV budgets constrained to free-return Earth-Moon trajectories; *i.e.,* the lunar vehicle would swing around the Moon and return to Earth without propulsion in the lunar vicinity.

(Note: it is probably the case that very small plane changes leaving or returning to Earth orbit would alleviate the time constraints.)

Mission Modes

Mission modes were screened using the rough ΔV budget presented in Table 1. The rough budget is compared to typical calculated values from the computer analysis. Base buildup and resupply needs are summarized in Tables 2 and 3. It is clear that delivery capabilities of about 13 metric tons of payload (unmanned) for base hardware deliver and 5 tons (manned) for resupply operations are needed. The resupply scenario presumed a lunar base crew of 12, lunar staytime of 164 days, and resupply mission with 4 crew exchanged every 55 days. A conceptual design analysis of the lunar transfer crew cab requirements concluded that an adequate crew cab could be derived from a space station common module and its subsystems and that the mass including crew will be approximately five tons.

Table 2.   Base Equipment Delivery Requirements

Flights 1 and 2	Habitat modules
Flights 3 and 4	Laboratory/Work modules
Flight 5	Construction equipment
Flight 6	Nuclear power plant
Flight 7	Solar/Regen fuel cell emergency power supply
Flight 8	Scientific equipment
Flight 9	Mobile explorer
Flight 10	Lunar oxygen production plant

Each flight has cargo capacity of 13 tons (28,500 pounds).

Table 3. Base Resupply Summary

Crew staytime 180 days
Exchange interval 4 crew/60 days
Resupply (Per 60 Days)

Food, water and atmosphere	3.30 Tons
EVA	0.27 Tons
Science	0.22 Tons
Equipment and Subsystems	1.21 Tons
	5.00 Tons (11,000 pounds)

Crew transport module (4 people)—5.0 tons
Net delivered—10
Net returned—5

We selected mission modes that could use orbital transfer stages similar in size and general configuration to the reusable orbit transfer vehicles (ROTVs) now under study for geosynchronous operations, while at the same time providing useful payload for lunar base support and buildup operations. A number of modes were considered, and those shown in Fig. 9 were selected for further analysis. The cargo and independent lunar surface sortie (ILSS) modes are appropriate to base buildup and early operations, as they do not depend in any way on lunar resources. If useful quantities of liquid oxygen can be produced on the lunar surface, the surface-refueling and orbiting lunar station (OLS) modes are attractive. These latter modes are fully reusable; the others are not.

The direct cargo mode offers payload adequate for base buildup with a simple mode that does not depend on lunar resources but that expends a stage. The ILSS mode is the most direct way of providing a crew round trip without lunar resources. The ILSS mode also expends a stage, the lunar lander, and its crew cab. The stage could be recovered, but this would require additional propellant to be delivered to lunar orbit and the propellant to be transfered to the lander after its return from the lunar surface. If the mode were to be used extensively, stage recover might be cost effective.

The surface refuel mode uses the lunar surface rather than an orbit as a staging point, and it is not constrained by lunar orbit selection factors. Its lunar stage is a somewhat more complex design, requiring lunar landing legs and an aerobrake on the same stage. The OLS mode is by far the most efficient as measured by payload versus delivery mass to low-Earth orbit. It does require propellant transfer operations in zero g, as the lander stage must be loaded with hydrogen from Earth and the Earth return stage loaded with lunar oxygen for the TEI and EOI maneuvers.

A nominal rocket engine specific impulse of 460 s (jet velocity of 4511 m/s) was used to compute stage sizes and propellant quantities for each burn of the mission profiles. Aeroassist was assumed for Earth return. Stage propellant loads cluster around 25 metric tons of hydrogen and oxygen at a mixture ratio of 6. This compares favorably with the 20–25 tons typically chosen for a GEO orbit ROTV. These stages are compatible with

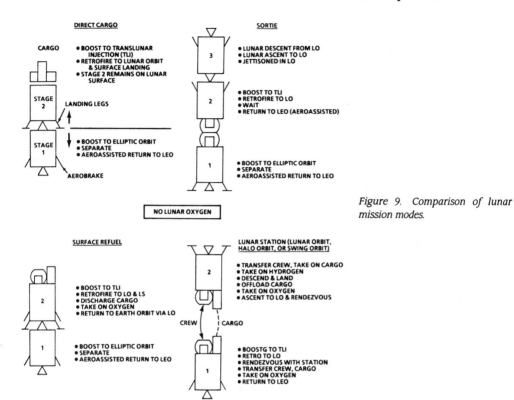

Figure 9. Comparison of lunar mission modes.

space station ROTV accommodation concepts, although the assembled vehicles may not be compatible unless provisions for handling such assemblies are provided. A summary of results from the mode analysis is presented in Table 4. The cost indicators in Table 4 include costs for Earth-to-orbit transportation and use/expenditure of lunar transport hardware but not indirect costs such as mission control operations. Benefits of lunar oxygen and of using a shuttle-derived cargo vehicle to deliver propellant to the space station for lunar operations are apparent.

Mode Performance and Sensitivities

The ILSS mode and three-stage vehicle system might be the first used to validate potential basing sites by manned visits before delivery of base hardware. It might also be used for base resupply operations until lunar oxygen production is validated and operational. Accordingly, performance and sensitivity analyses were made for this mode.

Two crew cabs are used, one for the lunar orbit and one for the lander. This avoids disrupting interfaces between a crew cab and its propulsion stage. For these calculations, the lunar lander and its crew cab were assumed jettisoned in lunar orbit to reduce Earth departure mass.

Table 4. Lunar Mission Mode Performance and Cost Screening Summary

Mode Name	Booster Stage Propellant Load (Metric Tons)	Lunar Orbit State Propellant Load	Lander State Propellant Load	Payload Delivered/ Returned	Propellant from Earth (Metric Tons)	Lunar Oxygen (Metric Tons)	Stages Expended	Crew Cabs Expended	Shuttle Transport Cost at 85m/Flt 26T- Payload	CLV Transport Cost at 75m/Flt (60T- Payload)	Lunar Transport Vehicle Cost	Total Mode Cost With Shuttle	Total Mode Cost With CLV
Direct cargo	24	Not req'd	25	13/0	50	None	1	0	218	83	55	273	138
ILSS	25	23	12	1/0.5 + Crew round trip	59	None	1	1	236	91	115	351	206
Orbiting lunar station (OLS)	Not req'd	28	20	5/0.5+ Crew round trip	28	18	0	0	106	41	20	126	61
Surface refuel	24	Not req'd	24	5/0.5+ Crew round trip	48	8	0	0	191	75	15	206	90

CREW CAB = 5 TONS
LANDED PAYLOAD = 1 TON EXCEPT NOTED

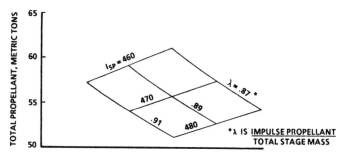

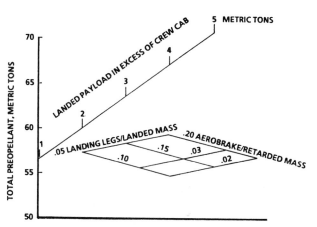

Figure 10. Sensitivities of ILSS modes.

Approximate sensitivities of the aeroassisted ILSS mode are shown in Fig. 10 for delivered payload, stage inert mass (defined in terms of propellant fraction), rocket specific impulse, and landing leg and aerobrake mass fraction. The mode is not highly sensitive; the others are even less so. "Not highly sensitive" means that the total propellant and mass variations are comparable to the uncertainties in design variables; the mode does not "go over a cliff" in the presence of uncertainties in typical preliminary design estimations.

Aeroassisted ILSS mode performance was also calculated by attaching a vehicle-sizing model to the lunar mission ΔV model described earlier. In these coupled models, vehicle performance and sizing calculations are performed for each ΔVs. Results are shown in Fig. 11. The total mass is greater than that shown in Fig. 10 because (1) delivered payload was 5000 kg; (2) accurate ΔVs are slightly higher than the rough values used for screening comparison; and (3) vehicle flight performance reserves were included. Results gave added confidence that the ILSS mode is a good candidate for an early manned return to the Moon.

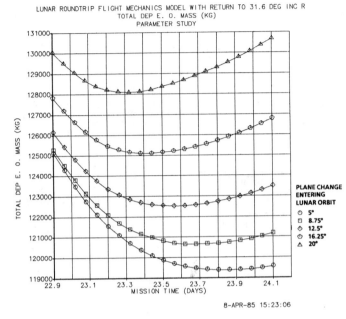

LUNAR ROUNDTRIP FLIGHT MECHANICS MODEL WITH RETURN TO 31.6 DEG INC R
TOTAL DEP E. O. MASS (KG)
PARAMETER STUDY

Figure 11. Earth orbit departure mass.

PLANE CHANGE
ENTERING
LUNAR ORBIT
- 5°
- 8.75°
- 12.5°
- 16.25°
- 20°

8-APR-85 15:23:06

CONCLUSIONS

Contemplation of a permanent manned lunar base in our space future should not be inhibited by the $80-billion costs of the Apollo program in present-year dollars. Few new developments are needed, and these are consistent with current paths of technology evolution. The transportation operations cost of supporting a modest lunar base can evolve to less than a billion per year in 1985 dollars.

Synergistic cost benefits accrue from cryogenic space propulsion, aeroassisted ROTV operations (the benefit for return from the Moon is greater than for return from GEO orbit), the space station as an intermodal transportatoin complex, a heavy-lift shuttle, and oxygen production on the Moon.

Mission designs can accommodate flight mechanics constraints in such a way as to permit repetitive operations between particular Earth and lunar orbits with negligible performance penalties.

Efficient mission modes can use stages in the ROTV-size range; these can be ROTVs or simple derivatives thereof. The space station can provide practical spaceport operations services for lunar operations.

A logical evolution exists from site exploration through buildup to support operations. All can use common transportation hardware and technology. An evolutionary step to very efficient resupply operations through lunar-produced oxygen is a natural beginning for the practical use of extraterrestrial resources.

IMPACT OF LUNAR AND PLANETARY MISSIONS ON THE SPACE STATION

G. R. Babb, H. P. Davis, P. G. Phillips, and W. R. Stump

Eagle Engineering, Inc., 711 Bay Area Blvd., Houston, TX 77598

The report examines the impact of several advanced planetary missions and a populated lunar base on the growth space station. Planetary missions considered include a Mercury Orbiter, a Saturn Orbiter with Multiple Titan Probes, and sample return missions from Mars, the comet Kopff, and the main belt asteroid Ceres. A manned lunar base buildup scenario is defined—from preliminary lunar surveys through 10 years of construction, to the establishment of a permanent 18-person facility with the capability to produce oxygen to be used as propellant. For the lunar base, the space station must hangar at least two OTVs, store 100 metric tons of cryogens, and support an average of 14 OTV launch, return, and refurbishment cycles per year. An average of 630 metric tons per year must be launched to the space station for lunar base support during the 10 years of base construction. Approximately 70% of this cargo from Earth is OTV propellant. An Unmanned Launch Vehicle (ULV) capable of lifting 100 metric tons net useful payload is considered necessary to deliver this propellant. An average launch rate of one shuttle every two months and one ULV every three months to the growth space station will provide the required 630 metric tons per year. Planetary sample return missions require a dedicated quarantine module.

INTRODUCTION

NASA and contractors are now working on the conceptual design of the Low Earth Orbit (LEO) space station. For the Initial Operational Capability (IOC) space station configuration to evolve smoothly, the designers must consider the requirements of the turn of the century "growth" space station. This study, performed by Eagle Engineering, Inc., for the Johnson Space Center (JSC) Solar System Exploration Division, examines the effects of advanced lunar and planetary missions upon the growth space station.

Mass estimates were developed for a lunar base with an emphasis on science, which, in its later phases of development, uses oxygen produced on the Moon in a transportation system sized to land its elements on the lunar surface. A 10-year flight schedule was developed, including weights, propellants, crew size, *etc.* The impact of such an undertaking upon the space station was then estimated.

Similarly, five advanced planetary missions were examined, as were three sample return missions, and two orbiter/probe missions. Weight statements and trajectories for each of the missions were determined. For this effort a NASA-developed standard OTV (Orbit Transfer Vehicle) was used. Propellant loads, configurations, and mission plans were determined. Some of these missions were found to require two OTVs in a two-stage stack. The effect on the space station of conducting these missions was then estimated. The final report of the study (Babb *et al.*, 1981) provides details and the full methodology.

Ground Rules and Assumptions

The OTVs used in this study are based upon a report by Scott *et al.* (1985). They are sized to deliver 9 metric tons one way or 6 metric tons round trip to Geosynchronous Orbit (GEO) using a single stage and to deliver 18 metric tons of lunar surface payload plus a lander to lunar orbit using two OTV stages in tandem. Both OTV stages then return to the space station.

For this study a mature aerobraking technology is assumed. All stages are to use LO_2/LH_2 with the exception of expendable ascent stages, which use storables.

LUNAR MISSIONS

The transportation operations required for buildup and support of a lunar base were examined using the space station and a fleet of large OTVs as the major transportation elements.

Lunar Base Description

The lunar base model examined was supplied by NASA/JSC (B. Roberts, personal communication, 1984) as an example of a research installation with an emphasis on science and some lunar materials utilization. With this base model, lunar oxygen production starts in the fourth year. The oxygen is provided for a Reusable Lunar Lander/Launcher (R–LEM) so that only hydrogen fuel needs to be brought from Earth.

At the end of the first 10 years, the lunar base staffed by 18 permanent personnel includes:

Five habitability modules
Five research units: a geochemical laboratory, a chemical/biological laboratory,
    a geochemical/petrology laboratory, a particle accelerator, and a radio
    telescope
Three production plants (preceded by pilot plants): oxygen, ceramics, and
    metallurgy
Two shops
Three power units
One earthmover/crane
Three mobility units and trailers

Lunar Base Support Requirements

This study focuses on the base buildup, starting in the year 2005. The eight years preceding the buildup of the base are devoted to unmanned exploration and mapping of the Moon. One landing of a roving vehicle per year is required and could be flown directly from the shuttle with a modified Centaur-class vehicle.

There are 25 base elements that average 17.5 metric tons each. Figure 1 shows delivery of a habitability module. They are delivered over a period of 10 years for a total of 465 metric tons. An additional 233 metric tons of miscellaneous cargo is delivered

Figure 1.   Unloading module on lunar surface.

during the manned resupply missions. Launch manifest and lunar mission schedules were derived for the entire 10-year buildup period. Table 1 shows the detailed schedule for the first year (2005).

Lunar Transportation System

The elements of the transportation system are as follows.

1. Aerobraking Orbital Transfer Vehicle (AOTV)—A 49 metric ton gross mass LO_2/ LH_2 propulsion stage (42 metric tons of propellants).

2. Expendable Lunar Lander (E-Lander)—A LO_2/LH_2 landing stage with 13.6 metric tons of propellant that will land 17.5 metric tons.

3. OTV Manned Module (OMM)—A 5.5 metric ton orbit-to-orbit reusable crew transport module with four personnel to be carried on the OTV.

4. Lunar Landing Manned Module (LLMM)—A 3.25 metric ton expendable module for temporary life support of four crew members during lunar landing and launching. It is attached to a 7.6 metric ton expendable launcher.

5. Reusable Lunar Lander/Launcher (R-LEM)—A 5 metric ton LO_2/LH_2 single-stage vehicle using lunar-produced O_2 for propellant.

6. Reusable Lunar Landing Manned Module (R-LLMM)—A 5 metric ton, six man, lunar-based crew compartment for the R-LEM. This crew compartment is maintained and stored at the lunar base.

Table 1. Detailed Launch Manifest and Lunar Mission Schedule

Month	Launch No.	Type	Cargo Manifest	Cargo Wt.*	Lunar Flight No.	Flight Type	Space Station Tasks for Flight	LOX/LH2 at Depot*
							Year 2005	
Jan	5-1	SD-ULV	LOX/LH2 propellant supply unit	100				100
Feb	5-2	STS-ACC	E-lander, +base element #1	21	L5-1	Unmanned Delivery	Prepare (Check out & fuel) 2 OTVs and E-lander; Check out cargo; & mate stack; -(2 OTVs, E-lander, and base element)	0
March	5-3	SD-ULV	LOX/LH2 propellant supply unit	100				100
April	5-4	STS-ACC	E-lander, +base element #2	21	L5-2	Unmanned Delivery	Prepare (Check out & fuel) 2 OTVs and E-lander; Check out cargo; & mate stack; -(2 OTVs, E-lander, and base element)	0
May	5-5	SD-ULV	LOX/LH2 propellant supply unit	100				100
	5-6	STS-ACC	E-lander, +base element #3	21	L5-3	Unmanned Delivery	Prepare (Check out & fuel) 2 OTVs and E-lander; Check out cargo; & mate stack; -(2 OTVs, E-lander, and base element)	0
July	5-7	SD-ULV	LOX/LH2 propellant supply unit	100				100
Aug	5-8	STS-ACC	E-lander, +base element #4	21	L5-4	Unmanned Delivery	Prepare (Check out & fuel) 2 OTVs and E-lander; Check out cargo; & mate stack; -(2 OTVs, E-lander, and base element)	0
Sept	5-9	SD-ULV	LOX/LH2 propellant supply unit	100				100
Oct	5-10	STS-ACC	E-lander, +E-LLMM/Ascent, +OMM, +4 crew +2 ton PL	20	L5-5	Manned Sortie	Prepare 2 OTVs and E-lander; Check OMM, E-LLMM/Ascent; mate stack -(2 OTVs, OMM, E-LLMM/Ascent, & E-lander); And transfer crew to OMM	100
Dec	5-12	STS-ACC	E-lander, +E-LLMM/Ascent, +OMM, +4 crew +2 ton PL	20	L5-6	MS-(4M +2t for 14 days)	Prepare 2 OTVs and E-lander; Checkout OMM, E-LLMM/Ascent; mate stack -(2 OTVs, OMM, E-LLMM/Ascent, & E-lander); And transfer crew to OMM	0

Year 2006

Month		Vehicle	Payload	Mass*		Mission	Operations	
Jan	6-1	SD-ULV	LOX/LH2 propellant supply unit	100				100
Feb	6-2	STS-ACC	E-lander, +base element #5	21	L6-1	Unmanned Delivery	Prepare (Check out & fuel) 2 OTVs and E-lander; Check out cargo; & mate stack; -(2 OTVs, E-lander, and base element)	0
March								
April	6-3	SD-ULV	LOX/LH2 propellant supply unit	100				100
May	6-4	STS-ACC	E-lander, +E-LLMM/Ascent, +OMM, +4 crew +2 ton PL + 4 tons of AOTV elements	20	L6-2	MS-(4M +2t for 14 days)	Prepare 2 OTVs and E-lander; Check OMM, E-LLMM/Ascent; mate stack -(2 OTVs, OMM, E-LLMM/Ascent, & E-lander); And transfer crew to OMM	0
June								
July	6-5	SD-ULV	LOX/LH2 propellant supply unit	100				100
Aug	6-6	STS-ACC	E-lander, +base element #6	21	L6-3	Unmanned Delivery	Prepare (Check out & fuel) 2 OTVs and E-lander; Check out cargo; & mate stack; -(2 OTVs, E-lander, and base element)	0
Sept								
Oct	6-7	SD-ULV	LOX/LH2 propellant supply unit	100				100
Nov	6-8	STS-ACC	E-lander, +E-LLMM/Ascent, +OMM, +4 crew +2 ton PL + 4 tons of AOTV elements	20	L6-4	MS-(4M +2t for 14 days)	Prepare 2 OTVs E-lander; and OMM; Checkout E-LLMM/Ascent; Mate Stack -(2 OTVs, OMM, E-LLMM/Ascent, & E-Lander); And transfer crew to OMM	0

*in million tons

7. Large OMM—An enlarged, 8 metric ton, reusable crew transport module for six personnel, carried on an OTV.

8. H_2 Transfer Tank—A 1 metric ton expendable container for carrying 4 metric tons of LH_2 to the lunar surface as fuel for the R-LEM.

9. Orbital Maneuvering Vehicle (OMV)—A small, 3 or 4 metric ton, remotely operated propulsion stage to provide controlled close-in operations at the space station.

Earth Launch Vehicles

The Earth launch vehicles include the following.

1. Space Shuttle (STS)—The space shuttle can launch 25 metric tons of crew and payloads to the space station (about 50% improvement over STS capabilities cited in the recent NASA literature).

2. Shuttle/Aft-Cargo Carrier—Shuttle with a cargo compartment on aft end of the External Tank. This allows the launch of oversized vehicles. It is used for launching E-Landers.

3. Shuttle-Derived Unmanned Launch Vehicle (SDULV)—Unmanned launcher, which was designed using shuttle elements, can deliver 100 metric tons of LO_2/LH_2 to the space station propellant depot. It is used for launching all cryogenic propellants.

Figure 2 shows a pair of OTVs with an E-Lander and an unmanned base element as they debark from the space station orbit. An OMV is shown returning to the space station after having moved the "stack" to a safe distance.

Figure 2. OTV departing space station with lander and module.

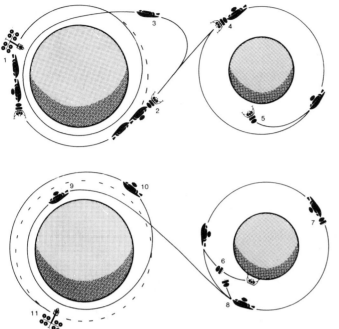

Figure 3. Manned lunar mission scenario. 1-Stack departs space station; 2-trans-lunar injection burn; 3-first stage returns to space station; 4-second stage, lander, and manned module insert into lunar orbit; 5-lander descends; 6-ascent stage departs lunar surface; 7-ascent module rendezvous with second stage; 8-second stage returns to Earth with OMM, ascent module discards; 9-aerobraking; 10-circularization above space station orbit; 11-rendezvous with space station.

Figure 3 depicts the flight scenario for a manned lunar flight. The unmanned flights are similar except that no cargo elements are left in orbit with the OTV, and none return from the lunar surface. The OTV returns to Earth empty.

The Earth launch tonnage requirements to the space station from Earth for the lunar base over the 10-year buildup are shown in Fig. 4. The LO_2/LH_2 tonnage requirements shown include the propellant for both the OTV and the lunar lander. Such propellant is only the portion (of the total required) that would be launched by the SD-ULV to the Low Earth Orbit propellant depot.

IMPACT OF THE LUNAR MISSIONS ON THE GROWTH SPACE STATION

The space station must provide propellant storage and transfer facilities (propellant depot), the capability for assembly of the mission stack, facilities for payload checkout and integration into mission stacks, maintenance and checkout of vehicles stored on-orbit (OTVs, OMVs, OMMs), flight control (rendezvous, proximity operations, and docking), personnel billeting, and temporary payload storage.

Hardware required to be added to the growth space station includes:

1. Permanent basing (hangars, storage, and shops) for four OTVs, three OMMs, and two OMVs.

2. Gantrys for preparing mission stacks of up to 40 m length of two OTVs, plus a lunar lander, plus various manned and unmanned lunar cargo elements.

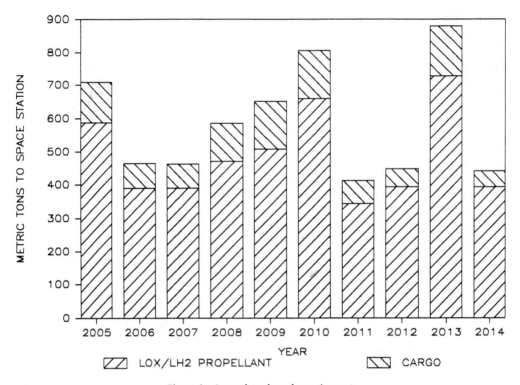

Figure 4.   Lunar base launch requirements.

3. A propellant depot for cryogenic LO_2/LH_2 propellant with the capacity of at least two tanker units of 100 metric tons each.

4. A propellant transfer capability to perform a measured propellant transfer from the depot to various vehicles in the mission stack at the assembly docks. A rate of 5 tons per hour is required to complete transfer in one 24-hour period.

5. Temporary storage for lunar vehicles and 20–30 tons of lunar payload.

6. An additional habitat module for housing the additional space station crew and temporary billeting of 4–6 transient lunar base personnel.

7. An estimated 20 kW of continuous *additional* power with appropriate heat rejection. This power budget breaks down into 10 kW for depot cryogenic refrigeration, 5 kW minimum for the extra habitat, and 5 kW or more for gantrys.

Manpower requirements identified at the space station are 14 man-weeks per lunar sortie. Five of these man-weeks will be required for general OTV turnaround and maintenance (Maloney *et al.*, 1983). An estimated five man-weeks will be required for stacking the lander and the manned module and for fueling the lander. It is assumed that an extensive checkout will be performed on the completed stack. One man-week will be dedicated to manned module refurbishment outfitting and checkout. Finally, three man-weeks are estimated for traffic control and OMV operations. These operations include one shuttle arrival for cargo and crew transfer, one ULV arrival for propellant, one ULV propellant tank disposal, one stack departure, and two separate OTV arrivals. These operations require a minimum extra crew complement of two persons and possibly another pair for other required tasks not yet identified.

Since an average of one lunar sortie will be required every eight weeks, two dedicated crew members will probably be required. Less complex unmanned lunar and planetary missions will require less manpower.

All of the required space station capabilities need to be online at the beginning of the lunar base buildup in 2005. Capabilities will have to be developed earlier and procedures learned from geosynchronous mission preparations or some of the more difficult unmanned planetary missions.

A brief sensitivity study showed that reasonable changes in I_{sp} and inert weight do not alter the scale of the operation to first order. An I_{sp} change of 20 seconds combined with a 20% decrease in vehicle inert weights can reduce the weight in Low Earth Orbit (LEO) from 10–20% (mostly LO_2/LH_2 propellant), a reduction of about one unmanned launch of propellant per year.

A brief examination of planetary mission launch from lunar orbit using lunar produced oxygen for propellant found no great advantage over launch from LEO.

PLANETARY MISSIONS

A set of missions was examined for impact to the growth space station. They were chosen to show how the space station with reusable OTVs might enable more ambitious planetary exploration and to see how the use of this infrastructure for planetary exploration would affect the growth space station. This set of missions is an example set and not a proposed addition to the NASA Plantary Exploration Program (1983).

Table 2.   Planetary Missions—Performance Summary

	C3 (km/s^2)	Type of OTV*	Payload out of LEO[†]	LEO Total Departure Mass[†]	OTV Propellant Load[†]	Propellant + Payload (Lift Req.)[†]
Mars Sample Return	9.0	1 Stage Reusable	8.89	44.03	27.76	36.65
Kopff Sample return	80.7	2 Stage, 1st Stage Returns	8.38	92.49	71.51	79.89
Ceres Sample Return	9.9	2 Stage, 1st Stage Returns	43.57	131.59	75.47	119.04
Mercury Orbiter	18.7	1 Stage Reusable	5.63	41.62	28.90	34.53
Titan Probes/Saturn Orbiter	50.5	1 Stage Expendable	6.34	53.54	41.81	48.15

*Isp = 455.4 seconds, all stages have a total propellant capacity of 42 metric tons. A = 3,731 kg, B = .0785. Stages that do not return have the aerobrake removed.
[†]In metric tons

All of the example missions used the delta V tables and detailed weight statements for spacecraft to the subsystem level that can be found in Babb *et al.* (1981). For the planetary missions considered, Table 2 lists the C3's and gives the weight breakdowns in terms of payload out of LEO, LEO OTV propellant load, and propellant plus payload (life required).

Mars Sample Return (MSR) Mission (November 1996 Launch)

The Mars sample return mission was the most complex one studied. The general mission plan and orbital mechanics data were taken from Bourke *et al.* (1984, 1985) and Feingold *et al.* (1982).

The space-based aerobraked OTV will load 27.76 metric tons of LO_2/LH_2 into its 42 metric ton capacity tanks and return to the space station after releasing the spacecraft. Figure 5 shows the spacecraft within its aeroshell during mating with the OTV. Figure 6 depicts the mission sequence. The returned sample is to be retrieved from Earth orbit with an OMV and brought to the space station.

Table 3 summarizes the impact of the various planetary missions on the space station/ OTV system. The major impact is that the Quarantine Module is now required to provide a place to receive the returned samples. Each of the sample return missions is assumed to use the Quarantine Module, which is discussed in more detail following the sections on individual missions.

Figure 5. Mars sample return mission OTV mating.

Figure 6.   OMV returns sample to quarantine module.

Comet Kopff Sample Return Mission (July 2003 Launch)

This mission was complex, but not as well studied as the MSR mission. The general mission plan was taken from Feingold *et al.* (1983) and Draper (1984). As Table 2 shows, a stack of two standard OTVs is required to take this Mars sample return size payload to a C3 of 80.7 $(km/s)^2$. In the mission scenario depicted in Figure 7, only the first stage returns to the space station. The Mariner Mark II (MMII) spacecraft carries the samplers out and back (Feingold *et al.*, 1983; Draper, 1984). The returned sample uses aerobraking to enter Earth orbit and is taken to the Quarantine Module by an OMV.

This mission requires the Quarantine Module, and in addition, the capability to stack and checkout two OTVs. The aerobrake is removed from the second stage, which does not return. An older OTV, near the last of its estimated 10 or so missions might be used for this stage.

Ceres Sample Return Mission (October 1994 Launch)

This mission is similar to the Kopff mission previously discussed. The mission sequence, depicted in Figure 7, is essentially the same as that of the Kopff sample return mission. The trajectory uses a double Mars–gravity–assist outbound and is ballistic inbound. Both of these missions use the MMII spacecraft for the outbound and inbound legs. The returning Mariner spacecraft and the large delta V required for a ballistic return on the Ceres mission leads to a particularly large weight penalty. A substantial reduction in Earth launch weights

Table 3.   Planetary Missions—Impacts on the Space Station

Requirements	Mars Sample Return	Kopff Sample Return	Ceres Sample Return	Mercury Orbiter	Titan Probes/ Saturn Orbiter
Space Station Hardware					
No. of OTV's expended (not returned)	0	1	1	0	1
No. of OTV refurb. kits	1	2	2	1	1
Gantry to stack two stages		yes	yes		
Check out equip. for two stage stack		yes	yes		
Quarantine module	yes	yes	yes		
Additional power, KW	5	5	5		
Additional thermal control, no. of standard modules	1	1	1		
Space Station Manhours					
OTV refurbishment	52	103	103	52	52
Aerobrake removal		21	21		21
OTV/payload integration & C/O	11	21	21	11	11
Fuel, release, and launch	24	36	36	24	24
Rendez./retrieve OTV using OMV	12	12	12	12	
Shuttle rendez./payload removal	3	2	12	2	2
ULV fuel delivery	7	17	18	7	10
Sample retrieval using OMV	8	8	8		
Sample analysis and shipment	24	16	16		
Total mission manhours	138	236	247	106	119

might be achieved by designing dedicated spacecraft for these missions. On the other hand, expending an old OTV, particularly if the capability to stack two stages already existed for other purposes, might be more cost effective. More study of this mission is required to choose the most reasonable solution.

The impact of the Ceres sample return mission on the growth space station configuration, shown in Table 3, is essentially the same as for the Kopff sample return mission.

Mercury Orbiter Mission (June 1994 Launch)

The Mercury Orbiter mission uses a MMII (Draper, 1984) and a dual Venus swingby trajectory (Friedlander *et al.*, 1982). One reusable 42 metric ton capacity OTV uses 28.90 metric tons to launch this mission and return.

One standard mission cycle (payload integration/OTV fuel/checkout/launch/ refurbishment) is required of the space station.

Saturn Orbiter/Multiple Titan Probes Mission (April 1993 Launch)

The Saturn orbiter with multiple Titan probes mission, patterned after the one described by Swenson *et al.* (1984) uses the MMII spacecraft as a carrier and an Earth-gravity-

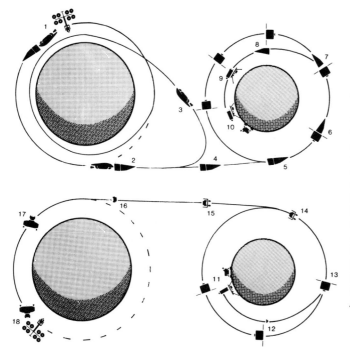

Figure 7. Mars sample return scenario. 1–Stack leaves space station; 2–trans-Mars injection; 3–first stage returns; 4–trans-Mars voyage; 5–aerocapture and Mars orbit insertion; 6–jettison MOV aeroshell; 7–lander and orbiter separate; 8–lander enters Mars atmosphere; 9–landing on martian surface; 10–collect samples; 11–launch from Mars; 12–Mars rendezvous vehicle injection into Mars orbit; 13–Mars orbiter vehicle maneuvers to rendezvous with MRV; 14–trans-Earth injection; 15–trans-Earth voyage; 16–Earth orbit capsule insertion into Earth orbit; 17–OMV rendezvous with EOC; 18–OMV returns EOC with sample to space station quarantine module.

assist trajectory. To make the mission different from the others in terms of its impact on the growth space station, additional probes were added. One standard 42 metric ton propellant capacity OTV can launch the mission, but does not have enough fuel to return. If the expended OTV is an old one and the assembly of a new OTV is not charged exclusively to this mission, then the impact on the growth space station (Table 3) is less than for the Mercury Orbiter. However, the man-hours required to remove the aerobrake from the Saturn Orbiter OTV more than make up for the space station man-hours gained from no retrieval.

QUARANTINE MODULES

Given the existence of a space-based reusable OTV and the capability to stack two of them (which is assumed for this study) the major design impact to the growth space station of planetary missions is the requirement for a Quarantine Module for the sample return missions.

There are several ways to handle returned samples. De Vincenzi and Bagby (1981) discuss a dedicated space station. Another attractive proposal is to environmentally separate one module of the space station from the rest with its own life support, with the module being rigidly attached but not interconnected by pressurized passageways, yet permitting use of station power and cooling. An OMV would deliver a returned sample to an airlock on this module (see Fig. 8) where it would be repackaged, perhaps with a glove box,

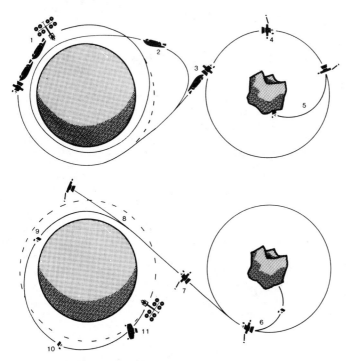

Figure 8. Ceres or Kopff sample return scenario. 1–Stack departs space station; 2–first stage burn, separation and return to space station; 3–second stage burn, trans-Ceres/Kopff voyage; 4–spacecraft rendezvous and asteroid/comet survey; 5–lander on surface, spacecraft in orbit; 6–spacecraft recovers samplers and departs for Earth; 7–trans-Earth voyage; 8–carrier and Earth orbit capsule separate; 9–EOC aerocapture for Earth orbit insertion; 10–circularization above space station orbit; 11–OMV rendezvous with EOC and return to space station quarantine module.

into a "superbox" capable of withstanding any conceivable re-entry accident without releasing material. A small amount of the material might also be examined at the space station. The "superbox" would then be returned to Earth for processing in a facility similar to the present Center for Disease Control (CDC) facility.

SUMMARY

Construction of a lunar base has a major impact on the growth space station. The space station would become much more important as a transportation node than current plans indicate. The planetary missions studied have significantly less impact, with the exception of the addition of a Quarantine Module to the space station. Other options for handling quarantine of planetary samples are now being studied.

REFERENCES

Babb G. R., Phillips P. G., and Stump W. R. (1981) *Impact of Lunar and Planetary Missions on the Space Station, Final Report.* Eagle Engineering Report No. 84–85D. Eagle Engineering, Houston. 132 pp.

Bourke R. D., Blanchard D. P., and DeVries J. P. (1984) *Mars Sample Return Missions.* JPL Publication No. D-1698. Jet Propulsion Laboratory, Pasadena. 75 pp.

Bourke R. D., Friedlander A. L., DeVries J. P., Norton H. N., and Blanchard D. P. (1984) *Mars Sample Return, Complete Report.* JPL Quarterly Review, Advanced Planetary Studies. Jet Propulsion Laboratory, Pasadena. 115 pp.

De Vincenzi D. L. and Bagby J. R. (1981) Orbiting Quarantine Facility. NASA SP-454. NASA/Washington. 134 pp.

Draper R. F. (1984) *The Mariner Mark II Program.* AIAA Paper No. 84-0214. AIAA, New York. 33 pp.

Feingold H., Friedlander A. L., Hoffman S., Limperes D., Niehoff J., Schaefer K., Soldner D., Spadoni D., and Wells B. (1983) *Comet Nucleus Sample Return Mission.* Science Applications Rept. No. 83/1152. Science Applications, Albuquerque. 144 pp.

Friedlander A. L., Soldner J. K., and Niehoff J. (1982) *Performance Assessment of Planetary Missions as Launched from an Orbiting Space Station.* Scientific Applications Report No. SAI 1-120-340-T19. Science Applications, Albuquerque. 109 pp.

Maloney J., Peuna L., Siden L., Hanson J., and Bianchi T. (1983) *Definition of Technology Development Missions for Early Space Station's Orbit Transfer Vehicle Servicing. Phase III, Task 1 Space Station Support of Operational OTV Servicing.* No. GDC-SP-83-067. General Dynamics, San Diego. 107 pp.

NASA Planetary Exploration (1983) *Planetary Exploration Through the Year 2000.* NASA Publication. U.S. Govt. Printing Office, Washington. 29 pp.

Scott C. D., Roberts B. B., Nagy K., Taylor P., Gamble J. D., Cerimele C. J., Knoll K. R., Li C. P., and Reid R. C. (1985) *Design Study of an Integrated Aerobraking Orbital Transfer Vehicle.* NASA TM-58264. 40 pp.

A MOON BASE/MARS BASE
TRANSPORTATION DEPOT

Paul W. Keaton

Los Alamos National Laboratory, MS D434, Los Alamos, NM 87545

Placement of the next space outpost, after the low-Earth-orbit space station, will strongly affect the evolution of future space programs. The outpost will store rocket fuel and offer a haven to space workers, as well as provide a transportation depot for long missions. Ideally, it must be loosely bound to the Earth, easy to approach and leave, and available for launch at any time. Two Lagrange equilibrium equilibrium points, $L_1(SE)$, between the Sun and the Earth, and $L_2(EM)$, in the Earth-Moon system, have excellent physical characteristics for an outpost; for example, less than 2% additional rocket propellant is required for docking at $L_1(SE)$ on the way to lunar bases or Mars bases. We apply the rocket problem, the two-body problem, and the three-body problem in discussing alternative locations for space depots. We conclude that Lagrange point halo orbits are the standard by which alternative concepts for transportation depots must be gauged.

INTRODUCTION

An evolutionary manned space program will put outposts along routes to places with economic, scientific, and political importance. These outposts will be "filling stations" for storing rocket fuel, warehouses for holding bulk shielding material, assembly plants for building large structures, and transportation depots for connecting with flights to other destinations. Some outposts may produce oxygen and hydrogen from raw materials obtained elsewhere. Each outpost can provide a refuge from solar flare radiation, a hospital for emergencies, and an oasis to those whose missions call for prolonged space travel.

The obvious initial choice for such an outpost is a space station in low-Earth orbit (LEO). LEO marks the first reasonable resting spot in climbing out of the deep potential well of the Earth's gravitational field, for leaving behind the aerodynamic drag of the Earth's atmosphere. And LEO is still within the protection of the Earth's magnetic field so that galactic cosmic rays and lethal solar flares are not a life-threatening hazard to unshielded occupants. A LEO space station will also provide early opportunities to perfect life support systems and conduct physiological experiments. The knowledge gained will promote a better understanding of the problems of engaging people in long-duration space activities.

This first step, the LEO space station, is the largest. Placement of the second step will affect future space programs, including lunar bases, Mars bases, and manned access to Earth's geosynchronous orbit (GEO). The purpose of this paper is to discuss the physics of how to decide where that second outpost in space should be.

LOCATING A TRANSPORTATION DEPOT

Any space habitat beyond the protection of the Earth's magnetic field will require some radiation shielding. The annual biological dose from galactic cosmic rays is about 50 rem (Silberberg *et al.*, 1985); 5 rem per year has been allowed for radiation workers on Earth. In addition, several solar flares per 11-year sun spot cycle would be lethal to astronauts without a radiation "storm cellar" of some type. Although the first few hours of a solar flare may be unidirectional, the radiation, which lasts for a day or two, soon becomes isotropic, so shielding is required on all sides. A transportation depot, therefore, should have easy access to some extraterrestrial source of bulk shielding material such as lunar regolith. The cost of lifting inert material from the surface of the Earth would thus be avoided.

In general, we must consider where extraterrestrial resources will be obtained. If, for example, the main source is the Moon, it will be reasonable to have lunar manufacturing plants and remove only finished products from the surface. If much of the traffic is to and from the Moon, it may be sensible to have a transportation depot there also. However, there is strong evidence that Mars' two moons, Phobos and Deimos, have compositions similar to a carbonaceous chondrite—a type of meteorite that is rich in water and organics (Carr, 1981, p. 200), so we may find that the resources of the Moon, which are quite dry and contain only traces of carbon, and those of Mars' moons will complement the needs of a growing space program. Furthermore, there is a reasonable chance that one of the 73 catalogued Earth-crossing asteroids (Lau and Hulkower, 1985) could supply valuable materials. The same amount of rocket propellant is required to send unmanned freighters from LEO to the surface of Mars' outer moon, Deimos, as to the surface of the Moon, and about 10% less propellant is needed to reach asteroid 1982DB. This argues against placing a transportation depot on any body of substantial gravity, such as the Moon, because each trip to and from the body surface will extract an expenditure of rocket propellant at least equal to that needed to achieve escape velocity.

The many trips to and from a transportation depot will waste fuel and diminish its usefulness if it is poorly situated in space. Consider GEO, for example. Because of its operational significance, a space platform is needed at GEO. However, GEO is about the worst possible place for a transportation depot. More propellant mass is required to insert a rocket payload into circular geosynchronous orbit than to escape the Earth's gravitational field entirely (Cornelisse *et al.*, 1979, p. 396). In addition, the return trip to the Earth from GEO requires more propellant mass than from nearly any other orbit radius (Taff, 1985). Furthermore, the geosynchronous radius of 42,240 km (6.63 Earth radii) is at the outer edge of the Van Allen radiation belt and at the inner edge of the geomagnetic tail (Gosling *et al.*, 1984, p. 46), a location that may necessitate considerable radiation shielding for people stationed there. Looking beyond GEO itself and toward lunar bases, Mars bases, and products derived from extraterrestrial resources, we find no wisdom in placing a transportation depot at the Earth's geosynchronous orbit.

The velocity v of a transportation depot relative to its local gravitational center is also an important consideration for establishing its location in space. The faster an object is moving, the smaller will be the velocity increase, Δv, required to make a given kinetic energy increase. This is easy to see; in non–relativistic mechanics, the kinetic energy T of a mass m is given by $mv^2/2$. For small increments of velocity, we may differentiate T with respect to v. Thus, the increase in kinetic energy is $\Delta T = (mv)\, \Delta v$, so that the larger v becomes, the smaller will be the Δv required to bring about a given ΔT. Remembering that the larger the Δv, the greater the rocket propellant mass required to accelerate the rocket, we can see the considerable savings in fuel if a rocket starts for Mars from LEO rather than from a much higher, slower orbit.

This last point leads to seemingly contradictory criteria for locating a transportation depot: it should not be tightly bound to a massive planet or moon because every encounter is high in fuel cost, and yet it should be capable of producing large velocities, which come from trajectories near massive bodies. One compromise is to establish highly elliptical orbits that give large velocities at perigee while requiring smaller binding energies to Earth than low, circular orbits. This compromise, which has many disadvantages (repeated passings through the Van Allen radiation belt, for example) will not be considered further here.

The ideal location for the second transportation depot, after the LEO space station, would be a spot that can be reached from LEO with no more than escape velocity, that requires no fuel to stay there, and that has an infinite launch window. Also, it would be easy to coast near the Earth from the ideal location; the rocket could, thereby, achieve a high velocity, leaving open options for igniting the engines to initiate interplanetary travel or aerodynamic maneuvering in the Earth's upper atmosphere. An added bonus would be obtained if the spot had velocity relative to the Moon so that lunar gravitational assists ("slingshots") could be used for increasing or decreasing a rocket's velocity. The ideal location for a transportation depot is depicted schematically in Fig. 1.

This logic leads us to discuss the subject of Lagrange points as candidates for locating transportation depots. If the Earth and Moon were fixed in space, there would be one point between them where the attraction toward the Earth would just equal the attraction toward the Moon. Because the Earth and Moon are not fixed in space but revolve around

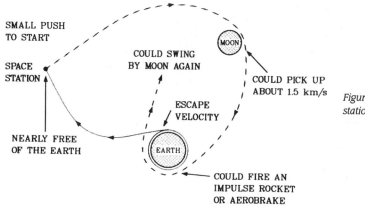

Figure 1. *Ideal place for a space station.*

each other, there are, instead, five such equilibrium points—Lagrange points, or libration points. These equilibrium points exist for other revolving two-body systems as well. The Earth-Moon Lagrange point, L_2(EM), and the Lagrange point between the Sun and Earth, L_1(SE), are particularly favorable locations for a transportation depot. They can be reached from LEO with escape velocity, they are easy to approach and leave, and from them it is not difficult to swing close by the Earth to initiate a high-velocity Δv firing before going to other planets. They also afford easy access to the surface of the Moon. Their "halo" orbits can be maintained with almost negligible fuel year after year. This has been verified experimentally by the International Solar Earth Exploration (ISEE-3) satellite launched in 1978, which was maintained at L_1(SE) for four years with a station-keeping Δv expenditure of 10 m/s per year (Farquhar *et al.*, 1984). In 1982, ISEE-3 was moved to measure the Earth's geomagnetic tail and, in late 1983, with gravitational assists from the Moon, was moved on to rendezvous with the Giacobini-Zinner comet in September, 1985. The name of the satellite has been changed to International Comet Expedition (ICE). We will return to the subject of Lagrange points in discussing the three-body problem.

Figure 2 helps us put in perspective some of the important points of this paper: the amount of Δv required to reach LEO from the Earth's surface and the cumulative Δv necessary to reach GEO, the L_1(SE) Lagrange point, and other places. This is not a potential energy diagram, so Δv depends on the path taken. However, it is correct to imagine a rocket leaving the Lagrange point and picking up velocity as it "slides down" the curve to LEO, where it ignites its engines for a trip to Mars. Lunar gravitational assists can be included in the mission profile. Thus, the propellant needed to get to the Lagrange point is not wasted, and L_1(SE) can be thought of as the first stage of a multi-stage rocket trip from LEO to Mars.

In the rest of this paper, we deduce the findings presented in this section from the laws of classical mechanics.

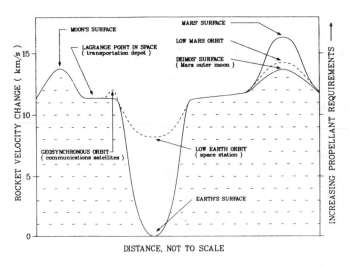

Figure 2. Delta-v budgets for special locations.

THE ROCKET PROBLEM

The fundamental rocket equation equates the instantaneous change in momentum of a rocket of mass m and velocity v, with the instantaneous change in momentum of the exhaust propellant of velocity c relative to the rocket. It may be written

$$THRUST = m\dot{v} = -\dot{m}c \tag{1}$$

where the dot indicates a derivative with respect to time. For constant c, (1) may be integrated exactly. If we remember that the initial rocket mass m_i is the sum of the final rocket mass m_f and the propellant mass m_p, then (1) leads to

$$m_p/m_i = [1 - \exp(-\Delta v/c)] \tag{2}$$

which shows the propellant mass required to change the velocity of a rocket by a given amount. The exhaust velocity of the main shuttle engine, which burns hydrogen and oxygen, is about 4.5 km/s. To illustrate, a Δv of 3.9 km/s is required to place a communications satellite into circular GEO from a circular shuttle orbit of 250 km altitude. If the exhaust velocity is c = 4.5 km/s, then from (2), $m_p/m_i = 0.58$. That is, as it leaves the shuttle, 58% of the mass of the satellite and associated rocketry is propellant mass.

THE TWO-BODY PROBLEM

Historically, interest in the two-body problem with central forces arose from astronomical considerations of planetary motions. Here we will use the results of that earlier work to show how people might travel to the planets. Two powerful formulas follow easily from the conservation of energy E, written as

$$T + V(r) = E \tag{3}$$

where, as before, T is the kinetic energy of a mass m with velocity v, and $V(r) = -GMm/r$ is the potential energy of m at a distance r from the center of a spherically symmetric mass M, and $G = 6.67 \times 10^{-11}$ $m^3/kg - s^2$ is the universal gravitational constant in MKS units. If an object that rests on the Earth (of mass $M = 5.976 \times 10^{24}$kg and average radius $r_0 = 6371$ km) is given a velocity, v_{esc}, just large enough to escape to infinity, then the total energy of the system is E = 0. It follows from (3) that $v_{esc}^2(r_0) = 2GM/r_0 = (11.19$ km/s$)^2$. Generalizing, we may write the first fundamental equation as

$$v^2(r) = v_{esc}^2(r) + v_0^2 \quad \text{(unbound)} \tag{4}$$

where $v_{esc}^2(r) = 2GM/r$, v_0 is the velocity m would have after escaping from M (called the hyperbolic velocity) and v(r) is the velocity needed at r to achieve a hyperbolic velocity of v_0. In this case, with $E = mv^2/2 > 0$, m is said to be unbound, and the trajectories trace hyperbolas, with the limiting case of $E = 0$ being a parabola. On the other hand, when a rocket of hyperbolic velocity v_0 encounters a planet, the rocket velocity increases according to (4). If it does not collide or fire its engines, the rocket reaches its largest velocity closest to the planet surface (a point called periapsis) and then leaves the planet in a different direction, losing speed until it again reaches its former hyperbolic velocity. If, instead, the rocket is to be captured into orbit around the planet, it can retrofire its engines at periapsis, slowing itself until the velocity is less than escape velocity so that it cannot escape the planet. The change in velocity, Δv, will determine the high point (apoapsis) of the resulting orbit. If the planet has a sufficient atmosphere, as do Venus, Mars, Earth, and Jupiter, the periapsis can occur low enough to permit atmospheric drag to slow the rocket (to aerobrake it) below escape velocity. These processes, which decrease the velocity below escape velocity, result in a negative value of E in (3).

If $E < 0$, m is said to be bound to M (we always assume that $m \ll M$), and the trajectories are elliptical orbits. If a is the semi-major axis of the ellipse, it can be shown that $E = -GM/(2a)$ (Goldstein, 1950, p. 79). Substituting this into (3), we arrive at the second fundamental equation,

$$v^2(r) = v_{esc}^2(r)[1 - r/(2a)] \qquad \text{(bound).} \qquad (5)$$

A particular case of interest occurs when the satellite is in a circular orbit so that the radius is always equal to the semi-major axis. Setting $r = a$ in (5) shows that the circular velocity, v_{cir}, is always equal to the escape velocity divided by $2^{1/2}$. Thus, the escape velocity from a 500-km LEO is calculated to be 10.77 km/s; the circular velocity is calculated to be 7.62 km/s. Likewise, the Earth, traveling in a nearly circular orbit around the Sun of mass $M = 1.989 \times 10^{30}$ kg at a distance of 149.6×10^6 km, travels at an average circular velocity of 29.78 km/s.

The simplest example of orbital transfers from a circular radius of r_1 to a circular radius of r_2 can be worked out with (5). The so-called least-energy transfers, or Hohmann transfers, are obtained by directing the rocket thrust tangent to the orbit at r_1 so that the velocity is increased by Δv_1, just enough to coast on an elliptical path and reach its apoapsis at r_2 after traveling 180° around the dominant mass. Then, a second tangential rocket thrust will increase the velocity by an amount Δv_2 to insert it into the new circular orbit at r_2. The major axis of the elliptical transfer orbit is $2a = r_1 + r_2$, which completely determines $v(r_2)$ in (5). Then $\Delta v_1 = v(r_1) - v_{cir}(r_1)$, $\Delta v_2 = v_{cir}(r_2) - v(r_2)$, and the total $\Delta v = \Delta v_1 + \Delta v_2$. Using (4) and (5) repeatedly, we find that the Δv necessary for traveling from LEO to a rendezvous with Deimos is 5.5 km/s. For comparison, the Δv necessary to go to the Moon from LEO amounts roughly to the escape velocity from LEO, 3.2 km/s, plus the escape velocity from the Moon, 2.4 km/s, a total of 5.6 km/s.

In the case of Hohmann transfers, if we set $R = r_2/r_1$ and $S(R) = (\Delta v_1 + \Delta v_2)/v_{cir}(r_1)$, it follows from (5) and the above discussion that

$$S(R) = \left[\frac{2R}{1+R}\right]^{\frac{1}{2}} - 1 + \frac{1}{R^{\frac{1}{2}}}\left[1 - \left(\frac{2}{1+R}\right)^{\frac{1}{2}}\right] \tag{6}$$

(Taff, 1985, p. 159). Figure 3 shows $\Delta v_1 + \Delta v_2$ for orbital transfers from a 500-km LEO. The various distances in Fig. 3 are given to indicate a scale. Of course, the Sun's gravity would dominate orbits beyond the Earth's sphere of influence, which extends out from the Earth to about 930,000 km. The most interesting feature of S(R) is that it has a maximum, which is near R = 15.58. This means that more Δv, and hence more fuel, is required to place a payload from LEO into circular orbit at 100,000-km altitude than into a circular orbit at higher altitudes. Or, putting it differently, less Δv is required to go from LEO to infinity and back into the Moon's orbit around the Earth than to initiate a direct Hohmann transfer. Further details on this and bielliptic transfers are given in Taff (1985, p. 160).

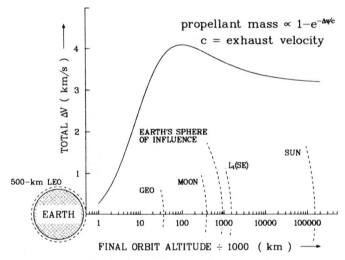

Figure 3. *Delta-v versus circular orbit altitude above the Earth.*

A more realistic Hohmann orbit transfer will include changing the orientation of the plane by an angle $\Delta\theta$. The optimum maneuver executes a small angular change $\Delta\theta_1$ at the lower orbit and a larger angular change $\Delta\theta_2$ at the higher orbit, so that the total angular change is $\Delta\theta = \Delta\theta_1 + \Delta\theta_2$. In that case, (6) generalizes to

$$S(R,\Delta\theta) = \left[\frac{1+3R}{1+R} - \cos(\Delta\theta_1)\left(\frac{8R}{1+R}\right)^{\frac{1}{2}}\right]^{\frac{1}{4}} + \frac{1}{R^{\frac{1}{4}}}\left[\frac{3+R}{1+R} - \cos(\Delta\theta - \Delta\theta_1)\left(\frac{8}{1+R}\right)^{\frac{1}{2}}\right]^{\frac{1}{4}} \tag{7}$$

where again $R = r_2/r_1$ and $\Delta\theta_1$ is fixed so as to minimize $S(R,\Delta\theta)$. Figure 4 shows the total $\Delta v = \Delta v_1 + \Delta v_2$ necessary to transfer from a 500-km LEO to higher circular Earth

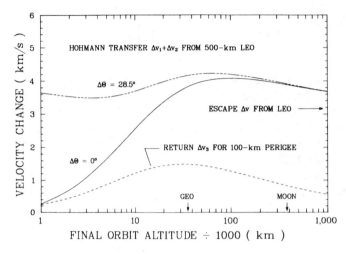

Figure 4. Earth orbital transfer missions.

orbits with a 28.5° plane change. The latitude of the Kennedy Space Center is approximately 28.5°, so that is a typical LEO to GEO transfer angle. Figure 4 also shows the necessary Δv_3 (dashed line) for returning to a 100-km perigee, from which aeromaneuvering is assumed feasible. An important feature of Δv_3 is that it reaches a broad maximum at 32,000-km altitude, which is close to the 36,000-km altitude of GEO. It therefore requires more propellant mass to return from GEO to the Earth than from almost any other circular orbit.

THE THREE-BODY PROBLEM

Analytic expressions for the orbits have not been found for the three-body problem, which is more complicated than the two-body problem. Ultimately, trajectories are calculated by numerical methods on digital computers. For a clear and succinct elementary treatment of the three-body problem, the reader is referred to Deslodge (1982, chapter 62). Here we are concerned only with "halo orbits" around one of the Lagrange points.

Consider two masses, which are a distance D apart and revolving around each other in perfect circles with an angular velocity ω under the influence of gravitational forces. It follows from the two-body problem that $\omega^2 = G(m_1 + m_2)/D^3$. The center of mass will be between m_1 and m_2, a distance αD from m_1, where $\alpha = m_2/(m_1 + m_2)$. We assume that $m_1 \leq m_2$. We establish a right-handed coordinate system with its origin at the center of mass and rotating with an angular velocity ω such that m_1 and m_2 are always stationary on the x-axis. For example, designating coordinates as (x,y,z), we find the coordinates of m_1 are $(-\alpha D,0,0)$ and of m_2 are $[(1-\alpha)D,0,0]$. We place the y-axis in the plane of rotation and the z-axis along the angular velocity vector. Now consider a third body of mass m that is so small it does not perturb the orbits of m_1 and m_2. These conditions describe the restricted three-body problem. Although it is specialized, this is an important problem because it describes reasonably well the situation of a rocket of mass m traveling in the Earth-Moon system or the Sun-Earth system.

The influence of m_1 and m_2 on m at some position (x,y,z) is described in the restricted three-body problem by a potential-like function,

$$U(x,y,z) = -\omega^2 D^2 \left[\frac{(x^2 + y^2)}{2D^2} + \frac{1 - \alpha}{(S_1/D)} + \frac{\alpha}{(S_2/D)} \right]$$

$$S_1 = [(x + \alpha D)^2 + y^2 + z^2]^{1/2} \qquad\qquad \text{and} \qquad (8)$$

$$S_2 = \left\{ [x - (1 - \alpha)D]^2 + y^2 + z^2 \right\}^{1/2}$$

where s_1 is the distance between m and m_1, and s_2 is the distance between m and m_2. The equations (Deslodge, 1982) of motion of m are

$$\ddot{x} - 2\omega\dot{y} = -U_x , \qquad\qquad\qquad 9(a)$$

$$\ddot{y} + 2\omega\dot{x} = -U_y , \qquad\qquad\qquad 9(b)$$

$$\ddot{z} = -U_z , \qquad\qquad\qquad \text{and} \quad 9(c)$$

$$\ddot{\vec{r}} + 2\vec{\omega} \times \dot{\vec{r}} = -\vec{\nabla} U \qquad\qquad\qquad 9(d)$$

where $U_x = \partial U/\partial x$, etc., and (9d) expresses (9a)–(9c) in vector notation. Because of the $\dot{x}$ and $\dot{y}$ terms in (9a) and (9b), U is not an ordinary potential function. In fact, particles can be trapped near maxima and saddle points of U, as well as near minima. The Lagrange equilibrium points can be found by setting the gradient of U, which is the effective force on m, to zero. In doing so, the equilibrium points are seen to be in the plane of rotation, with $z = 0$. There are three collinear Lagrange points, L_1, L_2, and L_3, with $y = 0$, and two equilateral points, L_4 and L_5, with $s_1 = s_2 = D$. The five Lagrange points for the Earth-Moon system are shown in Fig. 5, along with L_1 and L_2 for the Sun–Earth system.

A three-body analogue to (5), which relates the velocity magnitude of m at a point in space to its current orbital parameters, may be deduced from (9). Taking the inner

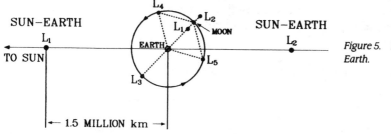

EARTH–MOON SYSTEM

SUN–EARTH L_1 TO SUN

SUN–EARTH L_2

L_4 L_2 MOON L_1 EARTH L_5 L_3

← 1.5 MILLION km →

Figure 5. Lagrange points near the Earth.

product of $\vec{v} = \dot{\vec{r}}$ and (9), we find both sides of the resulting equation are perfect derivatives with respect to time, so their algebraic sum is a constant. Therefore,

$$\tfrac{1}{2}v^2 + U(\vec{r}) = E = \text{constant} \tag{10}$$

expressing also a kind of energy conservation in the rotating system. To illustrate how (10) may be used, imagine that a satellite is known to have zero velocity when it is near the Sun-Earth L_1, which is located at $\vec{r} = (0.9900D,0,0)$. Substituting these coordinates into (10), we find that $E = U(\vec{r}) = -3.001(\omega D)^2/2$. If then, with no thrust added, we later find that the satellite is at a 500-km perigee on the opposite side of the Earth from L_1 (this formula tells nothing about the trajectory necessary to arrive there), its potential must be $U = -3.131(\omega D)^2/2$, from which it follows that $v^2/2 = (3.131 - 3.001)(\omega D)^2/2$. That is, $v = (0.13)^{1/2}(\omega D)$. Because $(\omega D) = 29.78$ km/s, $v = 10.74$ km/s, which is very close to the escape velocity for LEO calculated from the two-body problem. This shows that the velocity necessary in LEO to reach L_1(SE) is between 10.74 km/s and 10.77 km/s.

Consider now the conditions necessary to establish a satellite around the Lagrange point L_1. At L_1, $(U_x)_1 = (U_y)_1 = (U_z)_1 = 0$ by definition, where the notation $(U_x)_1$ indicates $\partial U/\partial x$ evaluated at L_1, etc. Making a first-order Taylor expansion of the gradient U about L_1, setting $X = x - x_1$, and substituting into (9), we have the equations of motion near L_1

$$\ddot{X} - 2\omega\dot{y} = -(U_{xx})_1 X; \quad (U_{xx})_1 = -\omega^2[1 + 2f^2] \tag{11(a)}$$

$$\ddot{y} + 2\omega\dot{X} = -(U_{yy})_1 y; \quad (U_{yy})_1 = -\omega^2[1 - 2f^2] \tag{11(b)}$$

$$\ddot{z} = -(U_{zz})_1 z; \quad (U_{zz})_1 = +\omega^2 f^2 \qquad \text{and} \quad 11(c)$$

$$f^2 = \frac{1-\alpha}{(s_1/D)^3} + \frac{\alpha}{(s_2/D)^3} \tag{11(d)}$$

where all other terms in the Taylor expansion vanish. The partial derivatives in (11) are given in Deslodge (1982). To the first-order expansion, motion in the z direction is simple harmonic and independent of motion in the xy plane. Because L_1 is a mathematical saddle point, motion in the plane contains exponentially diverging solutions as well as periodic solutions, and therefore L_1 represents an unstable equilibrium. However, with the proper initial conditions, the diverging amplitudes can be set to zero, and bound periodic orbits, or halo orbits, result:

$$X(t) = X_o \cos(b\omega t); \quad b = [1 - f^2/2 + (f/2)(9f^2 - 8)^{1/2}]^{1/2} \tag{12(a)}$$

$$y(t) = -(X_o\gamma) \sin(b\omega t); \quad \gamma = (1 + b^2 + 2f^2)/(2b); \qquad \text{and} \quad 12(b)$$

$$z(t) = z_o \cos(f\omega t). \tag{12(c)}$$

Note that the satellite is traveling on an elliptical path in a direction around L_1 opposite to that of the Earth around the Sun when the initial conditions are correct for a periodic orbit in the xy plane. The orbital parameters are $f^2 = 4.061$, $b = 2.086$, and $\gamma = 3.229$, depending only on the ratio of m_2/m_1, and $\omega = 2\pi$ rad/yr. Because $\gamma > 1$, the semi-major axis is always along the y-axis, perpendicular to the Sun-Earth line. Unlike elliptical orbits in the two-body problem, the period of revolution is independent of the size of the ellipse. The angular velocity in the plane is $b\omega$; in the z direction, it is $f\omega$. With $L_1(SE)$, $f = 2.02$ and $b = 2.09$, so the halo orbit will have a period of about 6 months. These observations are all in agreement with the halo orbit of ISEE-3 described by Farquhar *et al.* (1984). For the Moon, $L_1(EM)$ is located at $x_1/D = 0.8369$, where $D = 384,400$ km, and ω may be calculated by observing that the sidereal month is 27.32 days. From (11) and (12), we see that $f = 2.27$, $b = 2.33$, and $\gamma = 3.59$. The period of rotation of a satellite around $L_1(EM)$ is therefore about 12 days. Equations (11) and (12) are valid for any of the collinear Lagrange points, provided that proper values of s_1 and s_2 are substituted into (11d) to find f^2. For example, $x_2/D = 1.0100$ and 1.1557 for L_2 of the Sun-Earth and Earth-Moon systems, respectively, so the corresponding values of f are 1.99 and 1.79. The period of a halo orbit around $L_2(SE)$ is not significantly different from that of $L_1(SE)$, but for $L_2(EM)$, $b = 1.86$, and the period of the halo orbit is higher, about 15 days.

In exploring the basic physics of placing a space station at one of the Lagrange points, one last question remains. How large is the Δv push needed to make a rocket return from $L_1(SE)$ to low Earth orbit? We attempt here only a heuristic estimate. Because $L_1(SE)$ and the Earth are both revolving around the Sun with the same angular velocity ω, and because they are different distances from the Sun, the Earth is traveling faster than $L_1(SE)$ in a heliocentric inertial frame of reference. This amounts to a difference of about 0.01×29.78 km/s $= 298$ m/s, which can be taken as an initial estimate of the Δv necessary to eject from $L_1(SE)$ and fall toward the Earth with essentially no angular momentum barrier. The number agrees well with the numerically calculated value of 279 m/s given in Farquhar and Dunham (1985). If, on the other hand, the rocket is orbiting $L_1(SE)$ with a semi-major axis of about $X_0\gamma = 650,000$ km, as was the ISEE-3, calculating y_{max} from (12b) shows that the rocket reaches velocities relative to $L_1(SE)$ as high as $(X_0\gamma b/D)\omega D = 0.009 \times 29.78$ km/s $= 268$ m/s. This occurs when the rocket is moving parallel with the Earth, so only an additional $298-268 = 30$ m/s of Δv would appear to be needed to return to LEO from the halo orbit. This estimate is close to the ISEE-3 experiment that required $\Delta v = 36.3$ m/s for insertion into the halo orbit in November, 1978. Subsequently, in June, 1982, only $\Delta v = 4.5$ m/s was required to eject ISEE-3 from halo orbit and back into the geomagnetic tail. For our purposes, we will define "very low Δv" to be less than 50 m/s.

A drawback of these very low Δv injections and ejections is that the transit times take months. The transit times can be reduced to weeks while still keeping Δv under 100 m/s. Perhaps the very low Δv encounters with $L_1(SE)$ will be useful for only unmanned cargo ships carrying large masses. In addition, because the halo orbit period is six months, a space station will be at the proper place to receive freighters with very low Δv injections only twice a year. For each opportunity to dock, we estimate that the launch window will exist for about three weeks for very low Δv-class missions. These limitations will not pose serious problems when freight can be parked anywhere in halo orbit and boarded later by people at the appropriate time.

In addition to L_1(SE), other Lagrange points can be considered for a transportation depot. Although L_2(SE) shares all of the same kinematic advantages, it lies in the Earth's geomagnetic tail and may not be suitable because of the radiation environment. Each of the five Earth-Moon Lagrange points could be a candidate for a transportation depot, but estimates of the injection velocities, such as those described above, indicate that the points require a Δv in the range of 1000 m/s. However, for L_2(EM) this can be overcome to a great extent by a clever maneuver. It has been shown that, using a retrograde lunar gravitational assist, the Δv necessary to enter a halo orbit at L_2(EM) is about 300 m/ s instead of the 1230 m/s necessary for direct insertion from LEO (Farquhar and Dunham, 1985; Farquhar, 1972). We conclude that there are two places in the Earth's vicinity that are well qualified for hosting a transportation depot—L_1(SE) and L_2(EM).

Looking beyond the Earth's vicinity once the technology is developed to establish a manned space station around L_1(SE), we can use that technology at other places in the solar system. All of the major planets and moons have L_1 points that can sustain halo orbits. For example, Lagrange points of the Sun-Mars system have already been mentioned in connection with a Mars mission (Farquhar, 1969). An important factor in considering L_1(SM) at Mars is that the first manned missions there will carry return propellant, heavy shielding for radiation protection in transit, engines, and other things not needed at Mars. The less tightly bound this equipment is to Mars, the less fuel will be needed to break it away later on the return trip. A ship going to Mars would save propellant by slowing at periapsis to just under escape velocity and coasting to L_1(SM). All of the equipment for returning to Earth would be placed in a halo orbit and left there during the descent to Mars. Later, reversing the procedure, the crew could return to Earth and waste very little fuel in the process. Eventually, a transportation depot at

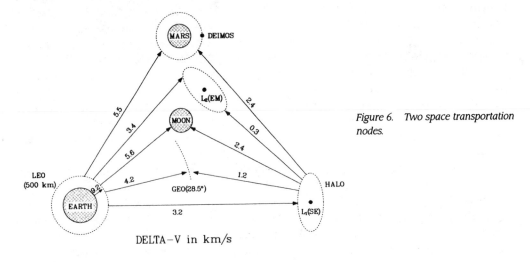

Figure 6.   Two space transportation nodes.

L_1(SM) would require its own space station, especially if water were found at Phobos or Deimos. Here, however, we are focusing on the second transportation depot, the one to follow the LEO space station.

Figure 6 emphasizes the "nodal" aspects of space stations at LEO and in a halo orbit around L_1(SE). It shows the Δv in kilometers per second required to move from each of the two nodes to GEO, Moon, L_2(EM), and Deimos. Notice that it costs very little extra total Δv to stop over at L_1(SE) on the way to one of these places from LEO. This makes L_1(SE) an important staging area for rocket propellant, water, bulk shielding material, and vehicle assembly for Moon and Mars expeditions. Freighters can go to L_1(SE) in months and take advantage of the very low Δv transfers into halo orbit. People can be carried there in a shorter time by paying a higher Δv expenditure in small orbital transfer vehicles (OTVs).

SUMMARY

The key to a successful evolutionary space program is the placement of effective transportation nodes in the supporting infrastructure. Such outposts have always been important in opening frontiers. For the settlement of space, a Lagrange equilibrium point between the Sun and Earth has the nearly ideal physical characteristics of a transportation depot: it is very lightly bound in the Earth's gravitational well; it can be reached with essentially escape velocity; the launch window is always open; it can accommodate a wide range of plane angles for LEO space stations; its halo orbits require only 10 m/s per year of station-keeping propellant to remain stable; and from there it is easy to leave and pass near the Earth at essentially escape velocity—affording several options. Rockets going to and from L_1(SE) can obtain free acceleration and braking by passing near the Moon's surface.

An in-depth study is necessary to determine the best place to put the next transportation depot after the LEO space station, but the laws of nature will not change. Lagrange point halo orbits are the present standard by which any alternative concept for a transportation depot must be gauged.

Acknowledgments. *When the author was invited by the Lunar Base Steering Committee to express his views on this subject, he was reluctant to write a paper because many of the ideas were not original. However, he was persuaded by the argument that a tutorial format might provide background and establish the framework for future deliberations. It is a pleasure to acknowledge several helpful conversations with Robert W. Farquhar, of NASA's Goddard Space Flight Center, during the course of this work.*

REFERENCES

Carr M. H. (1981) *The Surface of Mars.* Yale University Press, New Haven. 232 pp.

Cornelisse J. W., Schoyer H. F. R., and Wakker K. F. (1979) *Rocket Propulsion and Space Flight Dynamics.* Pitman, New York. 505 pp.

Deslodge E. A. (1982) *Classical Mechanics, Vol. I and II.* Wiley and Sons, New York. 991 pp.

Farquhar R. W. (1969) Future missions for libration-point satellites. *Astronaut. Aeronaut. J., 7,* 52–56.

Farquhar R. W. (1972) A halo–orbit lunar station. *Astronaut. Aeronaut. J., 10*, 59–63.

Farquhar R. and Dunham D. (1985) Libration–point staging concepts for Earth–Mars transportation. *Proc. Manned Mars Mission Workshop*. In press.

Farquhar R., Muhonen D., and Church L. C. (1984) *Trajectories and Orbital Maneuvers for the ISEE-3/ICE Comet Mission*. AAS-84-1976, American Insittute of Aeronautics and Astronautics, New York. 9 pp.

Goldstein H. (1950) *Classical Mechanics*. Addison–Wesley, Reading, ME. 399 pp.

Gosling J. T., Baker D. N., and Hones E. W. Jr. (1984) Journeys of a spacecraft. *Los Alamos Science, 10*, 32–53.

Lau C. O. and Hulkower N. D. (1985) *On the Accessibility of Near-Earth Asteroids*. AAS-85-352, American Institute of Aeronautics and Astronautics, New York. 27 pp.

Silberberg R., Tsao C. H., Adams J. H. Jr., and Letaw J. R. (1985) Radiation transport of cosmic ray nuclei in lunar material and radiation doses. This volume.

Taff L. G. (1985) *Celestial Mechanics*. Wiley and Sons, New York. 520 pp.

ACHROMATIC TRAJECTORIES AND THE INDUSTRIAL-SCALE TRANSPORT OF LUNAR RESOURCES

T. A. Heppenheimer

Center for Space Science, 11040 Blue Allium Avenue, Fountain Valley, CA 92708

Large-scale transport of lunar material can be accomplished by launches in small payloads by mass-driver along unguided trajectories. Achromatic trajectories then overcome the problem of dispersions due to launch velocity errors. These trajectories incorporate a focusing effect, which reduces the dispersions by four orders of magnitude from their expected values. The mass-driver is to be located close to the lunar equator at 33.1° E longitude. A mass-catcher then can maneuver near the L_2 Lagrangian point, intercepting the payloads. For an optimized catching trajectory, the catcher requires $\delta V = 187$ m/s per month, peak thrust = 0.1414 newtons per ton of loaded catcher mass, and for a reference propulsion concept, the Rotary Pellet Launcher, peak power of 0.316 kilowatts per ton. Caught material can be transferred with $\delta V = 60$ m/s to a processing facility (space colony) in stable high Earth orbit. This transport system can accommodate 10^5 tons per month of lunar material, with an appropriately sized mass-catcher.

INTRODUCTION

One of the principal rationales for a permanent lunar base is that it can serve for the mining and transport of lunar resources. It is appropriate, then, to look beyond the initial establishment of this lunar base and to consider its use in a resource-transport program, particularly one of large scale. In particular, it has often been proposed that resources be launched into space with a mass-driver or electromagnetic catapult. The existing literature on mass-drivers is largely included in Grey (1977a,b, 1979, 1981).

There is also the question of the operational use of a mass-driver, within an overall system for resource transport. Payloads launched by mass-driver follow an unguided, ballistic flight that is therefore subject to potentially large miss distances, at a target, due to errors at launch. During the 1970's, in connection with the concept of space colonization, I developed a theory for operational mass-driver use (Heppenheimer and Kaplan, 1977; Heppenheimer 1978a,b,c; Heppenheimer 1979; Heppenheimer *et al.*, 1982a,b). The point of departure for this work was the concept of achromatic trajectories (Heppenheimer and Kaplan, 1977; Heppenheimer, 1978a; Heppenheimer, 1979), *i.e.*, trajectories incorporating a focusing effect. Payloads launched along such trajectories still will reach their target despite errors in even the most sensitive component of launch velocity. It was demonstrated that achromatic trajectories provide a framework within which the entire problem of operational use of a mass-driver is readily solvable. In what follows, I give a brief summary and overview of this existing literature and its concepts.

ACHROMATIC TRAJECTORIES

Let a payload be launched by mass–driver, initially tangent to the lunar surface with velocity V_T. It arrives at the target with velocity V_i. If the payload experiences a velocity error (ϵ) at launch, with what error will it miss the target? Calling this error δy, a naive analysis based upon two–body dynamics gives the approximation

$$\delta y = (V_T/V_i)\ (d_m/V_i)\ (\epsilon)$$

where d_m is the lunar diameter, 3476 km. An attractive location for the target is the L_2 Lagrangian point, some 64,000 km behind the Moon (Johnson and Holbrow, 1977). Then, approximately, $V_T = 2400$ m/s, $V_i = 260$ m/s; hence $\delta y = 1.23$ km per cm/s of ϵ.

One must anticipate serious difficulties in controlling V_T to the inferred accuracy. As a result, the attractiveness of mass–drivers at first appears open to question. Achromatic

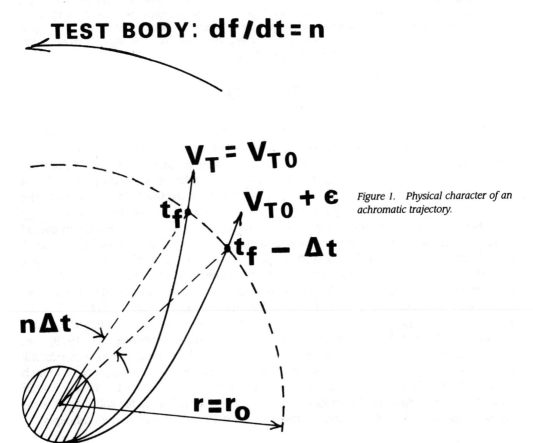

Figure 1. *Physical character of an achromatic trajectory.*

trajectories come to the rescue at this point, reducing this sensitivity by four orders of magnitude.

Figure 1 shows how this occurs. A test body is in orbit, at distance $r = r_o$ moving with angular velocity $df/dt = n$, where f is true anomaly. Its orbit need not be circular. We wish to hit this body with a projectile launched by mass-driver. Nominally, this calls for launch with velocity $V_T = V_{TO}$; the flight time is t_f. Now suppose that the launch velocity is actually $V_{TO} + \epsilon$. The trajectory will be flatter and the flight time reduced to $t_f - \delta t$. Yet, if we can take advantage of the flattening in the trajectory to reduce the transfer angle from nominal by $n\delta t$, then the projectile still hits the target and $\delta y = 0$.

Hence, in a coordinate system rotating with the target, the two trajectories cross. This crossing, due to the rotation of coordinate systems, provides the focusing effect. Figure 2 shows the geometry of such crossings. Here δy is the product of two quantities:

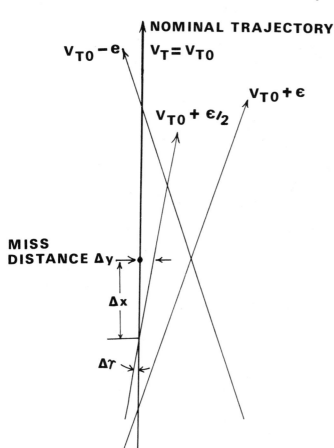

NOMINAL TRAJECTORY

$V_{TO} - e$ $V_T = V_{TO}$

$V_{TO} + \epsilon$

$V_{TO} + \epsilon/2$

MISS DISTANCE Δy

Δx

Δr

Figure 2. Focusing of achromatic trajectories in a rotating coordinate system.

δx, the location of the crossing point along the nominal trajectory; and $\delta\gamma$, the angle at crossing. But $\delta x \propto \epsilon$, $\delta\gamma \propto \epsilon$, so $\delta y \propto \epsilon^2$. More precisely, for ϵ in cm/s and for the achromatic trajectory passing through the L_2 Lagrangian point, $\delta y = 0.1901\epsilon^2$ meters (Heppenheimer, 1979). This compares with the naively expected value cited earlier, $\delta y = 1234.1\epsilon$ meters.

One may further consider a family of neighboring trajectories, originating with slightly different V_T and having the crossing geometry of Fig. 2. Such a family possesses an envelope (Fig. 3), any point of which is associated with a specific value of V_T. This envelope is called a focus locus. If the target maneuvers so as to always stay on the focus locus, then there always exists a V_T that permits reaching the target via an achromatic trajectory.

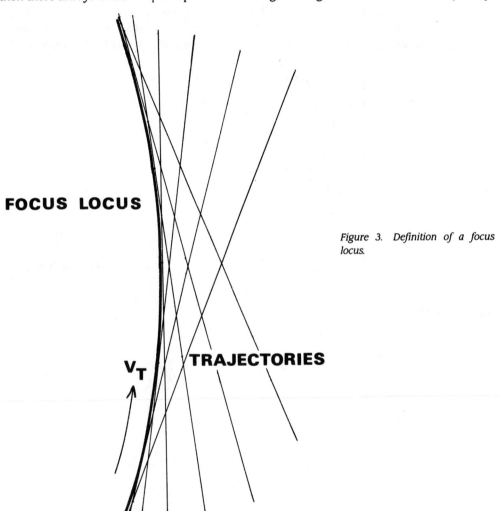

FOCUS LOCUS

V_T

TRAJECTORIES

Figure 3. Definition of a focus locus.

The mass-driver is a large, fixed lunar installation. At any time, a single, well-defined focus locus is associated with its location. But its effective location in the reference coordinate system is not constant because the Moon undergoes physical librations in longitude and latitude. During any one month, these simultaneous oscillations trace out an ellipse. The focus locus then oscillates in position, always lying on a cylinder whose cross-section is essentially this ellipse. Thus, there is a family of focus loci, associated with the different effective values of launch longitude and latitude due to the librations of the Moon. The problem is then constrained by requiring the mass-catcher, the target, to maneuver in space so as to always lie on the focus locus associated with the mass-driver's location.

Simple methods exist for determining focus loci (Heppenheimer, 1979). These involve numerical integrations of trajectories, carried out within the circular restricted three-body problem (wherein the Moon's orbit about the Earth is regarded as circular) or, for greater accuracy, the elliptic restricted problem. In the former, the lunar librations must be put in "by hand," but in the latter they arise naturally. Figure 4 illustrates focus loci calculated in the elliptic problem for a mass-driver located at 33.1° E longitude on the lunar equator

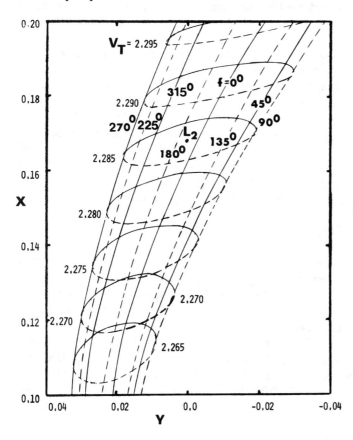

Figure 4. Focus loci in the elliptic restricted three-body problem.

Table 1. Normalized Variables in Restricted 3-Body Problem

Quantity	Variable	Unit Value	Significance
distance	x, y	Lunar orbit radius (mean = 384,410 km)	L_2 coordinates fixed
time	t	104.362 hours	Moon's rotation by l radian
velocity	V	1023.17 m/s	Mean lunar orbit velocity
acceleration		0.00273 m/s^2	(V/t)

(Heppenheimer and Kaplan, 1977). Such a location provides a focus locus through L_2, in the circular problem.

Here, and in subsequent discussion, x and y are normalized distances, given as fractions of the variable Earth-Moon distance (Table 1). With such coordinates, the location of L_2 remains fixed. Launch velocities V_T are also normalized, being given in terms of the mean lunar orbital velocity. The variable f is lunar true anomaly, which defines its orbital position; it is evident how the focus locus swings from one side to the other as the Moon revolves during a month.

From such focus-locus computations, one derives data sufficient to characterize the required mass-catcher maneuvers. In particular, these data permit derivation of the Catcher Equations (Heppenheimer, 1978a; Heppenheimer, 1979), describing the mass-catcher motion under the following restrictions:

(a) The catcher must always maneuver so as to be on the focus locus as it shifts location.
(b) The catcher varies in mass, owing to its interception of the stream of payloads from the mass-driver, and is under continuous propulsion.
(c) The payload stream imposes a momentum flux upon the catcher.

THE MASS-CATCHER

With the Catcher Equations, one derives optimal catcher trajectories, which minimize propellant expenditure subject to the above conditions. The optimization procedure employs such standard methods as the Maximum Principle (Heppenheimer, 1978b). It is useful to constrain the motion by limiting the allowed velocities at the beginning and end of a catching cycle. We impose the condition that, initially, the catcher velocity must nearly match that of the payload stream. As a result, the catcher is not immediately subject to potentially damaging high velocity impacts. Instead, there is time for a protective layer of caught material (a regolith) to build up. Also, the final x-component of velocity is constrained to a (normalized) value of -0.01, some 10 m/s, to permit gentle injection of the caught payloads onto a transfer trajectory to a proposed space processing facility (Heppenheimer, 1978c).

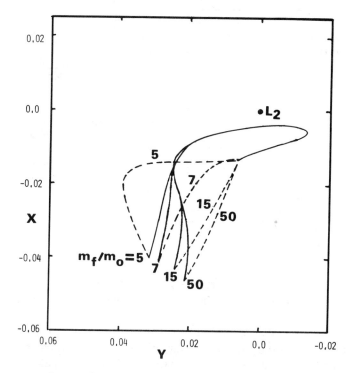

Figure 5. Some optimal mass-catcher trajectories in the plane.

The catcher is assumed to fill uniformly with time, starting with an initial mass m_o and finishing with total mass m_f. After disposing of the accumulated payload, the catcher moves propulsively back to the initial point. The entire cycle takes approximately one month. For convenience, the catcher is assumed to fill over a fraction of the cycle given by $(m_f - m_o)/m_f$ and returns to initial focus point during the remaining fraction, m_o/m_f. Figure 5 illustrates several optimal trajectories as a function of m_f/m_o. The motion is generally clockwise, tracking the shifting location of the focus locus (Fig. 4). The ballistic return trajectories are dashed.

From such computations, one derives data on the operational use of a mass–catcher and mass–driver, shown in Fig. 6 for $m_f/m_o = 15$. Here δ_0 is a constant phase angle (270°) chosen to minimize catcher propulsion requirements, representing the time of the month when catching begins. The catcher mass increases uniformly from $m_f/15$ at the curves' left-hand sides to m_f at the right-hand sides. The catcher follows the corresponding optimal trajectory of Fig. 5.

In a situation where the payload stream is interrupted, or the catcher drifts off the focus locus, the catcher follows its nominal path pending reacquisition of the payload stream. "Emergency acceleration" defines the thrust required to maintain the nominal motion in the absence of the payload momentum flux. Maximum thrust is thus given by the condition that a loaded catcher be capable of an acceleration of 0.052 in normalized units; this corresponds to a thrust of 0.1414 newtons per ton (1000 kilograms) of loaded

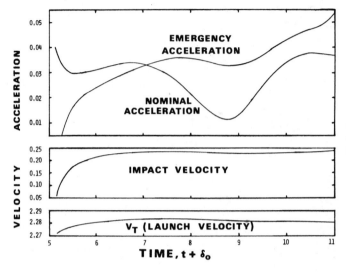

Figure 6. Some operational characteristics of an optimized catcher trajectory.

catcher mass. "Nominal acceleration" presents the thrust required to follow the nominal path of Fig. 5. This curve shows that the nominal thrust varies between 0.2 and 0.8 of the maximum, approximately.

The time integral of the curve, "nominal acceleration," gives the δV required to follow the nominal trajectory, 169.5 m/s. If V_e is the exhaust velocity of the catcher's thrusters, then $(169.5/V_e)$ is the fraction of loaded catcher mass required for thruster propellant. In addition, there is a δV of 260.7 m/s required for the return maneuver (dashed line) of Fig. 5. But the catcher mass then is $m_O = m_f/15$; hence the effective δV for a complete catching cycle increases by only $260.7/15 = 17.38$ m/s, to 186.88 m/s.

Figure 6 also gives the impact velocity of payloads striking the catcher and the launch velocity V_T as functions of time. Payloads launched by mass–driver are considered to be essentially simple bags of lunar soil. Thus, as they accumulate in the catcher, they build up a layer of such soil, the catcher regolith. There is much data on impacts into regolith, resulting from work on geological problems of lunar and planetary cratering. Heppenheimer (1978b) and Heppenheimer (1979) give elements of an analytic theory of catching, based on cratering and regolith theory. The catching theory predicts that mass loss from impacting payloads can be reduced to less than 1% by designing the catcher as a large open conical bag, rotating at several rpm. The rotation provides more than enough centrifugal force to hold caught mass in place. It also provides a differential velocity at the impact site, which ejecta must overcome in order to attain the free flight that can produce mass–escape from the catcher.

ENGINEERING APPLICATIONS

The achromatic trajectory passing through L_2 originates at equatorial longitude 33.1° E, location (1) in Fig. 7. Examination of lunar orbital photography shows that this site

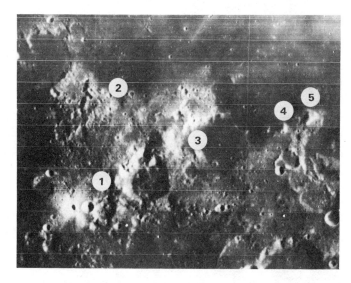

Figure 7.   Geography of a preferred mass-driver launch site.

is unsuitable because it is very mountainous. The nearby site (2) on level mare terrain is much better. Moreover, it is adjacent to a boundary between mare and lunar highland. Since mare rocks are rich in iron or titanium while highland rocks are richer in aluminum, a variety of ore types may be available.

Site (3) is similar but offers an additional advantage. The flight path can readily pass either or both of two mountains, (4) and (5), on which it is possible to set up downrange corrector stations (Heppenheimer *et al.*, 1982a,b). These adjust the flight of the payloads subsequent to launch. Heppenheimer *et al.* (1982b) argues that the payloads' trajectories can be controlled so as to achieve an accuracy, at L_2, of ±1.5 meters.

Figure 8 gives a reference design concept for the mass-catcher (Heppenheimer, 1978b, 1979). A suitable structural material is Kevlar–49, with yield strength $\sigma = 36.2$ gigapascals

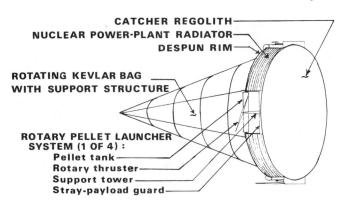

Figure 8.   Mass-catcher concept sized for 10^5 tons total mass.

50 meters

a

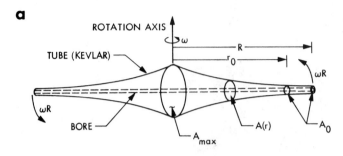

b

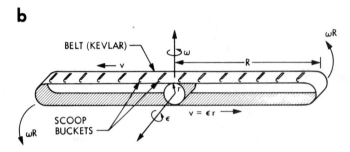

Figure 9. Design concepts for the Rotary Pellet Launcher.

= 525,000 psi and density $\rho = 1.45$ g/cm^3 (Johnson and Holbrow, 1977). For a uniform catcher regolith and stability in rotation, the cone angle must be greater than $\tan^{-1} 1/3 = 18.5°$. Propulsion is provided by Rotary Pellet Launchers, which rotate rapidly to eject pellets formed from lunar material. Their associated rotation velocity is of order $(2\sigma/\rho)^{1/2} = 2234$ m/s, which is the order of the pellet ejection velocities. It was earlier noted that for the catching mission, $\delta V = 187$ m/s and thrust $= 0.141$ newtons/ton. Hence the catcher must eject the order of $(187/2234) = 8.4\%$ of its loaded mass as propellant. Thruster power required is of the order of $(0.141 \times 2234) = 316$ watts/ton. A nuclear powerplant is advantageous; it can readily be protected from damage by stray payloads.

Figure 9 illustrates two concepts for the Rotary Pellet Launcher. Concept (a), discussed in Johnson and Holbrow (1977), maximizes the tip velocity by tapering the tube. With a tube that thickens exponentially toward the root, tip velocities can well exceed $(2\sigma/\rho)^{1/2}$. However, pellets then must roll or slide down the tube, possibly producing rapid abrasion. This difficulty may be overcome by the variant concept, (b). It features a rapidly rotating conveyor belt that carries pellets in scoop buckets out to the ends. Its tip velocity is less than $(2\sigma/\rho)^{1/2}$, but if the load–bearing belt has mass much greater than that of the pellets and scoop buckets it carries, then this velocity may be approached as a limit.

The operational use of the catcher involves a proposed "catching cycle" (Fig. 10) (Heppenheimer, 1979). Catching is initiated following the end of the catcher's return trajectory, shown as the dashed line of Fig. 5. Initially, catching is done under a constraint of low impact velocity, to protect the bag. This continues for two to three days, until some 1/10 of the final catcher load has been caught, forming a regolith. The catcher

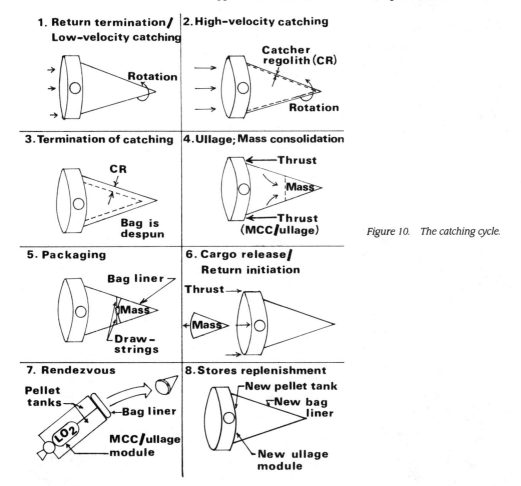

Figure 10.   The catching cycle.

then is sufficiently massive to require an optimal trajectory rather than one constrained by a requirement for low impact. For the next three weeks, it follows the loop of Fig. 5, with payloads impacting at some 250 m/s. The bag and its caught material are maintained in rotation by using the thrusters to despin the rim while torque is applied to the bag.

When the catcher is full, the thrusters serve to despin the bag. The mass within then is consolidated via an ullage maneuver. Part of the required thrust comes from the thrusters; extra thrust can be provided from compressed O_2, which is available from a remote space facility for ore processing. The consolidated mass may be packaged within a detachable bag liner simply by pulling drawstrings to close the mouth of the bag. With the cargo packaged, the catcher backs away, leaving the cargo to enter free flight on a trajectory to the space manufacturing facility (Heppenheimer, 1978c).

The catcher initiates a return maneuver, along the dashed curve of Fig. 5. While on this return leg, it can rendezvous with a supply of stores sent along trajectories that

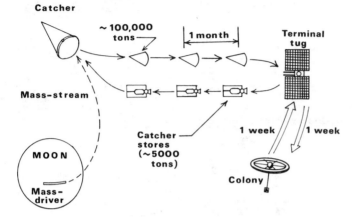

Figure 11. Overall system for transport of lunar resources.

are the reverse of those followed by the released cargo. Resupply can include a new O_2 module, tanks full of pellets for the thrusters, and a new bag liner. During the stores' transit, the O_2 module can provide midcourse corrections. Because it serves both for ullage and for such midcourse corrections, in Fig. 10 it is referred to as the MCC/ullage module.

Finally, Fig. 11 shows the overall transport operation. The traffic back and forth, in both cargo and stores, follows similar trajectories. Along the trajectory near the space facility, the cargo is intercepted by a terminal tug (Heppenheimer, 1979). It is based at the space colony and executes transfers between the colony orbit and the cargo orbit. Its basic mission resembles that of the catcher: to accelerate large cargoes through a small δV, at accelerations of some 10^{-4} m/s^2. Its propulsion systems thus will closely resemble those of the catcher, but it can be solar powered, since it does not face damage from stray payloads.

The catcher thus is to operate as an autonomous facility, resupplied from the colony. It need not tap into its caught material during catching operations. Instead, like an oil tanker, the catcher simply carries and contains its cargo, as a bulk mass. This mass, in turn, could run to 10^5 tons/month, or even more (Heppenheimer, 1978b).

REFERENCES

Grey J. (editor) (1977a) *Space Manufacturing Facilities*, Vol. 1. American Institute of Aeronautics and Astronautics, New York. 266 pp.

Grey J. (editor) (1977b) *Space Manufacturing Facilities*, Vol. 2. American Institute of Aeronautics and Astronautics, New York. 356 pp.

Grey J. (editor) (1979) *Space Manufacturing Facilities*, Vol. 3. American Institute of Aeronautics and Astronautics, New York. 574 pp.

Grey J. (editor) (1981) *Space Manufacturing Facilities*, Vol. 4. American Institute of Aeronautics and Astronautics, New York. 455 pp.

Heppenheimer T. A. (1978a) Achromatic trajectories and lunar material transport for space colonization. *J. Spacecraft and Rockets, 15*, 176–183.

Heppenheimer T. A. (1978b) A mass-catcher for large-scale lunar material transport. *J. Spacecraft and Rockets,* *15,* 242–249.

Heppenheimer T. A. (1978c) Steps toward space colonization: colony location and transfer trajectories. *J. Spacecraft and Rockets, 15,* 305–312.

Heppenheimer T. A. (1979) Guidance, trajectory and capture of lunar materials. In *Space Manufacturing Facilities, 3* (J. Grey and C. Krop, eds.), pp. 473–490. American Institute of Aeronautics and Astronautics, New York.

Heppenheimer T. A. and Kaplan D. (1977) Guidance and trajectory considerations in lunar mass transportation. *AIAA J., 15,* 518–525.

Heppenheimer T. A., Ross D. J., and Hannah E. C. (1982a) Electrostatic velocity adjustment of payloads launched by lunar mass-driver. *J. Guidance and Control, 5,* 200–209.

Heppenheimer T. A., Ross D. J., and Hannah E. C. (1982b) Precision release and aim of payloads launched by lunar mass-driver. *J. Guidance and Control, 5,* 291–299.

Johnson R. D. and Holbrow C. (editors) (1977) *Space Settlements, A Design Study.* NASA SP-413, NASA, Washington, D.C. 185 pp.

A LUNAR-BASED PROPULSION SYSTEM

Sanders D. Rosenberg

Aerojet TechSystems Company, P.O. Box 13222, Sacramento, CA 95813

As activities in cis- and trans-lunar space and on the Moon increase during the 21st century, the use of a lunar-based propulsion system, refueled by propellants manufactured from lunar resources, may offer large cost savings when compared with a space-based propulsion system refueled from the Earth. Oxygen/hydrogen (LO_2/LH_2) bipropellant propulsion appears to be attractive because of its estimated high delivered specific impulse, *i.e.*, 485 s. However, difficulties associated with the long-term storability and low density of LH_2 detract from this performance. Other bipropellant combinations may have advantages in this context. The potential utility of the oxygen/silane (LO_2/$LSiH_4$) bipropellant combination for use in a lunar-based propulsion system and the potential for the on-site manufacture of lunar oxygen and silane are considered in this paper. It appears that oxygen and silane can be produced from common lunar mare basalt in an integrated facility. The carbothermal process uses lunar materials efficiently to produce oxygen and silane-precursors with minimum terrestrial resupply. The production of silane from lunar materials may require a key, lunar-produced intermediate, magnesium silicide (Mg_2Si). Mineral acid or water terrestrial resupply will be required to produce silane by this synthesis. It appears that the propellant properties of oxygen and silane are more than adequate to support the development of a lunar-based propulsion system. Silane is stable and storable in space and lunar environments and has properties that are compatible with those of oxygen. Using standard Aerojet-JANNAF procedures, the estimated delivered performance of the propulsion system is 340–350 s at a mixture ratio of 1.50 to 1.80. Penalties normally associated with pressure-fed propulsion systems may be minimized in the lunar environment, *i.e.*, 1/6 g. A pressure-fed propulsion system may prove to be quite competitive with a pump-fed system.

INTRODUCTION

The Moon is made of oxygen. That is, the Moon is rich in minerals from which oxygen can be manufactured. Unfortunately, the Moon does not possess adequate, easily exploited sources of hydrogen- or carbon-containing minerals. This poses a particular problem for a designer of a lunar-based propulsion system, as an adequate supply of both oxidizer and fuel is required to power the system. One solution to the problem is to transport the required fuel from the Earth, or from low Earth orbit, to the Moon.

Hydrogen is one candidate fuel because it offers excellent performance with oxygen. Unfortunately, hydrogen has an extremely low density and is very difficult to store as a liquid. Monomethylhydrazine ($CH_3N_2H_3$) is another candidate fuel since it offers satisfactory density, storage properties, and performance. It would, however, have to be transported from the Earth for use in a lunar-based propulsion system.

Perhaps there is a middle position in regard to the manufacture of a suitable fuel on the Moon. There may be a satisfactory fuel that can be manufactured by using lunar resources and some chemicals resupplied from the Earth, namely, silane.

LUNAR RESOURCES FOR PROPELLANT MANUFACTURE

The minerals required to manufacture oxygen on the Moon are abundantly available. Olivine [$(Mg,Fe)_2SiO_4$], pyroxene [$(Ca,Mg,Fe)SiO_3$], and ilmenite ($FeTiO_3$) are particularly attractive raw materials for lunar oxygen manufacture. The major minerals, such as olivine, pyroxene, and the plagioclase feldspars $[(Ca,Na)Al_2Si_2O_3]$ occur in concentrations approaching 100%. The minor minerals generally occur at concentrations of less than 2%; however, some, particularly ilmenite, occur at concentrations of up to 20%.

The chemistry of the lunar minerals of interest has been confirmed by the analysis of Apollo samples (Williams and Jadwick, 1980). In regard to lunar oxygen manufacture,

Table 1. Apollo Sample Analyses

Compound	Mare, wt %	Highland, wt %
Analyses of Typical Lunar Olivine		
SiO_2	37.36	37.66
TiO_2	0.11	0.09
Cr_2O_3	0.20	0.15
Al_2O_3	<0.01	0.02
FeO	27.00	26.24
MnO	0.22	0.32
MgO	35.80	35.76
CaO	0.27	0.16
	<0.01	<0.01
Total	100.97	100.40
Analyses of Typical Lunar Pyroxenes		
SiO_2	47.84	53.53
TiO_2	3.46	0.90
Cr_2O_3	0.80	0.50
Al_2O_3	4.90	0.99
FeO	8.97	15.42
MnO	0.25	0.19
MgO	14.88	26.36
CaO	18.56	2.43
Na_2O	0.07	0.06
Total	99.73	100.39
Analyses of Typical Lunar Ilmenite		
SiO_2	0.01	0.21
TiO_2	53.58	54.16
Cr_2O_3	1.08	0.44
Al_2O_3	0.07	<0.01
FeO	44.88	37.38
MnO	0.40	0.46
MgO	2.04	6.56
ZrO	0.08	0.01
V_2O_2	0.01	<0.01
Na_2O	<0.01	0.13
Total	102.16	99.37

the focus is on the concentrations of silicon dioxide and iron oxide. The concentration of magnesium oxide is important for the manufacture of lunar silane. Note the comparatively high concentrations of these oxides in lunar olivine, Table 1. On balance, lunar olivine appears to be the mineral of choice for the manufacture of the propellants required for a lunar-based propulsion system.

PROPELLANT MANUFACTURE FROM LUNAR RESOURCES

Ilmenite, which is essentially ferrous titanate, can be reduced directly with hydrogen to form water, iron, and titanium dioxide. Electrolysis of the water yields oxygen, the required oxidizer, and hydrogen, which is recycled in this process (Kibler *et al.*, 1984; Gibson and Knudsen, 1984).

$$2 \text{ FeTiO}_3 + 2 \text{ H}_2 \longrightarrow 2 \text{ H}_2\text{O} + 2 \text{ Fe} + 2 \text{ TiO}_2 \qquad (1)$$

$$2 \text{ H}_2\text{O} \xrightarrow{\text{electrolysis}} 2 \text{ H}_2 + \text{O}_2 \qquad (2)$$

In the ideal case, 32 g of oxygen is obtained from 320.2 g of mare ilmenite, Table 1. However, *at best*, only 20% of the lunar material that must be processed is ilmenite. Therefore, 1,601 g of raw material must be processed to obtain 32 g of oxygen, a yield of only 2.0%.

It is apparent that the lunar raw material will have to be enriched in ilmenite before hydrogen reduction, *i.e.*, (1), to reduce the size of the chemical plant. Enrichment studies are currently being conducted (Agosto, 1984). Note, however, that while efficient enrichment reduces the size of the processing plant required for the hydrogen reduction step, it does not reduce the amount of lunar raw material that will have to be mined and transported. In the *ideal* case, 100.0 kg of mare material will have to be processed to manufacture 2.0 kg of oxygen.

The other major constituent oxides cannot be reduced directly with hydrogen to form water and the elemental metal. Silicon dioxide, as contained in olivine and pyroxene, can be reduced using the carbothermal process to form oxygen, silicon, and magnesium oxide. The ferrous oxide is reduced as well to form oxygen and iron (Rosenberg *et al.*, 1964a,b, 1965a,b, 1985).

$$\underset{\text{olivine}}{\text{Mg}_2\text{SiO}_4} + 2 \text{ CH}_4 \xrightarrow[\text{reduction}]{\text{carbothermal}} 2 \text{ CO} + 4 \text{ H}_2 + \text{Si} + 2 \text{ MgO} \qquad (3)$$

or

$$\underset{\text{pyroxene}}{\text{MgSiO}_3} + 2 \text{ CH}_4 \xrightarrow[\text{reduction}]{\text{carbothermal}} 2 \text{ CO} + 4 \text{ H}_2 + \text{Si} + \text{MgO} \qquad (4)$$

$$2 \text{ CO} + 6 \text{ H}_2 \xrightarrow[\text{reduction}]{\text{catalytic}} 2 \text{ H}_2\text{O} + 2 \text{ CH}_4 \qquad (5)$$

$$2 \text{ H}_2\text{O} \xrightarrow{\text{electrolysis}} 2 \text{ H}_2 + \text{O}_2 \tag{6}$$

In the ideal case, 32 g of oxygen is obtained from 123.56 g of mare olivine (see Table 1), a yield of 25.9%. Thus, 3.86 kg of olivine will have to be processed to manufacture 1.0 kg of oxygen, providing a 13-fold advantage over ilmenite in the ideal ilmenite case.

While the Moon has abundant resources for oxygen manufacture, fuel sources are scarce indeed. There are no known sources of free or bound water or carbonaceous minerals. The concentration of solar wind implanted hydrogen of approximately 50 to 200 ppm in lunar regolith (Carter, 1984; Friedlander, 1984) appears to be too low to be of practical value for large scale fuel manufacture.

Silane is an article of commerce in the United States today, as it is used in the manufacture of chips for our electronics industry. Silane is usually manufactured by a process that may prove to be too complex for lunar application, *e.g.*, the reduction of silicon tetrachloride with lithium aluminum hydride.

It may be possible to simplify the manufacturing process while minimizing dependence on terrestrial resupply by the use of the reaction between hydrochloric acid and dimagnesium silicide (Sneed and Maynard, 1947).

$$2 \text{ MgO} \xrightarrow{\text{electrolysis}} 2 \text{ Mg} + \text{O}_2 \tag{7}$$
lunar derived
e.g., carbothermal process

$$2 \text{ Mg} + \text{Si} \longrightarrow \text{Mg}_2\text{Si} \tag{8}$$
lunar
derived,
e.g., carbothermal process

$$\text{Mg}_2\text{Si} + 4 \text{ HCl} \longrightarrow 2 \text{ MgCl}_2 + \text{SiH}_4 \tag{9}$$
terrestrial
resupply

In a more complex cyclic electrolysis process, it may be possible to electrolyze the magnesium chloride to form the required magnesium and hydrogen chloride, and oxygen as a byproduct. In this synthesis, water would be resupplied from Earth rather than hydrochloric acid [see also (8) and (9) above].

$$2 \text{ MgO} + 4 \text{ HCl} \longrightarrow 2 \text{ MgCl}_2 + 2 \text{ H}_2\text{O} \tag{10}$$
lunar derived

$$2 \text{ MgCl}_2 + 2 \text{ H}_2\text{O} \xrightarrow{\text{electrolysis}} 2 \text{ Mg} + 4 \text{ HCl} + \text{O}_2 \tag{11}$$
terrestrial
resupply

The resupply of water from the Earth may be less troublesome than that of hydrochloric acid. Water transport should be a standard item in the support of permanent lunar bases. Note that the oxygen that occurs as a byproduct in the manufacture of silane, *i.e.*, (7) and (11), can be used to support propulsion needs. The byproduct water, *i.e.*, (10), would be put to good use as well.

The technology for the manufacture of silane from lunar resources has not been studied. Process research and demonstration are required.

PROPELLANT PROPERTIES

The use of liquid oxygen is well established. It is used today to power the Space Shuttle Main Engine and the Centaur Upper Stage RL-10 Engine. It will be used in the future to power propulsion systems for advanced orbit transfer vehicles, becoming an integral part of the United States Space Station (Davis, 1983a,b; Babb *et al.*, 1984).

Comparative physical property values for oxygen, silane, and methane are presented in Table 2. Note that silane has a broader liquidus range than methane, which is a great benefit to the propulsion system and rocket engine designer. Silane is hypergolic with oxygen, which is an additional benefit. Methane and hydrogen are not hypergolic with oxygen. This gives additional complexity to the engine system.

The propellant properties of silane have not been defined adequately. However, it appears that silane has the potential to be an adequate space storable propellant.

Table 2. Physical Properties of Potential Lunar Propellants

Propellant	Melting Point °F	°C	Boiling Point °F	°C	Specific Gravity (liquid)
Oxygen, O_2	−361	−218.4	−297	−183	1.142 (−297°F)
Silane, SiH_4	−301	−185	−169.4	−111.9	0.68 (−301°F)
Methane, CH_4	−296.7	−182.6	−258.3	−161.3	0.46 (−296.7°F)

1. SiH_4 is thermally stable to *ca.* 800°F.
2. O_2/SiH_4 is hypergolic.
3. SiH_4 is a liquid at the nbp of O_2.

A LUNAR-BASED PROPULSION SYSTEM

Pressure-fed liquid bipropellant engines and propulsion systems are attractive because of their comparative simplicity. Their disadvantages are lower performance and the need for heavier weight tanks and a tank pressurization system. This results in heavier propulsion systems, *i.e.*, increased propellant and hardware weights.

A pressure-fed engine in the STS Orbiter Maneuvering System Engine format seems to be appropriate for use with the $LO_2/LSiH_4$ bipropellant combination, Fig. 1. This Aerojet TechSystems engine, which delivers 6,000-lbF thrust and operates with the $N_2O_4/CH_3N_2H_3$

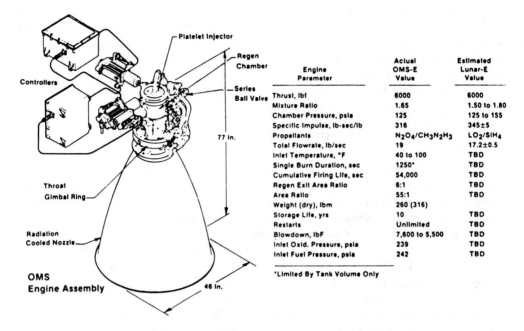

Engine Parameter	Actual OMS-E Value	Estimated Lunar-E Value
Thrust, lbf	6000	6000
Mixture Ratio	1.65	1.50 to 1.80
Chamber Pressure, psia	125	125 to 155
Specific Impulse, lb-sec/lb	316	345±5
Propellants	$N_2O_4/CH_3N_2H_3$	LO_2/SiH_4
Total Flowrate, lb/sec	19	17.2±0.5
Inlet Temperature, °F	40 to 100	TBD
Single Burn Duration, sec	1250*	TBD
Cumulative Firing Life, sec	54,000	TBD
Regen Exit Area Ratio	6:1	TBD
Area Ratio	55:1	TBD
Weight (dry), lbm	260 (316)	
Storage Life, yrs	10	TBD
Restarts	Unlimited	TBD
Blowdown, lbF	7,600 to 5,500	TBD
Inlet Oxid. Pressure, psia	239	TBD
Inlet Fuel Pressure, psia	242	TBD

*Limited By Tank Volume Only

Figure 1.   *The Space Shuttle OMS-Engine provides a format for the lunar engine.*

bipropellant combination at a mixture ratio of 1.65, has a regeneratively cooled thrust chamber. The OMS-E has a delivered specific impulse of 316 s.

If the regeneratively cooled thrust chamber can be adequately cooled with silane, and there is every reason to believe it can because of silane's thermal stability (Table 2), the development of a suitable long-life, reusable engine based on LO_2/SiH_4 appears to be quite achievable. Such an engine should have a delivered specific impulse of approximately 345 s.

The pressure-fed propulsion system penalties associated with an Earth-launched system would have to be reassessed in the light of a Moon-launch system. In the 1/6 g environment of the Moon, the apparent disadvantages of a pressure-fed system may be less, and a pressure-fed engine system operating at a higher chamber pressure may be appropriate. Detailed analysis would have to be performed to address these and other issues associated with the selection of a pressure-fed propulsion system.

Eagle Engineering has studied the impact of lunar-produced silane upon lunar oxygen production logistics in the context of the transfer of the lunar-produced oxygen from the Moon to low Earth orbit. A comparison was made between the use of a pump-fed LO_2/LH_2 propulsion system (Isp_v 480 s, MR 6.0) and a pressure-fed $LO_2/LSiH_4$ propulsion system (Isp_v 345 s, MR 1.80). Lunar-produced silane was used as the fuel in the propulsion system in place of Earth-supplied hydrogen. A small gain, *i.e.*, 2.5% in mass of oxygen transferred was derived by the substitution of silane for hydrogen. The $LO_2/LSiH_4$ propulsion system offers a modest benefit despite the 135 s difference in assumed specific impulse values.

In a postscript to this report, Eagle Engineering indicated that the use of a 5.0% higher mass fraction for the stage and a 1.45% higher specific impulse for the propulsion system resulted in a 29% increase in the mass of lunar-produced oxygen delivered to orbit on each ascent (Davis, 1983b). These improvements could be attained by the development of a pump-fed $LO_2/LSiH_4$ propulsion system.

Much work remains to be done to put such propulsion systems in place. However, the work required appears to be straightforward. No inventions are required to demonstrate technology readiness and enter development and production.

CONCLUSIONS

Whether or not a lunar-based propulsion system, more particularly, one based on $LO_2/LSiH_4$, will ever be developed will depend upon many factors that are beyond the scope of this paper. However, the technology to manufacture oxygen and silane on the Moon and also to develop a lunar-based propulsion system based on the bipropellant $LO_2/LSiH_4$ combination is within our reach. The development of a lunar-based propulsion system can be accomplished within the time frame under consideration, *i.e.*, by the start of the 21st century. All that is required are the need, the will, and the funds.

Acknowledgments. The author wants to acknowledge the support that he received from Hubert P. Davis and his colleagues at Eagle Engineering in Houston, Texas and Richard E. Walker and his colleagues at Aerojet TechSystems in Sacramento, California.

REFERENCES

Agosto W. H. (1984) Electrostatic concentration of lunar soil ilmenite in vacuum ambient (abstract). In *Papers Presented to the Symposium on Lunar Bases and Space Activities of the 21st Century*, p. 24. NASA/Johnson Space Center, Houston.

Babb G. R., Davis H. P., Phillips P. G., and Stump W. R. (1984) Impact of lunar and planetary missions on the space station (abstract). In *Papers Presented to the Symposium on Lunar Bases and Space Activities of the 21st Century*, p. 53. NASA/Johnson Space Center, Houston.

Carter J. L. (1984) Lunar regolith fines: A source of hydrogen (abstract). In *Papers Presented to the Symposium on Lunar Bases and Space Activities of the 21st Century*, p. 27. NASA/Johnson Space Center, Houston.

Davis H. P. (1983a) Lunar oxygen impact upon STS effectiveness. *Eagle Engineering Rpt. No. 8363*. Eagle Engineering, Houston. 78 pp.

Davis H. P. (1983b) Lunar silane impact upon lunar oxygen production logistics. *Eagle Engineering Rpt. No. 8370*. Eagle Engineering, Houston. 125 pp.

Friedlander H. N. (1984) An analysis of alternate hydrogen sources for lunar manufacture (abstract). In *Papers Presented to the Symposium on Lunar Bases and Space Activities of the 21st Century*, p. 28. NASA/Johnson Space Center, Houston.

Gibson M. A. and Knudsen C. W. (1984) Lunar oxygen production from ilmenite (abstract). In *Papers Presented to the Symposium on Lunar Bases and Space Activities of the 21st Century*, p. 26. NASA/Johnson Space Center, Houston.

Kibler E., Taylor L. A., and Williams R. I. (1984) The kinetics of ilmenite reduction: A source of lunar oxygen (abstract). In *Papers Presented to the Symposium on Lunar Bases and Space Activities of the 21st Century*, p. 25. NASA/Johnson Space Center, Houston.

Rosenberg S. D., Guter G. A., Miller F. E., and Jameson G. R. (1964a) *Catalytic Reduction of Carbon Monoxide with Hydrogen*. NASA CR-57. NASA, Washington. 64 pp.

Rosenberg S. D., Guter G. A., and Miller F. E. (1964b) The on-site manufacture of propellant oxygen utilizing lunar resources. *Chem. Eng. Prog., 62*, 228–234.

Rosenberg S. D., Beegle R. L. Jr., Guter G. A., Miller F. E., Rothenberg M. (1965a) *The Manufacture of Propellants for the Support of Advanced Lunar Bases*. SAE Paper No. 650835. Society of Automotive Engineers, New York. 16 pp.

Rosenberg S. D., Guter G. A., and Miller F. E. (1965b) The utilization of lunar resources for propellant manufacture. Post Apollo Space Exploration. *Adv. Astronaut. Sci., 20*, 665.

Rosenberg S. D., Guter G. A., and Miller F. E. (1965c) Manufacture of oxygen from lunar materials. *Ann. N. Y. Acad. Sci., 123*, 1106.

Sneed M. C. and Maynard J. L. (1947) *General Inorganic Chemistry*. Van Nostrand, New York. 708 pp.

Williams R. J. and Jadwick J. J. (1980) *Handbook of Lunar Materials*. NASA RP-1057. NASA, Washington. 120 pp.

LAUNCHING ROCKETS AND SMALL SATELLITES FROM THE LUNAR SURFACE

K. A. Anderson[1], W. M. Dougherty[2], and D. H. Pankow[2]

University of California, Berkeley, CA 94720
[1]Physics Department and Space Sciences Laboratory
[2]Space Sciences Laboratory

Scientific payloads and their propulsion systems optimized for launch from the lunar surface differ considerably from their counterparts for use on Earth. For spin-stabilized payloads, the preferred shape is a large diameter-to-length ratio to provide stability during the thrust phase. The rocket motor required for a 50-kg payload to reach an altitude of one lunar radius would have a mass of about 41 kg. To place spin-stabilized vehicles into low altitude circular orbits, they are first launched into an elliptical orbit with altitude about 840 km at aposelene. When the spacecraft crosses the desired circular orbit, small retro-rockets are fired to attain the appropriate direction and speed. Values of the launch angle, velocity increments, and other parameters for circular orbits of several altitudes are tabulated. To boost a 50-kg payload into a 100-km altitude circular orbit requires a total rocket motor mass of about 90 kg.

INTRODUCTION

The scientific investigation made possible by the Apollo project in the 1970s led to remarkable success in defining the Moon as a solar system planetary body. The synthesis of this great quantity of geochemical and geophysical data has, among other achievements, led to a theory of the Moon's origin that is more consistent with these data than previous theories have been.

Much was learned during the Apollo days about the Moon's gravity and magnetic fields, its atmosphere, and its interaction with the solar wind. However, much has been left unknown about these important topics, due in part to the restriction of the Apollo landing sites and in-orbit investigation to a fairly small range of latitudes about the Moon's equator. A powerful new attack on unsolved lunar scientific questions would be made possible by manned bases on the Moon's surface. With such resources, sounding rockets and even small orbiting vehicles carrying scientific payloads could be brought to the Moon and launched into a variety of interesting trajectories. In this paper, we attempt to define some of the technical characteristics such vehicles would have. We illustrate the use of rockets and satellites launched from the lunar surface with a few scientific experiments that are of current interest. We recognize this study to be a highly preliminary one and that the flight of a lunar polar orbiter before manned bases are established could greatly change what scientific experiments would be done. In any case, detailed planning and selection of the scientific experiments would be a prerequisite to any such program of rocket and satellite launching from the Moon's surface.

DISCUSSION

The launching of rockets or orbiting vehicles from the Moon's surface is quite different from the launching of such vehicles from Earth's surface for the following two reasons: (1) there is no atmosphere to produce aerodynamic forces on the vehicle; (2) the surface gravity is low. The acceleration due to gravity at the Moon's surface is about 1.63 m/s^2.

These environmental factors have several consequences to the design of the rocket motors and payloads:

(1) A nose cone to protect the payload from aerodynamic forces is not required. The advantage gained in this way is that total mass of the system is lowered. The reduction in mass results not only from absence of the nose cone but elimination of the nose cone ejection mechanism as well. For some payloads under some launch conditions,

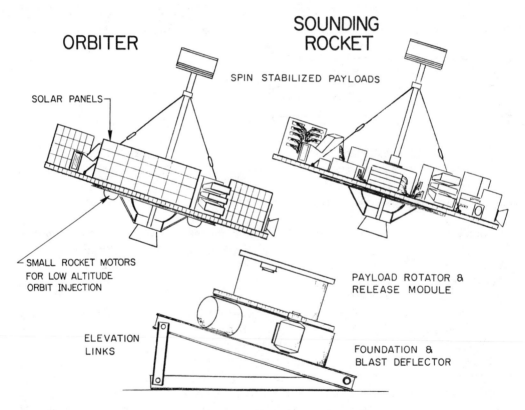

Figure 1. Possible configuration for a small scientific payload and propulsion system suitable for launch from the lunar surface. For use as an orbiting vehicle, the sounding rocket must be augmented by solar panels and small rocket motors to provide a second velocity increment. A 50-kg spin-stabilized payload can be put in a low altitude circular orbit by about 90 kg of rocket motors and fuel mass.

protection from dust blown off the surface by the rocket motor might be necessary; then a dust cover and means of opening it would be required. However, such a device would be simpler and less massive than the nose cone and ejection mechanism used in launches from Earth's surface.

(2) The absence of aerodynamic forces means that the diameter of the payload may be made as large as desired. A large diameter would result in a much simplified mechanical design and ease of assembly of the subsystem on the payload platform. For example, the added space would make layout and cabling much easier. The greater accessibility results in ease of troubleshooting and ease of replacement of subsystems or experiments. Figure 1 shows how such a payload platform might appear.

The large diameter payload has one other great advantage. The moment of inertia about the thrust axis of the vehicle, $I_{\parallel}$, could be made much larger than the moment of inertia about axes perpendicular to the thrust axis of the rocket motor, $I_{\perp}$. Thus, simple spin-stablized vehicles would have large $I_{\parallel}/I_{\perp}$ ratios, the requirement for "flywheel stability." This situation is highly desirable since there is much less tendency for large-angle precession of the thrust axis about the direction of flight. Rockets with a small $I_{\parallel}/I_{\perp}$ ratio, the condition for most Earth-launched rockets, have a tendency to go into a flat precessional motion rather than remaining nose up.

(3) The low gravitational force on the Moon's surface greatly reduces the total impulse requirement in order to achieve a given peak altitude or to attain low altitude orbital speed. In the case of chemical rocket motors, this means that the mass of the motor plus fuel is greatly reduced compared to the terrestrial situation. A further significant mass reduction follows from the absence of atmospheric drag. Because there is no need to keep the cross section of the vehicle small, the motors may be made spherical, thereby attaining a more favorable fuel-to-total-motor-mass ratio than is attainable from thin, cylindrical motors. Table 1 gives specifications of two rocket motors, one that will take a 50-kg payload to altitude of 100 km and the other will take the same payload to an altitude of 1738 km (one lunar radius). The term payload includes everything on the vehicle except the motor and its fuel.

An interesting scientific use of a sounding rocket capable of taking a plasma, particle, and field diagnostic payload to an altitude of 1738 km above the Moon's surface is indicated

Table 1. Specifications of Solid Fuel Rocket Motors Suitable for Launching Small Scientific Payloads From the Lunar Surface

	Length	Diameter	Mass at Ignition	Propellant Weight	Mass of Empty Motor
Motor A	29.4 cm	19.6 cm	13.9 kg	11.1 kg	2.8 kg
Motor B	50.5 cm	33.7 cm	41.1 kg	32.9 kg	8.2 kg

Motor A will take a 50-kg payload to an altitude of 100 km above the Moon's surface; Motor B will take the same payload to an altitude of one lunar radius (1738 km). The specific impulse of the propellant was taken to be 296 s.

in Fig. 2. When the Moon is in the solar wind, a void is formed behind it, and an expansion fan forms to fill the void. A limb shock is likely to be present, and the void and expansion fan can be expected to have interesting physical properties as well. The rocket just mentioned above would be able to sample all these regimes and their boundaries and provide a

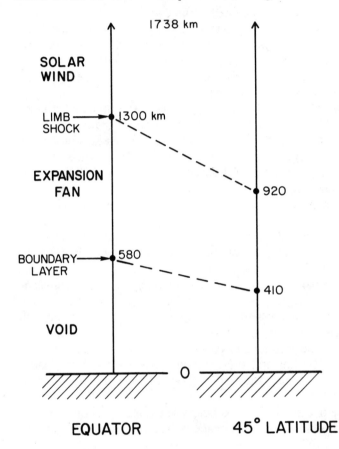

Figure 2. A sounding rocket launched vertically upward from the Moon's surface would encounter several distinctive regimes in the interaction of the solar wind with the Moon. The void, expansion fan, and solar wind would all be sampled as would two thin structures, the boundary layer between the void and expansion fan, and the limb shock.

detailed description of the boundary layer and shock structure. Figure 3 indicates where these boundaries would appear on the rocket trajectory for typical solar wind conditions and for two different launch latitudes, the equator and 45°.

We have also briefly considered the use of electromagnetic launchers (EML) to inject scientific payloads into a variety of trajectories from the lunar surface. There are several disadvantages to this method. For example, the acceleration a required to attain low altitude orbital speed of 1.686 km/s is given by

$$a = \frac{1.41 \times 10^6}{S} \text{ m/s}^2 \qquad (1)$$

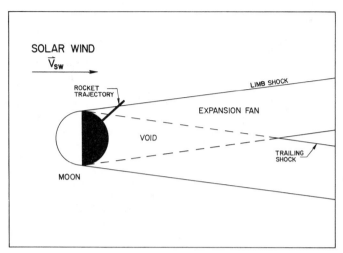

SOLAR WIND

$\vec{V}_{SW}$

ROCKET TRAJECTORY

LIMB SHOCK

EXPANSION FAN

VOID

TRAILING SHOCK

MOON

Figure 3. Altitudes at which the different features of the solar wind interaction with the Moon would appear during the rocket flight shown in Fig. 2. The peak altitude of the rocket is one lunar radius, its flight time about 45 minutes. To achieve this trajectory, a payload of 50 km requires a rocket motor plus fuel mass of about 41 kg.

where S is the length of the EML. Even a launcher 100 m in length would subject the payload to an acceleration of 14,100 m/s^2 or 1439 times Earth's gravitational acceleration. While scientific payloads might be designed to withstand such acceleration, it would be more expensive to do so. Another disadvantage of the EML is that almost certainly magnetometers would become severely contaminated by the very large magnetic fields generated by this type of launcher. Finally, an EML dedicated to launching scientific payloads would be more expensive than launch set-ups for chemically propelled vehicles (see Fig. 1). This is especially true if a variety of launch locations, azimuths, and elevations are desired.

We have also investigated launch of spin-stabilized spacecraft utilizing solid fuel motors into circular orbits of specified altitude without reorientation of the spin axis (*i.e.*, the spin axis remains inertially fixed during the injection sequence). Of course, attitude control systems could be used, but these add mass to the vehicle and increase the cost. Circular orbits can be achieved from vehicles initially at rest on the Moon's surface by using two velocity increments. The simple launch platform used for sounding rocket launchers could also be used for launch into circular orbits (Fig. 1). The first event in the launch sequence is to set the azimuth and elevation angles of the platform. Azimuth would be determined by experiment requirements. The elevation angle required to achieve a circular orbit of desired altitude is calculated by a method described below. For circular orbit altitudes in the range 50–100 km, the angle is about 3.5°. The spacecraft is then placed in the platform and spun up. The main motor is ignited and burns for about 11 seconds. A speed close to the desired low altitude circular orbit speed is reached at burnout. This velocity increment, ΔV_1, is about 1830 m/s. At burnout of the main motor, the spacecraft is approximately 8 km downrange at an altitude of 0.4 km. The second velocity increment, ΔV_2, is applied where the elliptical orbit crosses the desired circular orbit after aposelene. We have found values of ΔV_2 that result in nearly circular orbits of 50–100 km altitude. This sequence of events is illustrated schematically in Fig. 4.

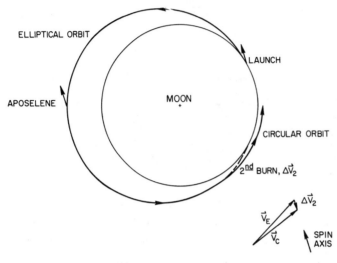

Figure 4. Illustration of the injection sequence of a spin-stabilized spacecraft into a circular orbit about the Moon. The spin axis remains inertially fixed over the elliptical trajectory. Orbit altitude is controlled mainly by the time to second burn. $\vec{V}_E$ is the orbit velocity just prior to the second burn; $\vec{V}_C$ is the resultant circular orbit velocity. The second velocity increment is much smaller than the first one.

Table 2 gives the launch angle and velocity increment necessary to achieve circular orbits for several altitudes. The second velocity increment can be achieved by small rocket motors whose thrust is directed opposite to the main motor. There is little advantage to jettisoning the main motor after burnout since its mass is so small. The mass of the small motors needed to provided the ΔV_2 would be less than 10 kg. Ignition of the small motors would be accomplished by a timer set to close the firing circuits a preset time after launch. These times are also given in Table 2.

A computer program tests Keplerian ellipses that intersect the Moon's surface and cross the desired circular orbit. The spin axis of the rocket, fixed at launch, is tangent to the ellipse where it enters the Moon. A velocity increment is made when the ellipse crosses the desired orbit. This increment is calculated to change the direction of motion to that of a circular orbit. If this increment results in the proper speed, as well, the computer prints the appropriate data. If not, another ellipse is tested. For the examples in the Table, the original eccentricity was arbitrarily fixed at $\epsilon = 0.20000$, and the aposelenes were all about 840 km. The launch angles for these ellipses are about 3.5° above the local

Table 2. Injection Parameters for Low Altitude Circular Orbits About the Moon.

Nominal Orbit	Launch Angle, γ	ΔV_1	ΔV_2	Time to Injection	Final Orbit		
					τ_{max}	τ_{min}	Eccentricity
50 km	3.6°	1835.0 m/s	−233.9 m/s	137.10 min	61.2 km	48.4 km	0.00356
75 km	3.5°	1835.6 m/s	−249.0 m/s	129.51 min	86.4 km	72.8 km	0.00374
100 km	3.2°	1836.6 m/s	−261.1 m/s	128.37 min	100.3 km	98.9 km	9.00035

A payload mass of 50 kg is assumed.

horizontal direction. Shallow launch angles would present difficulties only for launch sites near lunar mountains. Circular orbit insertion would take place about 310° around the Moon from launch and requires thrust from small motors directed opposite to the main motor (see Fig. 1).

The rocket will travel 8.3 km during the 11 s primary burn, but lunar gravity will deflect it into a slightly different ellipse than that calculated above. The velocity and altitude at main motor burnout are used to determine the length of the major axis of the new ellipse. The eccentricity and orientation of the major axis are determined numerically.

The most important difference between the original ellipse and the actual trajectory is the time the rocket crosses the circular orbit. The time between launch and orbit crossing is found by numerical inversion of Kepler's equation. Errors introduced by neglecting the duration of the second burn (2 seconds) are negligible. The unadjusted angle and magnitude of the velocity increment are not far from the ideal, and the final orbit is nearly circular.

Accuracies better than those obtained here are quite possible, and the quality of the orbit may be limited only by rocket motor performance and perturbations due to lunar gravity anomalies. If the required tolerances for the total impulse of the primary motor cannot be assured, the second burn could be initiated by radar altimeter, accelerometer, or Moon-based tracking.

We have not carried out analyses of the dispersion of the orbit altitude achieved because this depends on a detailed system design and engineering analysis. Nor have we attempted to determine the orbit parameters and launch times that will maximize the lifetime of near-circular low-altitude orbits.

For spacecraft having orbit lifetimes of weeks or months, a solar-cell power source would be required, as well as a set of batteries to operate the system throughout the shadow interval. Since the spacecraft is spinning, the solar cells should be mounted vertically on the periphery of the experiment platform (see Fig. 1). The angle, β, that the solar rays make with respect to a vertical line drawn from the solar cell surface is given by

$$\beta = | \lambda - \gamma | \qquad (2)$$

where λ is the latitude of the launch site. For launch sites in the latitude range 20–30°, the sun's rays lie close to the equator for the spacecraft. When the launch site for polar orbiting spacecraft is the Northern hemisphere, the launch direction should be toward the north; from Southern hemisphere bases, the launches are toward the south.

CONCLUSION

We conclude with a brief list of scientific investigations that could be carried out from low altitude lunar circular orbits:

(1) A low altitude polar orbit would permit a global survey of the Moon's small scale magnetic fields by both flux gate magnetometry (Russell et al., 1975) and electron reflection magnetometry (Howe et al., 1974; Anderson et al., 1977). Field strengths at

the surface of only 3 x 10^{-7} Gauss (0.03 nT) and scale sizes of a few hundred meters or less could be determined. The flux gate magnetometry will provide accurate directional information on the more extended magnetic fields and the electron reflectance can provide some information on direction. When combined, the two methods of magnetometry can also yield information on the size and depth of the magnetized lunar rock. One goal of such studies is to determine if the Moon once possessed an active dynamo (Runcorn, 1978) and if not, to determine the probable cause of the ancient magnetic fields. Another objective is to correlate magnetic features with surface geological features.

(2) The small scale features of the Moon's gravity field, including mascons, could be determined by tracking the low altitude orbiter (Sjogren *et al.,* 1974). Transmitters suitable for this purpose, and of sufficient power to be received by Earth stations, could be included on low altitude orbiter flights.

Acknowledgments. We thank the Advanced Projects Group of Bristol Aerospace Limited, Winnipeg, Canada, for design information on solid fuel rocket motors. The work at Berkeley was supported by NASA Grant NGL-05-003-017.

REFERENCES

Anderson K. A., Lin R. P., McGuire R. E., McCoy J. E., Russell C. T., and Coleman P. J. Jr. (1977) Linear magnetization feature associated with Rima Sirsalis. *Earth Planet. Sci. Lett., 34,* 141–151.

Howe H. D., Lin R. P., McGuire R. E., and Anderson K. A. (1974) Energetic electron scattering from the lunar remanent magnetic field. *Geophys. Res. Lett., 1,* 101–104.

Runcorn S. K. (1978) The ancient lunar core dynamo. *Science, B199,* 771–773.

Russell C. T. Coleman P. J. Jr., Fleming B. K., Hilburn L., Ioannides G., Lichtenstein B. R., and Schubert G. (1975) The fine-scale lunar magnetic field. *Proc. Lunar Sci. Conf. 6th,* pp. 2955–2969.

Sjogren W. L., Wimberly R. N., and Wollenhaupt W. R. (1974) Lunar gravity via the Apollo 15 and 16 subsatellite. *Moon, 9,* 115–128.

4 / LUNAR SCIENCE

T HE GOAL OF PROJECT APOLLO, to place a man on the surface of the Moon and return him safely to Earth, was a task for engineering and technology. The resultant transportation system was so robust in its capability that a rich scientific harvest also was gathered in the process. The visibility and magnitude of the Apollo program left the impression, even within the scientific community, that the major lunar scientific questions had been answered. Since the Surveyor project, NASA's unmanned planetary program has ignored the Moon, preferring to concentrate its admittedly limited resources on other bodies in the solar system.

In 1972, lunar scientists proposed launching a remote-sensing satellite into lunar polar orbit. Carrying a small number of geochemical and geophysical sensors, the Lunar Polar Orbiter (LPO) was designed to expand to global coverage the limited Apollo orbital science data set. Remote-sensing information from the orbiting Service Modules of Apollos 15, 16, and 17 had been invaluable in revealing the scale and extent of the planetary processes whose nature and timing were decoded in analyses of the returned samples. Low orbital inclination and limited time at the Moon resulted in tantalizingly incomplete results, often hinting at new geological insights just beyond the orbital coverage. Although the LPO had unquestioned scientific value, the mission concept fell victim to a scarcity of funds and a planning philosophy emphasizing complex exploration missions with great public appeal.

In the current planetary exploration program, a Venus Radar Mapper and a Mars Observer have been approved as new starts. A Comet Rendezvous and Asteroid Flyby has been proposed. The Soviets have announced an LPO mission to be launched before 1990, but lunar scientists in the United States are uncertain as to the availability and completeness of the future Soviet data. However, the announcement has caused some NASA program managers to suggest that a U.S. lunar mission should be put off for another ten years until the Soviet results have been evaluated.

Arguments for or against a near-term orbiting lunar survey satellite are couched in terms of purely scientific priorities within a fiscally restricted unmanned program. If an orbiter were seen as a precursor mission to a lunar base program, then one could argue from a different perspective that its entire cost would be a fraction of 1% of the manned program over a decade. It also becomes a long lead item because approximately eight years elapse from the declaration of a mission new start until a mature analysis of the results. Of course, hypothetical justifications by a lunar base program are irrelevant to real policy debates because no long range plan for manned lunar activities exists within NASA at the present time. In addition, some lunar base scenarios, such as a return to an Apollo site, require no precursor mission. However, new data from a low cost scientific satellite serves other purposes.

If we assume that NASA will continue historical behavior and propose a new major program only after the current one is complete or operational, then a lunar base will be planned no sooner than about 1995, after completion of the LEO space station. Scientists active in the Apollo data analysis will be at retirement; younger researchers will have no experience in lunar problems unless new lunar data sources are developed. A Lunar Observer mission, to use current nomenclature, is an inexpensive investment in expertise for the space activities of the next century.

The scientists who have contributed papers to this section are, for the most part, veterans of the Apollo scientific program and are well steeped in the critical scientific questions facing contemporary and future investigators. Taylor reviews the major scientific questions of planetary evolution and discusses the role of manned surface exploration in searching for the answers. Haskin *et al.* consider several field geological investigations in the vicinity of a base to address problems of geochemistry and petrology of the lunar surface. Vaniman *et al.* describe volcanism on the Moon and point out the resource potential of volcanic features. Cintala *et al.* present an extended geological traverse of 4000 km length around the Imbrium Basin. Although an engineering challenge, an advanced exploration of this type would yield invaluable scientific data on lunar stratigraphy. Friesen develops an observational strategy for detecting any volatile emissions from the lunar interior, while Wilhelms gives a thorough discussion of the scientific payoffs from a Lunar Observer orbital satellite. Hood *et al.* elaborate on the geophysical information to be gained from orbital studies, especially when combined with a surface network of geophysical stations. Strangway recalls the distinctive lunar geophysical properties and discusses key measurements needed to study them. Ander elaborates on the geophysical exploration techniques that are particularly suited for use in the lunar environment.

These manuscripts form an unusually complete survey of the state of lunar science and, by extension, our understanding of the terrestrial planets. The Moon, whose geological evolution ground to a halt so early in the history of the solar system, preserves an invaluable record of those times. In contrast, the Earth manifests a collection of dynamic geological processes seen only in part elsewhere among the planets. Studies of these two bodies in concert extend understanding of our environment both to the past and to the future.

THE NEED FOR A LUNAR BASE: ANSWERING BASIC QUESTIONS ABOUT PLANETARY SCIENCE

G. Jeffrey Taylor

Institute of Meteoritics and Department of Geology, University of New Mexico, Albuquerque, NM 87131

Information derived from manned and unmanned exploration of the Moon has allowed us to develop an outline of lunar origin and evolution. Nevertheless, many questions await answers. These questions will be answered in part by data obtained from future unmanned missions such as the Lunar Geoscience Observer, but complete answers (and the new questions that will inevitably accompany them) will come only from additional manned missions. Although robot spacecraft can perform remarkable feats, they cannot match human ability to do field work, to react to unexpected discoveries, to do sophisticated preliminary analyses, and to deploy certain complicated equipment. One drawback to people living on the Moon is that they will affect the pristine lunar environment. Consequently, the first thing we must do when (and possibly before) the base is established is to obtain uncontaminated samples from around the Moon. This global contingency sample could be stored on the Moon for future study.

INTRODUCTION

The Moon has barely been explored, although there has been some impressive preliminary work. Six manned Apollo landings returned 382 kg of rock and soil. Three unmanned Luna spacecraft sent by the Soviet Union brought back 300 g of lunar soil from areas other than those sampled by Apollo. Geochemical and geophysical instruments onboard Apollo Command Modules and on subsatellites released from them made important measurements from orbit. Before Apollo, a series of unmanned Ranger and Lunar Orbiter spacecraft radioed back thousands of photographs. Several unmanned Surveyor spacecraft landed on the Moon, giving us a glimpse of the Moon's surface and an indication of its chemical composition and compositional variability. From this information we have been able to reconstruct an outline of lunar history. Nevertheless, gaping holes exist in our knowledge of the Moon's nature and evolution. The greatest accomplishment of lunar exploration so far is to show us how little we know.

This paper has five purposes: first, to point out that our ignorance about the Moon is all the more unfortunate because it is such an ideal place to study important planetary processes and to enlighten ourselves about early planetary history; second, to recount briefly what we have learned from the first phase of lunar exploration; third, to offer a list of unanswered questions about the Moon and its history and to explain why we must return to the Moon to answer them; fourth, to point out that people must return, not only unmanned machines; and fifth, to warn that pristine samples must be collected and preserved prior to or soon after a lunar base is established, because base activities could compromise the scientific integrity of the lunar environment.

THE MOON'S CENTRAL PLACE IN PLANETARY SCIENCE

In a sense, the Moon is the cornerstone of planetary science. The intensity and duration of its geologic activity have left behind a record of its earliest history, including clues to its origin; yet, at the same time, the rocks and surface features have recorded information about the processes operating in and on the Moon as it evolved. Consequently, the Moon contains information about planet formation (though admittedly cryptically inscribed), its early differentiation into core, mantle, and crust (written more clearly, but still far from translated), its subsequent chemical evolution (clearer), and its impact history (scrawled on its cratered surface, but not yet understood in detail). These are key events in solar system history whose traces have been annihilated on Earth by vigorous terrestrial geologic processes. To understand Earth's early history, we must study the Moon.

The Moon is a natural laboratory for studying significant processes such as impact cratering and volcanism. Vaniman *et al.* (1985) discuss the importance of understanding lunar volcanism. Cratering has shaped the surfaces of the terrestrial planets, the satellites of Jupiter and Saturn, and the bodies (presumably asteroids) from which meteorites come. Studies of terrestrial impact craters have led to some understanding of the process, but even the youngest impact craters on Earth have been substantially eroded, and none of the huge early basins remain. On the Moon, craters up to hundreds of kilometers across are preserved and could be studied in detail (*e.g.*, Cintala *et al.*, 1985).

The Moon also has the virtue of being accessible; it's close to us. No doubt many other bodies throughout the solar system contain information about planetary origin, evolution, and processes, but they are not as easily reached as is the Moon. We should continue to explore them with unmanned spacecraft, including sample returns and manned reconnaissance missions, but detailed manned exploration can start most easily on the Moon.

AN OUTLINE OF LUNAR HISTORY

To set the stage for the litany of unanswered questions posed in the next section, I first outline our current understanding of the Moon's history. Detailed reviews of lunar rocks and the Moon's origin and evolution can be found in S. R. Taylor (1982), Norman and Ryder (1979), Warren (1984), and Hartmann *et al.* (1985); recent non-technical sketches are given by Rubin (1984) and G. J. Taylor (1984, 1985a,b).

The Moon's origin has puzzled scientists for hundreds of years. The three traditional theories of lunar origin—fission from Earth, capture by Earth, and forming with Earth as a double planet system—all have serious flaws. The probability of capturing a fully formed Moon is so tiny that most dynamacists consider it impossible. The capture hypothesis cannot adequately explain why the Moon has a small iron core, in contrast to Earth and the other inner planets. The fission theory accounts for the difference in iron between Earth and Moon by requiring the Moon to spin off of the primitive Earth after core formation. However, there are problems explaining how the Earth was spinning fast enough (once every 2.5 hours) to fling off a blob of material from which the Moon formed. Studies of the dynamics of planetary accretion indicate that it is unlikely that Earth was ever rotating that rapidly. In fact, it is hard to model accretion and account for the observed amount of angular momentum possessed by the Earth-Moon system, let alone substantially more of it. There are also problems with the double planet idea in accounting for the angular momentum of the Earth and Moon and in explaining why material around Earth stayed in orbit, rather than falling to Earth.

A new story for the Moon's origin has recently blossomed. Growing from seeds planted by Hartmann and Davis (1975), the new theory depicts a cataclysmic birth of the Moon: nearing the end of Earth's accretion, after its core had formed and while it was still molten, an object the size of Mars (one-tenth Earth's mass) smashed into it at an oblique angle. The resulting monumental explosion deposited large quantities of vaporized and molten impactor and Earth into orbit, and the primitive Moon formed from this material. This story seems capable of explaining the compositions of Earth and Moon and the total angular momentum of the system. Based on our knowledge of planetary accretion, the large impact is not dynamically far-fetched (Wetherill, 1985).

From analysis of lunar samples, we know that when the Moon formed 4.6 b.y. ago, it was substantially molten. The lunar highland crust contains exceptional quantities of plagioclase feldspar. Plagioclase constitutes more than 90% of many samples (called "anorthosites") and averages about 75% in the explored highlands. This large abundance of one mineral led to the suggestion (Wood et al., 1970) that when the Moon formed, it was enveloped by a huge magma system, commonly known as the lunar "magma ocean." Plagioclase, which has a relatively low density, floated to the top of the ocean as it crystallized, forming the original crust. The ocean had solidified by 4.4 b.y. ago.

Near the end of the magma-ocean epoch, magmas generated inside the Moon began to intrude into the crust and probably erupt onto its surface. These are represented by the "Mg-suite" rocks of the lunar highlands. They contain less feldspar and more olivine and pyroxene than do the anorthosites, and they constitute a diverse set of samples. This activity continued until about 4.1 b.y. ago, at which time basaltic magmas rich in elements such as potassium (chemical symbol K), the rare earth elements (REE), and phosphorus (P) rose to the surface and intruded into the crust. The characteristic enrichments in these elements has earned such basalts the nickname KREEP.

As this magmatism was constructing the lunar crust, the Moon continued to be bombarded by planetesimals (Fig. 1). The prominent circular basins now seen on the Moon, which are hundreds of kilometers across, formed around 3.9 b.y. ago, though numerous others had formed prior to 4 b.y. This bombardment transformed the upper 2–10 km of the lunar crust into a rubble pile. The nature of highland rocks reflects this: most are breccias, rocks composed of fragments of other rocks. Most of the fragments are themselves mixtures.

Beginning around 3.9 b.y. ago, lavas that formed by partial melting inside the Moon began to fill the basins with mare basalts, which contain less feldspar and more olivine, pyroxene, and ilmenite (an oxide of iron and titanium) than do highland rocks. Ages of lunar samples demonstrate that most of the mare volcanism ended around 3.0 b.y. ago, but photogeologic studies indicate continued activity to about 2.0 b.y. (Boyce et

Figure 1. A view of the farside of the Moon taken during the Apollo 16 mission. The abundance of craters attests to the fierce bombardment the Moon experienced. The dark-floored crater (upper center), Kohlschutter, is 60 km in diameter and appears to have been flooded by mare basalts, one of the rare occurrences of such lavas on the farside. The smooth horizon is interrupted by Mare Muscoviense, which also contains basalt flows. The pole protruding on the right is the boom for the gamma-ray spectrometer carried on the Apollo 16 Command Module. Apollo photo number AS16-0729(M).

al., 1974), or younger (Schultz and Spudis, 1983). Lavas with compositions like those of mare basalts had erupted prior to 4.0 b.y. ago, but repeated impacts as the Moon was bombarded during that time erased much of the evidence (see Vaniman *et al.*, 1985).

UNANSWERED QUESTIONS AND MISSING INFORMATION

Detailed though it may seem, the above account of lunar history is little more than an outline. Many questions remain unanswered. What is the Moon's bulk composition? Does the Moon have a metallic core? How large is the core? Was there really a magma ocean? If so, how did it form? How deep was it? Was the Moon totally molten when it formed? How does the crust vary in thickness and composition laterally and vertically? How are the major lunar rock types related to one another? Have we identified all the major rock types? What were the effects of giant impacts on the Moon? Over what period of time did the large basins form? What does the presence of volatiles in lunar volcanic glasses imply about the inventory of such elements in the Moon? How old are the oldest rocks? What are the ages of the youngest rocks?

Studies of our present collection of lunar samples will continue to provide insight into these questions, as will improved understanding of how chemical processes operate inside planets. However, poor geographic sampling of the Moon limits what we can learn from the samples collected during the first round of lunar field work. The Apollo and Luna missions sampled rock and soil from the central nearside only, at a mere nine localities. The farside (Fig. 1) is completely unsampled. Furthermore, spectral observations of the Moon from Earth-based telescopes (Pieters, 1978) indicate that we have sampled less than half of the types of mare basalts present on the nearside. Such telescopic observations also show that the types of highland rocks that are rare among lunar samples, such as those rich in olivine and high-Ca pyroxene, occur on the floors and walls of large craters (Pieters, 1982, 1983). Obviously, we have not sampled all the types of rocks present on the Moon. This is the first priority when sampling begins again.

We have scant knowledge about the lunar interior. Three seismic stations set up by Apollo crews provided a glimpse inside the Moon, but they were located close together, on the nearside only, and in an area where the crust might be anomalously thin. The Apollo seismic network detected a change in seismic velocities 60 km beneath the surface, interpreted as the boundary between the crust and mantle, and another at a depth of about 1000 km, which seems to be a region that is partly molten. However, we do not know how the crust's thickness varies around the Moon or whether subtle contrasts exist in either the crust or the mantle. We have no direct evidence for an iron-nickel core like the Earth's. The presence and size of a metallic core has great bearing on theories for the Moon's origin and on the source of the unexplained magnetization of lunar rocks. Clearly, when we return to the Moon, an extensive network of seismometers must be set up and perhaps active seismic measurements carried out to probe the structure of the lunar interior.

We do not know enough about the composition of the Moon's crust or about how the composition varies laterally and vertically. The most compelling evidence for the lunar magma ocean and one of the best indicators of its depth is the exceptional abundance of plagioclase feldspar in the highlands crust (Warren, 1984), but we do not know how the abundance of this mineral (reflected chemically in the concentration of aluminum) varies around the Moon or with depth in it. Spectral reflectance measurements (Pieters, 1982, 1983) show that rocks rich in olivine or high–Ca pyroxene occur in the central peaks of some craters, whereas other central peaks appear to be pure anorthosite. Because central peaks and crater floors represent deep materials brought up by impacts, it appears that there are complex variations with depth in the lunar crust. Sampling these areas is essential when lunar exploration begins in earnest from a lunar base.

How are the major types of lunar rocks related to one another? On Earth, we can directly observe the kinship among different lithologies. For example, we can study successive lava flows and know for a fact that they erupted from one volcano. In contrast, no lunar rocks were collected from discernible rock layers. Even the largest boulders were simply large rocks strewn on the lunar surface. However, photographs from orbit (Fig. 2) show that layers of rock exist in the walls of large craters; these could be sampled. Similarly, layers of rock in rilles, such as Hadley Rille where Apollo 15 landed, almost certainly represent successive lava flows. Furthermore, although the lunar crust has been devastated by impacts, careful field work might reveal huge chunks of igneous bodies that had differentiated into a series of related rocks. Such fragments of igneous intrusions might be found in the walls of large craters (Fig. 3). Tilted structural blocks may also provide access to layered crustal rocks (Haskin *et al.*, 1985).

The huge impacts that excavated basins such as Imbrium, Serenitatis, and Orientale hurled material across the Moon and dredged up rocks from great depth. We do not

Figure 2. Layers of basalts adorn the wall of the crater Bessel (17 km diameter), which punctured south-central Mare Serenitatis. These layers of basalts could be sampled, as could layers visible in rilles. No mare basalts have been collected from an identified lava flow. Apollo photo number AS15-9328(P).

understand the details of impact events of this magnitude. By studying some basins in detail we could assess how the formation of basins has obscured primary crustal compositional variations by lateral mixing and determine how deeply they excavated.

Figure 3. Close view of the walls of the crater Copernicus (90 km diameter), the most prominent crater on the lunar nearside. The walls of this and similar craters might contain huge chunks of intrusive bodies that could help us understand how various kinds of lunar highland rocks are related to each other. Sampling would be a challenge: the walls are steep and rise 4 km above the crater's floor. Smooth areas are probably impact melt deposits. Lunar orbiter photo number V-M155.

A traverse across the large lunar basin, Imbrium, is described vividly by Cintala *et al.* (1985). Because such basins have formed on numerous other planets and satellites, research on lunar craters has implications beyond the Moon itself.

The study of lunar basalts and pyroclastic deposits provides information about the Moon's interior and thermal history. Our present knowledge about lunar volcanism is as incomplete as the other topics discussed in this section. The gaps in our knowledge and suggestions for filling them are discussed by Vaniman *et al.* (1985) in this volume.

PEOPLE ARE NEEDED

There is no doubt that we have a tremendous amount to learn about the Moon and its history, but why send people? Why not return to the Moon with unmanned spacecraft? The answer: good idea! We *should* send unmanned spacecraft to study the Moon before we actually go back in person. A logical first step is a polar orbiter equipped with assorted remote-sensing devices to map the Moon's surface, as has been suggested for one of NASA's Observer missions (the Lunar Geoscience Observer). We ought also to have some unmanned sample returns before the lunar base is established to help define specific research programs at the permanent lunar facility. The returned samples will also help keep lunar science vital until the base is established. The current cadre of lunar scientists know much more than they could ever convey in publications. Their experience is an irreplaceable asset that must be preserved, and a new generation of lunar experts must be trained. Today's students will be the scientists who answer the questions—and ask the next ones.

Unmanned machines have performed remarkable feats, but they will never match a trained geologist in doing field work. As pointed out by Spudis (1984) in a talk at the Lunar Base Symposium and emphasized by Cintala *et al.* (1985) and Haskin *et al.* (1985), field work requires people. Field work is iterative because scientists need time to think about what they have seen before going further. Most importantly, creative minds can react to the inevitable unexpected discoveries; pre-programmed machines cannot, so we lose opportunities to sample the unexpected in detail. Also, it is not possible to program a machine with all the experience and perspective that a veteran geologist possesses. Machines cannot select samples as intelligently as can people.

Once samples are collected from outcrops or plucked from the lunar surface, people are required to do a reasonable preliminary analysis on the Moon. Unmanned probes do remarkable chemical analyses of soil or rock chips grabbed by a robot arm, and scanning electron microscopes with energy-dispersive detectors are being designed to fly on comet rendezvous missions. Nevertheless, these do not compare to the information gleaned by an experienced petrographer peering into a petrographic microscope or to the chemical data obtained by neutron activation or X-ray fluorescence analysis of carefully chosen subsamples. People are essential for good preliminary analysis of samples collected on the Moon.

In addition to their observational and decision-making abilities, people are needed to deploy and repair complex equipment. For example, can an automated spacecraft really plant a heat-flow probe or a seismograph properly? Lay out a set of seismic receivers to deduce local stratigraphy? Study the walls of a trench dug in the fragmental lunar regolith? Probably not.

Finally, a disclosure statement is in order. Although I am obsessed with anything related to finding out how the Moon formed and evolved, I am also enthralled by the concept of people going into space. I have been thrilled by the exploits of Yuri Gagarin and John Glenn, of Neil Armstrong and Sally Ride. They are our representatives in space. Machines give us information; people give us impressions and inspiration.

PEOPLE CAN BE A PROBLEM

Many of the scientific investigations described in this volume require the unique lunar environment, *e.g.*, the amazingly low atmospheric pressure (10^{-12} torr) and the virtual absence of H_2O, CO_2, and nitrogen compounds. For example, an important project will be to search for indigenous lunar volatiles (Friesen, 1985). Consequently, environmental changes must be monitored continuously after a lunar base is constructed. Most importantly, as noted by Duke *et al.* (1984), it seems that some environmental degradation will occur as a result of lunar base activities, so the lunar surface should be sampled thoroughly and characterized during (or even before) the initial lunar base operation. In short, we ought to obtain a global "contingency sample." Many samples could be stored in sealed containers on the Moon for future reference. This would ensure the preservation of pristine, uncontaminated, pre-base lunar samples.

Acknowledgments. *This paper and my talk at the Lunar Bases Symposium benefitted from discussions with Klaus Keil, Ed Scott, and Horton Newsom. The paper has been improved by comments from Carle Pieters, Dave Vaniman, and the esteemed editor of the book, Wendell Mendell. This work was funded in part by NASA Grant NAG 9-30.*

REFERENCES

Boyce J. M., Dial A. L., and Soderblom L. A. (1974) Ages of nearside light plains and maria. *Proc. Lunar Sci. Conf. 5th*, pp. 11-23.

Cintala M. J., Spudis P. D., and Hawke B. R. (1985) Advanced geologic exploration supported by a lunar base: A traverse across the Imbrium-Procellarum region of the Moon. This volume.

Duke M. B., W. W. Mendell, and P. W. Keaton (Compilers) (1984) *Report of the Lunar Base Working Group.* LALP 84-43, Los Alamos National Laboratory, Los Alamos. 41 pp.

Friesen L. J. (1985) Search for volatiles and geologic activity from a lunar base. This volume.

Hartmann W. K. and Davis D. R. (1975) Satellite-sized planetesimals and lunar origin. *Icarus, 24*, 504-515.

Hartmann W., Phillips R. J., and Taylor G. J. (editors) (1985) *Origin of the Moon.* Lunar and Planetary Institute, Houston. In press.

Haskin L. A., Korotev R. L., Lindstrom D. J., and Lindstrom M. M. (1985) Geochemical and petrological sampling and studies at the first Moon base. This volume.

Norman M. N. and Ryder G. (1979) A summary of the petrology and chemistry of pristine highlands rocks. *Proc. Lunar Planet. Sci. Conf. 10th*, pp. 531-559.

Pieters C. M. (1978) Mare basalt types on the front side of the Moon: A summary of spectral reflectance data. *Proc. Lunar Planet. Sci. Conf. 9th*, pp. 2825-2849.

Pieters C. M. (1982) Copernicus Crater central peak: Lunar mountains of unique composition. *Science, 215*, 59-61.

Pieters C. M. (1983) Composition of the upper lunar crust: Preliminary results from near infrared reflectance data (abstract). In *Lunar and Planetary Science XIV*, pp. 608-609. Lunar and Planetary Institute, Houston.

Rubin A. E. (1984) Whence came the Moon? *Sky & Telescope, 68*, 389-393.

Schultz P. H. and Spudis P. D. (1983) Beginning and end of lunar lunar mare volcanism. *Nature, 302*, 233-236.

Spudis P. D. (1984) Lunar field geology and the lunar base (abstract). In *Papers Presented to the Symposium on Lunar Bases and Space Activities of the 21st Century*, p. 6. NASA/Johnson Space Center, Houston.

Taylor G. J. (1984) Moon rocks. *The Planetary Report, 4*, 4-6.

Taylor G. J. (1985a) Lunar origin meeting favors impact theory. *Geotimes, 30*, 16-17.

Taylor G. J. (1985b) Earth's Moon: Doorway to the solar system. In *The Planets* (Byron Preiss, ed.). Bantam, New York. In press.

Taylor S. R. (1982) *Planetary Science: A Lunar Perspective.* The Lunar and Planetary Institute, Houston. 481 pp.

Vaniman D. T., Heiken G., and Taylor G. J. (1985) A closer look at lunar volcanism from a base on the Moon. This volume.

Warren R. W. (1984) The magma ocean concept and lunar evolution. *Ann. Rev. Earth Planet. Sci., 13*, 201-240.

Wetherill G. W. (1985) Occurrence of giant impacts during the growth of the terrestrial planets. *Science, 228*, 877-879.

Wood J. A., Dickey J. S., Marvin U. B., and Powell B. N. (1970) Lunar anorthosites and a geophysical model of the Moon. *Proc. Apollo 11 Lunar Sci. Conf.*, 965-988.

GEOCHEMICAL AND PETROLOGICAL SAMPLING AND STUDIES AT THE FIRST MOON BASE

Larry A. Haskin, Randy L. Korotev, David J. Lindstrom, and Marilyn M. Lindstrom

Department of Earth and Planetary Sciences and McDonnell Center for the Space Sciences, Washington University, St. Louis, MO 63130

Strategic sampling appropriate to the first lunar base can advance a variety of first-order lunar geochemical and petrological problems. Field observation and collection of samples would be done on the lunar surface, but detailed analysis would be done mainly in terrestrial laboratories. Among the most important areas of investigation for which field observations can be made and samples can be collected at the initial base are regolith studies, studies of mare and highlands stratigraphy, and a search for rare materials such as mantle nodules. Since the range of exploration may be limited to a radius of about 20 km from the first lunar base, locating the base near a mare-highlands boundary would enable the greatest latitude in addressing these problems.

INTRODUCTION

The number of problems in lunar geoscience that can be studied from a base on the Moon in conjunction with laboratory studies on Earth is nearly overwhelming to contemplate (*e.g.*, G. J. Taylor, 1984). Such problems range across studies of the petrological products of the Moon's geochemical differentiation, of large-scale displacements of segments of the Moon's crust, of processes of local regolith development, and of ancient solar wind, just to name a few broad areas. Many studies are of the most detailed sort and might at first seem to pertain to second-order problems. In fact, the understanding of such apparently trivial phenomena as mixing and heterogeneity in local regolith can have profound effects on interpretation of sampled materials in terms of all the areas of study mentioned above. Without such information, interpretations of data on Apollo samples must remain more speculative than desired, an especially serious situation when the importance of the Moon as an analog for other cratered bodies in the solar system is considered. This paper discusses briefly some projects for lunar geochemistry and petrology that could be carried out at the first lunar base. Proper exploration of the Moon will require extensive mobility and capability to sample material at depth (*e.g.*, Cintala *et al.*, 1985) that are not considered in this relatively simple scenario. Rather, we assume a multi-purpose base with mainly constructional equipment, relatively short tours of duty (three to six months), and scientists whose tasks include general assistance in setting up and operating the base. The emphasis is on field observation, sample collection, initial characterization, and selection and preparation of samples for detailed study in laboratories on Earth.

It does not seem realistic for the first base on the Moon to have sophisticated instrumentation for mineralogic and chemical characterization of collected materials. The hand lens and binocular microscope should suffice for most purposes, plus perhaps the ability to make some thin sections and examine them under a petrographic microscope. A small X-ray fluorescence spectrometer of the type based on radioisotope beta-ray excitation might prove useful for preliminary characterization of rock powders (see also the discussion by Lugmair, 1984). It does not seem plausible to provide more sophisticated apparatus for detailed laboratory study at a Moon base until tours of duty are extended to periods of years. It is difficult to imagine the rich variety of techniques available in terrestrial laboratories becoming available on the Moon for a very long time. The astronaut-petrologists and –geochemists will be busy enough designing and carrying out strategies for mapping and sample collection that they need not be burdened with more sophisticated study *in situ*. However, the scientist-astronauts will probably be confined to habitable quarters during the two-week lunar nights and will have time to do careful preliminary characterization of sampled materials. In fact, the scientists will probably be confined to working quarters underground during most days as well as nights because of the relatively high radiation levels on the lunar surface. Thus, outdoor activities may be mainly teleoperated with infrequent excursions to the surface, carefully planned for efficient scientific yield (see also the discussion by Spudis, 1984). The observational power of human geologists is not likely to be matched soon by robotic substitutes, however, and direct observation will be essential for many studies.

Here, we assume a base with dirt-moving equipment (expected to be necessary for emplacement of habitats, preparation of landing pads, assistance in providing feedstock for processing of lunar soil, *etc.*), rover capability over at least a 20-km range from the base headquarters, and adequate habitat and working area to enable preliminary characterization of collected samples as indicated above. We also assume the continuous presence of one or more geoscientists dedicated to careful observation and sampling of the local environment, working in partnership with other geoscientists in terrestrial laboratories. The analog is the field study and sampling followed by laboratory study of collected studies here on Earth.

Below, we describe some projects important to lunar science that could be done under the conditions outlined above. Some projects are tailored to Apollo sites. These are not meant to indicate that a return to any of those sites is essential, but to show specifically how individual projects could be designed. All projects mentioned are offered as examples in a generic sense, with selection and design dependent on the particular site chosen for the base.

REGOLITH CHARACTERIZATION

Most or all of the materials sampled at a lunar base will be regolith materials. Only in mare terrain do we know for certain that we can find bedrock in place. Thus, to some extent, most sampling strategies will contribute to regolith characterization, whether that is the main purpose of the study or not. What we refer to as "regolith characterization" in this section consists of observations and collection of samples to characterize the shallow (1–2 m) regolith at the base site. It will be necessary to collect some of the samples for this characterization early, before activities associated with establishment of the base have disturbed the material.

The scientific problems to be studied are extensions of those approached in the shallow coring and trenching done at Apollo and Luna sites. What different igneous rock types contributed to the regolith? What fraction of the material is derived locally and what fraction is exotic? Can rays from distant craters be traced? What are the patterns of mixing and bedding? What marker horizons can be found and how far can they be traced? Can buried ancient soils be located, and what can we learn from them (McKay, 1984a)? Such questions are fundamental to understanding the Moon's surface history.

If mobility extends for a radius of 20 km from the base site, sampling strategy should be developed to understand the shallow regolith over the entire accessible area. This can be done mainly by sampling with drive tubes or coring devices similar to those used for Apollo missions, with collection of bulk samples from the upper few centimeters at each sampled location. Two or three regions, a few kilometers distant from each other, should be intensively sampled, with sufficient cores spaced 1–10 m apart to enable definition of small-scale structure and to determine dips of marker horizons over short distances. Cores at more widely spaced intervals can then be used to characterize the broader scale features of the regolith, showing changes in relative amounts of components over the entire area and enabling the tracing of crater ejecta. Such a study can provide directional indications for ejecta from distant craters; work of this type has led to tentative ages for the fresh craters Tycho (Arvidson *et al.*, 1976) and Copernicus (*e.g.*, Eberhardt *et al.*, 1973), based on assumed sampling of ray material. Particularly interesting would be the discovery of abundant ray material 65 m.y. old, the age of the terrestrial Cretaceous-Tertiary boundary (recently suggested to have resulted from impact by a large meteoroid), perhaps as one manifestation of periodic bombardment of the Earth-Moon system (Hörz, 1984, 1985; McKay, 1984b). By combining petrographic and geochemical information (identification of rock types, mineral chemistry, major- and trace-element analysis, isotopic studies, nuclear particle track measurements, *etc.*) for well defined regolith strata, we can learn much about local and more distant regions.

These core samples would be studied only in terrestrial laboratories. At least one hundred cores would be necessary for initial characterization of the local regolith. More rapid means of core dissection than those used on Apollo cores would be required. Only obvious horizon boundaries in opened cores would need to be noted and only pebbles removed selectively during dissection. Continuous-strip thin sections could be provided for petrographic analysis. However, since petrographic analysis of lunar fines is difficult, primary characterization could be done geochemically, using methods such as X-ray fluorescence and instrumental neutron activation to show compositional discontinuities and using ferromagnetic resonance to show changes in soil maturity, as means of identifying horizon boundaries (*e.g.*, Korotev *et al.*, 1984). Following initial characterization, materials could be selected for more specific studies.

Shallow trenches were dug at Apollo sites, but they yielded little stratigraphic information. With dirt-moving equipment, a much more interesting set of experiments can be done. Trenches meters deep can be dug and visual and photometric studies of texture and bedding made. Observed horizons can be sampled for comparison with the core samples. The cohesiveness of the soils should make it possible to produce relatively deep trenches for scientific study, 10–20 m, by stepping them in terraced intervals of 10–20 cm vertical dimension. Advantage should be taken continuously of excavations done for other purposes such as habitat construction, landing pad preparation, and acquisition of feedstock for materials processing.

CRUSTAL STRATIGRAPHY

Knowledge of crustal stratigraphy is crucial to understanding the fundamental geochemical and petrological nature and evolution of the Moon, once thought to be composed of average non–volatile solar system material that had not undergone chemical differentiation to form an internally zoned planet like Earth. The nature of the first lunar samples collected laid to rest any notions that the Moon was chemically undifferentiated. For a summary of the history of evolution of models for the development of the lunar crust, see the article by Walker (1983). Observation of the mineral plagioclase feldspar in materials from the lunar regolith led to the hypothesis that its outer regions were initially molten, forming a global magma ocean (e.g., Wood et al., 1970). On cooling, such a magma ocean would precipitate dense ferromagnesian minerals (olivine, pyroxene, ilmenite), which would sink, and less dense plagioclase, which would remain suspended. The plagioclase, it is speculated, would float to form a crust of anorthosite tens of kilometers thick. Anorthosite is a type of rock that consists of over 90% plagioclase feldspar. It is found as hand-specimen–sized samples among the materials collected from the Apollo 15 and 16 highlands and appears to be present as the central peaks in some lunar craters (C. Pieters, personal communication, 1984). However, it is not found among the large fragments of rocks collected from the Apollos 14 or 17 highlands, nor do most of the remotely sensed highlands surfaces have the composition of anorthosite. Different varieties of anorthosite are known, suggesting different origins (Hubbard et al., 1971; Dowty et al., 1974; Warren et al., 1983; Lindstrom et al., 1984).

To produce such a variety of rock types, crystallization of the hypothesized magma ocean would have to have been very complex. If it produced mainly anorthosites, how did the rest (e.g., Taylor, 1975) of the sampled and remotely sensed rock types form? A few may have been lavas extruded onto the surface, but most seem to be of intrusive origin (i.e., formed beneath the surface, as anorthosites themselves are formed on Earth). These intrusive rocks (troctolites, norites, dunites) contain ferromagnesian minerals. Plagioclase did not separate from the ferromagnesian minerals when the abundant, feldspar-bearing troctolites and norites formed. Since crystals of plagioclase have a density only slightly lower than that of the liquid suspending them, it might be expected that substantial quantities of the liquid would be entrapped among the suspended crystals as they floated to form the crust. This liquid would crystallize to produce ferromagnesian minerals along with the plagioclase. If so, then the lunar crust might have a composition characteristic of noritic anorthosite or anorthositic norite, more in line with the composition inferred from soil data and remote sensing data (e.g., Taylor, 1975; Korotev et al., 1980). Norites are fairly common among the least altered Apollo highland samples, but noritic anorthosite is not; most sizable rocks collected of that composition are breccias, mixtures of precursor igneous rocks. If the anorthosite crust is as thick as has been suggested (20–60 km), why are intrusive rocks with ferromagnesian minerals so common among the collected highlands samples? Was an original anorthosite crust invaded by later magmas that crystallized to produce the norites and troctolites in analogy with the intrusive complexes found on Earth? Was the original crust mainly noritic anorthosite with only isolated "rockbergs" of anorthosite? Mapping and sampling of stratigraphically located highland materials would enormously improve our knowledge.

Tilted Blocks

It is not known whether any coherent blocks of crust that preserve original stratigraphy survived the heavy cratering of the lunar highlands. However, impact cratering and mixing processes were insufficient to homogenize the lunar highlands, and fragments of igneous rock at least as large as hand specimen size survived, so there is hope. Some blocks of original lunar crust may have become tilted, exposing to view a vertical cross section of their original horizontal layering. Perhaps the most promising example currently known is Silver Spur, about 20 km from the Apollo 15 landing site. Based on astronaut observations and photography, Silver Spur appears to be a tilted, coarsely layered block that exposes some 800 m of originally buried layers (Swann *et al.*, 1972). The layering could be igneous, as desired for the study outlined here. It could also be a series of overlapping ejecta blankets, still of interest but for regolith studies. It does not appear to be an artifact of lighting conditions such that the layering is merely apparent, as observed in other regions of the Apennine front. Assuming that the layering is real, one argument in favor of its being igneous layering is that the breccias collected at the nearby Apennine front are relatively simple, consisting of one type of impact melt rock plus fragments of coarse-grained igneous rocks (Ryder, 1976; Ryder and Bower, 1977). This allows hope that impacts in this particular region did not destroy entirely the original igneous materials. To map and sample the steep face of Silver Spur would require at least two traverses down the face (assuming that the layering proves to be real) with local removal of obscuring regolith and sampling of the rocks in each stratum. Efficient transportation would be needed up the shallow slope of the top stratum to the summit, presumably by a rover. Then an astronaut in some sort of winch-driven sling would have to be lowered safely down the face, sweeping, mapping, and sampling as he went. An automated hammering or coring device might be needed for efficient sample collection. The project would require at least several days and perhaps weeks of sampling effort, depending on the coherence of the rocks and the depth of obscuring regolith. The rocks would undergo preliminary examination at the lunar base, then selected portions would be shipped to Earth for further characterization and study.

Craters

Impacts on the lunar surface excavate craters whose walls can reveal the strata into which the craters were punched. Whether that stratigraphy pertains mainly to layering of highlands or mare igneous rocks or mainly to sheets of ejecta (including frozen melt pools) depends on the size of the crater and the material penetrated. The internal stratigraphy is of interest in either case and, for craters under a few kilometers in diameter, could be mapped and sampled with the aide of teleoperated dirt-moving equipment. Although there are craters at all sites, including the Apollo 15 site from which most of the examples in this discussion are drawn, to emphasize that these proposed studies are generic, we discuss as an example North Ray Crater at the Apollo 16 site. That crater is regarded as having penetrated into the highlands Descartes formation. Astronauts Duke and Young (Muehlberger, 1972) collected some 30 samples of light and dark breccias from its southern rim and described and photographed its northern wall. Initially, Ulrich (1973) attempted to match the breccias to light and dark materials seen in the crater wall and in the obscuring talus. He suggested that the crater penetrated through a blanket of Orientale ejecta into a blanket of Imbrium ejecta. Geochemical and petrographic studies (Lindstrom and Salpas, 1981, 1983), however, found many rock types among the breccias, so that the characterization as dark or light is misleadingly simple. Dark materials include three to five types of impact melt rocks; light materials include anorthosite, three types of granulite, and very KREEP-rich and KREEP-poor materials, indicating complex layering of the strata excavated. The Apollo 16 site was the site most nearly typical of lunar highlands; as determined from remote sensing data the significance of these rock types is Moon-wide. Other geological models for the Apollo 16 landing site have been developed (Spudis, 1984), but better knowledge of rock types and interrelationships is essential to understanding the nature and history of the region. The radius of North Ray Crater is about 1-km, and its depth is 230 m (Ulrich, 1981). The slope of its walls is only about 30°, shallow enough to be entered by dirt-moving equipment. Such a device of even modest capacity (10 m^3/h) could cut a trench 10 m wide and 2 m deep from the rim to the center of the floor in a lunar day (about 328 hours). Of course, just how much dirt would have to be removed is speculative, but this estimate demonstrates the possibility for determining the stratigraphy of at least the upper two thirds or so of the rim. Most of the excavation could be done by remote control, with an astronaut visiting the site only periodically to map trench walls and collect samples. The information gained, in conjunction with that from samples taken from the rim and radially outward, would yield first-order insight into the petrologic character of the materials of the Descartes formation, into the mechanics of impact cratering, and into regolith mixing.

Basalts

Our understanding of lunar volcanism is far from adequate to answer key questions about the Moon's interior composition and melting processes (*e.g.*, Vaniman *et al.*, 1984, 1985). Knowledge of stratigraphic relationships is vital to proper sampling of mare basalts as well as of materials from the lunar highlands. Several types of basalts and several different flows seem to have been sampled by the Apollo 11, 12, 15, and 17 missions (*e.g.*, Taylor, 1975), and further types were collected during the Apollo 14 and Luna 16 and 24 missions. Moreover, each site yielded a perplexing variety of basalt compositions. Spectral data from ground-based remote sensing indicate the presence of yet other types, some of which appear to be more prevalent than those already sampled (*e.g.*, Pieters, 1978).

Without geological control over sampling, spatial relationships among the types at a given site and compositional variability within single flows cannot be established. Complete characterization of a mare basin may require traverses across the basin (Spudis *et al.*, 1984; Cintala *et al.*, 1985) to collect samples from all types of basalts that cover the surface and from craters that have penetrated beneath the uppermost flows to provide samples of earlier basalt types. All of the basalt samples collected by the Apollo and Luna missions, except perhaps one from the Apollo 15 mission, were collected from the regolith, and it is unknown from what flows or exact locations they derived. A lunar base near a rille offers the opportunity to sample at least several different flows along the rille wall.

Hadley Rille, visited by the Apollo 15 mission, is some 400 m deep and 1.5 km wide. Its original walls and floor are largely obscured by talus and debris from nearby impacts, but *in situ* bedrock was observed and photographed on the wall opposite the landing site. Layered strata were observed to outcrop discontinuously in the upper 60 m of the opposite rim. The more massive layers are only 1–3 m thick (Swann *et al.*, 1972); some layers are observed to "weather out" differently from others, suggesting that the layers represent different flows or, in some cases, bands of intervening regolith. How many different flows might be represented is unclear, however, as several layers could derive from a single flow. In addition to the flows possibly represented in the vertical face at one location, portions of the rille farther "downstream" could expose a different set of flows. Even the intervening regolith, if it exists, can provide information about the eruption sequence of different basalt types if there was substantial change in the types of material on the impacted surfaces from which the ejecta were produced. Sampling and more detailed study of the rille strata are required before we can know how many flows are represented and what the characteristics of flow tops and bottoms are.

Further study of mare basalts *in situ* is clearly important to understanding lunar volcanism. Detailed study of rille strata can provide observational information on the characteristics of flows, which presumably resemble the least viscous terrestrial pahoehoe flows of Earth's rift zones and oceanic island volcanoes such as those of Iceland and Hawaii. What is the typical thickness of lunar basalt flows? What are the characteristics of flow tops and flow bottoms; are they vesicular as on Earth? If so, what are the volatile components responsible for the vesicularity (a problem not yet solved from study of the samples of vesicular basalt acquired by the Apollo missions)? Are successive flows similar in character or do the different lava types known alternate in the eruption sequence? These and other questions can be answered by studying the walls of Hadley Rille and collecting samples. This requires equipment such as that discussed above for sampling the face of a raised, tilted fault block because sides of rilles are too steep, at least where seen by the Apollo 15 astronauts, to traverse on foot or by rover. The opposite wall of a rille, while an important source of supplementary information about the structure and stratigraphy of the rille, is too far away at 1.5 km for detailed observation. Thus, astronauts who study the stratigraphy will have to do so mainly on the face they cannot see from the side they are on prior to descent. A power coring device or hammer is likely to be essential for proper sampling because lunar basalts, like their terrestrial counterparts, are coherent and tough. Several traverses down the face of the rille, each requiring one to a few days, would be needed for this study. It may be difficult to determine from field evidence and microscopic examination how many different varieties of basalt have been sampled, so for this study an X-ray fluorescence unit at the Moon base might be of particular value.

RARE MATERIALS

The presence of certain materials can be inferred from the nature of the known lunar rock types. These include possible lava lakes and mantle xenoliths in mare regions and concentrations of relatively volatile elements (*e.g.*, sulfur, chlorine, arsenic, zinc) in mare pyroclastic deposits and in products of metamorphism in the highlands. It is to be expected that other, unanticipated rare materials will be discovered in the vicinity of the lunar base. These unanticipated materials will deserve sampling and study to understand their origins and natures, and they may also serve as horizon markers or as tracers of specific volcanic or regolith-modifying events. Lava lakes are important in demonstrating the nature of magmatic crystallization and as a potential source of ores. Their analogs on Earth have been studied, for example, on the Hawaiian volcano Kilauea. Magma becomes trapped in a pit crater or other depression, then forms an insulating crust by cooling against its container and against the convecting atmosphere (Wright *et al.*, 1976). (On the Moon, radiation of heat into space would have to produce the upper crust.) Such events account for only a small percent of the exposed lavas on Kilauea and depend on the availability of depressions for their formation. They can thus be expected to be rare on the Moon as well, although flooded impact craters may offer added sites of a type not found on Hawaii.

Some varieties of lunar basalt are rich in ilmenite, a titanium mineral of possible interest for the manufacture of oxygen and of iron and titanium metals. Conceivably, that mineral or a chromium ore such as chromite (Taylor, 1984) might crystallize selectively in a lava lake to produce a useful as well as scientifically interesting ore. A lava lake might appear, dissected, on the wall of a crater or a rille. A lava lake might have been included in the target of an impact event so that coarse-grained products of mare volcanism might be found as rock chunks in the regolith. Such chunks would be relatively rare and might be discovered only by observation by an alert astronaut searching basaltic rubble.

The counterpart of the mare lava lake should also have occurred in the highlands. In addition to lava lakes born from magmas seeping from the Moon's deep interior, magma lakes may have formed in the lunar highlands within melt sheets from impact craters; some might have undergone analogous chemical differentiation by crystal settling.

One of the main sources of information about materials of Earth's mantle is mantle xenoliths, fragments of material broken from deep rocks and transported to the surface during volcanism. These are rare and tend to accompany more explosive types of volcanism, perhaps because they have high enough velocities of flow to suspend them against the force of gravity. On the Moon the mare lavas have low viscosity, which works against the raising of mantle fragments, but there is also lower gravity, which makes it easier than on Earth. Xenoliths have not been observed among the lunar basalts collected so far, but, if they are as rare as on Earth, none would be expected in such a small sampling. On the lunar surface, where thousands of fragments of lunar basalt can be examined, there is a much better chance of finding them. If found, they would provide direct evidence of the nature of the Moon's interior and would be invaluable in constraining ideas about its mineralogy and composition. Perhaps mantle nodules are best sought in regions where the most explosive form of volcanism is believed to have occurred on the Moon. These may be near the so-called dark halo craters, from which lava fountains have been inferred to come and of which the orange glass found at the Apollo 17 site may be a product.

Most highlands rocks at the lunar surface are breccias, fragments of earlier rocks broken up and consolidated by impact processes. Within many of the breccias, however, are recognizable fragments of highlands igneous rocks. In many cases, these fragments show alteration from the impact and brecciation processes that made them. A few, however, show features characteristic of types of metamorphic processes found on Earth. For example, Apollo 16 breccia 67016 (Norman, 1981) contains fragments of igneous anorthositic norite, most of which are sulfide free but some of which are metamorphosed with veins of iron sulfide (troilite) and regions in which a ferromagnesian silicate mineral (pyroxene) has been partially altered to iron sulfide. It appears that a sulfur-bearing fluid had invaded that rock and begun to metamorphose it. Such a fluid could also extract rare elements from large quantities of rock, in analogy to the formation of hydrothermal ores on Earth. It is not evident whether the metamorphism accompanied or preceded the impact event that produced the breccia as sampled. Impact energy may have been the ultimate driving force for the metamorphism, but the style seems to be that of a percolating fluid.

Sulfur is also enriched in lunar pyroclastic glasses (e.g., Butler and Meyer, 1976); this sulfur is presumably of igneous origin. Sulfur is not the only apparent agent of metamorphic alteration. Some metamorphosed intrusive rocks, including anorthosites, are rich in phosphate and numerous trace elements. The origin of this material is unknown, but igneous processes as currently understood do not seem capable of producing it. The observed enrichment in trace elements is reminiscent of that found in the enigmatic lunar material called KREEP, but the concentrations of the major constituents are quite different, and the concentrations of the trace elements in the phosphate-rich regions are higher than those found in KREEP (e.g., Warren and Wasson, 1979). Additional evidence that the apparent metamorphism may not have been caused by the immediate impacts producing the breccia samples but may be a more regional event is the nature of the highlands breccias sampled at Apollo 15. Those breccias contain clasts of igneous rocks, some of which show signs of metamorphism, but most of which retain their igneous textures and do not appear to have undergone numerous impacts as the breccias from Apollos 14 and 16 seem to have done. A search for rocks with sulfide mineralization could be carried out in conjunction with other observing and sampling activities. The pattern and frequency of occurrence of such rocks could indicate whether metamorphic processes are widespread, with the possibility of locating specific regions affected. A promising site for such observations would be the Silver Spur project outlined above.

CONCLUSIONS

The projects outlined above are but a few of many that would constitute important scientific research at a lunar base. They address restricted aspects of geochemistry and petrology, broad areas that are themselves only a subset of the geological sciences. They are designed to optimize the opportunities in the immediate vicinity of the base and to take advantage of the equipment that will be needed for construction and materials processing as well as for scientific study. Thus, they are suited to an initial, multipurpose base. They are also designed to take advantage of a very substantial terrestrial laboratory effort, with the astronaut–geologist or –geochemist mainly responsible for observation, details of experiment design, and sample collection during the tour of duty on the Moon. On returning from the Moon, the astronaut would then be an invaluable colleague to the numerous laboratories investigating a broad variety of lunar problems. The types of science outlined above are important for first-order understanding of the nature of the Moon and for constraining our ideas about its history.

Acknowledgments. This work was supported in part by the National Aeronautics and Space Administration under grant NAG-956.

REFERENCES

Arvidson R., Drozd R., Guiness E., Hohenberg C., and Morgan C. (1976) Cosmic ray exposure ages of Apollo 17 samples and the age of Tycho. *Proc. Lunar Sci. Conf. 7th*, pp. 2817–2832.

Butler P., Jr., and Meyer C., Jr. (1976) Sulfur prevails in coatings on glass droplets: Apollo 15 green and brown glasses and Apollo 17 orange and black (devitrified) glasses. *Proc. Lunar Sci. Conf. 7th* , pp. 1561–1581.

Cintala M. T., Spudis P. D., and Hawke B. R. (1985) Advanced geologic exploration supported by a lunar base: A traverse across the Imbrium-Procellarum region of the Moon. This volume.

Dowty E., Prinz M., and Keil K. (1974) Ferroan anorthosite: a widespread and distinctive lunar rock type. *Earth Planet. Sci. Lett., 24*, 15–25.

Eberhardt P., Geiss J., Grögler N., and Stettler A. (1973) How old is the crater Copernicus? *Moon, 8*, 104–114.

Hörz F. (1984) Mass extinctions and impact cratering (abstract). In *Papers Presented to the Symposium on Lunar Bases and Space Activities of the 21st Century*, p. 16. NASA/Johnson Space Center, Houston.

Hubbard N. J., Gast P. W., Meyer C., Nyquist L. E., and Shih C. (1971) Chemical composition of lunar anorthosites and their parent liquids. *Earth Planet. Sci. Lett., 13*, 71–75.

Korotev R. L., Haskin L. A., and Lindstrom M. M. (1980) A synthesis of lunar highlands chemical data. *Proc. Lunar Planet. Sci. Conf. 11th*, pp. 395–429.

Korotev R. L., Morris R. V., and Lauer H. V. (1984) Stratigraphy and geochemistry of the Stone Mountain core (64001/2). *Proc. Lunar Planet. Sci. Conf. 15th*, in *J. Geophys. Res., 89*, C143–C160.

Lindstrom M. M., Knapp S. A., Shervais J. W., and Taylor L. A. (1984) Magnesian anorthosites and associated troctolites and dunite in Apollo 14 breccias. *Proc. Lunar Planet. Sci. Conf. 15th*, in *J. Geophys. Res., 89*, C41–C49.

Lindstrom M. M. and Salpas P. A. (1981) Geochemical studies of rocks from North Ray Crater, Apollo 16. *Proc. Lunar Planet. Sci. 12B*, pp. 305–322.

Lindstrom M. M. and Salpas P. A. (1983) Geochemical studies of feldspathic fragmental breccias and the nature of North Ray Crater ejecta. *Proc. Lunar Planet. Sci. Conf. 13th*, in *J. Geophys. Res., 88*, A671–A683.

Lugmair G. W. (1984) The early geochemical evolution of planetary crusts: A detailed study of the crustal evolution of the Moon from a lunar base (abstract). In *Papers Presented to the Symposium on Lunar Bases and Space Activities of the 21st Century*, p. 10. NASA/Johnson Space Center, Houston.

McKay D. S. (1984a) A search for ancient buried lunar soils at a lunar base (abstract). In *Papers Presented to the Symposium on Lunar Bases and Space Activities of the 21st Century*, p. 8. NASA/Johnson Space Center, Houston.

McKay D. S. (1984b) Can a record of the 65 million year terrestrial mass extinction event be found on the Moon? (abstract). In *Papers Presented to the Symposium on Lunar Bases and Space Activities of the 21st Century*, p. 17. NASA/Johnson Space Center, Houston.

Muehlberger W. R. *et al.* (1972) Preliminary geologic investigation of the Apollo 16 landing site. *Apollo 16 Preliminary Science Report*, NASA SP-315, pp. 6-1 to 6-81. NASA, Washington, D.C.

Norman M. D. (1981) Petrology of suevite lunar breccia 67016. *Proc. Lunar Planet. Sci. 12B*, pp. 235–252.

Pieters C. M. (1978) Mare basalt types on the front side of the Moon. *Proc. Lunar Planet. Sci. Conf. 9th*, pp. 2825–2849.

Ryder G. (1976) Lunar sample 15405: Remnant of a KREEP basalt-granite differentiated pluton. *Earth Planet. Sci. Lett., 29*, 255–268.

Ryder G. and Bower J. F. (1977) Petrology of Apollo 15 black-and-white rocks 15445 and 15455—fragments of the Imbrium impact melt sheet? *Proc. Lunar Sci. Conf. 8th*, pp. 1895–1923.

Spudis P. D. (1984) Lunar field geology and the lunar base (abstract). In *Papers Presented to the Symposium on Lunar Bases and Space Activities of the 21st Century*, p. 6. NASA/Johnson Space Center, Houston.

Spudis P. D., Cintala M. J., and Hawke B. R. (1984) A geological traverse across the Imbrium Basin region (abstract). In *Papers Presented to the Symposium on Lunar Bases and Space Activities of the 21st Century*, p. 7. NASA/Johnson Space Center, Houston.

Swann G. A. *et al.* (1972) Preliminary geologic investigation of the Apollo 15 landing site. *Apollo 15 Preliminary Science Report*, pp. 5-1 to 5-112. NASA SP-289, NASA, Washington, D.C.

Taylor G. J. (1984) Lunar science from a Moon base: Answering basic questions about planetary science (abstract). In *Papers Presented to the Symposium on Lunar Bases and Space Activities of the 21st Century*, p. 71. NASA/Johnson Space Center, Houston.

Taylor L. A. (1984) Layered intrusives on the Moon: A source of chromium deposits? (abstract). In *Papers Presented to the Symposium on Lunar Bases and Space Activities of the 21st Century*, p. 15. NASA/Johnson Space Center, Houston.

Taylor S. R. (1975) *Lunar Science: A Post-Apollo View.* Pergamon, New York. 372 pp.

Ulrich G. E. (1973) A geologic model for North Ray Crater and stratigraphic implications for the Descartes region. *Proc. Lunar Sci. Conf. 4th*, pp. 27–39.

Ulrich G. E. (1981) Geology of North Ray Crater. In *Geology of the Apollo 16 Area, Central Lunar Highlands* (G. E. Ulrich, C. A. Hodges, and W. R. Muehlberger, eds.), pp. 45–81. U.S. Geol. Surv. Prof. Paper 1048.

Vaniman D., Heiken G., and Taylor G. J. (1984) A closer look at lunar volcanism (abstract). In *Papers Presented to the Symposium on Lunar Bases and Space Activities of the 21st Century*, p. 13. NASA/Johnson Space Center, Houston.

Vaniman D., Heiken G., and Taylor G. J. (1985) A closer look at lunar volcanism from a base on the Moon. This volume.

Walker D. (1983) Lunar and terrestrial crust formation. *Proc. Lunar Planet. Sci. Conf. 14th*, in *J. Geophys. Res., 88*, B17–B25.

Warren P. H., Taylor G. J., Keil K., Kallemeyn G. W., Rosener P. S., and Wasson J. T. (1983) Sixth foray for pristine nonmare rocks and an assessment of the diversity of lunar anorthosites. *Proc. Lunar Planet. Sci. Conf. 14th*, in *J. Geophys. Res., 88*, B151–B164.

Warren P. H. and Wasson J. T. (1979) The origin of KREEP. *Rev. Geophys. Space Phys., 17*, 73–88.

Wood J. A., Dickey J. S., Jr., Marvin U. B., and Powell B. N. (1970) Lunar anorthosites and a geophysical model of the Moon. In *Proc. Apollo 11 Lunar Sci. Conf.*, pp. 965–988.

Wright T. L., Peck D. L., and Shaw H. R. (1976) Kilauea lava lakes, natural laboratories for study of cooling, crystallization, and differentiation of basaltic magma. In *The Geophysics of the Pacific Ocean Basin and Its Margin, AGU Monograph 19*, American Geophysical Union, Washington D.C.

A CLOSER LOOK AT LUNAR VOLCANISM FROM A BASE ON THE MOON

D. T. Vaniman and G. Heiken

Earth and Space Sciences Division, MS D462, Group ESS-1, Los Alamos National Laboratory, Los Alamos, NM 87545

G. J. Taylor

Institute of Meteoritics, University of New Mexico, Albuquerque, NM 87131

The American Apollo and Soviet Luna missions provided much information on lunar volcanism, but major questions remain. Field work associated with a manned lunar base will be indispensable for resolving the scientific questions about ages, compositions, and eruption processes of lunar volcanism. From a utilitarian standpoint, a better knowledge of lunar volcanism early in the lunar base program will yield profitable returns in improved lunar construction methods (*e.g.*, exploitation of rille or lava-tube structures) and in access to materials such as volatile elements, pure glass, or ilmenite for lunar industry.

INTRODUCTION

Volcanism is a fundamental planetary process. Lavas probably flowed across Mercury's lunar-like surface, erupted to build huge shield volcanoes on Mars, and even oozed across the surfaces of some asteroids. Glowing lavas still make their way to the surfaces of Earth, Jupiter's satellite Io, and probably Venus, too. A major part of the Moon's landscape was formed by basaltic lava flows, which now constitute the dark areas called maria.

Volcanism is much more than a surface process, however. Lavas originate inside a planet when rocks undergo partial melting. Consequently, lavas serve as probes of the mineralogy and chemical composition of otherwise inaccessible planetary interiors. Much of the research done on samples of lunar volcanic rocks has been geared to understanding the varieties and origins of their source regions in the Moon's interior. Volcanic rocks also contain information about a planet's thermal history, which is intimately related to its origin and evolution. Did it form cold and heat up? Did it form hot and then cool? Did it heat up, cool, and then heat up again? Thus, although covering only 17% of the Moon's surface and constituting a tiny 1% of its crustal volume (Head, 1976), lunar volcanic rocks give us a broad picture of its geochemical processes.

Although a broad range of volcanic rock types probably occurs on the Moon (Head, 1976) the dark, low-silica rocks known as basalts constitute the only volcanic samples returned from lunar missions. Basalts of the lunar maria contrast with the light-colored highlands areas that are so strikingly visible when one looks at the Moon. Basalts of the lunar highlands do occur, but only a few small samples have been found within impact-generated breccias. However, the concept of "highland volcanism" may stretch the limits of our previous conceptions of what lunar volcanism is.

The term "volcanism" implies eruption of a melt derived from solid material at depth. It has been suggested that some very old volcanic rocks on the Moon could have crystallized from liquid remnants of a very large magma system ("magma ocean") that enveloped the Moon at some time prior to 4.4 b.y. ago (Wood, 1975). This is one explanation (Warren and Wasson, 1979) for some lunar magmatic compositions that are rich in the otherwise rare elements potassium, rare earths, and phosphorus (immortalized in the acronym "KREEP"). Such an origin as a result of crustal cooling is conceptually very different from volcanism caused by deep internal heating, though the actual extrusion may be very similar. There is also a possibility that some lavas could have been brought to the surface by impact excavation of crustal magma chambers (Schultz and Spudis, 1979). Although there is a slight semantic problem in calling such lavas "volcanic," it is intriguing to consider the chances of finding these rocks on the Moon, since such lavas have not been preserved on Earth. Because of our very limited knowledge of these other magma types, this paper emphasizes speculations drawn from the larger data base on mare basalts.

Lunar mare basalts differ markedly from terrestrial basalts. Among major chemical constituents, titanium content is, in general, considerably higher, whereas aluminum, potassium, and sodium contents and magnesium-to-iron ratios are much lower. Potassium contents are in fact so low in mare basalts that this element is often listed as a trace constituent (less than one tenth of one percent). Conversely, chromium is often listed among the major constituents in mare basalts (about 0.5% Cr_2O_3), whereas it is strictly a trace element in terrestrial basalts. The higher chromium content of lunar mare basalts, along with their high to very high titanium content, suggests that useful concentrations of chromite and ilmenite may occur on the Moon. Metal oxides may well be concentrated by separation of heavy crystals from mare basalt magmas (Taylor, 1984). But the definition of an ore (*i.e.*, a profitably mined geologic commodity) may lead to a classification as "ore" for some deposits that would be mundane on Earth. A key feature of the Moon is its depletion in volatile elements and compounds; in particular, mare basalts are depleted in water and carbon-oxygen compounds in comparison with terrestrial basalts. Sulfur, however, is as abundant or more abundant in mare basalts than it is in terrestrial basalts. Could accessible sulfur concentrations become an ore for lunar base exploitation? Mare basalt resources may also encompass the forms resulting from volcanic processes, such as lava tubes (Hörz, 1985; Khalili, 1984). Operations from a lunar base must be as aware of the resource potential of lunar volcanic rocks as of the outstanding scientific questions to be answered from studies of lunar volcanism.

VOLCANIC ORIGINS, AGES, AND COMPOSITIONS

Mare-flooding basalts that have been dated range from 3.8 to 3.1 b.y. old. Clasts of mare basalts within breccias extend the time of such volcanism back to 4.2 b.y. (Taylor *et al.*, 1983). Volcanic rocks younger than 3.1 b.y. have not been found among the lunar samples but photogeologic and remote sensing studies suggest that there are high-titanium basaltic lavas as young as 2.5 b.y. (Head, 1976) or younger (Boyce *et al.*, 1974; Schultz and Spudis, 1983). The answers to several important questions of lunar history are tied to volcanic processes: when they began, when they ceased, and how much activity occurred.

The oldest lunar volcanism may be difficult to decipher because of the widespread impact disruption of samples older than 3.9 b.y. Model isotopic ages derived from lunar basalts indicate that the source regions for magma production were established at 4.4 b.y.; the Moon may eventually provide volcanic rocks as old as this, or we may even find "pseudovolcanic" residual liquids that predate these source regions. It is encouraging that, even with our limited sampling of the Moon, we have obtained volcanic rocks that are almost as old as the Moon's initial differentiation. More detailed sampling might provide a map of chemical compositions and times of eruption that will record the ancient separation of lunar crust and mantle. From the experience we have had so far, it appears that the best place to look is within soils and breccias where small volcanic clasts occur. The oldest breccias exposed deep within the walls of large basins accessible from a lunar base should provide useful samples of these rocks.

The youngest lunar volcanism should be much easier to find, for these rocks will be the least impact-disrupted of lunar surface materials. Photogeologic and remote sensing studies have already pointed to volcanic flows in Mare Imbrium that may be as young as 2.5 b.y. (Schaber, 1973) or younger (Boyce *et al.*, 1974; Schultz and Spudis, 1983). Spectral data indicate that these flows consist of high-titanium basalt (Pieters *et al.*, 1974; Basaltic Volcanism Study Project, 1981), an interesting observation when it is considered that high-titanium basalts are also among the oldest (3.6–3.8 b.y.) radiometrically dated samples from lunar mare. The age-composition progression among mare basalt samples has already revealed a complex lunar mantle history that goes far beyond single-event melting of differing mantle source regions; closer study is required to evaluate what really went on. The history of lunar interior melting will also be an important parameter for determining lunar cooling history and the inventory of heat-producing radioactive elements in the lunar interior.

The volume of lunar volcanic rocks and variation of eruption rates through time are important parameters for determining a coupled history of lunar heat transfer and mantle evolution. A speculative summary of lunar eruption rates over time (Basaltic Volcanism Study Project, 1981) proposes two extreme scenarios: (1) very little volcanism prior to 3.9 b.y., or (2) abundant volcanism prior to 3.9 b.y. The total volumes of volcanic rocks estimated for these two scenarios differ by a factor of about two: around 10^7 km^3 with minimal early volcanism and about 2×10^7 km^3 with abundant early volcanism. The difference in implications for lunar evolution, however, is greater. If there are large volumes of older basalt hidden beneath the lunar highlands, then large-scale melting of the lunar mantle may have followed fast on the heels of the "magma ocean" crystallization. The resolution of this question lies in the discovery of how much old volcanic material underlies the highlands. Dark-haloed craters in the highlands have been pointed out as promising sites for such an investigation because they expose low-albedo basaltic materials that apparently underlie the high-albedo (non-basaltic) highlands debris (Schultz and Spudis, 1979). The dark-haloed craters are excellent examples of a type of lunar deposit that has not yet been sampled, where a closer look is vital to our understanding of the Moon.

Mare Basalt Compositions

The presently available samples of lunar basalts have been studied in great detail, though not yet exhaustively. Data from these samples show that mare basalts fall into three main categories, based on their content of TiO_2: high-Ti (8.5-13.0 wt % TiO_2), low-Ti (2-5 wt %), and very low-Ti (<1 wt %). In addition, some low-Ti basalts are relatively enriched in Al_2O_3 and are called aluminous mare basalts. However, remote sensing data indicate that there are at least twenty different types of mare basalts (Basaltic Volcanism Study Project, 1981). Moreover, as discussed below, our present collection of lunar basalts may not represent the full variety of lava types present at a given site. For review of mare basalt chemical and mineralogical compositions, see Papike et al. (1976), chapter 1.2.9 in the Basaltic Volcanism Study Project (1981), and chapter 6 in S. R. Taylor (1982).

Sample analyses, geochemical calculations, and experiments at high temperatures and pressures suggest that most mare basalts formed by remelting of deep rocks (at least 100 km) that had originally formed from the "magma ocean." We do not know, however, how much assimilation of surrounding rock and how much fractional crystallization took place as the basaltic magmas migrated to the lunar surface and then flowed across it. Both these processes affect conclusions about the nature and origin of the mantle source rocks. Detailed sampling of mare lava flows accessible from a lunar base will help to test the effects of assimilation and fractional crystallization.

Sampling a Larger Range of Basalt Types

Based on analyses of lunar samples, mare basalts seem to form distinct groups with respect to a number of chemical discriminants. Extremes of TiO_2 content have been recognized, but remote sensing data obtained by telescopic observations of the Moon and by gamma ray and x-ray fluorescence experiments flown on the Apollo 15 and 16 command modules indicate that a full range of intermediate TiO_2 contents might be present among mare basalts. Moreover, a detailed analysis of spectral properties, summarized in chapter 4 of the Basaltic Volcanism Study Project (1981), suggest that the Apollo and Luna missions have sampled at most only one third of the basalt types exposed on the Earth-facing side of the Moon. A map of known, similar-to-known, and unknown basalt types based on this study is shown in Fig. 1. Figure 1 also summarizes some of the data from remote sensing using three of the critical signals used for this map: TiO_2 content, aluminum-to-silicon ratios, and K_2O content. An important feature is that the remote sensing data actually map the *soils* on top of the mare, and when we compare the soils with actual underlying basalt samples, the match is not good because of the mixing that goes along with lunar soil formation. Clearly we have much to learn about lunar basalts. Consequently, expeditions from a lunar base must sample all types of mare basalts that remote sensing data suggest are present beneath the veil of soil. A major prerequisite for doing this sampling intelligently is a more thorough photographic and spectral coverage of the Moon. This coverage can be obtained from unmanned polar-

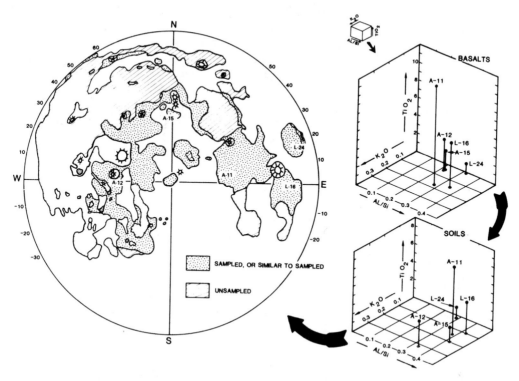

Figure 1. *Nearside map of the Moon based on remote sensing studies (Basaltic Volcanism Study Project, Chapter 4, 1981) showing areas of mare basalt that have been sampled or appear similar to samples in hand, contrasted with those areas that appear to be distinctive and unknown. The two perspective plots compare three compositional parameters measured by remote sensing (K_2O, TiO_2, and Al/Si compositions) used in generating the map. The perspective plots show that the map is based on the spectral reflectance properties of surface soils that mask and distort the true underlying basalt compositions, particularly in their K_2O and Al/Si compositions.*

orbiting lunar missions such as those tentatively planned by the United States, Japan, and the Soviet Union.

Sampling Problems at a Given Site

When working on Earth, geologists take great pains to collect samples from discernible rock units (*e.g.*, from a single lava flow). No lunar basalt was collected in this manner. All were pieces of rock chipped loose by impacts and strewn about on the surface. As a result, we do not know how many individual lava flows were sampled at each landing site. We can make intelligent guesses from chemical and mineralogical data, but we have no definitive field data relating one basalt sample to another. Considering also that the chemical variability in mare basalt flows may be high (Haskin *et al.*, 1977), we may have scarcely sampled the lunar maria at all.

Good sampling requires thorough field work, in which samples are taken from identifiable rock units such as those exposed in the walls of rilles and craters. At this time, we have visited only two sites where the volcanic deposits were relatively undisturbed: (1) the edge of Hadley Rille where a basalt flow was exposed (Apollo 15 site, Fig. 1) and (2) an overturned section of pyroclastic rocks within a crater rim in the Valley of Taurus Littrow (Apollo 17). Where such features are not exposed, samples can be obtained by drilling or by digging deep, wide trenches in the lunar surface (Korotev, 1984). Construction operations at a lunar base could also be integrated into the sampling of volcanic units, particularly where drilling or digging is involved.

Lateral Variations in Single Flows

Mare basalt lavas were much more fluid than terrestrial basalts (Murase and McBirney, 1970) and flowed easily across the Moon's surface. As they flowed, it is quite likely that they crystallized with heavier crystals concentrating near the flow bottoms. How much crystallization took place before a given flow reached an area sampled by an Apollo or Luna mission is anyone's guess. It is also possible that as lavas flowed across the Moon they reacted with the underlying regolith, assimilating a variety of chemical components. As a result, few if any of the basalts returned to us are likely to be "primary magmas" that maintained chemical integrity from their origins inside the Moon until sampled billions of years later. This is especially true if crystallization and assimilation took place inside the Moon before the lavas erupted.

It is difficult to obtain field data about processes operating inside a planet, but we can test the extent to which lavas crystallized or reacted as they flowed on the surface. Mapped flows could be sampled where they emerged (vents) and at intervals to their distal ends. It is possible that deposits of heavy minerals, including potentially valuable ones such as ilmenite ($FeTiO_3$), could be found by this type of exploration. Exploration on this scale will only be feasible where prolonged operations are supported from a lunar base.

VOLCANIC PROCESSES

Carefully documented stratigraphic studies of igneous rocks will provide information not only on the Moon's thermal history, but also on the processes involved in the filling of the mare basins. To understand lunar volcanic processes, it is important to find certain features of the mare basins and study the processes that formed them:

(1) Dikes: Determine dike locations, orientations, and ages to study the migration of dike systems as the basins filled.

(2) Products of explosive volcanism: Determine the extent and means of deposition of pyroclastic deposits; find source vents for these deposits and determine their shape, location, number, and size; find the lava flows that erupted contemporaneously with explosive activity.

(3) Lunar lava flows: By mapping and sampling determine the eruption rates, sources, processes of crystal concentration and volatile loss during flow over

long distance, and mode of transport.

(4) Physiographic features interpreted as volcanoes: Determine if they *are* volcanoes; determine whether the variety of landforms, such as domes, shields (Fig. 2), cones, and fissure mounds imply a variety of magma types and eruption processes that we have not sampled. Studies of vent areas will test ideas concerning cold traps that may have concentrated volcanic volatile phases as sublimates in and near the fissures. Subsidence of ponded lavas may also provide information on degassing of these ponds or on backflow of lavas into vents during waning stages of the eruptions.

(5) Products of eruption processes that we have not studied on Earth: (a) rille formation is very important on the Moon (Fig. 3) but is still one of the great mysteries that challenge the volcanologist. Are the rilles a result of high eruption rates and temperatures, 1/6 g, or lack of an atmosphere? (b) Explosive eruptions into a vacuum have been modeled, and we have representative samples of their products, but we have not explored the extent and nature of these pyroclastic deposits.

This list of features is obviously not complete, for volcanism on the Moon is very different than that on Earth. Even on Earth we have found that generalizations about volcanoes are misleading. As we study more and more terrestrial volcanoes we find that each is different. The same might be said about the Moon's volcanoes as we move through the first phases of exploration from a lunar base.

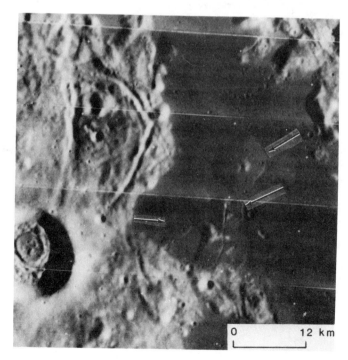

0   12 km

Figure 2.   Mare Veris, located at the base of the Rook Mountains scarp, eastern Mare Orientale. The rings of this large basin are only partly filled with lavas. Many of the basaltic volcanoes, the sources of these lava flows, are visible. In this Lunar Orbiter image, there are three lava shields (arrows); the topmost has a summit crater and the lower-right has a fissure vent crossing the shield.

Figure 3. Oblique photograph of Hadley Rille, eastern edge of Mare Imbrium. The rille (and the Apollo 15 landing site) lie at the base of the Apennine Mountains (bottom and right edge of the photograph). The rille begins at the cleft in lower left, traverses a lava-filled graben parallel to the Imbrium Basin rim, and becomes shallow, disappearing under mare lavas.

0 5 10 15 km

VOLCANIC RESOURCES

Lunar volcanic features and volcanic rocks can be used to provide shielding, shelter, construction materials, and bedrock for anchoring structures. Specific uses of natural mare structures such as lava tubes or lava caves for shelter are considered in this volume (Hörz, 1985; Khalili, 1984). Soils of the mare basins provide materials of higher average atomic number than lunar highland soils. This property makes mare soils more attractive for shielding against solar and cosmic radiation. The iron-rich mare soils also melt at lower temperatures (~1200°C) than the highland soils (~1400°C), bringing them more readily within the range of construction processes that rely on fused brick or melted soil techniques. In addition to their lower melting temperatures, the mare soils also have a low melt viscosity (~10–100 poises) that makes them useful for thermal construction in which flowage is desired (Khalili, 1984).

Beyond the construction phase of lunar base operations, the profitable use of lunar resources must be explored for space-based industries and to support the development of space-based civilization. Lunar mare volcanic rocks are known to be sources of ilmenite ($FeTiO_3$) and chromite ($FeCr_2O_4$), minerals with metallic ore value on Earth that may likely be found as ores in mare regions of the Moon (Taylor, 1984). Such ores can be formed by gravitational separation of crystals from cooling magma of low viscosity, a process likely to be the most important one for forming metallic ores on the Moon. However, the volatile elements associated with some mare basalt deposits might also be considered as lunar ores.

Lunar basalts are known to be markedly depleted in water and carbon-oxygen compounds that are important volatile constituents in terrestrial volcanism. This fact is often construed to suggest that no volatile elements can be found on the Moon in abundances comparable to Earth. Such is not the case, for sulfur has an abundance (about 0.2 wt % in high-Ti basalts) that exceeds the sulfur abundance in most terrestrial basalts. Based upon samples from the Apollo 11 and 17 sites and on some good remote sensing studies, "dark mantle" deposits of the Moon's nearside appear to be glasses of volcanic origin and may serve as unique resources for sulfur and sulfide-associated elements that are of great use for some manufacturing processes. Sublimates on surfaces of these glass particles, the residue of gases extruded during explosive volcanism, may provide the best source of volatiles on the Moon (Fig. 4). Furthermore, close study of these deposits may shed light on the source of volatile elements within the Moon, which has implications for theories of lunar origin (Delano, 1982). Volatile elements associated with lunar pyroclastic

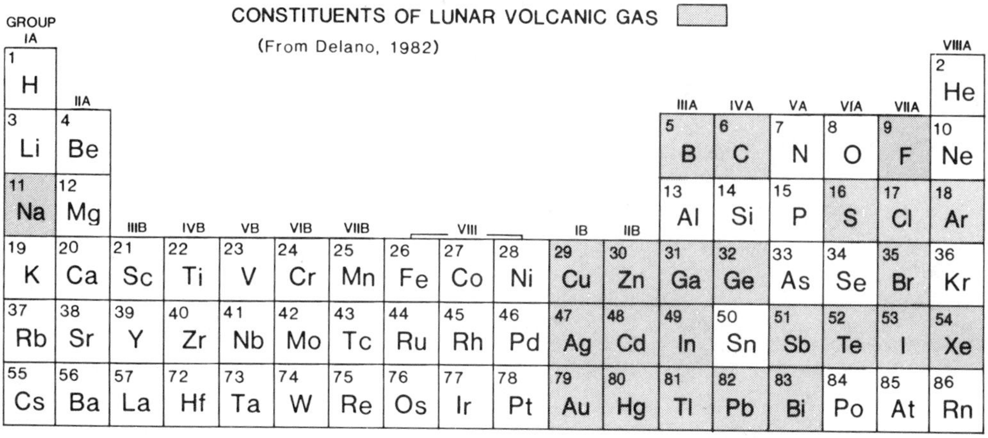

Figure 4. Constituents of lunar volcanic gas, based on Delano's (1982) synthesis of volatile elements correlated with lunar volcanic glasses. In abundance, sulfur predominates and has an abundance (about 0.2 wt % in high-Ti basalts) that exceeds the sulfur abundance in most terrestrial basalts. Water and CO-CO_2 are notably deficient in lunar basalts.

Table 1. A Closer Look at Volcanological Processes on the Earth's Moon—Summary

Processes to be Studied	Scientific Significance	Pragmatic Significance	Exploration
Filling of the maria. Eruption rates and basin structure.	a. History of lunar volcanism. b. Vent locations and relation to basin structure. c. Variations in eruption processes in stages of mare filling and lava flowback. d. Deformation of mare surfaces during filling.	a. Search for volatiles associated with vents. b. Lava tubes and rilles for structures that must be buried.	a. Visit mare basins of different ages, levels of fill, and with a variety of volcanic landforms. *At each location*: determine the volcanic stratigraphy in crater and rille walls or by coring. Following field studies, obtain a complete complement of laboratory studies such as age dates, petrology, and chemistry. b. Begin exploration with the rings of Mare Orientale (Fig. 2)
Explosive volcanic activity.	a. Evolution of lunar magmas through time with regard to volatile phases. b. Size, shape, and extent of vents responsible for mare volcanism. c. Relations of lava flows and pyroclastic deposits. d. Dark mantle deposits—a few vents with widespread deposits or many vents with small deposits?	a. Locate concentrations of volatiles (sublimates in vent "cold traps"). b. Major deposits of fine-grained glass for lunar resource development.	a. Visit vent areas identified. b. Map the extent and stratigraphic variations of the dark mantle deposits. Look for exposures in rilles and crater walls. c. Visit the variety of volcanic landforms seen on satellite photos of the Aristarchus Plateau and Marius Hills. d. Gravity and active seismic surveys of suspected vents.
Magma migration within the Moon.	a. Rise of lunar magmas—dike formation. b. Search for mantle and crustal xenoliths.		a. Visit dike and vent localities.

glasses are indicative of the combined fruits of a closer look at lunar volcanism in which both utility and science may be served.

CONCLUSIONS: A PROGRAM OF EXPLORATION

As part of geologic exploration from a lunar base (Table 1), the evolution of mare basins could be studied by visiting maria of different ages and with different levels of lava fill. A series of landings and traverses within the rings of the Orientale basin (Fig. 2) would provide an opportunity to visit shield volcanoes and fissure vents that may have erupted only enough lava (and ash?) to fill the bottoms of ring depressions. Visiting progressively older mare basins for field studies will eventually provide the observations and samples necessary for reconstruction of the history of lunar mare volcanism. Older highlands volcanism will be considerably more difficult to study because of intense cratering of those surfaces; reconstruction of this activity may be limited to breccia sample studies.

Another major phase of exploration will be to visit volcanic vents. Maps can be made of pyroclastic rocks, their associated sublimates, and the stratigraphy of those deposits along with the relation between clastic rocks and lavas. Deposits near vent areas may also contain xenoliths (foreign fragments of deep crustal or mantle rocks included in lavas). The search for lava tubes should begin near those vents identified by photogeologic mapping. The association of lava tubes and volatile-rich pyroclastic deposits makes locales such as these particularly enticing for the support of both science and utilization. Volcanic stratigraphy and the changes in basalt composition with time may be studied in rille walls, crater walls, and, if no outcrops are visible, by coring from the mare surface. Primary sites for these studies are along basin margins such as eastern Mare Serenitatis and Mare Imbrium and in what appear to be volcanic plateaus such as the Marius Hills and the Aristarchus Plateau. The possible terrains are many, and they provide a large range of sites accessible to a lunar base.

Acknowledgments. We are grateful to B. Hahn for preparation of the manuscript, to A. Garcia for drafting of figures, and to D. Eppler, G. Ryder, and W. Mendell for helpful reviews.

REFERENCES

Basaltic Volcanism Study Project (1981) *Basaltic Volcanism on the Terrestial Planets.* Pergamon, New York. 1286 pp.

Boyce J. M., Dial A. L, and Soderblom L. A. (1974) Ages of the nearside light plains and maria. *Proc. Lunar Sci. Conf. 5th*, pp. 11–23.

Delano J. W. (1982) Volatiles within the Earth's Moon (abstract). In *Papers Presented to the Conference on Planetary Volatiles*, pp. 27–28. Lunar and Planetary Institute, Houston.

Haskin L. A., Jacobs J. W., Brannon J. C., and Haskin M. A. (1977) Compositional dispersion in lunar and terrestrial basalts. *Proc. Lunar Planet. Sci. Conf. 8th*, pp. 1731–1750.

Head J. (1976) Lunar volcanism in space and time. *Rev. Geophys. Space Phys., 14*, 265–300.

Hörz F. (1985) Lava tubes: Potential shelters for habitats. This volume.

Khalili E. N. (1984) Magma and ceramic structures created *in situ* (abstract). In *Papers Presented to the Symposium on Lunar Bases and Space Activities of the 21st Century*, p. 82. NASA/Johnson Space Center, Houston.

Korotev R. L. (1984) The geologic study of lunar volcanic rocks as supported by a permanent lunar base (abstract). In *Papers Presented to the Symposium on Lunar Bases and Space Activities of the 21st Century*, p. 14. NASA/ Johnson Space Center, Houston.

Murase T. and McBirney A. R. (1970) Viscosity of lunar lavas. *Science, 167*, 1491–1493.

Papike J. J., Hodges F. N., Bence A. E., Cameron M., and Rhodes J. M. (1976) Mare basalts: crystal chemistry, mineralogy and petrology. *Rev. Geophys. Space Phys., 14*, 475–540.

Pieters C., McCord T. B., Charette, M. P., and Adams J. B. (1974) Lunar surface: Identification of the dark mantling material in the Apollo 17 soil samples. *Science, 183*, 1191–1193.

Schaber G. G. (1973) Lava flows in Mare Imbrium: Geologic evidence from Apollo orbital photography. *Proc. Lunar Sci. Conf. 4th*, pp. 73–92.

Schultz P. H. and Spudis P. D. (1979) Evidence for ancient mare volcanism. *Proc. Lunar Planet. Sci. Conf. 10th*, pp. 2899–2918.

Schultz P. H. and Spudis P. D. (1983) Beginning and end of lunar mare volcanism. *Nature, 302*, 233–236.

Taylor L. A., Nyquist L. E., and Laul J. C. (1983) Pre-4.2 AE mare-basalt volcanism in the lunar highlands. *Earth Planet. Sci. Lett., 66*, 33–47.

Taylor L. A. (1984) Layered intrusives on the Moon: A source of chromite deposits? (abstract). In *Papers Presented to the Symposium on Lunar Bases and Space Activities of the 21st Century*, p. 15. NASA/Johnson Space Center, Houston.

Taylor S. R. (1982) *Planetary Science: A Lunar Perspective*. Lunar and Planetary Institute, Houston. 481 pp.

Warren P. H. and Wasson J. T. (1979) The origin of KREEP. *Rev. Geophys. Space Phys., 17*, 73–88.

Wood J. A. (1975) Lunar petrogenesis in a well-stirred magma ocean. *Proc. Lunar Sci. Conf. 6th*, pp. 1087–1102.

ADVANCED GEOLOGIC EXPLORATION SUPPORTED BY A LUNAR BASE: A TRAVERSE ACROSS THE IMBRIUM-PROCELLARUM REGION OF THE MOON

Mark J. Cintala

NASA/Johnson Space Center, Advanced Research Projects Office, Code SN12, Houston, TX 77058

Paul D. Spudis

U. S. Geological Survey, 2255 North Gemini Drive, Flagstaff, AZ 86001 and Department of Geology, Arizona State University, Tempe, AZ 85287

B. Ray Hawke

Planetary Geosciences Division, Hawaii Institute of Geophysics, University of Hawaii, Honolulu, HI 96822

Inherent with the existence of a permanent manned presence on the Moon should be the ability to conduct extended geological explorations. Not only would a wide variety of features become accessible to scientists, but the sophisticated investigations performed in the field would make the spectacular Apollo efforts pale in comparison. An example of such a traverse is presented here, with the Imbrium Basin and its environs the region selected to be studied. A field crew of six to eight members would travel a total distance of almost 4000 km as they visited 29 separate localities in an attempt to characterize the processes involved in the formation and evolution of a variety of major lunar features. Among the sites chosen in this expedition would be the Apennine Mountains, the Apennine Bench, Mare Imbrium, the Aristarchus Plateau, Oceanus Procellarum, the extreme western highlands, and Eratosthenes, Copernicus, and Aristarchus Craters.

INTRODUCTION

The Apollo missions to the Moon were able to acquire data that without question revolutionized the planetary sciences. A multitude of other disciplines, however, also reaped major gains. Among these other branches of science are those dealing with particles and fields, solar-planetary relationships, astronomy, astrophysics, isotopic studies, biology, and the interplanetary medium. This list is certainly not exhaustive, nor does it include topics in engineering or other disciplines affected by "spinoff," which would be much too extensive to address here. Although Apollo was inarguably undertaken as a politically motivated project, the sheer volume of the scientific return and analytical innovations was probably unsurpassed by that of any other single effort on record. Thus, if history can be taken as a suitable guide, it would require no gift of prophecy to foresee that the existence of a lunar base, however limited in initial extent, would provide the opportunity to expand our knowledge of the Moon—and of planets in general—by orders of magnitude.

The scientific importance of maintaining a permanent or semi-permanent manned presence on the Moon is perhaps best illustrated by considering the time spent on or in the vicinity of the Moon by Apollo astronauts. Apollo command/service modules spent a total of 716 hours and 2 minutes in closed lunar orbit, while the total stay-time on the surface was 299 hours and 44 minutes. Of the time on the surface, only 81 hours and 9 minutes saw the crews outside of their lunar modules (Baker, 1981). More correctly, at least one crew member was outside of the landing vehicle for that period of time. Generously assuming that both astronauts were performing science-related extravehicular activities (EVA) for the entire period of surface operations, a total of 162.3 man-hours are found to have been spent on surface science during the entire Apollo program. By way of comparison, this amounts to just under seven days for a two-person field team working 12 hours a day, which would comprise the beginnings of a reconnaissance effort in a terrestrial field-geology context. To contend that the Moon is a well-studied object from a geological standpoint would be severely optimistic.

Separate from these time limitations were others that were just as confining. The transportation capability on the surface during the final three missions, while outstanding in comparison with the walking EVAs of the first three flights, nevertheless left much to be desired. The lack of timely rescue capabilities levied the requirement that the astronauts could never be farther from the lunar module than their consumables (oxygen and cooling water) would permit them to walk, should the roving vehicle have failed; this cast an imposing shadow over the traverse planning teams (e.g., Muehlberger et al., 1980) and, of course, on the geological exploration itself. The quantity of scientific equipment that could be delivered to the surface of the Moon was relatively small, owing to the limited payload capacity of the lunar modules. (This is not to detract from the capabilities of the spacecraft or equipment itself, which were marvels of engineering. It is instead an indication that the variety of scientific experimentation was determined, for the most part, by engineering constraints that were in turn fixed by the available technology.) At the other end of the lunar visits, the quantity of lunar samples to be returned to Earth was likewise preordained by vehicle performance guidelines. All of these criteria were non-negotiable; the very success of the Apollo missions under these and other anticipated but unspecified contingencies serves as a testament to the ingenuity and dedication of all those involved in the construction of the flight plans and field activities.

The new generation of field geology made possible by the existence of a lunar base would suffer from few of the difficulties or limitations cited above. As an example, the quantity of samples available for study would be limited only by the ability of the scientists to find them, and the quality of those chosen would be enhanced by the time available to the field scientists in making their selections. The easy access to laboratory facilities would permit any number of investigations without, what would be in retrospect, the unreasonable restrictions caused by vehicle capabilities. The nature of the study—not time constraints, solar flares, or utter hopelessness of rescue— would dictate the duration of a stop at any particular locality. Should the time spent at a given locality be greater than that planned, the ability to rest, resupply, and repair malfunctioning equipment would minimize the impact on subsequent studies at different sites.

This contribution indulges in some speculation regarding a model traverse that might be undertaken by a field-geology team supported by a lunar base and its facilities. It is not intended to be a definitive study, but it might instead serve as an example of the sort of exploration made possible by the new capabilities, as well as to illustrate the requirements and rationale behind an extended scientific traverse across the lunar surface.

FIELD SUPPORT

It would be an exercise in futility to attempt to predict the equipment available to the next generation of lunar explorers, not only in terms of its sophistication, but also its quantity. Therefore, in the interest of a succinct contribution, a number of items and capabilities will be assumed. It is hoped that this list will not be unreasonable and that the reader will allow the unheralded appearance of these devices and instruments for the sake of the exercise. Without them or their counterparts, the traverse described herein would certainly be impossible; indeed, without the technology necessary to construct such hardware, the lunar base itself might be equally improbable.

Fundamental Requirements

In terms of both duration and distance, the expedition to be described here will be long by any standards. It could indeed be stressful to a single crew while presenting taxing demands on the technology used to develop vehicle-power and mobile life-support systems. A myriad of operational scenarios could be devised that would invoke unmanned, automated, and/or teleoperated segments of the traverse, such as those between scientific sites. Evaluation of the interaction betwen man and machine at that level, however, is well beyond the scope and intent of this contribution. A fully manned operation will be assumed for the duration of the exploration.

The team will consist of scientists and support technicians; should the geologic traverse vehicle (GTV) be manned during its movement between sites, the technicians would probably double as GTV operators and mechanics. The number of scientists is difficult to suggest, but some requirements help in refining the estimate. As discussed below, it would be extremely desirable to have relatively short-range excursion capability from the "base camp" (as defined by the location of the GTV). As envisioned here, these "rovers" would require a crew of two for reasons of efficiency and safety. Thus, should two rovers be allowable, and with two scientists left at the base camp, the number of scientists becomes six for the purposes of this exercise. Thus, including two technicians/transport crewmembers, the size of the field team is suggested to be six to eight, depending on the number of rovers.

Transportation and Logistical Support

The shape and details of the GTV are unknown and of little concern to this paper. Its mode of transportation is also problematic, although it will more likely be a wheeled or tracked vehicle than a rocket-powered one for reasons of economy. While the flying version would be desirable in the sense that it would be faster and might double as a remote-sensing platform, lunar-orbiting spacecraft could provide such support. Therefore, the GTV assumed here will be a ground vehicle capable of negotiating steep (>30°) slopes of poorly consolidated material. It will be required to include or provide the following:

- Shelter and consumables
- First-order analytical equipment (described below)
- Navigational equipment and communications
- Sample-collecting equipment
- Reusable geophysical instrumentation (*e.g.*, magnetometers, gravimeters, *etc.*)
- Multi-spectral cameras
- Pressure suits for all crewmembers, as well as spares

The two rover vehicles would be carried or towed by the GTV; each should have a nominal traverse distance of at least 50 km between rechargings or refuelings. As it would be extremely desirable for the scientists to study the surroundings as they moved between stops, the rovers should be ground vehicles in order to maximize the scientific return of the forays from the base camp. It would be very useful if they were also pressurized, offering sufficient room for sleeping and other activities while the independent excursions were taking place. Clearly, it would not be efficient to suffer untimely returns to the base camp when dictated by consumables or crew fatigue. Thus, each would be, in effect, a scaled-down, more agile version of the GTV in some respects, but without the extensive scientific support instrumentation. Multiple EVAs would be possible from each rover before they would return to the base camp for recharging.

Resupply of the consumables required by the team, as well as delivery of samples to the better equipped main complex would be accomplished by periodic ferry flights originating at the lunar base. The frequency of these flights would be set by a number of factors to be determined elsewhere.

Analytical Instrumentation

It should go without saying that the more instrumentation capable of being carried inside the GTV, the more "bits per buck" will be obtainable. On the other hand, it is unlikely that a complete geochemical/petrologic laboratory could be included in the vehicle. With this in mind, it would be highly desirable to have the following instrumentation:

- microscopes (binocular, petrographic) and thin-sectioning equipment
- equipment for first-order chemical analysis (primarily to obtain whole-rock compositional information)
- computer support

It would probably be more efficient for the rovers to have payload capacity relegated to samples than to analytical instrumentation. It would be useful, however, if they would include binocular microscopes and small X-ray flourescence units with supporting hardware. This would provide the crews with enhanced sample characterization capabilities, thus permitting decisive sample selection in the field.

Field Equipment

Personal field equipment similar to that carried on Apollo would be used by all crewmembers, but a major advantage of this sort of exploration is the ability to perform investigations on a much larger scale than had been done previously. Thus, deep drill-coring operations (*i.e.*, more than a few hundred meters) would be within the realm of possible field operations. Deep coring, however, is an extremely time-intensive proposition even on Earth with all of the requisite materials readily at hand. The Moon will present an environment that is much more inimical to such activities; lubricating fluids for the core tubes and bits, for instance, will present a severe challenge to geological engineers. A new drilling technology, the foundations of which might well have been laid already (Rowley and Neudecker, 1984), would be required in order to make such important studies possible. Given the deep-coring capability, a core-extrusion unit will be necessary for use in the field: 100 m of core-tubing, 10 cm in diameter and full of rock and regolith would possess a mass of roughly 2 metric tons. It is obvious that the field team would have to break down the core in the field, sampling it at strategic intervals. In this way, the bulk of the core could be left at the site for future retrieval, if desired, while representative samples could be returned to laboratories better equipped than that aboard the GTV.

The use of a backhoe or similar device would permit regolith studies on a scale that would minimize statistical extrapolations, setting stratigraphic studies on a firm basis. Comparatively deep trenches could be excavated with little effort, yielding a spatially extensive cross-section of regolith stratigraphy. Indeed, removal of regolith to the basal bedrock in some mare areas could become commonplace, a capability that cannot be overemphasized in terms of regolith science and studies of solar history.

These two devices will increase the scientific capabilities of the field team commensurate with the magnitude of the entire expedition. Indeed, they would provide a considerable incentive for the very concept of the extended traverse.

THE TRAVERSE

Before the traverse itself is described, a few points should be made. First and foremost, the route is presented only as an example and should be treated as such. The area covered, however, was chosen for a number of reasons.

The Imbrium Basin and its environs have been geological favorites for many years, not only for the basin's prominence on the lunar nearside, but also for the wide diversity of geologic formations and other features found in that lunar quadrant. A very important aspect of this region of the Moon, however, is the fact that it contains features that were crucial in the development of the lunar stratigraphic system (Shoemaker and Hackman, 1962). This classification scheme has been the vehicle for describing the geologic history of the Moon as it is presently understood (*e.g.*, Mutch, 1970; Wilhelms and McCauley, 1971; Wilhelms, 1985); its only major drawback is the lack of established ages for a range of specific features, which would define an absolute chronology. One of the most important contributions of the proposed exploration thus would be the "calibration" of the lunar relative time-scale in terms of absolute ages, which would occur upon sampling the key formations to provide fodder for the various age-dating techniques. In addition to establishing an absolute time-scale for the lunar nearside, the traverse would provide the opportunities to examine a number of critical processes, which are described below.

Multi-Ring Basin Formation

Perhaps the most important process operating during the early histories of the terrestrial planets was the formation of the huge multi-ring basins characteristic of all large, solid bodies studied to date (Baldwin, 1949, 1963; Hartmann and Kuiper, 1962; Stuart-Alexander and Howard, 1970; Hartmann and Wood, 1971; Moore *et al.*, 1974; Head *et al.*, 1975; and many others). The impact events responsible for their formation on the Moon not only created major topographic features (*i.e.*, gigantic crater-like structures that measure thousands of kilometers in diameter), but they also rearranged millions of cubic kilometers of lunar crust (see, for example, Moore *et al.*, 1974; Head *et al.*, 1975), created vast quantities of shock-melted material (Head, 1974a; Moore *et al.*, 1974), generated sources of seismic energy that modified pre-existing terrain (Schultz and Gault, 1975a,b), and provided topographic "traps" for large volumes of subsequently erupted volcanic materials (see the review of Head, 1976). Thus it should come as no surprise that an understanding of these features and the mechanisms that were active during and after their formation is very high on the priority list for lunar geologists.

Large Crater Formation

Aside from some of the mare basalts, virtually every lunar sample that has been studied exhibits signs of shock damage (G. Ryder, personal communication, 1984), which is a consequence of crater formation by impact. Even a casual look at a telescopic photograph of the Moon is sufficient to demonstrate the importance of large craters in shaping the landscape and affecting the evolution of the lunar crust. By analogy, the same must be true of the other solar system bodies that possess high densities of craters. Thus, the mechanisms involved in the formation of large craters—indeed, craters of all sizes—are among the most important in the evolution of planetary surfaces. In this light, the study of craters, especially large ones (tens of kilometers across), will also receive considerable attention on the traverse.

Volcanism

Insofar as mare basalts cover more than one-sixth of the lunar surface (Head, 1975), volcanism played a highly visible role in the development of the surface and in the evolution of the interior of the Moon. These deposits are extremely diverse in composition, both on the basis of remote-sensing geochemistry (*e.g.*, Pieters, 1978; Bell and Hawke, 1984) and analysis of returned samples (see the review of Papike *et al.*, 1976), as well as in the ages of their emplacement (*e.g.*, Boyce *et al.*, 1974; Schultz and Spudis, 1983). These characteristics imply that the lunar interior (the source of the basalts) underwent a highly complex evolution during and after the period of early lunar bombardment. The origins of various lunar volcanic features are still problematical but have the potential of yielding important information on the factors governing the different styles of lunar volcanism. Among such structures are mare rilles and ridges, domes, dark mantles and dark halo craters, cones, individual flows, and the intricate volcanic complexes exemplified by the Marius Hills and Aristarchus Plateau (*e.g.*, Whitford–Stark and Head, 1977). Since the Imbrium Basin is flooded with mare basalts, substantial emphasis will be placed on the study of volcanic deposits and related features.

The suggested route, which is approximately 4000 km in length, is illustrated in Fig. 1. In reality, the starting and ending points would be governed strongly by the location of the lunar base itself, although it is easily conceivable that another leg could be added

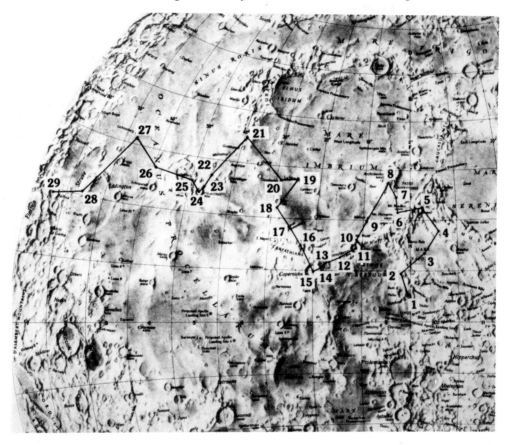

Figure 1. The proposed traverse is illustrated here, somewhat schematically, on a National Geographic (Lambert Equal Area) base map. While the individual segments of the trip are drawn here as straight lines, they would, in reality, be much more sinuous: orbital remote-sensing data would undoubtedly perturb the path taken by the field team in their quest for the widest variety of data possible. With this in mind, it is also important to note that the indicated stations represent only the prime field sites—many shorter, less complex stops will occur between the numbered locations.

to the trip from the base to a suitable starting site. The following paragraphs give an abbreviated description and rationale for the major stops, which are keyed to the numbers in the figure. At the end of each paragraph are the distance between that site and the previous site and letters indicating the principal purposes for studying that site. They are as follows:

A—Aristarchus Plateau, an impact-volcanic complex

B—Basin structure and stratigraphy, usually applied to the Imbrium Basin on this traverse

C—Large crater deposits and/or structure

V—Volcanic features and/or deposits.

Murchison Crater Floor (Stop 1)

An old, degraded crater, Murchison possesses a floor with an unusual morphology, being partially covered with either impact melt or volcanics. In addition, the walls and surroundings of the crater are mapped as Fra Mauro Formation, while parts of the floor are classified as Cayley Formation (Wilhelms, 1968). Both of these unit types were visited by Apollo missions, and their origins are still intensely debated. Finally, a ray from the young crater Triesnecker crosses the center of Murchison's floor, which provides an opportunity to establish a date for that crater's formation. (B, V; 0 km)

Fra Mauro Formation/Bode Dark Mantle (Stop 2)

The Fra Mauro Formation on the backslope of the Imbrium Basin will be sampled again on a line radial to the basin center (this technique will be employed throughout the traverse; since deeper materials should have been ejected to shorter overall distances, any radial variations in composition should, in theory, be related to vertical inhomogeneities in the target before the impact). A "dark mantle" of probable pyroclastic origin, associated with the sinuous rille Rima Bode and shown to be bluish in color by Earth-based spectral observations (Pieters *et al.*, 1973), occurs in the "backwaters" of Sinus Aestuum. Early basalts to the east of these deposits are also bluish, indicating that they could be related (Head, 1974b). (B, V; 110 km. This single leg of the trip is more than 15% longer than all of the Apollo traverses combined.)

Mare Vaporum (Stop 3)

The Vaporum basalts were emplaced in a pre-Imbrian impact feature that was about 200 km in diameter. Work at this site will concentrate on the geophysical study of the Vaporum Basin's structure as well as on the Vaporum basalts themselves. (B, V; 150 km)

Ina (Stop 4)

First noticed on Apollo 15 panoramic photography, this feature has been interpreted to be a caldera (El-Baz, 1972; Strain and El-Baz, 1980). It is remarkable in that it has virtually no superposed impact craters, an observation that indicates a very young age for this feature and the probable volcanic process that created it. Nearby Imbrium Basin deposits will also be visited. (B, V; 150 km)

Conon Crater (Stop 5)

A crater 21 km in diameter and, more importantly, 3 km deep represents a substantial excavation into its target terrain. Insofar as Conon is located in the backslopes of the Apennine Mountains, which form a portion of a ring surrounding the Imbrium Basin, it represents an important "window" into the stratigraphy of Imbrium ejecta, as well as possible pre-Imbrian materials. Its ejecta should contain exciting clues regarding the crustal structure of the Moon before Imbrium was formed. (B, C; 110 km)

Apennine Scarp/Possible Imbrium Impact-Melt Pool (Stop 6)

The basin-facing side of the Apennine Mountains probably represents a major fault zone; if so, the stratigraphy sought at Conon Crater might also be exposed here. A multi-spectral survey of the scarp from the GTV should provide information on any such layering. A number of small "pools" of possible melt generated by the Imbrium Event are also in this region; study of these rocks would yield very useful data in deciphering the nature of the Imbrium-forming impact. (B; 115 km)

Apennine Bench Formation (Stop 7)

This region, just inside the Apennine Front south of Archimedes Crater, has been something of an enigma to geologists. While it has superficial resemblances to impact melt deposits such as those found inside the Orientale Basin (Head, 1974a; Moore *et al.*, 1974), most recent interpretations give it a relatively old volcanic origin (Hackman, 1966; Hawke and Head, 1978; Spudis, 1978). If such were the case, it would represent the largest recognized deposit of non-mare volcanics on the Moon. (B, V; 90 km)

Montes Archimedes (Stop 8)

Orbital gamma ray detectors found abnormally high thorium concentrations in the rugged area just south of Archimedes Crater (*e.g.*, Metzger *et al.*, 1979). This area is also very red in a spectral sense (Malin, 1974; see, for example, McCord *et al.*, 1976 for a description of lunar spectral types). The origins of these "red spots," which are scattered across the nearside, are uncertain (*e.g.*, Malin, 1974). Sampling the rocks exposed at Archimedes would be a significant step in unraveling this mystery. [B, V(?); 80 km]

Wallace Crater (Stop 9)

Wallace is an old, flooded, unremarkable crater whose position in the Imbrium Basin brings it more attention than it deserves of its own accord. Rays and secondary-crater fields from Copernicus Crater occur in this area; thus, this site will mark the beginning of the radial sampling process for that classic crater. Relatively young basalts unsampled by Apollo are also abundant at this site, and a geophysical study will shed light on pre-mare basin stratigraphy. (B, C, V; 240 km)

Eratosthenes Crater Ejecta/Southwest Apennines (Stop 10)

Eratosthenes Crater is the type area for the definition of the Eratosthenian System of the relative lunar time-scale and, as such, will be studied in some detail. Examination of its ejecta will shed light not only on the emplacement dynamics of ejecta from large craters, but also on the azimuthal variation of Imbrium Basin deposits—the projectile that formed Eratosthenes impacted the southwestern portion of the Apennine Mountains. Deposits from the pre-Imbrian Aestuum Basin appear to have been excavated by this impact, so there is a chance that these materials could also be collected at this stop. (B, C; 100 km)

Eratosthenes Crater Interior (Stop 11)

This site is important not only from a scientific standpoint, but it will also provide a good test for the terrain-handling capabilities of the GTV in that local slopes of up to 30° will be encountered. The interior of the crater will be sampled, with special emphasis on the impact melt on the floor; it will be used to establish the time of formation of the crater. A multi-spectral panorama of the crater interior will be very useful, as will a detailed study of the central peaks and the geophysical profiling of the crater subsurface. (C; 40 km)

Copernicus Crater Rays and Secondaries (Stop 12)

The radial sampling of Copernicus will continue with a stop closer to the crater but still in its discontinuous deposits. In addition to the crater's ejecta, the "background" basalts will be collected, since the two are undoubtedly well mixed at this distance from the crater. These samplings will aid in deciphering the dynamics of ejecta emplacement, which are only partially understood. (B, C, V; 80 km)

Copernicus Crater Continuous Ejecta (Stop 13)

This radial sampling stop is located in the "continuous ejecta deposit" of the crater, which is most likely a combination of crater ejecta and local material that intermixed as the ejecta impacted (Oberbeck, 1975). One of the major goals at this stop is to determine the relative proportions of actual crater ejecta and local material in the deposit. The mare basalts are very thin here, leading to the probability that much of the Copernicus ejecta will consist of Imbrium ejecta, which is much older; thus, azimuthal sampling of Imbrium material should also continue at this site.

Copernicus Crater Rim Materials (Stop 14)

The view into the 93-km crater at this site should help to mollify the rigors endured by the crew to this point in the trip. Ejecta from deep in the crust should be present at this location, as are large concentrations of impact melt (Howard and Wilshire, 1975; Hawke and Head, 1977). Samples of the ejecta will aid in the reconstruction of the effects of the Imbrium impact event, and the melt will be used in dating Copernicus, which will also define the beginning of Copernican time in the lunar stratigraphic system. This site is tailor-made for a panoramic multi-spectral survey. (B, C; 30 km)

Copernicus Crater Central Peaks (Stop 15)

Recent Earth-based observations suggest that the central peaks of Copernicus are composed largely of olivine (Pieters, 1982), suggesting the presence of uncommon lunar rock types. The material in the peaks probably came from a significant depth (on the order of several kilometers), and a sampling effort here would be very informative. The thick impact-melt deposits on the floor will be sampled for compositional and textural variations to be compared with those examined at the rim. A geophysical survey of the crater is very high in priority. (C; 30 km, downhill)

Montes Carpatus/Copernicus Crater Ejecta (Stop 16)

The Apennines grade into the Carpathian Mountains on the extreme southern edge of Imbrium; a stop is planned here to continue the azimuthal study of Imbrium stratigraphy. The on-going examination of Copernicus ejecta will also profit from this locality, and possible pyroclastic deposits in the area will be sampled. (B, C; 140 km)

Tobias Mayer Rilles/Copernicus Crater Ejecta (Stop 17)

A muted volcanic complex exists to the northwest of Copernicus near the crater Tobias Mayer. Associated with this complex are a number of small rilles and potential calderas. This area will be studied in some detail, both geochemically and geophysically. The final opportunity to sample Copernicus ejecta in any significant concentration will probably occur at this stop. This will be the first of a series of mare sites located well into Mare Imbrium and Oceanus Procellarum. (B, C, V; 110 km)

Euler Crater (Stop 18)

This 28-km crater is surrounded by fairly young basalt flows (e.g., Schaber, 1973), which are the principal targets of this leg of the exploration. The continuous ejecta deposits of Euler, which probably contain pre-mare materials, will also be sampled. (B, C, V; 145 km)

Mons La Hire (Stop 19)

The Imbrium Basin is so thoroughly flooded that few remnants of its inner ring structure remain exposed. La Hire is one of those unburied massifs, and it, too, is spectrally red (Malin, 1974; Head and McCord, 1978). It will be studied both as a segment of an inner ring of Imbrium and as a spectrally distinct feature. (B; 180 km)

Eratosthenian Flows West of Mons La Hire (Stop 20)

In mid- to late-Eratosthenian time, a series of eruptive events covered a large portion of the Imbrium Basin from the southern border with Oceanus Procellarum to the northern reaches of Mare Imbrium (e.g., Schaber, 1973). Remote-sensing data suggest a composition for these lavas that is not represented in the samples returned by Apollo astronauts or Luna spacecraft (Whitaker, 1972; Etchegaray-Ramirez et al., 1983). These flows thus represent a significant, relatively late volcanic episode in the Moon's history. (B, V; 30 km)

Gruithuisen Domes (Stop 21)

Two separate groups of "red spots"—the Mairan and Gruithuisen Domes—are detectable with Earth-based instrumentation near the northwestern boundary of the Imbrium Basin. They are domical features, and have been interpreted to be the result of a late stage of non-mare extrusive volcanism (Head and McCord, 1978). Whether they are indeed volcanic or simply represent more red ring massifs (such as Mons La Hire) is still debated. Sampling these distinctive features will help to settle the question and perhaps provide data on a potentially important process. A number of spectrally differentiable basalt types (Pieters, 1978) will also be sampled during the trip to the domes. (B, V; 410 km)

Prinz Rilles/Aristarchus Crater Ray (Stop 22)

The Aristarchus Plateau and its immediate surroundings represent an area of remarkable diversity, in terms of both geology and the processes that were involved in its evolution (e.g., Zisk et al., 1977). Five full sites will therefore be dedicated to the exploration of this region of the Moon. This stop will be utilized to investigate the Prinz Rilles, which comprise a series of valley-like depressions in obviously volcanic terrain. A ray from Aristarchus Crater extends across this location; collection of material in that area will begin the radial sampling of this 40-km crater. Highland samples in this area will also be collected as the opportunities arise. (A, B, C, V; 300 km)

Aristarchus Crater Rim (Stop 23)

If it could be possible, the view here should be even more impressive than its equivalent was at Copernicus, since Aristarchus is less than half the diameter of the former, but almost as deep. The crater's continuous ejecta blanket will be examined at this site, while the rim structure and stratigraphy will be probed. The abundant impact-melt flows and ponds in this area will also be sampled for comparison with that on the crater's floor. (A, C, V; 75 km)

Aristarchus Crater Floor (Stop 24)

The melt sheet and central peaks will be high-priority sampling objectives here. In addition, a multi-spectral panorama will be acquired, and a geophysical profile will also be made. Special emphasis will be placed on an attempt to establish the stratigraphy of the crater's northwestern wall, which cuts through the volcanic plateau. The GTV will receive another workout during these investigations. (A, C, V; 25 km)

Vallis Schröteri/Aristarchus Plateau Dark Mantle and Basalts (Stop 25)

An effort as intensive as any on the traverse will be made at this site. The dark mantle materials, which probably represent pyroclastic eruptives, will be on an equal sampling priority with the flow basalts in the area, and a number of geophysical profiles will be taken across the Plateau. It is anticipated that the GTV-defined base camp will be rather mobile during this leg of the exploration, because the largest sinuous rille on the Moon, Schröter's Valley, will also be an object of extensive scrutiny. It is likely that Aristarchus ejecta is fairly common across much of the Plateau, so the radial sampling effort might well continue. [A, C(?), V; 60 km]

Schiaparelli Basalts (Stop 26)

Remote-sensing data show the basalts to the northeast of Schiaparelli Crater to be titanium-rich, but they appear to be very young on the basis of superposed-crater abundances. Since the vast majority of high-titanium basalts returned by the Apollo missions are very old (e.g., Taylor, 1982), samples of these flows will be very interesting for modelers of the evolution of the lunar interior. (V; 180 km)

Lichtenberg Crater and Basalts (Stop 27)

Not only is Lichtenberg Crater interesting because it excavated pre-mare material in northern Oceanus Procellarum, but the basalt flows that embay its ejecta deposits appear to be the youngest recognized lava flows on the Moon (Schultz and Spudis, 1983). (C, V; 300 km)

Struve L Crater (Stop 28)

Struve L is about 14 km in diameter and is a very good candidate for an Orientale Basin secondary crater. It also possesses a floor that is unusual in the sense that it might be comprised predominantly of impact melt from the Orientale Basin (Schultz, 1976a). If not, it would still offer a very good chance at obtaining Orientale ejecta. This stop is in a region of the Moon (the western "shores" of Oceanus Procellarum) that is teeming with diverse features and formations, many of which will be sampled between sites. (B, C; 370 km)

Balboa Crater (Stop 29)

Large craters with fractured floors are not uncommon on the Moon. While their origins are not certain, the leading hypothesis to account for their morphology is the intrusion of magma below the crater into the material that was disaggregated by the impact (Schultz, 1976b). Balboa represents one such crater, and a geophysical survey of its interior should provide some answers to the questions regarding the responsible processes. (C; 190 km)

CONCLUSIONS

This traverse would be an ambitious undertaking, but the scientific dividends it would yield are equal in magnitude to the challenge. Many variations of the route can be proposed, especially in the Oceanus Procellarum region. In fact, much difficult debate was involved in planning the path that is presented above. Some might view this as a sign of uncertainty or the lack of a clear goal for the exploration. On the contrary, the prospects opened by such capabilities are overwhelming, particularly in light of the pressures and limitations under which all similar planning had occurred in the past. When suddenly confronted with the profoundly exciting ability to travel over such great distances with highly sophisticated support equipment, it is almost unfair to ask that one's composure be maintained. The prospects are magnificent, and much remains to be done. . .

Acknowledgments. *The authors would like to thank Fred Hörz, Wendell Mendell, Jeff Taylor, and Dave Vaniman for very helpful—not to mention entertaining—reviews of this paper. Arthur C. Clarke and Stanley Kubrick, ahead of their time as usual, provided no small incentive for thinking big. (We might, however, respectfully suggest a site different from Clavius Base. . .)*

REFERENCES

Baker D. (1981) *The History of Manned Spaceflight*, Crown Publishing Co., New York. 544 pp.

Baldwin R. B. (1949) *The Face of the Moon*, University of Chicago Press, Chicago. 239 pp.

Baldwin R. B. (1963) *The Measure of the Moon*, University of Chicago Press, Chicago. 488 pp.

Bell J. F. and Hawke B. R. (1984) Lunar dark-haloed impact craters: Origin and implications for early mare volcanism. *J. Geophys. Res., 89,* 6899–6910.

Boyce J. M., Dial A. L., and Soderblom L. A. (1974) Ages of the lunar nearside light plains and maria. *Proc. Lunar Sci. Conf. 5th,* pp. 11–23.

El-Baz F. (1972) New geological findings in Apollo 15 lunar orbital photography. *Proc. Lunar Sci. Conf. 3rd,* pp. 39–612.

Etchegaray-Ramirez M. I., Metzger A. E., Haines E. L., and Hawke B. R. (1983) Thorium concentrations in the lunar surface: IV. Deconvolution of the Mare Imbrium, Aristarchus, and adjacent regions. *Proc. Lunar Planet. Sci. Conf. 14th,* in *J. Geophys. Res., 88,* A529–A543.

Hackman R. J. (1966) *Geologic map of the Montes Archimedes region of the Moon.* U. S. Geol. Survey Misc. Geol. Inv. Map I-463.

Hartmann W. K. and Kuiper G. P. (1962) Concentric structures surrounding lunar basins. *Comm. Lunar Planet. Lab., 1,* 51.

Hartmann W. K. and Wood C. A. (1971) Moon: Origin and evolution of multiring basins. *Moon, 3,* 3–78.

Hawke B. R. and Head J. W. (1977) Impact melt on lunar crater rims. In *Impact and Explosion Cratering* (D. J. Roddy, R. O. Pepin, and R. B. Merrill, eds.), pp. 815–841. Pergamon Press, New York.

Hawke B. R. and Head J. W. (1978) Lunar KREEP volcanism: Geologic evidence for history and mode of emplacement. *Proc. Lunar Planet. Sci. Conf. 9th,* pp. 3285–3309.

Head J. W. (1974a) Orientale multiringed basin interior and implications for the petrogenesis of lunar highland samples. *Moon, 11,* 327–356.

Head J. W. (1974b) Lunar dark-mantle deposits: Possible clues to the distribution of early mare deposits. *Proc. Lunar Sci. Conf. 5th,* pp. 207–222.

Head J. W. (1975) Lunar mare deposits: Areas, volumes, sequences, and implication for melting in source areas (abstract). In *Origins of Mare Basalts and Their Implications for Lunar Evolution,* pp. 66–69. Lunar Science Institute, Houston.

Head J. W. (1976) Lunar volcanism in space and time. *Rev. Geophys. Space Phys., 14,* 265–300.

Head J. W. and McCord T. B. (1978) Imbrian-age highland volcanism on the Moon: The Gruithuisen and Mairan Domes. *Science, 199,* 1433–1436.

Head J. W., Settle M., and Stein R. S. (1975) Volume of material ejected from major lunar basins and implications for the depth of excavation of lunar samples. *Proc. Lunar Sci. Conf. 6th,* pp. 2805–2829.

Howard K. A. and Wilshire H. G. (1975) Flows of impact melt at lunar craters. *J. Res. U. S. Geol. Survey, 3,* 237–251.

Malin M. C. (1974) Lunar red spots: Possible pre-mare materials. *Earth Planet. Sci. Lett., 21,* 331–341.

McCord T. B., Pieters C., and Feierberg M. A. (1976) Multispectral mapping of the lunar surface using ground-based telescopes. *Icarus, 29,* 1–34.

Metzger A. E., Haines E. L., and Etchegaray-Ramirez M. I. (1979) Thorium concentrations in the lunar surface: III. Deconvolution of the Apenninus region. *Proc. Lunar Planet. Sci. Conf. 10th,* pp. 1701–1718.

Moore H. J., Hodges C. A., and Scott D. H. (1974) Multiringed basins—Illustrated by Orientale and associated features. *Proc. Lunar Sci. Conf. 5th,* pp. 71–100.

Muelhberger W. R., Hörz F., Sevier J. R., and Ulrich G. E. (1980) Mission objectives for geological exploration of the Apollo 16 landing site. In *Proc. Lunar Highlands Crust* (J. J. Papike and R. B. Merrill, eds.), pp. 1-49. Pergamon Press, New York.

Mutch T. A. (1970) *Geology of the Moon,* Princeton University Press, Princeton, N. J. 391 pp.

Overbeck V. R. (1975) The role of ballistic erosion and sedimentation in lunar stratigraphy. *Rev. Geophys. Space Phys., 13,* 337-362.

Papike J. J., Hodges F. N., and Bence A. E. (1976) Mare basalts: Crystal chemistry mineralogy, and petrology. *Rev. Geophys. Space Phys., 14,* 475-540.

Pieters C. M. (1978) Mare basalt types on the front side of the Moon: A summary of spectral reflectance data. *Proc. Lunar Sci. Conf. 9th,* pp. 2825-2849.

Pieters C. M. (1982) Copernicus crater central peak: Lunar mountain of unique composition. *Science, 215,* 59-61.

Pieters C. M., McCord T. B., Zisk S., and Adams J. B. (1973) Lunar black spots and nature of the Apollo 17 landing area. *J. Geophys. Res., 78,* 5867-5875.

Rowley J. C. and Neudecker J. W. (1984) *In situ* rock melting applied to lunar base construction and for exploration drilling and coring on the Moon (abstract). In *Papers Presented to the Symposium on Lunar Bases and Space Activities of the 21st Century,* p. 77. NASA/Johnson Space Center, Houston.

Schaber G. G. (1973) Lava flows in Mare Imbrium: Geologic evaluation from Apollo orbital photography. *Proc. Lunar Sci. Conf. 4th,* pp. 73-92.

Schultz P. H. (1976a) *Moon Morphology,* University of Texas Press, Austin, TX. 626 pp.

Schultz P. H. (1976b) Floor-fractured lunar craters. *Moon, 15,* 241-273.

Schultz P. H. and Gault D. E. (1975a) Seismic effects from major basin formations on the Moon and Mercury. *Moon, 12,* 159-177.

Schultz P. H. and Gault D. E. (1975b) Seismically induced modification of lunar surface features. *Proc. Lunar Sci. Conf. 6th,* 2845-2862.

Schultz P. H. and Spudis P. D. (1983) Beginning and end of lunar mare volcanism. *Nature, 302,* 233-236.

Shoemaker E. M. and Hackman R. J. (1962) Stratigraphic basis for a lunar time scale. In *The Moon* (Z. Kopal and Z. K. Mikhailov, eds.), pp. 289-300. Academic Press, London.

Spudis P. D. (1978) Composition and origin of the Apennine Bench formation. *Proc. Lunar Planet. Sci. Conf. 9th,* pp. 3379-3394.

Strain P. L. and El-Baz F. (1980) The geology and morphology of Ina. *Proc. Lunar Planet. Sci. Conf. 11th,* pp. 2437-2446.

Stuart-Alexander D. E. and Howard K. A. (1970) Lunar maria and circular basins— A review. *Icarus, 12,* 440-456.

Taylor S. R. (1982) *Planetary Science: A Lunar Perspective,* Lunar and Planetary Institute, Houston. 481 pp.

Whitaker E. A. (1972) Lunar color boundaries and their relationship to topographic features. *Moon, 4,* 348-355.

Whitford-Stark J. L. and Head J. W. (1977) The Procellarum volcanic complexes: Contrasting styles of volcanism. *Proc. Lunar Sci. Conf. 8th,* pp. 2705-2724.

Wilhelms D. E. (1968) Geologic map of the Mare Vaporum quadrangle of the Moon. *U. S. Geol. Survey Misc. Geol. Inv. Map I-548.*

Wilhelms D. E. (1985) The geologic history of the Moon. *U. S. Geol. Survey Prof. Paper 1348.* In press.

Wilhelms D. E. and McCauley J. F. (1971) Geologic map of the nearside of the Moon. *U. S. Geol. Survey Misc. Inv. Map I-703.*

Zisk S. H., Hodges C. A., Moore H. J., Shorthill R. W., Thompson T. W., Whitaker E. A., and Wilhelms D. E. (1977) The Aristarchus-Harbinger region of the Moon: Surface geology and history from recent remote-sensing observations. *Moon, 17,* 59-99.

SEARCH FOR VOLATILES AND GEOLOGIC ACTIVITY FROM A LUNAR BASE

Larry Jay Friesen

McDonnell Douglas Technical Services Co., Mail Code 7G, 16055 Space Center Boulevard, Houston, TX 77058

A lunar base can be used as a central point from which to monitor lunar seismic activity and to search for lunar volatile emissions. This project could be carried out in a twofold effort with (1) a network of instrument packages placed over the lunar surface by manned rover vehicles, and (2) a polar-orbiting satellite. Each instrument package would include a seismometer plus mass spectrometers capable of detecting and analyzing gas species. The polar satellite would carry spectrometers and photometers operating at ultraviolet, visible, and infrared wavelengths, a mass spectrometer, and an alpha spectrometer to look for surface radon enhancements. The primary purposes for this project are to obtain a detailed analysis of lunar seismic activity and internal structure, to determine whether lunar gas venting occurs and, if so, with what intensity, and to locate the sources of any volatiles found.

INTRODUCTION

A lunar base can be used as a central point from which to monitor lunar geologic activity and to search for lunar volatile emissions. There are two primary purposes for such an investigation. The first purpose is to determine whether gas venting does in fact take place on the Moon, and to learn whether there are reservoirs of volatiles anywhere on or within the Moon. This interest is in part scientific curiousity and in part economic. If the Moon contains any substantial volatile reservoirs, and if these contain any substances of value such as water or carbon dioxide, the logistical problems of supporting a base on the Moon might be considerably eased. The second purpose is to monitor lunar seismic activity to determine its temporal and spatial patterns and to study the internal structure of the Moon. This sort of investigation was done during the Apollo program (Lammlein *et al.*, 1974; Latham *et al.*, 1978), but with a larger number of detectors placed more widely over the lunar surface, and by observing for a longer period of time, we can expect to gain a more precise and detailed picture of the lunar structure and seismic activity patterns.

Existing evidence suggesting that gas venting occurs on the Moon takes three primary forms:

    1. Unusual and unexplained brightenings or obscurations have been observed by astronomers from Earth and are often referred to as lunar transient events (LTE). Middlehurst (1967, 1977) and Cameron (1977) have provided extensive and detailed summaries of locations, times, and activity patterns for reported LTE.

2.   Apparent venting was observed instrumentally during the Apollo program, both at the surface [including the water vapor reported by the Apollo 14 Suprathermal Ion Detector Experiment (SIDE) (Freeman *et al.*, 1972)] and from lunar orbit (Hodges *et al.*, 1973).

3.   Greatly enhanced concentrations of ^{222}Rn and its decay daughter ^{210}Po were indicated by alpha spectrometers carried on board the Apollo 15 and 16 service modules at certain locations on the lunar surface, as compared with the surrounding regions (Gorenstein *et al.*, 1973, 1974). In many of these locations, the decay rates of ^{222}Rn and ^{210}Po were not in equilibrium with each other, indicating a change of emission rate for radon at these locations that was rapid compared to the half-life of ^{210}Pb (21 years) for ^{210}Po excesses or ^{222}Rn (3.8 days) for ^{222}Rn excesses.

METHOD OF INVESTIGATION

The search and monitoring effort proposed would be carried out in a twofold effort: with a network of instrument packages placed over the lunar surface and with a polar-orbiting lunar satellite monitored from the base.

It is very probable that scientific work conducted from a lunar base will involve geologic exploration and field work (Spudis, 1984) including traverses to considerable distances from the base, such as the one proposed by Spudis *et al.* (1984) for the Imbrium basin. It should not require much additional effort to carry along instrument packages on such traverses and to deploy the packages at appropriate locations on the lunar surface. This paper assumes that the first lunar base will be on the Earth-facing hemisphere of the Moon, as will the early geologic traverses. Since LTE reports and ^{210}Po enhancements appear to favor edges of maria (Middlehurst, 1967; Bjorkholm *et al.*, 1973), it would be worthwhile to place instrument packages in several of the major basins. Additional instrument packages would very likely be placed at other locations where there is cause to suspect, from LTE reports or Apollo alpha spectrometer observations, that gas venting may occur; for example, at the craters Aristarchus (Bjorkholm *et al.*, 1973; Middlehurst, 1977) and Alphonsus (Kozyrev, 1963). In order to follow up on the suggestion of Arnold (1979) that volatiles may be frozen in shadowed regions at the lunar poles, at least one instrument package should be placed as far north or south as feasible.

A set of proposed instrument locations is shown in Table 1. The proposed locations are subject to change; they will depend, for example, on what geological traverses are selected. The entire network would probably not be put in place at once; rather it would be placed station by station, as traverse opportunities permit. The most important factors in site selection are that ultimately one wants a network of many stations and even more important, these stations should be widely separated. Three or four instrument stations separated by thousands of kilometers would be worth more than ten or fifteen separated by only tens or hundreds. Indeed, one would eventually like to extend this network to the farside of the Moon.

Table 1. Proposed Locations for Lunar Monitoring Instrument Stations

Mare or Crater Location	Approximate Lunar Latitude and Longitude
Northern Imbrium	45°N, 15°W
Aristarchus	24°N, 47°W
Southeast Serenitatis, near Tranquillitatis	20°N, 25°E
Crisium	17°N, 58°E
Grimaldi	6°S, 68°W
Alphonsus	14°S, 3°W
Nectaris	14°S, 34°E
Humorum	23°S, 40°W
Newton	75°S, 4°W

A few words about the vehicles needed to conduct geologic traverses and emplace the instrument packages: these would be pressurized vehicles, probably the size of a camper or large van, with a normal crew of two. For safety's sake, the vehicles should set out in pairs, and each should be capable of providing life support for all four people in the event of a mishap to the other vehicle. Also for safety's sake, the vehicles should be furnished with digging equipment, not only for geological sampling, but also so that they can bury themselves rapidly in the event a solar flare occurs while they are far from the base.

The most promising power source for such vehicles appears at present to be hydrogen-oxygen fuel cells. Nuclear power was considered, but for man–rated vehicles it appears to require an unacceptably large shielding mass (French, 1984). Preliminary calculations based on information in McCormick and Huff (1984) and in *Fuel Cells for Transportation Applications* (Huff, 1981) indicate that fuel cell powered vehicles could have very good range. For excursions of more than one or two thousand kilometers from the lunar base, the vehicles might need to tow special fuel trailers or borrow the technique from mountain climbers and early polar explorers of setting up caches of consumables along the route.

The base would maintain communication with the vehicles (and the instrument packages, once emplaced) via a communication satellite that could be emplaced at the L1 Lagrange point between the Earth and the Moon.

Each instrument package should include seismometers to monitor for moonquakes and impacts, plus mass spectrometers capable of detecting and characterizing any gas species that may be present. The information needed from the mass spectrometers will be the chemical species and incoming energy of any gas atoms, molecules, or ions detected. If the mass spectrometers can be designed to distinguish the direction of arrival of the incoming gas, that information will also be useful. To gain the maximum scientific return from this program, it would be useful to monitor the sun as well as the Moon. A long-term monitoring of the solar wind would be a valuable tracer of solar activity. For this purpose, the mass spectrometers should be designed to be able to measure the composition and energy of solar wind ions, as well as to monitor any native lunar gas that may be present. Other instruments that might be included in the packages, for still other investigations, would be cosmic ray detectors and radio receiver elements for long baseline interferometry.

In addition to the instrument packages, a number of small gas reservoirs should be placed on the Moon. Once the instrument network was in place, these reservoirs would calibrate the network by providing gas releases of known location, time, duration, quantity, and composition.

Just as a lunar base program is likely to include geological traverses, it is also likely to include polar-orbiting lunar satellites for various types of mapping. In fact, a polar satellite such as the proposed Lunar Geoscience Orbiter is likely to go into operation before a lunar base is constructed, and data from it may well influence base site selection. I suggest that one or more lunar polar satellites continue operating into the manned phase of lunar base operation, and that the instruments for a polar satellite include some capable of detecting lunar activity, if any occurs. Such instruments would include spectrometers and photometers operating in the ultraviolet, visible, and infrared (useful for mapping as well as for monitoring activity) wavelengths. There should also be a mass spectrometer to look for lunar gases and an alpha spectrometer to look for surface concentrations of radon and polonium. In addition to providing an additional data point and being able to monitor surface radioactivity, a lunar polar orbiter's optical instruments may be able to detect one type of lunar activity that the surface stations cannot: large-scale dust motion. This will be significant if, as Geake and Mills (1977) have suggested, some LTE involve dust clouds.

A polar-orbiting satellite has the limitation that it can only pass directly over any given lunar location once every two weeks. To alleviate this, the satellite could be placed in as high an orbit as is consistent with its other scientific objectives, and could be given a capability to look a significant angle away from vertical. It might be necessary for the satellite's instruments to scan from one side of its ground track to the other as it passed over the Moon.

To summarize, the equipment required for this study includes: (1) a lunar base, (2) rover vehicles, (3) instrument packages, (4) a lunar polar-orbiting satellite, (5) a communication satellite at the L1 orbit, and (6) gas reservoirs to calibrate the instruments.

If any events are detected by this network of surface instruments and polar satellite, how will they be interpreted? For seismic events, each seismic station should record the time, duration, and intensity of any observation. From this information, geophysicists will attempt to determine the spatial and temporal patterns of lunar seismic activity and to deduce as much as possible about the internal structure of the Moon, in a manner very similar to what was done during the Apollo program.

If any gas venting is observed, each surface station will record for each observation the start time, the duration, the intensity as a function of time, and the composition as a function of time. The polar-orbiting satellite will obtain information similar to that provided by the ground stations. In addition, the polar satellite will make note of its orbital position throughout any observation, the surface locations being observed at the time, and any associated surface phenomena observed, for example: radon enhancements, optical brightenings, or obscurations. From the times and intensities for each venting observed at each ground station and by the satellite, the personnel at the base will try to deduce the time and location at which that event originated. Researchers will also look for coincidences between the times and places of venting events and those of seismic events.

It would greatly aid this investigation if professional and amateur astronomers on Earth can be enlisted. A network of astronomers who are prepared to observe the Moon on short notice is needed. There might be an existing astronomical organization enlisted, or one may be especially set up for this purpose as was done to look for LTE in the 1960s in Operation Moon Blink (Cameron and Gilheany, 1967). Some means of rapid notification would be established—perhaps a telephone tree. Whenever the lunar surface mass spectrometers or the polar-orbiting satellite report a strong burst of activity, the Moon base personnel would immediately relay the information to the Earth-based astronomers, who would observe the Moon and report any unusual phenomena they detect.

CONCLUSIONS: WHAT CAN BE LEARNED FROM A STUDY OF LUNAR VOLATILE AND GEOLOGIC ACTIVITY?

For seismic activity, we wish to determine in an much detail as possible its patterns in location and time. We also wish to determine the internal structure of the Moon in as much detail as possible. This would follow the seismic work done during the Apollo program in greater precision and detail.

For gas venting, we seek answers to four primary questions:

1. Does gas venting actually take place on the Moon? If so, does it involve sufficient amounts of material and energy to account for LTE observed from Earth? Also, does gas venting coincide in location or time with any seismic activity?

2. Do significant reservoirs of volatiles exist anywhere within the Moon, or is the entire Moon as depleted in volatiles as the surface locations we have examined so far? If reservoirs exist, where are they, how deep, and how great are the quantities of volatiles they contain? Do venting locations tell us where the volatiles are, or only where channels to the surface are?

3. What information or constraints does the presence (or absence) of volatiles provide regarding the origin, history, and present state of the Moon? Does this provide any clues about conditions elsewhere in the early solar system, especially in the region of the terrestrial planets and asteroids?

4. Do volatiles exist in sufficient quantity, and in locations accessible at low enough cost, to constitute a resource usable by a lunar base?

Acknowledgments. I would like to thank J. R. Huff for information he generously supplied about fuel cell power systems and likewise thank J. R. French, who was equally generous with information about nuclear power systems. I would also like to thank D. Heymann and R. Reedy for valuable suggestions.

REFERENCES

Arnold J. R. (1979) Ice in the lunar polar regions. *J. Geophys. Res., 84,* 5659–5668.

Bjorkholm P. J., Golub L., and Gorenstein P. (1973) Distribution of ^{222}Rn and ^{210}Po on the lunar surface as observed by the alpha spectrometer. *Proc. Lunar Sci. Conf. 4th,* pp. 2793–2802.

Cameron W. S. (1977) Lunar transient phenomena (LTP): manifestation, site distribution, correlations, and possible causes. *Phys. Earth Planet. Inter., 14,* 194–216.

Cameron W. S. and Gilheany J. J. (1967) Operation moon blink and report of lunar transient phenomena. *Icarus, 7,* 29–41.

Freeman J. W., Hills H. K., and Vondrak R. R. (1972) Water vapor, whence comest thou? *Proc. Lunar Sci. Conf. 3rd,* pp. 2217–2230.

French J. R. (1984) Nuclear powerplants for lunar bases (abstract). In *Papers Presented to the Symposium on Lunar Bases and Space Activities of the 21st Century,* p. 116. NASA/Johnson Space Center, Houston.

Geake J. E. and Mills A. A. (1977) Possible physical processes causing transient lunar events. *Phys. Earth Planet. Inter., 14,* 299–320.

Gorenstein P., Golub L., and Bjorkholm P. J. (1973) Spatial features and temporal variability in the emission of radon from the moon: an interpretation of results from the alpha particle spectrometer. *Proc. Lunar Sci. Conf. 4th,* pp. 2803–2809.

Gorenstein P., Golub L., and Bjorkholm P. (1974) Radon emanation from the Moon, spatial and temporal variability. *Moon, 9,* 129–140.

Hodges R. R., Hoffman J. H., Johnson F. S., and Evans D. E. (1973) Composition and dynamics of the lunar atmosphere. *Proc. Lunar Sci. Conf. 4th,* pp. 2855–2864.

Huff R. (compiler) (1981) *Fuel Cells for Transportation Applications.* Los Alamos Progress Report LA-9387-PR, Los Alamos National Laboratory, Los Alamos. 62 pp.

Kozyrev N. A. (1963) Volcanic phenomena on the Moon. *Nature, 198,* 979–980.

Lammlein D. R., Latham G. V., Dorman J., Nakamura Y., and Ewing M. (1974) Lunar seismicity, structure, and tectonics. *Rev. Geophys. Space Phys., 12,* 1–21.

Latham G. V., Dorman H. J., Horvath P., Ibrahim A. K., Koyama J., and Nakamura Y. (1978) Passive seismic experiment: a summary of current status. *Proc. Lunar Planet. Sci. Conf. 9th,* pp. 3609–3613.

McCormick J. B. and Huff J. R. (1984) Fuel cell propulsion system for lunar surface vehicles (abstract). In *Papers Presented to the Symposium on Lunar Bases and Space Activities of the 21st Century,* p. 113. NASA/Johnson Space Center, Houston.

Middlehurst B. M. (1967) An analysis of lunar events. *Rev. Geophys., 5,* 173–189.

Middlehurst B. M. (1977) A survey of lunar transient phenomena. *Phys. Earth Planet. Inter., 14,* 185–193.

Spudis P. D. (1984) Lunar field geology and the lunar base (abstract). In *Papers Presented to the Symposium on Lunar Bases and Space Activities of the 21st Century,* p. 6. NASA/Johnson Space Center, Houston.

Spudis P. D., Cintala M. J., and Hawke B. R. (1984) A geological traverse across the Imbrium basin region (abstract). In *Papers Presented to the Symposium on Lunar Bases and Space Activities of the 21st Century,* p. 7. NASA/Johnson Space Center, Houston.

UNMANNED SPACEFLIGHTS NEEDED AS SCIENTIFIC PREPARATION FOR A MANNED LUNAR BASE

Don E. Wilhelms

U.S. Geological Survey, Astrogeology, MS 946, 345 Middlefield Road, Menlo Park, CA 94025

Additional knowledge of the Moon's geology, geophysics, and geochemistry is required to maximize the scientific return from a future program of manned lunar exploration. Relatively simple and inexpensive unmanned missions could provide the necessary new data if they are targeted on the basis of knowledge obtained from the first round of lunar exploration. Polar orbiters, sample returners, and seismic probes are required.

INTRODUCTION

Much has been learned about the Moon since the Soviet Union's Lunas 2 and 3 began the era of spacecraft exploration in 1959. The origin of the maria (volcanic basalt) and of most craters (impact) is settled. The approximate compositions of the Moon's crust and mantle and of many geologic units are far better known than before the space age. Representative deposits exposed at the surface have been dated on both the relative and absolute time scales, and the antiquity of the Moon's face has been established. Any comparison of pre-1960 and current lunar literature quickly shows how far our knowledge has advanced (see reviews by Taylor, 1975, 1982; Basaltic Volcanism Study Project, 1981; Wilhelms, 1984, 1985).

Nevertheless, many important scientific questions remain unanswered. The Moon's mode and place of origin are still unknown. Its subsurface structure is poorly known even in relatively well explored areas. The pre-mare impact record is too poorly calibrated to establish such fundamental issues as the time of crustal solidification and the origin and lifetime of large solar system projectiles. The timing and volume of volcanism before 3.8 and after 3.2 aeons ago are uncertain (1 aeon = 10^9 years). The relation among composition, age, source depth, and extrusion site of the mare basalt flows is hypothetical. The compositions and ages of most farside maria are unknown. Terra (highland, upland) compositions are known only approximately from a few spot samples and from low resolution orbital measurements that covered a small percentage of the surface. The ages of most of the rayed craters are uncertain within broad limits. The origin of central peaks and shallow floors of complex craters is uncertain. The origin of basin rings and even the position of the boundary of basin excavation—central questions in studies of impact mechanics, lunar petrology, and stratigraphy—are frustratingly elusive. Lunar remote studies and direct exploration, now almost quiescent, have not completed their task.

Some of these matters can probably be settled by continued experimental and field study on Earth. The origin of complex craters and basin rings might yield to further study of terrestrial craters, laboratory and large-scale explosive experiments, and physical theory.

Other gaps in our knowledge, such as the distribution of mare basalt compositions, can be partly filled by continued geologic mapping, crater-frequency counts, telescopic spectral studies, and petrologic theory. Still other pieces of the lunar puzzle may be found by examining or reexamining the large and still incompletely exploited Apollo sample collection (Ryder, 1982).

Investigation of most of the remaining geologic questions, however, requires resumption of lunar spaceflights. This volume concerns a proposed manned lunar base. Such a base can be neither effectively sited nor productively exploited scientifically without additional preparatory exploration by unmanned spacecraft. This paper explains how three types of unmanned missions can address the questions that remain to be answered by the second round of lunar exploration. Two classes of missions, orbital surveys and sample returners, have already proved their value. Seismometers, which were not successfully included on the unmanned precursors to Apollo, will also be needed

POLAR ORBITER

Lunar Orbiters 4 and 5 provided indispensable photographic coverage of most of the Moon from their near-polar (85°) orbits. Our current knowledge of the Moon would be appallingly incomplete without these missions, which were originally intended for detailed landing-site studies in the equatorial belt and not for reconnaisance purposes. Lunar Orbiter, however, carried no geochemical or geophysical instruments. Except for gravity data obtained by doppler tracking, the only geochemical and geophysical data obtained by any orbiting mission were those obtained at relatively low spatial resolutions by Lunas 10 and 11 (flown by the Soviet Union in 1966) and by Apollos 15–17 (1971–1972) from parts of the belt between 30° N and S latitudes. A new lunar global orbiter could gather important data concerning at least six major topics.

Mare Compositions

At present, remote sensing data useful for extrapolating to large areas the compositional data obtained from returned samples are available only for the small area overflown by Apollos 15–17 and, for the nearside, from telescopic remote sensing (Pieters, 1978). A polar orbiter could readily obtain data in several wavelengths that could be used to determine the compositional variability of all the lunar maria. The estimated compositions of the basalts could then be correlated with their ages and with the sizes and inferred depths of the containing basins. The volumes, depths, and thermal histories of the mantle source zones of the basalts could be partially inferred from these correlations.

Terra Compositions

Terrae cover 83% of the Moon and terra materials constitute almost all of the lunar crust. Ignorance of terra compositions is thus even more serious than ignorance about the volumetrically minor mare basalts. The terrae have been directly sampled at only five spots (Apollos 14–17 and Luna 20), and, because most terra materials are mixtures of the original igneous rock types, the remote sensing data so far available have not closely specified the compositions of the unsampled remainder. This ignorance severely limits the petrologist, geochemist, and cosmogonist attempting to learn the origin of the Moon and the solar system.

Gravity

Except for a few small areas overflown at low altitudes, regional and local lunar gravity fields are poorly known. Two major problems are the mass balance of basins and the nature of the offset of the Moon's center of mass from its center of figure. Modeling has indicated, probably correctly, that mascons (positive gravity anomalies) are due mostly to mare basalt (Solomon and Head, 1980); however, whether the gravity field following basin formation and before filling by the basalt was positive, negative, or neutral is not known. Another unanswered but even more fundamental question is whether the mass offset results from a first-order heterogeneity in the Moon's crust, mantle, or core, or from the gravity anomaly created by a giant nearside basin.

Topography

Gravity modeling, geodesy, estimates of geologic units' thicknesses, spaceflight engineering and operations, and other important scientific and technical tasks depend on knowledge of a planet's topography. For the Moon, the heights of basin rims and other rings and the elevations of the contained maria are particularly significant. Yet the topography of none of the basins has been completely determined. That of the Orientale basin, which is the model for others because it is relatively young and large, is known only within wide limits. Accurate photogrammetry can be performed only for the illuminated parts of the Apollo groundtracks. Refinements of the rest of the nearside's topography still depend on telescopic selenodesy and radar modeling. The topography of the non-overflown parts of the farside and polar regions is almost completely unknown. The need for these basic data is obvious.

Magnetism

Another problem only partly approachable with the limited existing data is the origin of the remanent magnetism found in lunar samples and from orbit. Several alternative hypotheses for the origin of the fields and the way the magnetism was acquired are still viable; different alternatives may apply at different stages of lunar history. Global measurements of the remanent magnetism from orbit will help determine whether the Moon possesses a core, a central question in considerations of lunar composition, thermal history, and origin.

Stratigraphy

Determining the post-accretional history of a planet and extrapolating geochemical and geophysical data depend on knowledge of the stratigraphy of its near-surface rocks. The lunar stratigraphy has been worked out to a good approximation on the nearside and central farside. However, the photographs necessary for stratigraphic analysis have not been obtained at adequate resolution for the poles, most of the limb regions on both hemispheres, the farside at latitudes greater than about 40° N and S, or a zone between 100 and 120° W. The latter gap is particularly severe because it includes part of the Orientale basin. The east limb includes the large and puzzling Crisium basin, and the other poorly photographed zones also include basins whose stratigraphic sequence and ring structure should be examined. Thus, any future polar orbiter should include an imaging system.

UNMANNED SAMPLE RETURN

Solution of other problems requires additional samples from the Moon itself. Lunas 16, 20, and 24 proved the value of unmanned samples returned from targets that, apparently, were not selected in advance except within broad selenographic limits. Despite the small sample size, they provided two absolute ages in the maria and one in the terra, added two types to the list of mare basalt compositions, and sampled typical terra material at an outlying point not reached by Apollo. A relatively inexpensive program of unmanned sample-returning spacecraft could yield significant advances if our current knowledge of lunar geology is applied to the selection of landing point sites. The following five categories of geologic questions are most important. Table 1 lists 19 sites or groups of sites, in approximate order of descending priority, where sampling missions could address these objectives. Data from each site could be extrapolated to larger areas by means of currently available or future orbital sensing. Each probe is considered capable of returning a single sample of regolith randomly selected from within the designated area. Other geoscientists could augment and amend the list.

Absolute Ages
Several more ages are needed to calibrate the lunar stratigraphic column, particularly for basins (items 1, 6, 10, 15–17), young maria (items 2, 13, 14), and young craters (items 7, 8). Only the age of the Imbrium basin (about 3.85 aeons) and a questionable age of the Nectaris basin (3.92 aeons; James, 1981) are available to date the old part of lunar history. The highest priority is given here to dating the Nectaris basin, whose relative age is well known, and which, if securely dated, would therefore provide the needed calibration for the pre-mare cratering rate (Wilhelms, 1985). Dating of events in the last 3 aeons of lunar history is similarly imprecise. Maria younger than 3.16 aeons (Apollo 12) have been dated relatively (Boyce, 1976; Schultz and Spudis, 1983) but not radiometrically, and the Apollo 12 unit is hard to date on the relative time scale.

Compositions and Rock Textures
Although the emplacement mechanism of most lunar geologic units is now known to a good approximation, that of many plains and domelike hills is still questionable. These units should be sampled to decide once and for all whether non-mare volcanism has occurred on the Moon (items 3, 4, 17). Also, some impact phenomena need to be further explored, notably the relative volumes of impact-melt rock and clastic ejecta (items 15–17, 19).

Table 1. Potential Landing Sites for Future Unmanned Lunar Sampling Missions

Item	Stratigraphic Unit	Landing Area	Objective
1	Nectaris basin	(a) Ejecta near 35° S, 42° E (b) Plains (impact melt?) near 22° S, 41° E	(a) Absolute age (b) Composition
2	Copernican mare	Southeast of Lichtenberg, near 31° N, 67° W	(a) Absolute age (b) Composition of source
3	Terra plains	(a) Albategnius (b) Ptolemaeus	Nonmare volcanism or buried mare basalt flows?
4	Terra domelike landforms	(a) Gruithuisen delta or gamma (b) Hansteen alpha	Nonmare volcanism?
5	Farside mare	(a) Floor of Tsiolkovskiy (b) Mare Ingenii	Composition of source
6	Maunder Formation (Orientale-basin impact melt)	South of Mare Orientale	(a) Age of Orientale basin (b) Composition of crust
7	Crater Copernicus	Impact melt on floor	(a) Absolute age (b) Composition of crust
8	Crater King	Impact melt on rim or floor	(a) Composition of farside crust (b) Absolute age
9	Ancient crust	Near 30° N, 160° E	(a) Composition (b) Absolute age
10	South Pole–Aitken basin massifs	South of Korolev, 21.5° S, 160° W	(a) Composition of farside crust (b) Absolute age
11	Pre-Late Imbrian mare (?) basalt	(a) Center of Schickard (b) North of Balmer	(a) Absolute age (b) Composition
12	Early Late Imbrian mare	Mare Marginis, in Ibn Yunus	(a) Absolute age (b) Composition (KREEP–rich?)
13	Eratosthenian mare	(a) Southwestern Mare Imbrium (b) Surveyor 1 region, near 2.5° S, 43.5° W	(a) Absolute age (b) Calibration of color spectra
14	Central Mare Serenitatis	Between Bessel and Dawes	(a) Calibration of color spectra (standard spectrum) (b) Absolute age (near Imbrian–Eratosthenian boundary)
15	Orientale-basin lobate ejecta	Near 53° S, 79° W	Impact melt or other ejecta?
16	Alpes Formation (knobby Imbrium basin ejecta)	Southeast of Vallis Alpes, near 45° N, 5° E	(a) Impact-melt/debris content (b) Composition of deep ejecta
17	Apennine Bench Formation (planar deposit in Imbrium basin)	Near 27° N, 8° W	(a) Impact melt or KREEP–rich volcanic materials? (b) Absolute age (c) Calibrate orbital geochemical data
18	Reiner Gamma Formation (irregular bright patch on mare)	Near 7.5° N, 59° W	(a) Absolute age (b) Magnetism involved in origin?
19	Fissured crater floor deposits	Floor of Murchison, 1° W, 5° N	Ejected Imbrium basin impact melt?

Terra-crust Composition

The remote sensing data discussed above need to be calibrated by "ground truth" at points of known stratigraphic context (items 1, 6–10, 16). Such extrapolation from small to large areas has proved to be the most efficient use of lunar sample data.

Mantle Compositions

Similar remarks apply to the mare basalts; extrapolations from samples of currently unsampled color and age units could readily calibrate the existing and future remotely sensed properties (items 2, 5, 12–14).

Compositions and Ages of Premare Volcanic Basalt

Basalts appear to have formed in abundance before and near the end of the disruption by the early impact barrage (Schultz and Spudis, 1983), but their extent, compositional variability, and age spread are not known. This is a problem for directed sampling of certain breccias and thin plains units (items 3?, 11, 17?).

SEISMIC PROBES

Knowing the average and local thickness of the lunar crust is a prerequisite to assessing gravity models, the composition of the Moon, the depths of basin excavation, the elevation of the source regions of mare basalt and KREEP, and many other important facets of lunar geoscience. Yet the crustal thickness is known only in the limited area reached by the Apollo ALSEPs; in other areas it must be determined by extrapolation and modeling that are highly model-dependent (Basaltic Volcanism Study Project, 1981, section 4.5.2). The important problem of the core could also be resolved by good seismic data. Additional seismic data are therefore among the most urgently needed returns from renewed lunar exploration. Passive seismometers like those of the Apollo ALSEPs could record large meteorite impacts and greatly improve knowledge of the Moon's third dimension. Because the ALSEPs were shut down in 1977, new instruments are required. Also, a much greater spatial spread than that of the ALSEPs, including the farside, is needed.

CONCLUSION

If geoscience is to play a major role in mankind's return to the Moon, more scientific data are needed as preparation for that venture. The most severe gap is in knowledge of the subsurface structure and of surface composition and stratigraphy outside the near-equatorial nearside. Near-polar unmanned orbiters, unmanned samplers, and geophysical instruments emplaced on the surface can substantially improve on the present data base. These relatively simple and inexpensive instruments can be targeted intelligently on the basis of our current understanding of the Moon, incomplete as it is. Even if no manned missions will take place, these unmanned missions would add enormously to our dawning understanding of the Moon's makeup and history.

Acknowledgments. Paul D. Spudis contributed to Table 1 and reviewed the manuscript. Richard J. Baldwin, Michael B. Duke, and Peter H. Schultz also contributed valuable suggestions.

REFERENCES

Basaltic Volcanism Study Project (1981) *Basaltic Volcanism on the Terrestrial Planets.* Pergamon, New York. 1286 pp.

Boyce J. M. (1976) Ages of flow units in the lunar nearside maria based on Lunar Orbiter IV photographs. *Proc. Lunar Planet. Sci. Conf. 7th,* pp. 2717–2728.

James O. B. (1981) Petrologic and age relations of the Apollo 16 rocks: Implications for the subsurface geology and the age of the Nectaris basin. *Proc. Lunar Planet. Sci. 12B,* pp. 209–233.

Pieters C. M. (1978) Mare basalt types on the front side of the moon: A summary of spectral reflectance data. *Proc. Lunar Planet. Sci. Conf. 9th,* pp. 2825–2849.

Ryder G. (1982) Why lunar sample studies are not finished. *EOS, Trans. AGU, 63,* 785–787.

Schultz P. H. and Spudis P. D. (1983) The beginnning and end of lunar mare volcanism. *Nature, 302,* 233–236.

Solomon S. C. and Head J. W. (1980) Lunar mascon basins: Lava filling, tectonics, and evolution of the lithosphere. *Rev. Geophys. Space Phys., 18,* 107–141.

Taylor S. R. (1975) *Lunar Science: A Post-Apollo View.* Pergamon, New York. 372 pp.

Taylor S. R. (1982) *Planetary Science: A Lunar Perspective.* Lunar and Planetary Institute, Houston. 481 pp.

Wilhelms D. E. (1984) Moon. In *The Geology of the Terrestrial Planets* (M. H. Carr, ed.), pp. 107–205. NASA SP-469. NASA, Washington, D.C.

Wilhelms D. E. (1985) *The Geologic History of the Moon.* U.S. Geological Survey Prof. Paper 1348. In press.

THE NEXT GENERATION GEOPHYSICAL INVESTIGATION OF THE MOON

L. L. Hood and C. P. Sonett

Lunar and Planetary Laboratory, University of Arizona, Tucson, AZ 85721

C. T. Russell

Institute of Geophysics and Planetary Physics, University of California, Los Angeles, CA 90024

Primary deficiencies of current lunar geophysical data sets include: (1) uncertainties in the seismic velocity structure of the mantle, especially at depths > 500 km; (2) inadequate global coverage of high-resolution gravity field, topography and, to a greater extent, magnetic field data; (3) sparsity of surface heat-flow measurements; (4) lack of definitive constraints from either seismic or magnetic field data on the existence and radius of a metallic core; and (5) lack of ground-truth samples over most of the accessible surface. This paper discusses how these deficiencies may be removed in the course of preparing for and conducting the next set of manned lunar missions. A polar orbiting geoscience survey could complete lunar potential field and topographic mapping and obtain indirect global heat-flow measurements, while simultaneously providing basic geochemical and geophysical data for long-term manned site selection and resource evaluation. Later deployment of autonomous surface geophysical stations at widely spaced locations during manned exploration and utilization activities would provide the improved seismic and electromagnetic sounding and direct surface heat-flow data needed for a more accurate characterization of the deep interior. The result would be a significantly improved understanding of lunar origin and internal evolution.

INTRODUCTION

The primary aim of planetary geophysics is to determine the structure, composition, and state of a given body and the relationship between internal processes and surface tectonic features. The proximity of the Moon, and its small size implying relatively low internal pressures and temperatures, make it an obvious initial case for application of geophysical techniques to bodies other than the earth. The major goals are to distinguish among lunar origin models and to test theories of lunar (and hence planetary) evolution. The purpose of this paper is to outline some of the lunar geophysical studies that are incomplete, how they can be completed in our future exploration strategy, and what they might tell us.

Some initial progress toward geophysical exploration of the Moon was made as a result of the Apollo manned landing series and associated scientific investigations. For example, it was established that a plagioclase-rich crust with a thickness of 50–60 km beneath Oceanus Procellarum is present, implying differentiation of at least the outer few hundred kilometers of the early Moon. Based on analyses of gravity and topography data, it was shown that the farside crust is either less dense or thicker than the nearside crust; the latter asymmetry is sufficient to explain the observed center-of-figure to center-of-mass offset of about 2 km and may also account for the predominance of the nearside maria (see, for example, Taylor, 1982). Both seismic velocity measurements and analyses of mare basalt samples have been used to infer approximate limits on the composition of the lunar upper mantle (depths < 400–500 km). The lunar mean density and moment-of-inertia have been determined with excellent and fair accuracy, respectively. However, many fundamental issues remain unresolved, and their lack of resolution at present handicaps attempts to geophysically distinguish among either lunar origin models or models of lunar internal evolution.

Major unresolved issues include: (1) the depth of initial melting and differentiation; (2) the structure, thermal state, and bulk composition below the upper mantle; (3) the existence and radius of a possible metallic core; (4) the origin(s) of lunar paleomagnetism; (5) the value of the globally averaged surface heat flow; and (6) the importance of subsolidus convection in the present-day Moon.

Numerous additional aspects of the structure, composition, and history of the crust and uppermost mantle are also not well constrained. Included are the depths of mare fill in circular and irregular maria, the thickness of basin and crater ejecta deposits, the geometry and vertical extent of fault surfaces at depth, the existence and geometry of subsurface igneous intrusive bodies, and the state of stress in subsurface bedrock. The latter question is of particular importance for understanding the support of lunar mass concentrations (mascons) and for estimating the extent of global contraction. Further, near-surface geophysical studies of diverse phenomena including impact cratering physics, near-Earth meteoroid flux history, and the history of solar wind and cosmic ray ion bombardment are generally incomplete. After a brief burst of investigation during the Apollo program these investigations ceased, leaving much undone.

It is reasonable to expect that the next major throughput of geophysical data pertinent to the questions noted above will occur in the process of preparing for and carrying out more extensive manned exploration and utilization of the Moon, possibly involving establishment of a permanent base. Independent of any future manned lunar activity, an orbiting Lunar Geoscience Observer (LGO) mission may obtain expanded orbital geochemical and geophysical data sets during the 1990s as part of the planned Planetary Observer series (Randolph, 1984). Resulting measurements may then play a basic role in allowing a more knowledgeable manned site selection process and assessment of lunar resource potential prior to future landings. Thus both orbital geophysical and manned mission planning objectives may be simultaneously realized. However, several of the most basic geophysical issues such as the composition, structure, and thermal state of the deep interior can be adequately addressed only with the acquisition of long-term surface seismic and electromagnetic sounding and direct surface heat-flow measurements. While unmanned landers may provide some useful new measurements, manned emplacement of sensitive geophysical stations followed by periodic calibration checks and maintenance over at least a decade may be necessary to accurately resolve properties of the deep interior. In the following, we briefly discuss the present status of major lunar geophysical data sets, their relationship to the issues cited above, and the extent to which future orbital surveys and surface measurements may resolve these issues. Emphasis is intentionally placed on measurements pertaining to global scale properties that are of most interest in evaluating lunar origin and evolution models.

SEISMIC DATA

Analyses of lunar seismic signals afford a primary means of probing the structure, composition, and thermal state of the interior. Derived inferences concerning the composition of the middle and lower mantle and the radius of a possible metallic core are of basic interest in distinguishing among lunar origin models (e.g., Newsom, 1984). Because of the small lunar seismic energy release and the occurrence of intense scattering in a near-surface brecciated zone, a long-term data base is required for accurately deducing the velocity structure of the lunar mantle and deep interior. Unfortunately, the reception of transmitted data from the four stations of the Apollo seismic network was terminated on September 30, 1977, as a cost-saving measure. The resulting marginal quantity of measurements and the limited selenographic separation of the seismometers has made interpretations difficult. A general summary of final conclusions based on the available data has been given by Nakamura et al. (1982).

Sources of lunar seismic signals include artificial and natural impacts, weak repetitive deep-focus moonquakes triggered by tidal stresses (Toksöz et al., 1977), and more energetic but sparse, shallow moonquakes that are most probably entirely tectonic in origin (Nakamura et al., 1979). The generally low signal-to-noise ratio reduces the accuracy of identified wave arrival times despite the application of various signal processing techniques. This factor combined with the poor areal distribution of the Apollo array and the temporally limited data base has resulted in large uncertainties in the seismic velocity structure of the middle and lower mantle. In the upper mantle (depths < 400–500 km), P- and S-wave velocities are relatively well determined and imply a Mg/(Mg+Fe) ratio for mantle olivines and pyroxenes in the range 0.70–0.85 (Nakamura et al., 1974). Such a range is in agreement with independent estimates for mare basalt source regions at comparable depths based on petrologic analyses of surface samples (Ringwood and Essene, 1970; Morgan et al., 1978). In the middle mantle (500–1000 km depth), the most recent analysis of the complete arrival time data set (Nakamura, 1983) has yielded only one-standard deviation limits on the mean P- and S-wave velocities i.e., the probability that actual velocities are within the derived range is only about 68%. The derived velocity limits would be consistent with an increase in the Mg/(Mg+Fe) ratio of middle mantle olivines and pyroxenes if the garnet abundance is not too large. Such an increase would in turn suggest that initial melting and differentiation extended down to at least 1000 km depth (Nakamura, 1983). In the lower mantle (> 1000 km depth) essentially no seismic velocity limits or compositional constraints are available.

Although an early interpretation of a P-wave arrival from a single farside meteoroid impact suggested the presence of a small low-velocity core (Nakamura et al., 1974), this interpretation was never confirmed with additional data during the active operation period of the seismic network, and it is now regarded as undefinitive (e.g., Goins et al., 1981). In principle, mantle density models consistent with seismic velocity limits and other constraints such as moment-of-inertia ($I/MR^2 = 0.3905 \pm 0.0023$; Ferrari *et al.*, 1980) provide an alternate means of evaluating the likelihood of the existence of a metallic core. In particular, a more magnesium-rich (lower-density) middle and lower lunar mantle as suggested by Nakamura's (1983) seismic velocity model would require a substantial iron core with radius > 360 km in order to match the mean density and moment-of-inertia values (Hood and Jones, 1985). However, current uncertainties in mantle seismic velocities and their interpretation preclude deduction of the existence and radius of a metallic core on this basis alone. Additional independent constraints on the core radius such as may be provided by magnetic field data (see below) or by interpretations of laser-ranging data (Yoder, 1981) are needed.

In general, analyses of existing lunar seismic data have established the existence of a crust, an elastic upper mantle, and an attenuating lower mantle indicating temperatures much cooler than the solidus for depths less than about 500 km and suggesting a close approach of the selenotherm to the solidus at depths > 1100 km. However, compositional inferences based on seismic velocity structure needed for resolving fundamental issues such as depth of initial differentiation, bulk composition of the middle and lower mantle, and the existence of a metallic core are much more uncertain. Additionally, other questions of lunar seismicity including the nature of lateral heterogeneities and the focal depths, source mechanisms, and precise epicenters of the tectonically important shallow moonquakes remain largely unanswered. A significant change in this situation is likely to occur only with the acquisition of an improved data set involving emplacement of more widely separated surface seismometers and compilation of continuous measurements over a decade or more. Incorporation of a stored-mode operating system with relatively brief downlink transmission at regular intervals rather than continuous (24-hr) transmission, as was the case for the Apollo seismometers, would minimize the cost of long-term data acquisition. Finally, it should be noted that an effort to record the moon's free oscillations using improved seismic sensor design and emplacement (*e.g.*, in deep boreholes below much of the scattering layer) may lead to significantly improved constraints on the density and velocity profile of the deep interior.

ELECTROMAGNETIC SOUNDING DATA

An alternate but equally difficult approach toward investigating the existence of a metallic core and the thermal state of the interior involves determining limits on the electrical conductivity as a function of depth via measurements of induced magnetic fields. Two primary techniques were developed in the Apollo era. The first used a single orbiting satellite, such as one of the Apollo 15 and 16 subsatellites, during periods when the Moon was in the quasi-vacuum conditions of the geomagnetic tail. After a period of five to ten hours, the geomagnetic tail magnetic field can diffuse through the poorly conducting lunar mantle so that any residual distortion of the field is due to a more highly conducting core. This distortion is marginally measurable by a spacecraft magnetometer and results in an estimate for the residual induced dipole moment. Measurements of the latter were obtained indicating the probable existence of a core with radius > 400 km and conductivity ≥ 10 S m^{-1} (Goldstein *et al.*, 1976; Russell *et al.*, 1981). However, the core conductivity limit is consistent with either a partially molten silicate core (conductivity ~ 10 S m^{-1}) or a metallic core (conductivity $\sim 10^5$ S m^{-1}). Thus definitive evidence for a metallic core was not obtained. Also, the available Apollo data base was limited because of the finite lengths of the subsatellite missions and the need to eliminate periods when measurable ambient plasma densities were present. Confirmation of these measurements via future orbiters such as LGO would provide improved evidence for a highly conducting lunar core.

A second electromagnetic sounding technique employed both orbital (Explorer 35) and surface (Apollo 12) magnetometers to measure the external forcing field and the output field (forcing field plus induced field), respectively, as a function of frequency. These measurements together with a suitable theoretical model allowed limits to be established on the mantle electrical conductivity profile. The technique was applied both for times when the Moon was in the geomagnetic tail (Dyal *et al.*, 1976), and for times when the moon was in the supermagnetosonic solar wind (Sonett *et al.*, 1972; Hood *et al.*, 1982a; see the review by Sonett, 1982). Results are in reasonable (order-of-magnitude) agreement and show that the conductivity rises from 10^{-4} to 10^{-3} S m^{-1} at depths of a few hundred kilometers to 10^{-2}–10^{-1} S m^{-1} at depths of 1000–1200 km. The same technique can be applied to put bounds on the radius of a highly conducting (partially molten silicate or metallic) core. For example, solar wind response measurements have been found to imply an upper bound of 435 km on the radius of such a core (Hobbs *et al.*, 1983). However, uncertainties resulting from accuracy of the theoretical model used to interpret the data and possible intercalibration errors between the surface and orbiting magnetometers suggest that a more conservative upper limit on the core radius from time-dependent induction studies is near 500 km. When an array of surface magnetometers is again deployed on the lunar surface, steps must be taken to eliminate the possibility of instrument drift and intercalibration differences in order to ensure that maximally accurate limits on the core radius are obtained.

A further application of derived lunar mantle conductivity profiles is to empirically limit the range of allowed radial temperature profiles using laboratory measurements of conductivity versus temperature for relevant minerals. Results show reasonable consistency with thermal history model calculations and experimental solidus curves but are sensitive to compositional assumptions (Hood *et al.*, 1982b). Nevertheless, the results are encouraging, and significant experimental constraints on the form of the selenotherm may be possible when mantle composition models are more refined as a result of improved seismic data.

HEAT-FLOW DATA

Heat-flow probe measurements were successfully obtained at only two of the Apollo landing sites, Apollo 15 and 17, yielding final estimates of 21 and 16 mW/m^2, respectively (Langseth *et al.*, 1976). Derivation of globally representative averages from these isolated values is necessarily difficult, and values ranging from 11 mW/m^2 (Warren and Rasmussen,

1984) to 18 mW/m^2 (Langseth *et al.*, 1976) have been suggested. Assuming nominal radioactive element ratios and a steady-state balance between heat production and loss, the latter rates would imply bulk Moon uranium abundances of 29 and 46 ppb, respectively, compared to 14 ppb for C1 chondrites and 18 ppb for the bulk Earth. However, a steady-state balance may be inappropriate if the interior heat production is declining with time (Schubert *et al.*, 1979; Hsui, 1979), so that the above average heat flow values may be consistent with as little as 20 and 35 ppb, respectively. The larger of these two bulk Moon uranium abundances is significantly larger than terrestrial and chondritic values, supporting the view that the Moon is generally enriched in refractory elements (Kaula, 1977; Taylor, 1982, 1984). However, the lower of these abundances would not imply a significant lunar enrichment. Thus establishment of the true globally averaged heat-flow value would be a useful step toward the construction of more accurate lunar bulk composition models, which may ultimately distinguish the mode of lunar origin. Lateral heat-flow variations are also of interest in the context of a number of geological problems including the energy partitioning during impact cratering; measurements within and near comparatively recent impact craters would be useful in this regard. In addition it should be self-evident that the observed heat flow and inferred radioactive element abundances impose primary constraints on lunar thermal evolution models (*e.g.*, Toksöz *et al.*, 1978).

In the absence of new surface measurements, the best near-term method for determining the actual globally averaged heat flux involves orbital measurements of subsurface radio brightness temperature at a series of wavelengths near 10 cm (Minear *et al.*, 1977). Such a remote heat-flow instrument is being proposed for inclusion on the planned LGO mission (D. Muhleman, personal communication, 1985) and may result in a much improved assessment of lateral heat-flow variations and their dependence on physiography. Later surface measurements in different (non-mare) locales than were sampled during Apollo would provide ground-truth controls on the orbital data, thereby further increasing the accuracy of the global heat-flow determination.

GRAVITY/TOPOGRAPHY DATA

A primary application of combined gravity and topography data is the estimation of lateral crustal thickness and/or density variations, thereby allowing extrapolations from sparse seismic measurements. Resulting estimates for the volume and mass of the crust are important for constraining lunar bulk composition, depth of differentiation, and overall radial density models.

Precise measurements of the lunar gravitational field using observed accelerations of artificial satellites have thus far been limited by lack of direct line-of-sight Doppler tracking data on the farside and by an incomplete set of satellite orbital parameters (*e.g.*, Michael and Blackshear, 1972). Although harmonic analysis techniques have been developed to deduce approximate long-wavelength gravity fields over the entire Moon from nearside tracking data (Ferrari and Ananda, 1977; Bills and Ferrari, 1980), the available high-resolution coverage needed for detailed geophysical modeling studies of individual crustal structures is limited to low-latitude portions of the nearside. Similarly, lunar topographic data are absent for large regions of the farside and polar zones, so that global spherical harmonic models are of reduced accuracy and are limited to wavelengths representative of features of basin-scale or larger (Bills and Ferrari, 1977a).

The observed crustal gravity field is generally mild and is dominated by the mascon anomalies associated with major basins (Muller and Sjogren, 1968). Analyses of combined topographic and gravity data indicate that isostatic compensation is nearly complete for pre-Imbrium highland topography but is incomplete for post-Imbrium structures such as the Apennine Mountains (Ferrari *et al.*, 1978), indicating global cooling and thickening of the elastic lithosphere since about 4 Ae. From detailed studies of several nearside basins, including Grimaldi and Mare Serenitatis, over which high resolution measurements were obtained, it has been provisionally concluded that the mascons are most probably due mainly to rapid mantle rebound at the times of basin-forming impacts combined with a small additional mass excess due to later mare basalt filling (*e.g.*, Taylor, 1982). Assuming that lateral density variations are entirely the result of thickness variations of a nearly constant-density crust (Airy isostatic compensation), available gravity and topography data have been applied to show that (1) the average farside crustal thickness would be greater than that of the nearside and (2) crustal thickness would range from 30–35 km beneath the mascons to 90–110 km beneath the highlands (Bills and Ferrari, 1977b). However, it is also possible that a significant Pratt compensation component (lateral density variations in a constant thickness crust) is present, implying a somewhat different mean crustal thickness and density (Solomon, 1978). Either model would in principle explain the observed center-of-figure to center-of-mass offset. In order to establish the degree of Airy versus Pratt compensation, seismic crustal thickness determinations at a greater variety of surface locations are needed. Existing seismic data obtained near the Apollo 12 and 14 landing sites on Oceanus Procellarum indicate a thickness of 50–60 km (Nakamura *et al.*, 1974; Toksöz *et al.*, 1974) with possible evidence for a somewhat greater thickness beneath the Apollo 16 Descartes highlands site (Goins *et al.*, 1977). More than these three control points are needed. In addition, more complete high-resolution gravity and topography coverage would allow firmer conclusions about the completeness of isostatic compensation in different physiographic regions. More accurate estimates of crustal mean density, thickness, and lateral variations needed for bulk composition modeling would follow, as well as an improved understanding of the evolution of the early lunar lithosphere.

PALEOMAGNETIC DATA

Prior to the Apollo missions, the Moon was regarded as magnetically inert because previous flybys and orbiters had detected no global magnetic field and the low mean density required a small or non-existent core. Unexpectedly, both surface traverses and orbital surveys detected widespread crustal magnetic fields with scale sizes up to about 100 km, demonstrating a pervasive magnetization of lunar surface materials. If this magnetization is ascribable in whole or in part to the existence of a former core dynamo, by analogy with the terrestrial case, then the presence of a substantial metallic core would be required. Such an inference would be a valuable supplement to direct geophysical measurements, which are currently not definitive.

The dominant ferromagnetic carriers in lunar materials are metallic iron grains typically produced in impacts by reduction of pre-existing iron silicates. Repeated heating of returned samples necessary for Thellier-Thellier paleointensity determinations resulted in alteration of the magnetic properties of these grains, so that accurate paleointensities have been difficult to obtain. The few available Thellier-Thellier determinations, together with measurements by other more approximate paleointensity methods, indicate maximum lunar surface paleofields of the order of 1 Gauss during the 3.6–3.9 Ae period with much lower fields outside of this period (Cisowski et al., 1983). The latter authors conclude that the inferred high-field epoch is most directly interpreted in terms of a temporary core dynamo. On the other hand, significant paleointensities exist for some young samples; in particular, 70019, an Apollo 17 impact glass sample dated as < 100 m.y. old, has yielded a Thellier-Thellier paleointensity of 0.025 G, considerably larger than present-day surface magnetic fields (Sugiura et al., 1979). This fact, combined with the observation that many orbital anomalies are correlated with basin ejecta units and swirl-like albedo markings that may have impact-related origins (Hood et al., 1979), has led to suggestions that impacts themselves may generate transient fields that could have contributed substantially to the paleomagnetism (e.g., Hood and Vickery, 1984). However, other observations, including the correlation of one anomaly with Rima Sirsalis, an extensional graben-like feature south of Grimaldi (Anderson et al., 1977; Srnka et al., 1979), and evidence that surface field amplitudes on the nearside maria correlate with surface age (Lin, 1979) are supportive of the core dynamo hypothesis.

One means of evaluating the evidence for a former core dynamo involves the determination of approximate bulk directions of magnetization for major anomaly sources using orbital magnetometer data. Sources of comparable age should ideally be magnetized along field lines oriented in the shape of a dipole centered in the moon if a steady core dynamo was responsible for their magnetization. Unfortunately, the coverage of orbital magnetometer data is presently limited to narrow equatorial bands on the nearside plus a few farside areas (Hood et al., 1981) so that a total of only 28 sources with widely varying probable ages can be studied in this manner. Consequently, the directional evidence from these sources for a core dynamo is currently extremely difficult to evaluate, with different analysts drawing very different conclusions (Hood, 1981; Runcorn, 1983).

Two major steps can be taken to further understand the paleomagnetism and evaluate its implications for lunar internal evolution and history. First, a global vector magnetic anomaly map with sufficiently high resolution would lead to much more definitive evidence for or against the former existence of a large-scale lunar magnetic field. Determination of both anomaly correlations with surface geology and directional properties of the magnetization as described above would be possible in much greater detail than was the case using Apollo data. Such an anomaly map can be obtained by the planned LGO spacecraft if periapsis altitudes of < 50 km are achieved and are distributed reasonably well around the Moon. Second, the return of samples from at least some major anomaly sources, including oriented "bedrock" samples, preferably from dated mare basalt flows, would greatly facilitate the interpretation of both the orbital data and sample paleointensity determinations. Although automated landers may be capable of returning surface regolith samples, it is unlikely that the return of oriented specimens undisturbed by meteoroid impact and comminution will be achieved without manned landings. Thus a complete resolution of the paleomagnetism enigma may come only in the course of future manned exploration and utilization missions.

Acknowledgments. *Valuable reviews by S. C. Solomon and M. Ander are appreciated. Supported in part at the University of Arizona by grant NSG-7020 from the National Aeronautics and Space Administration.*

REFERENCES

Anderson K. A., Lin R. P., McGuire R. E., McCoy J. E., Russell C. T., and Coleman P. J., Jr. (1977) Linear magnetization feature associated with Rima Sirsalis. *Earth Planet. Sci. Lett., 34,* 141–151.

Bills B. G. and Ferrari A. J. (1977a) A harmonic analysis of lunar topography. *Icarus, 31,* 244–259.

Bills B. G. and Ferrari A. J. (1977b) A lunar density model consistent with topographic, gravitational, librational, and seismic data. *J. Geophys. Res., 82,* 1306–1314.

Bills B. G. and Ferrari A. J. (1980) A harmonic analysis of lunar gravity. *J. Geophys. Res., 85,* 1013–1025.

Cisowski S. M., Collinson D. W., Runcorn S. K., Stephenson A., and Fuller M. (1983) A review of lunar paleointensity data and implications for the origin of lunar magnetism. *Proc. Lunar Planet. Sci. Conf. 13th, in J. Geophys. Res., 88,* A691–A704.

Dyal P., Parkin C. W., and Daily W. D. (1976) Structure of the lunar interior from magnetic field measurements. *Proc. Lunar Planet. Sci. Conf. 7th,* pp. 3077–3095.

Ferrari A. J. and Anada M. P. (1977) Lunar gravity: A long-term Keplerian rate method. *J. Geophys. Res., 82,* 3085–3097.

Ferrari A. J., Nelson D. L., Sjogren W. L., and Phillips R. J. (1978) The isostatic state of the lunar Apennines and regional surroundings. *J. Geophys. Res., 83,* 2863–2871.

Ferrari A. J., Sinclair W. S., Sjogren W. L., Williams J. G., and Yoder C. F. (1980) Geophysical parameters of the earth-moon system. *J. Geophys. Res., 85,* 3939–3951.

Goins N. R., Dainty A. M., and Toksöz M. N. (1977) The deep seismic structure of the moon. *Proc. Lunar Planet. Sci. Conf. 8th,* pp. 471–486.

Goins N. R., Dainty A. M., and Toksöz M. N. (1981) Lunar seismology: The internal structure of the moon. *J. Geophys. Res., 86,* 5061–5074.

Goldstein B. E., Phillips R. J., and Russell C. T. (1976) Magnetic evidence concerning a lunar core. *Proc. Lunar Planet. Sci. Conf. 7th,* pp. 3321–3341.

Hobbs B. A., Hood L. L., Herbert F., and Sonett C. P. (1983) An upper bound on the radius of a highly electrically conducting lunar core. *Proc. Lunar Planet. Sci. Conf. 14th, in J. Geophys. Res., 88,* B97–B102.

Hood L. L. (1981) Sources of lunar magnetic anomalies and their bulk directions of magnetization: Additional evidence from Apollo orbital data. *Proc. Lunar Planet. Sci. 12B,* pp. 817–830.

Hood L. L., Coleman P. J., Jr., and Wilhelms D. E. (1979) The moon: Sources of the crustal magnetic anomalies. *Science, 204,* 53–57.

Hood L. L., Herbert F., and Sonett C. P. (1982a) The deep lunar electrical conductivity profile: Structural and thermal inferences. *J. Geophys. Res., 87* 5311–5326.

Hood L. L., Herbert F., and Sonett C. P. (1982b) Further efforts to limit lunar internal temperatures from electrical conductivity determinations. *Proc. Lunar Planet. Sci. Conf. 13th, in J. Geophys. Res., 87,* A109–A116.

Hood L. L. and Jones J. (1985) Lunar density models consistent with mantle seismic velocities and other geophysical constraints (abstract). In *Lunar and Planetary Science XVI,* pp. 360–361. Lunar and Planetary Institute, Houston.

Hood L. L., Russell C. T., and Coleman P. J., Jr. (1981) Contour maps of lunar remanent magnetic fields. *J. Geophys. Res., 86,* 1055–1069.

Hood L. L. and Vickery A. (1984) Magnetic field amplification and generation in hypervelocity meteoroid impacts with application to lunar paleomagnetism. *Proc. Lunar Planet. Sci. Conf. 15th*, in *J. Geophys. Res., 89*, C211–C223.

Hsui A. T. (1979) The bounds of the heat production rate within the moon. *Proc. Lunar Planet. Sci. Conf. 10th*, pp. 2393–2402.

Kaula W. M. (1977) On the origin of the moon, with emphasis on bulk composition. *Proc. Lunar Planet. Sci. Conf. 8th*, pp. 321–331.

Langseth M. G., Keihm S. J., and Peters K. (1976) Revised lunar heat-flow values. *Proc. Lunar Planet. Sci. Conf. 7th*, pp. 3143–3171.

Lin R. P. (1979) Constraints on the origins of lunar magnetism from electron reflection measurements of surface magnetic fields. *Phys. Earth Planet. Int., 20*, 271–280.

Michael W. H., Jr. and Blacksheat W. T., (1972) Recent results on the mass, gravitational field, and moments of inertia of the moon. *Moon, 3*, 388–402.

Minear J. W., Hubbard N., Johnson T. V., and Clarke V. C., Jr. (1977) *Mission Summary for Lunar Polar Orbiter*. Jet Propulsion Laboratory Document 660–41, California Institute of Technology, Pasadena, CA. 36 pp.

Morgan J. W., Hertogen J., and Anders E. (1978) The moon: Composition determined by nebular processes. *Moon and Planets, 18*, 465–478.

Muller P. M. and Sjogren W. L. (1968) Mascons: Lunar mass concentrations. *Science, 161*, 680–684.

Nakamura Y. (1983) Seismic velocity structure of the lunar mantle. *J. Geophys. Res., 88*, 677–686.

Nakamura Y., Latham D., Lammlein D., Ewing M., Duennebier F., and Dorman J. (1974) Deep lunar interior inferred from recent seismic data. *Geophys. Res. Lett., 1*, 137–140.

Nakamura Y., Latham G. V., Dorman H. J., Ibrahim A. K., Koyama J., and Horvath P. (1979) Shallow moonquakes: Depth, distribution, and implications as to the present state of the lunar interior. *Proc. Lunar Planet. Sci. Conf. 10th*, pp. 2299–2309.

Nakakmura Y., Latham G. V., and Dorman H. J. (1982) Apollo lunar seismic experiment—final summary. *Proc. Lunar Planet. Sci. Conf. 13th*, in *J. Geophys. Res., 87*, A117–A123.

Newsom H. E. (1984) The lunar core and the origin of the moon. *Eos, Trans. AGU, 65*, 369–370.

Randolph J. E. (1984) *Planetary Observer Planning FY 1984 Final Report*. Jet Propulsion Laboratory Document D-1846, California Institute of Technology, Pasadena. 313 pp.

Ringwood A. E. and Essene E. (1970) Petrogenesis of Apollo 11 basalts, internal constitution and origin of the moon. *Proc. Apollo 11 Lunar Sci. Conf.*, pp. 769–799.

Runcorn S. K. (1983) Lunar magnetism, polar displacements and primeval satellites in the earth-moon system. *Nature, 304*, 589–596.

Russell C. T., Coleman P. J., Jr., and Goldstein B. E. (1981) Measurements of the lunar induced magnetic moment in the geomagnetic tail: Evidence for a lunar core. *Proc. Lunar Planet. Sci. 12B*, pp. 831–836.

Schubert G., Cassen P., and Young R. E. (1979) Subsolidus convective cooling histories of terrestrial planets. *Icarus, 38*, 192–211.

Solomon S. C. (1978) The nature of isostasy on the moon: How big a Pratt-fall for Airy models? *Proc. Lunar Planet. Sci. Conf. 9th*, pp. 3499–3511.

Sonett C. P. (1982) Electromagnetic induction in the moon. *Rev. Geophys. Space Phys., 20*, 411–455.

Sonett C. P., Smith B. F., Colburn D. S., Schubert G., and Schwartz K. (1972) The induced magnetic field of the moon: Conductivity profiles and inferred temperature. *Proc. Lunar Planet. Sci. Conf. 3rd*, pp. 2309–2336.

Srnka L. J., Hoyt J. L., Harvey J. V. S., and McCoy J. E. (1979) A study of the Rima Sirsalis lunar magnetic anomaly. *Phys. Earth Planet. Int., 20*, 281–290.

Sugiura N., Wu Y. M., Strangway D. W., Pearce G. W., and Taylor L. A. (1979) A new magnetic paleointensity value for a young lunar glass. *Proc. Lunar Planet. Sci. Conf. 10th*, pp. 2189–2198.

Taylor S. R. (1982) *Planetary Science: A Lunar Perspective*. Lunar and Planetary Institute, Houston. 481 pp.

Taylor S. R. (1984) Tests of the lunar fission hypothesis (abstract). In *Papers Presented to the Conference on the Origin of the Moon*, p. 25. Lunar and Planetary Institute, Houston.

Toksöz M. N., Dainty A. M., Solomon S. C., and Anderson K. (1974) Structure of the moon. *Rev. Geophys. Space Phys., 12*, 539–567.

Toksöz M. N., Goins N. R., and Cheng C. H. (1977) Moonquakes: Mechanisms and relation to tidal stresses. *Science, 196,* 979–981.

Toksöz M. N., Hsui A. T., and Johnston D. H. (1978) Thermal evolutions of the terrestrial planets. *Moon and Planets, 18,* 281–320.

Warren P. H., and Rasmussen K. L. (1984) Megaregolith thickness, heat flow, and the bulk composition of the moon (abstract). In *Papers Presented to the Conference on the Origin of the Moon,* p. 18. Lunar and Planetary Institute, Houston.

Yoder C. F. (1981) The free librations of a dissipative moon. *Philos. Trans. Roy. Soc. London, Ser. A, 303,* 327–338.

GEOPHYSICS AND LUNAR RESOURCES

D. Strangway

Department of Geology, University of Toronto, Toronto, Ontario, Canada

The Apollo program provided a wealth of information about the nature of the Moon's upper 1–2 km. As a result, our knowledge of the physical properties of this material gives a sound basis for planning the geophysical aspects of a resource mapping and exploration program when a lunar base is established. Possible approaches to exploration include gravity, magnetic, electromagnetic, and seismic methods.

INTRODUCTION

The Apollo program taught us a great deal about the physical properties of the lunar surface. This information is of importance when considering the question of the role that geophysical methods will play in connection with the establishment of a lunar base and the resulting opportunity for exploration. This paper will not consider the question of establishing the global properties of the Moon and the nature of the deep interior, although that is an important topic in its own right. This paper considers only the question of more local and regional exploration, either for the purpose of geophysical mapping or for the assessment of resources. During the Apollo program we learned a great deal about gravity fields on the Moon and about magnetic fields and their ancient past. Seismic exploration methods were also employed at several sites and gave information on the nature of the regolith. The Moon is especially suited to the use of electromagnetic methods since, with the lack of any fluid, the Moon is quite transparent to most electromagnetic signals in a manner quite different from most terrestrial environments.

Arnold and Duke (1978) reviewed and laid out the types of resources that are expected to be of significance. The first of these was quite simply large quantities of material suitable to be used as collected for shielding materials, either on the lunar surface or for long duration missions for protection from the radiation environment. The second was rather more sophisticated and related to the use of large volumes of material for the extraction of specific materials such as titanium, aluminum, magnesium, silicon, or iron. The third was to consider the possibility of first detecting and then exploiting specific concentrations of unusual minerals or materials such as chromite, or ice in the polar region.

GEOPHYSICAL METHODS APPLIED TO LUNAR RESOURCE STUDIES

Magnetic

The magnetic method for mapping geophysical information and for the detection of magnetic ores of the various iron oxides or of the various sulphides is commonly used on Earth. It is now common to use airborne magnetic methods for mapping of very large areas. This is often the first step, for example, in establishing a reference for exploration programs in third-world countries. In general, the distribution of iron oxide in the form of magnetite (Fe_3O_4) and its solid solution series with titanium is mappable due to the magnetization induced by the Earth's magnetic field.

The Moon, on the other hand, has no modern-day magnetic field; furthermore, essentially no magnetite has been detected. This means that lunar magnetic mapping and exploration is entirely different in concept than terrestrial mapping. The lunar soils contain small particles of metallic iron or iron-nickel resulting largely from the reduction that takes place upon impact melting in the lunar vacuum (Strangway et al., 1977). Most soils have about 0.5% metallic iron. The source rocks, either basalts or anorthosites from which the soils were derived, also contain metallic iron but in much lower quantities (typically 0.1% for basalts). Since there is no ambient field, no magnetic anomalies arise from this source. The particles are, however, very small, and many of them are in particle sizes that are close to the superparamagnetic single-domain threshold (approximately 150 Å). This means that the phenomenon of magnetic relaxation that is a function of both grain size and iron content (Carnes et al., 1975) is very strong. Magnetic relaxation or viscosity refers to the property by which a magnetic material exposed to a magnetic field does not lose the magnetization as soon as the field is turned off; rather, it decays away over a period of time. It would be quite straightforward to develop an active magnetic mapping system on a traversing vehicle that could measure iron content and give some information on the grain-size distribution of metallic iron in the regolith. This information could be of use in selecting high-Fe soils or for the design of magnetic separation techniques.

Many observations have shown that the lunar samples carry some permanent magnetization. The source of the field that caused this magnetization is not well understood, but it could be due to an early lunar field with a value of a few thousand nanoTesla. Anomalous lunar magnetic fields have been detected from orbital magnetometers, from surface magnetometers (fixed stations and portable), and from an orbital electron reflection experiment. Significant volumes of the lunar outer layers carry this remanence and indicate that bombardment has not completely broken up and randomized the source layers. There is strong evidence that the greater anomalies are associated with breccias that are enriched in fine-grained iron just as the soils are (Strangway et al., 1975; Hood et al., 1978). Uniformly layered units do not give magnetic anomalies, but lateral variations do give rise to anomalies. Thus, magnetic mapping is particularly useful for detecting lateral changes and boundaries. It would be of extreme interest to obtain a complete satellite map of magnetic anomalies as part of a map of magnetic fields over large areas of the lunar surface. A lunar base, supported by a roving vehicle, could also provide much information on lunar magnetic history and be used to select sites for detailed study.

Electromagnetic Sounding

Electromagnetic sounding has been extensively used on the Earth for mapping near-surface properties as well as for sounding to considerable depth. Both ground and airborne systems have been useful in the search for conducting minerals—in particular for massive sulphide deposits. The Moon is known to be extremely dry and the electrical resistivities of lunar rocks and soils are extremely high, often as resistive as some of our best commercial insulators (Strangway *et al.*, 1977; Strangway and Olhoeft, 1977). As a result, electromagnetic energy propagates very readily into the Moon; it is possible to consider the concepts of reflection and refraction rather than the diffusion process usually considered on the Earth. Electromagnetic depth penetration can be considerable and can be accompanied by high resolution. The extreme resistivity of the surface layer means that the Moon is not amenable to the more conventional electrical resistivity methods that require current to be driven into the ground (see for example, Linlor, 1970).

A surface experiment confirmed the electrical properties of the surface at the Apollo 17 site (Strangway *et al.*, 1977). Frequencies from 1–32 Mhz were used to penetrate the surface and were recorded on the rover as it traversed away from a fixed transmitter. The results encountered were very similar to those found over ice sheets. Penetration was great at frequencies of 1, 2, and 4 Mhz, and reflections from depths of over 1 km could be detected. In addition, it was possible to measure the dielectric constant of the surface layers and its increase with depth. At the higher frequencies, the regolith proved to be a strongly scattering medium in which the contrast of dielectic constant between soil (approximately 3) and rock (approximately 7–8) caused many reflections. This experiment specifically confirmed that high-resolution mapping of the subsurface was possible with electromagnetic methods. A radar-like experiment used in a manner completely analogous to radar sounding of ice sheets would give much information on stratigraphy in the regolith. Such methods could be set up to function on a traversing vehicle. This would yield information on layering in the regolith, on the soil-rock interface, and on regolith thickness. It could also locate regions of high conductivity (radar-reflective) materials. In addition, the electrical losses are sensitive to the amount of ilmenite present. This approach would be useful in many aspects of assessing the character of the regolith, the distribution of ilmenite, or the presence of massive concentrations of metallic conducting phases such as chromite or other magmatic segregations.

Seismic

This paper does not review the need for a lunar-wide seismic observatory but does consider the use of seismic methods for assessing the character of the regolith. In the dry lunar vacuum, seismic propagation is very different than it is in normal terrestrial materials. The regolith is comprised of finely ground powder and pieces of breccia, basalt, and other igneous or metamorphic boulders with a wide range of sizes. At depths of a few meters to tens of meters there may be bedrock, as inferred, for example, from the walls of Hadley Rille at the Apollo 15 landing site. Nevertheless, these units are also heavily fractured. Since there is no fluid, the lithostatic load is supported entirely by particle-to-particle compression, and there is very little loss of seismic energy as it propagates through the rock. The net result is that the Moon has very high Q, and seismic energy is heavily scattered (Cooper *et al.*, 1974). This has meant that the active seismic experiments carried on several of the Apollo missions showed a straightforward velocity increase with depth following a fairly simple law (Johnson *et al.*, 1982). There is some argument about the precise nature of the relationship, but it is nevertheless consistent with a soil in which seismic velocity increases uniformly with depth. The large array used at the Apollo 17 site shows a sharp increase in velocity at a depth of 1.4 km that gives velocities typical of solid rock. There is a suggestion of a discontinuity at a depth of 32 m, perhaps corresponding to a stratigraphic horizon (ejecta blanket or lava flow) in the regolith.

It is clear that the application of the shallow seismic method on the lunar surface will be very different than on Earth. Nevertheless, mapping of the stratigraphy may be possible using a combination of shallow reflection and refraction techniques. Very small energy sources can give useful results, and one could imagine a rover-mounted impulse transmitter and receiving device.

Gravity

There has been only a small amount of gravity work done on the lunar surface, although orbital tracking has revealed the presence of large gravitational anomalies in the form of mascons (Muller and Sjogren, 1968). Because of the strength of the lunar crust, even very old features such as craters filled by dense mare basalts have not relaxed since they were formed nearly 4 billion years ago. Apollo 17 landed in a mare basalt-filled valley in the highlands of the Moon, and enough stations were measured to give a profile across the valley (Talwani *et al.*, 1973). The gravity values were high in the middle of the valley over the basalts and dropped by as much as 50 mGal as the highland massifs were approached. The massifs are rich in anorthositic highland materials of lower density. A simple model of basalts with a density contrast of 0.8 g/cm^3 and a thickness of 1 km explains these results adequately.

Thus, gravity methods could be used to assess the character of basaltic rocks and would of course be quite useful in exploration for local concentrations of heavy minerals that might be expected as products of magmatic segregation.

Discussion

The experience of the Apollo program has given information on the physical properties of regolith, basalt, and breccia in the lunar vacuum condition. In this environment, many physical properties are very different than those on Earth. The Apollo experiments using electromagnetic, magnetic, seismic, and gravity measurements give a clear idea of how these measurements can be carried out at a lunar base. These geophysical methods can and should play a role in the mapping that will be done in association with a lunar base as expeditions are mounted from this base. Geophysical methods can also play a role in assessing the character of the regolith materials for exploitation, especially if these methods are combined with geochemical techniques such as γ-ray and x-ray studies. Drilling also will be of considerable use for outlining stratigraphy between the holes and for confirming the geophysical interpretations. If there is a concentrated effort to search for ore deposits, many of these methods will be even more effective than they are on the Earth because of the Moon's more uniform background signals and the fact that electromagnetic energy can penetrate the dry surface much more readily than on Earth. The use of these methods for long traverses or for more local surveys seems clear, both for stratigraphic mapping and to search for and understand lateral changes. If sites are selected for deeper drilling, the use of geophysical methods in preliminary site surveys is essential.

Acknowledgments. The author's work is supported by the Natural Science and Engineering Research Council of Canada.

REFERENCES

Arnold J. B. and Duke M. B. (editors) (1978) *Summer Workshop on Near-Earth Resources.* NASA CP-2031. NASA, Washington, D.C. 107 pp.

Carnes J. G., Strangway D. W., and Gose W. A. (1975) Field-strength dependence of the viscous remanent magnetization in lunar samples. *Earth Planet. Sci. Lett., 26,* 1–7.

Cooper M. R., Kovach R. L., and Watkins J. S. (1974) Lunar near-surface structure. *Rev. Geophys. Space Phys., 12,* 291–308.

Hood L. L., Russell C. T., and Coleman P. J. (1978) The magnetization of the lunar crust as deduced from orbital surveys. *Proc. Lunar Planet. Sci. Conf. 9th,* pp. 3057–3078.

Johnson D. M., Frisillo A. L., Dorman J., Latham G. W., and Strangway D. W. (1982) Compressional wave velocities of a lunar regolith sample in a simulated lunar environment. *J. Geophys. Res., 87,* 1899–1902.

Linlor W. I. (1970) *Electromagnetic Exploration of the Moon.* Mono , Baltimore. 246 pp.

Muller P. M. and Sjogren W. L. (1968) Mascons lunar mass concentrations. *Science, 161,* 680–684.

Strangway D. W. and Olhoeft G. R. (1977) Electrical properties of planetary surfaces. *Philos. Trans. Roy. Soc. London, Ser. A, 285,* 441–450.

Strangway D. W., Rylaarsdam J. C., and Annan A. P. (1975) Magnetic anomalies near Van de Graaf Crater. *Proc. Lunar Sci. Conf. 5th,* pp. 2975–2984.

Strangway D. W., Pearce G. W., and Olhoeft G. R. (1977) Magnetic and dielectric properties of lunar samples. In *The Soviet American Conference on Cosmochemistry of the Moon and Planets,* (J. H. Pomeroy and N. J. Hubbard, eds.), Part 1, pp. 417–431. NASA SP-370, NASA Washington, D.C.

Talwani M., Thompson G., Dent B., Kahle H. G., and Buck S. (1973) Traverse gravimeter experiment, Apollo 17. In *Apollo 17 Preliminary Science Report,* 13-1 to 13-13. NASA SP-330, NASA, Washington, D.C.

SURFACE ELECTROMAGNETIC EXPLORATION GEOPHYSICS APPLIED TO THE MOON

Mark E. Ander

Geophysics Group, Los Alamos National Laboratory, Los Alamos, NM 87545

With the advent of a permanent lunar base, the desire to explore the lunar near-surface for both scientific and economic purposes will arise. Applications of surface exploration geophysical methods to the Earth's subsurface are highly developed. This paper briefly addresses some aspects of applying this technology to near-surface lunar exploration. It is noted that both the manner of application of some techniques, as well as their traditional hierarchy as assigned on Earth, should be altered for lunar exploration. In particular, electromagnetic techniques may replace seismic techniques as the primary tool for evaluating near-surface structure.

INTRODUCTION

Application of geophysical techniques to the Moon will provide information in two areas: scientific and economic. Scientific data will provide a more detailed understanding of the lunar interior, which, when coupled with geologic data, will give us better insight into the creation and evolution of the Moon. It can also aid in mapping of structure and lithology and provide information on the physical properties of rocks.

Exploration data will assist in prospecting for mineral concentrations and for water, an even more precious commodity. We are primarily interested in emphasizing surface exploration geophysics as applied to the upper few kilometers of the Moon's crust.

There are several important lunar characteristics that differ from terrestrial conditions and, therefore, alter the manner of application of geophysical methods, as well as their traditional hierarchy as assigned on the Earth. The absence of water precludes secondary alteration, mineral concentration, erosion, and deposition by hydrologic means. It also causes lunar materials to have an extremely high electrical resistivity. The only active geologic processes known to be occurring on the Moon at present are crater formation and subsequent regolith deposition via meteorite impacts.

Although some thin layering has been observed associated with ejecta blankets, continuous sedimentary beds as they occur on the Earth are unknown on the Moon. Other features affecting geophysical studies include the extensive impact structures, with attendant lithologies and shock metamorphism, and also lack of a core-generated magnetic field.

Standard geophysical exploration techniques employed on the surface of the Earth include gravity, magnetics, seismics, electrical, and electromagnetics. Although these techniques will be discussed separately, our experience on Earth has repeatedly proven the necessity for integrating data from as many geophysical and geological sources as possible to derive reasonable models and interpretations.

There are several other surface exploration geophysical techniques that will not be discussed in this paper because they are not widely used but have specific terrestrial applications, *e.g.*, radiometrics used in uranium exploration, and radon and other gas detectors used in geothermal exploration. Also, there are a wide variety of below-surface (well logging) and above-surface (remote sensing, potential field mapping, etc.) techniques that can be used either in planning or in conjunction with a surface exploration program, but these are not within the scope of this paper and will not be discussed here. Furthermore, electromagnetic techniques will be emphasized because of their expected importance in lunar near-surface exploration relative to other geophysical exploration techniques.

GRAVITY

Gravity is typically the first technique applied in terrestrial prospecting because it is relatively inexpensive and often provides important constraints on possible exploration targets. Gravity exploration could easily be undertaken on the surface of the Moon with little change in application other than that expected because the gravitational field of the Moon is approximately one sixth that of the Earth's field. Gravity exploration will probably play an important role in mapping gross structure in the subsurface, and it might be useful in the search for large buried fragments of meteoritic material.

Density information is needed for constraints on modeling or inversion of gravity data. This information is usually obtained from measuring specific gravity of rock samples obtained from the surface or from cores or cuttings obtained from drill holes. Density may also be inferred from gamma-gamma or seismic velocity drilling logs once these logs have been calibrated for lunar materials. Gravity data reduction to obtain residual Bouguer anomalies will be similar in principle to the procedure used on Earth. It will differ in practice only in that a lunar reference equipotential surface must be used rather than an Earth reference spheroid and that a "standard" density must be agreed upon for regolith material in order to perform a Bouguer correction. It will be necessary to obtain the lunar reference equipotential surface to as high a precision as possible using satellite data. Initial models of global lunar gravity have been obtained by fitting sixteenth-degree harmonic series to data collected from the Lunar Orbiter and Apollo missions (Wong *et al.*, 1971; Ferrari and Ananda, 1977; Bills and Ferrari, 1980). Lunar gravity modeling and inversion techniques will be identical to those applied on Earth.

MAGNETICS

Magnetics has traditionally been used as a major tool for detection of ore bodies on the Earth. Magnetic exploration on the Moon will be primarily affected by the lack of a large ambient lunar magnetic field such as that occurring on Earth. The small remanent field on the Moon generated by metallic iron disseminated in the lunar regolith would only represent noise to magnetic exploration. Strangway (1976) has already pointed out that, in order to use magnetics for exploration, an artificial magnetic source in the form of a large coil must be placed on the surface to induce magnetization in the rocks.

The addition of such an artificial magnetic source modifies the magnetic technique to such an extent that it essentially becomes a special case of the active inductive technique, which will be discussed later.

SEISMICS

Seismics has traditionally played the prominent role in terrestrial prospecting because of its extensive application in hydrocarbon exploration. Since it is not likely that petroleum fields exist on the Moon and because seismics is not typically useful for minerals exploration, it will probably not be very useful as a lunar minerals prospecting tool. Data from the Apollo Seismic Net (Toksöz et al., 1974) indicates that lunar seismograms are not at all like Earth seismograms. In particular, lunar seismograms have a small first arrival, the amplitude builds up slowly, the signal reverberates for an extremely long time, and the amplitude envelope has a very long rise time. The long reverberating seismic wave train is explained by strong scattering and very low velocity attenuation in the near-surface regolith. Seismic energy appears to propagate through the lunar regolith by anisotropic diffusion. Thus, lunar seismic waves are very hard to interpret. They have been used during the Apollo Project to glean information about gross lunar structure. In particular, they may be used (with difficulty) to estimate the combined thickness of the regolith plus megaregolith. Unfortunately, from an exploration standpoint, the very high rate of seismic energy dispersion that occurs in the regolith will very sharply decrease the resolution and hence the usefulness of seismic waves in discerning shallow, subsurface structure. Toksöz et al. (1974) studied the structure of lunar seismograms by comparing them to synthetic seismograms. The application of both synthetic seismograms and seismic three-dimensional physical model scaling is developing very rapidly and may help to unravel the very complex structure of future lunar exploration seismic data.

ELECTROMAGNETICS

Electromagnetics, of which electrical methods are a subset, will probably replace seismics as the dominant technique for exploring the lunar near-surface. Application of electromagnetic techniques will be quite different from that on the Earth because of the very high resistivity, which may vary from 10^6–10^{15} ohm/m (Strangway et al., 1972; Olhoeft and Strangway, 1975; Strangway and Olhoeft, 1977) for lunar material, as opposed to 10^{-1}–10^5 ohm/m for terrestrial rocks. These exceptionally high lunar resistivity values greatly increase the usefulness of electromagnetic methods on the Moon. In terrestrial applications these techniques are mainly used to detect differences in salinity and amounts of water in the host rock; whereas, for lunar rocks, electromagnetics can be used to derive values for actual physical properties and also to detect lateral and vertical variations in these properties that would be most useful from the standpoint of prospecting.

Because externally generated electromagnetic fields diffuse into the ground, their skin depth of penetration (S) and hence their depth of investigation is a function of frequency (ω) and subsurface electrical resistivity (ρ). For a homogeneous Earth half-space, S =

$(2\rho/\mu\omega)^{\frac{1}{2}}$, where μ is the relative magnetic permeability. A given electromagnetic frequency will penetrate much deeper on the Moon relative to the Earth because the Moon's near-surface resistivity is much higher. Therefore, techniques with much higher frequency responses (possibly as high as 10^7 Hz) will be needed to obtain data to within a few meters of the lunar surface. With an increase in frequency comes an increase in resolution. This will be another major advantage of applying electromagnetic techniques to lunar exploration, relative to terrestrial exploration.

Unfortunately, with an increase in resistivity comes a decrease in frequency at which displacement currents become dominant. The presence of displacement currents substantially complicates electromagnetic exploration theory. Above about 10^4Hz, they become important in the Earth. Consequently, other electromagnetic techniques with frequency responses above 10^4Hz are seldom used in exploration because they are harder to quantify. In addition, on the Earth non-displacement current techniques exist that can sample to within a few meters of the Earth's surface, making the use of higher frequencies unnecessary in most cases.

Electromagnetic techniques can be subdivided into two categories: active and passive. Active techniques involve artificially generated and induced fields, whereas passive techniques utilize naturally induced fields. The active techniques include direct current (DC) resistivity, induced polarization (IP), conductive source electromagnetic (EM) depth sounding, inductive source EM depth sounding, and displacement current techniques such as ground penetrating radar (GPR) and very low frequency (VLF) soundings. The passive techniques include the self-potential (SP) technique, telluric current technique, geomagnetic depth sounding (GDS), magnetotellurics (MT), and audio-magnetotellurics (AMT). Currently, there are no passive techniques used on the Earth that operate in the frequency range dominated by displacement currents. Such techniques will have to be developed for lunar operation. For a discussion of each technique and their Earth applications, see Telford et al. (1976).

The IP and SP techniques are directly tied to the presence of ground water and therefore will have no lunar application. The separation of the various passive techniques is somewhat artificial, primarily involving the frequency response of the various types of magnetometers used to measure the magnetic fields, and, hence, each technique has a different depth range of investigation, given a particular subsurface electrical resistivity structure: $<10^{-4}$Hz for GDS, 10^{-4} – 10^2Hz for MT, and 10^1 – 10^4Hz for AMT. Telluric profiling is a subset technique of magnetotellurics in which only natural electric fields are measured. On the Moon, displacement current techniques will become very important. In order to properly interpret resulting data, sophisticated modeling schemes will have to be developed.

The sources and nature of passively induced electromagnetic fields located at the surface of the Earth and Moon are quite different. The Earth-atmosphere-ionosphere establishes a spherical wave guide where the ionosphere forms a conductor, the atmosphere is the dielectric, and the Earth is a "leaky" conductor. The motion of charged particles in the ionosphere, due to their interaction with the solar wind, induces EM fields into the waveguide that then propagate around the Earth. When these EM fields impinge upon the Earth, a small fraction of their energy diffuses or "leaks" into the Earth. This ionospheric source decreases in energy with increasing frequency and is overpowered by a second EM source above about 10–100Hz. The second EM source is due to worldwide lightning strikes (called spherics) that feed energy into the waveguide and drive that part of the natural spectrum that is above about 10Hz.

On the Moon there is no atmosphere, no ionosphere, and hence no lightning. There is also no large main magnetic field. Therefore, there is no shield, buffer, or modifier between the surface and extralunar fields. Extralunar fields, which consist mainly of solar wind and the terrestrial magnetic field, are free to directly induce EM fields beneath the surface. Extensive modification of passive EM exploration theory is required in order to take advantage of the passive exploration techniques on the Moon. So far, theoretical studies have been aimed at the response of the Moon to long wavelength, time-dependent external fields with the intent of investigating the gross electrical resistivity structure of the Moon. These studies are quite complex and a great deal of theoretical work must be performed before natural EM fields at sufficiently high enough frequencies can be utilized for near-surface exploration. For discussion of this work see Dyal *et al.* (1974), Sonett (1982), and Hood *et al.* (1982).

Active electromagnetics can be divided into conductive, inductive, and displacement current techniques. Active inductive techniques use a closed wire loop to generate a time-varying magnetic dipole that induces an electric current in the subsurface. Induction methods are highly sensitive to the presence of conductors, which leads to the supposition that such techniques could play a dominant role in exploring for metallic mineral concentrations beneath the lunar surface. Active conductive techniques use current passing through a grounded wire to generate an electromagnetic field. The DC resistivity technique is a special case of the conductive EM method. The high sensitivity of conductive methods to resistive targets implies that these techniques may be very important in mapping near-surface structure because of the highly resistive lunar surface. The contact resistance of the electrodes with the surface will be extraordinarily high and may pose a severe problem. Extremely high voltages, possibly as high as a megavolt, will probably be required to overcome the problem. Traditional equipment used for conductive electromagnetic exploration must be redesigned. Realistic sources for the increased power might be Cockroft–Walton or Van De Graaff generators that are presently used in particle physics applications.

Although active displacement current techniques have not been important for terrestrial exploration, they will be very important for lunar exploration. The primary active displacement current surface exploration techniques are the VLF prospecting technique and GPR. The VLF technique uses as a source radio transmission for communications with submerged submarines and long-range radio positioning at frequencies of 3–40 kHz. Such a system does not seem practical for lunar exploration. GPR is a relatively new exploration technique that provides high resolution but poor terrestrial depth of penetration. This technique operates by transmitting an electromagnetic pulse with a 100-MHz center frequency into the ground and receiving energy reflected by any changes in dielectric permitivity (Olhoeft, 1979; 1984). GPR will most probably be a very useful near-surface lunar exploration tool.

CONCLUSION

Although gravity and seismics may provide a great deal of information on the gross structure of the Moon and, in addition to magnetics, may provide some limited data on the lunar near-surface, the expected efficiency of electromagnetic techniques suggests that substantial efforts should be directed toward extending our abilities in applying electromagnetics to lunar exploration.

REFERENCES

Bills B. G. and Ferrari A. J. (1980) A harmonic analysis of lunar gravity. *J. Geophys. Res.*, 85, 1013–1025.

Dyal P., Parkin C. W., and Daily W. D. (1974) Magnetism and the interior of the moon. *Rev. Geophys.*, 12, 568–591.

Ferrari A. J. and Ananda M. P. (1977) Lunar gravity: A long-term Keplerian rate method. *J. Geophys. Res.*, 82, 3085–3097.

Hood L. L., Herbert F., and Sonett C. P. (1982) The deep lunar electrical conductivity profile: Structural and thermal inferences. *J. Geophys. Res.*, 87, 5311–5326.

Olhoeft G. R. (1979) Impulse radar studies of near-surface geological structures (abstract). In *Lunar and Planetary Science X*, pp. 943–945. Lunar and Planetary Institute, Houston.

Olhoeft G. R. (1984) Applications and limitations of ground penetrating radar (abstract). In *Society of Exploration Geophysicists Annual Meeting Expanded Technical Program Abstracts with Biographies*, pp. 147–148. Society of Exploration Geophysicists, Tulsa.

Olhoeft G. R. and Strangway D. W. (1975) Dielectric properties of the first 100 meters of the moon. *Earth Planet. Sci. Lett.*, 24, 394–404.

Sonett C. P. (1982) Electromagnetic induction in the moon. *Rev. Geophys. Space Phys.*, 20, 411–455.

Strangway D. W. (1976) Future geophysical techniques for probing beneath the regolith - prospecting objectives (abstract). In *Lunar Science VII, Special Session Abstracts*, pp. 132–136. Lunar Science Institute, Houston.

Strangway D. W. and Olhoeft G. R. (1977) Electrical properties of the planetary surfaces. *Phil. Trans. Roy. Soc. London*, A285, 441–450.

Strangway D. W., Chapman W. B., Olhoeft G. R., and Carues J. (1972) Electrical properties of lunar soil: Dependence on frequency, temperature, and moisture. *Earth. Planet. Sci. Lett.*, 16, 275–281.

Telford W. M., Geldard L. P., Sheriff R. E., and Keys D. A. (1976) *Applied Geophysics*, Cambridge University Press. 860 pp.

Toksöz M. N., Dainty A. M., Solomon S. C., and Anderson K. R. (1974) Structure of the moon. *Rev. Geophys. Space Phys.*, 12, 539–567.

Wong L., Buechler G., Downs W., Sjogren W. L., Miller P. M., and Gottlieb P. (1971) A surface layer representation of the lunar gravity field. *J. Geophys. Res.*, 76, 6220–6236.

5 / SCIENCE ON THE MOON

T HE EXPEDITION OF LEWIS AND CLARK, the voyages of Captain Cook, and the lunar landings of Project Apollo are but a few examples of the symbiosis between scientific investigation and historic human probings of frontiers. Obviously, exploration and study of the Moon itself will be part of lunar surface activities, but what other scientific opportunities arise from permanent human presence? The Lunar Base Working Group, which met in Los Alamos in the spring of 1984, addressed this question in a general way. Consideration was given only to experiments uniquely enhanced in the lunar environment. Gradual expansion of facilities and capability will lead to various experiments that become easy to perform, but special installations or laboratories should be planned only when the research cannot be readily performed on the Earth or in easily accessible locations in space.

Broadly speaking, candidate experiments will be those capitalizing on the unique elements of the lunar environment. These include low gravity, absence of a planetary magnetic field, access to the plasma environment of the solar wind and the Earth's geomagnetic tail, no atmosphere, absence of water and other volatiles, isolation from the terrestrial biosphere, easily created low temperature radiative environments, availability of laboratory volumes with very high thermal and seismic stability, and the easily achievable pointing stability for observations of all kinds. As a rule, the lunar surface is a candidate location for any experiment that suffers interference from noise of geologic, biologic, or human origin.

Astronomical observations from the Moon are both promoted and criticized. Observers appreciate lack of atmospheric absorption but are quick to counter that satellite astronomy offers the same advantage. However, the slow rotation rate of the Moon permits very long integration times (literally days) from a very stable pointing platform. When a lunar base becomes a going concern, ease of maintenance and changeout of equipment will increase the productivity of an observatory.

LUNAR BASES AND SPACE ACTIVITIES OF THE 21ST CENTURY (1985)

The papers on astronomical concepts predict significant increases in resolution and sensitivity from lunar-based observations. Bernard Burke's optical interferometer represents a technical challenge that would be impossible without the positional stability of the elements and the freedom from atmospheric absorption and dispersion. Burns takes advantage of the lunar distance from the Earth to create an extremely long baseline interferometer of unprecedented resolving power. Douglas and Smith describe a simply constructed radio interferometer capable of probing the universe at long wavelengths unobservable from the surface of the Earth. Haymes can address a number of important astronomical questions with observations at the high energy end of the spectrum.

Cosmic-ray astronomy, discussed by Adams and Shapiro, not only samples products of physical processes in the galaxy but also provides data on the radiation background that limits the time astronauts can spend unshielded on the lunar surface. Shapiro and Silberberg review the types of neutrino sources that might be observed to advantage from the lunar surface. Cherry and Lande expand on the subject of the background interferences in lunar-based and terrestrial-based neutrino detection and describe what a lunar detector might look like. Petschek evaluates the possibilities of detecting proton decay or neutrino oscillations with a lunar based experiment.

Astronomy and astrophysics dominates the discussion in this section but in no way exhausts the possibilities. For example, Hörz speculates that dating of lunar craters might give perspectives on the proposed periodic bombardments of the Earth associated with mass extinction events. The paper by Anderson *et al.*, in the section on Space Transportation mentions techniques to measure the space plasma environment around the Moon with small, lunar-based sounding rockets.

A large body of science on the Moon can be viewed as applied research. Agriculture will be part of advanced life support systems. Medical and physiological data collection will be part of the study of the adaptation of humans to the lunar environment and will complement work done in space on the zero gravity environment. Microbial engineering will be part of the closed lunar ecosystem and, in addition, may have application to material processing as discussed by White and Hirsch in the section on Lunar Materials. Civil engineering and construction technology will receive attention in the course of lunar development.

In the final analysis, the scientific discoveries enabled by a lunar base are not predictable. Even a reasonable list of possibilities will be assembled only when a large number of minds with a broad spectrum of knowledge are made aware of the research opportunities. One purpose of this book is to stimulate the imagination of the scientific and technical communities.

ASTRONOMICAL INTERFEROMETRY ON THE MOON

Bernard F. Burke

Mail Code 105-24, California Institute of Technology, Pasadena, CA 91125. Permanent address: Room 26-335, Massachusetts Institute of Technology, Cambridge, MA 02139

Optical interferometric arrays are particularly attractive candidates for a manned lunar base. The radio model already exists: the Very Large Array (VLA) of the National Radio Astronomy Observatory, situated on the plains of St. Augustine near Socorro, New Mexico. A Y-shaped array of 27 antennas, each arm being 20 km long, operates as a coherent array, giving 0.1 arcsecond resolution at 2 cm wavelength. An array of similar concept, but with optical elements, would therefore give angular resolution of nearly one microarcsecond resolution at optical wavelengths and would give an absolutely revolutionary new view of objects in the universe. It would not be built on the Earth's surface because the atmosphere damages the phase coherence too severely at optical wavelengths. It could be constructed in Earth orbit as an assemblage of station-keeping free flyers (proposals to do so have been put forward) but the technical problems are not simple, *e.g.*, controlling element position and orientation to 100 A in 20 km. If a permanent lunar base were available, an optical analog of the VLA would, in contrast, be a relatively straightforward project.

THE CASE FOR HIGH ANGULAR RESOLUTION

Galileo's telescope was the first step in improving the angular resolving power of the human eye; this thrust in astronomy continues in our own time. The atmosphere of the Earth has posed a barrier at about one arcsecond (perhaps one-third of an arcsecond at the best sites), but if optical instruments can be mounted in space there seem to be few fundamental difficulties in extending to the microarcsecond range. Most of the problems are of a practical nature, centered on structural stability, satellite station-keeping, instrument adjustment and control, and related technical questions; these problems are solvable in principle, but may turn out to be costly if conventional orbital concepts are followed. Although the surface of the Moon has not been seriously considered in the past, it appears that astronomical instruments of great power could make good use of a lunar location. A permanently occupied lunar base could play a key role in such a program.

Angular resolution can never be better than the diffraction limit λ/D, the wavelength divided by the aperture diameter, and at 5000 Å, a one-meter aperture gives one-tenth arcsecond resolution. Milliarcsecond and microarcsecond resolution will require interferometers of large size, but much wider classes of problems, all of great current interest, become accessible. These are illustrated in Fig. 1, which shows the approximate optical fluxes and angular sizes of a variety of stellar and extragalactic objects. Since it is the maximum flux and largest angular size that is indicated, objects in each class will generally fall along the locus indicated by the upward sloping arrows. An object ten times more distant than the closest member of its class lies at the tip of the arrow,

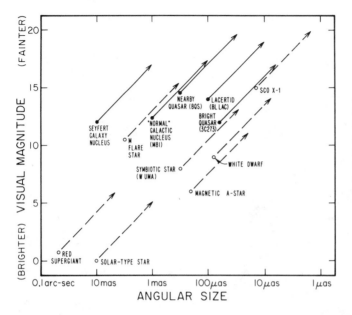

Figure 1. Magnitude and maximum angular sizes for a selection of stellar and extragalactic objects. The scales are chosen so that an object of a given class moves in the direction of the arrow, whose length corresponds to a factor of ten in distance.

for the given scale. The figure, therefore, gives the *largest* expected scale for each class of object.

For the various classes of stars, Dupree *et al.* (1984) have commented that measuring the size of a star is not enough, a conclusion that is generally valid for nearly all astronomical objects. Most interesting objects tend to be complex, and understanding the physical processes requires some detailed knowledge of the phenomena. For most stars, at least a factor of thirty resolution beyond the gross size is certainly needed (*i.e.*, about 100 pixels). Phenomena such as starspots, flares, and other analogs of solar processes will be interesting and, indeed, should be surprising. One is driven to the conclusion that every class of stellar object (except for the closest red supergiants) will demand an angular resolution of a milliarcsecond or better.

The extragalactic phenomena are still more demanding. The complexity of the processes is not known, since we do not have close analogs (such as the sun, for the stellar case) to guide us. The subject matter is of extraordinary interest, however: the physics of quasars, blacertids, and "ordinary" galactic nuclei press close to (or perhaps beyond) the limits of fundamental principles. It is clear that enormous energies are generated, both from radio and x-ray observations of these objects, and the indications are very strong that the energy source must be gravitational.

"Black holes," though not yet demonstrated in nature, may play a key role in these energetic processes. The optical study of the accretion processes and instabilities near the cores of the active extragalactic objects, with high angular resolution, should be as astounding as it has been in the radio case, where milliarcsecond resolution reveals velocities that appear to surpass the speed of light. Reference to Fig. 1 shows that only the broad-line regions at the nuclei of the closest Seyfert galaxies are accessible to an instrument of milliarcsecond resolution. The rest are smaller in angular size, and it is clear that an optical instrument having angular resolution in the 1–10 microarcsecond range would have truly extraordinary impact. None of the objects are brighter than the twelfth magnitude, and most are substantially fainter; an instrument having at least the collecting area of the Palomar 5-meter telescope is indicated. This challenge of obtaining angular resolution in the milliarcsecond to microarcsecond range, with a net collecting area of at least twenty to thirty square meters, is fully justified by the scientific rewards that would surely be gained.

APERTURE SYNTHESIS

Radio astronomers have, for the past several decades, circumvented the problem of obtaining high angular resolution by using interferometry, culminating in the concept that is called *aperture synthesis*. The methods were, ironically, developed by Michelson (1920) for measuring the diameters of stars at optical wavelengths, but the Earth's atmosphere hindered its quantitative use. The radio version of Michelson's stellar interferometer is illustrated in Fig. 2, which shows a pair of radio telescopes simultaneously receiving radiation from a distant source. There is a difference in arrival time, the geometrical time delay $\Delta\tau_g$, determined by the orientation of the source direction relative to the interferometer baseline. There is obviously no chance of interference if $\Delta\tau_g$ is larger than the coherence time t_c of the radiation, so a time delay must be inserted to compensate for this difference. Then, if the antennas are fixed and the source drifts through the reception

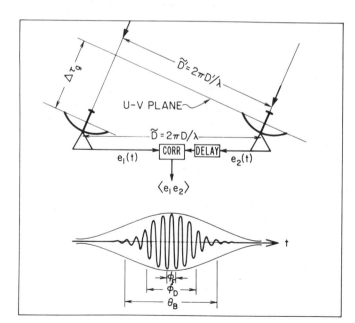

Figure 2. The Michelson stellar interferometer in its radio form. The output of the correlator, with the d.c. term removed, is shown for fixed apertures as a function of time; this is equivalent to variation with angle off-axis.

pattern, the product of the received signal amplitudes varies sinusoidally as the signals alternately interfere, constructively and destructively. These characteristic angular scales are important: the primary reception pattern of half-width θ_B, the fringe spacing ϕ_F, and the delay beam ϕ_D. The analysis is most straightforward if the antennas track the source, with the source itself being small compared to the primary beamwidth θ_B. The fringe spacing is determined by the projected baseline D′, which is the projection normal to the incoming radiation.

For the interferometer description, there is a third angle, the delay beam ϕ_D, that is determined by the receiving bandwidth or, equivalently, by the coherence time. If the time delay is set to match $\Delta\tau_g$ perfectly, the central fringe will have full amplitude, but as the time delay error grows, the interference conditions will be different at the upper and lower ends of the band. The interference effects cancel, and the fringe amplitude diminishes over an angle $\phi_D \sim 1/B\tau_B$, where B is the bandwidth and τ_B is the baseline length measured in light travel time. The number of fringes observed as a consequence is of the order of the inverse of the fractional bandwidth, an effect that has strong consequences for optical interferometry.

Given a two-element Michelson interferometer as illustrated in Fig. 2, the output is well-specified if the following conditions are met: the source under study must be small compared to both the primary resolution θ_B and the delay beam θ_D, and the delay compensation must approximate $\Delta\tau_g$ with an accuracy corresponding to a fraction of the fringe angle θ_F, or at least the error must be calibrated to that accuracy. The interferometer output is the convolution of its sinusoidal fringe pattern with the source brightness B(x,y) where x,y are angular coordinates on the sky. This means that the interferometer output is equal to the *Fourier transform* B(u,v) of the brightness distribution. The conjugate coordinates (u,v) are defined by the baseline and the source location as shown in Fig. 2: on a plane normal to the source direction, coordinates (u,v) are defined (North and East, for example) and the interferometer baseline D, measured in wave numbers $(2 \times D/\lambda)$, is projected onto that plane with the reference antenna (which can be chosen arbitrarily) at the coordinate origin. The plane is called the *u-v plane*, and the projected vector D′(u,v) defines the conjugate coordinates at which the Fourier transform B(u,v) is defined by the fringe amplitude and phase. If all interferometer baseline lengths and orientations are taken, the complete Fourier transform is determined, and performing Fourier inversion gives a true map B(x,y) of the source. In practice, of course, there is noise introduced by the apparatus; the coverage of the u-v plane is not complete; and due caution and knowledge must be exercised.

The process by which the Fourier transform is developed is known as *aperture synthesis*, and substantial literature has been developed for the radio case. The first complete description, in which the rotation of the Earth was used to move the interferometer baseline, was conceived by Ryle and Hewish (1960), and an authoritative summary of the two-element interferometer has been given by Rogers (1976). The most powerful aperture synthesis instrument is the radio array known as the VLA (the Very Large Array, operated by the National Radio Astronomy Observatory); it is described by Napier *et al.* (1983). The VLA probably provides the best model for a desirable optical instrument. Its 27 elements give 351 simultaneous baselines; this means that "snapshots" of fairly complex objects are nevertheless faithful representations if the target is not too complex, or if a dynamic range of a few hundred to one is sufficient. At the same time, for large fields of view and complex targets, its variable configuration and ability to use the rotation of the Earth to obtain more complete u-v plane coverage is vital. The size of the array, 20 km per arm of 35 km equivalent overall size, was set by the original operating requirement that it should equal conventional optical telescope resolution (1" at 20 cm, 0.3" at 6 cm). The same considerations will apply to an equivalent optical instrument. The discussion in the beginning of this paper, illustrated by Fig. 1, indicates that a mapping capability of ten microarcseconds would give a rich scientific return. At this angular scale, significant changes can be expected both for stars and active extragalactic objects within brief time spans. The system must therefore have a large number of elements, as in the case of the VLA. This gives two further advantages: a large number of objects can be studied in a short time because of the "snapshot" capability, and the more complete u-v plane coverage can yield maps of high dynamic range. If the optical array contains 27 elements, each element would have to have at least one m diameter to give a total collecting area comparable to the five m Palomar telescope. The instrument should cover the wavelength range 1216 Å (lyman-alpha) to 5 microns; for the mean wavelength of 5000 Å, this implies that an optical aperture-synthesis array should have a diameter of about 10 km.

One of the major considerations of any concept has to be the phase stability of the system. Incoherent and semicoherent interferometers (the Brown-Twiss interferometer is a brilliant example) suffer in signal-to-noise ratio and loss of phase information, and so must be rejected. For the complex objects of greatest interest, phase information is essential. This requirement exacts a price: control (or measurement) of the optical paths to $\lambda/20$ means that 250 Å precision is needed at $\lambda5000$, and proportionally tighter specifications are required as one goes to shorter wavelengths. The radio astronomers, in developing VLBI, have formulated a powerful algorithm, phase, and amplitude closure that eases the problem if there are enough receiving apertures. The technique has been applied to VLBI mapping problems with great success (Readhead and Wilkinson, 1978). If one has three elements, and hence three baselines, the instrumental phase shifts sum to zero; similarly, if there are four elements in an array, the instrumental perturbations to the amplitudes cancel. As the number of elements increase, the information recovery becomes more and more complete. For N antennas, a fraction $(N-2)/N$ of the phase information and $(N-3)/N-1$ of the amplitude information can be recovered. If N is 10 or more, the procedure appears to be thoroughly reliable. The phases must still be stable over the integration period; this means that the precision requirement on the optical paths must be held, but the *time* for which it is held is reduced. The desired sensitivity and the total collecting area therefore set the final stability specifications.

Up to the present time, two general classes of optical space interferometers have been proposed: station-keeping, independently orbiting interferometers, and structually mounted arrays. Examples of the first class are SAMSI (Stachnik *et al.*, 1984), in which pairs of telescopes are placed in near-Earth orbit, and TRIO (Labeyrie *et al.*, 1984), in which a set of telescopes are maneuvered about the fifth Lagrangian point in the Earth-Moon system. Among the structural arrays that have been proposed are COSMIC (Traub and Carleton, 1984), OASIS, a concept proposed by Noordam, Atherton, and Greenaway (unpublished data), and a variety of follow-on concepts to the Space Telescope being examined by Bunner (unpublished data). At the present time, all of these concepts hold promise for giving useful results in the milliarcsecond class, but when the number of elements grows to the order of 27 (or more) and when the spacings extend to 10 km (or even 100 km, for one microarcsecond resolution at $\lambda5000$) the solutions may prove to be expensive, perhaps prohibitively so.

A third class of optical array becomes feasible, however, if there is a permanently occupied lunar base. The Moon turns out to be a most attractive possible location for an optical equivalent of the VLA, capable of microarcsecond resolution.

A LUNAR VLA

Assuming that a lunar base has been established, the general outlines of a large optical array following the pattern of the VLA can be visualized with some confidence.

Figure 3. A schematic view of an optical aperture synthesis array on the Moon. The individual elements could assume forms very different from the versions shown.

A schematic form is shown in Fig. 3; a set of telescopes, suitably shielded, are deployed at fixed stations along a Y, each arm being 6 km long, for a maximum baseline length of 10 km. There is a fixed station that monitors the telescope locations by laser interferometers. The telescopes must be movable, but whether they are self-propelled (as shown in Fig. 3) or are moved by special transporters (as in the case of the VLA) is a technical detail. The received light signals are also transmitted to the central correlation station, but time delays must be inserted to equalize the geometrical time delays ($\Delta\tau_g$) illustrated in Fig. 2. These are not shown; a number of configurations are possible, probably in the form of laser-monitored moving mirrors.

The individual telescopes might well be approximately one m in diameter. The telescopes could be transported in disassembled form and, hence, they need not be extremely expensive since launch stress would not be a problem. A simple conceptual design indicates that each telescope might have a mass of 250 kg or less; the total telescope mass would then be about 7 tonnes for 27 telescopes plus a spare. The packing volume could be relatively small, since the parts would nest efficiently. The sketch in Fig. 3 shows each telescope being self-propelled, but if mass transportation to the Moon is a key consideration, one or two special-purpose transporters seems much more likely. Each might have a mass of about 200 kg.

The shielding of the telescopes is an interesting design problem. The simplest scheme would be to adopt the systems used on past telescopes in space, such as the International Ultraviolet Explorer (IUE), but the construction possibilities on the lunar surface may allow concepts that give dramatic improvements. Instead of mounting the shields on the telescopes themselves, the shields could be constructed as independent structures that sit on the lunar surface, free of the telescope itself. The shields might be very simple, low-tolerance, foil and foam baffles, keeping the telescope forever in the shade, radiatively cooled to a very low temperature, or perhaps kept at the average 200 K temperature of the lunar subsurface. It would appear that the thermal stresses might be kept very low by adapting the design to the lunar surface conditions.

Transmission of the received light from the telescopes to the central correlation station must proceed through a set of variable time delays as indicated earlier, and here there is a need for technical studies. For the 10 km maximum baselines proposed here, the maximum time delay rate would be 2.6 cm/s, which is not excessively high. The requirement of $\lambda/20$ phase stability is challenging: the motion should not have a jitter much greater than 100 Å/s rms, so a smoothness of something better than a part per million is needed; not an easy goal, but not beyond reason. The curvature of the lunar surface has to be taken into account unless a convenient crater can be found whose floor is suitably shaped. The height of the lunar bulge along a 6-km chord is 1.5 m and, hence, is not a serious obstacle. For the larger concept (60-km baseline, microarcsecond resolution at $\lambda5000$) the intervening rise of 150 m would be more serious, and suitable refraction wedges or equivalent devices would have to be arrayed along the optical path. The transmitted signal should probably be a quasi-plane wave; this translates to the requirement that the receiving aperture at the central correlator station should still be in the near field of the transmitting aperture of the most distant telescope. This specifies the diameter of the transmitted beam, which must have a diameter greater than 10 cm at $\lambda5000$, and 30 cm for 5 μ. If there were a desire to carry out aperture synthesis at 50 m (which there might well be), the transmitted beam would have to be at least a meter in diameter, a requirement that would still be easy to meet, since the tolerances would be relaxed.

The characteristics of the central correlator will depend on the results of detailed studies. Two general classes of optical systems can be projected: the "image plane" correlation geometry developed by Labeyrie (1984) for TRIO (a continuation of the traditional technique of Michelson), and the "pupil plane" correlation scheme generally used by radio astronomers, but realized in the optical regime by the astrometric interferometer of Shao et al. (1984).

One interesting advantage generally enjoyed by optical interferometry as compared to radio interferometry is the ease with which multi-banding circumvents the "delay beam" problem described earlier. Labeyrie (1980) has devised an ingenious dispersive system that efficiently eliminates the problem for most cases. The fringes are displayed in delay space and frequency space, but modern two-dimensional detectors such as CCDs (Charged Coupled Devices) handled the increased data rate easily.

The data rates are not excessive, being completely comparable to the data rates now being handled by the VLA. The 351 cross-correlations needed for a 27-element system (or 1404 if all Stokes parameters are derived) requires an average data rate of about 100 kilobaud for a 10-second integration period; future systems always require larger data rates, but even a projection of an order-of-magnitude increase does not seem to present formidable data transmission problems.

Finally, a word is in order concerning the use of heterodyne systems to convert the optical signals to lower frequencies. The technique is in general use in the radio spectrum, extending to wavelengths as short as a millimeter. Unhappily, the laws of physics offer no hope for astronomical use of heterodyne techniques at optical and ultraviolet frequencies. Every amplifier produces quantum noise, and the laws of quantum mechanics are inexorable: approximately one spurious photon per second per Hertz of bandwidth is produced by every amplifier. At radio frequencies, the quantum noise is swamped by the incoming signals since there is so little energy per quantum. Optical systems, with bandwidths of 10^{13} or 10^{14} Hertz, can afford no such luxury. The crossover in technology occurs somewhere between 100 and 10v. As infrared detectors improve, the shortest wavelength at which heterodyne detectors are practicable will be perhaps 50μ.

Except for these quantum limitations, the concepts developed for radio techniques carry over to the optical domain. The signal-to-noise analysis differs somewhat. The noise limits are determined by the Rayleigh noise of the system in the radio case, while the quantum shot noise of the signal itself determines the signal-to-noise ratio in an optical system. Otherwise, the extensive software armory developed for radio synthesis systems should be directly applicable to optical interferometers.

ARE THERE SERIOUS OBSTACLES?

Relatively little thought appears to have been given thus far to the advantages of the Moon as a base for astronomical instruments. There are a number of current misconceptions that seem to hold little substance.

1. *Does lunar gravity cause problems?* On the whole, the effects of lunar gravity appear to be beneficial. The relatively small (1/6 g) acceleration helps to seat bearings, locate contact points, and generally should provide a reference vector for mechanical systems. The lunar gravity removes dust from above the surface, keeping the density of light-scattering particles low.

Gravitational deflection for telescopes in the one-meter size range is completely negligible. Gravitational deflection does not depend upon the weight of a structure; elementary physics shows that the structural deflection s of a structure depends on the length l of the beam, Young's modulus Y, the density ρ, the gravitational acceleration g_m, and a dimensionless geometrical factor γ that decreases as the depth of the beam increases:

$$s \approx \gamma \left(\frac{\rho}{Y}\right) g_m \, l^2 \qquad (1)$$

On Earth, 4- and 5-meter telescopes have been built with mirror support systems that limit mirror deflection to a fraction of a wavelength of light under full gravity. A 1-meter mirror, located on the Moon but otherwise similar, would be stiffer than a terrestrial 4-meter mirror by a factor of about 100!

Deflection of the telescope structure can be controlled to high tolerances. Not only are superior materials like carbon-epoxy now available, but there are improved design methods such as the concept of homologous design (introduced by von Hoerner in 1978), in which a structure is designed that always deforms to a similar shape. In summary, gravitational deflection poses no problem.

2. *What about the thermal environment?* The Moon is an approximately 200 K blackbody subtending 2 π steradians on the underside of a lunar-based instrument. For a conventional satellite in low-Earth orbit (LEO), the Earth is an approximately 200 K blackbody subtending nearly 2π beneath the spacecraft; however, if the spacecraft is tracking a celestial object, the aspect is changing rapidly—on the order of 4° per minute. The telescope tracking a celestial source in the lunar environment is changing its aspect at about 0.01° per minute. When one considers the additional advantage of the natural lunar terrain for better thermal shielding to start with, and the ability to upgrade its quality at a permanent base, the lunar environment is almost certainly more favorable than LEO from the point of view of thermal stresses. The L5 case is different, since the elements would always be exposed to direct solar radiation.

3. *Is scattered light a problem?* Again, equipment in LEO has the Earth subtending nearly a hemisphere, but the Earth has high albedo and the Moon has low albedo. The lunar environment is strongly favored, and, as in the thermal case, one should be able to provide superior light shielding on the Moon.

4. *Is direct sunlight a problem?* The sun shines only half the time, and its direction changes slowly. Given the superior light baffling of the lunar-based telescopes, the lunar environment will probably turn out to be far superior to either LEO or L5, but thermal studies of real designs should be made.

5. *What about lunar dust?* The laser retro-reflectors have been in service for over a decade, with little performance degradation reported. Dust seems to be no problem,

probably because the Moon's gravity scavenges it rapidly. A very rare meteorite impact nearby might take one or two telescopes out of service, and the choice would have to be made to clean or replace the instruments.

6. *Is seismic activity a problem?* The Moon is far quieter than the Earth, with a low Q. At good seismic stations on the Earth, the seismic noise is less than one Å rms; the poor locations have high noise because of the effects of wind and surf. Lunar seismic activity is not a concern.

7. *Do the solid body tides of the Moon move the baselines too much?* Earth tides are routinely accommodated by geodesy groups conducting VLBI studies on Earth, where the motions amount to several wavelengths every 12 hours. The lunar tides are larger in amplitude, but they proceed so slowly that they can be compensated for. The 10 km maximum baseline of a lunar VLA is a smaller fraction of the lunar diameter than the 10,000 km VLB baselines are of the Earth's diameter, which diminishes the amplitude of baseline motion. The net tidal motion of the maximum baseline vector should be of the order of a few tenths of a millimeter. This is not a negligible motion, measured in wavelengths of light, but the slow lunar rotation leads to a manageable correction rate of the order of a few wavelengths per hour. The usual interferometric calibration routines should keep this error source under control.

8. *Can the baseline reference system be well defined?* The analogy with terrestrial VLBI is so close that the answer has to be affirmative. The errors can be controlled; the lunar soil is sufficiently competent to stably bear the load of a telescope; and, if necessary, hard points can be established to check on vertical motions. Interferometers are largely self-calibrating: there are enough quasi-stable reference points in the sky to allow the observations themselves to bootstrap the instrumental constants.

SUMMARY

A permanent lunar base can provide support for a variety of astronomical investigations. An optical interferometric array, perhaps of the general form of the VLA but designed for optical instead of radio wavelengths, would lead to a qualitative advance in our understanding of the universe. The Y configuration is well suited to expansion, and the VLA has demonstrated that it can make maps both rapidly (in its snapshot mode) and with high dynamic range (when multiple array configurations are used). Other configurations, such as maximum–entropy–derived circles, should certainly be examined.

A wide variety of scientific problems could be addressed by such an instrument. The stellar analogs of the solar cycle, the behavior of sunspots on other stars, the magnetic field configurations of other stars, and the behavior of dynamic plasma phenomena such as flares and winds, are all examples of star-related problems that ultimately would lead to both fundamental knowledge of how stars formed and evolve, and increase understanding of our own sun. A wide variety of extragalactic problems could be studied, including the fundamental processes associated with black holes and massive condensed objects as they are manifest in quasars, galactic nuclei, and other optically violent variables. There would surely be a number of dramatic surprises, both in stellar and extragalactic studies, and the instrument would certainly be at the forefront of astronomy from the time of its first use.

There seem to be no fundamental problems in building such an instrument. The total mass to be delivered to the lunar surface for the instrument would be 10–30 tonnes, roughly equivalent to one space station habitat module. The detailed system studies have not yet been made, but even a preliminary conceptual investigation indicates that the elements of the system are relatively straightforward. The presence of man is highly desirable for this particular instrument, in marked contrast to the free-flyer case in which the instruments are too easily perturbed by human presence.

How long would it take to build the instrument? The answer depends upon the timescale of development for a lunar base. Once a clear consensus exists to establish a base on the Moon, development of the components of a lunar VLA could be started and would be ready to be among the first large shipments of non-life-support systems to the Moon. Assembly and development time at the lunar base would depend on the details of the design and on the philosophy of lunar base operations.

Finally, it is clear that a large astronomical community would use the instrument. All the major astronomical facilities on Earth are heavily subscribed, and the VLA probably supports more users than any other astronomical instrument today. An interferometric array has many possible modes of operation: it can take brief snapshots; it can be broken into subarrays to serve multiple user groups simultaneously for specialized projects; and it can interweave long observing sequences with short projects in an efficient fashion. The VLA supports the observing programs of over a thousand scientists per year, and a lunar-based optical equivalent could be expected to do the same.

REFERENCES

Dupree A. K., Baliunas S. L., and Guinan E. F. (1984) Stars, atmospheres, and shells: Potential for high resolution imaging. *Bull. Am. Astron. Soc., 16,* 797.

Labeyrie A. (1980) Interferometry with arrays of large-aperture ground-based telescopes. *Proc. Optical and Infrared Telescopes for the 1990s* (A. Hewitt, ed.), p. 786. Kitt Peak National Observatory, Tucson.

Labeyrie A., Authier B., Boit J. L., DeGraauw T., Kibblewhite E., Koechlin L., Rabout P., and Weigelt G. (1984) TRIO: A kilometric optical array controlled by solar sails. *Bull. Am. Astron. Soc., 16,* 828.

Michelson A. A. (1920) An interferometer for measuring stellar diameter. *Astrophys. J., 51,* 257.

Napier P. J., Thompson A. R., and Ekers R. D. (1983) The VLA: A large aperture synthesis interferometer. *IEEE Proc., 71,* 1295.

Readhead A. C. S. and Wilkinson P. N. (1978) Phase closure in VLBI. *Astrophys. J., 223,* 25.

Rogers A. E. E. (1976) The two-element interferometer. *Methods of Experimental Physics, Vol. 12C: Astrophysics, Radio Observations* (M. L. Meed, ed.), p. 139.

Ryle M. and Hewish A. (1960) The synthesis of a large radio telescope. *MNRAS, 120,* 220–230.

Shao M., Colavita M., Staelin D., and Johnston K. (1984) The technology requirements for a small space-based astrometric interferometer. *Bull. Am. Astron. Soc., 16,* 750.

Stachnik R. V., Ashlin K., and Hamilton S. (1984) Space station SAMSI: A spacecraft array for Michelson spatial interferometry. *Bull. Am. Astron. Soc., 16,* 818.

Traub W. A. and Carleton N. P. (1984) COSMIC: A high-resolution large collecting area telescope. *Bull Am. Astron. Soc., 16,* 810.

A MOON-EARTH RADIO INTERFEROMETER

Jack O. Burns

Department of Physics and Astronomy, University of New Mexico, Albuquerque, NM 87131

In this paper, the logistical considerations and astronomical applications of placing a radio antenna(s) on the Moon as one element of a Moon-Earth Radio Interferometer (MERI) are described. Diffraction, interstellar scintillation, and Compton scattering are considered as processes that will influence the resolution of the interferometer. Compact sources with flux density less than 30 milliJanskys (mJy) can be observed at the optimum resolution of <30 microarcsec for wavelengths <6 cm. With such a resolution, one could perform fundamental astrometry experiments leading to a much improved value for the Hubble constant, or possibly map active regions on other stars and investigate with unprecedented linear resolution the nature of the "engine" at the center of the Milky Way and in active galaxies.

INTRODUCTION

The technique of radio interferometry and, in particular, Earth-Rotation Aperture Synthesis has proven to be enormously successful for ground-based radio astronomy. Operating radio interferometers include the MERLIN and 5-km Cambridge arrays in England, and the Westerbork Synthesis Radio telescope in the Netherlands. The most sophisticated aperture synthesis telescope is the Very Large Array (VLA) located in west-central New Mexico. It is composed of 27 individual antennas arranged in a Y-configuration (*e.g.*, Napier *et al.*, 1983). Each pair of radio antennas samples a particular Fourier component of the radio source brightness distribution at a given instant in time. As illustrated in Fig. 1, the turning of the Earth on its axis effectively synthesizes an aperture with resolution comparable to a single antenna with diameter equal to the maximum baseline between the outermost dishes. This is obviously a cheaper and more practical method for achieving high resolution mapping of extraterrestrial radio sources. The Fourier components gathered during a typical 12-hour integration are Fourier inverted with a computer algorithm (typically a gridded FFT) to produce a map of the sky brightness distribution. The accuracy of this map will depend upon the density of points in the Fourier transform plane (*i.e.*, how well the aperture was synthesized with the available antennas and the length of the integration). Typical maximum resolution for a high declination radio source using the VLA is 0.3 arcsec at 6 cm (35-km baseline) and the dynamic range (peak signal to RMS noise) can be thousands to one.

This technique has been extended to even longer baselines, termed Very Long Baseline Interferometry (VLBI), with individual antennas separated by entire continents. Data are recorded at each antenna on high speed videotape with accurate time markers as determined by hydrogen maser clocks. All the tapes are later brought together at a central computer processor and the data are correlated. Problems arise with the stability of the correlated phase due to differences in tropospheric and ionospheric refraction over the

APERTURE SYNTHESIS

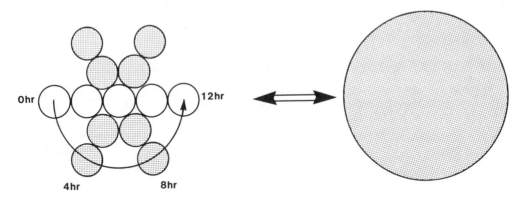

Figure 1. The principle behind Earth rotation aperture synthesis in radio interferometry. Imagine that an observer is stationed above the North Pole of the Earth looking down upon a linear alignment of five antennas. As the Earth rotates, the line sweeps out portions of a filled aperture. In 12 hours, the line has synthesized a circular aperture with diameter equal to the maximum baseline between the outermost antennas. This is equivalent to observing a radio source with a single very large antenna.

different telescope sites, and due to fluctuations in the local oscillator clocks. However, using closure phase and hybrid mapping techniques, it is now possible to recover an accurate map of the source brightness distribution and the relative positioning of radio features (*e.g.*, Pearson and Readhead, 1984). The accuracy and dynamic range depends upon the number and uniformity of antennas in the VLBI network. The proposed Very Long Baseline Array (VLBA), to be operated by the National Radio Astronomy Observatory, will consist of ten identical 25-m dishes located between Hawaii and Puerto Rico, and centered in New Mexico at the VLA. The resolution is expected to be at the submilliarcsecond level at centimeter wavelengths.

The maximum baseline for a ground-based VLBI is obviously limited to the diameter of the Earth. Also, both the European VLBI network and the American VLBA are oriented in a predominantly East-West direction, producing poor sampling of Fourier components in a North-South direction and poor sampling for southerly declination sources. An additional VLBI antenna in space could both increase the resolution and the image restoration accuracy when linked to a ground-based array. Recently, a European-American collaboration has developed a proposal for such a space-based antenna, called QUASAT (quasar satellite), which would have an elliptical orbit with a semi-major axis of 10,000 km and inclined 45° with respect to the equator (Schilizzi *et al.*, 1984). The resolution is expected to be about 350 microarcsec at 6 cm wavelength.

In principle, there is no reason why this VLBI technique could not be applied to baselines ranging between the Earth and the Moon. The major difference would be the observational technique used to synthesize the aperture. For a Moon-Earth Radio

Interferometer (MERI), the Moon's revolution around the Earth would provide the mechanism for the synthesis. Observations would be scattered over two-week intervals rather than the typical 12-hour continuous integrations that are now performed with the VLA and for VLBI. A first-generation antenna on the Moon could be quite simple and inexpensive, unfolding like a flower or an umbrella from the cargo bay of a Moon shuttle craft. Such an antenna, operating at centimeter wavelengths and possibly 10–15 m in diameter, will be needed almost immediately on the Moon for communications and data telemetry. Some time on this antenna could be initially "bootlegged" for tests of the MERI concept. This is an exciting prospect that is relatively inexpensive and promises to yield important new science. This paper examines the advantages of a telescope on the Moon as one element of a MERI, the practical considerations of wavelengths and receivers for setting up such an interferometer, and the scientific merits of this project.

ADVANTAGES OF A MERI

The most obvious advance over previously existing interferometers would be the improvement in resolution. At 6-cm wavelength, the resolution is 30 microarcsec, a factor of ten better than that proposed for QUASAT, 30 times better than the VLBA, and 10,000 times better than the VLA. More discussion on this point is given in the next section.

The thermal stability for an antenna on the Moon would be far better than that of an Earth-orbiting satellite dish. The MERI antenna would experience constant illumination from the sun in two-week intervals. Such thermal stability would be desirable to maintain constant pointing accuracy during a synthesis observation.

A larger antenna (say, 100 m) or a subarray of antennas on the Moon could be built out of relatively simple materials mined on the Moon. This is in keeping with the spirit of self-sufficiency for an advanced Moon colony. Once the mining and manufacturing techniques are developed, fabrication of dish antennas on the Moon will be far cheaper than transporting them from Earth.

Finally, one expects a long lifetime for the antennas on the Moon in comparison to that for a space satellite dish. The main advantage will be easy access to the Moon antenna(s) for repair and, particularly, for cryogen resupply. The lack of weather and low gravity on the Moon should minimize maintenance of the structure of the antennas.

CONSTRAINTS ON MERI

In considering the optimum wavelength at which the receivers will operate and the sensitivity required for interesting science, there are three constraints placed upon MERI. The first is the diffraction limit of the interferometer, which is simply given by

$$\theta_D = 5.47\lambda \tag{1}$$

where θ_D is the FWHM point response function of the instrument (microarcsec) and λ is the wavelength (centimeters). An average Earth-Moon baseline of 3.8×10^{10} cm is assumed.

The second constraint, involving scintillation of the interstellar medium (ISM), will also limit the resolution of MERI. Turbulence in the ISM (a plasma) within our galaxy effectively scatters radio radiation from distant sources, broadening the radio "seeing" disk in a manner somewhat analogous to the "twinkling" of stars produced by the passage of optical light through the turbulent atmosphere of the Earth. The predicted amount of scattering depends critically upon the galactic latitude of the radio source and the assumed model for electron density fluctuations. For sources high above the galactic plane and assuming a simple power-law spectrum for the turbulence following Rickett (1977) and Cordes *et al.* (1984), the scattering angle is given by

$$\theta_{ISM} = 0.60\lambda^{11/5} \tag{2}$$

where θ_{ISM} is again measured in microarcsec and λ is the wavelength in centimeters.

The final constraint is a theoretical limit placed upon the intrinsic sizes of compact radio sources, sometimes referred to as the Compton catastrophe or 10^{12} K brightness temperature limitation. The emission mechanism for active galaxies and quasars is believed to be incoherent synchrotron radiation at radio wavelengths. For compact sources, the densities of relativistic electrons and synchrotron photons are high. Inverse Compton scattering of the photons by electrons is plausible in this environment. Within a limiting angular distance of the core, the optical depth for this Compton process is so high that few, if any, synchrotron photons will escape from this region. Therefore, no information on the structure within this scattering disk can be retrieved. This angle depends upon both the wavelength and the flux density, S (measured in milliJanskys where 1 mJy = 10^{-29} watts/Hz/m²), such that brighter sources will have larger scattering disks. This limit is given by

$$\theta_{cc} \geq S^{1/2}\lambda \tag{3}$$

This limit will not apply to extended, optically thin sources.

A plot of these three effects is shown in Figure 2. One can see from the above equations that the Compton catastrophe limit is equal to the diffraction limit for radio sources with flux densities of 30 mJy. Furthermore, the ISM scintillation effects are negligible below a wavelength of about 6 cm. Therefore, to achieve the ideal resolution (*i.e.*, diffraction limit), one would like to observe compact sources (for astrometry) of <30 mJy at a wavelength of 6 cm or below.

The fortuitous combination of baselines and wavelengths makes the 6-cm band ideal for MERI (although one should not overlook the potential for operating at much shorter wavelengths). The 6-cm receivers currently available are some of the most sensitive used in radio astronomy, with system temeratures of <50 K. Such sensitivity will be very useful since we desire to observe relatively weak radio sources in the range of a few tens of mJy.

As noted above, we can achieve the diffraction limit for sources at high galactic latitudes with flux densities <30 mJy at 6 cm. This angular resolution of 30 microarcsec corresponds to the following linear dimensions:

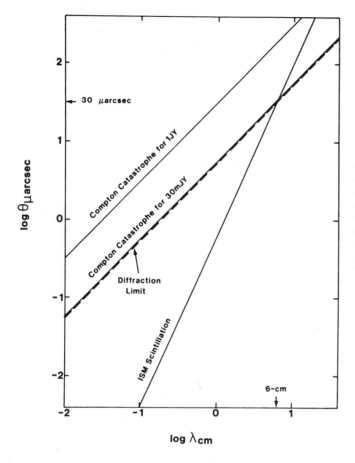

Figure 2. A plot of the minimum FWHM resolution of MERI vs. wavelength. Note that the optimum resolution of 30 microarcseconds occurs in the 6-cm band for compact sources of 30 mJy. ISM scattering effects are negligible for high galactic latitude sources observed at wavelengths <6 cm.

1.   At a distance of 300 pc (~240 times the distance to the nearest star), the disk of a solar-type star can be resolved. Studying active regions on such stars will be invaluable in understanding the solar–stellar connection.

2.   At the distance of the Galactic center, the linear resolution will be 0.15 astronomical units (1 au = average Earth-sun distance = 1.5×10^{13} cm). With such a resolution, one will be able to "look down the throat of the beast" that is responsible for prodigious amounts of electromagnetic radiation at all wavelengths. This linear dimension corresponds to 760 Schwarzschild radii for a 10^4 solar mass black hole.

3.   At the distance of the nearest radio galaxy, Centaurus A, the linear resolution is 450 au. Again one can seriously investigate the nature of the engine at the cores of active galaxies with such a resolution.

Can we achieve the sensitivity necessary to observe sources of a few tens of mJy? The integration time needed to obtain a signal-to-noise ratio, S/N, with radio antennas of diameter D, antenna temperatures T, antenna efficiencies ϵ, and bandwidths $\Delta\nu$ for a source with flux density S on a single Moon-Earth baseline is given by

$$t(s) = 3.1 \times 10^7 (S/N)^2 (T_m T_e) \, (\epsilon_m^2 \epsilon_e^2 D_m^2 D_e^2 S^2 \Delta v)^{-1} \tag{4}$$

Here, the subscript m refers to a Moon antenna and e to an Earth antenna. If we assume equal radio dishes on the Moon and the Earth, and S/N = 10, D = 25 meters (VLA and VLBA size dish), T = 50 K, ϵ = 0.65 (at 6 cm), Δv = 30 MHz, and S = 30 mJy, the integration time is only 30 minutes. If the Moon radio telescope is linked to one of the ground-based VLBI arrays, the integration time can be further reduced by a factor of N(N–1)/2, where N is the total number of antennas in the network. In either case, the integration time is quite reasonable.

To review, then, our analysis shows that the diffraction-limited resolution can be achieved for wavelengths less than 6 cm for compact sources at high galactic latitudes with flux densities <30 mJy. This requires a moderate aperture radio antenna (~25 meters) linked to a VLBI array on Earth using a wide bandwidth system. It is important to note that *no new technology* is required for MERI as outlined above. However, if aperture synthesis mapping of radio sources is to be seriously attempted at these resolutions, then additional antennas should be placed in orbits between the Earth and Moon.

SCIENTIFIC GOALS

There is a wealth of scientific data that could be collected with a MERI telescope that would significantly add to our knowledge of the local environment and the cosmos. These goals are divided into two parts that will depend upon the number of elements in MERI: astrometry and synthesis mapping.

Astrometry

First, the unprecedented relative position accuracy of MERI could be used to improve the celestial coordinate system, thereby improving celestial navigation and astronomical timekeeping.

Second, observations of point sources could potentially be used to accurately measure distances between the Earth and the Moon. In principle, baseline determinations with millimeter accuracy could improve by an order of magnitude those measured by laser ranging. Improved-stability maser clocks would be required, however.

Third, it may be possible to search for dark companion stars (black holes and neutron stars) or even planets around radio stars. One could simply look for perturbations of the radio star proper motions produced by the gravitational pull of a dark binary companion(s). The few tens of microarcsecond accuracy of MERI would allow very small perturbations (produced by planets of mass less than that of Jupiter) to be detected.

Fourth, one of the most exciting aspects of science with MERI is the fundamental cosmological experiments that could be performed. At present, the Hubble constant, which measures the rate of expansion of the universe at the current epoch, is not known to within a factor of two. Moran (1984) has shown that H_2O masers in our galaxy can be used as independent distance measures using classical proper motion and statistical parallax techniques. The impressive power output of these radio sources at discrete wavelengths makes them ideal for this purpose. The angular resolution of MERI could enable astronomers to extend this technique to other galaxies and to accurately determine their distances independent of the other less reliable assumptions currently invoked (Reid, 1984). A measure of the Hubble constant would follow from a well-defined statistical sample. Similarly, proper motions of radio galaxies in clusters combined with redshift measurements could allow statistical determinations of cluster distances and, therefore, the Hubble constant.

Synthesis Mapping

If a few radio antennas between the Earth and the Moon could be added to the initial single antenna on the Moon, then synthesis mapping at <30 microarcsec resolution becomes feasible. The following is a short list of the possible mapping projects in order of increasing distance and decreasing linear resolution.

First, one could potentially locate and map radio burst regions on other stars. The primary limitation here is sensitivity. One would probably need a 100-m antenna on the Moon to make this feasible.

Second, resolving regions in and around star formation nebulae would become possible with MERI. This could make significant impacts on our understanding of the early stages of star (and possible planetary) formation. The origins of the bipolar flows recently discovered in star formation zones could be explored.

Third, the center of the Milky Way contains a powerful source of electromagnetic energy. This energy, $\sim 10^{42}$ ergs/sec, emanates from a region of only a few light years across. Mapping this region at the resolution of MERI would almost certainly add new insights if not a definitive answer to our questions concerning the "engine."

Fourth, active galaxies and quasars are now known to possess "jets" of radio emission that appear to illuminate channels by which matter and energy are transported between the engine at the core of the galaxy (or quasar) and extended structures hundreds of thousands of light years out in intergalactic space. We do not understand how or why the radio jets are collimated as they are in two thin streams. The resolutions of current interferometers are simply too low to explore the nuclear regions where the initial collimation occurs. MERI will allow us to explore these regions in unprecedented detail.

Fifth, in a related vein, the engines themselves are not understood. Are galactic sources such as SS 433 and the Milky Way center simply scaled-down versions of those in the more powerful active galaxies? With MERI, we will be able to map the core regions at high enough resolution to address the nature of the engine.

Sixth, we could potentially test the fundamental physics of compact extragalactic sources. In particular, the Compton catastrophe predicts a minimum size for compact sources of a given flux density at a particular wavelength. We need to test this prediction. Observing compact sources at larger wavelengths and/or higher flux densities will provide the definitive test of this model.

MERI offers us an opportunity for a major leap forward in radio astronomy, both in terms of technique and science. As I have noted, the costs of a single lunar antenna are minimal but the science could be potentially quite promising. Therefore, I would hope that a single radio antenna linked as an interferometer to Earth-based radio telescopes would be one of the first "flowers" planted on mankind's return to the Moon.

Acknowledgments. *I wish to thank Pat Crane, Ed Fomalont, Steve Gregory, Frazer Owen, Marc Price, Peter Wilkinson, and Stan Zisk for useful input into the preparation of this paper. I would also like to thank the Lunar and Planetary Institute for travel support to the Symposium.*

REFERENCES

Cordes J. M., Ananthakrishnan S., and Dennison B. (1984) Radio wave scattering in the galactic disk. *Nature, 309,* 689.

Moran J. M. (1984) Masers in the nuclei of galaxies. *Nature, 310,* 270.

Napier P. J., Thompson A. R., and Ekers R. D. (1983) The Very Large Array: Design and performance of a modern synthesis radio telescope. *Proc. I.E.E.E., 71,* 1295.

Pearson T. J. and Readhead A. C. S. (1984) Image formation by self-calibration in radio astronomy. *Ann. Rev. Astron. Astrophys., 22,* 97.

Reid M. J. (1984) H_2O masers and distance measurements: The impact of QUASAT. In *QUASAT—a VLBI Observatory in Space,* p. 181. ESA Scientific Publication, The Netherlands.

Rickett B. J. A. (1977) Interstellar scattering and scintillation of radio sources. *Ann. Rev. Astron. Astrophys., 15,* 479.

Schilizzi R. T., Burke B. F., Booth R. S., Preston R. A., Wilkinson R. N., Jordan J. F., Preuss E., and Roberts D. (1984) The QUASAT project. In *VLBI and Compact Radio Sources* (R. Fanti, K. Kellerman, and G. Setti, eds.), p. 407. D. Reidel, Boston.

A VERY LOW FREQUENCY RADIO ASTRONOMY OBSERVATORY ON THE MOON

James N. Douglas and Harlan J. Smith

Astronomy Department, University of Texas, Austin, TX 78712

Because of terrestrial ionospheric absorption, very little is known of the radio sky beyond 10 m wavelength. We propose an extremely simple, low-cost Very Low Frequency (VLF) radio telescope, consisting of a large (approximately 15 × 30 km) array of short wires laid on the lunar surface, each equipped with an amplifier and digitizer, and connected to a common computer. The telescope could do simultaneous multifrequency observations of much of the visible sky with high resolution in the 10- to 100-m wavelength range, and with lower resolution in the 100- toward 1000-m range. It would explore structure and spectra of galactic and extragalatic point sources, objects, and clouds, and would produce a detailed quasi-three-dimensional mapping of interstellar matter within several thousand parsecs of the sun.

INTRODUCTION

The spectral window through which ground-based radio astronomers can make observations spans about five decades of wavelength, from a bit less than a millimeter to something more than ten meters. The millimeter cutoff produced by molecular absorption in the Earth's atmosphere is fairly stable, but the long wavelength cutoff caused by the terrestrial ionosphere is highly variable with sunspot-cycle, annual, and diurnal effects; scintillation on much shorter time scales is also present. Radio frequency interference imposes further limits, making observations at wavelengths longer than 10 m normally frustrating and frequently impossible.

Consequently, the radio sky at wavelengths longer than 10 m is poorly observed and is virtually unknown for wavelengths longer than 30 m, except for a few observations with extremely poor resolution made from satellites. Exploration of the radio sky at wavelengths longer than 30 m must be done from beyond the Earth's ionosphere, preferably from the farside of the Moon, where physical shielding completes the removal of natural and manmade terrestrial interference that the inverse square law has already greatly weakened.

THE LONG WAVELENGTH RADIO SKY

What may we expect in the long-wavelength radio sky (apart from the unexpected, which experience often shows to be more important)?

First, non-thermal radiation from plasma instabilities in solar system objects is present in rich variety, especially from the sun, Jupiter, Saturn, and Earth itself (which was unexpectedly discovered by telescopes flown for other purposes).

Second, the synchrotron radiation from the galaxy reaches a peak of intensity near 4 MHz (75 m), then drops off as absorption by ionized hydrogen becomes important, and possibly for other reasons, as well. This behavior has been seen by the low-resolution (20° beam) telescopes already flown.

Third, the plane of the Milky Way—already dimming at 10 m—becomes even more absorbed by ionized hydrogen, and many black blots of HII regions are seen in absorption against the bright radiation background. Such clouds, whose emission measure is too small to be noticed optically, would be obvious using a moderate to high resolution (1°–0.1°) VLF telescope. At longer wavelengths, our distance penetration becomes increasingly limited, decreasing with the square of the wavelength until, by 300 m, unit optical depth corresponds to only a few hundred parsecs.

Fourth, extragalactic discrete sources continue to be visible as wavelength increases (outside the gradually expanding zone of avoidance at low galactic latitudes), and their spectra can be measured, although their angular structure will be increasingly distorted by interstellar and interplanetary scattering. At these wavelengths, one is looking at the expanded halo parts of such objects, and turn-overs will be noted in the spectra of many. For wavelengths longer than about 300 m, HII absorption in our own galaxy will effectively prevent extragalactic observations, even at the galactic poles, and, at wavelengths longer than a kilometer or so, we will be limited to studying objects within a few tens of parsecs of the sun.

Finally, there are possible new features of the sky that can be studied only by a high resolution and high sensitivity telescope at long wavelengths. These include non-thermal emission from stars and planets or other such sources within a few parsecs (if any of these are significantly more powerful than the sun and Earth); radio emission of very steep spectra from new classes of galactic or extragalactic discrete sources that may have gone undetected to date in even the faintest surveys at short wavelengths, yet be detectably strong at 100 m; nearby and compact gas clouds, visible in absorption, whose presence has hitherto been unsuspected; and fine-scale structure in the galactic emission, which—given data at high-resolution and multiple low-frequencies—can be studied in depth as well as direction. In this connection the proposed telescope should provide a uniquely detailed and effectively three-dimensional map of interstellar matter in the galaxy out to distances of thousands of parsecs.

THE LUNAR VLF OBSERVATORY

As noted above, the low-frequency telescopes flown to date have had very poor resolution, although valuable for some studies on very bright sources, e.g., dynamic spectra of the sun, Earth, Jupiter, and the cosmic noise spectrum. Significant advances, however, will require high resolution (say 1°, corresponding to 15 km aperture at 300 m wavelength) and high sensitivity (many elements). A lunar base offers probably the best location in the solar system for constructing an efficient low-cost VLF radio telescope.

In contemplating any lunar-based experiment, the question must first be asked whether it is preferable to carry out the work in free space. For the proposed VLF observatory, the Moon offers a number of advantages:

1. It is a fine platform, able to hold very large numbers of antenna elements in perfectly stable relative positions over tens or even hundreds of kilometers separation (this would be excessively difficult and expensive to try to do in orbital configuration);

2. The telescope can begin modestly, though still usefully, and can continue to grow to include thousands of antenna elements added in the course of traverses of lunar terrain undertaken at least in part for other purposes;

3. The dry dielectric lunar regolith permits simply laying the short thin-wire antenna elements on the surface. No structures, difficult to build and maintain, are required;

4. Lunar rotation provides a monthly scan of the sky;

5. The lunar farside is shielded from terrestrial interference, although even the nearside offers orders-of-magnitude improvement over Earth orbit because of the inverse square law, and the much smaller solid angle in the sky presented by the Earth.

Limiting Factors

Various natural factors limit the performance of a lunar VLF observatory.

Long wavelength limits. (1) Interplanetary plasma at 1 AU has about 5 electrons/cm^3 corresponding to a plasma frequency (f_p) of 20 kHz, or a wavelength of 15 km. (2) The Moon may have an ionosphere of much higher density than the solar wind; 10^{-12} torr corresponds to about 40,000 particles/cm^3, if the mean molecular weight is 20. If such an atmosphere were fully singly ionized, f_p would be around 1.8 MHz, usefully but not vastly better than the typical values for the Earth of around 9 MHz. However, ground-based observations of lunar occultations suggest that N_e is actually less than 100. In this case, f_p would be less than 90 KHz (wavelength 4 km), and would set no practical limit to very low frequency lunar radio astronomy. It will clearly be very important for detailed planning of the lunar VLF observatory to have good measures of the lunar mean electron density and its diurnal variations.

Scattering. (1) The interstellar medium produces scattering and scintillation, and thus angular broadening of sources—e.g., the angular size of an extragalactic point source would be about 8 arcseconds if observed at 30 m wavelength. The size grows with wavelength to the 2.2 power, becoming 1/3° at 300 m. (2) Interplanetary scintillation is more important. Obeying essentially the same wavelength dependence as interstellar scintillation, it ranges from about 50 arcsec at 30 m to a few degrees at 300 m (1 MHz). However, it is still worthwhile designing the telescope with higher resolution than 2° at 1 MHz, since techniques analogous to speckle interferometry may recover resolution down to the limits set by interstellar scintillation, which will be relatively small especially for nearby sources in our galaxy.

Interference. (1) Solar: the intensity of the cosmic background radiation is on the order of 10^{-15} w m^{-2} Hz^{-1} $ster^{-1}$. The sun is already known to emit bursts stronger than this by an order of magnitude in the VLF range, so the most sensitive observations may have to be carried out during lunar night. (2) Terrestrial: Nearside location will always expose the telescope to terrestrial radiations. Consider two known types: auroral kilometric radiation is strong between 100 and 600 kHz; an extremely strong burst would produce flux density at the Moon of about $2(10)^{-15}$ w m^{-2} Hz^{-1}—far stronger than the cosmic noise we are trying to study. Fortunately it is sporadic, and limited to low frequencies. Also, it probably comes from fairly small areas in the auroral zones, so that its angular size as seen from the Moon will be small. Lunar VLF observations below 1 MHz will therefore be limited unless the telescope is highly directive with very low sidelobes, or built on the lunar farside.

Terrestrial radio transmitters may leak through the ionosphere in the short wavelength portions of the spectrum of interest. If we assume a 1-Mw transmitter on Earth with a 10 kHz bandwidth, the flux density at the Moon would be about $5(10)^{-17}$ w m^{-2} Hz^{-1} without allowing for ionospheric shielding. This would be a serious problem; much weaker transmitters with some ionospheric shielding would merely be an occasional nuisance. Again, this is an argument in favor of a farside location, particularly for frequencies above 4 MHz or so.

Considerations of Telescope Design

It would be futile to carry out a detailed telescope design at this point; however some general considerations can be addressed:

Frequency range. The telescope should be broadband, but capable of observing in very narrow bands over the broad range to deal with narrow-band interference. The upper limit of frequency should be around 10 MHz or 30 m. Even though this wavelength can be observed from the ground, it is extraordinarily difficult to do so. The initial normal lower limit should be about 1 MHz or 300 m, although the capability for extending observations with reduced resolution to substantially longer wavelengths should be retained.

Resolution. It is probably useless to attempt resolution at any given frequency better than the limit imposed by interstellar scintillation, e.g., about 1/3° at 1 MHz. A reasonable initial target resolution for the observatory might be 1° at 1 MHz. Although this is somewhat better resolution than the limit normally set by interplanetary scintillation, it is probably attainable using restoration procedures. This choice of target resolution implies antenna dimensions of 15 × 15 km for a square filled array, or of 30 × 15 km for a T configuration.

Filling factor. A 1° beam may be synthesized from a completely filled aperture (100 × 100 elements, for a total of 10^4) or by a T, one arm of which has 200 elements, the other 100, for a total of 300 elements—far less. Many other ways of filling a dilute aperture also exist, including a purely random scattering of elements over the aperture. The filled array has far greater sensitivity, but, what is also important in this context, it has much better dynamic range and a cleaner main beam. This will be of great benefit in mapping the galactic background, particularly in looking at the regions of absorption, which will be of such interest at these frequencies. The sensitivity of the filled array is also decidely better: a 1° beam produced by a filled aperture at 1 MHz with a bandwidth of 1 kHz and an integration of 1 min has an rms sensitivity of 1 Jy; the same sensitivity would require an integration of 1 day with the dilute array of 300 elements. The most sensible approach is probably to begin with a dilute aperture and work toward the filled one, the power of the system increasing as more antenna elements are set out.

Telescope construction. The telescope would be an array of many elements. Each element should be thought of as a field sensor rather than as an ordinary beam-forming antenna—in other words as a very short dipole. The inefficiency of such devices can be great before noise of the succeeding electronics becomes a factor, in view of the high brightness temperature of the cosmic background radiation. An A/D converter at each element would put the telescope on a digital footing immediately. The exceedingly low power requirement at each antenna element could be met with a tiny solar-powered battery large enough to carry its element through the lunar night.

Communication with the telescope computer at lunar base via radio or perhaps by individual optical-fiber links would bring all elements together for correlation. Bandwidth of the links need only be about 1 kHz per element if only one frequency is to be observed at a time, although maximum bandwidth consistent with economics will produce maximal simultaneous frequency coverage. In any event the central computer will produce instant images of a large part of the visible hemisphere with the 1° resolution, at one or many frequencies, which can be processed for removal of radio frequency interference (rfi) and bursts prior to long integrations for sky maps at various frequencies in the sensitivity range of the system.

Short wavelength operation. Operation at the short-wavelength boundary of the telescope range will be a different proposition. Element spacing for 1 MHz is very dilute indeed for 10 MHz; some portion will have to be more densely filled, and operated against the rest of the system as a dilute aperture. At 10 MHz the system would have a resolution of about 0.1°. In this way, an extremely powerful telescope for work both on extragalactic sources and on galactic structure would result.

ESTABLISHING THE VLF OBSERVATORY

The individual antenna elements—short wires—will probably weigh about 50 gm each. Their associated microminiaturized amplifiers, digitizers, transmitters, and solar batteries can all be on several tiny chips in a package of similar weight. Allowing for packaging for shipment to the Moon, the initial array should still weigh less than 50 kilograms! Materials for the entire filled array would only need about a ton of payload. If individual optical fiber couplings to the central computer are used, each of these should add only a few tens of grams to the total, not appreciably affecting the extraordinarily small cost of transporting the system to the Moon.

A powerful computer is of course required, to process continuously the full stream of digital information. Some on-base short-term storage of processed data is probably also desirable, but at frequent intervals this would presumably be dumped back to Earth. Again, with the increasing miniaturization yet steady growth in power of computer hardware over the next twenty years, the required computer facilities may also be expected to weigh less than a hundred kilograms. It thus seems clear that at least the initial, and quite possibly the ultimate, VLF observatory system could be carried to the Moon as a rather modest part of the very first scientific payload.

Laying out the initial system of several hundred antenna elements on the lunar regolith should require only a few days of work with the aid of an upgraded lunar rover having appropriate speed and range. (Such vehicles will be an essential adjunct of any lunar base for exploration, geological and other studies, and general service activities). The elements need not be placed in accurately predetermined positions, but their actual relative positions need to be known to a precision of about a meter. This can easily be done, as the layout proceeds, by surveying with a laser geodometer. The conspicuous tire marks produced by the rover vehicle will delineate the sites of each of the antenna elements for future maintenance or expansion of the system. A concentrated month using two vehicles each carrying teams of perhaps three workers would probably suffice to lay out the full proposed field of 100×100 elements.

These estimates, while necessarily rough at this preliminary stage of planning, strongly suggest that because of its extreme simplicity and economy, its almost unique suitability for lunar deployment, and its high scientific promise, the VLF observatory is a major contender for being the initial lunar observatory—perhaps even the first substantial scientific project that should be undertaken from Lunar Base.

LUNAR BASED GAMMA RAY ASTRONOMY

Robert C. Haymes

Space Physics and Astronomy Department, Rice University

INTRODUCTION

Gamma ray astronomy is the study of the universe through analysis of the information carried by the highest energy electromagnetic radiation. The energy of the photons it analyzes ranges upwards from about 0.1 MeV, the upper energy end of the x-ray band, through the MeV energies of nuclear transitions and radioactivity, up through the GeV energies of cosmic ray-matter interactions, to beyond the 10^3TeV energies radiated by the highest energy galactic cosmic rays. The celestial sources where gamma ray emission is a major fraction of the energy release are some of the most bizarre, energetic objects in the universe, including supernovae, neutron stars, black holes, galactic cores, and quasars.

SIGNIFICANCE

If the gamma ray photon fluxes are above the sensitivity threshold, then a variety of phenomena may be studied through their measurement. We currently believe that nucleosynthesis of the heavier elements takes place in *supernovae*, stellar catastrophic explosions that each rival a whole galaxy of stars in brightness. Supernovae are expected to emit specific gamma ray energy spectra from which the mode(s) of nucleosynthesis may be deduced (Ramaty and Lingenfelter, 1982).

Black holes are places where present-day physics is, at best, on shaky ground. Gamma ray astronomy permits us to study matter as it falls into a black hole, because the matter becomes heated to gamma ray temperatures as it does so.

Active galaxies generate energies comparable with their relativistic self energy, but the nature of their energy source(s) is unknown. The various theoretical possibilities suggested thus far all predict different gamma ray spectra.

Quasars may each generate as much energy as do ten billion stars, but they appear somehow to do it in a volume not much larger than that of only one star. Much of the quasar's radiation is in the gamma ray band, and our most important clues to the phenomenon may therefore come from such astronomy.

The sources of *cosmic rays* have long been mysterious. Whatever and wherever they are, the sources are likely to generate high-energy gamma rays because of the acceleration of the charged-particle cosmic rays. Because photons are uncharged and therefore travel in straight lines that are unaffected by magnetic fields en route, gamma ray astronomy uniquely offers the opportunity of at long last locating and studying the sources of this ubiquitous, extreme-energy particle radiation.

Quantized cyclotron emission may already have been detected from the presumed direction of one *neutron star*. If confirmed, this evidence would provide for the existence of magnetic fields seven orders of magnitude more intense than any generated in the laboratory. *Gamma ray bursters* are possibly associated with neutron stars and have been known for over a decade, but their nature is as mysterious as ever. Also not understood is the nature of *transient sources* of cosmic gamma radiation.

Over thirty steady sources of high energy gamma radiation have already been detected, sources that do not seem to have counterparts in other spectral bands (*e.g.*, the optical, radio, and x-ray). These sources therefore seem capable of somehow accelerating charged particles to the extreme energies required for gamma ray production, while also suppressing the usually copious radiation of lower energy photons.

In addition to acceleration of ultrarelativistic charged particles, the gamma radiation is produced in several ways. These include radioactivity and nuclear de-excitation, matter-antimatter annihilation, decays of elementary particles, and some effects of relativity. Through this branch of astronomy, qualitatively different and often uniquely available information will be acquired on the mechanisms, history, and sites of cosmic nucleosynthesis, the structure and dynamics of the Milky Way, the nature of pulsars, the sites and properties of intensely magnetized regions, the isotopic composition of the matter in the space surrounding black holes, the nature of the huge energy sources powering active galactic nuclei, and the universality, composition, and sources of very high-energy matter throughout the universe. We already know that, at least in some of these phenomena, the emissions are most luminous in the gamma ray part of the spectrum. Full understanding of these sources will require detailed study of their gamma radiation, in combination with study of the emissions in the other spectral bands.

SENSITIVITY LIMITATIONS FOR GAMMA RAY ASTRONOMY

In most situations, if a source radiates high-energy photons, it will radiate smaller fluxes of them than it will of lower energy photons. The sensitivity of observational gamma ray astronomy is limited by the small fluxes of gamma ray photons from celestial sources. It is also limited by the background.

Gamma ray astronomical background appears to have two components. One component is a sky (possibly cosmic) background. The other component has two sources. One arises from both ambient gamma radiation, and the other from radioactivity induced in the observing instrument by the particle radiation environment that exists in space.

Gamma ray astronomy is best conducted far from Earth, because the atmosphere is a source of gamma rays. Energetic particles continually bombard the atmosphere. Examples of such particles include cosmic rays and the high-energy protons that compose the Inner Van Allen belt; others are solar-flare accelerated ions. These particles produce gamma ray photons when they interact with the atmosphere. Some of the produced photons head out into space, forming a "gamma ray albedo." The ambient gamma radiation comprising the albedo has an intensity that exceeds, by several orders of magnitude, the fluxes of gamma ray photons from even the brightest cosmic sources.

Bombardment by high-energy particles also activates the materials composing a gamma ray detector, making the instrument itself a source of the very radiation one is attempting to measure from the cosmos (e.g., Paciesas et al., 1983). The brightness of this background component is dependent on the mass of target matter and the magnitude of the bombarding fluxes. It may be reduced to cosmic ray levels by observing from sites that are outside the geomagnetically trapped radiation and that are shielded from solar protons.

GAMMA RAY ASTRONOMY OBSERVATIONS

Earth's atmosphere is an absorber of cosmic gamma radiation. To avoid unacceptable attenuation of the small photon fluxes from celestial sources, gamma ray astronomy must therefore be conducted from outside most, if not all, of the atmosphere. Fragmentary data have been acquired with balloon flights near the top of our atmosphere. Almost all of the balloon flights have been restricted to durations of a day or less, and all were sporadic. Even though most of the major advances in understanding the phenomena encountered in other spectral bands were made when long-term measurements were undertaken, there has been little systematic, long-term observational gamma ray astronomy.

The HEAO-1 and HEAO-3 satellites scanned the sky and undertook some short-duration observations of selected discrete gamma ray sources. The major long-term efforts thus far have been the European COS-B mission and the groundbased monitoring of Cerenkov pulses in Earth's atmosphere. For about six years in the 1970s, the low altitude COS-B (Mayer-Hasselwander et al., 1982) satellite carried a relatively small spark chamber. The chamber measured the arrival directions and the energies of gamma rays in approximately the 0.1–5 GeV energy band. That mission has produced almost all of our present information on astronomical sources of gamma rays in that energy band.

All of our information on emissions in the 1–1000 TeV energy band, which may be providing our first direct looks at the sources of galactic cosmic rays and which could tell us how high in energy particle acceleration goes in the nuclear regions of active galaxies, has come from groundbased monitoring of nanosecond Cerenkov light pulses in the atmosphere. Some of these pulses are due to the interaction of very high energy gamma ray photons with the atmosphere (Samorski and Stamm, 1983). The relative frequency of such pulses increases when the gamma ray source transits the observer's meridian. These extreme-energy events are relatively rare in occurrence; several-year integration times are necessary for statistical validity. For absolute flux information, it is necessary to distinguish those showers produced by γ-ray photons from those due to cosmic ray nuclei. Such a distinction is now made on the basis of relative μ-meson richness in the showers.

In 1988, NASA plans to launch the Gamma Ray Observatory (GRO), the first full-fledged systematic investigation into gamma ray astronomy, into a low altitude orbit. The three-axis stabilized spacecraft will operate for two years. Data will be collected from selected targets for one to two weeks at a time. The GRO will carry four gamma ray astronomy experiments in the one-square-meter class. Described by Kniffen (written communication, 1981), each experiment has different scientific objectives.

One GRO experiment is a spark chamber. Its collecting area for photons is about ten times greater than the chamber that flew on COS-B, and it will have about 1° angular resolution. Since the observing times will be comparable with the COS-B times, the spark chamber experiment, called EGRET, will therefore extend the sensitivity of the COS-B observations of 100 MeV–5 GeV cosmic sources and the gamma ray background.

The second GRO experiment, called COMPTEL, is an imaging double Compton telescope for 1–30 MeV studies. Very little is presently known about this spectral region, most of which lies above the energies of nuclear transitions and radioactivity but below the energy where decay of neutral π-mesons into gamma rays is an important source of photons. Compton scattering is a major photon–matter interaction mechanism at these energies.

Double Compton telescopes consist of two layers of scintillation counters. A gamma ray that interacts in the first layer generates a pulse. The recoil Compton photon, if its direction is suitable, interacts with the second layer. Delayed coincidence between the two layers is required for a given event to be accepted as due to a gamma ray in the direction of the instrumental cone of acceptance. The time between these two pulses is given by the ratio of the separation distance to the speed of the recoil, which is the speed of light. COMPTEL's spacing is about 3 m, so the delay time is about 10 nanoseconds. This is an excellent way of rejecting background; particles and photons from other directions will not give the correct time signature. Double Compton experiments, however, are inefficient (the photon-detection efficiency is of order 10^{-5}), since the recoils have to be directed only in the direction of the second layer for an event to be counted. They also do not require total energy deposition, and reliance must therefore be placed on calculations of most probable energy loss as a function of energy, in order to convert the observed pulse-height spectrum to an energy spectrum. COMPTEL's imaging will be crude at best, since its angular resolution is several degrees.

Third is OSSE, the Oriented Scintillation Spectrometer Experiment, which consists of four independently pointable, equal-area actively collimated scintillation counters. Its goal is to conduct astronomical spectroscopy in the 0.1–10 MeV energy band. This band is characteristic of radioactivity and transitions of excited atomic nuclei.

Actively collimated counters almost completely surround the photon counter with a thick collimator that consists of an efficient scintillation counter whose output is connected in anticoincidence with the photon counter. Absence of coincidence between the scintillation and photon counters is required for a photon–counter event to be accepted. Total deposition of energy in the photon counter is imposed by this requirement. The coincidence requirement also rejects counts due to charged particles; a particle that caused a count in the photon counter most likely had to traverse the scintillation counter-collimator in order to reach the photon counter, and it would have generated a pulse from the scintillation counter in order to do so. The magnitude of the field of view of such experiments is defined by the size of the opening in the "active collimator" (*i.e.*, the surrounding scintillation counter). Although total energy-deposition is required and the photon-detection efficiency may be near 100%, activation of the crystals themselves by particle bombardment is a serious source of background. For example, the "collimator" is itself induced by the bombardment to become a radioactive source of gamma rays, which are not distinguished by the system from external cosmic gamma rays coming through the aperture. At high energies in the band, photon leakage through the finite thickness collimator is also a serious background source. To date, the noise (i.e., background)-to-signal ratios of actively collimated astronomical experiments have typically been 10 or more, even for the brightest cosmic sources. The two brightest cosmic sources, the central region of the Galaxy and the Crab Nebula supernova remnant, have photon fluxes at 1 MeV that are of order 10^{-3} photons/cm²-s. OSSE will attempt to reduce background effects by simultaneously measuring source and background, using its independently targetable 2000 cm² modules. OSSE has an angular resolution of about one degree and an energy-dependent energy resolution that is about 0.05 MeV at 1 MeV.

The fourth experiment, called BATSE, is primarily intended to measure cosmic gamma ray bursts with a lower fluence threshold than heretofore available. Its very large photon collection area also makes it a very sensitive detector of rapidly varying existing gamma ray sources, such as pulsars. BATSE consists of six one-square-meter scintillation counters that are each pointed in different directions from the stabilized spacecraft. They cover the entire hemisphere of sky. So far as is known, bursts seem to originate from all sky locations with equal probability; a burst that occurs anywhere on the hemisphere will illuminate all six counters differently. Each counter uses sodium iodide as the scintillator, for maximum light output and best energy resolution from a large-area detector. The ratios of their count rates will locate the burst on the sky to an accuracy of a degree or so. The time history of their count rates will measure the light curve of the event, and pulse height data from the six will provide some information on the energy spectrum of the emitted gamma radiation. Pulses from existing sources are part of BATSE's data stream. Sources that have known "signatures," such as pulsars that have known periods, may be sorted out from the other data and their energy dependence measured with good precision out to higher energies than previously done.

There can be little doubt that, if successful, the GRO will add greatly to our meager knowledge of the universe at gamma ray wavelengths. But, the two-year overall lifetime limits the time that may be devoted to a given source; variability information, which may be crucial to a correct understanding, will suffer. The low altitude orbit means that sensitivity will also suffer, since the materials composing the spacecraft and the instruments will be subjected to a continuous irregular bombardment by high energy particles, resulting in relatively high, time-varying backgrounds.

THE FUTURE AND THE NEEDS OF THE SCIENCE

It is vital, for progress in gamma ray astronomy, to establish small error boxes on the sky for the locations of the different sources. Good locations will make deep searches practical in other wavelength bands, such as the radio and the optical, for counterparts of the gamma ray sources. With optical and other identifications made, progress in understanding the sources is likely. Good positional accuracy for small-flux sources may be obtained with increased photon collecting area, or by increased observing time, or by a combination of the two. The importance of long observing times for variability studies has already been noted. As the next step in the post-GRO era, instruments in the 10–100 m² class appear to be in order.

For missions beyond the GRO, there is the low Earth orbit, long-duration space station. The space station offers the opportunity to conduct first-time measurements of the variability of gamma ray sources. The low altitude orbit, however, also means that a sensitivity limitation will again be imposed, because of the comparatively intense bombardment by Van Allen and South Atlantic Anomaly particles.

What is ideally needed for support of gamma ray astronomy conducted from within Earth's gravitational sphere of influence is an indefinite-duration observatory that is (a) capable of orienting large instruments; (b) located far from large masses; (c) at worst, bombarded by high energy particle fluxes no greater than cosmic ray fluxes; and (d) operated such that its instrumentation may be updated as technology advances.

An observatory located at one of the Earth-Moon Lagrangian Points best fits this ideal. If other considerations rule out such a location, it appears that the surface of the Moon itself would be an acceptable alternative site.

A LUNAR OBSERVATORY

From Apollo data, the Moon's surface is already known to be a low radioactivity environment, compared with Earth's surface or atmosphere. Background radiation from the surroundings will be lower on the lunar surface than it will be in a satellite in orbit about Earth.

The Moon is a satellite orbiting at 60 Earth radii. Instruments on the Moon therefore are in orbits well beyond the regions where the geomagnetically trapped particles exist. There will be no activation by the intense particle fluxes encountered in the South Atlantic Anomaly.

The monthly passage of the Moon through the plasmas in the geomagnetic tail is unlikely to present a problem for gamma ray astronomy, because the energies of the plasma particles in the far tail are all too low to generate gamma ray photons. Should a solar flare occur while the Moon crosses the tail and while solar-flare ions with high energies travel back "upstream" along the tail, the background will be increased. But the tail crossings are only about five days long, and the rest of each month should be free of such problems.

With certain modifications, gamma ray astronomy instrumentation resembling that of a scaled-up GRO payload is envisaged for the lunar observatory. In the context of gamma ray astronomy, a lunar observatory would have much the same goal as does GRO: increased sensitivity measurements over as wide an energy range as feasible.

One observatory instrument that would not require pointing is a spark chamber. In a spark chamber, the incident gamma ray is converted into a positron–electron pair of particles; the tracks of these two particles may be measured by the pattern of little sparks they cause as they move through high electric fields in the chamber's fill gas. The length of the tracks yields energy information, and their direction yields directional information. Post-GRO progress would require a spark chamber whose total sensitive volume would be 15 m (diameter) × 3 m thick, filled to one-atmosphere pressure with an inert gas such as argon. Its overall weight is likely to be nearly 100 tons on Earth. Given a GRO-like observing time for a given source, the sensitivity would be improved by a factor of 10 over that of EGRET. Data processing with so large a chamber would become a serious problem, because the hundred-fold increase in gamma ray detection rate would correspondingly increase the rates of non-background (i.e., non-single-track-events in the chamber's volume) events. Each such non-background event requires measurement of the length and direction of the two visible tracks in the chamber. This may be done in principle by humans with photographs; it is more likely to be automatically done with digitized TV pictures and a computer, in the observatory, or done non-photographically, as in the GRO, but with a hundred-fold increase in data rate. Lunar observatory-large spark chamber data rates are therefore likely to be continuously over one Mbit/s.

Continuous observations mean that gas leaks must be compensated for. A lunar observatory would require some ability to store replacement filling gas for the chamber.

Double Compton instruments also do not require active pointing; all sources on the visible sky would be simultaneously measured by a double Compton on the lunar surface. A scaled-up GRO double Compton with a ten-meter diameter and three-meter height would weigh perhaps 30 tons on Earth. It would have a sensitivity of 10^{-5} photons cm^{-2} sec^{-1}, given one week of observing.

Actively collimated astronomical instruments do require pointing, usually in an on-source, off-source sequence. Future instruments of this type are likely to need pointing accuracy and stability of one arcminute. A hypothetical instrument would have several square meters (total collecting area) for the photon counter, and the photon counter would be constructed of segmented hyperpure germanium solid-state radiation detectors for maximum energy resolution and lowest induced background. The active collimator used to define the field of view and impose a total energy deposition requirement and the active coded aperture used to define the angular resolution would most likely be comprised of thick scintillation counters. On Earth, the instrument would weigh perhaps 10 tons.

Scintillation counters operate well at room temperature. Their photomultiplier tubes or photodiodes require temperature stabilization. Germanium gamma ray counters require an operating temperature of below 100 K. If radiative cooling is not practical, cryostats will be necessary, requiring replenishment of the cryogen. The observatory must be able to resupply the cryogen (e.g., liquid nitrogen or solid carbon dioxide) as needed.

Finally, measurements of gamma ray bursts must be extended. The principal objective here is to attempt to identify the source(s) of the bursts. This means determining their positions on the celestial sphere with arcsecond accuracy, so that other wavelength identifications may be confidently made. Probably the most progress would not be made with a simple increase in the size of a BATSE derivative, but by significantly increasing the length of the baseline available for measurements of position. A network of burst detectors, each not much larger than the BATSE instrument and that have good (*i.e.*, microsecond or better) event timing accuracy over long times, should be operated simultaneously on the Moon and throughout as much of the solar system as feasible.

Each instrument would be sensitive to photons from all directions. Therefore, none of the members of the network would require pointing. The relative timing of the detection at the various sites of a given burst would provide high accuracy data on the angular coordinates of the burst site. Detectors located on the Moon and in low orbit about Earth could form the beginning of such a network of detectors; these would give a baseline 60 Earth radii long. Such a two-station network could locate bursts with one arcsecond accuracy (Chupp, 1976). Each detector should additionally have spectroscopic capabilities, at least in the 0.1–1.0 MeV spectral range and preferably beyond.

Large area and long observing time will place severe demands on the dependability of all the instruments. Large instruments tend to be more complex; they are likely to composed of more modules. Long times without failure are difficult to achieve, technically. The availability of a lunar observatory staffed with trained maintenance personnel makes this more practical. A not insignificant contribution to the increased practicality arises from the penetrating power of gamma radiation. Providing the walls of the observatory are at most a small fraction of a gamma ray mean free path in thickness, gamma ray astronomy instruments may be located *inside* a shirtsleeve environment, which facilitates maintenance and calibration. At an energy of 1 MeV, the gamma ray mean path in aluminum is 16 gm/cm^2, and it is 40 gm/cm^2 at 100 MeV photon energy.

Long-term measurements could run the risk of instrument obsolescence, if the instruments were not upgraded. The availability of a well equipped observatory staffed with trained scientific personnel and in good communication with Earth obviates this problem. Because of the observatory, investments in large instruments may be cost effective; modifications and improvements may be made as they develop.

REFERENCES

Chupp E. L. (1976) *Gamma-ray Astronomy*. D. Reidel, Boston. 195 pp.

Mayer-Hasselwander H. A., Bennett K., Bignami G. F., Buccheri R., Caraveo P. A., Hermsen W., Kanbach G., Lebrun F., Lichti G. G., Masnou J. L., Paul J. A., Pinkau K., Sacco B., Scarsi L., Swanenburg B. N., and Wills R. D. (1982) Large-scale distribution of galactic gamma radiation observed by COS-B. *Astron. Astrophys., 105,* 164.

Paciesas W., Baker R., Bodet D., Brown S., Cline T., Costlow H., Durouchoux P., Ehrmann C., Gehrels N., Hameury J., Haymes R., Teegarden B., and Tueller J. (1983) A balloon-borne instrument for high-resolution astrophysical spectroscopy in the 20–8000 keV energy range. *Nucl. Instrum. Methods, 215,* 261–276.

Ramaty R. and Lingenfelter R. E. (1982) Gamma-ray astronomy. *Annu. Rev. Nucl. Part. Sci., 32,* 235–269.

Samorski M. and Stamm W. (1983) Detection of 2×10^{15}–2×10^{16} eV gamma-rays from Cygnus X-3. *Astrophys. J. Lett., 268,* L17.

IRRADIATION OF THE MOON BY GALACTIC COSMIC RAYS AND OTHER PARTICLES

James H. Adams, Jr.

E. O. Hulburt Center, Naval Research Laboratory, Washington, DC 20375

Maurice M. Shapiro

Max Planck Institut für Astrophysik, 8046 Garching bei München, Federal Republic of Germany

Men and sensitive instruments on a lunar base can be profoundly affected by the radiation environment of the Moon. The ionizing radiation incident upon the lunar surface is comprised of the galactic cosmic rays (GCR) and energetic particles accelerated in the solar neighborhood. The latter consist mainly of solar energetic particles (SEP) from flares and of other particles energized in the heliosphere. The cosmic radiation bombarding the Moon consists overwhelmingly of relativistic and near-relativistic atomic nuclei ranging in energy from 10^8–10^{20} eV, approximately 98.6% of which consists of hydrogen and helium. The remainder spans the rest of the periodic table, with conspicuous peaks in abundance at C, O, Ne, Mg, Si, and Fe. The GCR composition is roughly similar to that of the sun, with some notable differences. Differential energy spectra and composition of cosmic rays as well as the intensities, composition, and the spectra of SEP and particles accelerated in the heliosphere are reviewed. We also summarize the analytic models developed by the Naval Research Laboratory (NRL) group to describe the energy spectra and elemental compositions of the various components.

THE LUNAR RADIATION ENVIRONMENT

The Moon is constantly bombarded by galactic cosmic rays (GCR). Figure 1 (from Simpson, 1983) shows a sampling of the data available on the differential energy spectra of the most prominent particle types in galactic cosmic rays. The intensity of this highly penetrating particle radiation varies in response to solar activity. In a way that is not yet fully understood (Fillius and Axford, 1985), the out-flowing solar wind modulates the cosmic ray intensity so that it is anti-correlated with the general level of solar activity. This causes the average intensity of cosmic rays with energies greater than 10 MeV/amu to increase 2.5 times from the maximum to the minimum of the 11-year solar activity cycle. Low energy cosmic rays are affected more strongly than the higher energy ones. Figure 2 (from Simpson, 1983) compares the elemental compositions of galactic cosmic rays and the solar system, normalized at silicon. The two compositions are comparable for the most abundant elements. The odd elements, in general, and Li, Be, B, F, Sc, Ti, V, Cr, and Mn, in particular, are overabundant in galactic cosmic rays. This difference is the result of the propagation of galactic cosmic rays through approximately 7 g/cm^2 of interstellar gas, on average, before reaching Earth (Shapiro and Silberberg, 1970).

During periods of minimum solar activity, additional components can be observed at low energy. One component is constantly present. This component, discovered by Garcia-

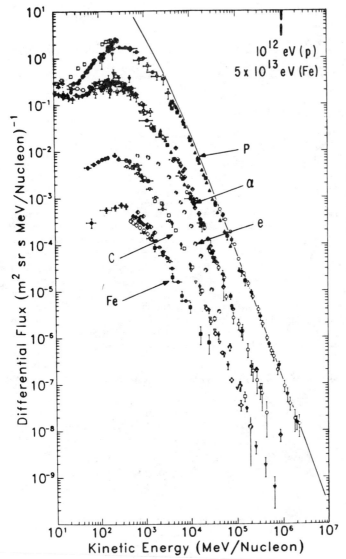

Figure 1. The differential energy spectra for the elements (from the top) hydrogen (P), helium (α), carbon (C), and iron (Fe). Also shown is the electron spectrum (labeled e). The solid curve shows the hydrogen spectrum extrapolated to interstellar space by unfolding the effects of modulation. The turn-up of the helium spectrum below about 60 MeV/nucleon is due to the contribution of the anomalous component of helium. This figure was taken from Simpson (1983).

Munoz *et al.* (1973), is called the anomalous component because of its unusual nature. Figure 3 shows the spectra of H, He, C, and O in the interplanetary medium. The anomalous component is the broad peak in the low energy oxygen spectrum and the absence of a dip in the helium spectrum at 10 MeV/amu that causes the helium flux to exceed the hydrogen flux in this energy range. A second component is sometimes accelerated in regions where fast and slow moving solar wind streams collide (Gloeckler, 1979). These co-rotating energetic particles will sometimes cause modest increases in the 1–10 MeV/

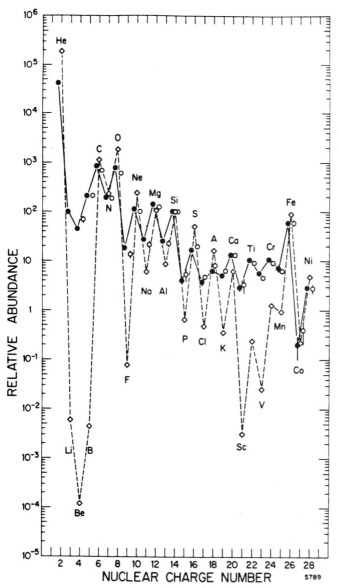

Figure 2. The cosmic ray element abundances (He-Ni) measured at Earth compared to the solar system abundances. The two abundances are normalized at silicon. The diamonds represent the solar system abundances, while the open circles are cosmic ray measurements at high energies in the 1000–2000 MeV/nucleon range. Hydrogen, not shown, is about 20 times more abundant in the solar system than in cosmic rays, using silicon as the normalization. This figure was taken from Simpson (1983).

amu hydrogen and helium fluxes bombarding the Moon. Because these components exceed the cosmic ray background only at low energies, their contribution to the total particle intensity bombarding the Moon is small.

Occasionally there are major increases in the radiation intensity at the Moon due to solar energetic particle (SEP) events. These events last from hours to days and range in size from the limit of detection to an intensity more than 70,000 times that of galactic

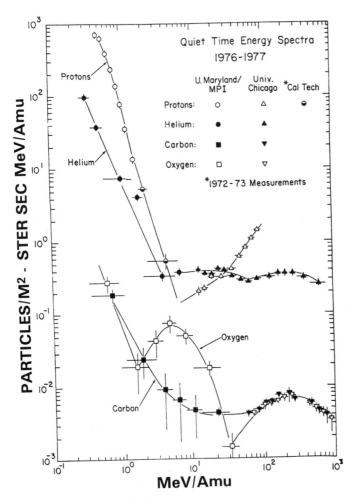

Figure 3. Differential energy spectra of hydrogen, helium, carbon, and oxygen observed in the interplanetary medium near the Earth during the solar minimum in 1976–77 during quiet times. The anomalous cosmic ray component appears between 2 and 30 MeV/ nucleon and is characterized by the large overabundance of helium and oxygen compared to hydrogen (protons) and carbon, respectively. This figure was taken from Gloeckler (1979).

cosmic rays. So large are the largest of these events that they determine the particle fluence at the lunar surface over a solar cycle. It is usually true that half the particles to strike the Moon in an 11-year solar cycle arrive in less than a day and are the result of one, or at most a few, large SEP events. This striking feature will make a flare watch an important part of any future lunar expedition, as it was during the Apollo program.

McGuire et al. (1983) show the record of SEP events for the last three solar cycles. Their results are reproduced in Fig. 4. As this figure shows, the frequency of SEP events varies with the overall level of solar activity as gauged by the smoothed Zurich sunspot number. McGuire et al. find the solar-cycle-averaged hydrogen fluxes above 10 MeV for cycles 19, 20, and 21 are 378, 93, and 65 particles/cm^2 s. These fluxes are 132, 33, and 23 times larger than the solar-cycle-averaged galactic cosmic ray hydrogen flux, respectively.

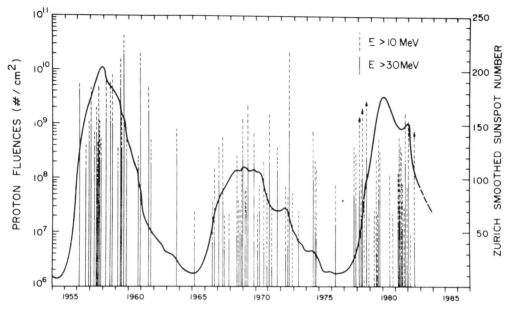

Figure 4. Hydrogen fluences above 10 and 30 MeV in Solar Energetic Particle Events during solar cycles 19, 20, and 21. The solid curve represents Zurich smoothed sunspot numbers. This figure was taken from McGuire et al. (1983).

The spectra of SEP events are much softer than the galactic cosmic ray spectrum. Even during the peak intensity of most SEP events, galactic cosmic rays are the principal source of particles above a few hundred MeV/amu. The largest observed flares have, at their peak, dominated the flux up to 10,000 MeV/amu.

The elemental composition of SEP events is very similar to the solar system composition (shown in Fig. 2) on average but can be highly variable from one event to the next. The composition even varies with particle energy in individual events (Chenette and Dietrich, 1984). SEP events that are enriched in one heavy ion tend to be enriched in the others as well. Dietrich and Simpson (1978) have shown that this systematic enrichment increases strongly with atomic number.

THE NRL CREME MODEL

A procedure has been developed at Naval Research Laboratory to characterize cosmic ray effects on microelectronics (CREME) used in spacecraft and aircraft (see Adams *et al.*, 1981; Adams *et al.*, 1983; and Tsao *et al.*, 1984). This procedure relies on a detailed numerical model of the near-Earth particle environment (Adams *et al.*, 1981), which is directly applicable to characterizing the radiation environment on the Moon. A set of formulas describes the differential energy spectra of each of the elements in galactic cosmic rays and how these spectra are modified by the contributions from the anomalous

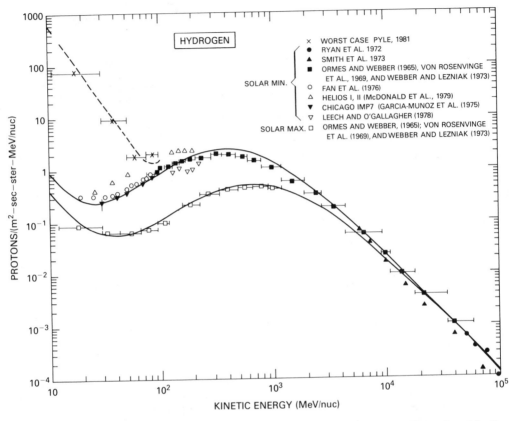

Figure 5. Hydrogen differential energy spectra (taken from Adams et al., 1981). The data are selected for the extremes of solar maximum and solar minimum. The solid curves are from the formulas to fit the cosmic ray spectra for solar minimum (upper curve) and solar maximum (lower curve). The dashed curve is from a formula constructed to give instantaneous flux levels so high at each energy that they are exceeded only 10% of the time.

component, co-rotating energetic particle streams, and from small flares. The model also contains formulas for the differential energy spectra of each of the elements in SEP events and formulas for calculating the probability of occurrence of such events.

Galactic cosmic rays were modeled by using all the available data to determine the shapes of the differential energy spectra of hydrogen, helium, and iron at the extremes of solar maximum and solar minimum. Figures 5, 6, and 7 show how the model (solid lines) fit the data for hydrogen, helium, and iron, respectively. The model spectra are for the extremes of solar maximum and minimum. Intermediate cases are interpolated with a sinusoidal solar modulation factor of $\sin[2\ (t-t_0)/10.9\ \text{years})]$, where $t_0 = 1950.6$. The other elemental spectra are obtained by multiplying either the helium or the iron spectrum by a constant or, in some cases, energy-dependent scale factor. By comparing

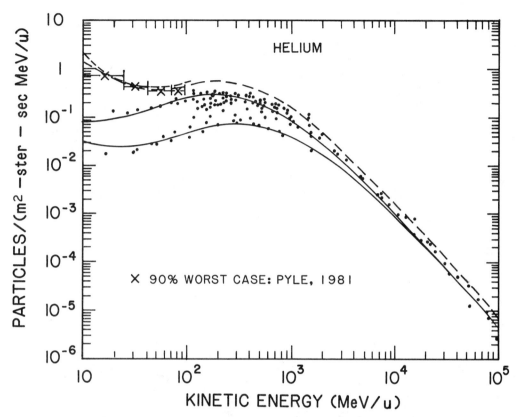

Figure 6. Helium differential energy spectra (taken from Adams et al., 1981). The solid curves are from the formulas to fit the cosmic ray spectra for solar minimum (upper curve) and solar maximum (lower curve). The dashed curve is from a formula constructed to give instantaneous flux levels so high at each energy that they are exceeded only 10% of the time.

this model with recent data, we find that it seems to predict the absolute cosmic ray flux to within a factor of two. The relative abundances are accurate to about 20%.

The contributions at low energies from co-rotating particle streams and small SEP events were accounted for along with the overall uncertainty by the 90% worst-case model. This model is shown as the dashed curves in Figs. 5–7. The contributions of the anomalous component to the helium, nitrogen, and oxygen spectra are modeled by Adams *et al.* (1981), who show how these may be combined with the cosmic ray model spectra to account for the contributions of the anomalous component at low energies.

Following the scheme of King (1974), Adams *et al.* (1981) divided all the large SEP events into ordinary large flares and anomalously large flares. A formula was fitted to the means of the log-normal distributions of the integral SEP flux above several energy thresholds and then differentiated to obtain the mean hydrogen differential energy spectrum for ordinary large SEP events. This procedure was repeated using values 1.28 standard

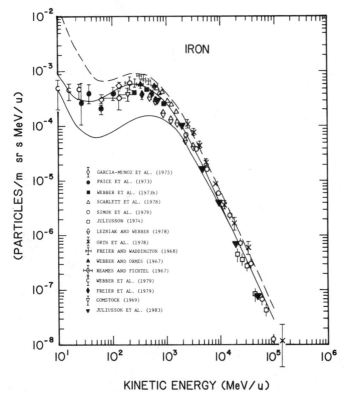

Figure 7. *Iron differential energy spectra (taken from Adams et al., 1981). The solid curves are from the formulas to fit the cosmic ray spectra for solar minimum (upper curve) and solar maximum (lower curve). The dashed curve is from a formula constructed to give instantaneous flux levels so high at each energy that they are exceeded only 10% of the time. This dashed curve has been constructed by comparison with helium, since iron data to establish this curve directly are lacking.*

deviations above the means of the log–normal distributions to obtain a spectrum for the 90% worst-case SEP event. Again, following King (1974), the SEP event of August 4, 1972 was used as the model for anomalously large events. These three model hydrogen spectra are shown in Fig. 8.

The composition of the SEP events is given by Adams *et al.* (1981) as elemental abundances relative to hydrogen for both the mean heavy ion composition and a 90% worst-case enrichment in the heavy elements. These two compositions, shown in Fig. 9, indicate the degree of variability in the SEP composition. Burrell distribution formulas are also provided by Adams *et al.* (1981) to calculate the probability of an SEP event during any time period.

RADIATION EFFECTS ON THE MOON

If a large permanent base is established on the Moon, the 5 rem/y exposure limit for Earth-based radiation workers might be a more appropriate standard for radiation protection. As Silberberg *et al.* (1985) show, adherence to this standard would make it necessary to bury a lunar habitat beneath several meters of lunar regolith and limit human activity on the lunar surface to "regular working hours," *i.e.*, about 1800 hours

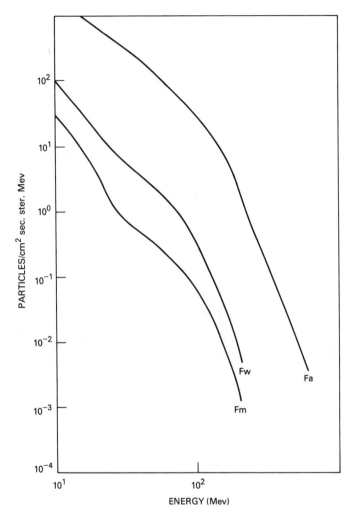

Figure 8. Hydrogen differential energy spectra (taken from Adams et al., 1981). These spectra are for the peak intensities of three model solar energetic particle (SEP) events. The curve labeled Fm is for the mean large SEP event (using the definition of King, 1974). A second model SEP event (curve Fw) has been constructed such that only one SEP event in 10 will have a peak intensity, at any energy, that is greater than predicted by this event. These two curves may be compared to get a feel for the range of flare sizes. The Fa curve is modeled after the peak of the SEP event of August 4, 1972. This is one of the most severe SEP events ever observed.

per year inside an enclosed vehicle. Extravehicular activity would have to be restricted. Even then, people working on the surface would have to remain near the habitat so that they could rush to shelter in case of a large SEP event. Long expeditions across the lunar surface would be risky unless shelters could be constructed a few hours travel time apart, or a means could be provided to quickly rescue the expedition and return its members to the safety of the buried lunar habitat.

The lunar radiation environment affects not only people, but electronic systems as well. It has long been known that electronic components are affected by the total radiation dose they have accumulated. This radiation damage produces changes in conductivity or shifts in device thresholds that cause a malfunction of the electronic circuit. Electronic components have been developed that can tolerate very large total doses, so that it is

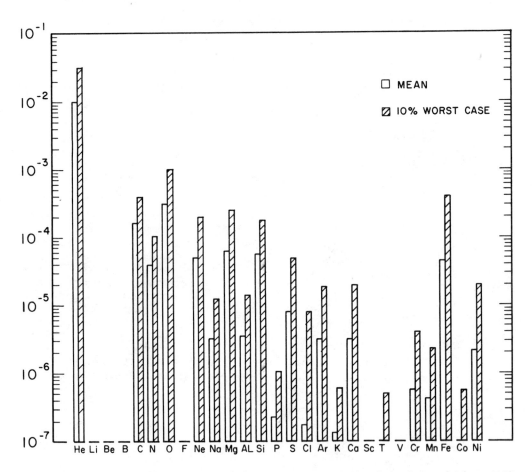

Figure 9. Elemental composition of Solar Energetic Particle (SEP) events (taken from Adams and Gelman, 1984). The mean SEP composition, normalized to hydrogen, is compared with a composition (10% worst case) constructed so that only one SEP event in 10 will be richer in any heavy ion. Comparing the two compositions, we can see that the iron to hydrogen ratio in a heavy ion-rich event may exceed the oxygen to hydrogen ratio for a typical event.

possible to design electronic systems for use on the Moon with operational lifetimes in excess of 10 years.

Recently, it has been discovered that single, intensely ionizing particles can produce a burst of hole–electron pairs so large that the resulting charge or current can change the logic state of a modern digital microcircuit (Adams *et al.*, 1981, 1983; Tsao *et al.*, 1984). This change of state damages not the electronic circuitry but the information stored in it. These events are therefore called "soft upsets."

The operational impact of a soft upset depends on the microcircuit affected. If the microcircuit is in the program memory of a computer, the program will no longer be operable and must be reloaded. If it is in the microprocessor's address registers or program counter, the actions the computer takes will be unpredictable. Soft upsets in control circuitry can also result in unplanned events such as thruster firings. It is clear that a single soft upset could cause the loss of equipment and personnel. Unlike total dose sensitivity, soft upset susceptibility is a fundamental feature of modern large-scale integrated circuits. It appears unlikely that such compact circuits can be made immune to soft upsets. The problem has been attacked at the system level instead, with redundancy, fault tolerance, and software checking. These methods reduce, but do not eliminate, the risks posed by soft upsets.

COSMIC RAY EXPERIMENTS FOR A LUNAR BASE

The Moon offers the possibility of doing cosmic ray experiments that would be difficult to carry out in Earth orbit. On the Moon, lunar regolith can be used for the massive absorbers needed in some large detector systems. The Moon also offers a site for the construction of large detector arrays beyond the protection of the Earth's magnetic field. Two possible experiments are discussed here.

The energy spectrum of cosmic rays has been measured directly up to 1 TeV/nuc (Watson, 1975). The only direct measurement above this energy is due to Grigorov *et al.* (1971). Therefore, most of what is known about cosmic rays above 1 TeV/nuc is based on indirect measurements that provide only total particle energy as determined from the shower of secondary particles produced in the atmosphere by an incident cosmic ray. Direct measurements at these higher energies will make it possible to establish the particle's charge and, hence, its velocity and magnetic rigidity. The latter quantity can be compared with the available data on galactic magnetic fields to determine whether the particles at these high energies could have come from our galaxy or must be extra-galactic.

The best device for measuring these high energies directly is an ionization calorimeter of the type developed at Goddard Space Flight Center (Balasubrahmanyan and Ormes, 1973). One square meter calorimeters could be constructed on the lunar surface, using regolith to replace the heavy iron plates in the Goddard design. A single calorimeter of this size would detect events up to 10,000 TeV/nuc in the first year of operation, and 100 such units could extend the spectrum to the interesting region above 100,000 TeV/nuc in a few years. A secondary benefit of such an experiment might be the opportunity to study elementary particle interactions at energies well above those achieved at accelerators. Such investigations could lead to new discoveries pointing the way for new particle physics experiments on Earth.

Experiments employing NASA's Long Duration Exposure Facility, presently underway and planned for the near future, are expected to establish the flux of actinide nuclei in galactic cosmic rays. This will tell us whether cosmic ray source material resembles the interstellar medium or is enriched in nuclei synthesized by the rapid neutron capture (r-) process. Whatever the nucleosynthetic origin of cosmic rays may turn out to be, these near-term experiments will not tell us how much time has elapsed since cosmic ray material was synthesized. To answer this question, it will be necessary to measure the abundances of the individual actinide nuclei. The relative abundances of Th, U, Np, Pu, and Cm tell us the elapsed time since the nucleosynthesis of cosmic rays in the range of 10^7 to 10^9 years. To measure them would require 2000 m^2 ster. years of collecting power. As suggested by Waddington (personal communication, 1984), this could be provided by a cylindrical array of scintillators 10 m in diameter and 10 m high. Such an apparatus could be placed on the lunar surface, and it would collect a suitable sample of events in less than 5 years. The array would use time of flight across the cylinder to measure velocity, so that the scintillator signals could be corrected for velocity to obtain the particle's charge.

CONCLUSIONS

The ionizing radiation environment on the lunar surface poses a hazard to men and sensitive instruments. Measures to protect crews from this environment can be expected to influence the design of lunar bases and the planning of lunar surface activities.

The lunar surface offers a site for large cosmic ray experiments to measure the abundances of rare elements and extremely high energy particles. The experiments that are possible on the Moon will provide new information on the origin of cosmic rays and possibly on the interaction of ultra-high energy particles with matter.

REFERENCES

Adams J. H. and Gelman A. (1984) The effects of solar flares on single event upset rates. *IEEE Trans. Nucl. Sci., NS-31,* 1212–1216.

Adams J. H., Silberberg R., and Tsao C. H. (1981) Cosmic ray effects on microelectronics, Part I: *The Near-earth Particle Environment.* NRL Memorandum Report 4506, Naval Research Laboratory, Washington, DC. 92 pp.

Adams J. H., Letaw J. R., and Smart D. F. (1983) *Cosmic Ray Effects on Microelectronics, Part II: The Geomagnetic Cutoff Effects.* NRL Memorandum Report 5099, Naval Research Laboratory, Washington, DC. 45 pp.

Balasubrahmanyan V. K. and Ormes J. F. (1973) Results on the energy dependence of cosmic-ray charge composition. *Ap. J., 186,* 109–122.

Chenette D. L. and Dietrich W. F. (1984) The solar flare heavy ion environment for single event upsets: A summary of observations over the last solar cycle, 1973–1983. *IEEE Trans. Nucl. Sci., NS-31,* 1217–1222.

Dietrich W. F. and Simpson J. A. (1978) Preferential enhancements of the solar flare-accelerated nuclei carbon to zinc from ~20–300 MeV nucleon. *Ap. J., 225,* L41–L45.

Fillius W. and Axford I. (1985) Large scale solar modulation of ≤500 MeV/Nucleon galactic cosmic rays seen from 1 to 30 AU. *J. Geophys. Res., 90,* 517–520.

Garcia-Munoz M., Mason G. M., and Simpson J. A. (1973) A new test for solar modulation theory: The 1972 May-July low-energy galactic cosmic-ray proton and helium spectra. *Ap. J., 182,* L81–L84.

Gloeckler G. (1979) Compositions of energetic particles populations in interplanetary space. *Rev. Geophys. Space Phys., 17,* 569–582.

Grigorov N. L., Gubin Yu V., Rapaport I. D., Savenko I. A., Akimov V. V., Nesterov V. E., and Yakovlev B. M. (1971) Energy spectrum of primary cosmic rays in the 10^{11}–10^{15} eV energy range according to the data of proton-IV measurements. *Proc. 12th Intl. Cosmic Ray Conf., 5,* pp. 1746–1751.

King J. H. (1974) Solar proton fluences for 1977–1983 space missions. *J. Spacecraft Rockets, 11,* 401–408.

McGuire R. E., Goswami J. N., Jha R., Lal D., and Reedy R. C. (1983) Solar flare fluences during solar cycles 19, 20, and 21. *Proc. 18th Intl. Cosmic Ray Conf., 44*, pp. 66–69.

Shapiro M. M. and Silberberg R. (1970) Heavy cosmic ray nuclei. *Ann. Rev. Nucl. Sci., 20*, 323.

Silberberg R., Tsao C. H., Adams J. H. Jr., and Letaw J. R. (1984) Radiation doses and LET distributions of cosmic rays. *Rad. Res., 98*, 209–226.

Silberberg R., Tsao C. H., Adams J. H. Jr., Hulburt E. O., and Letaw J. R. (1985) Radiation transport of cosmic ray nuclei in lunar material and radiation doses (abstract). This volume.

Simpson J. A. (1983) Elemental and isotopic compositions of the galactic cosmic rays. *Ann. Rev. Nucl. Part. Sci., 33*, 323–381.

Tsao C. H., Silberberg R., Adams J. H. Jr., and Letaw J. R. (1984) Cosmic Ray Effects on Microelectronics, Part III: Propagation of Cosmic Rays in the Atmosphere. NRL Memorandum Report 5402, Naval Research Laboratory, Washington, D.C. 87 pp.

Watson A. A. (1975) Energy spectrum and mass composition of cosmic ray nuclei from 10^{12} to 10^{20} eV. In *Origin of Cosmic Rays* (J. L. Osborne and A. W. Wolfendale, eds.), pp. 61–96. Reidel, Dordrecht.

CELESTIAL SOURCES OF HIGH-ENERGY NEUTRINOS AS VIEWED FROM A LUNAR OBSERVATORY

Maurice M. Shapiro

Max-Planck Institut für Astrophysik, 8046 Garching bei München, West Germany

Rein Silberberg

Hulburt Center, Naval Research Laboratory, Washington, DC 20375

The detection of high-energy (HE) cosmic and solar flare neutrinos near the lunar surface would be feasible at energies much lower than for a terrestrial observatory. At these lower energies ($\geq 10^9$ eV) the neutrino background is drastically reduced below that generated by cosmic rays in the Earth's atmosphere. Because of the short mean free path (< 1m) of the progenitor pi and K mesons against nuclear interaction in lunar rocks, the neutrino background would be quite low. At 1 GeV, less than 1% of the pions would decay; at 10 GeV, 0.1% would decay. Thus, if the neutrino flux to be observed is intense enough and its spectrum is steep enough, then the signal-to-noise ratio is very favorable. The reduction in cross section at lower energies would not cancel the advantage of enhanced flux. The observation of HE neutrinos from solar flares would be dramatically enhanced, especially at lower energies, since the flare spectra are very steep. Detection of these neutrinos on Earth does not appear to be feasible. Moreover, higher-energy neutrinos ($> 10^{12}$ eV) that could, in principle, be detected are virtually absent from solar flares. A remarkable feature of solar flares as viewed in HE neutrinos from a lunar base is that the entire surface of the sun would be "visible." Indeed, flares on the *far* side of the sun would be producing more neutrinos moving toward the detector than those on the near side. Diffuse sources of HE neutrinos, such as the galactic disc (especially from the galactic center), would be detectable at energies between, say, 10^9 and 10^{11} eV. On Earth, they are swamped by the overwhelming atmospheric background.

INTRODUCTION

The advantages of a lunar observatory for neutrino astronomy were discussed some years ago by F. Reines (1965). In the present paper, we suggest that the investigation of neutrinos from astrophysical sites at energies between 1 and 10^3 GeV can be better carried out on the Moon than on the Earth. In the dense lunar materials, competition between nuclear interactions of pions and their decay suppresses the frequency of decay. In the tenuous upper atmosphere of the Earth, on the other hand, the decay of pions (and of their muon progeny) does generate neutrinos. Hence, the flux of neutrinos near the surface of the Moon is about 10^{-3} of that on the Earth at energies between 1 and 10^2 GeV, and about 10^{-2} at 10^3 GeV. Only the background due to prompt neutrinos from the decay of charmed particles in the atmosphere is not suppressed.

At energies below 1 GeV, however, the path length of pions against decay diminishes as the Lorentz factor approaches unity, and pion decay is no longer suppressed, even on the Moon. Furthermore, due to the absence of magnetic shielding on the Moon, the

— flux of low-energy cosmic rays incident on the lunar surface is much higher than the average flux at the top of the Earth's atmosphere. This further enhances the low-energy neutrino intensity ($E < 1$ GeV) on the Moon. [The suppression of neutrino background was quantitatively explored by Cherry and Lande (1984) in a paper presented at this conference.]

Accordingly, a lunar base is probably an unsuitable site for observing the low-energy neutrinos (~10 MeV) from stellar gravitational collapse. Moreover, it is not competitive for recording neutrinos at very high energies ($E > 10^3$ GeV); this can be done more readily with Cerenkov light detectors in a large volume of sea water (some 10^8 m^3) near the bottom of the ocean. Such an array—DUMAND (a Deep Underwater Muon and Neutrino Detector)—will be emplaced in the waters near Hawaii in the near future (Peterson, 1983).

CRITERIA FOR CANDIDATE NEUTRINO SOURCES TO BE EXPLORED ON THE MOON

What types of neutrino sources are likely to be observable between 1 and 10^3 GeV? This is the energy interval for optimum detection by a neutrino observatory under the lunar surface (about 100 m below). The sources should emit neutrinos much more copiously above 1 GeV than above 1 TeV, so as to permit the construction of a neutrino observatory significantly smaller than DUMAND. An important constraint is imposed by the interaction cross section of neutrinos, which increases linearly with energy between 1 and 10^3 GeV. As a result, the observation of lower-energy neutrinos becomes more difficult. This cross section is given by

$$\sigma_{\nu N} = (0.7 \text{ or } 0.8) \times 10^{-38} \, E_\nu \text{ cm}^2 \qquad (1)$$

and

$$\sigma_{\bar{\nu} N} = 0.3 \times 10^{-38} \, E_{\bar{\nu}} \text{ cm}^2 \qquad (2)$$

for neutrinos and anti-neutrinos, respectively. Let the energy spectrum of the neutrinos be

$$\frac{dJ}{dE_\nu} = KE_\nu^{-\alpha} \qquad (3)$$

Then the event rate is proportional to

$$\int_{E_0}^{E_{max}} \sigma(E_0) K E_\nu^{-\alpha} \, dE_\nu \qquad (4)$$

i.e., it is proportional to

$$E_0^{-(\alpha-2)} - E_{max}^{-(\alpha-2)} \qquad (5)$$

Thus, one criterion for significant source strength in the energy interval between 1 and 10^3 GeV is a steep neutrino spectrum, with the exponent α appreciably greater than 2.

SOME PROMISING CANDIDATE SOURCES

Solar flares generate particles having steep energy spectra, with $\alpha = 4\text{-}7$ at proton energies above 1 GeV. Erofeeva *et al.* (1983) explored the use of a deep underwater detector of 10^6 tons for observing neutrinos from solar flares. They did not investigate the neutrino background in their paper. We estimate that the background rate is about 10^3 per day. If the neutrinos are emitted in about 20 minutes, as are the gamma rays from a flare, then the background rate is down to 10 for the duration of the flare. If, moreover, an angular resolution of 1 steradian is obtained, then the background is down to ~1 event for the duration of the flare.

For observation of neutrinos from very large flares, such as occur about once per solar cycle, a *terrestrial* underwater observatory of 10^6 tons seems adequate. However, for larger observatories, $>10^6$ tons, the neutrino background on Earth becomes prohibitive. Thus, for observing fine-time structure or neutrino energy spectra of very large flares, or for recording somewhat smaller flares, a lunar observatory of $>10^6$ tons provides an opportunity to carry out studies of flares that are not possible on the Earth. Even flares on the remote side of the sun become observable, since neutrinos with energies $<10^{11}$ eV can traverse the solar diameter. In fact, for a given size of flare, neutrinos should reach the detector in greater numbers from the far side than from flares on the near side. This is due to the favorable rate of production of pions (hence, of daughter neutrinos) that move toward the observer, when the progenitor protons or other energetic nuclei— on the far side—are directed toward the solar surface.

Another, more diffuse source of neutrinos with a fairly steep energy spectrum $\alpha = 2.7$ is that from the central annulus of the galactic disk, $\pm 60°$ in longitude and $\pm 5°$ in latitude about the galactic center. Stecker *et al.* (1979) explored the detectability of these neutrinos at 10^3 GeV with a DUMAND array of 10^9 tons (having an effective detection volume of some 10^{10} tons). The estimated rate of neutrino events to be expected was 130 per year, swamped by 1.8×10^4 background events per year. At $E > 1$ GeV, the event rate is about 100 times higher, so that even in a smaller detector of $\sim10^7$ tons, the event rate is about 10 per year, with the signal exceeding the background in a lunar observatory.

In addition, there are many interesting discrete candidate sources of neutrinos: accreting neutron stars (including pulsars) in binary systems, active galactic nuclei with accretion disks from which matter drifts into ultra-massive black holes (Silberberg and Shapiro, 1979), and the expanding shells around young pulsars (Berezinsky, 1976; Shapiro and Silberberg, 1979). However, the energy spectra of neutrinos from these sources are as yet unknown.

Presented here are the results of a sample calculation for SS433, which appears to be one of the most promising candidate sources in our galaxy, at a distance of about

3 Kpc. This object is probably an accreting black hole in a binary system; it has two relativistic jets and other remarkable features. Its estimated power output is 3×10^{39} ergs/s (Grindlay *et al.*, 1984), but values that are higher by an order of magnitude have also been proposed (Eichler, 1980). If we assume that a power input of 3×10^{39} ergs/s yields protons of energy ≥ 10 GeV and that these protons suffer nuclear collisions, a detector of 10^6 tons would permit the observation of about 30 neutrino events per year. With 10^7 tons, several different sources of neutrinos become detectable.

CONCLUSIONS

We conclude that a neutrino detector of $\geq 10^6$ tons on the Moon—*i.e.*, one considerably more compact than the proposed DUMAND array—would open up a new window of neutrino astronomy, making possible the study of neutrinos at 1–10^3 GeV*. The effort must probably await the establishment of a substantial lunar colony; because of its large size, the detector would probably have to be locally constructed, perhaps of glass fabricated from lunar materials.

Acknowledgments. One of the authors (MMS) expresses his appreciation to Professors R. Kippenhahn and W. Hillebrandt for their hospitality at the Max-Planck Institut für Astrophysik in Garching. He thanks Professor F. Reines for stimulating his interest in this problem.

REFERENCES

Berezinsky V. S. (1976) Ultra HE neutrinos and detection possibilities by DUMAND. In *Proc. 1976 DUMAND Workshop* (A. Roberts, ed.), pp. 229–255. Univ. Hawaii, Honolulu.

Cherry J. L. and Lande K. (1985) Proposal for a neutrino telescope on the Moon (abstract). In *Papers Presented to the Symposium on Lunar Bases and Space Activities of the 21st Century*, p. 50. NASA/Johnson Space Center, Houston.

Eichler D. (1980) SS433: A possible neutrino source. In *Proc. of 1980 DUMAND Symposium, Vol. 2*, pp. 266–271. Hawaii Dumand Center, Honolulu.

Erofeeva I. N., Lyotov S. I., Murzin V. S., Kolomeets E. V., Albers J., and Kotzer P. (1983) Detection of solar flare neutrinos. In *Proc. 18th Intl. Cosmic Ray Conf.*, pp. 104–107. Tata Institute for Fundamental Research, Bombay.

Grindlay J. E., Band D., Seward F., Leahy D., Weisskopf M. C., and Marshall F. E. (1984) The central x-ray source in SS433. *Ap. J., 277*, 286.

Peterson V. Z. (1983) Deep underwater muon and neutrino detection. In *Composition and Origin of Cosmic Rays* (M. M. Shapiro, ed.), pp. 251–268. Reidel, Dordrecht.

Reines F. (1965), as reported in Shapiro M. M. (1965) Galactic cosmic rays. In *NASA 1965 Summer Conference on Lunar Exploration and Science*, pp. 317–330, NASA SP-88. NASA, Washington.

Shapiro M. M. and Silberberg R. (1979) Neutrinos from young supernova remnants. In *Proc. 16th Intl. Cosmic Ray Conf., Vol. 10*, pp. 363–366. Univ. Tokyo, Kyoto.

Silberberg R. and Shapiro M. M. (1979) Neutrinos as a probe for the nature of and process in active galactic neutrinos. In *Proc. 16th Intl. Cosmic Ray Conf., Vol. 10*, pp. 357–362. Univ. Tokyo, Kyoto.

Even a smaller detector of 10^4 tons could detect a giant solar flare like that of Feb. 23, 1956. The pulse is likely to be of such short duration (< 20 min) that the atmospheric background would not degrade terrestrial observation.

Stecker F. W., Shapiro M. M., and Silberberg R. (1979) Galactic and extra-galactic HE neutrinos. In *Proc. 16th Intl. Cosmic Ray Conf., Vol. 10*, pp. 346–351. Univ. Tokyo, Kyoto.

A LUNAR NEUTRINO DETECTOR

M. Cherry and K. Lande

Department of Physics, University of Pennsylvania, Philadelpha, PA 19104

The major experimental difficulty in neutrino astronomy lies in the fact that expected event rates are exceedingly small (typically 10^{-4} or fewer neutrinos per year per sr per ton of detector). A detector must therefore be extremely massive and must be located in a very low background environment. Over the energy range 1 GeV–10 TeV, the neutrino background on the Moon is lower than on the Earth (at some energies by as much as 10^{-3}–10^{-4}). At both lower and higher energies, the lunar background is just as high as that on Earth, but in the proper energy range, the Moon may be the only possible site for neutrino astronomy. We review the properties of terrestrial neutrino detectors located deep underground or underwater, discuss the calculated and measured backgrounds, and demonstrate the improvement to be obtained with a lunar location. In addition, we briefly discuss a possible design for a 10^{6} ton lunar neutrino detector.

INTRODUCTION

Neutrinos are produced as a result of collisions of cosmic rays with the ambient material in astronomical sources, the Earth's atmosphere and surface, and the lunar surface. Neutrinos are produced directly when cosmic rays interact to produce secondary charmed mesons, when π and K mesons decay to μ mesons and electrons, and when the μ mesons decay to electrons. Gamma rays are produced from neutral $\pi°$ meson decays resulting from the same cosmic ray interactions, and the numerous satellite, balloon, and ground-based measurements of astronomical gamma ray sources make it clear that such cosmic ray interactions occur in a great variety of sources and that large numbers of neutrinos must also be emitted. Neutrinos have a much greater interaction mean free path than do gamma rays, however ($\lambda_{\nu_\mu} / \lambda_\gamma \sim 10^{11}$ at 100 GeV), so that if the neutrinos can be detected, they can be used to probe to far greater depths in dense media. The significance of astronomical neutrino observations and the relationship to gamma ray astronomy have been discussed previously by, among others, Berezinsky and Zatsepin (1970), Lande (1979), Fichtel (1979), Stecker (1979), Shapiro and Silberberg (1983), Silberberg and Shapiro (1983), and Lee and Bludman (1985).

Due to the low detection rates and high terrestrial background, the Moon may provide the best (and perhaps the only) location for observations of astronomical neutrinos at energies 1 GeV–10 TeV. In the following section, we briefly review the estimates of neutrino detection rates from astronomical sources and the Earth's atmosphere. We list the existing large underground detectors and discuss some of the experimental difficulties associated with the low anticipated astronomical event rates and the high atmospheric background. In a later section, we describe the lower neutrino background expected for a detector on the Moon; in the last section, we discuss how a lunar neutrino detector might look.

PRESENT EXPERIMENTAL STATUS OF NEUTRINO ASTRONOMY

If observed high energy gamma rays are primarily nucleonic in origin (*i.e.*, they result from cosmic ray interactions and the resulting π° meson decays) and if the bulk of the emission originates in source regions whose thickness (t) is small compared to the gamma ray attenuation length (*i.e.*, $t \lesssim 10^2 g\ cm^{-2}$), then the gamma ray and neutrino emission must be comparable. Unfortunately, since the neutrino interacts so weakly, it can be detected only very inefficiently. Fichtel (1979) has estimated neutrino detection rates based on observed gamma ray fluxes; for the quasar 3C273, the active galaxy Cen A, the central region of our own galaxy, γ–ray pulsars, and the galactic γ–ray source Cyg X–3, he finds expected interaction rates of only 3×10^{-9} detected neutrinos per yr per kiloton of detector above 1 TeV and 3×10^{-10} neutrinos per yr per kiloton above 10 TeV, corresponding to neutrino fluxes of 10^{-12} and $10^{-13} cm^{-2}\ s^{-1}$, respectively. At lower energies, Cherry and Lande (unpublished data, 1985) expect rates of 10^{-5} neutrinos $ton^{-1}\ yr^{-1}\ sr^{-1}$ from the Crab and $10^{-4}\ ton^{-1}\ yr^{-1}\ sr^{-1}$ from the galactic center above 100 MeV.

These small rates of neutrinos interacting in a detector must be visible in the presence of a nearly isotropic background of neutrinos produced locally by cosmic rays interacting in the Earth's atmosphere. The fluxes of atmospheric neutrinos have been calculated. The recent analytic calculation of Dar (1983) and the detailed Monte Carlo calculation of Gaisser *et al.* (1983) are in substantial agreement with each other and with the measured interaction rates in the underground IMB proton decay detector (Bionta *et al.*, 1983). In this and other large underground detectors, it is impossible to measure energies as high as 1 TeV; rather, one measures the integral flux of all neutrinos with energies above a relatively low threshold (~200 MeV for the IMB detector). The rates measured deep underground ($10^{-2}\ T^{-1} yr^{-1}$) are in substantial agreement with the calculated interaction rate of atmospheric neutrinos (Gaisser and Stanev, 1984).

Gaisser and Stanev (1984) have also considered the case of neutrinos traveling upward through the Earth, interacting in the rock beneath the detector, and producing muons that then continue upward through the detector. If the detector is located sufficiently deep underground that the flux of penetrating downward–moving cosmic ray muons is sharply reduced, then the small upward flux can be measured. The resulting measured upward fluxes (~2 – 7 $\times 10^{-13} cm^{-2}\ s^{-1}\ sr^{-1}$) are also in substantial agreement with the calculations of atmospheric background.

It appears, then, that the atmospheric neutrino background is reasonably well understood, and one can compare the calculated atmospheric neutrino spectrum to predicted astronomical spectra. An example of such a comparison is shown in Fig. 1. The neutrino flux from the Earth's atmosphere is based on the analytic approach described by Dar (1983). The line labeled "Galactic" gives an estimate of the diffuse neutrino flux expected from the region of the galactic center. This estimate agrees substantially with the detailed calculation of Stecker (1979). One can see that the atmospheric spectrum is steeper than the galactic spectrum and falls below the galactic spectrum for neutrino energies $E \lesssim 10$ TeV. It is for this reason that many previous discussions of neutrino astronomy have typically emphasized neutrino energies above several TeV.

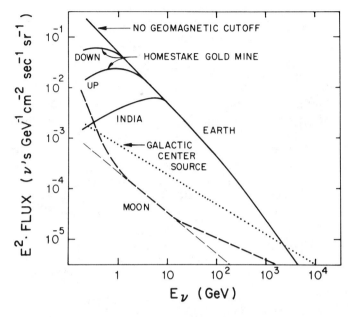

Figure 1. *Flux of muon neutrinos and anti-neutrinos. The terrestrial flux is shown for the case of no geomagnetic field, for vertically downward and upward neutrinos at the Homestake and IMB sites, and for downward neutrinos in India in the direction of maximum geomagnetic shielding. The lunar flux is calculated (light line) for π and K meson decays; the heavy line takes into account the effect of low-energy K^+ and high-energy charm decays.*

The detectors envisioned for these observations have typically been massive instruments located deep underground or underwater in order to reduce the background. For example, the proposed DUMAND detector (Stenger, 1984) involves instrumenting a 50 Mt 5×10^7 m^3 volume of the Pacific Ocean at a depth of 4.7 km; the IMB detector is an 8 kT water detector at a depth of 600 m in the Morton Salt Mine near Cleveland, Ohio; and a 1–5 kT liquid scintillation detector (Cherry *et al.*, 1983) is eventually planned at a depth of 1500 m in the Homestake Gold Mine in Lead, South Dakota. The largest underground detector so far proposed is approximately 30 kT of water (A. Mann and

Table 1. Operating Underground Detectors

Detector	Location	Mass (tons)	Type
< 100 Tons			
Soudan I	Minnesota, U.S.	30	Fe Calorimeter
100–1000 Tons			
Kolar	India	140	Fe Calorimeter
Homestake	South Dakota, U.S.	140	Liquid Scintillator
NUSEX	Italy–France	150	Fe Calorimeter
Baksan	USSR	330	Liquid Scintillator
Frejus	Italy–France	800	Fe Calorimeter
HPW	Utah, U.S.	800	Water Cerenkov
> 1000 Tons			
Kamioka	Japan	3000	Water Cerenkov
IMB	Ohio, U.S.	8000	Water Cerenkov

B. Cortez, private communications, 1985). A list of existing large underground detectors is given in Table 1. A number of detection techniques have been utilized and shown to be effective. In each case, though, the cost of the experiment is typically $1-5 million per kiloton.

Unfortunately, even for massive detectors located deep underground where the remnant cosmic ray background fluxes are low, neutrino and neutrino–induced muon rates are still meager—both in terms of the absolute number of events per year and compared to the atmospheric background. As an example, the calculated rates of high-energy muons penetrating from the Earth's surface to a depth of 1500 m and the rate of lower-energy neutrino-induced muons at the same depth are shown in Fig. 2 for the case of the Homestake Large Area Scintillation Detector (Cherry et al., 1985). The penetrating muons must have a minimum energy of $E_\mu \sim 2.6$ TeV at the Earth's surface in order to penetrate to the depth of the detector; the minimum energy detectable at the detector is $E_\mu \sim 2$ GeV. By contrast with the rates of Fig. 2, the galactic center spectrum of Fig. 1 would give 3 neutrino-induced muon events per year per sr above 2 GeV and 2 $yr^{-1}sr^{-1}$ above 100 GeV. With such extremely low event rates, it is absolutely essential to minimize the background as much as possible.

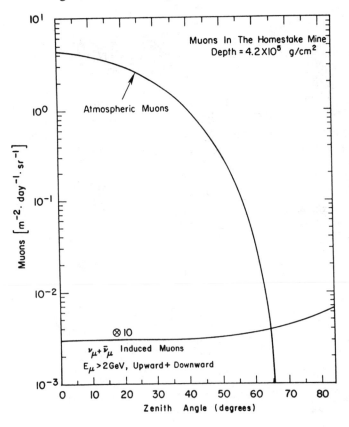

Figure 2. Predicted muon fluxes in the Homestake Gold Mine (Dar, written communication, 1984).

LUNAR AND TERRESTRIAL BACKGROUNDS

Over a large range of energies, the background problem can be alleviated by placing the detector above the atmosphere—for example, on the Moon. In the Earth's tenuous upper atmosphere, cosmic ray primaries have interaction lengths of 80 g cm^{-2}; the secondary pion interaction lengths are 120 g cm^{-2}. The total thickness of the atmosphere (1020 g cm^{-2} from the top of the atmosphere down to sea level) is many interaction lengths, so that even the Earth's surface is exceptionally well shielded from the primary cosmic rays and the hadronic components of the cosmic ray showers. In order to obtain the same hadron shielding, a lunar detector should be buried at a depth of at least 10^3g cm^{-2}, or 3 m.

Once a π (or K) meson is produced in the Earth's upper atmosphere, it can either interact or decay. Neutrinos are produced from the decays in Table 2. If the initial meson interacts, then neutrinos are produced by the decays of later-generation mesons formed lower in the atmosphere. The contribution from these secondary interactions is relatively small, however (20%). The meson decay length is a function of the meson energy E: $\lambda_\pi^{decay} = \beta\gamma c\tau = 7.8\beta\gamma$ m, where $\beta = v/c$ is the pion velocity in units of the speed of light c, $\gamma = E/mc^2$ is the energy in units of the mass, and τ is the pion lifetime at rest. By contrast, the pion interaction length is nearly independent of energy: λ_π^{inter} = 120 g cm^{-2} ($\lambda_\pi^{inter} \sim 6$ km in the Earth's upper atmosphere, $\lambda_\pi^{inter} \sim 0.4$ m in lunar rock). The probability of decay depends on the relative values of λ_π^{decay} and λ_π^{inter}.

$$\frac{\lambda_\pi^{inter}}{\lambda_\pi^{decay}} = \begin{cases} \dfrac{6000}{7.8\beta\gamma} \sim \dfrac{100}{E_\pi(GeV)} & \text{for the Earth} \\[4mm] \dfrac{0.4}{7.8\beta\gamma} \sim \dfrac{1}{110E_\pi(GeV)} & \text{for the Moon} \end{cases} \tag{1}$$

As long as this ratio is large, mesons will decay before they have a chance to interact, and the neutrino flux will be high. In the Earth's tenuous upper atmosphere, a large fraction of the mesons below 100 GeV decay; above 100 GeV, the decreasing decay probability suppresses neutrino production. In the dense lunar (or terrestrial) rock, the interactions occur long before the mesons have a chance to decay, and at all energies above 10 MeV, neutrino production is highly suppressed. The ratio of lunar and terrestrial neutrino fluxes from decay is roughly

$$\frac{\phi_\nu^\pi(Moon)}{\phi_\nu^\pi(Earth)} \sim \frac{\lambda_\pi^{inter}/\lambda_\pi^{decay}(Moon)}{\lambda_\pi^{inter}/\lambda_\pi^{decay}(Earth)} \sim \begin{cases} 1/110\, E_\pi & \text{for } E_\pi \lesssim 100 \text{ GeV} \\[2mm] 10^{-4} & \text{for } E_\pi \gtrsim 100 \text{ GeV} \end{cases} \tag{2}$$

We assume here that the production is the same in a lunar or terrestrial target of thickness 10^3 g cm^{-2}; we ignore the extra neutrinos produced in the terrestrial rock, since few mesons penetrate through the entire atmosphere; and we let the terrestrial ratio $\lambda_\pi^{inter}/$

Table 2. Most Probable Neutrino–Producing Decay
Modes of Non–charmed Mesons

π°, η°	$\rightarrow$	2γ
π^-, K_1	$\rightarrow$	$\mu^- \bar{\nu}_\mu$
π^+, K^+	$\rightarrow$	$\mu^+ \nu_\mu$

$$\mu^+ \rightarrow e^+ \nu_e \bar{\nu}$$

$$\mu^- \rightarrow e^- \bar{\nu}_e \nu_\mu$$

$$\pi^+ e^- \bar{\nu}_e$$

$$K^\circ \rightarrow \quad \pi^- e^+ \nu_e$$

$$\pi^+ \mu^- \bar{\nu}_\mu$$

$$\pi^- \mu^+ \nu_\mu$$

$\lambda_\pi^{\text{decay}}$ saturate at 1 for $E_\pi \lesssim 100$ GeV, where essentially all terrestrial pions decay. The neutrino flux from K mesons is similarly suppressed on the Moon.

The ratio ϕ_ν (Moon)/ϕ_ν (Earth) due to the competition between π and K meson interactions and decays is shown as the dashed curve in Fig. 3. At energies above 10 GeV the lunar suppression is reduced, however, by the effect of charmed meson production. The energy dependence of charm production is not yet well understood, but at Fermilab energies Ball *et al.* (1984) found cross sections for pN $\rightarrow$ D$\bar{\text{D}}$X and pN $\rightarrow$ ΛDX of 10–20μb/nucleon, about 10^3 times smaller than the corresponding pN $\rightarrow$ πX cross-sections. The lifetime of the charmed mesons is exceedingly short, however ($\tau_{\text{D}^\circ,\text{D}^\pm} \sim 4 - 9 \times 10^{-13}$ s, $\tau_{\text{F}\pm} \sim 2 \times 10^{-13}$ s), so that all charmed mesons decay promptly either on Earth or on the Moon. The effect of the Moon is therefore to suppress only the neutrinos from π and K mesons. The maximum suppression is given by the ratio (R) of prompt neutrinos from charm decay to neutrinos from π and K meson decay: R $\sim 10^{-3} - 10^{-4}$ (Elbert *et al.*, 1981; Inazawa and Kobayakawa, 1983). Above 10 GeV, the relative neutrino flux from charmed and π meson decays is given by the production ratio times the ratio of charmed meson to π meson decay probabilities:

$$\frac{\phi_\nu^c(\text{Moon})}{\phi_\nu^\pi(\text{Moon})} \sim 10^{-3} \times \frac{1}{1/110\, E_\pi} \sim \frac{E_\pi(\text{GeV})}{10} \tag{3}$$

The solid curve above 10 GeV shows this behavior.

Below about 1 GeV, the terrestrial neutrino flux is reduced by the effects of the Earth's geomagnetic field. At geomagnetic latitude (λ), zenith angle (θ), and azimuthal angle (φ)

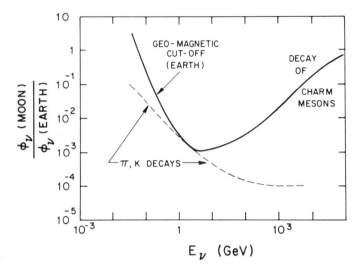

Figure 3. Ratio of lunar to terrestrial neutrino fluxes versus neutrino energy.

from magnetic North, the Earth's magnetic field effectively prevents primary cosmic rays with rigidity

$$p < 60 \cos^4\lambda \left(1 + \sqrt{1 + \cos^3\lambda \sin \theta \sin \varphi}\ \right)^2 GV \tag{4}$$

from reaching the atmosphere. For downward-moving protons at the IMB and Homestake detectors, this cutoff is near 2 GV; at the Kolar Gold Fields in India, the cutoff is at 16 GV. This geomagnetic suppression does not apply to the Moon, so $\phi_\nu(\text{Moon})/\phi_\nu(\text{Earth})$ increases at low energies; the details depend on the particular location and direction of cosmic ray incidence on the Earth, but the qualitative effect is shown in Fig 3.

Figure 1 shows the flux of neutrinos we calculate for the Earth without any geomagnetic field, for downward and upward trajectories at the Homestake detector, and for the direction of maximum cutoff rigidity (~60 GV) at the Kolar detector in India. The dashed line is the lunar flux under 4 m of rock. The dotted line is the flux expected from cosmic ray-matter interactions near the galactic center. The Moon offers a major suppression of the neutrino background between about 1 GeV and 1 TeV and, in particular, may make it possible to see the very interesting galactic center source as well as numerous other potential astronomical sources.

A LUNAR NEUTRINO DETECTOR

Since neutrinos can escape from dense sources that are opaque to gamma rays, it is quite possible that actual neutrino fluxes may turn out to be much larger than those predicted on the basis of the observed γ-rays. If one is to embark on a project as complex and costly as a lunar neutrino detector, however, one must probably adopt extremely

LUNAR NEUTRINO DETECTOR

CALORIMETER WITH PROPORTIONAL COUNTER
PLANES OR TUBES

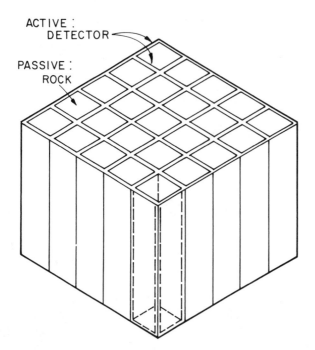

ACTIVE :
DETECTOR

PASSIVE :
ROCK

Figure 4. Schematic diagram of a possible lunar neutrino detector, consisting of active detector planes distributed through a large mass (10^6 tons) of lunar rock.

conservative design criteria. In particular, one must probably start from the very low neutrino fluxes estimated from the γ-ray measurements. For a flux level 10^{-5} ton^{-1}yr^{-1}sr^{-1}, one therefore needs a detector mass on the order of 10^6 tons. Since it is presumably unreasonable to transport 10^6 tons of water or liquid scintillator to the Moon, one must rather use local lunar material for the main detector mass. For example, a block of lunar surface material 30 m high × 100 m wide × 100 m long buried beneath several meters of soil would provide a well-shielded 900 kT neutrino target.

The detector might be instrumented as a calorimeter with planes of gas-filled drift chambers set out through the detector volume, as in Fig. 4. If the detector planes are separated by 1 m, then the detector threshold is given by the minimum muon energy required to penetrate 2 detector layers—about 1 GeV. The 30–100 m dimensions of the detector make muon energy measurements possible up to 100 GeV for some cases.

The detector would presumably have to be fabricated in a lunar laboratory. Wire, filling gas, and electronics could be supplied from Earth, but the main detector elements might be locally constructed gas-tight glass structures. Trenching, drilling, and digging of the detector volume would need to be done on site. Individual drift chamber dimensions might be 3 m $\times$ 3 m $\times$ 3 cm, requiring 3×10^4 elements. Although the size of such a detector is certainly mammoth by terrestrial standards, the electronic complexity is comparable to other high energy experiments.

The logistical, engineering, and financial problems associated with a lunar neutrino detector would be enormous. From the observational point of view, however, the low neutrino backgrounds would make lunar viewing conditions significantly better than anything possible on the Earth in the energy range 1–1000 GeV.

Acknowledgments. Funding for the University of Pennsylvania/Homestake program is provided by the United States Department of Energy. We have benefitted from many discussions with Drs. S. Bludman, A. Dar, T. Gaisser, H. Lee, and T. Stanev.

REFERENCES

Ball R. C. *et al.*, (1984) Prompt muon-neutrino production in a 400-GeV proton beam-dump experiment. *Phys. Rev. Lett., 51*, 743–746.

Berezinsky V. S. and Zatsepin G. T. (1970) Cosmic neutrinos of ultrahigh energy. *Sov. J. Nucl. Phys., 11*, 111–114.

Bionta R. M. *et al.* (1983) Results from the IMB detector. In *Fourth Workshop on Grand Unification* (H. A. Weldon, P. Langacker, and P. J. Steinhardt, eds.) pp. 46–68. Birkhauser, Boston.

Cherry M. L., Davidson I., Lande K., Lee C. K., Marshall E., Steinberg R. I., Cleveland B., Davis R., and Lowenstein D. (1983) Physics opportunities with the Homestake large area scintillation detector. *ICOMAN '83: International Colloquium on Matter Non-Conservation*, p. 133–143. INFN, Rome.

Cherry M. L., Corbato S., Kieda D., Lande K., Lee C. K., and Steinberg R. I. (1985) The Homestake large area scintillation detector and cosmic ray telescope. In *Solar Neutrinos and Neutrino Astronomy* (M. L. Cherry, K. Lande, and W. A. Fowler, eds.) pp. 32–49. American Institute of Physics, New York.

Dar A. (1983) Atmospheric neutrinos, astrophysical neutrinos, and proton decay experiments. *Phys. Rev. Lett., 51*, 227–231.

Elbert J. W., Gaisser T. K., and Stanev T. (1981) Possible studies with DUMAND and a surface air shower detector. In *DUMAND-80* (V. J. Stenger, ed.), pp. 222–228. Hawaii DUMAND Center, Honolulu.

Fichtel C. E. (1979) The significance of gamma-ray observations for neutrino astronomy. *Proc. 1978 DUMAND Summer Workshop* (A. Roberts, ed.) p. 289–312. DUMAND Scripps Institute of Oceanography, La Jolla, CA.

Gaisser T. K. and Stanev T. S. (1984) Interaction rates of atmospheric neutrinos. In *ICOBAN '84: International Colloquium on Baryon Non-Conservation* (D. Cline, ed.), Park City, UT. In Press.

Gaisser T. K., Stanev T., Bludman S. A., and Lee H. (1983) Flux of atmospheric neutrinos. *Phys. Rev. Lett., 51*, 223–226.

Inazawa H. and Kobayakawa K. (1983) The production of prompt cosmic ray muons and neutrinos. *Prog. Theor. Physics (Japan), 69*, 1195–1206.

Lande K. (1979) Experimental neutrino astrophysics. *Ann. Rev. Nucl. and Part. Sci., 29* (J. D. Jackson, H. E. Gove, and R. F. Schwitters, eds.), pp. 395–410. Annual Review, Palo Alto, CA.

Lee H. and Bludman S. A. (1985) Neutrino production from discrete high-energy gamma ray sources. *Astrophys. J., 290*, 28–32.

Shapiro M. M. and Silberberg R. (1983) Gamma rays and neutrinos as complementary probes in astrophysics. *Space Sci. Revs., 36*, 51–56.

Silberberg R. and Shapiro M. M. (1983) Sources of extragalactic cosmic rays: Photons and neutrinos as probes. In *Composition and Origin of Cosmic Rays* (M. M. Shapiro, ed.) pp. 231–244. Reidel, Dordrecht.

Stecker F. W. (1979) Diffuse fluxes of cosmic high-energy neutrinos. *Astrophys. J., 228*, 919–927.

Stenger V. J. (1984) The production of very high energy photons and neutrinos from cosmic proton sources. *Astrophys. J., 284*, 810–816.

NEUTRINO MEASUREMENTS ON THE MOON

Albert G. Petschek

Los Alamos National Laboratory, Los Alamos, NM 87545 and
New Mexico Institute of Mining and Technology, Socorro, NM 87801

Several possible neutrino experiments on the Moon are discussed in the light of the expected background rates. Observations of neutrino oscillations may be feasible.

Grand unified theories suggest a variety of novel physical processes, including nucleon decay, that is, the breakup of neutrons or protons into lighter particles, and neutrino oscillations, a process in which neutrinos with one type of weak interaction change back and forth into neutrinos with another interaction. In order to test these theories, a number of very large proton decay detectors with masses between 0.14 and 3 k tons have been built on Earth.

Protons are presumed to decay into various leptons and mesons, possibly $e^+ + \pi^0$. For a more complete list, see the review of particle properties (Particle Data Group, 1984). The detectors attempt to observe these decay products, or their daughters, by observing Cherenkov radiation in a large volume of water (Bionta *et al.*, 1983) or ionization in gas-filled proportional tubes (Peterson, 1983; Krishnaswamy *et al.*, 1983). A principal source of background in these detectors, none of which has observed a proton decay, is neutrino interactions. The neutrinos in question are generated by the interaction of cosmic rays with the upper atmosphere, producing relativistic pions and other particles that decay to produce neutrinos of high (Doppler shifted) energy before they interact with another nucleus in the rarefied upper atmosphere. Naively, one might expect a substantially lower background on the Moon, since interactions with the solid lunar material would take place before decay. Accordingly, the Lunar Base Working Group (Duke *et al.*, 1984) suggested that the Moon would be a suitable location for a proton decay detector sensitive to 10^{34} or 10^{35} years proton lifetime. Such a lifetime corresponds to between 0.06 and 0.006 decays per kiloton–year if all the nucleons, including those bound in nuclei, are active and fewer decays otherwise. Thus, exposures of 100 to more than 1000 k ton-years would be required, a massive undertaking on Earth, let alone on the Moon.

As is detailed by Cherry and Lande (1985) the neutrino background on the Moon is less than that on the Earth only in a limited energy band, one that just begins at the 1 GeV proton decay energy. The total energy of proton decay, including the rest energy of the decay products is, of course, the proton rest energy of 938 MeV, less than that at which the background is reduced. Higher backgrounds mean poor experiments; poor experiments do not get funded. Hence, it is neither worthwhile nor possible to move proton decay experiments to the Moon, and it will not be possible to piggyback neutrino detection on the huge decay detectors that would be required. Nevertheless, it is of interest to discuss what neutrino experiments might usefully be done on the Moon. Further discussion may be found in the paper by Shapiro and Silberberg (1985).

Another source of background, of interest in its own right, is astrophysical neutrinos. Their spectrum dominates atmospheric neutrinos even on Earth above 1000 GeV. These neutrinos originate in a variety of places. The ones with highest energy originate in interactions between cosmic rays and interstellar matter (Dar, 1983) or cosmic rays and the 3 K background radiation (Stecker, 1979). Lower energy neutrinos arise from stellar collapse (Burrows, 1984). This background should be the same on Earth as on the Moon, and the lower local background above 1 GeV will allow it to be studied to somewhat smaller energies on the Moon than on Earth.

Neutrinos produced in the Earth's atmosphere will also reach the Moon. Since the Earth subtends a solid angle of 5×10^{-5} of 4π at the Moon, the rate of interaction of atmospheric neutrinos with a lunar detector will be reduced from the terrestrial value of 100 per k ton year (Gaisser and Stanev, 1983) to 5×10^{-3} k ton^{-1} year^{-1}. This is a few during the 100–1000 k ton years of exposure required in the proton–decay experiment. If a really large detector were to be built, it might be possible to detect neutrino oscillations in a new range of Δm^2, the difference in the squares of the masses of the two neutrinos. The probability of transition to a second neutrino, for example, from the electron neutrino to the μ or τ neutrino, depends on the parameter $\Delta m^2 L/E$ (Boehm, 1983) where L is the flight path and E the neutrino energy. Experiments explore values of this parameter ~1 if Δm^2 is in eV^2, L is m, and E is MeV. Thus reactor experiments explore $\Delta m^2 = 1$ eV^2 or a little less (L is a few m, E a few MeV). Accelerator experiments use much higher energies and larger flight paths, but still explore a similar range of Δm^2. Neutrinos have been observed from the sun at well below the expected rate (Bahcall et al., 1982). Since the neutrinos emitted by the sun are electron neutrinos and the detection method (by the transmutation of Cl to Ar) detects only these same neutrinos, a possible explanation of the discrepancy between theory and experiment is that an oscillation into another neutrino has taken place. With this assumption the solar neutrino experiment can be viewed as an oscillation experiment with E a few MeV and L = 1.5×10^{11} m, corresponding to very small Δm^2.

Gaisser and Stanev (1984) have put limits on neutrino oscillations by observing the dependence of the intensity of upward–going neutrinos produced in the Earth's atmosphere on angle, that is, on path length to the detector from the source (the atmosphere). In this experiment the characteristic energy is 1000 MeV and the characteristic distance is an Earth diameter so that $\Delta m^2 \sim 10^{-4}$ eV^2 is explored, less than for accelerator or reactor experiments but much greater than for the solar experiment. If, as is suggested at the beginning of the preceding paragraph, these same neutrinos can be detected on the Moon, then another region of Δm^2 can be explored. This region corresponds to mass differences much larger than those of the solar experiments but smaller than those of Gaisser and Stanev. In contrast to the experiment with solar neutrinos, which must rely on a calculation of the source, the experiment on the Moon would derive its neutrino source from terrestrial measurements such as those used by Gaisser and Stanev.

In summary, a few interesting neutrino experiments can be contemplated on the Moon, but they require massive detectors. A body with no atmosphere but with a magnetic moment comparable to the Earth's to reduce the low energy neutrino background would be much more suitable.

REFERENCES

Bahcall J. N., Huebner W. F., Lubow S. H., Parker T. D., and Ulrich R. K. (1982) Standard solar models and the uncertainties in predicted capture rates of solar neutrinos. *Rev. Mod. Phys., 54,* 767–800.

Bionta R. M., *et al.* (1983) Results from IMB detector. In *Fourth Workshop on Grand Unification* (H. A. Weldon, P. Langacker, and P. J. Steinhardt, eds.), pp. 46–68. Birkhäuser, Boston.

Boehm F. (1983) Neutrino mass and neutrino oscillations. In *Fourth Workshop on Grand Unification* (H. A. Weldon, P. Langacker, and P. J. Steinhardt, eds.), pp. 163–173. Birkhäuser, Boston.

Burrows A. (1984) On detecting stellar collapse with neutrinos. *Astrophys. J., 283,* 848–852.

Cherry M. and Lande K. (1985) Proposal for a neutrino telescope on the Moon. This volume.

Dar A. (1983) Atmospheric neutrinos and astrophysical neutrinos in proton decay experiments. In *Fourth Workshop on Grand Unification* (H. A. Weldon, P. Langacker, and P. J. Steinhardt, eds.), pp. 101–114. Birkhäuser, Boston.

Duke M. B., Mendell W. W., and Keaton P. W. (1984) *Report of the Lunar Base Working Group.* LALP-84-43. Los Alamos National Laboratory, Los Alamos. 41 pp.

Gaisser T. K. and Stanev T. (1983) Calibration with cosmic ray neutrinos. *AIP Conference Proceedings #114* (M. Blecher and K. Gotow, eds.), pp. 89–97. American Institute of Physics, New York.

Gaisser T. K. and Stanev T. (1984) Neutrino induced muon flux deep underground and search for neutrino oscillations. *Phys. Rev. D 30,* 985–990.

Krishnaswamy M. R., Menon M. G. K., Mondal N. K., Narasimham V. S., Sreekantan B. V., Hayashi Y., Ito N., Kawakami S., and Miyake S. (1983) The K.G.F. nuclear decay experiment. In *Fourth Workshop on Grand Unification* (H. A. Weldon, P. Langacker, and P. J. Steinhardt, eds.), pp. 25–34. Birkhäuser, Boston.

Particle Data Group (1984) Review of Particle Properties, *Rev. Mod. Phys., 56,* S1–S304.

Peterson E. (1983) New results from the Soudan 1 detector. In *Fourth Workshop on Grand Unification* (H. A. Weldon, P. Langacker, and P. J. Steinhardt, eds.), pp. 35–45. Birkhäuser, Boston.

Shapiro M. M. and Silberberg R. (1985) High-energy neutrino astronomy from a lunar observatory. This volume.

Stecker F. W. (1979) Diffuse fluxes of cosmic high-energy neutrinos. *Astrophys. J., 228,* 919–927.

MASS EXTINCTIONS AND COSMIC COLLISIONS: A LUNAR TEST

Friedrich Hörz

Experimental Planetology Branch, SN4, NASA/Johnson Space Center, Houston, TX 77058

Chemical and physical evidence strongly indicates synchroneity between the Cretaceous/Tertiary mass extinctions and the collision of a large cosmic object with Earth some 65 m.y. ago. Statistical time series analysis of the marine extinction record for the past 200 m.y. reveals periodicity on the order of 30 m.y. Time series analyses on the formation ages of terrestrial impact craters may yield similar periodicities; extinction and cratering cycles may even be in phase. However, the crater analyses are somewhat ambiguous because of the small number of terrestrial craters that can be dated precisely. It is therefore suggested that additional formation ages of lunar craters be obtained such that statistically improved and sound time series analysis for collision events in the Earth-Moon system can be performed. If synchroneity between cratering rate and mass extinctions were confirmed, far-reaching implications for the evolution of life would result.

INTRODUCTION

Paleontologists, geochemists, planetologists, astrophysicists, and others are currently debating whether some or all major mass extinctions in the geologic record of Earth are caused by the collision of massive, cosmic objects (*e.g.*, Silver and Schultz, 1982; Holland and Trendall, 1984).

The fossil record abounds with evidence that major mass extinctions are followed by an explosive "blooming" of new life forms that rapidly occupy apparently empty habitats. Thus, the sudden disappearance of old life forms and the almost equally sudden radiation of new forms (or of a few survivors) are closely intertwined. The brevity of the time scales involved can be termed "sudden" only in a geologic context. Generally, an enormous diversity of both marine and continental life is affected. This argues very strongly for some "catastrophic" deterioration of the environment on global scales. As a consequence, modern paleontological views allow room for both catastrophism and orderly evolution; the term "spasmodic" evolution is frequently used (Raup, 1984).

As detailed below, a strong case can be made for the correlation between one major mass extinction and the impact of a massive cosmic object; tentative evidence exists for a second case. Speculative suggestions postulate a general, causal link between most (all?) mass extinctions and hypervelocity collisions. Large-scale collisions must have taken place throughout geologic time within the entire solar system because most planetary surfaces are pockmarked by hypervelocity craters. Many of these craters are 100 km in diameter and represent global catastrophies.

THE CRETACEOUS/TERTIARY MASS EXTINCTION

The entire debate started with a watershed discovery by Alvarez *et al.* (1980). They found that siderophile trace elements such as Ir and Pt are greatly enriched and sharply concentrated in exactly the same strata in which palenotologists had placed the boundary between the Cretaceous (K) and Tertiary (T) geologic time periods. The K–T boundary can be identified in some localities with centimeter precision by a sharp decline, if not total disappearance, of many Cretaceous life forms and the subsequent blooming of new, Tertiary species. It is estimated that at least 85% of all Cretaceous biomass became extinct and no land animal larger than 25 kg survived this catastrophe some 65 m.y. ago (Russell, 1979). Alvarez *et al.* (1980) proposed that the unusual concentration of siderophile elements in these strata must be remnants of a gigantic meteorite. Subsequent work by many geochemists demonstrated that this siderophile–rich layer is of global extent; importantly, it may be found in marine as well as continental sediments (Fig. 1). In addition, Bohor *et al.* (1984) discovered "shocked" quartz grains in the K–T layer. These grains display deformation features that are diagnostic for the passage of transient, high pressure shock waves, which in turn can only be generated in nature by high speed impact. Thus, there

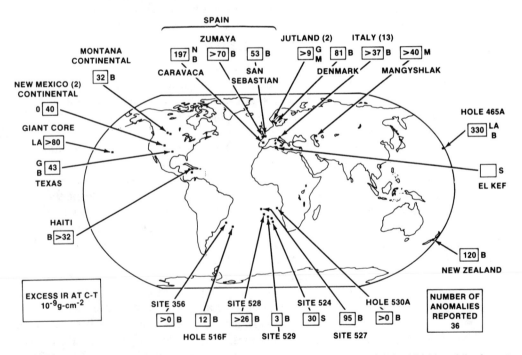

Figure 1. *Worldwide distribution of locations where 65-m.y.-old strata of anomalously rich siderophile element concentrations have been observed. The letters next to the boxes refer to major analytical groups specializing in such analyses, and the numbers in the box refer to the absolute concentration of the key element Ir in units of 10^{-9} g per cm^2 (from Alvarez et al., 1982).*

is chemical and physical evidence that the mass extinctions at the K–T boundary are indeed synchronous with a large collisional event.

The quantity of meteoritic material in the K–T layer requires an impacting body some 5–10 km across. Upon impact, such a body releases energies measured in millions of megatons of TNT (billions of Hiroshima bombs!), and it produces a crater some 50–100 km in diameter. These are minimum estimates; the crater could have been larger. Nevertheless, such a large, 65-m.y.-old crater is not presently known on Earth. This is a somewhat disturbing aspect of the collisional scenario, but by no means a fatal one, considering the dynamic nature of Earth. Also, very little is known about the actual kill mechanism(s), *i.e.*, what kind of environmental changes and specific stresses did the impact generate? Extreme temperature excursions can readily be envisioned as can changes of the pH of oceans and surface water. Direct poisoning by heavy metals and catastrophic floodings, if the impact occurred in the ocean, were also suggested (Toon *et al.*, 1982; O'Keefe and Ahrens, 1982; Hsu *et al.*, 1982; Gault and Sonett, 1982; and others). In all likelihood, a combination of these (and other?) effects resulted in short- and long-term imbalances in the environment and associated breakdown(s) of the food chain.

What is the probability of collision between a cosmic object some few kilometers across and the surface of Earth? Fortunately, there are three independent ways to estimate this probability: (1) The total number of impact structures on Earth is about 100 (see Fig. 2). This crater population may be used to calculate a minimum crater production rate for Earth (Grieve, 1982; Grieve *et al.*, 1985). (2) It was possible to date the ages of lunar volcanic lava flows returned by Apollo with isotopic methods. All craters that punched through these lava flows must therefore be younger than the flows themselves. Combining the observable number of such craters and the age of the target rock, lunar

TERRESTRIAL CRATERS

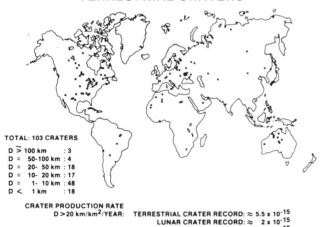

TOTAL: 103 CRATERS

D $\gtrsim$ 100 km   : 3
D = 50-100 km : 4
D = 20- 50 km : 18
D = 10- 20 km : 17
D = 1- 10 km : 48
D < 1 km   : 18

CRATER PRODUCTION RATE
D >20 km/km²/YEAR: TERRESTRIAL CRATER RECORD: $\approx$ 5.5 x 10⁻¹⁵
LUNAR CRATER RECORD: $\approx$ 2 x 10⁻¹⁵
ASTRONOMICAL OBSERVATIONS: $\approx$ 6 x 10⁻¹⁵

OR: $\approx$ 1 CRATER OF D =   10 km PER $\approx$ 10⁵ YEARS
$\approx$ 1 CRATER OF D =   20 km PER $\approx$ 4 MILLION YEARS
$\approx$ 1 CRATER OF D = 100 km PER $\approx$ 50 MILLION YEARS

Figure 2. Locations of known terrestrial craters (from Grieve, 1982) and their approximate size distribution. Note that three structures of 100 km or larger in diameter are known; however, none of them is close in age to the K-T boundary at 65 m.y. Approximately 50% of all craters are less than 10 km in diameter. While this inventory of craters contains a few older than 250 m.y., the crater production rates indicated are those calculated for the past 250 m.y.

crater production rates can be calculated (Neukum and Wise, 1976; Shoemaker, 1984). (3) Present-day astronomical telescope observations measure the frequency and orbits of small objects, generally less than 10 km in diameter, in the inner solar system. The collision probability of these objects with Earth may be computed (Wetherill and Shoemaker, 1982; Shoemaker, 1983). While the astronomical observations refer to the present, the terrestrial cratering record is an average over the last few hundred m.y. and the lunar cratering data integrate over the last 3 to 3.5 b.y., the age of lunar basaltic volcanism. Nevertheless, collision rates based on these different approaches agree to better than an order of magnitude, if not better than a factor of 5. Therefore, the probability for large-scale collisions on Earth may be calculated with considerable confidence (see Fig. 2). Large-scale collisions appear unavoidable in Earth's geologic record.

SYNCHRONEITY OF MASS EXTINCTIONS AND CRATERING RATE

From the above, it seems clear that the K–T collision was by no means unique. The question therefore arises whether mass extinctions may be associated with collisions in general. The search for siderophile element enhancement in paleontological boundary layers has just begun. Ganapathy (1982) found such an enhancement for the Eocene/ Oligocene extinctions some 34 m.y. ago. This evidence, however, is tentative and lacks confirmation on a worldwide scale. Searches at other paleontological boundaries were unsuccessful to date.

A different perspective, however, provides incentive to consider a general correlation between impact and mass extinctions. Raup and Sepkoski (1984) subjected the marine mass extinction record for the last 250 m.y. to statistical time series tests. They found distinct periodicity of about 27 m.y. in the occurrence of mass extinctions. Depending on the approach and on various weighting factors, others have subsequently arrived at similar periodicities ranging from 26 to 32 m.y. Simultaneously, the terrestrial cratering record was subjected to time series tests by Alvarez and Muller (1984) and by Rampino and Stothers (1984a,b). Surprisingly, maxima in the terrestrial cratering record seem to reveal similar periodicity of 30 m.y. Importantly, both cycles appear to be approximately in phase (see Fig. 3).

If we accept a causal link between mass extinctions, cosmic collisions, and a period of some 30 m.y. as dictated by the biological record, we must identify mechanisms that gravitationally perturb small solar system objects on such time scales to become Earth-crossing objects. The most populous objects are comets. They reside at the outer fringes of the solar system, and their number is estimated to be 10^{13}. The existence of comets in the inner solar system is proof that comets can be gravitationally deflected into orbits that come close to the sun and thus Earth (Weissmann, 1982). Periodic character of this process requires a massive object that disturbs the comet reservoir on a cyclic schedule. Such an object could be a small companion star of the sun itself (Whitmire and Jackson, 1984; Davis *et al.*, 1984), but a more probable cause is the passage of our solar system through the galactic plane, where encounters with gigantic and massive molecular clouds seem unavoidable (Schwartz and James, 1984). In principle, both suggestions appear valid,

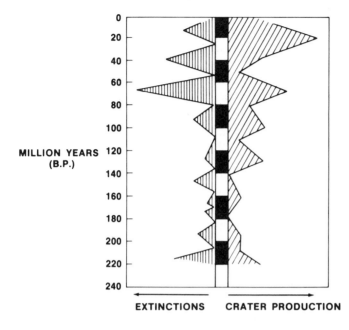

MILLION YEARS
(B.P.)

EXTINCTIONS   CRATER PRODUCTION

Figure 3. Periods of pronounced mass extinctions for the last 250 m.y. with peak height reflecting their magnitude (left-hand side; after Raup and Sepkoski, 1984). The right-hand side depicts maxima in the crater production rate as assembled by the author using the tabulation of Grieve (1982) for 35 crater ages (isotope ages and a few stratigraphic ages) of craters larger than 5 km.

although passage through the galactic plane is probably the more likely and less *ad hoc* mechanism (Clube and Napier, 1984; Hills, 1984; Torbett and Smoluchowski, 1984).

According to the above, statistical time series tests seem to yield a positive correlation between impact and mass extinctions. Plausible, if not compelling, mechanisms that may drive this 30 m.y. cycle appear to exist. Nevertheless, these generalizations are severely questioned by many.

Major criticism relates to the small number of terrestrial craters that can be subjected to statistical time series. Most craters are badly eroded. In many cases, erosion has totally removed the surficial impact melts. It is such melts, however, that are the choice material in isotopic age dating. Without impact melts, a crater formation time can generally not be determined exactly. Isotope ages (± 1%) exist for some 25 craters only. The formation times of all other craters are bracketed—at best—by stratigraphic evidence. The crater must be younger than its target rocks and older than the first sediments that backfilled the cavity. This stratigraphic bracketing yields on occasion relatively good ages, but it results mostly in age estimates that are useless for incorporation in the statistical time series tests.

Additionally, crater size must be considered. A 1 km diameter crater will certainly not produce a global catastrophy. Nevertheless, relatively small craters could be the sole survivors of impactor swarms, also referred to as "showers." The impact scars of the more massive shower members may have been totally obliterated by erosion or sedimentation, or the "large" impact may have happened in the ocean.

This brings us to a crucial point: The terrestrial cratering record is incompletely preserved as summarized by Grieve (1982) and Grieve *et al.* (1985). Erosion and sedimentation rates on Earth are highly variable on regional if not local scales. Craters may be obliterated by surface erosion; they may also be buried by sedimentation. Otherwise identical craters are subject to vastly different mechanisms and rates of erosion (or burial) depending on local, geological environment. For example, one of the best preserved large craters on Earth, the Ries Crater in Germany, 26 km in diameter, formed 15 m.y. ago and therefore at approximately the time when the Colorado River started (!) to carve the Grand Canyon. Some 1500 m of rocks were removed at Grand Canyon and yet the uppermost meters of ejecta are still preserved at the Ries. Due to fortuitous circumstances, the Ries was completely buried immediately after formation by 200 m of sediments; erosion and removal of these sediments started a few million years ago and has presently proceeded to a stage where the crater is being exhumed again. Thus, highly variable erosion and sedimentation environments rendered a fair number of craters unrecognizable during the past 200 m.y., and the terrestrial cratering record is incomplete.

From the above it follows that the terrestrial crater population may represent a subset of "survivors" only. For the purposes of statistical treatment, the formation ages of this limited sample may be biased. Second, choices have to be made as to which crater formation ages are appropriate, possibly introducing additional bias. Third, a minimum crater size applicable to the problem must be defined; the cut-off diameters used in the above time series tests are typically 5 or 10 km. Thus, considerable high-grading of the terrestrial crater population is required for time series analysis. Additional bias may be introduced in selecting this limited subset. Depending on personal preference and judgment, only some 20–40 craters may be suitable. This is a small statistical sample; inclusion or exclusion of a few craters may have significant effects as demonstrated by Grieve *et al.* (1985). Within permissible ranges of geological weighting factors, Grieve *et al.* defined a variety of sample sets and subjected them to time series tests. They obtained periodicities ranging from 18–30 m.y. Also, 30 m.y. cycles resulted that may or may not be in phase with the mass extinctions. Indeed, the possibility that the cratering ages are entirely random cannot be positively excluded.

As a consequence, statistical time series tests of the terrestrial cratering record yield ambiguos results. Statistical analyses of the small number of craters is simply not robust enough to demonstrate synchroneity with the mass extinctions. By the same token, none of the analyses demonstrate that such a correlation is inconsistent with the (limited) evidence at hand. It is therefore permissible to adopt a general, causal relationship between impact and mass extinction as a viable working hypothesis. We must subject this hypothesis to more rigorous tests.

Kyte (1984) analyzed chemically a deep ocean drill core that represented an essentially complete sedimentary record for the past 70 m.y. Enrichment of siderophile elements was only observed at the K–T boundary. Thus, this conceptually elegant test failed in detecting periodicity in Earth's collision record.

The most obvious solution for this important debate is a statistically sound data base. More craters need to be dated with the prerequisite precision. As stated above, isotope geochronology can only be performed on impact melts. Most terrestrial craters containing such melts are dated and already part of the data set. Some additional craters may be discovered, no doubt, as our abilities to recognize these impact scars improve; many of these new discoveries, however, will identify relatively eroded structures. Most well preserved structures that yield datable impact melts are probably part of the present data set because they are recognized with relative ease. Thus, in the author's intuitive view, the number of well dated terrestrial craters may be doubled, with luck.

THE LUNAR CRATERING RECORD

Within the solar system, Earth and its Moon occupy essentially one location; both bodies were subject to the same bombardment history. As a consequence, formation

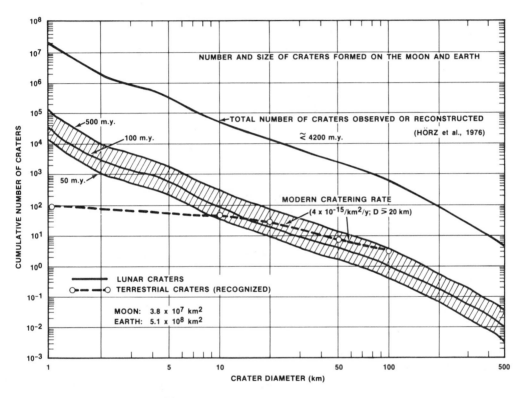

Figure 4. *Observed and calculated cumulative size frequencies of lunar craters 1 to 500 km diameter. The uppermost curve depicts the observable frequency for craters >50 km formed during the past 4 b.y.; however, the absolute number of craters <50 km are computed values (Hörz et al., 1976). These calculations are based on the well documented size frequency distribution of small craters on relatively young surfaces. The numbers of craters above a specific size that were produced during the past 500, 100, and 50 m.y.—bracketed by the hatched band—were calculated with Grieve's "modern" crater production rate for craters >20 km and assuming that the size frequency distribution of lunar craters remained constant through geologic time. The latter is a reasonable assumption; if anything, the size distribution of <50 km diameter craters is better known than that of very large craters for recent geological history. The terrestrial crater size distribution is depicted also; it becomes relatively flat at small crater sizes, indicating that a fair number of "small" craters were obliterated or have not yet been recognized on Earth. The absolute crater numbers refer to the entire surface areas of the Moon and Earth. The figure may therefore be used to estimate how many square kilometers need to be covered during manned exploration to sample a specific number of craters above a specific diameter.*

times of lunar collisions may be mixed with terrestrial crater ages in the statistical time series tests.

The heavily cratered lunar surface attests to a much better preservation of craters than on Earth. The Moon is geologically less active and lacks efficient erosive agents, such as water. The total number of craters larger than 100 km in diameter is 700 (as illustrated in Fig. 4). However, most large craters are remnants of an early bombardment phase with unusually high impactor fluxes (Wetherill, 1975; Neukum and Wise, 1976; Soderblom, 1977). The number of craters produced during the last 50, 100, and 500 m.y. is indicated in Fig. 4. Accordingly, a 100-km-diameter crater was formed once every 100 m.y., during which time the Moon also retained some 5 craters larger than 50 km or approximately 100 craters larger than 10 km. All of these craters are exceptionally well preserved by terrestrial standards and should yield proper impact melt samples for isotope geochronology. If craters as small as 5 km in diameter were included, the number of datable collisions could reach the hundreds. Clearly, the "modern" lunar cratering record has the potential to place any time series tests on statistically firm footing.

Figure 4 depicts the crater population of the entire Moon to illustrate that exploration, sampling, and ultimate dating of, for instance 100 lunar craters, requires systematic coverage of large fractions of the lunar surface. Long range sampling mobility of about 1000 km radius is necessary. Clearly, this amounts to a major undertaking, although sample acquisition spread over a 10 year period, possibly even longer, appears acceptable. Such time scales may be short compared to the efforts required in expanding the terrestrial cratering record to an adequate degree. Moreover, terrestrial studies include the risk of being only partly successful.

SUMMARY

In conclusion, synchroneity and causal relationship between mass extinctions and collisional events may be accepted as a viable working hypothesis, but not more. The hypothesis must be subjected to additional and more diagnostic tests. The most direct test demands that more craters are dated to a precision measured in small fractions of the proposed period, i.e., to a precision of ± 5 m.y. The number of datable terrestrial craters will always remain small compared to the inventory of relatively recent lunar craters. There is considerable doubt whether the terrestrial cratering record can ever be reconstructed with enough confidence to accept or reject a temporal and causal relationship between mass extinctions and collisions on Earth. A return to the Moon and extensive lunar exploration may therefore hold crucial clues in our quest to better understand how life evolved on Earth.

REFERENCES

Alvarez L. W., Alvarez W., Asaro F., and Michel H. V. (1980) Extraterrestrial cause for the Cretaceous-Tertiary extinction. *Science, 208,* 1095–1108.
Alvarez W., Alvarez L. W., Asaro F., and Michel H. V. (1982) Current status of the impact theory for the terminal Cretaceous extinctions. *Geol. Soc. Am. Spec. Pap. 190,* 305–316.

Alvarez W. and Muller R. A. (1984) Evidence from crater ages for periodic impacts on the Earth. *Nature, 308,* 718–720.

Bohor B. F., Foord E. F., Modreski J., and Triplehorn D. M. (1984) Mineralogic evidence for an impact event at the Cretaceous-Tertiary boundary. *Science, 224,* 867–869.

Clube S. V. M. and Napier V. M. (1984) Terrestrial catastrophism-Nemesis or galaxy. *Nature, 311,* 635–636.

Davis M., Hutt P., and Muller R. A. (1984) Extinction of species by periodic comet showers. *Nature, 308,* 715–717.

Ganapathy R. (1982) Evidence for a major meteorite impact on the Earth 34 million years ago: implications on the origin of North American tektites and Eocene extinctions. *Geol. Soc. Am. Spec. Pap. 190,* 513–516.

Gault D. E. and Sonett C. P. (1982) Laboratory simulation of pelagic asteroidal impact: atmosphere injection, benthic topography, and the surface radiation field. *Geol. Soc. Am. Spec. Pap. 190,* 69–92.

Grieve R. A. F. (1982) The record of impact on Earth: implications for major Cretaceous/Tertiary impact event. *Geol. Soc. Am. Spec. Pap. 190,* 25–38.

Grieve R. A. F., Sharpton V. L., Goodacre A. K., and Garvin J. B. (1985) A perspective on the evidence for periodic cometary impacts on Earth. *Earth Planet. Sci. Lett.* In press.

Hills J. G. (1984) Dynamical constraints on the mass and perihelion distance of Nemesis and the stability of its orbit. *Nature, 311,* 636–638.

Holland H. D. and Trendall A. F. (editors) (1984) *Patterns of Change in Earth Evolution.* Springer Verlag, Berlin. 431 pp.

Hörz F., Gibbons R. V., Hill R. E., and Gault D. E. (1976) Large-scale cratering of the lunar highlands: some Monte Carlo model considerations. *Proc. Lunar Sci. Conf. 7th,* pp. 2932–2945.

Hsu K. J., McKenzie J. A., and He Q. X. (1982) Terminal Cretaceous environmental and evolutionary changes. *Geol. Soc. Am. Spec. Pap. 190,* 317–328.

Kyte F. (1984) Cenozoic iridium sedimentation: Death of Nemesis? (abstract) *Geological Society of America Abstract with Programs, 16,* 567.

Neukum G. and Wise D. U. (1976) Mars: a standard crater curve and possible new time scale. *Science, 194,* 1381–1387.

O'Keefe J. D. and Ahrens T. J. (1982) The interaction of the Cretaceous/Tertiary bolide with the atmosphere, ocean and the solid Earth. *Geol. Soc. Am. Spec. Pap. 190,* 103–120.

Rampino M. R. and Stothers R. B. (1984a) Terrestrial mass extinction, cometary impacts and the Sun's motion perpendicular to the galactic plane. *Nature, 308,* 709–712.

Rampino M. R. and Stothers R. B. (1984b) Geological rhythms and cometary impacts. *Science, 226,* 1427–1431.

Raup D. M. (1984) Evolutionary radiations and extinctions. In *Patterns of Change in Earth Evolution* (H. D. Holland and A. F. Trendall, eds.), pp. 5–15. Springer Verlag, Berlin.

Raup D. M. and Sepkoski J. J., Jr. (1984) Periodicity of extinctions in the geologic past. *Proc. Natl. Acad. Sci. U.S.A., 81,* 801–805.

Russell D. A. (1979) The enigma of the extinction of dinosaurs. *Annu. Rev. Earth Planet. Sci., 7,* 163–182.

Schwartz R. D. and James P. B. (1984) Periodic mass extinctions and the Sun's oscillation about the galactic plane. *Nature, 308,* 712–713.

Shoemaker E. M. (1983) Asteroid and comet bombardment of the Earth. *Annu. Rev. Earth Planet. Sci., 11,* 15–41.

Silver L. T. and Schultz P. H. (editors) (1982) Geological implications of impacts of large asteroids and comets on Earth. *Geol. Soc. Am. Spec. Pap. 190,* 528 pp.

Soderblom L. A. (1977) Historical variations in the density and distribution of impacting debris in the inner solar system: evidence from planetary imaging. In *Impact and Explosion Cratering* (D. J. Roddy, R. O. Pepin, and R. B. Merrill, eds.), pp. 629–633. Pergamon, New York.

Toon O. B. (1984) Sudden changes in atmospheric composition and climate. In *Patterns of Change in Earth Evolution* (H. D. Holland and A. F. Trendall, eds.), pp. 41–61. Springer Verlag, Berlin.

Torbett M. V. and Smoluchowski B. (1984) Orbital stability of the unseen solar companion linked to periodic extinction events. *Nature, 311,* 641–642.

Weissmann P. R. (1982) Terrestrial impact rates for long and short-period comets. *Geol. Soc. Am. Spec. Pap. 190,* 15–24.

Wetherill G. W. (1975) Late heavy bombardment of the moon and the terrestrial planets. *Proc. Lunar Sci. Conf. 6th*, pp. 1539–1561.

Wetherill G. W. and Shoemaker E. M. (1982) Collision of astronomically observable bodies with the Earth. *Geol. Soc. Am. Spec. Pap. 190*, 1–14.

Whitmire D. P. and Jackson A. A. (1984) Are periodic mass extinctions driven by a distant solar companion? *Nature, 308*, 713–715.

6 / LUNAR CONSTRUCTION

THE FIRST HABITATS, LABORATORIES, and industrial plants to go to the lunar surface undoubtedly will be prefabricated and self-contained to the greatest extent possible. Masses and volumes will be constrained by the capabilities of the transportation network to the Moon. Current working concepts are very similar to the engineering sketches in 1971; *i.e.*, space station modules are placed on the surface and buried for protection from the galactic radiation flux. The base is expanded through the addition of more modules to an interconnected network.

For early engineering studies, emplacement of a fully configured module creates a realistic scenario for modeling transportation, surface operations, power, and other requirements. As a lunar surface facility evolves beyond an outpost or camp, expansion of operational capability must be weaned from dependence on expensive transportation from Earth. Large enclosed volumes will be needed for maintaining equipment, housing increasingly complex scientific apparatus, and providing comfortable living and working environments. Innovative architectural approaches using locally derived building materials and available tools will mark the beginning of true lunar habitation.

The space program has left behind the "man in a can" approach as human factors engineering has become an ever larger activity. Sophisticated computer-aided design is utilized in planning the interiors of habitation modules and laboratories in the LEO space station. As the durations of manned missions lengthen, the crew can no longer be expected to "adjust" to the situation. Psychological well-being and crew comfort become important components of productivity.

Although the design of a work station in space today embraces many more elements than the interior of a Mercury capsule, the human factors engineer still must operate within the constraints a prefabricated volume imported in the payload bay of the space shuttle. On a planetary surface, the presence of local resources enlarges the options for design. The human factors designer begins to assume a more familiar persona, that of an architect.

The environmental constraints and the available construction materials on the Moon will lead eventually to a lunar architectural style as recognizable as Gothic or Neoclassical. Land's paper presents an architect's eye view of lunar structures and speculates what types will be appropriate for lunar conditions. Kaplicky and Nixon concentrate on providing shielded volumes quickly and simply for protection of pressurized structures.

Lin notes that the abundance of calcium oxide in lunar minerals raises the possibility of concrete as a local construction material. If oxygen is produced on the Moon as a propellant for the Space Transportation System, then sufficient water should be available to combine with lunar-derived cement and regolith aggregate. Young discusses the versatility of concrete and cementitious material in a variety of structural contexts and identifies research needed to characterize lunar cement chemistry. Hörz reviews the possibility of using naturally occurring geologic structures for habitation.

The first humans to live and work on the Moon will be supported by an advanced technology. Yet, the basic incompatibility of human physiology with the environment will limit the flexibility of response to challenges of everyday existence. Our tools will be very sophisticated, but our actual resources will be limited initially. In many ways, the development of a lunar economic and social infrastructure will require the kind of adaptability and innovation seen in successful enterprises in the Third World. For this reason, Khalili's perspective on lunar architecture provides an interesting and thought-provoking contrast to "orthodox" scenarios.

The final two papers in the section touch on aspects of engineering and planning that are ubiquitous on Earth but badly neglected so far in lunar studies. The collection of data relevant to civil engineering was only an ancillary activity in the Apollo scientific investigations. As Johnson and Leonard demonstrate, a large body of lunar environmental data must be accumulated to properly design future lunar structures. As construction and habitation expand, the effects of human activity will impact the lunar environment. Briggs reviews some issues to be considered.

LUNAR BASE DESIGN

Peter Land

Illinois Institute of Technology, College of Architecture, Planning and Design, Chicago, IL 60616

A successful lunar base operation must have appropriately designed structures to house and facilitate the performance of its functions and personnel and to respond to the very special problems of the irradiated, near vacuum lunar environment. This paper on lunar base design proposes the concept of a radiation shield with pressurized enclosures underneath. It examines a range of factors related to base planning, including shielding considerations. Several ways of designing and building shields are described in detail, and the form and location of pressurized enclosures outlined. The paper also enumerates the related areas of needed research, development, and testing work upon which further progress will depend.

INTRODUCTION

Various impressions of extensive future "lunar cities" and base complexes have recently been proposed and illustrated. By terrestrial standards, most anticipate a large building operation, considerable quantities and weights of building materials, fairly heavy plant and equipment, and a sizable labor force. In view of the high cost of transport and the unknown performance of building materials and structures in the irradiated vacuum lunar environment as well as other factors, a different strategy is proposed.

An approach for the post–camp phase base is advocated that proceeds from particulars to generalities. Starting from a small building of simple configuration, it expands to testing and evaluation of materials and structural concepts. With this experience the base grows in stages to a larger installation, becoming more self-sufficient and using more lunar resources. An incremental approach contrasts with some earlier proposals that show sizable and finite arrays of structures built in one operation. It is doubtful that a large base is initially required or could even be built, and its design would probably be out of date before completion.

It is important to take time now to plan a long–range physical development strategy for the lunar base. This will guide the design thinking and be reflected in the initial shape of the complex. An evolutionary approach will probably generate a linear layout for the base, reflecting incremental growth, transportation, solar orientation, and excavation factors. The overall success of the lunar base operation will very much depend upon the building(s) and the structure(s) that will house a wide range of functions and processes.

LUNAR BASE CONCEPT

First generation structures of the post–camp stage would consist of two independent parts: pressurized enclosures under radiation shielding canopies. The size and shape of the enclosures will be determined by the dimensions of the operations they accommodate.

The height and extent of the shielding canopies will be influenced by the building system. Canopies will be heavy and, ultimately, made almost entirely from lunar resources. The pneumatic structures forming pressurized enclosures will be lightweight and packable into small volumes for transport and terrestrial manufacture.

In this concept, one main radiation shield consists of lunar regolith spread over a supporting structure and raised above the lunar surface. The shield can be expanded at the perimeter on one or more sides, where and when needed. Structurally independent pressurized enclosures of the required shape and volume are erected under the shield. Part or all of the shielded space can be pressurized. Different heights can be obtained under the shield by dropping the floor level, where required, by excavating with drag-line technique. Equipment such as antennae, heat exchangers, telescopes, *etc.*, can be mounted over the shield or conveniently placed in an equipment "park" on one side. This concept of the base aims at simplicity in general configuration, building technology, erection, and expansion. It reduces the chance of failure in building or maintenance and minimizes the need for heavy equipment in construction.

The design concept of a lunar base will be influenced by many factors, but two are of particular importance: cosmic radiation and maximum use of lunar materials. Data on radiation levels at the lunar surface indicate that 1.5–2.0 m of regolith would be required to provide shielding of sufficient density to block radiation to acceptable levels, such as dosages encountered by terrestrial x-ray workers. With this thickness of shielding, regolith on supporting structures is a viable concept for the base.

All lunar operations will, as far as possible, be carried out under the shield(s). Under the shield will be either a pressurized "shirtsleeve" or a non-pressurized "suited" environment. Servicing and assembly of large pieces of equipment would not necessarily require a pressurized environment but could be done under the shield, where operators would have radiation protection and would be suited for pressurization only.

RADIATION AND SHIELDING CONCEPT

These design proposals for radiation shields with structures supporting regolith are based upon generally accepted data for radiation levels and regolith density for shielding. However, fresh data (See R. Silberberg *et al.*, this volume) indicate that initial estimates of the regolith thickness for radiation protection may be too low. In particular, the hazard from secondary neutrons, generated within the shielding material by cosmic rays must be carefully evaluated. Additionally, "storm cellars" with very thick overburdens must be constructed for safe haven during occasional solar flares. If a great increase in regolith thickness were required from these considerations, the concept of supporting structure might become uneconomic and the greater thickness would be more economically provided by tunnels or caverns. Therefore, radiation levels and regolith density are pivotal questions affecting lunar base design.

RADIATION AND BASE LAYOUT

A range of tasks must be carried out by unshielded workers on the lunar surface. The duration of these activities for each person will be severely limited by radiation exposure, measured as accumulated dosage over a period of time. This will consist of high intensity, unshielded lunar surface exposure together with some very low intensity radiation under the shield. To maximize the permissible time that a person can work unprotected on the lunar surface, or indeed at the lunar base itself, radiation dosage, when not actually engaged on surface operations, must be minimized.

This will affect the design of the base in two ways. First, all parts of the base should be consolidated under one shield, as far as is practical. In this way, no unnecessary radiation dosage will be accumulated by personnel in moving between different installations and parts of the base, as would be the case with a fragmented based layout. Second, the effectiveness of the shielding should be maximized. Some parts of the base must be separated from the main installation for operational and safety reasons; connecting links must be shielded in those cases. Since the radiation flux is isotropic, the edges of the shield must also be protected by regolith mass to screen out horizontal infiltration. Entrances should be labyrinthine, with overlapping screen walls to effectively block radiation.

Several options are viable for the design of the shield support structure and are described as follows. The first bays of the support structure would need to be erected quickly to give a radiation-free work area; therefore, they would probably utilize entirely terrestrially manufactured components.

BASE STRUCTURES

Flat Shield, Pressurized Enclosures Beneath (Figs. 1,2)
The structure supporting the regolith consists of floors resting on deep lattice girders connected to columns and erected in sections. The bay dimension of the structure and

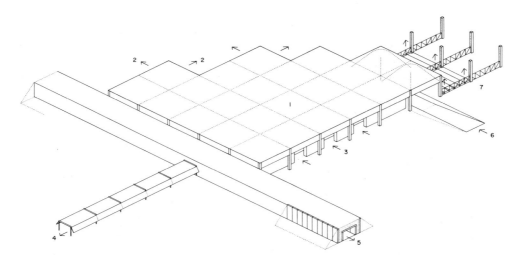

Figure 1. BASE CONCEPT I. Flat shield raised in sections, pressurized enclosures beneath. *Overall view of base. (1) Regolith shielding. (2) Perimeter expansion. (3) Base entry through overlapping radiation barrier walls, from lunar surface equipment and installations "park." (4) Solar shaded links to other parts of base. (5) Shielded links to other parts of base. (6) Ramp access to lower levels. (7) Initial erection sequence.*

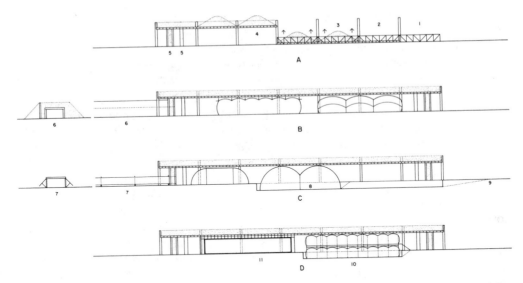

Figure 2. BASE CONCEPT I (see Fig. 1). *Section A showing erection sequence. (1) Aluminum lattice girders and columns in place. (2) Prestressed floors, employing moulded regolith components, attached at ends to girders. (3) Floor loaded with regolith raised by jacks up column guides. (4) Regolith leveled off. Sections B–D show range of pressurized enclosures shapes and types for different uses under shield. (5) Entry through overlapping radiation barrier walls. (6) Shielded link to other parts of base. Bermed or with walled sides. (7) Solar shaded link to other parts of base. (8) Large pressurized enclosures for big equipment maintenance. (9) Ramp access to lower levels. (10) Small pressurized enclosures for agriculture, etc. (11) Pressurized enclosure (to be developed later) using impervious membranes applied to interior surface of rectangular enclosure. Extra top and side reinforced panels to resist outward thrust of pressure.*

spacing of the columns will be influenced by the functions of the base and the pressurized envelopes underneath. Column spacing could be close on one axis, perpendicular to the floor span, and wide on the other, or it could be equally spaced on both axes. Flexibility in column spacing and relatively large spans are feasible. Because of the 1/6 gravity environment, the dead load of dry regolith is not high, and the lattice girders can be very deep in the thickness of the regolith.

The bays of this shield support structure would be raised by pneumatic jacks to the required level, at which point the ends of the beams would be permanently connected to the columns. The regolith overburden could be loaded either before erection of the structure or after. Placing regolith at ground level mainly involves pushing operations, with some leveling needed after the platforms are raised. Placing regolith on elevated floors requires lifting and reaching operations that require more energy and equipment.

Prestressed floors. The floors would consist of moulded regolith components, prestressed from end to end with stranded fibre-glass tendons by using small, portable hand jacks. Components would be assembled flat on a leveled lunar surface, with tendons inserted and prestressed to form narrow floor sections. All sections would be connected at their ends to the transverse girders.

Folded aluminum floors. Here, the floors use all aluminum lightweight components. Folded aluminum sheet material with a deep section for high strength/weight ratio can be fabricated with a profile to permit "nesting" for transportation. The maximum length of components is determined by payload bay space dimensions of the transport vehicle. This floor system of terrestrial manufacture would be used for the initial sections of the shield. Later, with a production plant installed, moulded regolith components would be produced for the floors.

Pneumatic component floors. These would employ inflatable beams, which have been successfully used for bridges in military application to carry trucks and tanks over gullies and craters. They consist of large-diameter inflated long tubes, smaller cross tubes, *etc.,* with an aluminum deck over all. This floor system of terrestrial manufacture would also be used for initial sections of the shield.

Low Arch Shield, Pressurized Enclosures Beneath (Figs. 3,4)

If the structure supporting the regolith is a low arch working in compression with no tensile stresses, then no reinforcement is required. Components of such an arch can be made of moulded regolith, assembled over a movable pneumatic support form. The arches are assembled in sections, each the width of the form, embracing several rings of components. After one section is in place and covered with regolith, the form is partially deflated and moved forward to assemble a new arch section.

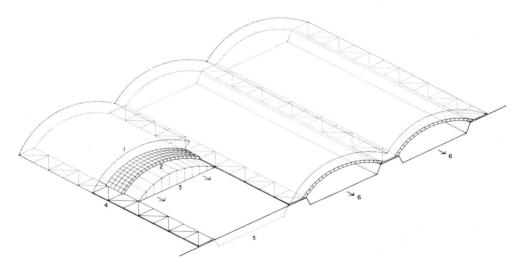

Figure 3. BASE CONCEPT II. Low arch shield using moulded regolith components assembled over temporary, movable pneumatic support form, pressurized enclosures beneath. *General view of base. (1) Regolith shielding. (2) Interlocking, moulded regolith arch components. All components identical dimensions. (3) Movable pneumatic form supporting arch assembly. (4) Aluminum lattice girders to accommodate outward thrust of arches. Girders assembled flat on surface with short components and anchored to surface with vertical pins or connected with transverse cables at convenient, widely space intervals. (5) Height increased where required by excavation. (6) Expansion.*

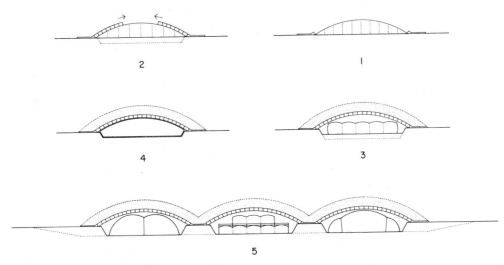

Figure 4. BASE CONCEPT II. (See Fig. 3). *(1) Inflated arch support form. (2) Interlocking, moulded regolith arch components laid over inflated form. (3) Regolith pushed over arch, pneumatic support form removed, excavated underneath where required by dragline scoop and pressurized enclosures erected. (4) Alternative pressurized enclosure using hermetic membrane applied to inner surface of shield. (5) Interconnected arch shields with range of pressurized enclosures.*

The thrust of the arch is horizontal and is accommodated by two aluminum lattice girders conveniently assembled from short sections flat on a level surface, one on each side of the arch. The girders are anchored by pins driven into the lunar surface or by transverse connecting cables. All the girders are conveniently deep, and cables can be widely spaced. As in the first concept, initial sections of the arch can be quickly assembled using terrestrially manufactured components to provide immediate radiation protection. These components could be of folded aluminum or plastic with a profile that permits "nesting" for compact and economic transportation.

The weight and size of the prestressed floors and compression arch shield structures using moulded regolith components will be determined by the lifting capacity of one or two persons or equipment, as well as the moulding technique. The design of the components will employ a thin-rib, deep-section configuration to maximize stiffness and minimize weight. Components will be interlocking both transversely and longitudinally and self-aligning under stress. They will be manufactured in a lunar plant from presorted regolith, formed accurately to the required configuration through moulding, either by firing to the necessary temperature for sintering and surface sealing using a direct solar or electrical furnace, or by using a double-mix epoxy or portland-type cement to bond regolith aggregates.

Low Arch Shield with Pneumatic Support Structure (Fig. 5)

In this concept, a pneumatic structure permanently supports the regolith. The deflated structure is laid on a leveled surface, regolith is pushed over, and the structure is inflated.

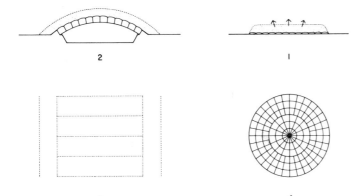

Figure 5. BASE CONCEPT III. Low arch shield with pneumatic support structure. *Deflated pneumatic hybrid structure flat on ground, regolith pushed over. (2) Structure inflated, raising the regolith, afterwards evened out or thickened where necessary. (3) Plan showing the concept applied in sections for a continuous low arch shield. (4) Plan showing concept applied in a low, domed shield.*

The raised regolith is evened out afterward or thickened where necessary. The upper surface of the structure is ribbed to anchor into the regolith. This concept can be applied in sections to form a continuous low arch or a single domed structure.

ARCHED AND DOMED SHIELD SUPPORT STRUCTURES

Arches or domes must be fairly flat since regolith cannot be placed on curved sides that rise too vertically. The dome form can be erected without supporting formwork for most, but not all, of its height, if the courses are raised equally all around the perimeter. However, this dome form must be almost a hemisphere with steep sides that are difficult to cover and that have excessive middle height, making it inconvenient to use. Another considerable disadvantage of the dome is that many components (a dome will have thousands) will have different dimensions, greatly complicating component moulding. In contrast, the low arch form would have only slightly inclined sides, so that regolith can be easily pushed over it, and all building components are dimensionally identical. Also, the arch form can be very conveniently expanded lengthwise.

PRESSURIZED ENCLOSURES AND PNEUMATIC STRUCTURES

Pneumatic structures under shielding canopies can be of three types: air supported, air inflated, and hybrid. Each would need to be evaluated for lunar application. The *air-supported* structure has one structural membrane supported by the push of internal pressure. The *air-inflated* structure has beams, columns, and arches that are independently pressurized and that support membranes between them. The two concepts are combined in *hybrid* structures making this type particularly attractive for lunar applications. Cable mesh containment technique gives the advantage of special shaping and additional membrane support for accommodating higher stresses in the lunar vacuum environment. Rigid elements could be incorporated in the membranes to obtain stiffening, flattening, curving, sealing, mounting, *etc.*

Pneumatic structures have good potential for lunar application in combination with shielding canopies, especially for the initial building thrust after the post–camp stage.

They are small in volume and light in weight, can be formed in a wide range of shapes, and can provide environments at a range of pressures. A great deal of design work and technical experience covering work done over more than 40 years is available in this specialized technology area. Since about 1950, thousands of small and large structures have been erected in many countries for many uses. Recent advances in flexible plastic material with very high strength/weight ratios make pneumatic structures particularly attractive for lunar application under radiation shielding.

SUNKEN AND BERMED STRUCTURES

These could accommodate smaller spans and spaces of the lunar base. To avoid the need for a heavy conventional excavator, lightweight equipment must be fully evaluated, particularly dragline techniques. In a lunar application a dragline would consist of continuous cables with attached scoops running over one motorized and one free vertical capstan. The dragline would run continuously with minimum attendance to excavate trenches of any depth or width. Shielding platforms of any of the types discussed would be erected in the trenches and the loose regolith pushed over to the required thickness. Pneumatic structures would afterwards be inflated beneath the platforms. The use of dragline technique would suggest a linear base arrangement, and the powered capstan could afterward be used as a transportation spine and system.

Each of these concepts has merits and weak points. The design that combines regolith directly on a pneumatic support structure has the disadvantage that if a reduction in pressure is experienced, the shield will drop and crush the contents underneath. There is the risk of failure in the other proposals, but independent, pressurized enclosures may protect their contents and support a failed shield.

SOLAR SHADING CANOPIES

Canopies are proposed to create partial or complete shade over walkways or vehicular driveways linking different parts of the base that for safety or functional reasons must be separated from the main base shield. A horizontal canopy would give total shade at lunar noon. Temperature in the canopy shade would depend upon the width of the canopy, since radiative thermal transfer or conduction via the ground will occur at the edges. Solar shade with low temperature means that personnel moving under the canopies by walking or vehicle need not be suited for cooling, but only for pressure.

Canopies might be perforated with small holes to permit the passage of some light as a fine pattern. This would slightly raise the temperature of any intercepting surface, the intensity depending upon the size of the perforations in the canopy.

The feasibility of lightweight, portable or mobile shading canopies must be studied. These could be placed on the lunar surface where and as required: for servicing the plant, vehicles, mining operations, construction, *etc.* Personnel working on these tasks could possibly have more freedom of movement and greater work range and duration with lighter suits.

SERVICE STATIONS

The distance a person can travel will be limited by radiation exposure time on the lunar surface. Any long distance travel by relatively slow moving vehicles is difficult to envisage. To undertake long distance movement, shielded service stations must be built at strategic spacing for radiation-free resting and sleeping environments, supplies, servicing, *etc.* They would also offer emergency shelter at the time of increased radiation that comes with solar flares, generally predictable in advance, for persons some distance from the main base. Ideally, surface vehicles for long distance travel must be developed with radiation shielding.

SOLAR ORIENTATION

This might be a very important determinant in the layout of the base, or parts of it. As the sunsets and sunrises are relatively long and low-angled, energy build-up on vertical or steeply inclined surfaces might be considerable; this problem should be studied in base layout and design. Entrances and external operational edges of the complex should be orientated away from the sun to minimize temperatures at these points. Also, vertical surfaces, perhaps in combination with horizontal ones, could be developed to provide shade where needed.

INTERIOR ENVIRONMENT

The psychology of interior space and treatment in sealed environments is an important aspect of the base design. The mental stability and vitality of base inhabitants is an essential factor and will be influenced by interior design. Experience from sealed environments, such as in submarines, some industrial complexes, tunnels, *etc.*, must be fully evaluated for possible application in the lunar context.

RESEARCH AREAS

If the lunar base is to be on line by 1995, research and development must be initiated in the near future. The main technology and engineering issues generated by the base concepts and for which terrestrial based testing, development, and research work must be done include the following:

1. Regolith should be tested for moulding building components using heat, sintering, and sealing. Lunar-based experiments are needed for a simple solar furnace.

2. Regolith moulding using bonding agents and cementitious materials such as portland cements and double mix epoxies needs testing in terrestrially based vacuum experiments.

3. Regolith potential for glass and ceramic building materials should be determined. Increased strength of materials in an anhydrous lunar environment should be evaluated.

4. Degradation of materials in a lunar environment should be studied, especially in such materials as plastic, including Kevlar, Teflon, and adhesives, and in metals, in particular, aluminum variants. There is a need for radiation/vacuum exposure experiments.

5. Physical movement of materials following wide diurnal temperature changes should be tested, as should regolithic ceramic-based components, plastics, *etc.*

6. The shape and size of components for compression arch shields and their assembly should be modeled and tested, along with interlocking joints and component profiles.

7. The shape and size of components for prestressed flat shields and their assembly should be modeled and tested.

8. Pneumatic, pressurized envelopes in a wide range of shapes and sizes using Kevlar, Teflon, and steel cable materials should be tested. Net as a structural element, containing and shaping an internal pressure membrane, could be researched. A technique for generating a range of shapes should be developed.

9. The initial stage of the base and community layout should be planned for expansion. Options should be diagrammed and analyzed.

10. The influence of transportation on community layout should be studied, with emphasis on a linear transport route for moving people and goods, shielded or unshielded, pressurized or not. Connections to other parts of the base could be diagrammed and options analyzed.

11. The influence of solar orientation on community layout should be tested with models and a solar simulator. Glare and thermal gain must be minimized.

12. Shaded canopies linking separate parts of the base should be modeled and tested with a solar simulator. Both vertical and horizontal shades should be studied; perforated shades should also be studied.

13. Trenching methods for excavation should be studied using dragline techniques: a rotating cable with scoops travelling around two or more surface capstans. Its influence on base layout and possible later use for transport should be considered.

14. Inside/outside air-lock/valve design for equipment and vehicles as well as individuals should be researched with attention to physical convenience, dust filtration, pressure leaks, and various sizes.

15. The psychological aspects of interior design should be studied, since emotional stability can be influenced by human-related dimensions, color, textures, *etc.*

FORM AND FUNCTION IN LUNAR BASE DEVELOPMENT

Functional considerations will determine the width, height, span, and areas of the various functional components of the lunar base (which need to be more precisely defined). As yet, we do not have specifications and dimensions for the range of anticipated base functions. A small base planning group should be formed to work in close collaboration with all specialized areas of the lunar base group to determine the dimensional characteristics of the base functions with their environmental and servicing needs. The functional inventory will influence the base design, but the ultimate design will also be affected by the building system and shape(s) decided upon.

As far as can be estimated, some large-span enclosures will be needed for the servicing and/or assembly of large pieces of equipment, including lunar surface and spacecraft, telescopes, *etc.* However, the major part of the base could probably be interconnected spaces of fairly small dimensions that would still be much larger than camp stage modules and more economic to erect and maintain. Therefore, both large- and small-area units must be considered.

Although the post-camp stage objective is ultimately to develop an entirely lunar-based construction capability, this would be difficult in the beginning. Structures for radiation shielding will use lunar regolith, but pressurization will generate tensile forces that cannot be handled by first generation regolith processing, such as the fabrication of relatively simple ceramic/glass block components. Nevertheless, structures with floor areas larger than those provided inside imported camp stage modules must be erected as soon as possible.

At a later phase in the evolution of the lunar base, shielding and pressurization might be combined in structures entirely fabricated from regolith. However, this would occur after the base has a fairly large combined floor area to house the necessary plant with workshop capability. A second generation building system combining pressurization with shielding, using regolith in a rigid system, must be able to accommodate tensile stresses generated by pressurization and temperature movement. Speculation on second generation structures suggests a flat or almost flat upper surface to support regolith shielding. Curved forms can better handle pressurization, but loose regolith cannot be heaped onto the steeply inclined surfaces of some curved forms. Outer skirt walls under flat or arched shields must be able to accommodate tensile stresses resulting from the outward push of pressurization. The way this will be done depends upon the manner in which the regolith is used. Skirt wall panels using interlocking ceramic components can be prestressed vertically; panels using cements or epoxies as bonding agents with regolith aggregates could also employ short filament glass fibres for reinforcement within the mix. Sealing between panels could be provided by adhesive tapes applied internally and kept in position by outward pressure.

As mentioned at the beginning of this paper, impressions of some lunar bases presented over the past few years suggest enormous technical problems. The designs and options presented by this author indeed reveal technical unknowns. However, the required information and solutions can be obtained through research and development, some main areas of which have been outlined. These indicate that an integrated program of planning, research, and design can lead to the building of a lunar base within an optimum time frame.

A SURFACE-ASSEMBLED SUPERSTRUCTURE ENVELOPE SYSTEM TO SUPPORT REGOLITH MASS-SHIELDING FOR AN INITIAL-OPERATIONAL-CAPABILITY LUNAR BASE

Jan Kaplicky and David Nixon

Future Systems Consultants, 1103 South Hudson Avenue, Los Angeles, CA 90019

The early deployment of a lunar base in the form of a manned research outpost and habitat could be aided by potential savings of time and money achieved through the use of direct derivatives of module types being developed for the space station. These would be grouped as a complex at lunar surface level. To achieve solar flare radiation protection, the complex would require shielding, which could be provided by an elevated superstructure envelope that would support the required depth of regolith mass. This preliminary design concept examines a typical configuration for a simple and economical superstructure envelope as a prelude to detailed investigations into different design options, their characteristics, and performance criteria.

PURPOSE

The deployment and occupation of a manned lunar base following selection of a suitable lunar surface location may well be the next logical step beyond the space station in terms of both scientific research and space commercialization. The chances of an early manned presence on the lunar surface might be improved if the extensive technology developed for the space station could also be used to develop an Initial-Operational Capability (IOC) lunar base in advance of permanent and fully manned facilities. Such a base may be operated initially as a Phase II facility, gradually evolving into a Phase III facility (Duke *et al.*, 1985). One area where direct transfer of technology is possible is in the provision of crew living quarters and working facilities. The design of these could be based on the habitation, logistics, and laboratory modules developed for the space station with modifications for lunar surface application, including fitting out for one-sixth gravity operation. This approach could save considerable time and money, allowing attention and funding to be focused on the development of an Earth-to-Moon space transportation system that will be needed for advanced base construction, manning, and operation.

DESIGN CONCEPT

A program for a lunar base, which might initially aim to provide a research outpost and living habitat for up to six persons, might take the form of surface deployment and linkage of a series of modules grouped into a complex. This complex would comprise habitation, logistics, and laboratory modules derived from the three generic types being developed for the IOC space station in 1993. A pressurized construction workshop module and unpressurized pilot oxygen production plant module would also be required.

A major obstacle to the utilization of space station modules for an early base appears to be the difficulty of providing rapid and permanent protection against solar flare radiation and mini/micrometeoroid impact if the modules are deployed on the surface. The concept of protecting modules by burying them beneath 2 m of lunar regolith would not be feasible, since the superimposed loading of regolith on outer module surfaces would be likely to reach figures in excess of 5000 N/m^2 (taking into account regolith volume weight in one-sixth lunar gravity and minimum 2 m depth of material required).

This figure would almost certainly be incompatible with the proposed thin, pressure-hull construction of the space station common module identified in the Space Station Reference Configuration Description issued by NASA (Space Station Program Office, 1984). Were it to be possible, once buried, module exteriors would be inaccessible for inspection and maintenance. Base complex reconfiguration and modification—which might involve module, utility, or environmental control, and life-support system repositioning, repair, or upgrading—would also become extremely difficult.

To solve these problems and avoid the need for extensive or deep excavations, it would be possible to develop a simple, manually deployable, superstructure envelope to enclose the module complex at lunar surface level and provide the necessary shielding by means of a "stand-off" layer of regolith deposited over the upper surface of the envelope. This would provide full protection to modules and any external environmental control and life-support system equipment while leaving space around the modules clear for access, circulation, or additional growth. Such an envelope system could be manufactured from a standardized and simplified "kit-of-parts" and transported to the lunar location in stowed form by an Earth-to-Moon transportation system for on-site construction by mission specialists.

Several design configurations for an envelope of this type are possible. This outline design concept describes a typical configuration based on a simple, rectangular plan capable of providing protection for six modules, each nominally 10 m in length and 4.5 m in diameter. The concept is schematic and is intended as a prelude to detailed investigations into complex grouping alternatives, structural design, load characteristics, and material properties.

STRUCTURE AND MATERIALS

The overall envelope is configured as a shallow, flat-topped mound of loose regolith supported by a continuous tension membrane connected to a regular grid of telescopic columns and tapered beams beneath. The column and beam grid delineates the volume occupied by the module group. Physical access would be provided at both ends, which would be left open with suitable allowance for protective overhangs. Beams and columns would be based on an orthogonal grid of structural bays of 5×2 m-size each with the 5-m beam span straddling the girth of the modules, and with an internal clear height of 5 m. A series of high-tensile, fine-mesh membranes would be stretched between the beams to provide support for the regolith mass above, with mesh bays experiencing controlled convex billowing under load. The mesh would be made of woven graphite fiber, sized to allow deposit of approximately 1 mm regolith grain-size upwards. Regolith would be deposited over the previously erected envelope by a manually remote-controlled mobile conveyer system designed to deliver loosely compacted material on a bay-by-bay basis.

Vertical columns and angled beams would be fabricated from advanced composites, which would be derived from graphite/epoxy or similar technology developed for the space station's deployable truss structure. Each column would be connected to a circular footpad to spread the superimposed load over the ground surface, estimated to be in the region of 55,000 N compressive load per column, assuming a 5 × 2 m clear bay size (the minimum feasible bay size that would work with module complex group dimensions). Columns would be designed as telescopic tubes to facilitate low preassembly of all beams to columns at node points and mesh membrane captive attachment to all beam edges. With main beams spanning 5 m in a lateral direction (side-to-side), short 2-m length struts would be needed to provide support to the frames at right-angles to the beam lines in a longitudinal direction (end-to-end). These would typically be spaced at 2.5-m intervals. Cross-bracing would also be required to provide longitudinal stiffening.

Assuming that a Lunar Base IOC would require the provision of shielding for up to six modules (together with access, circulation, and some exterior-mounted environmental control and life-support system equipment), the total plan area of the envelope would amount to approximately 546 m². This plan assumes that the minimum feasible protected area comprises a rectangular plan of 26 m end-to-end and about 21 m side-to-side, allowing the emplacement of two rows of three modules each, with modules located side-to-side and pointing in the longitudinal direction toward the open ends of the envelopes, as shown in Figure 1. This configuration would require a basic component schedule that would include 56 telescopic columns, 84 footpads, 70 main beams, 143 lateral struts, and 65 woven mesh membrane panels. Preliminary (and very approximate) estimates suggest a diameter of about 125–150 mm for the outermost tube of the telescopic columns, 300 × 75 mm describing the mean section profile of main beams, 75 mm diameter struts and cross-braces, and 500–750 mm diameter footpads. The actual footpad diameter would be determined by the ground-bearing strength condition in the selected location; it is based on the assumption that loosely compacted surface regolith would be removed to expose densely compacted material capable of achieving a reasonable bearing pressure.

MANUFACTURE, TRANSIT, AND ASSEMBLY

The manufacture and evaluation of the superstructure envelope system would take place in a 1 g environment on Earth, destined for a one-sixth g environment on the Moon. This would enable the system to be fully preassembled and load-tested in a simulated lunar setting using a dry sand-based aggregate mix to represent the lunar regolith shielding.

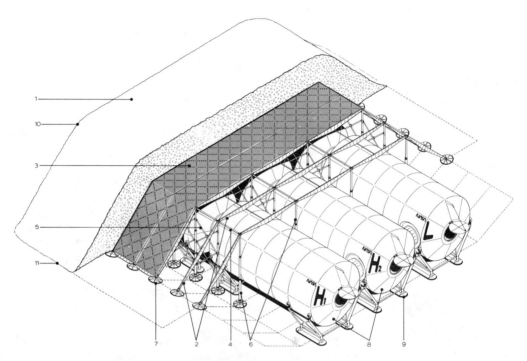

Figure 1.   Cutaway illustration of superstructure envelope system: 1—regolith mass shielding; 2—main tapered beams; 3—graphite fiber mesh; 4—longitudinal struts; 5—longitudinal bracing; 6—telescopic tubular columns; 7—circular footpads; 8—linked habitat/laboratory/workshop modules; 9—module ground support cradles; 10—crest of slope; 11—base of slope.

As with the Apollo missions, it would be possible to test most lunar surface operations in a terrestrial environment beforehand. This would apply to the entire construction sequence of the envelope system: unpacking, layout, assembly, hoisting, leveling, tie-down, regolith deposition, and possibly even module insertion and interconnection.

The fully stowed envelope system, complete with tools and accessories, would be delivered to low-Earth orbit by the shuttle, with transit from low-Earth orbit to lunar surface by means of an Earth-to-Moon transportation system. Preliminary outline estimates of material/component weights indicate that the complete system would amount to 5000–7500 kg launch weight and be capable of being stowed in a volume equivalent to a cylinder measuring 5.5 m long by 2.5 m in diameter, thus enabling its conveyance in a single mission. Another mission would convey a lunar tractor (or similar vehicle), which would be required to move the lunar base modules from their soft-landed locations to the selected lunar base site, as well as the mechanical conveyer system required to transfer surface regolith from the lunar base environs to the upper surface of the envelope canopy. As with the proposed space station assembly, each manned module would require a single dedicated mission.

Once all equipment were brought to the lunar base location, the superstructure envelope would be erected in a predetermined sequence by a team of mission specialists. This would proceed after loose surface regolith clearance and excavation. All beams, struts, columns, and footpads would first be assembled at "shoulder" level (*i.e.*, at the height determined by the top of the outermost telescopic column tube prior to extension), followed by attachment of the mesh membrane panels. Each structural bay would then be raised to the correct height, working from one end of the envelope to the other, using manually operated screws and crank handles. After leveling, tightening, and anchoring down, regolith deposition would be carried out, and the superstructure envelope would be ready for module insertion, interconnection, and subsequent lunar base operation.

CONTINUATION OF RESEARCH

Essentially, the concept outlined in this paper represents a simple "elevated-bunker" approach to providing mass shielding for a lunar research outpost and habitat, as shown in Figure 2. Several design variations are possible, depending on module complex grouping, local ground conditions, and logistics considerations.

The central point, however, is that a system of this type could provide a rapid and efficient means of surface protection for an early manned presence and is therefore well

Figure 2.   View of the lunar base from the surface, showing system construction.

worth further investigation. Although the use of modified space station modules may be the most logical and economical solution to providing habitable facilities, an elevated envelope system of the type discussed in this paper could be applicable to other lunar base habitat options, including an advanced lunar base community covering a large surface area. These reasons clearly point to the need for further research on the design concept. In the future, the authors plan to address the following issues in particular. (1) Investigation of the range of static and dynamic (seismic/impact) loads to which the structural system would be subject, based on known lunar regolith mass properties. (2) Analysis of the range of forces and stresses imposed on all structural members comprising the system, in accordance with preliminary element and component detailed design and physical material properties. (3) Organization of data generated in (1) and (2) into a representative criteria set that could form guidelines for a preliminary lunar structural design code. (4) Detailed investigation into activities and phasing implicit in the entire logistics sequence including, in serial order, the following: system prototyping, manufacture, testing and simulation, stowage, Earth-to-Moon transportation, lunar surface retrieval, unpacking and layout, low-assembly, hoisting, leveling and tightening, anchoring and tie down, and, finally, regolith deposition.

REFERENCES

Duke M. B., Mendell W. W., and Roberts B. B. (1985) Towards a lunar base programme. *Space Policy, 1,* 49–61.

Space Station Program Office (1984) *Space Station Reference Configuration Description.* JSC 19989, NASA/Johnson Space Center, Houston. 798 pp.

CONCRETE FOR LUNAR BASE CONSTRUCTION

T. D. Lin

*Construction Technology Laboratories, The Portland Cement Association,
5420 Old Orchard Road, Skokie, IL 60077*

Prior to the establishment of lunar scientific and industrial projects envisioned by the National Aeronautics and Space Administration, suitably shielded structures to house facilities and personnel must be built on the Moon. One potential material for the construction is concrete. Concrete is a versatile building material, capable of withstanding the effects of extreme temperatures, solar wind, radiation, cosmic rays, and micrometeorites. This paper examines data published by NASA on lunar soils and rocks for use as concrete aggregates and as possible raw materials for producing cement and water, and investigates the technical and economic feasibility of constructing self-growth lunar bases. A hypothetical 3-story, cylindrical concrete building with a diameter of 210 ft (64 m) was analyzed for conditions of a vacuum environment, lunar gravity, lunar temperature variations, and 1 atmosphere internal pressure. The advantages of concrete lunar bases are subsequently discussed.

INTRODUCTION

A project proposed within the National Aeronautics and Space Administration to build permanent lunar bases after the turn of the century has drawn tremendous interest from scientific and engineering communities across the nation. Lunar bases will enable mankind to extend civilization from Earth to the Moon. Ample solar energy, low gravitational force, and abundant minerals on the Moon will provide excellent conditions for the development of scientific and industrial space activities. Prior to the establishment of these activities, suitably shielded lunar structures must be built to house facilities and to protect personnel from the effects of solar wind, radiation, cosmic rays, and micrometeorites.

As a material capable of withstanding these effects, concrete is proposed for construction and can be produced largely from lunar materials. This discussion covers the process of making cement from lunar material, concrete mixing in a lunar environment, physical properties of concrete at lunar surface temperatures, and structural design suitable to lunar conditions.

CEMENTITIOUS MATERIALS

Cements used in construction on Earth are made basically with raw materials such as limestone, clay, and iron ore. A burning process transforms the raw materials into primarily calcium-silicate pebbles called clinker. The clinker is then ground into micron-sized particles known as cement. A wide variety of cements are used in construction. The chemical compositions of these cements can be quite diverse, but by far the greatest amount of concrete used today is made with portland cements. A typical portland cement

(Mindess and Young, 1981) consists of about 65% calcium oxide (CaO), 23% silica (SiO_2), 4% alumina (Al_2O_3), and small percentages of other inorganic compounds. Among these constituents, calcium oxide is the most important in the cement manufacture.

Other types of cements produced with lower calcium oxide content are available, e.g., slag cement, expansive cement, alumina cement, and low calcium silicate cement. High alumina cement has 36% calcium oxide while low calcium silicate cement has only 30% (Takashima and Amano, 1960). Theoretically, a cementitious material can be made with any proportion of $CaO:SiO_2:Al_2O_3$ that falls within the calcium–silica–alumina phase diagram.

LUNAR MATERIALS

Information from Apollo lunar soils and rocks indicates that most lunar materials consist of sufficient amounts of silicate, alumina, and calcium oxide for possible production of cementitious material. Table 1 shows the chemical compositions of some selected lunar samples (Morris, 1983; Ryder and Norman, 1980; Fruland, 1981). It appears that the content of calcium oxide in lunar material is relatively low in comparison with other major cement ingredients; our discussion, therefore, will center around the calcium oxides.

Table 1.   Chemical Compositions of Selected Lunar Samples

| Element | Major Elements, wt % | | | | |
	Mare Soil (10002)	Highland Soil (67700)	Basalt Rock (60335)	Anorthosite Rock (60015)	Glass (60095)
SiO_2	42.16	44.77	46.00	44.00	44.87
Al_2O_3	13.60	28.48	24.90	36.00	25.48
CaO	11.94	16.87	14.30	19.00	14.52
FeO	15.34	4.17	4.70	0.35	5.75
MgO	7.76	4.92	8.10	0.30	8.11
TiO_2	7.75	0.44	0.61	0.02	0.51
Cr_2O_3	0.30	0.00	0.13	0.01	0.14
MnO	0.20	0.06	0.07	0.01	0.07
Na_2O	0.47	0.52	0.57	0.04	0.28

A review of available literature on Apollo lunar samples reveals that a typical mare soil has a CaO content of nearly 12% by weight, highland soil 17%, basalt rocks 14%, and anorthosite rocks, a calcium-rich plagioclase in the feldspar group, almost 19%. A rock type with 19% CaO content is a good candidate for lunar cement production. Lunar sample 60015, a coherent, shock-melted anorthosite rock, is an example. The rock is approximately $12 \times 10 \times 10$ cm and is largely coated with a vesicular glass up to 1 cm thick, as shown in Fig. 1. The glass layer has been interpreted as a quenched liquid derived from melting the surface layer of the anorthosite rock. Quenched glass generally is an amorphous substance and represents a potential cementitious material if ground to fine particle size.

Figure 1. Pristine cataclastic anorthosite glass-coated sample 60015.

Glasses are common in lunar soils. Table 2 shows the averaged glass contents in lunar samples brought back by the Apollo missions. Note that samples taken from Shorty Crater rims have glass content as high as 92.3%. The chemical compositions of glass could possibly be similar to glass sample 60095 shown in Table 1.

Table 2.   Average Glass Content of Lunar Samples

Mission	Average Glass Content, %
Apollo 11	6.6
Apollo 12	18.0
Apollo 14	12.2
Apollo 15	29.4
Apollo 16	10.6
Apollo 17	31.1

PROCESS METHODS

Figure 2 shows condensation temperatures of various elements in basalt rocks (Wood, 1975). Interestingly, all cementitious elements including Ca, Al, Si, Mg, and Fe have condensation temperatures about 1400 K, at least 200° higher than those of non-cementitious elements. This unique physical property may enable us to separate cementitious elements from non-cementitious ones in the process of cement manufacture.

However, a temperature of 3000 K or higher will be needed for the elemental evaporation in the process. This may cause some degree of difficulty in finding suitable material for containment use.

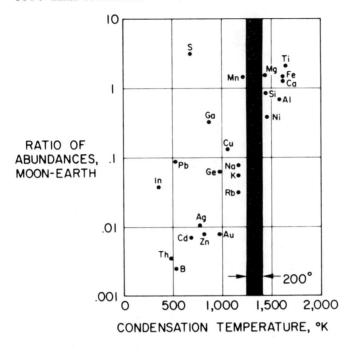

RATIO OF ABUNDANCES, MOON-EARTH

CONDENSATION TEMPERATURE, °K

Figure 2. Condensation temperature of basalt minerals.

Figure 3 shows residual fractions of multicomponent melt consisting of FeO, MgO, SiO_2, and Al_2O_3 of solar elemental abundances during the evaporation process at temperatures up to 2000°C (Hashimoto, 1983). The complete evaporation of FeO in Stage I may be utilized for metallic iron beneficiation. The remaining residues in Stage IV have high concentrations of CaO and Al_2O_3 and a small amount of SiO_2. Calculated $CaO:Al_2O_3:SiO_2$ proportions of the combined residues at lines A and B of Fig. 3 fall in the stoichiometric range of commercial high alumina cement (Lea, 1971).

AGGREGATES

Aggregates generally occupy about 75% by weight of the concrete and greatly influence concrete properties. Aggregates, according to American Standard for Testing and Materials (ASTM), are not generally classified by mineralogy. The simplest and most useful classification is based on specific gravity. Lunar soils and rocks all have specific gravities higher than 2.6 and are believed to be quality material for aggregate use. To produce concrete on the Moon, lunar rocks can be crushed to suitable coarse aggregate size, and the abundant lunar soils can be sieved to good gradation of fine aggregates.

Glassy soils used as aggregates may develop alkali–aggregate reactions that could cause the concrete to crack or spall. The lunar materials have never been exposed to oxygen and water, and the chemical and physical stability of these materials when exposed to water are not yet fully known. Research on lunar soil for possible aggregate application is indeed important.

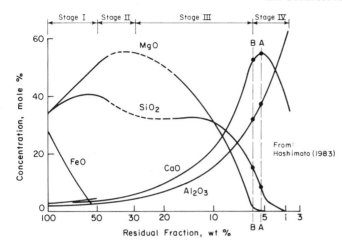

Figure 3. Change of composition of residual molten oxide material as a charge is vaporized away into a vacuum, at 2073 K.

CHEMICAL COMPOSITION, WEIGHT PERCENT			
Compound	Solar Elemental Abundances		Alumina Cement
	@ A-A	@ B-B	
CaO	42.7	40.0	36–42
Al$_2$O$_3$	52.3	48.8	36–51
SiO$_2$	5.0	11.0	4–9

WATER PRODUCTION

There have been studies on oxygen and metal production using lunar materials. Proposed methods include an alkali–hydroxide–based scheme (Cutler, 1984), hydrogen reduction of ilmenite (Agosto, 1984), and others. The ilmenite reduction reaction yields iron and water.

$$FeTiO_3 + H_2 \rightarrow TiO_2 + Fe + H_2O$$
ilmenite iron water

Hydrogen is not readily available on the Moon and may have to be imported from Earth. The terrestrial hydrogen can be transported in the form of liquid hydrogen (H_2), methane (CH_4), or ammonia (NH_3) (Friedlander, 1984). In considering the need for carbon and nitrogen for life support and the higher boiling points of methane at –322°F (–161°C) and ammonia at –91°F (–33°C) than liquid hydrogen at –486°F (–252°C), it may be more advantageous to import methane and ammonia to the Moon rather than liquid hydrogen.

REINFORCED CONCRETE

Concrete is basically a mixture of two components: aggregate and cement paste. The paste, comprised of cement and water, binds the aggregate into a rock–like mass as it hardens.

The flexural strength of plain concrete is generally low, about one-tenth of its compressive strength. However, concrete reinforced with either steel or glass fibers has increased flexural strength, strain energy capacity, and ductility. Test data reveal that concrete reinforced with 4% by weight of steel fibers possesses nearly twice the flexural strength of plain concrete (Hanna, 1977). These fibers act as crack arresters, that is, the fibers restrict the growth of microcracks in concrete.

STRUCTURAL DESIGN

Design of structures for a lunar base differs from design of structures on Earth. First, there are no wind and earthquake loads on the Moon. Second, the lower lunar gravity, one–sixth that of Earth, could permit an increase in the span length of a flexural member to 2.4 times, based on the flexural theory of a simply supported beam.

Figure 4 shows a proposed three–level concrete structure with a diameter of 210 ft (64 m). The structure is assumed to be subjected to 1 atmosphere pressure inside

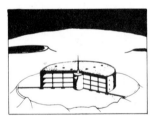

a) Concrete Lunar Base

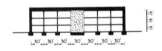

b) Elevation

Figure 4. Proposed three-level concrete lunar base.

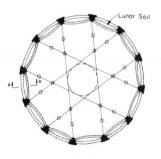

c) Plan View

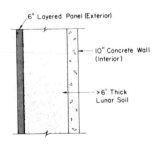

d) Cross Section A-A

and vacuum outside. The cylindrical tank at the center of the system serves as a safety shelter for inhabitants in case the system suffers damage or air leak. It could also serve as a "storm cellar" during solar flares. The roof will be covered with lunar soil of suitable thickness 6–18 ft (2–6 m) (Arnold and Duke, 1978) to protect personnel and facilities from harmful effects of cosmic radiation.

Plain concrete is normally weak in tension but strong in compression. Conceivably, the major demand on the structural system will be the high tensile stresses in the wall resulting from the internal pressure. To solve the problem, use of circular panels facing outward and supported by columns will change the tension into compression (Fig. 4c). Steel tendons can then be used to secure the columns into position. For effective use, these tendons could be placed around the cylindrical tank, stressed to provide hoop forces on the tank, and then anchored to columns at the opposite side. The 6-inch-thick (15 cm) layered panels at external faces of the wall (Fig. 4d) are non–load-bearing units. They are used to contain the soil between the internal and external panels. A layered system that is free to expand can minimize the thermal stresses due to extreme temperature changes on the Moon.

The proposed concrete lunar base structure has 90,000 ft^2 (8,360 m^2) of usable area. Approximately 250 tons of steel and 12,200 tons of concrete would be needed for the construction. That much concrete requires approximately 1,500 tons of cement and 490 tons of water. All these materials can be obtained on the Moon except hydrogen. The needed hydrogen from Earth is about 55 tons.

ADVANTAGES OF A CONCRETE LUNAR BASE

Concrete lunar bases offer the following advantages:

1. *Economic.* Table 3 compares energy requirements for four major construction materials (Mindess and Young, 1981). To produce 1 m^3 of aluminum alloy requires 360 GJ energy; 1 m^3 of mild steel requires 300 GJ; 1 m^3 of glass requires 50 GJ; and 1 m^3 of concrete requires 4 GJ. The energy ratio between aluminum alloy and concrete is 90:1. Less energy requirement in the production can be translated into lower cost.

Table 3. Typical Properties of Construction Materials

Materials	α (10^{-6}/°C)	k (W/m K)	E_{rq}' (GJ/m^3)
Aluminum Alloy	23	125	360
Mild Steel	12	50	300
Glass	6	3	50
Concrete	10	3	3.4
			(4.0)*

*H$_2$O is made from ilmenite.

2. *Compartmentalization.* One major advantage of concrete is that it can be cast into any monolithic configuration. A lunar structure could be compartmentalized to prevent catastrophic destruction in case of any local damage.

3. *Concrete Strength.* Lunar surface temperature may vary from –250°F (–150°C) in the dark to +250°F (120°C) facing the sun. Figure 5 shows that the strength of heated concrete is practically unaffected by 250°F (120°C) (Abrams, 1973). Concrete maintained at 75% relative humidity and temperature of –150°F (–100°C) increases in strength two and one–half times that of concrete maintained at room temperature, and two times at –250°F (–150°C). Concrete that has 0% relative humidity neither gains nor loses the strength in the course of the cooling period, down to –250°F (–150°C) (Montage and Lentz, 1962).

4. *Heat Resistance.* Concrete is thermally stable up to 1100°F (600°C). The low thermal conductivity as shown in Table 3 and high specific heat make concrete an excellent heat resistant construction material.

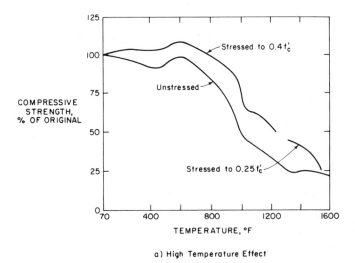

a) High Temperature Effect

Figure 5. Effect of temperature on compressive strength of siliceous aggregate concrete.

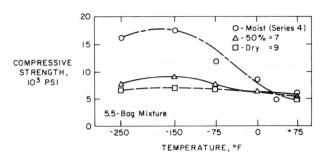

b) Low Temperature Effect

5. *Radiation Shielding.* Most radiation energy will be converted into heat energy in the course of attenuation in an exposed body. In general, a hardened concrete consists of 95% aggregate and cement and 5% water by weight. Both aggregate and cement are non-metallic and inorganic, and are excellent materials for absorbing gamma-ray energy. Water is the best substance for absorbing neutron energy (U. S. Scientific Laboratory, 1950).

6. *Abrasion Resistance.* Micrometeorites can strike the Moon with relative speeds up to 25 miles/s (40 km/s). These microparticles may abrade the surface of the lunar structures. Concrete possesses high abrasion resistance, which increases proportionally with concrete strength.

7. *Effect of Vacuum.* Exposed to the lunar environment, the free moisture in concrete may eventually evaporate, but the chemically bonded water will not. Again, Fig. 5 shows that the loss of free moisture, which generally takes place around 212°F (100°C) has no adverse effect on concrete strength.

A pressurized concrete structure may not be completely airtight. To solve this problem, an epoxy coating, or another sealant that hardens without oxidation, can be applied on the internal surface.

CONCLUSION

Reinforced concrete has many material and structural merits for the proposed lunar base construction. The attractiveness of this proposal lies in the fact that most of the components of the concrete can be produced simply from lunar materials. The scenario for the self-growth lunar base is as follows:

1. Materials: Cement could be obtained by high-temperature processing of lunar rocks. Aggregates would be obtained by physical processing of lunar rocks and soils. Lunar ilmenite would be heated with terrestrial hydrogen to form water, while the residual iron could be processed into fibers, wire, and bars for reinforcement.

2. Concrete: Casting and curing chambers for concrete could be developed from empty shuttle fuel tanks. Temperature and humidity control, as well as controlled drying and recycling of excess water, are vital parameters because water is an expensive commodity in space.

3. Construction: The evaluation of the most suitable structural design must include considerations of constructability. It is possible to optimize the concrete properties in order to achieve the most suitable design, both for ease of construction and for maintenance-free service.

Conceivably, concrete lunar bases will be essential facilities for the scientific and industrial developments on the Moon. Perhaps concrete will provide the ultimate solution to the colonization of outer space. The task of constructing lunar bases is a great challenge to scientists and engineers in this fascinating space age.

REFERENCES

Abrams M. S. (1973) *Compressive Strength of Concrete at Temperatures to 1600° F.* RD016.01T. Portland Cement Association, Skokie, IL. 11 pp.

Agosto W. N. (1984) Electrostatic concentration of lunar soil ilmenite on vacuum ambient (abstract). In *Papers Presented to the Symposium on Lunar Bases and Space Activities of the 21st Century*, p. 24. NASA/Johnson Space Center.

Arnold J. R. and Duke M. B. (editors) (1977) *Summer Workshop on Near-Earth Resources.* NASA Conf. Public. 2031. NASA, Washington. 95 pp.

Cutler A. H. (1984) An alkali hydroxide based scheme for lunar oxygen production (abstract). In *Papers Presented to the Symposium on Lunar Bases and Space Activities of the 21st Century*, p. 21. NASA/Johnson Space Center, Houston.

Friedlander H. N. (1984) An analysis of alternate hydrogen sources for lunar manufacture (abstract). In *Papers Presented to the Symposium on Lunar Bases and Space Activities of the 21st Century*, p. 28. NASA/Johnson Space Center, Houston.

Fruland R. M. (1981) *Introduction to the Core Samples from the Apollo 16 Landing Site.* NASA Curatorial Branch Public. 58. NASA/Johnson Space Center, Houston. 45 pp.

Fruland R. M. (1982) *Catalog of the Apollo 16 Lunar Core 60009/60010.* NASA Lunar Curatorial Branch Public. 61. NASA/Johnson Space Center, Houston. 155 pp.

Hanna A. N. (1977) *Steel Fiber Reinforced Concrete Properties and Resurfacing Applications.* RD049.01P, Portland Cement Association, Skokie, IL. 16 pp.

Hashimoto A. (1983) Evaporation metamorphism in the early solar nebula—Evaporation experiments on the melt $FeO-MgO-SiO_2-CaO-Al_2O_3$ and chemical fractionations of primitive materials. *Geochem. J., 17,* 111.

Lea F. M. (1971) *The Chemistry of Cement and Concrete.* Chemical Publishing, New York. 497 pp.

Mindess S. and Young J. F. (1981) *Concrete.* Prentice-Hall, New York. 671 pp.

Monfore G. E. and Lentz A. E. (1962) *Physical Properties of Concrete at Very Low Temperatures.* RD 145. Portland Cement Association, Skokie, IL. 39 pp.

Morris R. V. (1983) *Handbook of Lunar Soils.* NASA Planetary Materials Branch Public. 67. NASA/Johnson Space Center, Houston. 914 pp.

Ryder G. and Norman M. D. (1980) *Catalog of Apollo 16 Rocks.* NASA Curatorial Branch Public. 52. NASA/Johnson Space Center, Houston. 1144 pp.

Takashima S. and Amano F. (1960) Some studies on lower calcium silicates in portland cement. *Reviews of 74th General Meeting of Cement Association of Japan.* Yogyo Cement Co., Osaka. 44 pp.

U. S. Scientific Laboratory (1950) *The Effects of Atomic Weapons.* Combat Forces Press, Washington. 730 pp.

Wood J. A. (1975) The Moon. *Sci. Amer., 233,* 93–102.

CONCRETE AND OTHER CEMENT-BASED COMPOSITES FOR LUNAR BASE CONSTRUCTION

J. Francis Young

Departments of Civil Engineering and Ceramic Engineering, University of Illinois at Urbana-Champaign, 208 N. Romine, Urbana, IL 61801

The use of concrete and other materials based on a cementitious matrix is evaluated for the construction of a manned space station on the Moon. Consideration is given to the most recent developments in the science and technology of cementitious materials and the feasibility of *in situ* construction using lunar materials. It is concluded that concrete construction on the Moon should be technically feasible.

INTRODUCTION

Recently it has been suggested (Lin, 1984) that concrete could be successfully used in the construction of a lunar base. This exotic use for such a mundane material seems at first sight impracticable. We associate concrete with massive structures involving the use of structural elements with large cross-sections that require large volumes of concrete. When we look at typical concrete properties (Table 1) we are still more convinced of the inappropriateness of concrete on the Moon. It is relatively weak, even in compression, and also fails in a brittle manner, so that tensile steel is required as reinforcement. It suffers from dimensional instabilities and various durability problems, as any highway traveller can testify.

The wide popularity of concrete, however, indicates that it has important advantages that outweigh the problems cited above. A major advantage is versatility. Concrete can be cast in almost any shape imaginable: witness the famous Sidney Opera House in Australia, or the Bahai temple now under construction in India (Anonymous, 1984). Furthermore, by a suitable choice of constituent materials and their proportions, concrete can be manufactured to exact and unique specifications for any given application. Another dominant advantage is economy: large masses of concrete can be produced cheaply

Table 1. Typical Properties of Conventional Concrete.

	Normal Strength	High Strength
Compressive Strength	5000psi (35MPa)	12,000psi (85MPa)
Flexural Strength	800psi (6MPa)	1300psi (9MPa)
Tensile Strength	500psi (3.5MPa)	800psi (6MPa)
Modulus of Elasticity	4×10^6 psi (28GPa)	5×10^6 (35GPa)
Strain at Failure	0.002	0.003
Drying Shrinkage	0.05–0.1%	0.05–0.1%

because concrete is made from inexpensive ingredients obtained primarily from local sources. This was one of the major assumptions of Lin's (1984) proposal: that concrete can probably be made from lunar materials, thereby avoiding the costly alternative of transporting construction materials from Earth.

Economy is relative, however. If large quantities of concrete are required, the costs of mining, processing, and fabrication on the Moon may exceed the costs of ferrying lightweight prefabricated components made from steel or plastics from Earth. Continuing developments in concrete technology have led to high strength concrete two to three times stronger than conventional concrete. Such concretes can be used to produce structures of lower mass, and they also have better overall performance, since strength is an indicator of the quality of the concrete. Concrete with 14,000 psi (100 MPa) compressive strength can now be produced under field conditions (Burge, 1983; Radjy and Loeland, 1985), while special cementitious systems have been developed whose strengths are at least double the above figure (Hirsch *et al.*, 1983; Young, 1985). Since these materials do not use large-sized aggregates (particulate fillers) they cannot be considered concretes in the conventional sense, so I refer to them as cement-based materials. They not only have the potential to make a large impact on the economics of future lunar construction, but are also likely to be more suitable for the lunar environment.

Lunar base construction is an interesting materials problem for which there are several possible solutions. Lin (1984) sought to highlight the potential for concrete-based lunar structures, and in this paper I will examine the feasibility of his approach in light of the most recent developments of cement-based systems. I will be concerned only with the materials aspects; further development of structural designs is a separate problem whose solution depends on the decisions made with regard to a set of optimum material properties.

NEW CEMENT-BASED MATERIALS

Dr. Lin's structural design, while quite an elegant solution to the problem, assumes conventional concrete properties; he is considering a brittle material with only modest strength (up to 10,000 psi [70 MPa] in compression). His preliminary design called for 10-in thick precast sections, which represents quite a large mass in the lunar structure. Although transportation from Earth is eliminated by using lunar resources wherever possible, it must be assumed that mining and processing will still be expensive. Thus, reduction in mass would be desirable, but is possible only with a greater improvement in concrete performance.

As mentioned above, high performance cement-based composites have been developed during the last few years and are now entering the market place. These composites are attaining compressive strengths in the range of 30,000–40,000 psi (200–300 MPa), and, when reinforced with fibers, they can exhibit good ductility. Two distinct approaches have been taken, leading to MDF and DSP cements, as described below. Further evaluation of their properties and improved versions of these materials can be anticipated over the next few years.

Table 2. Typical Mechanical Properties of High Performance Cement-Based Materials.

	MDF Cement Paste	DSP Cement Paste
Compressive Strength, MPa	300	250
Flexural Strength, MPa	150	4
Modulus of Elasticity, GPa	50	40
Strain at Failure	—	0.003
Fracture Energy, J/m^2	200	40
Critical Stress Intensity Factor, $Mpa.m^{1/2}$	3	0.3

MDF (Macro–Defect–Free) cements were developed and named by Birchall and co-workers (Birchall et al., 1982; Birchall, 1983; Alford and Birchall, 1985; Kendall et al., 1983; Kendall and Birchall, 1985). The approach is to apply new processing techniques to the creation of the paste (cement + water) that forms the binding matrix in the composite. A water-soluble polymer is added as a processing aid, and only small quantities of water are used. The Earth-dry formulation is mixed under high shear, during which the polymer prevents excessive entrapment of air by cavitation and provides cohesiveness and plasticity. The blended dough can now be roll-mixed (calendered) to eliminate residual entrapped air bubbles (the macro-defects). The plastic dough is readily molded into the desired shapes by extrusion, pressing, or other conventional plastics processing operation. The material is moist-cured at 80°C followed by air drying; a combination of cement hydration and polymer dehydration during this regime densifies the matrix. The properties of MDF pastes are impressive, as can be seen in Table 2, although as yet we do not fully understand the fundamentals of the material. The ratio of compressive strength to flexural strengths strongly suggests that the polymer is not simply a processing aid, but is actively contributing to the engineering properties. Portland cement has been successfully replaced by calcium aluminate cement (which actually appears to perform better).

DSP cements, which were conceived by Bache (1981), represent a quite different approach, since they were designed to be castable, like concrete. This is done by control of the particle size distribution of the cement to minimize the void space between cement grains, which must be filled with water to make the system castable. As in the case of the MDF systems, the idea is to minimize the amount of water used in order to minimize the porosity of the final product, since material properties are always very strongly porosity dependent. Bache added a very finely divided silica (particle size 0.1 μm), a by-product of silicon and silicon alloy production, to Portland cement in order to provide the void-filling capabilities. This material has the added advantage of reacting chemically with the cement paste to become an integral part of the cementitious matrix. A dispersing surfacant is also necessary to achieve castability. Later approaches to the DSP concept have used more than one addition for particle size control (Roy et al., 1985; Wise et al., 1985). The properties of DSP materials are comparable to MDF pastes (see typical values in Table 2) except that they are much more brittle due to the absence of a polymer phase. DSP cements can also be formulated with calcium aluminate cements.

The structures of the paste matrix formed by both systems are quite similar at the micron level. Residual unhydrated grains of cement act as a "micro-aggregate" embedded in a matrix of hydration (reaction) products. The unhydrated cement makes up quite a large volume fraction of the paste, in contrast to cement paste in conventional concrete. We do not yet know the implications of this observation with regard to service performance. Fibers or fillers can be added to either matrix to achieve desired properties, such as ductility, abrasion resistance, etc.

These new materials are beginning to be used commercially as useful substitutes for metals or reinforced plastics. MDF composites can be drilled, tapped, or machined. They have been used (Alford and Birchall, 1985) in turntables and speaker cabinets for stereo systems, since the polymer imparts good acoustic properties. They are also being considered for ballistic protection and electromagnetic radiation screening. DSP composites are being used for press tools and molds in the aerospace and automotive industries (Wise et al., 1985) where they replace metals and for machinery parts where abrasion resistance is of prime concern (Hjorth, 1983, 1984). DSP-based molds have been successfully used in vacuum forming (Wise et al., 1985). Such molds have been found to have lower vacuum leak rates than the aluminum molds that they replace.

It is clear from the above discussion that these advanced materials have the potential to provide structures of low mass (reduced cross-sections) to act as an air-tight radiation shield and with sufficient ductility and abrasion resistance to resist meteorite impacts. However, at present little is known about the long-term properties of these materials: their response to large temperature fluctuations, strong drying, impact loadings, fatigue, prolonged vacuum, etc. An extensive evaluation program will be necessary to obtain this kind of information, although no doubt it will be gradually developed. Studies will need to be made on a variety of composites made with different combinations of cements, aggregates, and fibers.

LUNAR PRODUCTION OF MATERIALS

The potential for utilizing lunar resources for the constituents of cement-based composites is an attractive scenario. However, the manufacture of cement, which is a crucial step, will not be a straightforward problem. Both Portland cement and calcium aluminate cement contain over 60 wt % of CaO, whereas lunar rocks and soils are relatively low in CaO (<20 wt %). Limestones do not occur on the Moon. It is possible that a CaO enrichment scheme could be developed using differential vaporization (Agosto, 1984), but the economics may be prohibitive. Perhaps digestion by molten alkali hydroxides to break down minerals into their oxides (Cutler, 1984) might be feasible.

However, one should not be constrained by conventional cement chemistry. One could attempt to manipulate the existing chemical composition to provide a reactive material by fusion-recrystallization processes. Perhaps reactive glasses with a reactive solution, such as carbonic acid (Young et al., 1974) or a polyelectrolyte (Wilson, 1978,

1979) could be used, either of which can initiate very rapid reactions. Development of a suitable cementitious system may require some innovative approaches but should not be an insurmountable problem. It remains to be seen to what extent the strategies of the DSP and MDF systems could be successfully implemented if a new cement chemistry is adopted.

The question of suitable aggregates is much less critical. Processed Moon rocks should be satisfactory provided they are not excessively weak and there is a reasonable thermal match with the cementitious matrix. This is necessary to avoid the creation of internal stresses that could cause internal microcracking and loss of properties (*e.g.*, vacuum tightness, abrasion resistance, *etc.*). Lack of moisture eliminates most durability problems encountered with aggregates on Earth, such as alkali-aggregate reactions or freeze-thaw distress. Whether lunar soil can be successfully used as aggregate will depend primarily on its fineness and particle characteristics. Since weathering does not occur, clay minerals should be absent, helping to reduce the potential water demand. It may be possible to use lunar soil as a densifying fraction in a DSP-type formulation.

Inorganic fibers—rock wool or glass, for example—could probably be made *in situ*, as could steel fibers using iron extracted from rocks. It might be economical to bring lightweight organic fibers (polypropylene, Kevlar, or carbon) from Earth, since the weight fractions needed to enhance ductility and resistance to cracking are quite low. Similarly organic compounds used in small quantities as processing aids might also be brought to the Moon economically.

The final ingredient is water, which must either be shipped in from Earth or synthesized on the Moon. It will therefore be a scarce and expensive commodity. This is an advantage technically, in that the manufacture or high performance concrete (or other cement-based composites) will now become the most attractive choice economically because less water would be used. One needs to carefully examine alternate hardening strategies that might not use water as the sole reactant, such as carbonation curing (Young *et al.*, 1974; Goodbrake *et al.*, 1979). Carbon dioxide could be obtained from the organic refuse of human activity. The actual water needed could be less than one quarter of that used in conventional concrete, especially if the water removed during drying is collected and recycled. Terrestrial hydrogen burnt in oxygen extracted from Moon rocks (Friedlander, 1984) is the most likely source of water, although it may be possible to obtain some hydrogen on the Moon (Carter, 1984).

FABRICATION OF CONCRETE

Concrete will have to be formed into precast elements under controlled conditions of humidity and temperature. This is, of course, the key to the hardening process, as well as to maintaining the high level of quality control that will be essential. The "curing" process will involve not only promoting the chemical reactions required for strength development, but also subsequent drying to equilibrium with the lunar atmosphere. This involves much stronger drying than is normally encountered on Earth, but laboratory studies tell us what to expect. Drying shrinkage will be about 2–3 times the usual values and, unless developed very slowly, will cause cracking. Controlled drying may well be the most crucial part of the whole "curing" process and may dictate the choice of hardening strategies. Use of heat and concomitant carbonation would not only increase the rate of strength development, but would also provide a more dimensionally stable material.

CONCLUDING REMARKS

The goal of making concrete on the Moon is an intriguing challenge, one that can almost certainly be met technically. Concrete is truly a versatile material: when the tight economic restraints of terrestrial construction are removed, it will be possible to take advantage of methods and strategies that most engineers are not yet aware of. The basic principles that would guide development are well established, although considerable research and development will still be needed to generate the necessary specific information.

REFERENCES

Agosto W. N. (1984) Lunar cement formulations (abstract). In *Papers Presented to the Symposium on Lunar Bases and Space Activities of the 21st Century*, p. 81. NASA/Johnson Space Center, Houston.

Alford N. M. and Birchall J. D. (1985) The properties and potential applications of Macro–Defect–Free cement. In *Very High Strength Cement-Based Materials, Mater. Res. Soc. Symp. Proc., 42.* pp. 265–276. Materials Research Society, Pittsburgh. In press.

Anonymous (1985) Temple for Bahai faith uses God's blueprints. *Eng. News Rec., 213,* 34–35.

Bache H. H. (1981) Densified cement/ultrafine particle based materials. *Papers Presented to the 2nd Intl. Conf. on Superplasticizers in Concrete.* Aalborg Portland, Aalborg, Denmark. 36 pp.

Birchall J. D. (1983) Cement in the context of an energy expensive future. *Phil. Trans. Roy. Soc. Lond., A310,* 31–42.

Birchall J. D., Howard A. J., and Kendall K. (1982) Strong cements. *Proc. Brit. Ceram. Soc., 32,* 25–32.

Burge T. A. (1983) 14,000 psi in 24 hours. *Concr. Intern., 7,* 39–40.

Carter J. L. (1984) Lunar regolith fines: A source of hydrogen (water) (abstract). In *Papers Presented to the Symposium on Lunar Bases and Space Activities of the 21st Century*, p. 27, NASA/Johnson Space Center, Houston.

Cutler A. H. (1984) An alkali hydroxide based scheme for lunar oxygen production (abstract). In *Papers Presented to the Symposium on Lunar Bases and Space Activities of the 21st Century*, p. 21. NASA/Johnson Space Center, Houston.

Friedlander H. N. (1984) An analysis of alternate hydrogen sources for lunar manufacture (abstract). In *Papers Presented to the Symposium on Lunar Bases and Space Activities of the 21st Century*, p. 28. NASA/Johnson Space Center, Houston.

Goodbrake C. J., Young J. F., and Berger R. L. (1979) Reactions of hydraulic calcium silicates with carbon dioxide and water. *J. Am. Ceram. Soc., 62,* 488–491.

Hirsch P., Birchall J. D., Double D. D., Kelly A., Moir G. K., and Pomeroy C. D. (editors) (1983) *Technology in the 1990s: Developments in Hydraulic Cements.* The Royal Society, London. 207 pp.

Hjorth L. (1983) Development and application of high–density cement-based materials. *Phil. Trans. Roy. Soc. Lond., A310,* 167–173.

Hjorth L. (1984) Silica fumes as additions to concrete. In *Characterization and Performance Prediction of Cement and Concrete* (J. F. Young, ed.), pp. 165–175. Engineering Foundation, New York.

Kendall K., Howard A. J., and Birchall J. D. (1983) The relationship between porosity, microstructure, and strength, and the approach to advanced cement-based materials. *Phil. Trans. Roy. Soc. Lond., A310,* 139–154.

Kendall K. and Birchall J. D. (1985) Porosity and its relationship to the strength of hydraulic cement pastes. In *Very High Strength Cement-Based Materials, Mater. Res. Soc. Symp. Proc., 42.,* pp. 143–148. Materials Research Society, Pittsburgh.

Lin T. D. (1984) Concrete structures for lunar base construction (abstract). In *Papers Presented to the Symposium on Lunar Bases and Space Activities of the 21st Century,* p. 79. NASA/Johnson Space Center, Houston.

Radjy F. F. and Loeland K. E. (1985) Microsilica concrete: A technological breakthrough commercialized. In *Very High Strength Cement-Based Materials, Mater. Res. Soc. Symp. Proc., 42.,* pp. 305–312. Materials Research Society, Pittsburgh.

Roy D. M., Nakagawa Z., Scheetz B. E., and White E. L. (1985) Optimized high strength mortar: Effects of particle packing and interface bonding. In *Very High Strength Cement Based Materials, Mater. Res. Soc. Symp. Proc., 42.,* pp. 133–142. Materials Research Society, Pittsburgh.

Wilson A. D. (1978) The chemistry of dental cements. *Chem. Soc. Revs., 7,* 265–296.

Wilson A. D. (1979) Glass ionomer cements—ceramic polymers. *Cements Res. Progr., 1984,* 279–310.

Wise S., Satkowski J. A., Scheetz B., Rizer J. M., MacKenzie M. L., and Double D. D. (1985) The development of a high strength cementitious molding/tooling material. In *Very High Strength Cement-Based Materials, Mater. Res. Soc. Symp. Proc., 42.,* pp. 253–263. Materials Research Society, Pittsburgh.

Young J. F., Berger R. L., and Breese J. (1974) Accelerated curing of compacted calcium silicate mortars on exposure to CO_2. *J. Amer. Ceram. Soc., 57,* 394–397.

Young J. F. (editor) (1985) *Very High Strength Cement-Based Materials. Mater. Res. Soc. Symp. Proc., 42.,* Materials Research Society, Pittsburgh. 317 pp.

MAGMA, CERAMIC, AND FUSED ADOBE STRUCTURES GENERATED *IN SITU*

E. Nader Khalili

Southern California Institute of Architecture, 1800 Berkeley Street, Santa Monica, CA, 90404

The accumulated human knowledge of "universal elements" can be integrated with space-age technology to serve human needs on Earth; its timeless materials and timeless principles can also help achieve humanity's quest beyond this planet. Two such areas of knowledge are in earth architecture and in ceramics, which could be the basis for a breakthrough—in scales, forms, and functions—in low gravity fields and anhydrous-vacuum conditions. With the added missing link of the element of fire (heat), traditional earthen forms can be generated on other celestial bodies, such as the Moon and Mars, in the form of magma structure, ceramic structure, and fused adobe structure. Ceramic modules can also be generated *in situ* in space by utilizing lunar or meteoritic resources.

TIMELESS MATERIALS—TIMELESS PRINCIPLES

The traditional techniques of building without centering, *i.e.*, leaning-arches, corbelling, and dry-packing can have greater applications in lower gravity fields, as well as higher material strength, than in the restricted conditions of these techniques' terrestrial origins. At the same time, the "high-tech" heat-obtaining skills of solar heat, plasma, microwave, and melting penetrators can provide ceramic-earth shelters and appropriate technology for both developed and underdeveloped nations. Through understanding and utilizing the principles of "Yekta-i-Arkan"—unity of elements—integration of tradition and technology in harmony with the laws of nature is possible at many levels of microcosm and macrocosm.

MAGMA STRUCTURE

Lunar base structures can be generated and cast, based on the natural space formations created by magma-lava flow such as tubes and voids. By utilizing existing lunar contours or by forming mounds of lunar soil to desired interior spaces, structures can be cast *in situ* with the generated magma. Either way, the upper layers of the mounds and the apex, consisting of unprocessed lunar resources, can generate magma flow with focused sunlight (Criswell, 1976).

Ceramic-glass (Grodzka, 1976) and/or other lunar fluxes may be added to the main composite for lowering the melting temperature. Basalt melting point, 900° to 1200°C, can be lowered to glass composites' melting point with added lunar flux. As the molten composite flows with the low gravity crawl, the lava crust can be formed in spiral, circular, or multi-patterned rib troughs on the mound. A controlled flowing magma can cast single- or double-curvature monolithic shell structures. The underlying loose soil mound can then be excavated and packed over the monolithic shell for radiation/thermal/impact shielding (Carrier, 1976). Since high depth of necessary soil coverage over the structure is detrimental to both architectural flexibility and harmonious interaction of inner and outer space environments, the variable magma viscosity can be utilized to reduce the estimated 2-m thickness (Land, 1984) of the packed soil protections depending on material composites and attained temperature degree/time parameters.

The viscosity of the generated magma and the packed regolith can counterbalance internal atmospheric pressure, and the semi-glazed interior can provide an airtight membrane. The pliability of the magma medium can present new dimensions in the creation of sculptured interiors for the ultimate functional utilization of the generated spaces. It also offers an aesthetic dimension, since the molded forms conform to human generic non-angular tendencies. The organic material of magma and the possibilities for ceramic glazing of the interior will open a new era in integration of the arts to scales unattainable for humans under the limits of terrestrial conditions.

Magma materials, basaltic in particular, have produced agricultural soils and with suitable atmospheric conditions have proved to produce vegetation. Plant successions have taken place in magma-lava metamorphosis in terrestrial lava tubes and voids. Many examples of flora can be seen in old lava beds of the volcanic regions of the world. Similar conditions will be present in lunar magma structures when the temperature-moisture ambient exists for a life-supporting environment. Thus, common spaces of lunar bases could be designated as mini-agricultural zones that could both generate suitable atmosphere to sustain human life and provide supplemental nutrition resources.

Natural lava structures, such as Craters of the Moon National Monument, can provide case studies in the design development stages. Research is needed to determine material composites, magma crust formation patterns, and span limitations.

PREFABRICATED MAGMA MEMBERS

Conventional structures can be built with magma in lunar base complexes by prefabricating structural members. Beams, columns, panels, and connections can be prefabricated with generated magma composed of unprocessed lunar resources fused with solar heat. Magma-lava solidified structural members can be reinforced with fibers or reinforcing mesh produced from lunar resources. The precast panels and members can be post-tensioned by tendons or fused with spot mortar composed of similar magma materials. Precast magma and ceramic members can be shaped to fit desired forms and functions. Lunar soil troughs and fused regolith layer form work can be utilized for casting systems.

CERAMIC STRUCTURE

The use of shielding ceramic tiles on the space shuttle points to the potential of ceramic materials for lunar and space applications. Ceramic structures of limited spans can be cast *in situ* on lunar sites; they can also be generated in space. On lunar sites, a centrifugally gyrating platform—a giant potter's wheel—featuring adjustable rims with high flanges can be utilized for the dynamic casting of ceramic and stoneware structures. A mass of lunar resources can be "thrown" in the stationary center zone of the platform and melted by focused sunlight to flow to the periphery rotating zone and cast desired shapes. Known lunar resources can also be spun on the same platform to create tensile fiber; by integrating the two operations, monolithic ceramic structures with tensile fiber reinforcing layers can be generated. Double-shell ceramic structures sandwiched with space and/or packed with insulating materials can provide radiation, thermal, and impact shielding. Such units can be used singularly for lunar camps or combined around a common hub and/or spine to form a lunar base complex.

The centrifugal platform system with its adjustable rim flanges can be utilized for lunar base infrastructure parts: pipes, ducts, and tunnel rings. Prefabricated sections for utility sheds can also be formed in single- or double-shell modules.

In space, a centrifugally gyrating platform moving in three dimensions can create more variations of ceramic structured modules than is possible in terrestrial or gravity fields. Attached to a space station, the gyrating platform can generate ceramic modules *in situ*. The resources for ceramic structures can either be of lunar or martian origin or, in space, from captured meteoroids.

FUSED ADOBE STRUCTURE

Lunar base structures can be constructed *in situ* utilizing lunar adobe blocks produced from unprocessed lunar soil or the by-products of industrial mining operations. Lunar adobe blocks can be formed by the fusion of lunar resources with solar heat. It is anticipated that vacuum conditions and the essentially zero-moisture content of lunar soils should significantly reduce thermal diffusity (Rowley, 1984). Lunar adobe blocks can be used to build structures without form work, employing the earth-architecture techniques of dry-packing, corbelling, and leaning-arches (Khalili, 1986). The low gravity field and vacuum conditions, which allow for a smaller angle of repose and enhance lunar soil cohesion (Blacic, 1984), will give greater opportunity, in the case of the leaning-arch technique, for larger spans and shallower vaults and domes. The same advantages will cause the soil-packed covering to follow desirable contours for more flexible interaction of interior and exterior space and solar orientation. Fused spot-mortar or lunar dust sprayed at fusion point temperature can be used to bond the blocks in medium and large span structures. Arches, domes, vaults, and apses can be constructed to fit the contours of the moonscape; these curved surfaces can create sun and shade zones that are functionally desirable.

For functional or aesthetic reasons, total or partial interior ceramic glazing of lunar adobe structures can be done with lunar resources containing glass (Heiken, 1976) and other fluxes by solar heat fusion or plasma technology. The difficulty of mechanical separation of lunar dust can be solved by the bulk use of the soil at its powder stage, involving pre-heating the dust and guniting it on the structure at the point of fusion.

The techniques of earth-architecture and the human skills that have evolved to deal with natural materials and to meet the historic challenges of harsh environments and terrestrial gravity can put future men and women in direct touch with the lunar world. Discovering suitable dimensions of blocks, techniques of construction, and appropriate material composites while developing their own sense of unity with the lunar entity can be the start of human independence from Mother Earth, creating shelters in the heavens. The organic growth of lunar architecture, with its own materials and equilibrium of elements, can be used to initiate an indigenous and ecologically balanced human environment without damaging the heavenly body.

On Earth, one of the main tasks of architects, engineers, and builders has historically been nothing but winning the fight against gravity; now and in the future, the chance for victory on the Moon will be six times as great as it has been here on Earth.

INITIAL *IN SITU* CONSTRUCTION

Locating a lunar lava tube may well be one of the first stages of setting up a lunar base site. Lava tubes can provide the most expedient and economical way of starting an indigenous lunar architecture. Terrestrial lava tubes are the best design model for exploring the development of appropriate life-supporting environments in lunar lava tubes. Either at the initial stage or in the following phases of lunar base construction, locating and utilizing lava tubes can be of great value.

An immediate construction system for the lunar base, after the initial camp setup, can utilize unprocessed lunar resources in a non-mechanized construction system. This system uses existing rocks of different sizes and dry-pack techniques. The low gravity field and higher rock fracture strength give added advantages for larger spans of corbelling and leaning-arch earth-structure systems. Meteoroid and/or indigenous rock structures covered with lunar soil for radiation and thermal shielding can provide immediate, non-life-supporting shelters. Structures built with the same techniques can be fitted with an airtight fabric mesh for human habitation (Blacic, 1984).

PAVING AND LUNAR DUST STABILIZATION

The lunar soil, with a particle size of about 70 microns, which adheres to everything and churns up with vehicular traffic, needs to be stabilized (Carrier and Mitchell, 1976). Fusion of the top layers of lunar soil with focused sunlight can form a magma-lava crust to arrest unstable lunar dust. Spacecraft landing pads, vehicular traffic roads, and pedestrian walkways can be paved with solar heat by on-spot fusion of the top layers, penetrating to desirable depth. Unprocessed lunar soil can be fused by solar energy via a manual or automatic and remote control "paving" vehicle. Inappropriate regolith areas can be topped with a layer of appropriate lunar soil before its fusion. For low temperature fusion, lunar fluxes can be sprayed on top of the soil prior to introducing solar heat. Paving surfaces of heavier traffic areas can be constructed from composites fused to ceramic and stoneware consistency with desired colors and textures.

As a general rule, it is the use of the universal principles of the terrestrial element of fire (heat)—the solar rays—that must be thought of at the forefront of mediums and materials for planetary base design and construction. Adhering to the philosophy of the use of local resources, human skills, and solar energy, we can achieve our quests on the Moon, Mars, and beyond.

We must learn from the accumulated human knowledge of earth-architecture, which has sheltered humans in the harshest conditions. Each person going to the Moon, regardless of his or her work, must be aware of these fundamental principles and techniques to participate in creating an indigenous architecture to form their communities, not only because of economic benefit but also because of spiritual reward. As an old Persian saying goes, "Every man and woman is born a doctor and a builder—to heal and shelter himself."

Acknowledgments. *The Geltaftan Group, consisting of of Manouchehr Sedehi, Mahmoud Hejazi, Ezzatollah Salmanzadeh, Ali Gourang, Ostad Asghar, and A. A. Khorramshahi, supported my work in earth-and-fire developments. Eyal Perchik, Alessandra Runyon, Tsosie Tsinhnahjinnie, Steven Haines, Ellwood Pickering II, Barclay Totten, students at the Southern California Institute of Architecture, have helped advance my research work.*

REFERENCES

Blacic J. D. (1984) Structural properties of lunar rock materials under anhydrous, hard vacuum conditions (abstract). In *Papers Presented to the Symposium on Lunar Bases and Space Activities of the 21st Century,* p. 76. NASA/Johnson Space Center, Houston.

Carrier W. D. III and Mitchell J. K. (1976) Geotechnical engineering on the Moon (abstract). In *Lunar Science VII, Special Session Abstracts,* pp. 92–95. Lunar Science Institute, Houston.

Criswell D. R. (editor) (1976) *Lunar Science VII, Special Session Abstracts (on Lunar Utilization),* pp. iii–vi. Lunar Science Institute, Houston.

Grodzka P. (1976) Processing lunar soil for structural materials (abstract). In *Lunar Science VII, Special Session Abstracts,* pp. 114–115. Lunar Science Institute, Houston.

Heiken G. (1976) The regolith as a source of materials (abstract). In *Lunar Science VII, Special Session Abstracts,* pp. 48–52. Lunar Science Institute, Houston.

Khalili E. N. (1986) *Ceramic Houses.* Harper and Row, San Francisco. In press.

Land P. (1984) Lunar base design (abstract). In *Papers Presented to the Symposium on Lunar Bases and Space Activities of the 21st Century,* p. 102. NASA/Johnson Space Center, Houston.

Rowley J. C. (1984) In-situ rock melting applied to lunar base construction and for exploration drilling and coring on the moon (abstract). In *Papers Presented to the Symposium on Lunar Bases and Space Activities of the 21st Century,* p. 77. NASA/Johnson Space Center, Houston.

LAVA TUBES:
POTENTIAL SHELTERS FOR HABITATS

Friedrich Hörz

Experimental Planetology Branch, SN4, NASA/Johnson Space Center, Houston, TX 77058

Natural caverns occur on the Moon in the form of "lava tubes," which are the drained conduits of underground lava rivers. The inside dimensions of these tubes measure tens to hundreds of meters, and their roofs are expected to be thicker than 10 meters. Consequently, lava tube interiors offer an environment that is naturally protected from the hazards of radiation and meteorite impact. Further, constant, relatively benign temperatures of -20°C prevail. These are extremely favorable environmental conditions for human activities and industrial operations. Significant operational, technological, and economical benefits might result if a lunar base were constructed inside a lava tube.

INTRODUCTION

This paper addresses the existence of natural caverns on the Moon in the form of "lava tubes," and it suggests that they could provide long-term shelter for human habitats and industrial operations.

The origin of lava tubes is genetically related to the formation of "sinuous rilles," which represent flow channels of molten lava. Such channels generally form at high extrusion rates of low viscosity magmas. Sinuous rilles are abundantly observed on lunar basalt surfaces (*e.g.*, Oberbeck *et al.*, 1969, 1971; Greeley 1971, 1972; Cruikshank and Wood, 1972; Head, 1976). The distribution of sinuous rilles on the lunar front side was mapped by Guest and Murray (1976).

Lava tubes are well known from basaltic volcanic terranes on Earth (Ollier and Brown, 1975; Greeley, 1971, 1972, 1975; Cruikshank and Wood, 1972; Hulme, 1973; Peterson and Swanson, 1974). A number of processes may contribute to their formation: (1) radiative cooling may cause surface crystallization and crusting–over of the liquid lava. (2) Commonly, such relatively thin crusts break apart and collapse because the melt below continues to flow. Solid but relatively hot chunks of this crust will raft on the lava river and may coalesce into larger and larger aggregates until a solid roof forms. (3) Radiative cooling takes place at the sides of such lava flows, leading to crusting and aggregation of solids and ultimately to the buildup of pronounced levees, which in turn increase channelled melt flow. Additional aggregation from these levees, aided by spattering of lava splashes, can lead to the formation of solid roofs.

Commonly, low viscosity magmas are also very hot. Hulme (1973) and Peterson and Swanson (1974) present field observations that lava tube cross sections may be modified and enlarged by thermal erosion, *i.e.*, by remelting of the tube's ceiling, walls, and floor.

Typical heights and widths of terrestrial lava tubes are generally measured in a few meters; cross-sectional dimensions in excess of 10 m are rare. The length of lava tubes on Earth may reach 10 to 20 km, but most lava tubes are only 1–2 km long. Greeley (1975) points out that the frequency of such underground lava conduits on Earth may have been underestimated in the past and that they are indeed relatively common around terrestrial shield volcanoes such as those in Hawaii.

LUNAR LAVA TUBES

High extrusion rates and extremely low viscosities characterize lunar basaltic volcanism (Moore and Schaber, 1975), conditions very conducive to the formation of lava channels and tubes. Open channels in the form of sinuous rilles are very abundant on lunar basalt surfaces. Their widths and depths are typically hundreds of meters, and they are commonly a few tens of kilometers long. They are, thus, much larger than their terrestrial analogs (*e.g.*, Oberbeck *et al.*, 1969, 1971). Indeed many of the above studies address the problem of how to properly scale the dimensions of terrestrial lava channels and tubes to their much larger counterparts on the Moon. Hulme (1973) argues for increased turbulence and increased thermal erosion during lunar basalt flow. In detail, the highly meandering nature of many lunar rilles is also not observed to the same degree in terrestrial analogs. Increased meandering is probably best explained by reduced gravity and extremely shallow flow gradients.

In contrast to numerous open flow channels in the form of sinuous rilles, bona fide lava tubes are rarely observed on the Moon; they could indeed be rare geologic features. On the other hand, they are subsurface and will therefore generally not show up in lunar surface imagery. The only lava tubes that can be recognized from lunar surface photos are those that have partially collapsed roofs. Thus, little can be surmised about the absolute frequency and global distribution of lunar lava tubes. They may well be more common than can be demonstrated at present. The important point and the crux of this paper is, however, that they do exist on the Moon.

Figure 1 shows a lava tube with large segments of collapsed roof. A modest topographic ridge forms the crest of the tube as pointed out by Oberbeck *et al.* (1969). The elongated depressions must be caved-in portions of this ridge system. Their elongated plane view and the lack of any raised rims distinguishes these depressions from circular impact craters. Note also the highly braided nature of the elongated depressions, in stark contrast to the random distribution of circular impact features dotting the surroundings. This observation in particular lends additional credence to the interpretation that the entire linear feature is a partially collapsed lava tube. Figure 2 represents another feature interpreted by Cruikshank and Wood (1972) as a partially collapsed lava tube. This tube seems to be unusually straight. The width of the open rille is approximately 200 m, and the uncollapsed roof segments are a few hundred meters long. Note the size of impact craters that were suffered by the seemingly intact roof segments.

What do we know about the roof thicknesses of lunar lava tubes, and are these roofs sufficiently massive and structurally stable to provide long-term shelter against

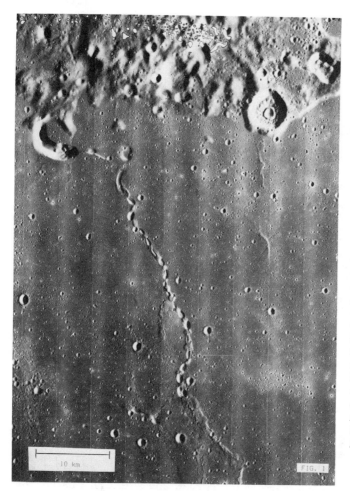

Figure 1. One of the most prominent lunar lava tubes, first described in detail by Oberbeck et al. (1969). The lava tube is approximately 40 km long and up to 500 m wide. Note that some sections of the roof are uncollapsed and that the tube continues underground toward the south (at bottom of picture). Also, note that slopes leading into the rille may be of different steepness; the flatter ones might be negotiated with ease. Uncollapsed sections of the tube are on the order of a few hundred meters long, particularly in the northern part. Dimensionally, these lava tubes would be more than adequate to serve as receptacles for modular habitats and a variety of machinery. Note that this lava tube happens to be within a few kilometers of a highland contact, and it is not inconceivable that access to different raw materials may be possible from a single lava tube (Lunar Orbiter 5, frame 182. Northern Oceanus Procellarum).

radiation and meteoroid bombardment? According to Oberbeck *et al.* (1969), the ratio of roof thickness (T_R) of terrestrial lava tubes relative to typical dimensions of tube cross sections (T_C) ranges from 0.25 to 0.125. Oberbeck *et al.* (1969) also use simple structural beam modeling to calculate that basalt "bridges" spanning a few hundred meters are possible on the Moon provided they are at least 40–60 m thick. These estimates happen to agree with the terrestrial T_R/T_C ratios. Importantly, these estimates are also in good agreement with the following observations: uncollapsed roofs of lava tubes display impact craters a few tens of meters across (see Fig. 2), occasionally as large as 100 m. The diameter/depth ratio of small lunar craters is approximately 4 to 5 (Pike, 1977). Thus, crater excavation depths approaching 20 m can be demonstrated. Using ballistic penetration mechanics (*e.g.*, Gehring, 1970) and associated spallation processes at the rear surface (the roof's ceiling) of a slab–like impact target, one can estimate conservatively that the

Figure 2. Lunar lava rille with uncollapsed roof sections that measure hundreds of meters. Note that mountains are close by which certainly differ in chemistry and mineralogy from the relatively flat basalt surfaces. This rille was extensively described by Cruikshank and Wood (1972) (Lunar Orbiter 5, frame M-191).

roof thickness must be at least two times larger than any crater depth; otherwise complete penetration of the slab (roof) would have occurred. Following these arguments, the maximum crater sustained by an uncollapsed roof yields a minimum measure of roof thickness. The thicknesses of some lunar lava tube roofs are thus a few tens of meters. In principle, the minimum roof thickness of specific lava tubes could be assessed using the above crater-geometry relationships.

While impact craters indicate that initial roof thickness must have been substantial, the cratering process has also contributed to the erosion and structural weakening of such natural basalt bridges. Judging from the thickness of lunar soils on representative basalt surfaces, the uppermost 5–10 m of solid bedrock (lava tube roofs) are totally comminuted into fine-grained lunar soil. Penetrative cracks associated with this average regolith depth are a factor of 3–5 deeper on the basis of seismically disturbed areas below terrestrial craters (Pohl *et al.*, 1977). Importantly, the above-mentioned spallation process occurs at the tube's ceiling even for impact events that are not penetrative; *i.e.*, it occurs as long as the stress amplitude of an impact-triggered shock wave exceeds the basalt's tensile strength (*e.g.*, Hörz and Schaal, 1981). This spallation–induced thinning and weakening is of more concern for the structural integrity of a given roof than surficial erosion. A few relatively large craters (>50 m) may have done more structural damage than the cumulative effects of many <20-m-diameter craters. Because meteorite impact is a stochastic process, it is difficult to predict the structural integrity and exact thickness of a lava tube roof with great precision. Nevertheless, rough estimates can be made via photogeologic techniques related to crater geometry as outlined above. An obvious strategy would be to select roofs or roof segments that have suffered relatively small cratering events only. Such segments are clearly safer than areas close to, if not directly below, a relatively large impact crater.

What do we know about cross-sectional dimensions of lunar lava tubes, and how do their interiors look? As indicated above, the linear dimensions of sinuous rilles and lava tubes are significantly larger on the Moon than on Earth; collapsed portions of some lunar lava tubes indicate correspondingly large tube interiors (see Figs. 1, 2). There is little doubt that lunar lava tubes have large enough cross sections to house most any habitat. Restrictions and enlargements of cross sections occur in terrestrial lava tubes, but on relatively modest scales. The surface relief of terrestrial lava tube interiors can be highly variable, ranging from relatively smooth to very rough and knobby. However, this variability occurs on relief scales that are extremely small compared to cross-sectional dimensions. We can only assume that lunar lava tubes display similar relief. In addition, the above-mentioned spallation products will have accumulated on the floor; they may possibly make initial trafficability cumbersome until removed or leveled (using readily available lunar soil as fill).

LUNAR BASE INSIDE A LAVA TUBE

Based on the foregoing, it appears that natural caverns of suitable sizes to house an entire lunar base exist on the Moon. Roof thicknesses in excess of 10 m will provide safe and long-term shelter against radiation and meteorite collisions. Creation of similarly shielded environments will constitute a significant and costly effort for any lunar base located at or close to the lunar surface. Substantial operational advantages for a lava tube scenario emerge as outlined below.

The primary suggestion advocated by this report is to use lava tubes merely as receptacles for prefabricated, modular habitats, either imported from Earth (initially?) or fabricated from lunar resources, if not in place (at later stages?). We do not suggest that the lava tube itself may be suitably modified to serve as the primary habitat. There are too many uncertainties related to detailed geometry of the cross section and to the surface roughness of the walls and floors. Indeed, lava tube interiors may be too large, at least initially. Furthermore, penetrative cracks in the roof may exist, which would make it extremely difficult, if not impractical, to render the enclosed volume airtight. Modest site preparation inside the lava tube would consist of leveling the floor with lunar soil, an earth-moving operation similar in scale to site preparation on the surface. The lava tube would then be ready to act as a receptacle for self-enclosed habitats as well as for a large number of industrial operations, all safely protected from radiation and meteorite impact.

The primary advantage of housing the lunar base in a naturally sheltered environment is the potential to use extremely lightweight construction materials. None of the components would have to support any shielding mass whatsoever. Indeed, many components, such as a habitat shell, would not even have to support much of their own weight because they could be supported from the walls and ceilings of the lava tube. Habitats could even be inflatable, supported by air pressure only. In any case, construction and selection of materials would be entirely dictated by expected wear and tear. Widespread use of thin foil materials (metals, plastics?) is possible not only for the habitat itself, but also for a variety of ducts, storage tanks, *etc.* Any lunar base will include a variety of machinery located outside the man-rated, shirt-sleeve environment. Some of this gear will have to be protected against meteorite impact (*e.g.,* all life-support systems). Much of this equipment will also have to be visited occasionally by crews for monitoring, maintenance, and repair. Inside a lava tube, the layout of this equipment could resemble that of terrestrial operations with all components freely exposed and easily accessed for inspection and repair. This seems particularly convenient for a variety of duct work, pipes, valves, storage tanks, *etc.,* used to transfer gases and liquids. It is also possible to house some machinery inside lightweight shells to create an optimum environment for its operation (*e.g.,* bio-processing plant). Such lightweight shells and habitats are easily connected with each other, providing great flexibility for expansion of the lunar base as well as for specific environmental engineering inside individual enclosures and compartments. In summary, numerous structural and operational advantages would present themselves if a lunar base could be designed and constructed without continuous concern for the hazards of radiation and meteorite impact.

Lava tube interiors offer additional environmental differences compared to the lunar surface. These differences may be beneficial for a number of engineering tasks and operational aspects. Being underground and some tens of meters removed from the lunar surface, there is a relatively constant-temperature environment (estimated at -20°C; Mendell, personal communication, 1985). This contrasts with the diurnal temperature cycle of -180° to +100°C at the surface. Temperature management inside a lava tube appears significantly easier than at the surface, where complex thermal insulation and control systems appear unavoidable. Also, the selection of materials functioning properly over a wide range of temperatures is severely limited at the surface; in contrast, a wide range of common materials may be used at the more benign and constant temperatures prevailing inside a lava tube. Furthermore, inside a lava tube, all equipment is well shielded from IR and UV radiation. Materials (*e.g.,* certain plastics) that otherwise deteriorate if exposed to this radiation could be used indiscriminately inside a lava tube. In short, additional environmental differences of a subsurface location may allow widespread use of common materials that may not be suitable for use on the lunar surface.

Some additional advantages for siting a lunar base inside a lava tube come to mind. The front and rear entrances of the tube may be sealed off rather readily to keep a relatively dust-free environment for all operations; loose dust may be a nuisance for a fair number of operations on the surface. It is also possible to conceive of lightweight, highly fexible suits for crews venturing outside the man-rated habitats but remaining inside the tube; neither thermal insulation nor meteorite impact is of great concern for such suits. Heavy, vibrating machinery may be solidly anchored to firm bedrock (a rarity on the lunar surface). The lava tube may serve as convenient "hangar" or "garage" for all kinds of equipment that have low duty cycles and that must be kept in a protected environment.

A major operational drawback in utilizing lava tubes may be their difficult accessibility. Negotiation of perhaps steep slopes and the climbing in and out of a local "hole" appears cumbersome, possibly impractical. Relatively shallow sinuous rilles, somewhat flattened by impact craters, exist however. Also, the Apollo 15 crew visited the edge of Hadley rille and felt that their Lunar Rover could have negotiated the slopes of this rille (Irwin, personal communication, 1985).

Location of a lunar base at the bottom of a hole seems not very economical from an energy point of view, because mass will have to be lowered and especially raised when needed on the lunar surface and when being readied for export to LEO or GEO. These energy considerations are, however, a matter of degree, because most large-scale industrial operations rely heavily on gravity for material transport. Some modest elevation difference between the source of lunar raw materials and the processing plant is desirable even for such simple operations as sieving and magnetic separation. For this reason, a lunar base may be more functional if located at the base of some slope. Why not a sinuous rille/lava tube where chutes or pipes may be laid out such that they terminate inside the lava tube at exactly that station where the high-graded raw materials are needed?

The most serious drawback in the utilization of lava tubes relates, however, to the present status of lunar surface exploration. Only a few lava tubes are recognized. High resolution photography of the entire lunar globe is needed to improve the inventory of lunar lava tubes and to determine their spatial distribution. Detailed imagery appears at present to be the only means for an improved understanding of their dimensions, roof thicknesses, and global distribution. Furthermore, lava tubes are viable candidates for shelters only if desired raw materials are close by. The distribution of specific lunar resources is also largely unknown at present. It appears prudent to further explore the lunar surface and its resources via remote sensing from polar orbit. Lava tubes are viable candidates to house a lunar base if basaltic raw materials are desired. Lava tubes are, however, not excluded if non-basaltic resources were the ultimate choice. As illustrated in Figs. 1 and 2, lava tubes occur within kilometers of non-mare terrains with lithologies that differ substantially from the surrounding basalts.

CONCLUSIONS

Establishment of a lunar base, its construction, its layout, its diverse functions, and its ultimate location will be the compromise result of numerous scientific, technical, and economic considerations. Some of these considerations may be incompatible with housing a lunar base inside a lava tube. The simple purpose of this contribution is to remind everybody that natural caverns exist on the Moon. They provide a natural environment that is protected from meteorite impact, shelters against radiation, and is at a constant, relatively benign temperature. Such a natural environment allows widespread use of lightweight construction materials, great flexibility in the choice of such materials, and it results in improved operational capabilities. If a lunar base were emplaced on the lunar surface, a qualitatively similar environment would have to be engineered with great complexity and cost.

REFERENCES

Cruikshank D. P. and Wood C. A. (1972) Lunar rilles and Hawaiian volcanic features: Possible analogues. *Moon, 3,* 412–447.

Gehring J. W., Jr. (1970) Engineering considerations in hypervelocity impact. In *High Velocity Impact Phenomena,* (R. Kinslow, ed.), pp. 463–514. Academic Press, New York.

Greeley R. (1971) Observations of actively forming lava tubes and associated structures, Hawaii. *Mod. Geol., 2,* 207–223.

Greeley R. (1972) Additional observations of actively forming lava tubes and associated structures, Hawaii. *Mod. Geol., 3,* 157–160.

Greeley R. (1975) The significance of lava tubes and channels in comparative planetology (abstract). In *Papers Presented to the Conference on Origins of Mare Basalts,* pp. 55–55. Lunar Science Institute, Houston.

Guest J. E. and Murray J. B. (1976) Volcanic features of the nearside equatorial lunar maria. *J. Geol. Soc. London, 132,* 252–258.

Head J. W. (1976) Lunar volcanism in space and time. *Rev. Geophys. Space Phys., 14,* 265–300.

Hörz F. and Schaal R. B. (1981) Asteroidal agglutinate formation and implications for asteroidal surfaces. *Icarus, 46,* 337–353.

Hulme G. (1973) Turbulent lava flows and the formation of lunar sinuous rilles. *Mod. Geol., 4,* 107–117.

Moore H. J. and Schaber G. G. (1975) An estimate of the yield strength of the Imbrian flows. *Proc. Lunar Planet. Sci. Conf. 6th,* pp. 101–118.

Oberbeck V. R., Quaide W. L., and Greeley R. (1969) On the origin of lunar sinuous rilles. *Mod. Geol., 1,* 75–80.

Oberbeck V. R., Greeley R., Morgan R. B., and Lovas M. J. (1971) *Lunar Rilles—A Catalog and Method of Classification.* NASA TM X-62,088, 83 pp.

Ollier C. D. and Brown M. C. (1965) Lava tubes of Victoria. *Bull. Volcan. 25,* 215–229.

Peterson D. W. and Swanson D. A. (1974) Observed formation of lava tubes. *Speleology, 2,* 209.

Pike R. J. (1976) Crater dimensions from Apollo data and supplemental sources. *Moon, 12,* 463–477.

Pohl J., Stöffler D., Gall H., and Ernstson K. (1977) The Ries impact crater. In *Impact and Explosion Cratering* (Roddy D. J., Pepin R. O., and Merrill R. B., eds.), pp. 343–404. Pergamon Press, New York.

DESIGN OF LUNAR-BASED FACILITIES: THE CHALLENGE OF A LUNAR OBSERVATORY

Stewart W. Johnson

The BDM Corporation, 1801 Randolph Road, S. E., Albuquerque, NM 87106

Ray S. Leonard

The Zia Company, P.O. Box 1539, Los Alamos, NM 87544

This paper focuses on the development process and systems engineering needed to emplace and support an astronomical observatory on the Moon. Factors taken into account are the types of observations to be accomplished, the structures and systems needed to support and to protect the elements of the observatories associated with the science mission, the interaction of the lunar regolith with the structures and foundations, and maintenance requirements. The requirements for advanced scientific research can be met with an appropriate mix of automation, robotics, and suited astronaut intervention. It is suggested that a time–phased, multidisciplinary approach be initiated to determine the observations to be made, the best sites for use, and the technical requirements, as well as to resolve criticial technical issues relating to the planning, design, development, and placement of the observatory. Fourteen critical issues related to the development of observatory facilities are listed and discussed. These critical issues will be sharpened in the future as operational requirements become clearer. In the final analysis, the design and development strategy must flow down from requirements. Cost and risk must be elements in the decision process. The question is posed as to how the technologies of the Hubble Space Telescope, the Space Infrared Telescope Facility, and the Gamma Ray Observatory may be adapted to a lunar setting.

SCOPE AND APPROACH

An astronomical observatory on the Moon offers the potential advantages of emplacement on a stable platform in an environment unencumbered by atmospheric obscurations. A radio astronomy observatory on the farside of the Moon would avoid much of the electromagnetic "noise" associated with man's terrestrial activities. The long lunar night would provide extended periods for dark sky observation (Johnson and Leonard, 1984; *Astronomy and Astrophysics for the 1980s,* 1982). The approach in this paper is to focus on the development process and systems engineering needed to emplace and support an astronomical observatory on the Moon.

SOME HIGHLIGHTS OF PAST EFFORTS RELATING TO OBSERVATORIES IN SPACE

The idea of a telescope in space was mentioned in 1923, when H. Oberth, a German rocket pioneer, suggested an orbital telescope (Longair and Warner, 1979). He realized the advantage of observations in space where stars do not twinkle and where there is negligible absorption in the ultraviolet and infrared. Since the launch of Sputnik in 1957, in a period of less than 30 years, many significant contributions to astronomy have been made by use of orbiting instrumentation. There were OAO, SAS-I (Uhuru), Ariel, ANS, Copernicus Orbiting Observatory, Skylab, OSO-7, Solar Max, Explorer, IMP, and others that added to our scientific understanding and to our experience in space operations. The recent Infrared Astronomy Satellite (IRAS) was an enormous success in opening new vistas in the solar system and the universe. The Einstein Observatory (HEA02) in 1979 probed x-ray sources. Currently under development are the Hubble Space Telescope (ST) and the Gamma Ray Observatory (GRO). Further in the future are the Space Infrared Facility (SIRTF), which will be a free flyer, and the Advanced X-Ray Astronomy Facility (AXAF), which will continue work done by the Einstein Observatory mission of 1979. The ST, the GRO, AXAF, and SIRTF are complementary in that they span a range of wavelengths. Each of these instruments is built upon earlier successful orbiting observatories that led to enticing discoveries and suggested improvements in sensors and instrumentation.

Establishment of scientific requirements and development of conceptual designs for a space-based telescope is a lengthy and iterative process. The Space Telescope was first proposed in the early 1960s at a summer study (Longair and Warner, 1979). Meetings

Figure 1.   Radio astronomy from the Moon has three advantages over terrestrial observation: man-made, terrestrial-originating background noise is avoided (particularly on the farside); there is less gravitational pull to cause distortions in the structures; and there is a slower period of rotation relative to objects being observed (from Malina, 1969).

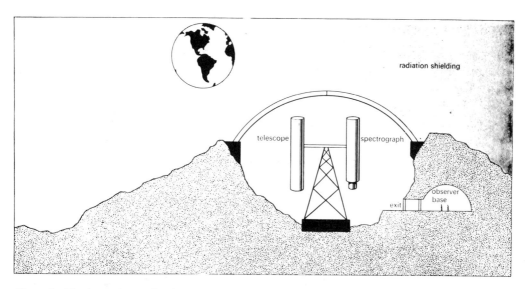

Figure 2. The Lunar International Symposium (LIL) of 1965 suggested this semi-permanent observatory in a small lunar crater. Radiation shielding lids of expanded foam materials are shown (Malina, 1969).

in 1967 and 1968 by an NAS *ad hoc* committee discussed how a space telescope could be used. A 1974 AIAA Symposium led to additional discussion of space telescope use. The NASA Space Telescope project was initiated by an advanced study (Phase A) activity in 1971 and 1972. During 1973–1976, Phase B scientific definition studies were carried out. Final design and development (Phases C and D) began in 1977, the year that Congress approved a 2.4-m space telescope. Launch of this Hubble Space Telescope is anticipated in 1986 or 1987. The telescope is to be maintained and refurbished in orbit at 2- or 3-year intervals and may be returned to Earth for major refurbishment at 5-year intervals. The operational life of the system may be 15 years. Might a similar instrument someday be located on the Moon?

Discussions of the scientific and engineering challenges facing a lunar surface telescope began over 20 years ago. In 1964, a Lunar International Laboratory (LIL) panel anticipated that a manned, permanent research center would begin operations on the Moon sometime in the period 1975–1985. At the International Academy of Astronautics Lunar International Laboratory Project Symposium in Athens in 1965, it was noted (Malina, 1969) that the Moon "represents...an ideal place to site an observatory for both optical and radio telescopes (p. 109)." Figures 1–3 illustrate concepts for lunar observatories suggested at the LIL Symposium (Malina, 1969).

The report of the astronomy working group (Hess, 1967) listed necessary measurements to be made on the Moon before the establishment of a lunar astronomical base. Designs of observatories for the Moon (e.g., for those illustrated in Figs. 1–3) will require extensive data on the lunar environment. The environmental characteristics (Table 1) to be determined at selected lunar sites may influence site selection. Sites suggested include the crater

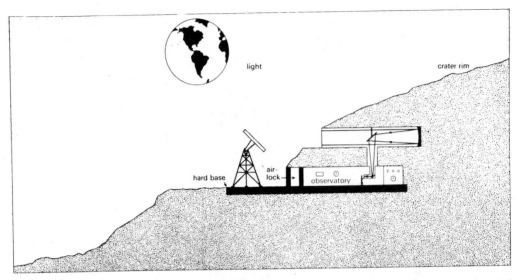

Figure 3.   This fixed-based observatory was also proposed at LIL such that sensitive equipment could be protected by a considerable thickness of lunar soil and rock (Malina, 1969).

Grimaldi because of its location near the equator and near the limb. Another option is a site slightly south of the equatorial plane to provide access to the Magellanic Clouds. The very large telescope was suggested at a site on the farside of the Moon to avoid earthlight.

A MIMOSA (1967) Program III that was to commence in 1971 and extend until 1988 involved 1-meter optical telescopes set up at the south pole and the center of the farside to evaluate the potential of lunar-based astronomy. A 12-man permanent

Table 1.   Characteristics to be Determined at Selected Sites*

1. Micrometeoroid environment
2. Radio frequency noise levels
3. Surface impedance and conductivity
4. Density and extent of lunar ionosphere (if it exists)
5. X-ray and γ-ray intensities, including zenith-angle distribution of intensities
6. Soil mechanics such as bearing strength and stability, depth profiles of temperature, seismic activity, and ionizing radiation
7. Thermal effects on astronomical instrumentation
8. Contaminants such as dust, spacecraft outgassing, spacecraft radio frequency interference, and astronaut seismic noises
9. Deterioration of precision optical surfaces
10. Evaporation rates for optical coatings

*from Report of Astronomy Working Group (Hess, 1967)

base in the crater Grimaldi was to use an array of radio, optical, and x-ray telescopes. MIMOSA was based on an upgraded Saturn V launch rate of three to four per year through the 1970s and six per year in the 1980s.

Successful operation of astronomical observatories on the Moon will necessitate the meeting of many stringent requirements for materials, structures, and controls. New proposals for observatories are arising in the scientific community. Some innovative concepts in interferometry (Burke, 1984) require precise knowledge of interactions between array elements, support structures, and the lunar regolith. Other, more traditional installations may require protected, pressurized environments. Each observatory instrument complex, from the simplest to the most elaborate, presents a challenge to soil mechanics, foundation engineering, structural material selection and design, and construction planning.

SOIL MECHANICS AND FOUNDATION ENGINEERING

Johnson *et al.* (1971) used Surveyor and Apollo mission results in an investigation of the lunar regolith as a site for an astronomical observatory. A telescope system was postulated involving a large reflector, and foundations were designed for cases of a deep regolith and a shallow regolith. It was noted that the lunar soil is fine-grained, relatively dense, and weakly cohesive. More information is needed on the lunar environment, including thermal cycling and behavior of the surface under dynamic loads. Previously, Johnson (1964) considered criteria for lunar base structures, taking into account gravitational, vacuum, and other effects.

There are known to be significant variations in the lunar soil both laterally and with depth as revealed by trenching and core tubes (Johnson and Carrier, 1971; Carrier *et al.*, 1972). In emplacing an observatory on the Moon, it will be necessary to have knowledge of soil and rock profiles and engineering properties at depth and to monitor soil and foundation behavior during observatory placement. The regolith may vary in thickness from 1–15 m and be underlain by perhaps jointed or fractured rock. It may be feasible to compact or stabilize the regolith. The wide range in lunar temperatures implies a thermal cycling (and expansion and contraction) of the regolith, suggesting that foundations should be placed below the depth of thermal cycling. Both total and differential settlements are to be controlled appropriately.

DESIGN AND CONSTRUCTION ALTERNATIVES

A variety of materials offers promise for use at a lunar observatory. The materials range from aluminum, graphite epoxy, and metal matrix composites to castings from lunar rock. The choice of materials will depend on progress in the development of the technology of the base and where the observatory will be sited. Early facilities and those away from the main base will probably be fabricated on Earth. Sensitive components will be shielded by burial in the lunar regolith. Air-inflated structures offer the possibility of providing mobile repair hangars that could be used at remote observatory sites.

If robots and automated construction equipment are to be used, consideration will have to be given to a myriad of design details. For example, connections and hookups (*e.g.*, for fluids) must take a positive connection with little adjustment required. Semi-autonomous construction equipment offers the possibility of providing tremendous cost savings in building and maintaining a lunar observatory. Developments on Earth are already validating concepts of semi-autonomous telecommanded systems of construction and exploratory vehicles and equipment for use in hazardous environments and in military contexts.

RELATIONSHIPS BETWEEN REQUIREMENTS AND CRITICAL ISSUES

An observatory on the Moon will be established to meet a set of pre-determined requirements. Requirements related to what missions it will perform (*e.g.*, radio astronomy, optical astronomy, x-ray astronomy, and gamma-ray astronomy), how to perform these missions (*e.g.*, mass and configuration on the Moon, energy needs, and shielding), and how well (resolution, available viewing time, data rate to Earth, etc.). From the requirements will flow a set of critical issues to be resolved. There will be an interplay between requirements and critical issues and technology development programs. Requirements will probably not be fixed until there is a definition of transport and logistics capabilities (mass transportable to the Moon) and the incremental costs to establish a type of observatory on the Moon.

Many critical issues must be resolved before an observatory can be established on the Moon. Some of the issues can be enumerated as follows.

(1) What is the operational role of the lunar astronomical observatory relative to in-orbit and Earth-based observatories? Various options are to be considered for prioritization of viewing and technical oversight management.

(2) What collectors and sensors are to be placed at the sites, and what is the desired time-phasing of placement? Can derivatives of ST, AXAF, SIRTF, GRO, and other systems be used to reduce development time and cost?

(3) What site or sites are to be used? What is the relationship of sensor or astronomical discipline to site, and how can a range of sites be utilized and developed in an optimal time-phasing to achieve scientific returns and be compatible with the lunar base infrastructure?

(4) What information is needed on potential sites, and how are the sites to be surveyed and screened (*e.g.*, for soil profiles and other properties)? Sites need proximity to support but distance from detrimental effects of mining and launch operations.

(5) How is the interface between observatory systems and the lunar regolith to be addressed (*e.g.*, selection of foundation elements to be used on the Moon)? Thermal profiles, conductance, impedance, presence of hot spots, and regolith strength and deformation characteristics need to be evaluated.

(6) How are sites to be preserved for the operational life of observatory systems (*e.g.*, leveling, excavation, dust mitigation, and pollutant control)? Some effort is suggested to establish protected areas for astronomy and to monitor site environments.

(7) How are required positioning tolerances to be met and controlled? Burke (1984) comments that for an optical interferometric array on the Moon, element position and orientation would be controlled to 100 Å in 20 km. What would be the trade-off between structural and foundation precision placement and position holding by closed loop automatic control?

(8) What materials should be used in support structures considering the lunar environment with its radiation, thermal cycling, vacuum, and meteoroid impacts? Are materials such as the metal matrix composites (with low coefficients of thermal expansion) required or are aluminum alloys feasible with appropriate control loops? The composite materials can be tailored to meet coefficient of thermal expansion and stiffness needs. Should structural elements formed from lunar soil be developed and used?

(9) How should logistics and construction needs be met with the appropriate mix of automation, robotics, vehicles, and suited man activity? Cost studies and evaluation of operational risks are needed to establish a desirable mix.

(10) Communications, data storage, and data transfer questions relative to the lunar base, lunar orbit, Earth orbit, and Earth's surface need to be investigated.

(11) System monitoring, protection, maintenance, upgrading, and refurbishment requirements need to be identified, and capabilities need to be planned and built into the system. Experience in Earth orbit should be helpful, particularly if derivatives of Earth-orbiting systems may be used.

(12) What separation distance restrictions are desirable in relation to other lunar activities anticipated such as the logistics base, mining and resources recovery, spaceport launches and landings, and habitat operation? A logical means of setting separation distances needs to be developed.

(13) Shielding for personnel and equipment against the radiation environment of space needs to be addressed. In the design of the observatory facilities, charged particles from the solar wind, solar cosmic rays, and high-energy galactic cosmic rays from outside the solar system will be of concern (Smith and West, 1983), as well as products of the interaction of primaries with shielding materials. Gammas and neutrons originating from power sources may also be present. In space, the proton dosage during a solar flare may be potentially lethal (Benton, 1984). Protection in "storm shelters" with heavy shielding is suggested. Microprocessors such as those used in experiments and life support equipment will have to be protected.

(14) Support requirements such as power, cooling, heating, and associated expendables need to be addressed. Included are cryogenics for cooling detector arrays and possibly advanced refrigeration (cryocooler) capabilities.

SYSTEM MODELING AND OPTIMIZATION

Resolution of these critical issues will be assisted by system models developed to facilitate the evaluation of various options before construction or fabrication of hardware. Analytical models may be constructed of flow of fluids, rate of heat rejection, power

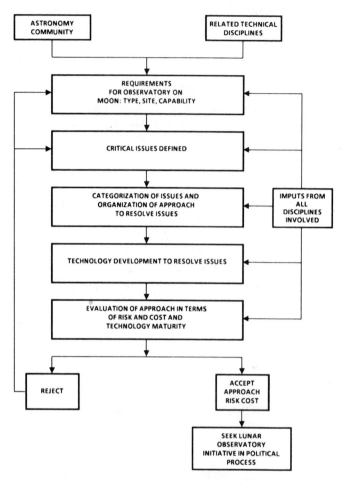

Figure 4. Illustrated are the generation of requirements for the observatory, the definition of critical issues, and how technology programs relate to ultimate decisions after risks and cost are taken into account.

consumption, process flow and performance, and costs. An important model from the standpoint of selecting alternatives is a life cycle cost (LCC) model. It is proposed that one of the first tasks of a lunar observatory working group should be the development of models complete with weighting factors for the various components and systems. Models will assist in calculating the effects of requirements (e.g., resolving power and data handling) on the cost of the observatory. In operations research, many techniques for formalizing the decision process have been developed. Some attributes that might be considered in formalized models to assist in the decision process are refurbishment requirements, power (quantity and quality) for experiments, lighting, heating, ventilation, air conditioning, environmental control, utilities (water, cryogenics, sewage, gas, and air), and thermal control. Models will also assist in trade-offs on structural requirements such as stiffness, natural frequency, thermal stability, precision of erection or alignment, damping, and passive and active control. Results of trade-off studies can be fed back into LCC models. Approaches to optimizing solutions of complex multi-attribute problems (Edwards, 1977) will be applied.

THE FUTURE EFFORT

The pathway to success in the development of a lunar astronomical observatory involves an early start and a multidisciplinary approach to state requirements and to clarify and resolve critical issues. Cost models based on life cycle cost and systems engineering methods will assist in planning and implementing a phased development of a lunar astronomical observatory. Mathematical modeling of the observatory system and its operation will help clarify critical issues and suggest improved approaches. Developmental programs will be initiated to resolve these issues and to reduce the risks involved in placing an observatory on the Moon. Figure 4 portrays a process whereby inputs from the multidisciplinary community are used to derive requirements. Perhaps the observatory will be an international facility built with inputs from many nations. If that is to be the case, then the breadth of input could be great and financial backing will be increased. Once it is determined what capabilities are desired and some allotted mass of equipment is known, issues will be defined. Technology programs will be developed to resolve the issues and to build and test the system.

Acknowledgments. *The authors acknowledge the support of the BDM Corporation and the Zia Company in the preparation of this paper.*

REFERENCES

Astronomy and Astrophysics for the 1980s (1982) *Vol. 1, Report of the Astronomy Survey Committee of the Assembly of Mathematical and Physical Sciences*, pp. 165–166, National Research Council, National Academy Press, Washington, D.C.

Benton E. V., Almasi J., Cassou R., Frank A., Henke R. P., Rowe V., Parnell T. A., and Schopper E. (1984) Radiation measurements aboard Spacelab 1. *Science, 225*, 224–228.

Burke B. F. (1984) Astronomical interferometry on the Moon (abstract). A Paper Presented at the Symposium on Lunar Bases and Space Activities of the 21st Century, NASA/Johnson Space Center, Houston.

Carrier W. D., III, Johnson S. W., Carrasco L. H., and Schmidt R. (1972) Core sample depth relationships: Apollo 14 and 15. *Geochim. Cosmochim. Acta, 3*, 3213–3221.

Edwards W. (1977) How to use multiattribute utility measurement for social decisionmaking. In *Systems Engineering: Methodology and Applications* (A. Sage, ed.), pp. 206–220. IEEE Press, New York.

Hess W. N. (editor) (1967) *Summer Study of Lunar Science and Exploration.* Report of Astronomy Working Group, pp. 369–390. NASA SP-157.

Johnson S. W. (1964) *Criteria for the Design of Structures for a Permanent Lunar Base.* Ph.D. Thesis, University of Illinois, Urbana, IL. 177 pp.

Johnson S. W. and Carrier W. D., III (1971) Lunar soil mechanics. *The Military Engineer, 63*, 324–328.

Johnson S. W. and Leonard R. S. (1984) Lunar-based platforms for an astronomical observatory. In *Proceedings of National Symposium and Workshop on Optical Platforms* (C. L. Wyman, ed.), pp. 147–158, volume 493. Society of Photo-Optical Instrumentation Engineers, Bellingham, Washington.

Johnson S. W., Rohloff K. J., Whitmire J. N., Pyrz A. P., Ullrich G. W., and Lee D. G. (1971) The lunar regolith as a site for an astronomical observatory. In *Proceedings of the Ninth International Symposium on Space Technology and Science* (M. Uemura, ed.), pp. 1059–1076. AGNE Publishing, Inc., Tokyo.

Longair M. S. and Warner J. W. (editors) (1979) *Scientific Research with the Space Telescope.* International Astronomical Union Colloquium, no. 54, Princeton, New Jersey. 327 pp.

Malina F. J. (1969) The lunar laboratory. *Sci. J.* (Special Issue: Man on the Moon), *5*, 108–113.

MIMOSA (1967) *Study of Mission Modes and System Analysis for Lunar Exploration, Final Report, Recommended Lunar Exploration Plan*, volume III. LMSC-A847942, Lockheed Missiles and Space Company, Sunnyvale, CA. 154 pp.

Smith R. E. and West G. S. (compilers) (1983) *Space and Planetary Environment Criteria: Guidelines for Use in Space Vehicle Development.* NASA TM-82478. 191 pp.

ENVIRONMENTAL CONSIDERATIONS AND WASTE PLANNING ON THE LUNAR SURFACE

Randall Briggs and Albert Sacco Jr.[1]

Worcester Polytechnic Institute, 166 South Street, Shrewsbury, MA 01545
[1]Chemical Engineering, Worcester Polytechnic Institute, advisor

Lunar operations in the near future will pose waste management judgments. Final decisions must be established to maintain a high degree of operational efficiency within technological and economical limits. The purpose of this study is to identify environmental considerations arising from a lunar manufacturing facility. Certain assumptions guide the setup and operation of the lunar base presented. The author does not suggest that such assumptions will become reality but, instead, presents an outline of a base that appears promising at the present time. The goal of this paper is to promote conversation and thought towards fundamental decisions that will need to be made before the building of lunar facilities begins. The following assumptions were taken: (1) the base will be manned by a crew of 15 workers; (2) the primary function of the base will be the extraction of oxygen from ilmenite ($FeTiO_3$) via a hydrogen reduction process; (3) nuclear power will form the central power supply on the surface; (4) a semi–closed life support system will be employed with the total resupply of food coming from Earth.

Many questions still exist as to what will be done with much of the refuse brought about by man's existence on the lunar surface. What will become of spent nuclear reactors? Will carbon and phosphate salts from man's waste be stored for later use? Will chemical transportation create a significant lunar atmosphere? These are some of the myriad of questions that confront planners of the lunar base. In order to achieve the best possible results from the operation, questions like these will soon have to be addressed.

INTRODUCTION

If NASA's tentative schedule is to be kept and the lunar facility is to be implemented before 2010, it is not too soon to begin contemplating the impact of man and his activities on the lunar surface and to discuss ways in which his presence and activities will not limit his possibilities in more distant times. A principal goal of any project such as the proposed lunar facility is to grant the most alternatives, economically and technically feasible at the time, to future planners. The paper identifies several wastes that promulgate future operational choices and several that limit more distant activities.

Table 1 presents a rough estimate of the oxygen needs that NASA might typically require at the time oxygen production begins on the lunar surface (oxygen needs and trip parameters were developed through personal communication with B. Roberts and W. Richards of the Johnson Space Center, 1984). A typical lunar oxygen mission can return a net amount of liquid oxygen (LOX) to low Earth orbit (LEO) totaling 54,400 kg (assuming an aerobrake weight that is 14.5% of entry weight, an overall trip I_{sp} of 480 s, and an OTV and lunar lander oxygen to fuel ratio of 7:1; mass payback ratio

Table 1. Typical Oxygen Needs of NASA in LEO as Lunar
Facility Becomes Operational (Yearly)

Mission from LEO	LOX per Mission (kg)	Total Missions	Total LOX (kg)
Lunar	77,200	4	309,000
Manned Geosynchronous	38,000	4	152,000
Other	—	—	91,000
		Total:	552,000

is 2.6). The production rate on the surface to get this amount of LOX to LEO is approximately 200,000 kg. The other 146,000 kg of LOX is used in the transport of cargo to LEO. This brings the lunar oxygen production rate to 200,000/54,400 or 3.7 kg LOX produced on the lunar surface per kg LOX free in LEO.

Referring to Table 1, a total of 552,000 kg LOX will be needed in LEO. Therefore the lunar oxygen production rate is set at 3.7 by 552,000, or 2.1×10^6 kg LOX per year.

MINING, BENEFICIATION, AND THE HYDROGEN REDUCTION OF ILMENITE

The reduction of ilmenite with hydrogen holds much promise for the production of oxygen on the lunar surface for several reasons. (1) The process employs only one significant chemical equation, $FeTiO_3 + H_2 \rightarrow Fe + TiO_2 + H_2O$. The water vapor is electrolyzed back to hydrogen, to be recycled, and oxygen. (2) The lunar maria is known to be 10% or more, by weight, ilmenite. (3) Beneficiation, with an electrostatic separator, is relatively easy and yields material of high purity for further processing.

Beginning with the production rate of 2.1×10^6 kg LOX and working under the assumptions of (a) 90–100% conversion in the reaction above, (b) lunar soil consisting of 10–12% ilmenite, (c) soil density of 1,800 kg/m³, and (d) no significant loss in the beneficiation process, the following approximate values are arrived at: 20,000 tons $FeTiO_3$ mined/year, 1.1×10^5 m³ bulk soil mined/year, and 18,000 tons $FeTiO_2$ generated/year.

LIFE SUPPORT

It is possible that the permanently manned outposts in space, at least initially, will have a life support system that is semi–closed (Spurlock and Modell, 1979). To achieve a greater payback from the lunar operation, components of water and air must be recycled to limit resupply transportation costs. The initial facility, however, may not be capable of supplying its own foodstuffs. The total resupply of food from Earth is therefore assumed.

Table 2.   Basic Requirements and Waste Generation for a
Lunar Crew of 15*

Requirements	Per Man, Daily (kg)	Total for Crew (kg)
Metabolic oxygen	0.9	13.5
Drinking water	3.6	54.0
Hygiene water	5.4	81.0
Food	0.6	9.0
Waste Production		
Carbon dioxide	1.0	15.0
Water vapor	2.5	37.4
Urine	1.5	22.5
Feces	0.16	2.4
Metabolic heat	12,660 kJ	189,900 kJ

*Sharpe (1969), p. 107.

Table 2 lists the material that a life support system must be capable of handling. Figure 1 displays the various units in a feasible life support system.

Central to the support system is the oxidation water reclamation unit. For many years, Robert Jagow has studied the wet oxidation process of waste recycle as it exists in industry and in its application to various other environments (Jagow, 1972, 1975). From this process, Jagow has shown that wash water, human waste, and trash can be oxidized at elevated temperatures and pressures to produce water, carbon dioxide, and a cake of phosphates and sodium salts. The water may be recycled for direct human use, and the carbon dioxide could be converted to oxygen and carbon. From Table 2, the carbon dioxide from respiration from the 15-member crew will be about 15.0 kg/day. About 1500 kg of pure carbon will be produced yearly.

POWER

A dependable, constantly operative power supply will be needed on the lunar surface. Survival will be dependent upon a few kilowatts of power to clean the air of the workers, regenerate the oxygen, and recycle the water. At the same time, large amounts of energy must be produced to accommodate the taxes imposed by the processing facilities. Nuclear power holds a promising key to these lunar needs.

Nuclear power is attractive for several reasons. Proposed units are easy to maintain and operate. It is anticipated that the unit can be dropped into place, started, and left alone until its productive life is over. Nuclear power units have a high energy density. Nuclear power is also capable of producing large amounts of both thermal and electrical energy—five to twenty watts thermal to every one watt electrically produced (French, personal communication, 1984).

Nuclear power units now in study under the program name SP-100 are being adapted to the zero-g environment. These reactors will have a lifetime of seven to ten years

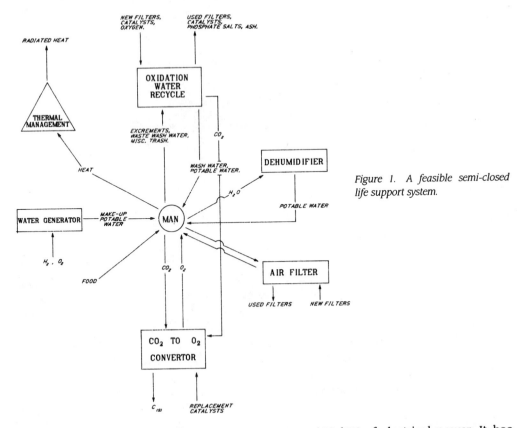

Figure 1. A feasible semi-closed life support system.

and will be capable of generating between 100 to 400 kW of electrical power. It has been proposed to conform such units to the lunar environment, employing native soil for shielding purposes (French, 1984).

TRANSPORTATION

Transport of man and materials to low lunar orbit (LLO) will conveniently be adapted from geosynchronous transport now being drawn up by NASA. Heavy lift vehicles (HLV) and the space shuttle will bring material to LEO, and orbital transfer vehicles (OTV) will travel to LLO, transferring the payload to lunar descent vehicles.

Initially it is conceivable that the lunar descent vehicles will be disposable. Until the production of oxygen begins, there will be no need to bring substantial material off the surface and, further, no ability to fuel vehicles after descent. Upon the production of oxygen, however, a great desire will arise to make these lunar surface transports reusable. Many flights will be needed to transport oxygen off the lunar surface for the trip back to LEO. If such flights were made with disposable transportation, operation costs would reduce the operation's payback.

NASA has estimated the total base mass to be approximately 181,500 kg at the time of oxygen production. Lunar descent platforms, each having a mass of 4,900 kg, have been estimated to be capable of transporting approximately 18,800 kg of cargo (Richards, personal communication, 1984). Thus, to transport the initial base to the Moon, 181,500/18,800 or about ten trips need to be made. This leaves 10 × 4,900 kg or 49,000 kg of scrap material on the lunar surface.

POTENTIALS AND PROBLEMS

Discussion of potentials to increase future operational alternatives and identification of problems that may arise from man's existence on the Moon can proceed from the general outline of the major components of the lunar base given above.

If man is to become self-sufficient on the lunar surface, the groundwork must be laid to close his life support system. Man must eventually develop food production on the lunar surface and cut the "umbilical cord" that will tie him to Mother Earth.

Current work, under the program name Controlled Ecological Life Support System (CELSS), is proceeding to achieve this goal. An experimental bio-regenerative life support system is being considered for attachment to the Space Station in 1998. It is possible that a similar system could be landed on the lunar surface. In this event, advantages exist to aid in the success of this project and reduce the economical cost incurred.

As discussed above, the life support system generates carbon from the conversion of carbon dioxide to oxygen and filters phosphate salts from the water reclamation unit. This material need not be deemed waste, since such rare and life-required elements could be employed in a pilot agricultural facility. It seems logical that these elements from the support system should be stored for later use. It has also been proposed that parts of the discarded transport platforms (up to 40%) be made of similar rare lunar materials (Babb et al., 1984). In this manner, what might previously be considered waste could be used to the advantage of man.

Nuclear power units have potential additional use. This opportunity arises from the large quantity of thermal energy that reactors generate. This excess heat can be radiated into space, if desired, and be considered waste. If, however, processing facilities such as the hydrogen reduction process are developed in a way as to be receptive to this alternative energy source, waste can again be used to the advantage of the base.

Several waste problems still remain unanswered. Will those operating the lunar base consider it economically feasible to refill the mining pits previously excavated? Many now working on this subject seem to believe that the idea of a moral duty to leave the Moon as it has been for eons is sheer "lunacy." However, Williams et al. (1979, p. 283) stated, "Although conveniently located craters can be refilled with wastes initially, back filling the mine site is the only feasible long term solution." Backfilling here would include the ilmenite-poor soil (90% of that is mined) and the reduced ilmenite, $FeTiO_2$, from the reactor.

What will become of the spent nuclear reactors? If those under design in the SP-100 program are to be employed, the energy rating of 100 kW and the high demands of the lunar facility will necessitate several reactors. To reduce the cost of transport, it has been proposed to strip the reactor of its shielding and, in its stead, employ native soil. This requires that the surrounding area be off-limits to personnel and overhead flights (French, 1984). The reactors will provide their services for about ten years, at which point they will be shut down. Two options exist at that time. One choice is to dig up the reactors and refuel them or move them to a different locality. This procedure, however, will require a strong surface infrastructure capable of safely handling such a technically difficult task; perhaps not developed fully enough for several decades after the initial setup. The second option is then forced upon the lunar base—to leave the reactors buried. Since the prospect of tens of discarded nuclear reactors will face the lunar facilities after several years of operation, the need of strict reactor management will arise to avoid limiting future operational plans. If not, the prospect of scattered sites, each being off-limits to lunar activities, will confront more distant lunar workers.

Will chemical transportation cause an impact in the formation of a restrictive lunar atmosphere? It has long been realized that the farside of the Moon provides astronomers the perfect locality for an observation post; the existence of a distinct lunar atmosphere will limit the options of these astronomers. It has been estimated that a long–lived atmosphere (with a life of several hundred years) could be created on the Moon with a mass of 10^8 kg or equivalently if an amount on the order of 60–100 kg/s entered for an extended period of time (Vondrak, 1974).

Calculations can be made for the scenario presented here. For each lunar descent and ascent a total of 98,000 kg exhaust vapor is formed (Richards, personal communication, 1984). Since each trip to the lunar surface returns 54,400 kg LOX to LEO, and a total of 552,000 kg of LOX is to be returned to LEO, about 11 trips must be made in the course of a year. In one year, then, 11 by 98,000 kg or 1×10^6 kg of vapor enter the lunar atmosphere. This is on the order of 0.01 kg/s.

One may conclude that although a long–lived atmosphere will not be developed in this case, a finite cover can be formed and may pose astronomical limitations. Further, if oxygen requirements increase in time and more flights on and off of the lunar surface occur, while some alternative form of transportation (*e.g.*, a mass driver) is not employed, the formation of a substantial lunar atmosphere is not out of the question.

What are the legal responsibilities of NASA and the United States when constructing and operating the base? In 1967, President Lyndon Johnson ratified the "Treaty on Principles Governing the Activities of States in the Exploration and Use of Outer Space, Including the Moon and Other Celestial Bodies," and pledged to abide by certain guidelines in outer space. The following passages are from this treaty.

Art. I, para. 1

"The exploration and use of outer space, including the moon. . . shall be carried out for the benefit and in the interest of all countries, irrespective of their degree of economic or scientific development, and shall be the province of all mankind."

Art. IX, para. 1

"State parties to the treaty. . . . shall conduct all their activities in outer space, including the moon. . . with due regard to the corresponding interest of all other State Parties to the treaty. State

Parties. . . . shall pursue the studies of outer space, including the moon. . . and conduct exploration of them so as to avoid their harmful contamination.

A State Party to the treaty which has reason to believe an activity or experiment planned by another state party in outer space, including the moon. . . . would cause potentially harmful interference with activities in the peaceful exploration and use of outer space. . . . may request consultation concerning the activity or experiment."

If NASA intends to operate processing facilities on the lunar surface, it must not alter the existing environment. If this is not done, it will surely elicit a response from other signatory nations. Intentions of operations and their impact must, in accordance with the treaty, be available to these nations. Therefore, NASA's lunar facilities must not cause other State Parties to question the environmental impact of base activites.

Many prospective lunar activities at this time have unanswered questions and pose many waste management problems. These problems arise from units and procedures that appear at this point promising to be included at the initial lunar base. The opportunity exists for several of the wastes to be utilized by the crew to increase opportunities and decrease cost of operation in the lunar environment. Table 3 summarizes the environmental considerations identified by this study. Questions and opportunities that are open to base planners must be addressed soon to deal with these waste problems and provide the maximum future operational options. It is not too soon to begin to address the impact of man and his activities on the lunar surface.

Table 3. Summary of Environmental Considerations Arising from Proposed Lunar Operations and Facilities

Waste Origin	Environmental Consideration	Potential Benefit
Transportation Hardware	Use of disposable descent platforms for pre-oxygen production transportation places hundreds of tons of scrap material scattered in the base vicinity.	Building platforms of materials beneficial to man on the Moon allows him to accumulate lunar deficient elements.
Chemical Propulsion	Traditional chemical transportation used at projected levels could lead to experimental and operational limitations.	None.
Life Support System (LSS)	Man's life processes produce bio-wastes. Most material is recycled; however initial LSS is not closed, giving rise to phosphate and carbon residue.	Life support residue is rare to the Moon. These materials can be utilized at a later time for lunar agriculture.
Nuclear Power	Reactors stripped of man-made shielding must employ native soil to limit radiation. Surrounding area will be restricted to personnel and overhead flights.	Thermal energy produced by reactors can be used in material processing.
Mining and Processing	Strip mining of topsoil produces large pits and piles on the surface. Unprocessed material is rejected by processing facilities.	None.

Acknowledgments.   This presentation was made possible by Worcester Polytechnic Institute, being adapted from a project under the guidance of Dr. Albert Sacco, Chemical Engineering.

REFERENCES

Babb G. R., Davis H. P., Phillips D. C., Stump M. R. (1984) Impact of lunar and planetary missions on the Space Station (abstract). In *Papers Presented to the Symposium on Lunar Bases and Space Activities of the 21st Century*, p. 53. NASA/Johnson Space Center, Houston.

French J. R. (1984) Nuclear powerplants for lunar bases. In *Papers Presented to the Symposium on Lunar Bases and Space Activities of the 21st Century*, p. 116. NASA/Johnson Space Center, Houston.

Jagow R. B. (1972) Design and development of a prototype wet oxidation system for the reclamation of water and the disposition of waste residues onboard space vehicles. NASI-9183, Lockheed, Sunnyvale, CA. 146 pp.

Jagow R. B. (1975) A study to evaluate the feasibility of wet oxidation for shipboard waste water treatment application. DOT-CG-23034-A, Lockheed, Sunnyvale, CA. 189 pp.

Sharpe M. R. (1969) *Living in Space*. Doubleday, Garden City, NY. 192 pp.

Spurlock J. and Modell M. (1979) Systems engineering overview for regenerative life-support systems applicable to space habitats. In *Space Resources and Space Settlements* (J. Billingham, W. Gilbreath, B. O'Leary, and B. Gosset, eds.) pp. 1-11. NASA SP-428, NASA, Washington, DC.

Vondrak R. R. (1974) Creation of an artificial atmosphere on the Moon. *Nature, 248,* 657-659.

Bunch T. E., William R. J., McKay D. S., and Giles D. (1979) Mining and beneficiation of lunar ores. In *Space Resources and Space Settlements* (J. Billingham, W. Gilbreath, B. O'Leary, and B. Gosset, eds.), pp. 275-288. NASA SP-428, NASA, Washington, DC.

7 / LUNAR MATERIALS AND PROCESSES

T HE AVAILABILITY OF LOCAL RESOURCES constitutes a qualitative difference between a space station and a planetary base. All consumables and construction materials for expansion must be imported to a space station. Because of the great expense of Earth–to–Moon transportation, judicious use of planetary surface materials ought to reduce the cost of maintaining and enlarging a surface facility, in principle. The next generation of launch vehicles may well be designed to lower launch costs, but no concepts on the horizon will result in enough savings to favor importation of terrestrial materials in bulk over utilization of lunar ones whenever possible.

Papers in this section focus on the exploitation of surface materials and on processes that might be important in that enterprise. The opening paper by Haskin is a thoughtful essay on general guidelines for developing lunar resources, bringing to bear his considerable expertise in lunar geochemistry. He identifies "those raw materials for use in manufacture that can most readily be obtained from lunar materials as we know them," always choosing the most conservative scenarios. Since the first step in processing lunar materials has to be some form of mining, Podnieks and Roepke outline research necessary for understanding lunar mining technology, drawing on their experience at the Bureau of Mines supporting Project Apollo. Agosto presents laboratory data on a beneficiation technique for the important lunar mineral ilmenite. The proposed electrostatic separation of the mineral from the bulk soil exemplifies the adaptation of technologies to the lunar environment.

Three papers from the Los Alamos National Laboratory also explore unusual technologies and material properties for lunar utilization. Rowley and Neudecker suggest a universal but power intensive technique for the many types of drilling, coring, and tunneling operations that can be envisioned on the lunar surface. Meek *et al.*, present results of testing microwave heating of lunar soil to extract implanted gases or to form ceramic materials by sintering and melting. Blacic shows that the anhydrous nature of the lunar environment is not all bad news. Silicates show a significant increase in strength in the absence of water; glasses, in particular, show an increase in tensile strength of an order of magnitude, making glass a prime candidate for structural uses.

Testing any of these ideas on actual lunar soils is difficult. The Apollo samples are allocated to scientific investigators only after extensive review of proposals. Final aliquots are usually in milligram quantities. Allton *et al.*, discuss the issue of simulants and explore the possibility that certain small portions of the collection with limited scientific value might be made available for critical testing in future programs.

Many processes in terrestrial manufacturing that are taken for granted must be reexamined at a fundamental level for application on the Moon. Pettit tries to picture what a lunar distillation column looks like. The parameters for this perfectly common process in chemical engineering have been developed over the years largely by trial and error in the terrestrial setting. An analysis of the lunar version must be theoretical and poses an interesting problem in the physics of the important processes taking place in the column. In a similar vein, Lewis looks at the design parameters of lunar machine tools. Such problems make interesting exercises for students because the basic assumptions behind the physics and engineering must be disinterred from the technical literature.

TOWARD A SPARTAN SCENARIO FOR USE OF LUNAR MATERIALS

Larry A. Haskin

Department of Earth and Planetary Sciences and McDonnell Center for the Space Sciences, Washington University, St. Louis, MO 63130

Many lunar materials have been proposed as raw materials for space manufacture. Only those that are abundant and extractable by relatively simple means may be feasible for use. These include mare and highland soils and very abundant rock types. Even the restricted set of materials and extraction processes considered can yield a good variety of raw materials.

INTRODUCTION

Visionaries seeking to show in general terms how our space environment can be used to our benefit have every justification for assuming the usefulness of the chemical elements available on the Moon or on other bodies in the solar system. Those proposing more specific uses of extraterrestrial materials should feel obliged to learn in detail the nature of the materials as found and to determine whether their strategies for use are reasonably based on material availability and probable economy of manufacture. In particular, it falls to the chemist, the chemical engineer, and the metallurgist to test the technological feasibility of individual stages of grander scenarios for product development from space resources.

If lunar ores and conditions resembled those on Earth, it would be straightforward to transplant terrestrial technology to the Moon's surface. Since they do not, it is necessary to consider whether it is better to try to adapt terrestrial technology to the Moon's very different conditions or to devise fresh technologies appropriate to them. This debate may continue for some time, as little research has been done so far to enable choices to be made. The best techniques for converting lunar materials to useful products may not even have been thought of or recognized yet, as we have little direct experience in coping with the space and lunar environments. We still tend to view most extraterrestrial conditions as obstacles and have not learned to react intuitively to them as the advantages that, for many purposes, they will surely be. At the very least, we owe it to ourselves to determine, at the laboratory bench scale, the basic characteristics of all the extraction and manufacturing processes we can conceive of during the next several years before choices must be made on whether to employ lunar matter for use in space and, if so, how.

Some proposed scenarios for manufacture seem to be based on misunderstanding of the nature of the starting materials or to demand for their execution outlandish layouts of factories, prepared reagents, and energy. It is better to use lunar materials instead of terrestrial ones only if there is significant gain in economy or time by doing so. There is, nevertheless, reason to be confident that the combination of those products that can be produced relatively simply from lunar materials for use in space will prove to be economical, will accelerate the growth of space activities, and will enrich human welfare and experience.

The purpose of the discussion below is to identify those raw materials for use in manufacture that can most readily be obtained from lunar materials as we know them. Availability of lunar materials has been considered by others (e.g., Williams and Jadwick, 1980; Williams and Hubbard, 1981; Arnold and Duke, 1977; Arnold, 1984). Information on the nature of the materials found on the Moon is summarized by Taylor (1975) and the details are found in the many volumes of the Proceedings of the Lunar and Planetary Science Conferences. The state of lunar exploration is incomplete, and there is reason to believe that concentrated ores may exist for elements that must now be regarded as rare or dispersed. The discussion here, however, describes minimum possibilities, based on the notion of making do with the least feasible amounts of separation and processing.

This ground rule will not necessarily lead to the most likely scenarios for space manufacture using indigenous materials. Utilization of lunar resources is complicated sufficiently just by the difficulties of transportation and living that any such manufacture is necessarily somewhat complex, and the best methods may need to be very complex to be economical. There nevertheless seems to be merit in asking what is most easily available. Besides, there is a certain satisfaction in imagining oneself to be a lunar homesteader trying to adapt to the Moon's environment with as few fancy tools and as little dependence on support from home as possible. As seen below, it appears that a substantial range of materials can be obtained with minimal separation and that scenarios based on the simplest technologies are not severely restrictive.

GROUND RULES

For this discussion, we limit ourselves to the most abundant materials observed at lunar sites visited by the Apollo missions. Excluded from consideration are less abundant materials already known, conceivable exotic ores, and hoped-for polar water. Materials are to be used directly for manufacture in "as found" condition or high-graded or separated by only the simplest means.

Although manufacturing processes are not considered here, corresponding constraints can be defined for them. For example, manufactured products should be used on the Moon or in space, not on Earth; however, any complex components would be imported. Steps in manufacture would have to be few, easy, and reliable.

These constraints, in turn, can be accompanied by ground rules for use of lunar products. Materials used in space will be those available and adequate, not necessarily those traditionally used, envisioned, or preferred on technical grounds for a given application. The quality of materials used will not necessarily match those of similar materials produced on Earth; sizes and quantities needed must be adjusted accordingly. Assembly and application of products of space manufacture must be simple and rapid and tolerances relatively loose.

TOOLS

The tools necessary to carry out any significant production of raw materials for manufacture cannot be regarded as simple, but merely as relatively simple. One item will have to be dirt-moving equipment, necessary for habitat excavation, as well as for transporting feedstocks to smelting or manufacturing sites and removing waste products and, perhaps, the products themselves. It is assumed that all processes, from gathering and production of raw materials through manufacture and assembly, will be substantially automated. Equipment probably will be tended, mainly from remote stations nearby. Equipment should be designed to allow straightforward repair, optimization to actual encountered conditions, and innovative adaptation by the operator to new feedstocks and conditions.

Electrical power, at least one megawatt, will have to be available for any significant processing or manufacturing activity. Initially, this probably will need to be nuclear power, although eventually solar power should be exploited to the fullest extent. Through the use of concentrating mirrors, solar power should be available at the outset for heating of materials to high enough temperatures to melt or even distill them.

LUNAR SURFACE CONDITIONS

The lunar surface is a source of high and unavoidable vacuum, both an inconvenience and a potential aid to manufacture. The acceleration of gravity is only one sixth that of Earth, which makes strengths of materials less of a problem for lifting and supporting, although not for withstanding changes in momentum. Also, the microgravity of orbiting factories can be used for processing lunar material, where appropriate. Half the time, the lunar surface is bathed in sunlight, without uncertainties from clouds; each day and night is about 328 hours long. Surface temperatures range from roughly -170°C in the shade to +120°C in direct sunlight. There is no medium for heat exchange as on Earth where abundant water and air are available, so waste heat must be dissipated by radiators. Sunlight is available nearly full time in orbit, and lunar bases at polar locations might also achieve full-time use of solar energy. The lunar surface abounds in fine dust, a convenient form of rock for some uses but a potentially serious problem for operation of equipment and for personnel comfort and health.

READILY AVAILABLE MATERIALS

Unprocessed Regolith

The material most readily available is unprocessed lunar soil (McKay, 1984). This consists mainly of fine rock flour (Papike, 1984), produced by past impacts of meteorites on the atmosphereless lunar surface. Some of the particles have been melted during impact and are present as glassy agglutinates. As encountered, this rock flour contains fragments ranging in size from clay to boulders. The minimum high-grading would consist of sieving out the larger fragments. Soils at mare sites are made mainly from crushed lunar basalt, similar in nature to common terrestrial basalt but drier and chemically more reduced. Fragments in mare soils are rich in the minerals plagioclase feldspar and pyroxene with ilmenite an important minor component. These minerals are not mainly present as separate grains, but are joined together as rock fragments or mixed and partly melted to make glassy agglutinates. From an elemental point of view, these soils are rich in oxygen (41%) and silicon (19%) (as are all known lunar soils), and in iron (13%), magnesium (6%), and, relatively, titanium, (up to 6%). Soils derived from highland rocks have lower abundances of iron and magnesium, but tend to be rich in aluminum (14%) and calcium (11%).

These elements do not occur in elemental form but in combination with oxygen. Their compositions are usually reported in terms of idealized oxides based on the probable oxidation state in the rock. It must be recognized that the concentrations as given are not intended to imply the presence of the oxides as such; the elements and their shares of oxygen are present as solid solutions, *i.e.*, as the minerals, mainly silicates, from which the rocks are made.

Most soils in mare regions appear to contain significant quantities of highland rocks, probably because many mare deposits are rather shallow and have allowed craters of intermediate size to penetrate into their highlands substrates. Soils distant from maria seem to have rather small percentages of mare components. Mixed soils, found mainly near mare-highland boundaries, have intermediate compositions.

In addition to the soils there are rocks. While perhaps not in as convenient a form to use as the already pulverized soils, they offer the possibility of more concentrated sources for some elements. Some rocks found on the Moon's surface are nearly monomineralic. These include anorthosite (nearly pure plagioclase feldspar, a calcium aluminosilicate) and dunite (nearly pure olivine, an iron-magnesium orthosilicate solid solution). Similarly, certain mare basalts are richer in titanium than are the soils derived from them. Not enough is known about the highlands to determine whether any monomineralic rock is present in sufficient abundance to serve as a convenient source of ore. Dunite fragments are rare at Apollo sites. Anorthosite fragments are abundant at the Apollo 16 site and were found also at the Apollo 15 site as isolated pieces in the regolith or as large clasts in breccia boulders. Breccias are rocks composed of broken fragments of prior rocks, compressed together to produced mixed rocks. Breccias are overwhelmingly the most common type of rock collected in the lunar highlands. Other rock types fairly common as clasts in breccias from the highlands are troctolite (olivine-plagioclase rocks) and norite (pyroxene-plagioclase rocks). Remotely sensed infrared spectra indicate that the central peaks of some large craters may be principally olivine and of other craters principally plagioclase (*e.g.*, C. Pieters, personal communication, 1984). This is unconfirmed by sampling since no such sites were visited by the Apollo or Luna missions. However, the soil at the station 11 site, Apollo 16, is highly enriched in anorthosite. By far the most common compositions observed by infrared remote sensing are those corresponding to norite, the same compositions that are typical of bulk or average samples of highland breccias (B. Hawke, personal communication, 1985)

Minimally Processed Regolith

There is some metallic iron, commonly about a tenth of a percent, in lunar soils. It is mainly present as fragments from meteorites that broke apart during crater-forming explosions and became mechanically mixed with lunar rock debris. It consists of alloys of iron that contain up to several percent nickel and some cobalt. This metal can be high graded by simply drawing a magnet through the soil. Pure metal will not be obtained, however, because many fragments are incorporated into agglutinates, which are glassy shards produced by melting of soil during meteorite impacts. Also, there are small amounts of indigenous metallic iron in lunar igneous rocks, a reflection of Moon's dearth of free oxygen. When bits of such rocks are incorporated into agglutinates, they contribute to their magnetism. A magnet gathers both magnetic agglutinates and metal fragments, and further processing to obtain pure metal would have to be done.

Within most lunar soils at least a few percent of the fragments are grains of single minerals. The percentage can be fairly high when the rocks from which the soils derived were coarse grained (*e.g.*, anorthosites) so that even sizeable fragments can be essentially monomineralic. It is lower but not negligible in soils derived from finer grained rocks (*e.g.*, basalt), because soils tend to be so finely pulverized that many of their grains are monomineralic fragments. All grains do not become monomineralic, because the micrometeorite–driven processes of partial melting and agglomeration to produce agglutinates work against continued communition of soil grains. Electrostatic processing (Agosto, 1984) offers the possibility for separating some minerals (*e.g.*, ilmenite) in highly concentrated form, potentially useful for providing special feedstock. Residues from this processing are also available and will have properties different from those of the bulk soil. Since these residues must be handled anyway, it is desirable to find uses for them and to control their properties as much as possible to provide for good secondary products. Most separation processes produce several different fractionated materials.

Melting of common lunar silicates can produce a variety of glasses with different properties. These can be cast or drawn, and under water-free lunar conditions may have great tensile strengths (Blacic, 1984). Speculation on possible products (Steurer, 1984; Khalili, 1984) and their properties is beyond the scope of this discussion of raw materials; glasses are mentioned here because they can be produced by such simple processing.

MAXIMUM "ALLOWED" PROCESSING

Herein, we offer a few examples of processing methods we regard as the most complicated allowable for serious discussion in realistic planning for an initial Moon base. Any of these processes will require substantial engineering and equipment to carry out. Along with the products of interest, we call attention to the residues, which may themselves have application.

Thermal Release of Gases

The simplest scheme for production of gases involves heating of lunar soil to release trapped solar wind. The most abundant implanted elements that can be extracted in this way are hydrogen, helium, nitrogen, and carbon. Since concentrations of these elements are low (mainly 50–100 parts-per-million by weight), large volumes of soil must be heated for a reasonable yield. The elements are tightly bound within surfaces of soil grains; soils must be heated to temperatures of 700–1100°C to release them. Provided that the tonnages of fines can be handled adequately, useful amounts could be obtained. Some indigenous, relatively volatile elements such as sulfur, chlorine, and noble gases (mainly argon) are similarly abundant in some soils and could also be extracted by heating. The products of heating of materials from indigenous and solar-wind sources would include water, hydrogen sulfide, carbon monoxide and dioxide, ammonia, and hydrogen cyanide, among others. For efficient handling, these gases may require oxidation (with lunar derived oxygen) to produce three readily separable fractions: water, carbon dioxide, and nitrogen plus noble gases. The available quantities of entrapped solar wind are sufficiently high and the economies of complete propellant production on the Moon so appealing that production of hydrogen deserves serious consideration (Carter, 1984; Friedlander, 1984; Rosenberg, 1984; Blanford *et al.*, 1984). The problem of heating large tonnages of lunar soil in a closed system to capture emitted gas has not received adequate consideration. The residue would be degassed lunar soil, or, if melted, glass.

Hydrogen Reduction of Ilmenite

The gaseous product that has received the most consideration so far is oxygen (*e.g.*, Rosenberg *et al.*, 1965; Driggers, 1976; Davis, 1983; Carroll *et al.*, 1983; Cutler, 1984a,b,c; Kibler *et al.*, 1984; Gibson and Knudsen, 1984; Waldron, 1984). Extraction of oxygen requires oxidation of that element from an oxide (*e.g.*, ilmenite) or from silicates (*e.g.*, mare basalt; separation of a pure silicate mineral does not seem necessary). The method receiving the most attention is one in which the water is electrolyzed and the hydrogen returned for further reaction (*e.g.*, Williams and Mullins, 1983). The ilmenite concentrate would have to be provided by highgrading of lunar soil as described in the previous section, and in that sense the "ilmenite process" is at least a two-stage process. The residue would be an intimate mixture of iron metal and unreduced iron oxide and titanium oxides, plus perhaps some silicate, depending on the purity of the ilmenite concentrate (Johnson and Volk, 1965). Electrolysis of molten silicate and other processes are also being considered for oxygen production (*e.g.*, Kesterke, 1971; Lindstrom and Haskin, 1979). None of the proposed processes is adequately understood yet at the laboratory bench scale.

Carbonyl Processing

The carbonyl process has been studied extensively, used industrially on Earth, and considered for use in space (*e.g.*, Lewis and Nozette, 1983; Lewis and Meinel, 1984). It should be relatively straightforward to use it to extract the metal concentrated magnetically from lunar soil and, in the same sense as the ilmenite process, would be part of at least a two-stage operation. Possibly, it could extract metal from unbeneficiated soil. The principal product would be high purity iron, and a secondary product would be high purity nickel. Cobalt and heavy noble metals will also be extracted. In such high purity, iron metal may attain remarkably high strength (Sastri, 1984). The residue would be mainly metal-free or at least metal-poor silicate. Carbonyl extraction of iron in the residue from hydrogen reduction of ilmenite has been demonstrated and required carbon dioxide pressures of 100–150 atmospheres and catalysts such as hydrogen sulfide for efficient separation (Vishnapuu *et al.*, 1973).

Electrolysis of Molten Silicate

Electrolysis of molten silicates to produce oxygen gas has the attraction that it requires, in principle, only sunlight for heat and electricity and lunar soil as its feedstock (*e.g.*, Lindstrom and Haskin, 1979). Experiments using simulated lunar basalt show that, while oxygen is being liberated at the anode, iron metal is simultaneously formed at the cathode, possibly as an alloy containing small amounts of chromium and manganese and with other impurities. Ilmenite-rich compositions yield iron metal alloyed with some titanium; titanium-poor compositions can yield iron alloyed with silicon. Since compositions of silicates can be significantly changed as a result of electrolysis, this method can leave residues with compositions different from those of indigenous lunar materials. It is perhaps not proper to regard the silicate residue as a residue, since the silicate melt might be the primary product of electrolysis.

Destructive Distillation

Solar furnaces are capable of producing very high temperatures. Some experiments have been carried out on meteoritic material (*e.g.*, King, 1982; Agosto and King, 1983), but no systematic study has been made on condensates or residues from the destructive distillation of simulated lunar silicates. The most refractory material formed probably would be calcium aluminate or, perhaps, calcium oxide. Extraction of gases, discussed above, is a relatively low-temperature form of destructive distillation of silicates. At higher temperatures, even perhaps during gas extraction, other useful substances such as alkali metal oxides might be concentrated by volatilization.

CONCLUSIONS

Unprocessed lunar soil can be used for radiation shielding both on the Moon and in space. Numerous glass products can be made, perhaps with special properties resulting from the dry lunar and space environment; these may become the principal structural materials for space. Iron and nickel can provide steel products, including electrical conductors. Ultra-pure iron may be as good as steel for many purposes. Oxygen and perhaps hydrogen gases can be produced for propellant and life support. This is a very good list of potential early raw materials for use on the Moon or in space. Much work still must be done to determine the conditions required for development of even simple procedures for obtaining and using these raw materials. The constraint of keeping things simple (relatively) does not seem to be an uncomfortably restrictive one.

Acknowledgments.   *Some of the ideas presented in this paper were developed during a NASA-sponsored summer workshop entitled Technological Springboard to the 21st Century, held in 1984 at the University of*

California in San Diego. Discussions at the workshop with M. Duke and S. Sastri are gratefully acknowledged. This work was supported in part by the National Aeronautics and Space Administration through grants NAGW-179 and NAG-9-56.

REFERENCES

Agosto W. N. (1984) Electrostatic concentration of lunar soil ilmenite in vacuum ambient (abstract). In *Papers Presented to the Symposium on Lunar Bases and Space Activities of the 21st Century*, p. 24. NASA/Johnson Space Center, Houston.

Agosto W. N. and King E. A. (1983) *In situ* solar furnace production of oxygen, metals, and ceramics from lunar materials (abstract). In *Lunar and Planetary Science XIV*, pp. 1–2. Lunar and Planetary Institute, Houston.

Arnold J. R. (1984) Lunar materials: Domestic use and export (abstract). In *Papers Presented to the Symposium on Lunar Bases and Space Activities of the 21st Century*, p. 100. NASA/Johnson Space Center, Houston.

Arnold J. R. and Duke M. B. (editors) (1977) *Summer Workshop on Near-Earth Resources*. NASA Conf. Public. 2031. NASA, Washington. 95 pp.

Blacic J. C. (1984) Structural properties of lunar rock materials under anhydrous, hard vacuum conditions (abstract). In *Papers Presented to the Symposium on Lunar Bases and Space Activities of the 21st Century*, p. 76. NASA/Johnson Space Center, Houston.

Blanford G., Børgesen P., Maurette M., and Moller W. (1984) On-line simulation of hydrogen and water desorption in lunar conditions (abstract). In *Papers Presented to the Symposium on Lunar Bases and Space Activities of the 21st Century*, p. 124. NASA/Johnson Space Center, Houston.

Carroll W. F., Steurer W. H., Frisbee R. H., and Jones R. M. (1983) Should we make products on the Moon? *Astronaut. Aeronaut., 21*, 80–85.

Carter J. L. (1984) Lunar regolith fines: A source of hydrogen (water) (abstract). In *Papers Presented to the Symposium on Lunar Bases and Space Activities of the 21st Century*, p. 27. NASA/Johnson Space Center, Houston.

Cutler A. H. (1984a) A carbothermal scheme for lunar oxygen production (abstract). In *Papers Presented to the Symposium on Lunar Bases and Space Activities of the 21st Century*, p. 22. NASA/Johnson Space Center, Houston.

Cutler A. H. (1984b) Plasma anode electrolysis of molten lunar minerals (abstract). In *Papers Presented to the Symposium on Lunar Bases and Space Activities of the 21st Century*, p. 23. NASA/Johnson Space Center, Houston.

Cutler A. H. (1984c) An alkali hydroxide based scheme for lunar oxygen production (abstract). In *Papers Presented to the Symposium on Lunar Bases and Space Activities of the 21st Century*, p. 21. NASA/Johnson Space Center, Houston.

Davis H. P. (1983) Lunar oxygen impact on STS effectiveness. *Eagle Engineering, Inc. Rept. #EEI 83-63*, Eagle Engineering, 44 pp. Houston.

Driggers G. W. (1976) Some potential impacts of lunar oxygen availability on near-Earth space transportation (abstract). In *Lunar Science VII, Special Session Abstracts*, pp. 26–34. The Lunar Science Institute, Houston.

Friedlander H. N. (1984) An analysis of alternate hydrogen sources for lunar manufacture (abstract). In *Papers Presented to the Symposium on Lunar Bases and Space Activities of the 21st Century*, p. 28. NASA/Johnson Space Center, Houston.

Gibson M. A. and Knudsen C. W. (1984) Lunar oxygen production from ilmenite (abstract). In *Papers Presented to the Symposium on Lunar Bases and Space Activities of the 21st Century*, p. 26. NASA/Johnson Space Center, Houston.

Johnson C. A. and Volk W. (1965) Reduction of ilmenite and similar ores. U. S. Pat #3,224,870.

Kesterke D. G. (1971) Electrowinning of oxygen from silicate rocks. *U. S. Bureau of Mines Rept. 7587*,. pp. 1–12. U. S. Bureau of Mines, Washington.

Khalili E. N. (1984) Magma and ceramic structures generated *in situ* (abstract). In *Papers Presented to the Symposium on Lunar Bases and Space Activities of the 21st Century*, p. 82. NASA/Johnson Space Center, Houston.

Kibler E., Taylor L. W., and Williams R. J. (1984) The kinetics of ilmenite reduction: A source of lunar oxygen (abstract). In *Papers Presented to the Symposium on Lunar Bases and Space Activities of the 21st Century*, p. 25. NASA/Johnson Space Center, Houston.

King E. A. (1982) Refractory residues, condensates and chrondules from solar furnace experiments. *Proc. Lunar Planet. Sci. Conf. 13th*, in *J. Geophys. Res., 87*, A429–A434.

Lewis J. S. and Meinel C. P. (1984) Carbonyls: Short cut from extraterrestrial ores to finished products (abstract). In *Papers Presented to the Symposium on Lunar Bases and Space Activities of the 21st Century*, p. 126. NASA/Johnson Space Center, Houston.

Lewis J. S. and Nozette S. D. (1983) Extraction and purification of iron group and precious metals from asteroidal feedstocks. *Adv. Astronaut. Sci., 53*, 351.

Lindstrom D. J. and Haskin L. A. (1979) Electrochemistry of lunar rocks. In *Space Manufacturing Facilities 3*, (J. Grey and C. Krop, eds.), pp. 129–134. American Institute of Aeronautics and Astronautics, New York.

McKay D. S. (1984) Evaluation of lunar resources (abstract). In *Papers Presented to the Symposium on Lunar Bases and Space Activities of the 21st Century*, p. 21, NASA/Johnson Space Center, Houston.

Papike J. J. (1984) Petrologic and chemical systematics of lunar soil size fractions: Basic data with implications for specific element extraction (abstract). In *Papers Presented to the Symposium on Lunar Bases and Space Activities of the 21st Century*, p. 122. NASA/Johnson Space Center, Houston.

Rosenberg S. D. (1984) A lunar-based propulsion system (abstract). In *Papers Presented to the Symposium on Lunar Bases and Space Activities of the 21st Century*, p. 29. NASA/Johnson Space Center, Houston.

Rosenberg S. D., Guter G. A., and Miller F. E. (1965) The utilization of lunar resources for propellant manufacture. *Adv. Astronaut. Sci., 20*, 665.

Sastri S. (1984) Role of iron in space (abstract). In *Papers Presented to the Symposium on Lunar Bases and Space Activities of the 21st Century*, p. 125. NASA/Johnson Space Center, Houston.

Steurer W. H. (1984) Use of lunar materials in the construction and operation of a lunar base (abstract). In *Papers Presented to the Symposium on Lunar Bases and Space Activities of the 21st Century*, p. 77. NASA/Johnson Space Center, Houston.

Taylor S. R. (1975) *Lunar Science: A Post Apollo View*, Pergamon, New York. 372 pp.

Vishnapuu A., Marek B. C., and Jensen J. W. (1973) Conversion of ilmenite to rutile by a carbonyl process. *U. S. Bureau of Mines Rept. 7719*. U. S. Bureau of Mines, Washington. 20 pp.

Waldron R. D. (1984) Diversity and purity: Keys to industrial growth with refined lunar materials (abstract). In *Papers Presented to the Symposium on Lunar Bases and Space Activities of the 21st Century*, p. 123. NASA/Johnson Space Center, Houston.

Williams R. J. and Hubbard N. (1981) *Report of Workshop on Methodology for Evaluating Potential Lunar Resource Sites*. NASA TM-58235. NASA, Washington. 120 pp.

Williams R. J. and Jadwick J. J. (1980) *Handbook of Lunar Materials*, NASA Ref. Public. 1057, NASA, Washington. 120 pp.

Williams R. J. and Mullins O. (1983) Enhanced production of water from ilmenite: An experimental test of a concept for producing lunar oxygen (abstract). In *Lunar and Planetary Science XIV*, pp. 34–35. Lunar and Planetary Institute, Houston.

MINING FOR LUNAR BASE SUPPORT

E. R. Podnieks and W. W. Roepke

U.S. Department of the Interior, Bureau of Mines, Twin Cities Research Center,
5629 Minnehaha Avenue South, Minneapolis, MN 55417

Mining and excavation technologies must be developed for the establishment of lunar bases that are capable of operating under the severe conditions of the lunar environment. The mining technology research needs for establishing a permanent lunar base utilizing lunar resources are outlined and briefly discussed. These requirements range from initial exploration needs to those required for gradually reaching full lunar base autarky. The extraterrestrial mining and excavation needs indicate that a multidisciplinary research effort must be initiated early in the lunar base program to provide timely research results for the subsequent development stages.

INTRODUCTION

The lunar base program NASA is proposing (Duke *et al.*, 1985) will require substantial rock and soil mechanics and excavation research and development before any self-sufficient lunar base will support extraterrestrial colonization. This ultimate goal of a self-supporting lunar base may require development of unorthodox mining and mineral extraction technology. This technology must include the ability to efficiently construct or excavate shelters for people and equipment to minimize thermal, radiation, and meteorite hazards, and to establish life support systems using lunar resources. These activities will require a knowledge of the characteristics and properties of lunar regolith and rocks, mining and excavation methods applicable to the lunar environment, and extraction processes for recovery of necessary resources involving material handling and special mining or metallurgical methods. Emphasis will need to be placed on making maximum use of lunar materials recovered with methods requiring the least amount of machinery. The development of lunar basing technology will establish basic concepts useful in any extraterrestrial regime. A substantial amount of research was conducted during the Apollo program on basic physical, geological, and chemical properties using simulated lunar materials from the rock suite developed by the Bureau of Mines (Fogelson, 1968). Engineering properties were determined for simulated lunar materials by the Bureau and others, using the Bureau's rock suite. The Bureau's research was oriented toward identification and definition of specific mining-related materials problems in the lunar environment. The research program consisted of:

1. Basic property and fragmentation studies on the effect of the lunar environment on physical, strength, elastic, and thermophysical properties affecting fragmentation (Atkins and Peng, 1974; Bur and Hjelmstad, 1970; Carpenter, written communication, 1969; Griffin and Demou, 1972; Heins and Friz, 1967; Lindroth, 1974; Lindroth and Krawza, 1971; Podnieks and Chamberlain, 1970; Podnieks *et al.*, 1968; Roepke and Schultz, 1967; Roepke, 1969; Roepke and Peng, 1975; Thirumalai and Demou, 1970; Willard and Hjelmstad, 1971).

2. Studies of lunar drilling, blasting, and novel fragmentation methods (Hay and Watson, personal communication, 1974; Paone and Schmidt, 1968; Roepke, 1975; Schmidt, 1969; Schmidt, 1970; Woo et al., 1967).

3. Research on handling and transportation of materials on the lunar surface (Crow and Bates, 1970; Johnson et al., 1973; Nicholson and Pariseau, 1971; Pariseau and Nicholson, 1972).

4. Lunar mineral processing studies (Fogelson et al., written communication, 1972; Fraas, 1970; Haas and Khalafalla, 1968; Kesterke, 1969; Khalafalla and Haas, 1970).

Although Apollo lunar landing results have not demonstrated serious friction and wear problems, this is considered by Bureau of Mines researchers to be due to localized lubrication in the immediate activity area by offgassing from equipment and astronaut space suits. The actual level of friction problems to be expected is best illustrated by the problems experienced with the lunar surface drill. Although core samples were obtained from the Apollo 15 site, the drill was never fully successful and never worked as easily as expected in loose regolith (J. Bensko, personal communication, 1971). No deep core samples could be obtained (Baldwin, 1972). Rock and soil properties can be quite different in a pure pristine lunar environment, as earlier Bureau of Mines work has indicated (Atkins and Peng, 1974; Lindroth, 1974; Podnieks and Chamberlain, 1970; Thirumalai and Demou, 1970). Results have shown that in uncontaminated hard vacuum the surface friction and strength of rocks are higher than in a terrestrial environment. This may be due to greater adhesive (cohesive) forces from higher surface energies and/or due to lack of more than monolayer coverage of water vapor. Adhesion on rock and tool surfaces soaked in ultrahigh vacuum to the point of minimal offgassing will create a problem with chip formation, removal, and clogging during any drilling process (Paone and Schmidt, 1968; Schmidt, 1969; Woo et al., 1967). An example of dust adhesion to space suits and tools used by the astronauts can be seen in Apollo 16 mission photographs.

Previous Bureau of Mines research in the tribology area (Roepke and Schultz, 1967; Roepke, 1969) with mineral–mineral and metal–mineral pairs using materials from the simulated rock suite (Fogelson, 1968) has shown that friction increases substantially in ultrahigh vacuum approaching the pristine lunar conditions. Even assuming that the surfaces still had at least a monolayer coverage of oxides and water vapor, the increase in friction ranged from 1.5 times to over 60 times. Since the lunar surface may consist of totally degassed pristine material, and since extended operations will permit extensive offgassing of equipment, the results suggest that an extremely high friction coefficient will be encountered during excavation and mining operations unless technology is established to minimize this friction.

Associated with the Bureau's work on friction, drilling experiments in ultrahigh vacuum were conducted with a drill designed for gas-free operation using polyimide bearings and gears impregnated with silver and tungsten disulfide. Laboratory results indicated that drilling operations can be performed in ultrahigh vacuum without the use of sealed bearings or any special lubricants (Roepke, 1975), but the earlier friction test results suggest that introduction of some gassy lubricant may be necessary for extended equipment use on the Moon.

Research on the use of explosives in vacuum was conducted by the Bureau of Mines to investigate the effects on explosive properties, including the blast wave pressure profiles (Hay and Watson, personal communication, 1974), and to perform small scale field cratering experiments in several simulated lunar material deposits to obtain indications of blasting effectiveness on the lunar surface. The preliminary results indicated that explosives and explosive components are available that can withstand the lunar environment without deterioration. The behavior of fragments from an explosion can be treated similarly to that in terrestrial explosions and should not present any problems, although lack of air currents and lower gravity will influence ejecta.

Substantial research was also directed toward developing novel methods of fragmentation involving mining equipment that minimizes moving parts. Work using electrothermal techniques for fragmenting simulated lunar rocks has shown some promising results (Thirumalai and Demou, 1970). Laser technology may also be useful for rock fragmentation purposes (Lindroth, 1974; Lindroth and Krawza, 1971).

Additional tests also were conducted using a simulated lunar soil to establish the range of cohesiveness and shear characteristics (Johnson et al., 1975) for near pristine materials. The size distribution of lunar soil recovered by Apollo 11 was used to develop simulated lunar materials for these tests using tholeiitic basalt. A knowledge of shear strength of soils in the outgassed pristine condition will be essential to all materials handling problems. The Bureau results suggested that compacted particulate materials will require greater energy input to induce flow in any materials handling situation, and this potential problem will be exacerbated by the low gravity condition. Although the Apollo results did not indicate any problems with lunar soils, the conditions were similar to those discussed earlier on friction between surfaces. The soils handled were not necessarily pristine and no flow characteristics were considered. If they are handled by equipment fully outgassed due to long soak time in lunar vacuum, and if the soils are handled in quantity, the increase in energy needed to induce and maintain flow may become a factor to be considered.

Although prior work has indicated that electrowinning of oxygen from silicate rocks is feasible, further studies are needed to establish the practical application for the lunar environment (Kesterke, 1969). Similar findings were obtained by using reduction of silicates with carbon (Haas and Khalafalla, 1968), of raw soil by fluorine, and ilmenite by hydrogen (Fogelson et al., 1972). These methods are not novel per se, but would be novel when used for obtaining oxygen in the lunar environment, since it will involve a transport of these catalysts to the Moon and their preservation by recycling.

The Bureau research effort during the Apollo years produced a variety of findings both in mining and resource extraction that provided ample proof of the need for a multidisciplinary mining systems development effort. Rock behavior is different in the lunar environment than on Earth; consequently, novel mining technologies may be required. The development of a lunar drill, a relatively simple device compared to the equipment necessary for extended base support, illustrated this point very well. Research and development of suitable mining technology will be in the critical path to establish shelters and maximize use of lunar resources for support of the lunar base.

LUNAR MINING NEEDS

Specific areas of research effort associated with lunar excavation and mining activities in support of lunar base activation must include both basic and applied research as well as prototype engineering development. An understanding of the materials being handled as well as environmental effects on both materials and mining equipment requires a systematic multidisciplinary approach.

The ultra high vacuum and the broad temperature range of the lunar surface will create a severe and restrictive working environment. Although much research was done during the Apollo period on materials properties, additional research is needed to specifically identify the problem areas of mining and handling the regolith materials and any hard rock substrata. From the mining perspective, it will be required that, at least initially, surface friction, strength, and thermal and dynamic (shock) rock properties need to be further investigated over the range of conditions expected to exist during lunar mining or excavation operations. The effects of a low pressure, low gravity environment on the failure mechanisms and on rock fragmentation products such as chips, flyrock, and dust must be established.

Knowledge of environmental effects on mining equipment and the operators will also be essential. Friction and wear of moving machine components may be a major equipment problem. Novel designs that minimize moving parts or that utilize solar or unconventional energy sources may be necessary.

ROCK MECHANICS RESEARCH

The early rock mechanics research in the Apollo program was conducted using a suite of 13 simulated lunar rocks selected by the Bureau of Mines (Fogelson, 1968) based on chemical analysis of the lunar surface by Surveyor V data. These were tested for basic behavior and properties such as fracture characteristics, strength, deformation, and friction (Atkins and Peng, 1974; Bur and Hjelmstad, 1970; Carpenter, written communication, 1969; Griffin and Demou, 1972; Heins and Friz, 1967; Lindroth, 1974; Lindroth and Krawza, 1971; Podnieks and Chamberlain, 1970; Podnieks *et al.*, 1968; Roepke and Schultz, 1967; Roepke, 1969; Roepke and Peng, 1975; Thirumalai and Demou, 1970; Willard and Hjelmstad, 1971) in a simulated lunar environment. These early preliminary studies need to be followed up with both real and simulated lunar rocks in order to extend the data base on materials properties needed for developmental research of the lunar base. A two–step approach is needed to further these studies: (1) determination of material behavior under simulated lunar environment in terrestrial facilities, and (2) using an orbiting space laboratory for long term degassing of materials and equipment to provide more realistic verification testing of drilling or other equipment. The results will provide the basic input needed to develop methods of excavation and construction of lunar shelters, and for mining processes providing life support or recovery of lunar resources.

ROCK FRAGMENTATION RESEARCH

The rock mechanics research should provide an extensive basis for research on the various engineering aspects of rock fragmentation in a lunar environment. Since previous lunar excursions have indicated a plethora of large boulders it seems very likely that fragmentation methods will be required to handle them and subsurface rock formations. More research is needed to help in identifying suitable fragmentation methods by using a methodology to classify lunar rock masses in terms of engineering parameters, rock strength and deformational properties, chemical and petrofabric descriptions, and index properties.

With any rock fragmentation or handling of regolith, dust is generated. It is important to determine dust generation characteristics for different methods of fragmentation and dispersion effects, as related to visibility, operator safety, and equipment use. The effects of low gravity and lack of surface winds must be considered on flyrock dispersion and airborne dwell time of dust clouds.

Use of explosives has been one of the primary methods of fragmentation in the terrestrial environment due to its relative simplicity and efficiency. Since minimal use of mechanical equipment is needed, explosives may be a good fragmentation method on the lunar surface. Initial work on blasting in vacuum has been done, but more fundamental knowledge of blasting in the lunar environment is needed to evaluate its feasibility, safety, and efficiency on the lunar surface.

Other fragmentation techniques already being studied by the Bureau include use of lasers, microwaves, and induction heating. These thermal systems cause the various minerals in the rock to heat at different rates dependent on their thermal conductivity, thus causing differential expansion and fragmentation.

MATERIALS HANDLING

Materials handling involves the removal of the regolith material, the fragmented rocks, and the transport from the site. Methods must be developed for effectively removing cuttings from drill holes during exploration or mining operation. Since there may be a synergism between the machine characteristics and the drilling media, the research must be done with specific design concepts adapted to the lunar environment.

Since materials handling is involved in a broad area of mining and excavation processes, the research needs to cover a variety of aspects.

EQUIPMENT DEVELOPMENT

One of the major problem areas associated with any exploration, excavation, or mining is the development of suitable equipment. The development of lunar-based excavation or mining equipment will require a major emphasis on tribology using designs that control friction and wear without hydrocarbon lubricants. It is unlikely that it will be possible to directly utilize terrestrial mine equipment designs or any direct extrapolation of them for application at a lunar base. Extensive revision of conventional technology or innovative new methods and designs will be required. Complete systems and design analysis will be required to identify and evaluate every step of the mining process in respect to the lunar base requirements. This analysis will identify equipment design needs and the necessary operating steps.

Human factors analysis will also play an important role, since it will be an essential part of developing a fully functional mining system in a lunar environment. The lunar miners are expected to be engineers and scientists who must become fully familiar and experienced with details of mining techniques unique to the lunar application. The special problems associated with mining in the lunar environment will amplify the many problems that any new operators will face. For these reasons, the system must be carefully designed for safety, ease of operation, and to satisfy the functional and operator needs while performing excavation or mining operations.

CONCLUSIONS

The Bureau of Mines past research with NASA on lunar resource utilization within the Apollo program has established a good foundation for more advanced and specific research efforts. The results indicate that on the lunar surface, mining or excavating must be performed using technology that is significantly different from terrestrial methodology. By developing a technology for lunar mining and excavating, a major step will be made toward further expansion of this concept to other planets or bodies in the solar system. The required technology will evolve from research results obtained on: (1) material properties, (2) environmental effects on these properties, (3) various energy mechanisms capable of fragmenting and removing material, and (4) relationships that govern the interaction of the mining and excavation mechanisms with lunar materials under specific environmental conditions.

Based on the conditions found on the surface of the Moon, the major excavation and materials handling will be in the regolith zone consisting of particulates and fine grained or blocky materials. However, the lunar surface also has large surface boulders and, by inference, large underground obstructions that must be fragmented. These tasks present an area of unknowns where research must provide answers generating engineering knowledge for the development of a lunar excavating, mining, and processing technology. The mining technology for extraterrestrial use most likely cannot be a simple adaptation of existing terrestrial methods, but will require novel approaches governed by the knowledge of the constraints of the lunar environment.

If the initial establishment of a lunar base is to occur in the first decade of the next century, it is essential that early research be undertaken in mining technology as part of the overall program effort. It will also be essential to do timely prototype development and testing of proposed lunar mining equipment. Past experience has shown that development of even the simplest exploration/mining tools for the lower gravity, gas-free lunar environment can be a lengthy effort.

Acknowledgments.   The authors wish to express their appreciation to R. E. Thill and D. E. Siskind for their technical input. They also wish to thank W. C. Larson for valuable comments and consultation during the preparation of this review of lunar mining research needs.

REFERENCES

Atkins J. O. and Peng S. S. (1974) *Compression Testing of Rock in Simulated Lunar Environment.* RI 7983, Bureau of Mines, Washington. 21 pp.

Baldwin R. R. (1972) Mission description. In *Apollo 15 Preliminary Science Report.* NASA SP-289, pp. 1-1 to 1-11. NASA/Johnson Space Center, Houston.

Bur T. R. and Hjelmstad K. E. (1970) Elastic and attentuation symmetries of simulated lunar rocks. *Icarus, 13,* 414–423.

Crow L. J. and Bates R. C. (1970) *Strengths of Sulfur-basalt Concretes,* RI 7349, Bureau of Mines, Washington. 21 pp.

Duke M. B., Wendell W. W., and Roberts B. B. (1985) Towards a lunar base programme. *Space Policy, 1,* 49–61.

Fogelson D. E. (1968) *Simulated Lunar Rocks.* In *Proceedings of the Sixth Annual Meeting of the Working Group on Exterrestrial Resources.* NASA SP-177, pp. 75–95. NASA, Washington.

Fraas F. (1970) *Factors Related to Mineral Separation in a Vacuum.* RI 7404, Bureau of Mines, Washington. 13 pp.

Griffin R. E. and Demou S. G. (1971) Thermal expansion measurements of simulated lunar rocks. *Proc. 3rd Amer. Inst. Phys. Conf. on Thermal Expansion,* pp. 302–311. American Institute of Physics, New York.

Haas L. A. and Khalafalla S. E. (1968) *The Effect of Physical Parameters on the Reaction of Graphite with Silica in Vacuum.* RI 7207, Bureau of Mines, Washington. 21 pp.

Heins R. W. and Friz T. O. (1967) *The Effect of Low Temperature on Some Physical Properties of Rock.* SPE 1714, American Institute of Mining Engineers, Denver. 8 pp.

Johnson B. V., Roepke W. W., and Strebig K. C. (1973) *Shear Testing of Simulated Lunar Soil in Ultrahigh Vacuum.* RI 7814. Bureau of Mines, Washington. 18 pp.

Kesterke D. G. (1969) Electrowinning of Oxygen from Silicate Rocks. In *Proceedings of the Seventh Annual Working Group on Extraterrestrial Resources.* NASA SP-229, NASA, Washington. 148 pp.

Khalafalla S. E. and Haas S. A. (1970) Carbothermal reduction of liquid siliceous minerals in vacuum. *J. High Temp. Sci., 2,* 95–109.

Lindroth D. P. (1974) *Thermal Diffusivity of Six Igneous Rocks at Elevated Temperatures and Reduced Pressures.* RI 7954. Bureau of Mines, Washington. 33 pp.

Lindroth D. P. and Krawza W. G. (1971) *Heat Content and Specific Heat of Six Rock Types at Temperatures to 1,000° C.* RI 7503, Bureau of Mines, Washington. 24 pp.

Nicholson D. E. and Pariseau W. G. (1971a) *Gravity Flow of Powder in a Lunar Environment. Part I: Testing of Simulated Lunar-basalt Powder for Gravity Flow.* RI 7543, Bureau of Mines, Washington. 34 pp.

Nicholson D. E. and Pariseau W. G. (1971b) *Gravity Flow of Powder in a Lunar Environment. Part II: Analysis of Flow Initiation.* RI 7577, Bureau of Mines, Washington. 20 pp.

Paone J. and Schmidt R. L. (1968) Lunar drilling. In *Proceedings of the Sixth Annual Meeting of the Working Group on Exterrestrial Resources.* NASA SP-177, pp. 107–117. NASA, Washington.

Pariseau W. G. and Nicholson D. E. (1972) Gravity flow bin design for lunar soil. Preprint, *2nd Symposium on Storage and Flow of Solids,* American Society of Mechanical Engineers, New York. 8 pp.

Podnieks E. R. and Chamberlain P. G. (1970) Method for rock property determination in ultrahigh vacuum. *Proc. 5th Space Simulation Conference, Special Publication 336,* pp. 209–223. National Bureau of Standards, Washington.

Podnieks E. R., Chamberlain P. G., and Thill R. E. (1972) Environmental effects on rock properties. *Proc. 10th Symposium on Rock Mechanics: Basic and Applied Rock Mechanics,* pp. 215–241. Society of Mining Engineers of AIME, Denver.

Roepke W. W. (1969) Friction tests in simulated lunar vacuum. In *Proceedings of the Seventh Annual Meeting of the Working Group on Exterrestrial Resources.* NASA SP-29, pp. 107–111. NASA, Washington.

Roepke W. W. (1975) *Experimental System for Drilling Simulated Lunar Rock in Ultrahigh Vacuum.* NASA CR-146419. NASA, Washington. 13 pp.

Roepke W. W. and Peng S. S. (1975) *Surface Friction of Rock in Terrestrial and Simulated Lunar Environments.* NASA CR-146359. NASA, Washington. 34 pp.

Roepke W. W. and Schultz C. W. (1967) Mass spectrometer studies of outgassing from simulated lunar materials in UHV. *Proc. 14th Nat. Vacuum Symp. of the Amer. Vacuum Soc.,* pp. 165–166. American Vacuum Society, New York.

Schmidt R. L. (1969) Developing a lunar drill: A 1969 status report. In *Proceedings of the Seventh Annual Meeting of the Working Group on Exterrestrial Resources.* NASA SP-229. NASA, Washington. 148 pp.

Schmidt R. L. (1970) Diamond flunk test on lunar hammer-drill. *Drilling, 32,* 34–35, 64.

Thirumalai K. and Demou S. G. (1970) Effect of reduced pressure on thermal-expansion behavior of rocks and its significance to thermal fragmentation. *J. Appl. Phys., 41,* 5147–5151.

Willard R. J. and Hjelmstad K. E. (1971) Effect of moisture and temperature on the fracture morphology of dacite. *Internat. J. Rock Mech. Min. Sci., 8,* 529–539.

Woo W. G., Bensko J., Lindelof L., and Paone J. (1967) A lunar drill concept. In *The Industrial Diamond Revolution,* pp. 257–274. Ind. Diamond Assn. of America, Chicago.

ELECTROSTATIC CONCENTRATION OF LUNAR SOIL MINERALS

William N. Agosto

Lunar Industries, P.O. Box 590004, Houston, TX 77259-0004

Estimates of the magnetic susceptibility of lunar ilmenite indicate that electrostatic separation may be preferable to magnetic separation as a beneficiation technique for concentration of the mineral in lunar soil. Single-pass electrostatic separations of terrestrial ilmenite from a lunar soil simulant in the 0.15–0.09 mm size range on a slide-type separator yielded higher grades in air or nitrogen (89 and 90 wt %, respectively) than in vacuum (78 wt %) due to the effects of air resistance, charging of the feed by gas ions, and the absence of vibratory feed in the vacuum runs. Single-pass concentrations on a 0.09–0.15 mm fraction of Apollo 11 soil 10084,853 increased the lunar ilmenite grade from 7 wt % to 51 wt % in dry nitrogen with the ferromagnetic agglutinates removed and to 29 wt % in vacuum (10^{-5} torr) with all agglutinates present. If all ilmenite-bearing soil phases are included, one-pass grades of lunar ilmenite increase to the mid 60s for the N_2 runs and to the mid 30s for the vacuum run. Soil ilmenite behaved like a conductor or semi-conductor and the agglutinates like non-conductors in all lunar runs. This behavior suggests that ilmenite grades and recoveries in the high 90s can be expected from comparable mare soil fractions with fully liberated ilmenite using multistage-multipass electrostatic concentrators of commercial design. However, redesign of the separator geometry is required for optimal performance in vacuum.

INTRODUCTION

Numerous authors (Andrews and Snow, 1981; Waldron and Criswell, 1982; Davis, 1983) have projected substantial economic advantages for lunar oxygen production as a major fuel and gas resource in extended space industrial operations. Several systems studies of lunar oxygen production (Davis, 1983; Gibson and Knudsen, 1985) have referenced a process utilizing ilmenite, the most abundant oxide in the returned lunar samples. Lunar ilmenite consists of 47 wt % FeO and 53 wt % TiO_2 stoichiometrically and is most heavily concentrated in the maria (McKay and Williams, 1979). Williams and Mullins (1983) have demonstrated the feasibility of oxygen extraction from terrestrial ilmenite by hydrogen reduction of the FeO component.

The paramagnetic properties of ilmenite at ordinary temperatures suggest magnetic separation as a method for concentrating it from mare soil. Accordingly, the magnetic susceptibility of lunar ilmenite was estimated in a 0.09–0.15 mm split of Apollo 11 soil 10084 using a Franz model L1 magnetic separator at the NASA/Johnson Space Center (JSC). The susceptibility was approximately 76×10^{-6} cgs mass units, similar to synthetic stoichiometric terrestrial ilmenite but orders of magnitude below that of natural terrestrial ilmenite, which averages $26,800 \times 10^{-6}$ cgs mass units (Carmichael, 1982). The low lunar ilmenite susceptibility probably reflects the virtual absence of Fe^{3+} compared with terrestrial varieties.

The susceptibility determination was made on the soil split after the high susceptibility agglutinates (approximately 75% of total sample agglutinates by microscopic count) were removed using a permanent magnet with a strength of approximately 1500 gauss. When the balance of the sample was run on the Franz separator, all paramagnetic splits that contained ilmenite also contained comparable abundances of agglutinates that had not been removed by the magnet. It was concluded that the lunar ilmenite may be difficult to separate magnetically from the population of soil agglutinates with comparable susceptibilities. Therefore, the alternate technique of electrostatic concentration of lunar soil ilmenite has been studied, and experimental results using electrostatic separators are reported here. The separators are based on industrial designs and were fabricated by Lockheed Engineering and Management Services Co. at NASA/JSC.

MINERAL ELECTROSTATIC SEPARATION

The primary commercial application of mineral electrostatic beneficiation is in processing beach sands and alluvial deposits for titanium minerals. All of the heavy mineral beach sand plants in Australia, and most in the United States (Florida), use electrostatic methods to separate rutile and ilmenite from zircon and monazite (Fraas, 1962; Kelly and Spottiswood, 1982). The most common electrostatic separator designs use a drum or slide configuration. In both designs, a high intensity electric field (several kV/cm) is established by a high voltage electrode spaced a few cm from the grounded drum or slide. Many separators use an additional ionizing electrode above the field electrode to charge the mineral feed with air ions and electrons before it enters the accelerating field (Fraas, 1962; Carpenter, 1970; Moore, 1973). The slide design (Figs. 1 and 2) was used for these experiments because its performance was superior to the drum design for small samples (0.1–5.0 g) (Agosto, 1983) and because it is simpler to operate in vacuum.

Minerals falling through the separating field commonly acquire charge by one or some combination of the following mechanisms: (1) Electrostatic induction; (2) Contact charging; (3) Ionic charging.

Electrostatic induction occurs primarily in conducting and semi–conducting grains which, because they are grounded by the slide, acquire charge opposite in polarity to the field electrode and are pulled toward it. As a result, conductor and semi–conductor particles preferentially report to the conductor bin or bins farthest from the foot of the slide.

Contact charging occurs when a material with a lower electronic work function gives up electrons and becomes positively charged by contact with a higher work function material. Since the rate of charging and discharging is an exponential function of the particle surface conductivity (Kelly and Spottiswood, 1982; Inculet, 1982), conductors lose excess charge rapidly on the slide. Given time to acquire sufficient contact charge, non–conductors tend to retain the charge and move toward the electrode of opposite polarity. Grains that acquire insufficient charge fall into the non–conductor bin or bins at the foot of the slide.

Mineral charging also occurs in the separator due to gas ion and electron bombardment generated by the ambient electric field. This mechanism was not introduced intentionally in these experiments, but evidence for the presence of ionic charging was noted in separation runs made in dry nitrogen.

The charging behavior of terrestrial varieties of the most common mare soil minerals (anorthite, ilmenite, olivine, pyroxene) in four size ranges <0.5 mm has been studied in slide-type electrostatic separators (Agosto, 1983). Ilmenite electrostatic concentration in air for soil-analog size fractions below 0.09 mm was found to be substantially poorer than in the larger size fractions. The reduced performance resulted from partial clumping of fines on the slide as well as billowing and dispersal of the falling fines, which were substantially retarded by the air. Virtually every investigator has reported these effects for particles under 0.075 mm (Fraas, 1962; Carta et al., 1964; Carpenter, 1970; Inculet, 1979; Kelly and Spottiswood, 1982). Since mean grain size for most lunar soils is under 0.1 mm, it appeared that vacuum operation of the separator might improve performance by eliminating the disruptive effects of air resistance on the flow of fines.

ILMENITE ELECTROSTATIC CONCENTRATION IN LUNAR SOIL SIMULANTS

Lunar soil simulants were prepared from mixtures of comminuted terrestrial anorthite (An_{90}), ilmenite (Quebec), olivine (Fo_{90}), and augite pyroxene ($Wo_{50}En_{33}$) in the weight ratio 4:1:1:4, respectively. Two size ranges (0.09–0.15 mm and 0.15–0.25 mm) were tested in air, and one size range (0.09–0.15 mm) was tested in nitrogen and in a vacuum of approximately 10^{-5} torr. Samples were weighed in quantities of 1–5 g and were washed and boiled in isopropyl alcohol until the supernatant fluid appeared clear. Prior to separation, all samples were dried, first on filter paper and then in glass receptacles in an air oven at 120°–130°C overnight.

Ilmenite electrostatic separation grades and recoveries from the simulants are reported in each of three ambients: air at one atmosphere; dry nitrogen at about one torr positive pressure; and in vacuum at 10^{-5} (±20%) torr. In all cases, grade is defined to be the cumulative concentration (wt %) of ilmenite in the conductor bins; recovery is the wt % of total sample ilmenite reporting to those bins. Ilmenite concentration was determined in each bin by microscopic count. The wt % of ilmenite was then calculated by weighing the contents of each bin and assuming that the ilmenite fraction had a density ratio of 1.5 relative to the balance of the simulant sample.

Samples were maintained at 100°–200°C in the separator prior to and during separation by heating both the vibratory feed hopper and the grounded slide to drive off adsorbed feed moisture and to enhance the conductivity of the ilmenite. Feed heating is a common practice in commercial electrostatic mineral separation operations (Fraas, 1962; Kelly and Spottiswood, 1982). Feed rate was approximately 6 g/min, and feed size ranged from a maximum of a few grams in the bench top runs in air to a few hundred milligrams in the nitrogen and vacuum runs.

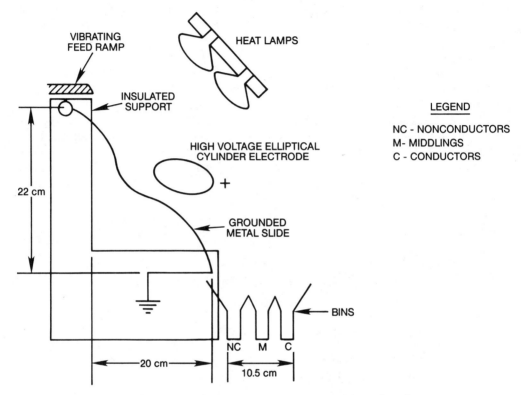

Figure 1. Mineral electrostatic separator: bench top slide configuration.

Maximum electric field strength was nominally +5 kV/cm DC but ranged between +2.5 and +7 kV/cm. Positive field electrode polarity was used throughout to make the ilmenite induced charge additive to the negative ilmenite contact charge on the aluminum slide (Agosto, 1984).

Four runs of ilmenite electrostatic concentration from simulants were made in air in a bench top apparatus (Fig. 1) where the feed hopper and slide were heated with infrared lamps. Separates were collected in three bins each 3.5 cm wide with bin 3 being the only conductor bin in the bench top setup. The feed size range was 0.15–0.25 mm.

Three other simulant runs in air and all simulant runs in N_2 and in vacuum were made in the slide configuration designed to fit the vacuum system (Figs. 2 and 3). The feed size range was 0.09–0.15 mm. The hopper and slide were resistance heated, and separates were collected in 7 bins each 2 cm wide. (The number of bins was increased to improve separates resolution for subsequent vacuum runs.) High concentrations of ilmenite reported to bins 2 through 7. Mean grade for all runs in air after one pass was 89±6 wt % (up from a mean ilmenite starting concentration of 8.8 wt %), and recovery was 51±9 wt %.

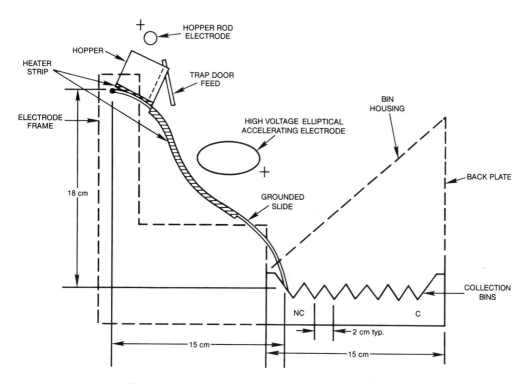

Figure 2. Mineral electrostatic separator: vacuum slide configuration.

For the nitrogen runs, the seven-bin collector was used in a glove box at a positive nitrogen pressure of about one torr. Ilmenite concentration data after one pass were collected in seven runs for four-component simulants in the 0.09–0.15 mm size range. High concentrations of ilmenite reported to bins 2 through 7. Mean grade after one pass was 90±7 wt % (up from 7.9 wt %), and recovery was 67±5 wt % for all runs in nitrogen.

For ilmenite electrostatic concentration in vacuum, the apparatus (Figs. 2 and 3) was scaled to fit a bell jar of approximately 70 liters volume. A trap door hopper was used instead of the the vibratory feed to reduce grain bounce and consequent random feed dispersal from the hopper mouth. Grain rebound is especially high in vacuum because both air resistance and sound energy radiation are greatly reduced (Fraas, 1970). A rod-shaped electrode was set above the hopper in an attempt to charge the feed sample by electrostatic induction prior to separation.

After exhausting the bell jar to 10^{-5} torr and heating the sample and slide to approximately 150°C, high voltage (15 kV) was applied to the hopper and the accelerating electrodes in parallel to obtain a field strength of approximately 3 kV/cm in the hopper for 10 minutes before separation. The trap door was then opened to drop the sample onto the slide and into the separating field of approximately 5 kV/cm. Material moved farther toward the back bins than in the air and nitrogen runs, and ilmenite concentrated

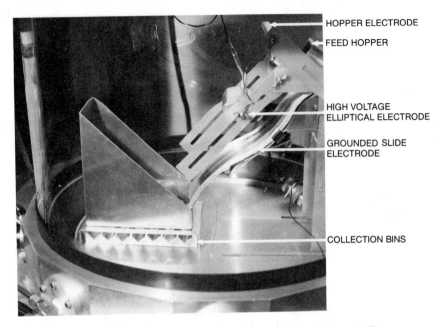

HOPPER ELECTRODE

FEED HOPPER

HIGH VOLTAGE
ELLIPTICAL ELECTRODE

GROUNDED SLIDE
ELECTRODE

COLLECTION BINS

Figure 3. Mineral electrostatic separator slide configuration in vacuum bell jar.

in bins 5, 6, and 7. More material fell outside the bin system as well. Grades of twelve single-pass runs in vacuum were calculated from the cumulative ilmenite concentration in bins 5, 6, and 7. Two recoveries were calculated: R_1, which excludes material that fell outside the bins, and R_0, which includes that material. The mean ilmenite grade after one pass of all the vacuum runs was 78±7 wt % (up from 9.8 wt %); $R_1 = 56±8$ wt % and $R_0 = 49±8$ wt %.

LUNAR ILMENITE ELECTROSTATIC CONCENTRATION

Table 1 lists lunar ilmenite grade and recovery data of four electrostatic concentration runs on a 0.09–0.15 mm sample of Apollo 11 soil 10084,853. Three of the runs were in dry nitrogen and one in vacuum. The lunar runs in nitrogen were made with the same apparatus and procedure as the simulant nitrogen runs. Similarly, the lunar vacuum run replicated the procedures of the simulant vacuum runs. Lunar ilmenite wt % was also calculated as in the simulant runs.

About 50 vol % of the sample were agglutinates before the first run, and the starting ilmenite content was about 10 wt %. The first nitrogen run in the table and the vacuum run were made without prior removal of agglutinates. However, before the other two nitrogen runs (#2 and #3), the ferromagnetic agglutinates were removed with a permanent magnet of approximately 1500 gauss. The ferromagnetic agglutinates comprised approximately 75 vol % of total sample agglutinates by microscopic count.

Table 1.   Lunar Soil Ilmenite Electrostatic Concentration (One Pass)

Apollo 11 Soil Sample 10084, 853
(grain size: 0.09–0.15 mm; sample charging time: 10 min.)

Run Number	Ambient	Starting Ilmenite Concentration (weight percent)	Sample Feed Temperature (°C)	Maximum Electrostatic Field Strength (kV/cm)	Ilmenite Grade (weight percent)	Recovery (weight percent)	Conductor Bins
1	Dry nitrogen	10	173	4.7	45	24	2–7
2a*	Dry nitrogen	7.3	—	3.0	51	48	2–7
2b†	Dry nitrogen	7.3	—	3.0	66	50	2–7
3a	Dry nitrogen	7.3	193	5.0	51	35	2–7
3b	Dry nitrogen	7.3	193	5.0	62	36	2–7
4a	Vacuum 1.5×10^{-5} torr	7.0	154	5.0	29	55	6,7,Tr**
4b	1.5×10^{-5} torr	70	154	5.0	37	60	6,7,Tr

* a excludes ilmenite bearing glass and polyphase components.
† b includes ilmenite bearing glass and polyphase components.
** Tr is the catch tray beyond the 7th bin.

Tables 2 and 3 give lunar soil component distributions among the seven collection bins after one pass of electrostatic concentration for the first nitrogen run (#1) and the vacuum run (#4), respectively, with all agglutinates present in both cases. Components microscopically discriminated and confirmed by SEM-EDS analysis were: agglutinate, anorthite, glass, ilmenite, polyphase, and olivine/pyroxene. Olivine and pyroxene counts were lumped together. The polyphase component consisted of lithic fragments that were about half ilmenite, and about half the dark glass grains had ilmenite compositions. The

Table 2.   Lunar Soil Ilmenite Electrostatic Concentration in Dry Nitrogen

Site	Weight/ Site (mg)	Aggl	An	Glass	Il	Poly	Px/Ol	Ot	Wt% Il	Wt Il (mg)
				Soil Component Distribution After One Pass (Run #1, Table 1)						
				(Modal Percent)						
Bin #1	61.2	31	11	11	5	18	22	1	7.3	4.5
2	2.4	11	1	14	24	30	17	3	32	0.8
3	1.1	11	–	9	40	27	14	–	50	0.6
4	0.4	–	–	–	82	14	4	–	87	0.3
5	<0.1	8	–	–	75	17	–	–	81	
6	<0.1	–	–	–	100	–	–	–	100	} 0.1
7	<0.1	–	–	–	100	–	–	–	100	
S+H+E	11.6	55	9	6	8	10	11	1	11	1.3
Tr	1.0	58	2	3	6	17	13	–	8.7	0.1
Total	77.7									7.7

Aggl-agglutinate; An—anorthite; E—field electrode; H—feed hopper; Il—ilmenite; Ot—other, Two of these grains were metal spherules; Poly—polyphase; Px/Ol—pyroxene/olivine, combined counts; S—grounded slide; Tr—catch tray under bins

Table 3. Lunar Soil Electrostatic Concentration in Vacuum (1.5×10^{-5} torr)

Site	Weight/ Site (mg)	Aggl	An	Glass	Il	Poly	Px/Ol	Wt% Il	Wt Il (mg)
				(Modal Percent)					
Bin #1	16.6	75	4	3	3	6	8	4.4	0.7
2	26.8	71	4	3	2	2	17	3.0	0.8
3	8.0	60	4	4	2	4	26	3.0	0.2
4	3.3	56	6	4	5	3	26	7.3	0.2
5	1.8	58	7	3	10	6	16	14.3	0.3
6	1.9	45	4	3	18	13	17	25.0	0.5
7	3.5	35	4	2	20	15	25	27.0	0.9
S+H+E	0.4	46	26	1	3	1	19	4.4	–
Tr	3.6	33	5	1	25	7	29	33.0	1.2
Total	65.9								4.8

The table header "Soil Component Distribution After One Pass (Run #4, Table 1)" spans the Aggl, An, Glass, Il, Poly, Px/Ol columns.

Aggl-agglutinate; An—anorthite; E—field electrode; H—feed hopper; Il—ilmenite; Poly—polyphase; Px/Ol—pyroxene/olivine, combined counts; S—grounded slide; Tr—catch tray under bins

discrepancy in glass, polyphase, and agglutinate counts between Tables 2 and 3 resulted from a confusion among those components when counting run #1.

In the first nitrogen run, with all the agglutinates present, approximately 12 wt % of the sample coated the field electrode and about 3 wt % clung to the slide and hopper that together comprise the electrical ground. By contrast, in a vacuum run that also contained all the agglutinates, only 0.5 wt % of the sample clung to the slide, hopper, and field electrode combined. In both cases, half this material was agglutinates. Nitrogen ion and electron charging of the agglutinate-rich sample of the first run probably account for the differences.

Lunar ilmenite behaved like a semi-conductor in all the runs. Although it reached 100% concentration in bins 6 and 7 of the first nitrogen run (Table 2) and 80 and 90 wt % in the other nitrogen runs, only a few grains (<1 wt % of the sample) appeared there. In the vacuum run, by contrast, about 14 wt % of the sample reported to bins 6 & 7 and beyond (into the catch tray) but only at a combined ilmenite grade of 29 wt % (up from 7 wt %). Recovery for the vacuum run was 55 wt % (Tables 1 and 3). Mean ilmenite grades and recoveries were not calculated for the lunar sample runs because of the small number of runs and their differing conditions. However, the identical grade (51 wt %) for runs #2a and #3a under the same conditions suggests reproducibility. Adjusted grades and recoveries that include half the glass and polyphase fractions in the conductor bins to reflect their ilmenite contributions are also reported for runs 2, 3, and 4 in Table 1. After adjustment, the grade of the vacuum run increases from 29–37 wt %, and runs #2 and #3 in nitrogen increase from 51 wt % to the mid 60s. The calculation was not done for run #1 because of uncertainty in the glass and polyphase counts.

Agglutinates behaved like non-conductors in all the lunar runs regardless of whether or not the ferromagnetic agglutinates were removed. Over 90% of sample agglutinates reported to the non-conductor bin or bins and the electrodes combined, and about 5% were spread over the remaining bins (Tables 2 and 3).

DISCUSSION

Significant one-pass concentrations of ilmenite were obtained in all the runs from the average eleven-fold increase for the simulant runs in nitrogen (7.9–90 wt %) to a four-fold increase in the lunar soil run in vacuum (7.0–29 wt %). Grades in nitrogen and air were significantly better than in vacuum probably for the following reasons:

1. Gas ionization products contributed to feed charging in air and nitrogen.

2. The absence of vibratory sample feed in vacuum may have reduced the efficiency of mineral contact and induction charging compared with the atmospheric runs.

3. Partial density segregation of ilmenite occurred as the feed fell in gas at atmospheric pressure.

4. Feed transit time in the separating field was greater in air and nitrogen than in vacuum because of air resistance.

5. The ilmenite concentration in the seventh bin of the vacuum runs was reduced due to two effects.

a. The horizontal component of acceleration imparted to feed grains by the slide (see Fig. 4) transported a few percent of the feed to the far bins in vacuum at zero electric field and led to a dilution of ilmenite separates reporting to the conductor bins with the field on. By contrast, none of the feed was transported mechanically to the far bins in air or in nitrogen at zero field.

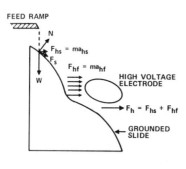

Figure 4. Force diagram of mineral grain on slide.

LEGEND

a_{hf}	—	HORIZONTAL COMPONENT OF GRAIN ACCELERATION DUE TO ELECTRIC FIELD
a_{hs}	—	HORIZONTAL COMPONENT OF GRAIN ACCELERATION DUE TO SLIDE
F_h	—	HORIZONTAL FORCE ON GRAIN
F_{hf}	—	HORIZONTAL FORCE ON GRAIN DUE TO ELECTRIC FIELD
F_{hs}	—	HORIZONTAL FORCE ON GRAIN DUE TO SLIDE
F_s	—	FORCE ON GRAIN PARALLEL TO SLIDE
m	—	MASS OF GRAIN
N	—	FORCE ON GRAIN PERPENDICULAR TO SLIDE
W	—	WEIGHT OF GRAIN

b. A back plate behind the seventh bin, for containment of material within the bin system (see Fig. 2), tended to deflect material striking it toward the seventh bin. As a consequence, more material was collected in the seventh than the fifth and sixth bins in the vacuum runs. That also may have contributed to dilution of the ilmenite concentrate and suggests that a more extended bin system might result in better grades on the slide apparatus in vacuum.

Recovery in the vacuum lunar soil run was higher than in the lunar sample runs in nitrogen and is in part due to the larger sample fraction that reached the conductor bins in vacuum. In the simulant runs, higher recoveries also tended to be associated with lower grades. Mean bin recovery (R_1) of the simulant runs in vacuum lies between the recoveries in air and nitrogen and probably reflects performance variability due to grain bounce within and outside the bin system in vacuum.

CONCLUSIONS

Significant concentrations of lunar soil ilmenite from 7.3–51 wt % have been obtained in one pass of an Apollo 11 soil sample through a slide-type electrostatic separator in nitrogen after the ferromagnetic agglutinates had been removed with a permanent magnet. Ilmenite was concentrated in same sample and in a similar apparatus (Figs. 2 and 4) from 7.0–29 wt % after one pass in vacuum with all the agglutinates present. Lunar soil ilmenite recovery was 55 wt % in the vacuum run and lower in nitrogen (ranging from 24–48 wt % in three separate runs).

It should be emphasized that the lower limit of the grain sizes tested in this study was 0.09 mm. Mineral electrostatic separation of fines in vacuum has not yet been investigated, presumably because there has been no terrestrial commercial application. However, Carta *et al.* (1964) reported improved one-pass grades and recoveries of a graphite ore (7.7% carbon) concentrated in air at reduced pressure (120 torr) compared with concentration at one atmosphere in two commercial drum electrostatic separators. Grades and recoveries were doubled in some cases even for grain size ranges less than 0.075 mm. Carta *et al.* attributed the improvement to several factors: (1) enhanced ionization charging of the feed due to enhanced corona current at the lower gas pressure, (2) improved contact of feed and grounded electrode (the drum), (3) increased uniformity of grain trajectories in all size ranges, and (4) reduced agglomeration of fines and dusting of the coarser grains with fines. Factors 2 through 4 are likely to be even more enhanced in vacuum. For that reason, mineral electrostatic separation of fines in vacuum is an important follow-up investigation to this study, especially for lunar field applications.

Electrostatic separators offer the advantages of low power consumption and mass and efficient high voltage generation in a vacuum environment such as the lunar surface. However, existing separator designs are intended for atmospheric operation and take advantage, either by design or by accident, of atmospheric effects that enhance separation performance for grain sizes >0.1 mm such as (1) mineral feed charging by air ionization in the separating field, (2) partial density segregation of mineral grains falling through air in the feed path of the separator, or (3) increased feed transit time in the separating field due to air resistance. Existing designs with minor modifications would probably work very well in a gas environment established on the Moon, but major modifications are required for efficient vacuum operation.

Multistage electrostatic separation systems like those used in all commercial electrostatic mineral concentrators should raise grades and recoveries to 90%+ levels (*e.g.*, Kelly and Spottiswood, 1982), provided the soil ilmenite is sufficiently liberated. Even without additional liberation, lunar soil ilmenite grades in the size range tested could reach the high 70s because combined glass and polyphase soil components that reported along with ilmenite to the conductor bins and beyond were comparable to ilmenite in abundance (Table 3) and are about half ilmenite in composition. Agglutinates are the major component of ilmenite bearing mare soils (Papike *et al.*, 1982), and their divergent electrical behavior to ilmenite is an indication that the two components are separable electrostatically. This work suggests that the best sequence for concentrating lunar soil ilmenite would be magnetic extraction of the ferromagnetic agglutinates followed by electrostatic concentration of ilmenite in the non–ferromagnetic soil fraction.

REFERENCES

Agosto W. N. (1983) Electrostatic separation of binary comminuted mineral mixtures. In *Space Manufacturing 1983, Advances in the Astronautical Sciences*, pp. 315–334. Amer. Astronaut. Soc., AAS 83-231, Univelt, San Diego.

Agosto W. N. (1984) Electrostatic separation and sizing of ilmenite in lunar soil simulants and samples (abstract). In *Lunar and Planetary Science. XV,* pp. 1–2, Lunar and Planetary Institute, Houston.

Andrews D. G. and Snow W. R. (1981) The supply of lunar oxygen to low earth orbit. In *Space Manufacturing Four*, pp. 173–179. American Institute of Aeronautics and Astronautics, New York.

Carmichael R. S. (1982) Magnetic properties of minerals and rocks. In *Handbook of Physical Properties of Rocks*, pp. 230–287. CRC Press, Boca Raton, FL.

Carpenter J. H. (1970) Electrical methods for the separation of minerals. *Min. Sci. Eng., 2*, 23–30.

Carta M., Ferrara G. F., Del Fa C., and Alfano G. (1964) Contribution to the electrostatic separation of minerals. *International Mineral Processing Congress, 1*, 427–446.

Davis H. P. (1983) *Lunar Oxygen Impact on STS Effectiveness.* Eagle Engineering Report No. 8363, Houston. 44 pp.

Fraas F. (1962) *Electrostatic Separation of Granular Minerals.* U.S. Bureau of Mines Bulletin 603. U.S. Bureau of Mines, Minneapolis. 155 pp.

Fraas F. (1970) *Factors Related to Mineral Separation in a Vacuum.* U.S. Bureau of Mines Report 7404. U.S. Bureau of Mines, Minneapolis. 13 pp.

Inculet I. I. (1979) Electrostatic separation of lunar soil. In *Space Manufacturing, 3*, pp. 109–111. American Institute of Aeronautics and Astronautics, New York.

Inculet I. I. (1982) Electrostatic beneficiation of coal. In *Physical Cleaning of Coal.* (Y. A. Liu, ed.), pp. 87–131. Marcel Dekker, New York.

Kelly E. G. and Spottiswood D. J. (1982) Introduction to Mineral Processing. Wiley and Sons, New York. 491 pp.

McKay D. S. and Williams R. J. (1979) A geologic assessment of potential lunar ores. In *Space Resources and Space Settlements* (J. Billingham, W. Gilbreath and B. O'Leary, eds.) pp. 243–255. NASA SP-428. NASA, Washington, DC.

Moore A. D. (1973) *Electrostatics and Its Applications.* Wiley and Sons, New York. 481 pp.

Papike J. J., Simon S. B., and Laul J. C. (1982) The lunar regolith: Chemistry, mineralogy and petrology. *Geophys. Space Phys., 20,* 761–826.

Waldron R. D. and Criswell D. R. (1982) Materials processing in space. In *Space Industrialization* (B. O'Leary, ed.) p. 97–130. CRC Press, Boca Raton, FL.

Williams R. J. and Mullins O. (1983) Enhanced production of water from ilmenite (abstract). In *Lunar and Planetary Science XIV, Special Session Abstracts*, pp. 34–35. Lunar and Planetary Institute, Houston.

IN-SITU ROCK MELTING APPLIED TO LUNAR BASE CONSTRUCTION AND FOR EXPLORATION DRILLING AND CORING ON THE MOON

John C. Rowley[1] and Joseph W. Neudecker[2]

Los Alamos National Laboratory, P.O. Box 1663, Los Alamos, NM 87545
[1]Geology and Geochemistry Group, Earth and Space Sciences Division, MS D462
[2]Analysis and Testing Group, Design Engineering Division, MS C931

An excavation technology based upon melting of rock and soil has been extensively developed at the prototype hardware and conceptual design levels for terrestrial conditions. Laboratory and field tests of rock melting penetration have conclusively indicated that this excavation method is insensitive to rock soil types, and conditions. Especially significant is the ability to form in-place glass linings or casings on the walls of boreholes, tunnels, and shafts. These factors indicate the unique potential for *in situ* construction of primary lunar base facilities. Drilling and coring equipment for resource exploration on the Moon can also be devised that is largely automated and remotely operated. It is also very likely that lunar melt glasses will have changed mechanical properties when formed in anhydrous and hard vacuum conditions. Rock melting experiments and prototype hardware designs for lunar rock melting excavation applications are suggested.

INTRODUCTION

The Los Alamos National Laboratory conducted a research and development project in excavation technology from 1960 to 1976 (Armstrong *et al.*, 1962; Hanold, 1973b, 1977; Smith, 1971). The project subsequently developed the potential advantages of a rock and soil melting excavation process for lunar applications (Rowley and Neudecker, 1980). Field and laboratory demonstrations of prototype rock and soil melting penetrator systems under terrestrial conditions have illustrated the unique features of this technology that may have application to lunar base facilities construction and exploration drilling and coring on the Moon. These basic features are detailed here. (1) The melting method is relatively insensitive to rock or soil types or conditions; (2) the technique can be automated for remote and untended operation; (3) the melting penetrators create a formed-in-place rock-glass structural lining (casing) for boreholes, tunnels, or shafts; (4) selective formation of debris (or "cuttings") as glass wool or glass pellets is possible; and (5) electrical energy is used for resistive heaters for the melting penetrators, although direct heating by nuclear power is possible for larger equipment. These results were obtained with soil and rock samples at terrestrial ambient conditions of moisture and partial pressures of oxygen.

It is anticipated that vacuum conditions and essentially zero moisture content of the lunar soils and rocks should have significantly reduced thermal diffusivity relative to terrestrial counterparts. Therefore, reduced heat losses could be expected for lunar applications. The absence of moisture and oxygen should reduce the corrosion rate of the refractory metal penetrators. The most important parameter in the rock and soil penetration process of excavation is the viscosity of the soils and rocks. This property for lunar soils and basalts, as reported in the literature (MacKenzie and Claridge, 1980), appears to be within the same range as terrestrial materials of roughly the same composition. In any event, tests and experiments at vacuum condition could be performed in order to extend and optimize the penetrator designs perfected in the previous Los Alamos work to lunar soils and rocks.

This paper summarizes the results of the previous Los Alamos research and development project with emphasis on those concepts: laboratory test hardware, field tests, and equipment designs as related to lunar uses. After recording what is known, the discussion turns to concepts and hardware that might be used for lunar base facility excavation and construction, especially primary structures. In addition, the potential for borehole and shaft "drilling" applications and for exploration core holes on the Moon is reviewed.

The final section deals with suggested research and development activities that could extend and optimize the rock and soil melting technology and hardware to lunar conditions. As indicated, these efforts would focus on a few basic experiments to determine lunar rock and soil melt properties at hard vacuum and anhydrous conditions; an effort to develop preliminary equipment design for projected needed lunar construction and drilling tasks; and most importantly, a study performed to evaluate the structural properties of lunar glasses (LG). In this latter area, we would like to strongly support the ideas of Blacic (1985) that suggest the LG may have very desirable structural properties. This prospect should be especially pursued because the advantages of formed-in-place linings or casings could be enhanced considerably.

The original work at Los Alamos termed the terrestrial excavation devices for soil and rock melting "the Subterrene." Perhaps for lunar applications a more appropriate term would be "Subselene."

PREVIOUS RESULTS

In the course of the previous Los Alamos subterrene research and development project, many different terrestrial soil and rock samples were melted under laboratory conditions to assess the performance of rock melting penetrator designs. Tables 1–3 illustrate the range of samples investigated, melting behavior, crush strengths, and two basalt compositions. The compositions of the two basalts recorded in Table 3 are especially relevant because of the close similarity to those cited by Blacic (1985) as the "average" for the lunar regolith. Boreholes were melted in an extremely wide variety of samples, both wet and dry: soils, sands, clays, shales, gravels, tuffs, basalts, and granites (Table 1). In all these experiments the melting penetrators formed competent glass walls on the borehole wall (Fig. 1a) or a separable, free-standing glass structure (Fig. 1b).

In the course of the research project several detailed evaluations were made of formed-in-place glass linings. One example is the Bandelier tuff rock-melt glass (Roedder, 1980)

Table 1. Typical Rock Melting Behavior

Material	Melting Temperatures, K		Remarks
	Start	Complete	
Bandelier Tuff	—	1750	Melt viscosity increased as quartz crystals were consumed.
Jemez basalt-1[*]	—	1570	Melts uniformly with some gas evolution.
Jemez basalt-2[†]	—	1510	
Dresser basalt	—	1570	
Charcoal granite[**]	—	1670	Dark phase melted first and then proceeded to consume the matrix.
Westerly granite	—	1760	
Sioux quartzite	—	1760	
Tennessee pink marble	—	—	Heated to 2270 K without melting; some decomposition.
Shale, Santa Fe County, New Mexico	1470	1560	Discrete phase melting accompanied by gas evolution. Viscosity increased as more material melted.
Caliche, Santa Fe County, New Mexico	1570	1850	
Green River Shale, Cuba, New Mexico	1550	1600	
Concrete	1620	1700	Localized melting. Less gas evolution than from shales or caliche.

[†]Started with rock fragments—1 to 3 mm
[*]Started with powder <1 mm
[**]Also called St. Cloud gray granodiorite

Table 2. Typical Crush Strength of Rocks and Rock Glasses

Item	Material	Average Crush Strength, MPa	Number of Specimens
1	Jemez basalt	44	10
2	Jemez basalt-glass[*]	108	4
3	Bandelier tuff	2.8	3
4	Bandelier tuff-glass from 51-mm-diam-hole wall		
	Axial	55	5
	Tangential	36	2
5	Bandelier tuff-glass from 114-mm-diam-hole wall		
	Axial (2.3 Mg/m^3)	126	4
	Axial (2.2 Mg/m^3)	110	3
	Radial (2.3 Mg/m^3)	115	3
	Tangential (2.3 Mg/m^3)	132	3

[*]Uniform glass prepared by Corning Glass Works

Table 3.   Typical Chemical Compositions of Rocks and Rock Glasses

Constituents	Composition, wt %			
	Dresser Basalt	Dresser Basalt-Glass	Jemez Basalt	Jemez Basalt-Glass
SiO_2	48.2	49.52	50.01	50.09
Al_2O_3	16.13	15.54	16.82	16.81
Fe_2O_3	7.65	8.19	2.83	4.38
FeO	5.41	4.68	7.60	6.53
MgO	6.25	6.50	6.70	6.69
CaO	8.69	10.05	9.62	9.68
Na_2O	2.54	2.47	3.94	3.40
K_2O	0.96	0.97	0.97	0.94
H_2O	0.38	0.004	0.14	0.003
TiO_2	1.45	1.66	1.38	1.46
P_2O_5	0.16	—	—	—
CO_2	0.048	0.003	0.02	0.003
B_2O_3	—	—	—	—
MnO	—	0.18	0.15	0.15

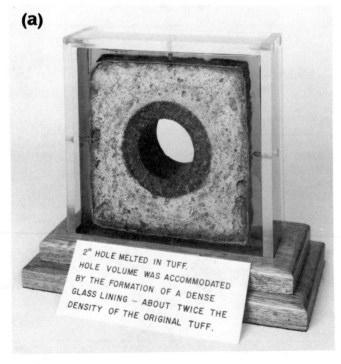

Figure 1.   Typical glass-lined holes: (a) Cross section of glass-lined hole (51-mm-diameter) melted in tuff rock; (b) Exterior view of glass-lined hole melted in loose soil and rock.

(b)

Figure 1. Typical glass-lined holes: (a) Cross section of glass-lined hole (51-mm-diameter) melted in tuff rock; (b) Exterior view of glass-lined hole melted in loose soil and rock.

where the properties of the glass-lining material were glass as summarized in Table 4. The study concludes that "the glass-lined hole formed by the penetrator is much more competent than an unlined hole and presents the possibility of engineering applications." The tuff-derived glass structures are somewhat less homogeneous than the glassy, basaltic melt glasses. Several reports on rock-melts, glass linings, and subterrene structures and the formation process were prepared by Nelson *et al.* (1975) and Krupka (1973, 1974).

The most basic conclusions reached from the laboratory research efforts were these. (1) Formed-in-place glass linings could be practically formed through proper handling, forming, and thermal processing (chilling) of the soil and rock melts (Lundberg, 1975; Stanton, 1974), and because these methods applied to all soils and rocks tested, a single penetrator design could be effectively used for virtually all natural terrestrial materials. (2) The melting process is quite insensitive to rapid variations in rock or soil types, void space, water content, or competence of the rocks or soils, and it is especially effective in consolidating "mixed ground" (*i.e.*, gravels or soils with rocks and cobbles). (3) A very uniform and precisely dimensioned borehole could be produced. (4) A high-temperature electric heater technology was perfected that used efficient low-voltage direct current resistance heaters (Armstrong, 1974; Krupka, 1972; Stark and Krupka, 1973). (5) Heat losses to the surrounding rocks or soils were low and predictable (Murphy and Gido, 1973; Cort, 1973; McFarland, 1974). (6) Low mass loss from the refractory metal penetrator

Table 4.   Properties of Tuff and Tuff-Glass

Property	Tuff[*]	Tuff Glass[†]
Density (Mg/m^3)	1.40	2.23
Grain Density (Mg/m^3)	2.54	2.40
Permeability (md)		
(a) No Confining Pressure	>100	2 to 5
(b) 50 MPa Confining pressure	—	0.1 to 0.3
Compressive Strength (MPa)	~4	~50
Moduli Average		
(a) No Confining Pressure		
E (GPa)	—	7
v	—	0.3
G (GPa)	—	2.6
(b) 50 MPa Confining Pressure		
E (GPa)	—	55
v	—	0.2
G (GPa)	—	7.0
Tensile Strength (MPa)	~1	~1

[*]A soft, friable, highly porous (41–45%) material.
[†]Grain size ~2 mm.

would lead to long equipment life (Stark and Krupka, 1975). Lastly, (7) materials, design methods, fabrication techniques, and analytical procedures were available to systematically construct and predict penetrator performance that scaled with size.

The subterrene project included a wide range of penetrator configurations (Fig. 2). The depicted shapes include nearly all concepts of hole making by melting. Figure 2a illustrates a "consolidating" penetrator (Neudecker, 1973) used in higher porosity materials; all the rock melted during formation of the hole will be densified, forming the glass lining. No debris removal is required. An alternate configuration for a melting penetrator, shown in Fig. 2b, is termed an "extruder" (Neudecker *et al.*, 1973). Pass-through port(s) allow the melt to flow back through the penetrator head into a device that chills the melt and forms "debris" (or "cuttings" or "muck," depending upon whether drilling or tunneling are considered). These solids can easily be formed as glass pellets, rods, or a glass wool-like material (Fig. 3). The core-consolidating mode of operation is shown in Fig. 2c, and cores with a glass encasement are possible (Murphy *et al.*, 1976). The final configuraton in Fig. 2d was not fabricated, but the knowledge and methods are all in hand to design and construct a kerf melting, coring extruding penetrator. This configuration might be the conceptual design for a large size tunneler. The cross section of the hole (tunnel) could be any (non-circular) geometry.

The project also developed and prepared the analytical tools (Lawton, 1974a,b) needed to perform design analyses and trade studies of the several excavation processes, *i.e.*, drilling and tunneling. These computer methods can be directly used to design soil and rock melting penetrators for lunar base application and exploring the Moon's subsurface structure and resources.

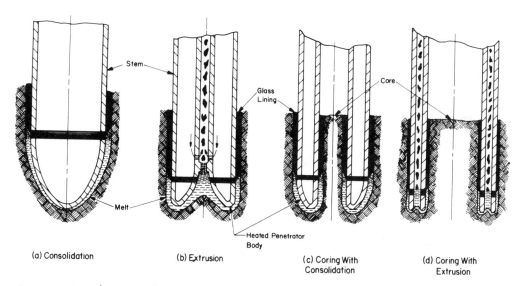

(a) Consolidation   (b) Extrusion   (c) Coring With Consolidation   (d) Coring With Extrusion

Figure 2. Schematic cross sections of different rock melting penetrators. (a) Consolidation of porous rock and soils, no debris produced. (b) Extruding of glass fiber or pellets to remove material in more dense materials. (c) A coring-consolidating configuration with glass-lined hole and core. (d) An extruding-coring combination mode of hole formation.

Figure 3. Typical debris or cuttings formed by extruder-type rock melting penetrators, glass pellets, and wool.

POTENTIAL APPLICATIONS ON THE MOON

It would be a straightforward process to develop design concepts and system designs for rock and soil melting in the lunar environment. Sensitivity, trade-off, and cost analyses can be performed as they were for geothermal well drilling (Altseimer, 1976) and tunneling equipment and operations, based on melting as the hole-making technology. The principal thrust would be to design glass-lined or stabilized openings, *i.e.*, to make structures or boreholes with in-place LG structural linings or casings. This would reduce the dependence on materials transported to or refined and fabricated on the Moon. To illustrate these concepts, two areas of excavation technology are outlined here: construction of subsurface primary structures, *i.e.*, tunnels or rooms, and drilling and coring.

The requirements for fairly deep burial as solar flare protection indicate that a tunneling procedure, in contrast to trenching and back-covering, may be more efficient. If the primary structural member of the tunnel walls can be formed-in-place LG, then a significant further advantage is achieved. If the LG surface can be sealed by continuous, direct vapor sputtering with a coating of metal (perhaps iron), then an airtight (or low leak rate) barrier may also be formed.

Figure 4 is a tunneler design (Hanold, 1973a; Altseimer, 1973) for loose soils or unconsolidated ground, and this design could be used on the Moon to produce such glass-lined tunnels in the regolith. The power source could be electric cables or a self-contained nuclear reactor. Such equipment will have only a few moving parts, chiefly

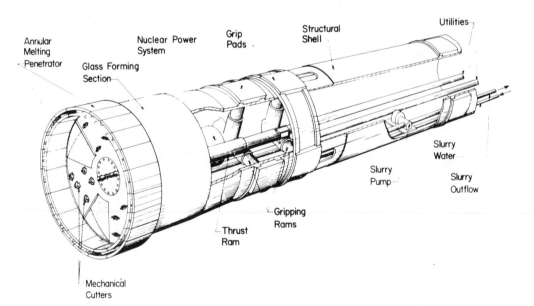

Figure 4. Large tunneler conceptual design for lunar base subsurface facilities construction. Glass lining is formed in place in the lunar regolith.

in the muck-handling system. Bulkheads of LG and LG fiber composites would provide extensive spaces for habitation, fabrication facilities, storage, and laboratories. A second concept for surface construction is illustrated in Fig. 5. A mound of regolith is prepared, and then a supporting LG "roof" is formed with a portable subselene system (shown schematically). The interior would then be excavated to form a room for use as a warehouse, vehicle storage, or large equipment housing, *etc.* The roof shape would be designed to support the overburden loads as well as side and edge reactions. A prototype of such

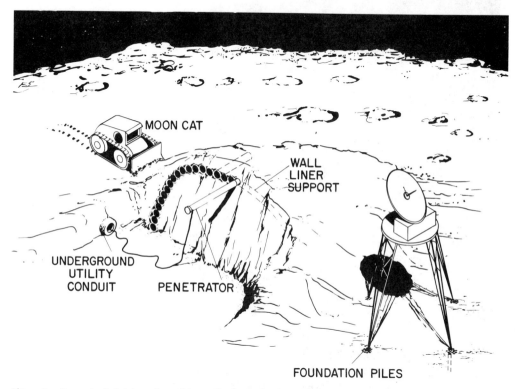

MOON CAT

WALL
LINER
SUPPORT

UNDERGROUND
UTILITY
CONDUIT

PENETRATOR

FOUNDATION PILES

Figure 5.   Conceptual sketches of possible applications of rock and soil melting for lunar base facility construction using melted-in-place glass lining.

a structure was built in the fashion described (Fig. 6), and though crude in appearance, it is quite adequate to support the overburden loading. Other applications are sketched in Fig. 5, such as horizontal holes (Sims, 1973) for utility installation.

A second area of application of lunar rock and soil melting is drilling equipment that could be adapted for lunar conditions, such as borehole or shaft drilling and coring applied to exploration (Fig. 7). It appears that vertical-hole-melting systems could be readily designed for use on the Moon, and the potential exists for essentially self-contained, remotely operated drill rig equipment (Altseimer, 1973). These possibilities result from

Figure 6. Photograph of prototype melted-in-place structure. Dimensions are 2 by 1 m at the opening, and 2 m deep. Interior rock and soil excavated after melting construction of arch and walls.

the automation potential of the melting methods and insensibility to subsurface conditions. An additional, inherent problem will be closed debris handling and separation systems.

A horizontal coring tool (Neudecker, 1974), termed a "geoprospector," has been devised for use on Earth. It is a mini-tunneler that is self-propelled, is remotely guided, and produces a continuous core. Such a device might be useful to explore along a proposed tunnel route.

RESEARCH ISSUES

Most of the basic concepts and methods to design subselene systems are available. However, a few areas of investigation are recommended if such equipment is to be developed. The tasks are primarily activities needed to extend or optimize the results available for terrestrial conditions to lunar environments. The recommended activities are as follows:

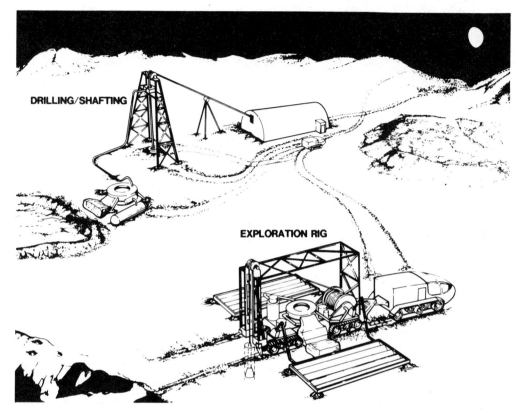

Figure 7.   Conceptual sketch of automatic, remotely operated drilling/coring and shafting equipment for exploration of the Moon.

(a) Measure lunar rock and soil melt properties such as viscosity, conductivity, and melting data for hard vacuum and anhydrous conditions simulating lunar conditions.

(b) Determine corrosion rates of penetrator material in lunar conditions by testing refractory metals in lunar rock melts under hard vacuum and very low moisture contents.

(c) Conduct conceptual prototype design, sensitivity, operation, and cost studies of lunar subselene hardware and equipment for both construction and drilling/coring functions.

(d) Produce subcomponents for the most promising reference designs and conduct laboratory and field tests. Automation and remote operation schemes should be explored.

(e) Create designs for debris-handling systems for drilling and coring applications; the approach should provide for a closed circulation of debris-handling fluids and a loss-free debris/fluid separator.

Power source designs and requirements would be an important element in such studies.

In addition, the very important aspect (outlined by Blacic, 1985) of determining the potential for very high strength LG should be investigated. The potential exists for obtaining even more efficient structures from melted-in-place lunar base structures and casings for boreholes or core holes.

SUMMARY

A major advancement in excavation technology that already exists appears to be ideally suited to lunar base facility construction chores for primary structures. Rock and soil melting technology is simple, and the basic hole-making process has no subsurface moving parts. A lunar glass structural lining or support can be formed in the melt chilling process. Tunneling machines based upon soil melting technology have been designed and could be extended and adapted to lunar use with modest effort. Another promising area of application is drilling, coring, and shafting. Here the benefits of remote and automated operations are potentially available. These uses will require some further development of debris-handling techniques and closed fluid and cooling systems.

The potential for direct use of glass derived from lunar materials for primary structures, linings, casings, and the like would reduce Earth-lift mass requirements, should reduce import costs from Earth significantly, and should require much less commitment and use of lunar-based structural materials refinement and fabrication facilities.

Acknowledgments.   A major portion of the past work on the subterrene summarized here was funded by a grant from the National Science Foundation (NSF). We would like to especially acknowledge the encouragement and guidance provided by William Hakala, the program manager at NSF. Support was also provided by the U.S. ERDA (Energy Research and Development Administration, now the U.S. Department of Energy). The authors would also like to recognize the contributions to the project by many individuals at Los Alamos and elsewhere. Robert Hanold and Charles Bankston contributed especially effective direction to the subterrene research effort as the project was completed in 1976.

REFERENCES

Altseimer J. H. (1973) *Systems and cost analysis for a nuclear subterrene tunneling machine.* Los Alamos Scientific Laboratory Report LA-5354-MS, Los Alamos, New Mexico. 22 pp.

Altseimer J. H. (1976) *Technical and cost analysis of rock-melting systems for producing geothermal wells.* Los Alamos Scientific Laboratory Report LA-6555-MS, Los Alamos, New Mexico. 37 pp.

Armstrong D. E., Coleman J. S., Berteus B., McInpeer B., Potter R. M., and Robertson E. S. (1962) *Rock melting as a drilling technique.* Los Alamos Scientific Laboratory Report LA-3243-MS, Los Alamos, New Mexico. 39 pp.

Armstrong, P. E. (1974) *Subterrene electrical heater design and morphology.* Los Alamos Scientific Laboratory Report LA-5211-MS, Los Alamos, New Mexico. 34 pp.

Blacic J. D. (1985) Mechanical properties of lunar rock materials under anhydrous, hard vacuum conditions: Applications to lunar-glass structural components. This volume.

Cort G. E. (1973) *Rock heat-loss shape factors for subterrene penetrators.* Los Alamos Scientific Laboratory Report LA-5435-MS, Los Alamos, New Mexico. 12 pp.

Hanold R. J. (1973a) *Large subterrene rock-melting tunnel excavation systems: A preliminary study.* Los Alamos Scientific Laboratory Report LA-5210-MS, Los Alamos, New Mexico. 35 pp.

Hanold R. J. (editor) (1973b) *Rapid excavation by rock melting—LASL Subterrene Program, December 31, 1972 to September 1, 1973*, Los Alamos Scientific Laboratory Report LA-5459-SR, Los Alamos, New Mexico. 35 pp.

Hanold R. J. (editor) (1977) *Rapid excavation by rock melting—LASL Subterrene Program, September 1973 - June 1976*, Los Alamos Scientific Laboratory Report LA-5979-SR, Los Alamos, New Mexico. 87 pp.

Krupka M. C. (1972) *Internal reaction phenomena in prototype subterrene radiant heater penetrators.* Los Alamos Scientific Laboratory Report LA-5084-MS, Los Alamos, New Mexico. 11 pp.

Krupka M. C. (1973) *Phenomena associated with the process of rock melting: Applications to the subterrene system.* Los Alamos Scientific Laboratory Report LA-5208-MS, Los Alamos, New Mexico. 10 pp.

Krupka M. C. (1974) *Selected physiochemical properties of basaltic rocks, liquids, and glasses.* Los Alamos Scientific Laboratory Report LA-5540-MS, Los Alamos, New Mexico. 24 pp.

Lawton R. G. (1974a) *The AYER heat conduction computer program.* Los Alamos Scientific Laboratory Report LA-5613-MS, Los Alamos, New Mexico. 27 pp.

Lawton R. G. (1974b) *PLACID: A general finite-element computer program for stress analysis of plane and axisymmetric solids.* Los Alamos Scientific Laboratory Report LA-5621-MS, Los Alamos, New Mexico. 31 pp.

Lundberg L. B. (1975) *Characteristics of rock melts and glasses formed by earth-melting subterrene.* Los Alamos Scientific Laboratory Report LA-5826-MS, Los Alamos, New Mexico. 13 pp.

MacKenzie J. D. and Claridge R. C. (1980) Glass and ceramics for lunar materials. In *Lunar Materials Processing*, (D. R. Criswell, ed.), p. 333. NSR09-051-001 (No. 24), Lunar and Planetary Institute, Houston.

McFarland R. D. (1974) *Numerical solution of melt flow and thermal energy transfer for the lithothermodynamics of a rock-melting penetrator.* Los Alamos Scientific Laboratory Report LA-5608-MS, Los Alamos, New Mexico. 15 pp.

Murphy D. J. and Gido R. G. (1973) *Heat loss calculations for small diameter subterrene penetrator.* Los Alamos Scientific Laboratory Report LA-5207-MS, Los Alamos, New Mexico. 17 pp.

Murphy H. D., Neudecker J. W., Cort G. E., Turner W. C., McFarland R. D., and Griggs J. E. (1976) *Development of coring, consolidating, subterrene penetrators.* Los Alamos Scientific Laboratory Report LA-6265-MS, Los Alamos, New Mexico. 39 pp.

Nelson R. R., Abou-Sayel A., and Jones A. H. (1975) *Characteristics of rock-glass formed by the LASL subterrene in Bandelier Tuff.* Terra Tek, Inc., Report No. TR-75-61, Terra Tek, Inc., Salt Lake City, Utah. 37 pp.

Neudecker J. W. (1973) *Design description of melting-consolidating prototype subterrene penetrators,* Los Alamos Scientific Laboratory Report LA-5212-MS, Los Alamos, New Mexico. 17 pp.

Neudecker J. W. (1974) *Conceptual design of a coring subterrene geoprospector.* Los Alamos Scientific Laboratory Report LA-5517-MS, Los Alamos, New Mexico. 14 pp.

Neudecker J. W., Giger A. J., and Armstrong P. E. (1973) *Design and development of prototype universal extruding subterrene penetrators.* Los Alamos Scientific Laboratory Report LA-5205-MS, Los Alamos, New Mexico. 16 pp.

Rowley J. C. and Neudecker J. W. (1980) Melted in-place lunar soil for construction of primary lunar surface structures. In *Lunar Materials Processing*, (D. R. Criswell, ed.), pp. 200–204. NSR09-051-001 (No. 24), Lunar and Planetary Institute, Houston.

Roedder E. (1980) Use of lunar materials in space construction. In *Lunar Materials Processing*, (D. Criswell, ed.), pp. 213–230. NSR09-051-001, (No. 24), Lunar and Planetary Institute, Houston.

Sims D. L. (1973) *A versatile rock-melting system for the formation of small-diameter horizontal glass-lined holes.* Los Alamos Scientific Laboratory Report LA-5422-MS, Los Alamos, New Mexico. 20 pp.

Smith M. C. (editor) (1971) *A preliminary study of the nuclear subterrene.* Los Alamos Scientific Laboratory Report LA-4547-MS, Los Alamos, New Mexico. 62 pp.

Stanton A. C. (1974) *Heat transfer and thermal treatment processes in subterrene-produced glass hole linings.* Los Alamos Scientific Laboratory Report LA-5502-MS, Los Alamos, New Mexico. 7 pp.

Stark W. A., Jr. and Krupka M. C. (1973) *Carbon receptor reactions in subterrene penetrators.* Los Alamos Scientific Laboratory Report LA-5423-MS, Los Alamos, New Mexico. 10 pp.

Stark W. A. and Krupka M. C. (1975) *Chemical corrosion of molybdenum and tungsten in liquid basalt, tuff, and granite with application to the subterrene penetrators.* Los Alamos Scientific Laboratory Report LA-5857-MS, Los Alamos, New Mexico. 21 pp.

MICROWAVE PROCESSING OF LUNAR MATERIALS: POTENTIAL APPLICATIONS

Thomas T. Meek[1], David T. Vaniman[2], Franklin H. Cocks[3], Robin A. Wright[4]

[1] *Materials Science and Technology Division, Los Alamos National Laboratory, Los Alamos, NM 87545*
[2] *Earth and Space Science Division, Los Alamos National Laboratory, Los Alamos , NM 87545*
[3] *Department of Mechanical Engineering and Materials Science, Duke University, Durham, NC 27706*
[4] *Department of Engineering Mechanics, Columbia University, New York, NY 10027*

The microwave processing of lunar materials holds promise for the production of water, oxygen, primary metals, or ceramic materials. Extra high frequency microwaves (EHF), between 100 and 500 gigahertz (GHz), have the potential for selective coupling to specific atomic species and a concomitant low energy requirement for the extraction of specific materials, such as oxygen, from lunar materials. The coupling of ultra high frequency (UHF) (e.g., 2.45 GHz) microwave frequencies to hydrogen–oxygen bonds might enable the preferential and low energy cost removal (as H_2O) of implanted protons from the sun or of adsorbed water that might be found in lunar dust in permanently shadowed polar areas. Microwave melting and selective phase melting of lunar materials could also be used either in the preparation of ceramic materials with simplified geometries (e.g., bricks) and with custom-tailored microstructures, or for the direct preparation of hermetic walls in underground structures. Speculatively, the preparation of photovoltaic devices based on lunar materials, especially ilmenite, may be a potential use of microwave processing on the Moon. Preliminary experiments on UHF melting of a terrestrial basalt and of an ilmenite-rich terrestrial rock show that microwave processing is feasible, particularly in ilmenite-bearing rock types.

INTRODUCTION

Many different suggestions have been made with regard to the potential use of lunar materials in space as well as in lunar applications. In broad terms, these applications may be classified as involving either structural uses or raw material supply. In either case it generally is true that thermal processing will be needed for most of these potential applications. For such heating the use of both ultra high frequency (UHF) 2.45 GHz microwaves as well as extra high frequency (EHF) microwaves between 100 and 500 GHz would appear to have very special advantages over other heating methods, as, for example, the use of focused sunlight. Some specific advantages of microwave heating are the following: a potential savings in processing energy for certain applications of at least an order of magnitude; the development, in particular geometries, of tailored microstructures; and the possibility of selectively heating only desired phases in rocks composed of many different minerals. Additionally, the use of microwave energy to process lunar materials offers the prospect of designing a continuous flow from the raw material to the finished product in a self-contained process. For example, large rocks could be fractured by coupling to particular phases that possess relatively large thermal expansion coefficients. Once fractured, the rocks could be melted and separated into other raw materials or used directly for fabrication into simple but useful geometries, *e.g.*, bricks. Moreover, if sufficiently high temperatures can be obtained, it would be possible to decompose lunar materials substantially into their constituent elements without the need for any chemical feedstocks or further electrochemical processing.

In the following section we discuss first the principle features of microwave heating of ceramic or mineral materials before proceeding to a specific discussion of the application of this method to lunar substances and the presentation of specific examples of microwave-melted terrestrial rocks and ceramic materials.

Microwave Processing

Microwave processing of ceramics can be accomplished in two ways: first, by using UHF radiation to couple radio-frequency energy to defects, impurities, and H-O bonds, and second, by coupling directly to the oxide lattice with EHF radiation. Depending on the composition of the lunar material, either or both of these radiation bands may be used. Additionally, lunar materials have been bombarded for geologic times by high energy photons and particles and as a result contain a high density of fossil radiation damage (Eddy *et al.*, 1980; McDougall *et al.*, 1971). It is quite possible that these radiation-induced defects will couple strongly to UHF radiation.

As an example of ceramic processing by microwaves, Fig. 1 shows the typical geometry of ceramic–glass–ceramic junctions fabricated using UHF radiation. Region II is the seal, while regions I and III are alumina substrates. Figure 2 shows a greatly magnified view of the same junction fabricated using conventional heating. Region I is the alumina substrate, and region II is the glass seal. Note the very different microstructures that are produced. The microwave fabricated geometry is diffusion-bonded, while the conventionally heated geometry is held together by surface wetting only. The energy required to form the geometry

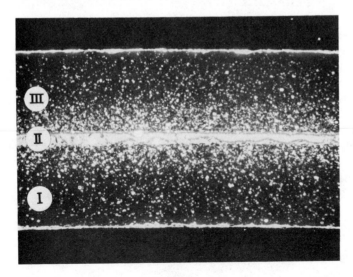

Figure 1. Microwave-formed seal between alumina substrates. (75X)

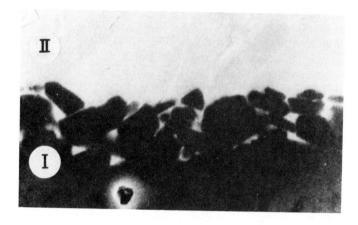

Figure 2. Conventional seal (interface along alumina substrate). (2000X)

of Fig. 1 was approximately 100 times less than that shown in Fig. 2, while the time required to complete the bonding shown in Fig. 1 was approximately 200 times less than that of Fig. 2. Microwave heating rates of up to approximately 80% of the fusion temperature have reached 32,000°C/hour. A rapid heating rate for a conventional heating cycle would be between 20° and 50°C per hour. The different microstructures evident when using microwave heating are due to high-temperature chemical reactions that occur so rapidly that the composition of the phases formed is not affected by the loss of reactants by evaporation.

Self-limiting temperatures have been observed in microwave heating. In Fig. 1, region II is the reacted interface between the alumina substrates and contains a fine grained region that is $ZrO-Al_2O_3$ (the small dark bands within the broader white band region between the alumina substrates.) The white band is a lead alumino silicate glass. Time-temperature data for this seal showed an initial rapid heating rate followed by a period when the maximum temperature of 1000°C was attained for about 45 minutes. After this time period temperature began to decrease with the microwave power still at the maximum setting. It is thought that temperature decreases due to the formation of phases with higher dielectric constant, and the time–temperature data may reflect high temperature reactions that resulted in the formation of these phases. It thus appears possible that the use of EHF radiation will enable selective coupling to be achieved by tuning to particular phases in the materials that have high coupling characteristics. In any case, it is evident that microwave heating can be used to heat ceramic materials with extreme rapidity and with very low power consumption, compared to conventional ceramic processing techniques.

Potential Applications

It is well known that UHF microwaves at 2.45 GHz couple strongly to water and that the lunar materials that have been examined to date appear to lack any water. The possibility exists, however, that some moisture may occur in permanently shadowed lunar polar regions (Watson *et al.*, 1961; Urey, 1967; Werner *et al.*, 1967), and a lunar polar-orbiting mission may be designed to detect such moisture by reflected radar or by other methods. Presumably polar moisture would be of an adsorbed nature and would not be actual ice. Assuming a water content of 10 ppm, heating an entire quantity of rock from -100°C to +100°C would require more than 10,000 times as much energy as that needed to heat the water alone. Neglecting the energy cost of the required material handling, and assuming both the coupling of 2.45 GHz to water and the generation of these microwaves to be 50% efficient or better, approximately 7 l of water could be obtained in 24 h from the power produced by 10 m^2 of solar cells having 10% efficiency. However, material handling costs would in fact be much smaller if the fossil hydrogen (and concomitant oxygen) were removed by a portable microwave unit and the resulting water vapor collected as ice after recondensation on a cold collecting plate.

It is also important to note that 40% by weight of lunar soil is, typically, oxygen (Mason and Melson, 1970). The extraction of this oxygen by microwave heating would be of enormous utility in the long-term support of any lunar base. The energy cost per unit weight of oxygen would be expected to be much greater than that for the extraction of water, due to the relatively high energy associated with metal-oxygen bonds. Such energies can be obtained with focused sunlight as well as by using microwave heating techniques (*i.e.*, by direct EHF coupling to the appropriate metal-oxygen bonds of a mineral lattice). It is also possible, however, that microwave heating might only be needed to reach temperatures high enough to cause melting, after which decomposition to produce both oxygen and metallics could be achieved by means of electrochemical decomposition of the molten rock. Focused sunlight methods of melting or decomposition could, of course, be usable only during the lunar day, whereas microwave melting, if powered from a nuclear source, would be continuously operable. As has been mentioned, it is to be expected that either UHF or EHF microwaves may preferentially couple to and selectively affect particular phases. In the making of brick-like materials in which only a small fraction of either crystalline or glassy phases need be liquified, such selective coupling could be expected to both lower the total energy requirement and aid in the development of tailored microstructures.

It is also possible to speculate that useful photovoltaic properties might be achievable using selected lunar minerals together with appropriate processing. The photovoltaic properties of elemental and compound semiconductors are of course well known. It is less widely known, however, that many minerals may also show certain semiconducting properties. In particular, mineral defect semiconductors can show well-defined, optical band gaps. The band gap of synthetic ilmenite has been measured to be 2.58 eV, for example. From band gap considerations alone, therefore, the theoretical efficiency of solar cells based on such ilmenite could be as high as 11% (Loferski, 1956). The very poor transport properties of such defect semiconductors would make it very unlikely that this level of efficiency could be reached. Nevertheless, the possibility of producing photovoltaic devices of low efficiency, but also low cost, using native lunar materials deserves mention. In such a process, the use of microwave heating with its possibility of coupling to selective phases could have useful advantages.

As an application of the foregoing concepts, UHF heating with 2.45 GHz microwaves has been utilized to produce brick-like materials using both a terrestrial basalt and an ilmenite-rich terrestrial rock. The basalt was not selected at random but was chosen because of its similarity in many components to lunar low-titanium olivine basalts, as shown in Table 1. This analogy is, of course, imperfect because of the considerably lower Fe and Cr contents and higher water, Na, K, and P content that is common to most terrestrial basalts in comparison to lunar mare basalts. The ilmenite-rich rock was a

Table 1. Similarities and Differences Between the Techado Mountain Basalt and Lunar Low-Ti Olivine Basalts

	Techado Mt. New Mexico	Average Apollo 12 Olivine Basalt	Average Apollo 15 Olivine Basalt
SiO_2	44.3	44.3	45.0
TiO_2	3.18	2.65	2.41
Al_2O_3	10.8	8.0	8.8
[†]FeO	12.0	21.1	22.4
MnO	0.18	0.28	0.30
CaO	10.1	8.6	9.8
Na_2O	2.10	0.22	0.28
K_2O	1.74	0.06	0.05
P_2O_5	0.88	0.08	0.08
Cr_2O_3	0.05	0.63	0.57

Data for the average Apollo 12 and Apollo 15 olivine basalts are from Papike *et al.* (1976).

*Alkaline basalt from Techado Mountain, New Mexico, normalized to 100% water-free composition for comparative purposes (sample contains 2.1% H_2O).

[†]Essentially all iron occurs as $FeO(Fe^{+2})$ in lunar basalts; Fe^{+3} also occurs in terrestrial basalts, but the Techado Mountain sample is reported as FeO only for comparative purposes.

Norwegian sample consisting of about 75% ilmenite with hematite exsolution, and a matrix of plagioclase with minor pyroxenes, pyrite, pleonaste spinel, biotite, and olivine. Since ilmenite is a defect semiconductor, its resistivity is relatively low (on the order of hundreds of ohm–cm) and it couples strongly to the UHF microwave field. It has been found in this present work that a mixture of 10 wt % of the ilmenite-rich rock and 90 wt % of the Techado Mountain basalt can be melted successfully with 2.45 GHz fields, whereas the Techado Mountain basalt will not melt by itself. The ilmenite acts, in this case, as a coupling agent and the resulting temperature rise is sufficient to cause the basalt to couple to this field also. These admixtures have also been melted using normal resistance furnaces. Inasmuch as the temperatures produced during microwave melting could only be measured by optical means, it is not certain what temperatures were reached in the center of the melt. The surface of the basalt/ilmenite melt, however, was found to reach approximately 1200°C.

Using a standard resistance furnace, the ilmenite rock was melted, then cooled at approximately the same rate as that used for the microwave melted material (air-quenching). Figures 3 and 4 show the microstructures found in each case. As may be

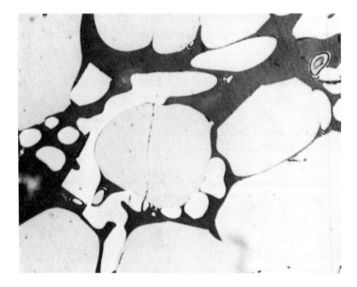

Figure 3. Resistance-furnace melt of ilmenite-rich Norwegian sample (Wards ilmenite 46W4115). Globular ilmenite and lesser amounts of globular hematite in silicate glass matrix. Long dimension of photograph is 1.1 mm.

seen, there appear to be dramatic structural differences between these two melts. The mineralogical examination of the normally melted material (Fig. 3) shows that it is dominated by a globular admixture of 75% ilmenite (without hematite lamellae) and 15% hematite, in a glassy silicate matrix. Titanomagnetite occurs, but is very rare (less than 1%). In contrast, the microwave melted material (Fig. 4) contains abundant (15%) cruciform or dendritic titomagnetite, along with anhedral to subhedral ilmenite (60%, also without hematite lamellae) in a silicate matrix. Hematite is rare (~1%). It is still unknown to what extent these differences are due to microwave versus normal melting and to what extent

Figure 4. Microwave melt of ilmenite-rich Norwegian sample (Ward's ilmenite 46W4115). Cruciform titanomagnetite, laths of ilmenite, and one of the rare hematites (lower left corner) are shown. Long dimension of photograph is 0.2 mm.

the increased oxidation of the mormally melted sample is due to slower heating and cooling while exposed to atmosphere. However, considerable differences between microwave and normal furnace-melted material might be expected, based on the differential selective coupling of the different phases present in these melts. In the present case, both normal melting and microwave melting produced a brick-like material that would clearly have utility as a structural material, particularly if produced *in situ* along tunnel walls.

Additionally, microwave melting could be used as a means of dust consolidation. In this case, outright melting would probably not be required because dust particle agglomeration via sintering could be expected to occur at temperatures well below melting. Of course, during such treatment any fossil hydrogen would be expected to be driven off with the concomitant production of water that could be collected via the use of cold collecting surfaces. Such heating of the entire mass of dust, by means of EHF microwaves, would consume far more energy than the selective coupling to the H-O bond by 2.45 GHz microwaves. In this way, however, the two-fold purpose of dust agglomeration and water production would be achieved in a single operation.

CONCLUSION

Microwave heating appears to have several potential applications in the processing of lunar materials. The ability of selected microwave frequencies to couple to specific bonds appears to be especially valuable, particularly in the selective removal of fossil solar hydrogen, possibly as water, from lunar soils.

In certain applications, dense, impermeable materials will be required, whereas in others (as for example, in extralunar radiation shielding applications), only a solid brick-like mass, whether or not gas-permeable, will be needed. In either case, it will certainly be true that the ability of microwave processing to interact selectively with one or more specific mineralogic constituents will be of immense help in preparing ceramic bodies with specific structural properties from lunar materials. Furthermore, the ability to heat, melt, or otherwise process lunar materials with a minimum expenditure in both energy and time are extremely valuable characteristics. From these considerations, it appears that the use of microwave heating for the processing of lunar and other materials has unique and important benefits.

REFERENCES

Eddy J. A., Merrill R. B., and Pepin R. O. (editors) (1980) *The Ancient Sun*, Pergamon, New York. 581 pp.

Loferski J. (1956) Theoretical considerations governing the choice of the optimum semiconductor for photovoltaic solar energy conversion. *J. Appl. Phys., 27*, 777–784.

MacDougall D. J., Lal D., Wilkening L. L., Bhat S. G., Liang S. S., Arrhenius G., and Tamhane A. S. (1971) Techniques for the study of fossil tracks in extraterrestrial and terrestrial samples. *Geochem. J., 5*, 95–112.

Mason B. and Melson W. G. (1970) *The Lunar Rocks*. Wiley-Interscience, New York, 117 pp.

Papike J. J., Hodges F. N., Bence A. E., Cameron M., and Rhodes J. M. (1976) Mare basalts: Chemistry, mineralogy, and petrology. *Rev. Geophys. Space Phys., 14*, 475–540.

Urey H. C. (1967) Water on the Moon. *Nature, 216*, 1094–1095.

Watson J., Bruce C., and Brown H. (1961) The behavior of volatiles on the lunar surface. *J. Geophys. Res., 66,* 3033–3045.

Werner M., Gold T., and Harwit G. (1967) On the detection of water on the Moon. *Planet. Space Sci., 15,* 771–774.

MECHANICAL PROPERTIES OF LUNAR MATERIALS UNDER ANHYDROUS, HARD VACUUM CONDITIONS: APPLICATIONS OF LUNAR GLASS STRUCTURAL COMPONENTS

James D. Blacic

Los Alamos National Laboratory, Geophysics Group, Division of Earth and Space Sciences, MS C335, Los Alamos, NM 87545

Lunar materials and derivatives such as glass may possess very high tensile strengths compared to equivalent materials on Earth because of the absence of hydrolytic weakening processes on the Moon and in the hard vacuum of free space. Hydrolysis of Si-O bonds at crack tips or dislocations reduces the strength of silicates by about an order of magnitude in Earth environments. However, lunar materials are extremely anhydrous, and hydrolytic weakening will be suppressed in free space. Thus, the geomechanical properties of the Moon and engineering properties of lunar silicate materials in space environments will be very different than equivalent materials under Earth conditions, where the action of water cannot be conveniently avoided. Possible substitution of lunar glass for structural metals in a variety of space engineering applications enhances the economic utilization of the Moon.

INTRODUCTION

The intent of this paper is to consider the effects of the environmental conditions of the Moon and free space on the mechanical properties of lunar rocks and materials derived from them. Mechanical properties of silicate materials are very different in the anhydrous, hard vacuum conditions of space compared to Earth due to the virtual absence of hydrolytic weakening processes there. The implications of this realization will be very important in the interpretation of geophysical measurements in investigating the structure of the Moon, in exploitation of lunar materials for construction of a lunar base, and in eventual space industrialization and habitation.

After documenting what is currently known about these environmental effects, I concentrate on the implications of "anhydrous strengthening" of an easily formed structural material derived from lunar regolith, namely, lunar glass. Although the importance of lunar-derived glass has been known for some time (Phinney *et al.*, 1977), the full implications of the potentially very great strength of lunar glass in the vacuum environment are not widely realized. In detailing some applications of lunar glass structural components, I support a philosophy that requires maximal utilization of common lunar materials with minimal processing before end use. It has become clear that large-scale exploitation of space is limited by the cost of Earth-lift of materials. Therefore, it is essential that every possible means be taken to utilize indigenous materials from the Moon and, eventually, the asteroids. In doing so, we should not fight the *in situ* environmental conditions (*e.g.,* low gravity, vacuum), or try to wedge Earth–derived processes into conditions for which they are not adapted; rather, we should attempt to take advantage of that which is given in new ways. It is in this sense that lunar glass can play a central role in easing full-scale entry into the new frontier of space.

HYDROLYTIC WEAKENING PROCESSES IN SILICATES

It has been known for some time that the fracture strength of brittle amorphous and crystalline silicates is determined in Earth environments by the damage state of surfaces and, most especially, the corrosive action of water in extending microcracks (Charles, 1958; Scholz, 1972). For example, the moisture sensitivity of glass is well known. Merely touching freshly formed glass rods will drastically reduce their tensile strengths, and less than one percent of the theoretical tensile strength of glass is normally realized in industrial practice (LaCourse, 1972). Similarly, the plastic strengths of crystalline silicates (*e.g.*, quartz and olivine) at elevated temperatures and pressures are strongly affected because trace amounts of water aid dislocation motion (Griggs,1967; Blacic, 1972). In both instances, the weakening mechanism is believed to involve the hydrolyzation of Si-O bonds (Griggs, 1967; Blacic and Christie, 1984; Charles, 1959; Michalske and Freiman, 1982). A schematic representation of one proposed mechanism is shown in Fig. 1 (Blacic and Christie, 1984). The great inherent strength of silicates is due to the strength of the network-forming silicon–oxygen bonds. However, it appears that the polar water molecule can easily hydrolyze these linking bridges by replacing the strong Si-O bond with a hydrogen-bonded bridge that is an order of magnitude weaker. This hydrolyzation can occur along dislocations, thereby increasing the mobility of dislocation kinks, or at highly stressed microcrack tips resulting in a lower applied stress to propagate the cracks. In both cases, the net result is a large weakening of the material when even very small amounts of water are present.

Whatever its detailed nature, the hydrolytic weakening mechanism is demonstrably a thermally activated rate process. Thus, the time- and temperature-dependent mechanical properties of silicates (brittle and plastic creep, static fatigue, subcritical crack growth) are dominated by moisture effects (Charles, 1958; Scholz, 1972; Blacic and Christie, 1984). As might be expected, these hydrolytic weakening processes are an important factor in such diverse areas as solid earth mechanics, geotechnology (drilling and mining), materials science (glass and ceramic technology), communications (fiber optics), national defense (high energy laser optics), and others, since, on Earth, it is practically impossible to avoid the presence of some water in the fabrication or use of materials, be they natural or synthetic. However, the case may be much different on the Moon and in free space.

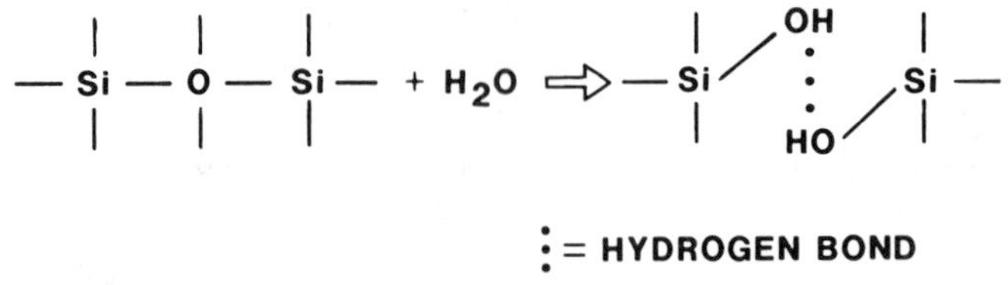

Figure 1. *Schematic representation of the Si-O bond hydrolyzation reaction.*

ANHYDROUS STRENGTHENING ON THE MOON

Although there is still hope that we may find some water preserved in the permanently shaded regions of the lunar poles (Arnold, 1979), a striking feature of all lunar materials examined so far is their almost total lack of water (Williams and Jadwick, 1980). The very small amount of water that is observed to evolve from heated lunar samples is likely due to either oxidation of solar wind–implanted hydrogen, present at about the 100 ppm level (Williams and Jadwick, 1980), or is the result of Earth contamination (Carrier *et al.*, 1973). There is no unequivocal evidence of native water in any lunar sample returned to date. This fact suggests that, in the hard vacuum of space, silicates derived from the Moon will not, if we can avoid contaminating them, exhibit the water-induced weakening that is so ubiquitous on Earth. In other words, lunar silicates may possess very high strengths due to an "anhydrous strengthening" effect relative to our common experience on Earth. This possibility has numerous implications for space industrialization, some of which are explored below.

There is some supporting laboratory evidence for the anhydrous strengthening phenomenon in lunar or lunar-simulant materials. The compressive strength of a mare-like simulant rock (basaltic intrusive) has been shown to increase by about a factor of two when samples are degassed and tested in a moderate laboratory vacuum compared to tests in 100% humid air (Mizutani *et al.*, 1977). Subcritical crack velocity measurements in a lunar analogue glass demonstrate many orders of magnitude reduction in crack velocity with decreasing partial pressure of water (Soga *et al.*, 1979). This suggests that static fatigue processes will be strongly suppressed or absent in lunar materials in a vacuum environment. Several investigators have found that very small amounts of water strongly affect the dissipation of vibratory energy (Q^{-1}) in lunar and terrestrial rocks (Pandit and Tozer, 1970; Tittmann *et al.*, 1980). These attenuation mechanisms are likely the result of the hydrolysis of crack surfaces with consequent reduction of surface energy in a manner similar to that shown in Fig. 1. The soil mechanics properties of Apollo samples and simulants have been shown to be strongly affected by atmospheric moisture contaminants in moderate and ultra high vacuum experiments (Carrier *et al.*, 1973; Johnson *et al.*, 1973). These latter results suggest that well-consolidated lunar regolith may be substantially stronger than similar materials on Earth with important implications for energy requirements for handling of lunar materials.

There are many additional examples, too numerous to document here, of research on the effects of water on the mechanical properties of terrestrial silicate materials. The main conclusion to be gained from all this work is that water, even in trace amounts, is all-important in explaining the great reduction in strength of silicates. However, in order to get a quantitative estimate of the possible increase in strength of lunar materials relative to their Earth counterparts, it is instructive to examine in some detail the elegant results of F. M. Ernsberger (1969) on glass.

In Ernsberger's experiments, etched glass rods are heated and deformed to produce entrapped bubbles in the form of oblate spheroids. The bubbles concentrate stress at the point of maximum curvature of the bubble-glass interface in a calculable way. In

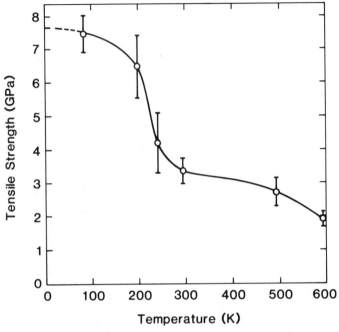

Figure 2. Tensile strength of Kimble R6 soda-lime glass in a relatively anhydrous environment as a function of temperature (Ernsberger, 1969).

addition, if care is taken, failure always occurs at the flaw-free bubble surface where the atmosphere is constant and relatively anhydrous. Using this technique, Ernsberger was the first to achieve controlled compressive failure of glass by shear fracture or densification. Scatter in tensile strength measurements was reduced compared to normal test methods; results are shown in Fig. 2 for soda-lime glass. The temperature dependence of strength shown in Fig. 2 is believed to be due to solid-state diffusion of weakening elements to the stress concentration, possibly sodium but more likely residual water dissolved in the glass. At reduced temperatures, the weakening element is immobilized and strength increases. The important aspect of this work, confirmed by other investigators for other glass compositions, is that the strength is about an order of magnitude higher in an anhydrous environment than it is for the same glass tested in a normal humidity atmosphere. This gives some idea of what might be expected for a lunar glass used in vacuum, although it probably represents only a minimum strength estimate because of the extremely anhydrous nature of lunar materials and the hard vacuum of space.

SOME POSSIBLE APPLICATIONS

Table 1 compares the mechanical properties of some structural metals likely to be produced from lunar regolith with estimates for lunar glass. Common soda-lime glass under Earth conditions is also listed for comparison. The range of tensile strength estimated for lunar glass is believed to be conservative, as discussed above, but even if only the low end of the range can be achieved, then one can see that lunar glass is very competitive

Table 1. Mechanical Properties of Lunar-Derived Materials

	T (GPa/10^6psi)	ρ	E (GPa/10^6psi)	T/ρ (GPa/10^6psi)	E/ρ (GPa/10^6psi)
Aluminum	0.17/0.02	2.7	70/10.2	0.06/0.009	25.9/3.76
Magnesium	0.20/0.03	1.7	45/6.5	0.12/0.017	26.5/3.84
Iron	0.28/0.04	7.9	196/28.4	0.04/0.006	24.8/3.60
Titanium	2.3/0.33	4.6	119/17.3	0.50/0.073	25.9/3.76
Alloy Steel	2.3/0.33	8.2	224/32	0.28/0.041	27.3/3.90
Soda-lime Glass (Earth Environment)	0.007/0.01	2.5	68/9.9	0.003/0.004	27.2/3.95
Lunar Glass (Space Environment)	0.007/0.01– 3.0/0.44 or greater?	2.8	100/14.5?	0.003/0.004–1.07/0.16	35.7/5.19?

T = ultimate tensile strength
ρ = specific gravity
E = Young's modulus

with—if not superior to—the metals obtainable from lunar materials with considerably more processing effort.

How can lunar glass be utilized? One obvious way is in the form of glass fibers in tensile stress situations. Although lunar glass will be very strong, it will still be a very brittle material, and therefore it makes sense to distribute the load over many small elements whenever possible. Thus, lunar glass fiber cloths (Criswell, 1977) and multiply stranded cords and cables should see wide application in a lunar base and large space structures such as solar power satellites (SPS). However, lunar glass fibers should always be coated with a metal such as Fe, Al, or Mg to protect the glass from inadvertant or purposeful exposure to water vapor. Otherwise, a highly stressed glass component might fail catastrophically due to water-induced stress corrosion. The metal coating could easily be incorporated into the production process and would also serve the desirable purpose of protecting the fibers from mechanical damage during production handling or use. This is commonly done in terrestrial fiber glass production in the form of organic sizing coatings.

Figure 3 schematically shows the elements I believe will be required in a lunar or space-based glass fiber production plant. I have assumed that sufficient electrical energy will be available [alternatively, direct solar melting could be used (Ho and Sobon, 1979)] and that there will be at least some minimal beneficiation of the feedstock. No lunar or space-based processing plant should be without some means of capturing the rare but highly valuable volatile elements in the lunar regolith. We also suggest in the figure that the relatively new Pochet-type furnace (Loewenstein, 1973) be investigated for use in lunar glass production because of the advantage it would seem to have in weight over traditional furnaces.

For applications requiring flexural, compressive, or mixed loadings such as for bulkheads in a habitat, or beams and columns in an SPS, fiber glass composites would be advantageous. Of the many types of composite materials seeing increasing terrestrial

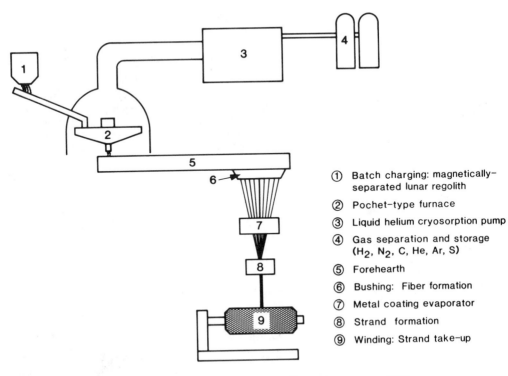

Figure 3.   Elements of a lunar-base glass fiber and gas recovery plant.

① Batch charging: magnetically-
   separated lunar regolith

② Pochet-type furnace

③ Liquid helium cryosorption pump

④ Gas separation and storage
   (H_2, N_2, C, He, Ar, S)

⑤ Forehearth

⑥ Bushing: Fiber formation

⑦ Metal coating evaporator

⑧ Strand formation

⑨ Winding: Strand take-up

usage, two would be especially attractive for space applications: metal matrix and ceramic matrix composites. Gas-tight metal matrix composites such as graphite–aluminum are now widely used in aerospace applications. If we follow the philosophy of minimal processing of lunar materials before end use, then lunar glass fiber (LG)–Fe matrix composites should be developed since native iron will be available from regolith beneficiation for fiber coating in any case. The lunar vacuum would make the diffusion bonding and liquid metal infiltration techniques (Davis and Bradstreet, 1970) of composite production advantageous. This lunar glass–metal matrix composite should be very useful in lunar base habitat construction. If a lighter weight composite is wanted (for example, for SPS applications), then silica fiber–Mg composites could be produced when a more sophisticated lunar processing capability becomes available.

Ceramic matrix composites offer some special advantages in certain applications. Large space structures such as antennas and support structures of an SPS are sensitive to the potentially large thermal strains associated with periodic eclipses. Table 2 lists thermal expansion coefficients for some structural materials. Note that glass generally has lower thermal expansivity than common structural metals, and also that some compositions derivable from abundant lunar materials (*e.g.*, titanium silicate glass) exhibit extremely low thermal expansion. If one were willing to import from Earth small amounts

Table 2. Thermal Expansion

	$\Delta L/L$ (10^{-6} °C^{-1})
Aluminum	24.0
Magnesium	25.0
Titanium	8.5
Iron	12.0
Steel	12.0
Invar	1.2
E-Glass	4.8
Corning #7971 Titanium Silicate Glass	0.03

of graphite fiber (which has a negative thermal expansivity), then composites having zero thermal expansion could be produced (Browning, 1982). Ceramic matrix composites exhibit one other desirable property. If the reinforcement fibers do not chemically bond to the ceramic matrix, but instead are held dominantly by frictional forces, then the composite exhibits an enhanced ductility and residual strength beyond the yield point to relatively large strains and is notch insensitive in a manner similar to metals (A. Evans, personal communication, 1984). Thus, I envision a composite in which Fe-coated LG fibers are imbedded in a lunar glass matrix. Such a material may have very desirable structural properties and may represent the best structural material that can be formed entirely from the most common lunar materials with the least amount of processing.

Finally, I would like to support the suggestion of Rowley and Neudecker (1984) that lunar habitats be formed by melting in-place, glass-lined tunnels using the "subterrene" (perhaps in the present context, as they note, better termed "subselene") technology. If the glass-lined tunnels were sputter-coated with a metal to protect the glass from water vapor, and if the LG fiber composites were used for bulkheads, etc., then extensive lunar habitats with more than adequate radiation shielding from the largest solar flare storms could be produced from 100% lunar materials. No doubt engineers and architects will find many more uses than we have thought of for a lightweight structural material with several hundred thousand psi tensile strength.

RESEARCH NEEDS

Most of what I have advocated concerning the possible high strength of lunar materials in hard vacuum environments has been based on research of terrestrial silicates under terrestrial or, at best, poorly simulated space conditions. Ultimately, our contentions must be proved at full-scale using actual lunar materials under *in situ* conditions. A lunar-based materials testing laboratory would seem necessary for this and should be an early, high priority lunar base facility. Until reoccupation of the Moon, however, much can be learned, and perhaps our basic contentions can be proved by experiments using lunar simulants formed and tested under ultra high vacuum laboratory conditions on Earth. This approach would seem initially preferable to LEO shuttle experiments because of the relatively poor vacuum environment of the shuttle resulting from the normally low orbits achieved and, perhaps more importantly, outgassing of the vehicle itself. Perhaps the free flying or tethered experimental platforms proposed in conjunction with the space station will improve this situation and will be needed to evaluate the effects of extended exposure to radiation and micrometeoroid fluxes, but for now ultra high vacuum experiments in Earth laboratories appear most appropriate. Most urgently needed are basic mechanical properties such as tensile and compressive strengths, fracture toughness, and thermal properties. With these results in hand, investigation of potential composite materials can proceed followed by bench top and proto-type engineering of the manufacturing facilities that will be required. Also, research and evaluation of the "subselene" approach to lunar habitat formation should proceed because of the advantages it would seem to have over imported structures.

SUMMARY

Although the apparent absence of water on the lunar surface makes it difficult to do many of the things we would like on the Moon, in at least one respect it may be a blessing. It appears that the anhydrous, hard vacuum environment and the inherently dry nature of lunar regolith materials down to the ppb level make possible the use of lunar glass for structural applications that would be impossible on Earth. In view of the fact that the initial cost of large-scale industralization and scientific exploitation of the space environment is dominated by Earth-lift requirements, the possible extensive use of lunar glass structural materials in a wide variety of applications offers promise of very large savings in Earth export expenses and thereby enhances the economics of utilizing the Moon. From a purely scientific point of view, it is likely that the anhydrous strengthening phenomenon will have numerous implications for a wide range of geological and other scientific investigations on the Moon in which mechanical properties play an important role.

Acknowledgments. *I would like to acknowledge helpful discussions with J. Clemens, R. Dick, E. Mathez, M. Rovetta, J. Rowley, and T. Shankland.*

REFERENCES

Arnold J. R. (1979) Ice in the lunar polar regions. *J. Geophys. Res.*, *84*, 5659–5668.

Blacic J. D. (1972) Effect of water on the experimental deformation of olivine. *Flow and Fracture of Rocks* (H. C. Heard, I. Y. Borg, N. L. Carter, and C. B. Raleigh, eds.), Amer. Geophys. Union Mongr. 16, 109–115.

Blacic J. D. and Christie J. M. (1984) Plasticity and hydrolytic weakening of quartz single crystals. *J. Geophys. Res.*, *89*, 4223–4239.

Browning D. L. (1982) Large space structures. In *Space Industrialization*, Vol. 2 (B. O'Leary, ed.), pp. 71–73. CRC Press, Boca Raton.

Carrier W. D., Bromwell L. G., and Martin R. T. (1973) Behavior of returned lunar soil in vacuum. *J. Soil Mech. Div., ASCE*, 979–996.

Charles R. J. (1958) Static fatigue of glass. *J. Appl. Phys.*, *29*, 1549–1560.

Charles R. J. (1959) The strength of silicate glasses and some crystalline oxides. In *Proceedings of the International Conference on Atomic Mechanisms of Fracture*, p. 225–249. MIT Press, Cambridge.

Criswell D. R. (1977) Appendix: Materials packaging. *Space-Based Manufacturing from Nonterrestrial Materials* (G. K. O'Neill and B. O'Leary, eds.), *Prog. Astronaut. Aeronaut., 57*, 97–123.

Davis L. W. and Bradstreet S. W. (1970) *Metal and Ceramic Matrix Composites*, pp. 73–100. Cahners, Boston.

Ernsberger F. M. (1969) Tensile and compressive strength of pristine glasses by an oblate bubble technique. *Phys. Chem. Glasses, 10*, 240–245.

Griggs D. T. (1967) Hydrolytic weakening of quartz and other silicates. *Geophys. J. Roy. Astron. Soc., 14*, 19–31.

Griggs D. T. and Blacic J. D. (1965) Quartz: Anomalous weakness of synthetic crystals. *Science, 147*, 292–295.

Ho D. and Sobon L. E. (1979) Exterrestrial fiberglass production using solar energy. In *Space Resources and Space Settlements* (J. Billingham and W. Gilbreath, eds.), NASA SP-428, pp. 225–232.

Johnson B. V., Roepke W. W., and Strebig K. C. (1973) Shear testing of simulated lunar soil in ultra high vacuum. *U. S. Bur. Mines, 7814*, 1–18.

LaCourse W. C. (1972) The strength of glass. In *Introduction to Glass Science* (L. D. Pye, H. J. Stevens, and W. C. LaCourse, eds.), pp. 451–543. Plenum Press, New York.

Loewenstein K. L. (1973) *The Manufacturing Technology of Continuous Glass Fibers*. Elsevier, Amsterdam, 280 pp.

Michalske T. A. and Freiman S. W. (1982) A molecular interpretation of stress corrosion in silica. *Nature, 295*, 511–512.

Mizutani H., Spetzler H., Getting I., Martin R. J., and Soga N. (1977) The effect of outgassing upon the closure of cracks and the strength of lunar analogues. *Proc. Lunar Sci. Conf. 8th*, pp. 1235–1248.

Pandit B. I. and Tozer D. C. (1970) Anomalous propagation of elastic energy within the Moon. *Nature, 226*, 335.

Phinney W. C., Criswell D., Drexler E., and Garmirian J. (1977) Lunar resources and their utilization. *Space-Based Manufacturing from Non-terrestrial Materials* (G. K. O'Neill and B. O'Leary, eds.), *Prog. Astronaut. Aeronaut., 57*, 97–123.

Rowley J. C. and Neudecker J. W. (1984) *In situ* rock melting applied to lunar base construction and for exploration drilling and coring on the Moon (abstract). In *Papers Presented to the Symposium on Lunar Bases and Space Activities of the 21st Century*, p. 77, NASA/Johnson Space Center, Houston.

Scholz C. H. (1972) Static fatigue of quartz. *J. Geophys. Res., 77*, 2104–2114.

Soga N., Spetzler H., and Mizutani H. (1979) Comparison of single crack propagation in lunar analogue glass and the failure strength of rocks. *Proc. Lunar Planet. Sci. Conf. 10th*, pp. 2165–2173.

Tittmann B. R., Clark V. A., and Spencer T. W. (1980) Compressive strength, seismic Q, and elastic modulus. *Proc. Lunar Planet. Sci. Conf. 11th*, pp. 1815–1823.

Williams R. J. and Jadwick J. J. (editors) (1980) *Handbook of Lunar Materials*, p. 113–116. NASA RP-1057. NASA/Johnson Space Center, Houston.

GUIDE TO USING LUNAR SOIL AND SIMULANTS FOR EXPERIMENTATION

J. H. Allton, C. Galindo Jr., and L. A. Watts

Northrop Services, Inc., P.O. Box 34416, Houston, TX 77234

The vision of a lunar base has stimulated experimentation needed for the planning and construction of lunar vehicles, habitats, and factories. The following discussion is a guide to facilitate the design and interpretation of technology experiments on lunar soil and lunar soil simulants. Lunar soil, once it is taken from the Moon for study in the laboratory, may not represent true *in situ* lunar conditions. The proposed simulated soils are different from genuine lunar soils in several important respects, mostly due to the effects of micrometeorites and solar wind on the Moon. However, these proposed simulants do replicate the lunar soil grain size distribution, gross mineralogy, and general chemical composition and are useful for studies of these properties. There are several reserves of lunar material that are suitable for tests requiring genuine lunar soil.

INTRODUCTION

Past studies have concentrated on unlocking scientific secrets of lunar soil. Extracts of scientific studies on chemistry and petrography of the 163 individual soils and an extensive bibliography are found in *Handbook of Lunar Soils* (Morris *et al.*, 1983). A review of lunar soil petrography is given by Heiken (1975). Although lunar soil chemistry is fairly well known, engineering properties and industrial reactions are not as well studied.

Mechanics and thermal information on *in situ* soil conditions was gathered by early researchers from television, surface photography, and measurements using a penetrometer and heat flow probe. Observations on the lunar surface include the Apollo lunar module descent engine blowing dust, depths of footprints on crater rims, the rover throwing dirt in the "grand prix," drilling, trenching, scooping, and raking. *In situ* properties are most relevant to the use of lunar soil for tunneling, heaping, and excavating, and as a substrate for buildings and vehicles. The properties of interest are the *in situ* bulk density and porosity. For surface activities similar to those conducted on the Apollo missions, these properties are probably known well enough.

Other properties, intrinsic to the soil grains, become important for those experiments where lunar soil is an active ingredient in a process. These properties include composition, rock form (crystalline, glassy), grain size, grain shape, grain strength, grain surface reactivity, dielectric constant, and magnetic susceptibility. Present interest includes experiments for extracting oxygen from the soil, melting or chemically reacting the soil for use as structural material, and growing organisms on the soil.

BRIEF DESCRIPTION OF LUNAR SOIL

Lunar soil can be described in familiar terrestrial terms as well-graded silty sands or sandy silts with an average particle size by weight between 0.040 and 0.130 mm (Carrier *et al.*, 1973). The density of *in situ* bulk lunar soil, as determined from large diameter core tube samples, is typically 1.4 to 1.9 g/cm^3. The bulk density increases with depth, and below 10–20 cm the soil is often at higher density than is required to support the overburden in lunar gravity (Carrier *et al.*, 1973). Spheres, angular shards, and fragile, reentrant, vesicular grains are among the diverse shapes found in most lunar soils. The most abundant particles composing the soil are igneous or breccia lithic grains, mineral grains, glass fragments, and the unique lunar agglutinates. Major lunar minerals are pyroxenes, anorthite, ilmenite, and olivine. Compositionally, the lunar soils fall into two broad groups: the highlands soils, which developed on anorthositic bedrock, and the mare soils, which developed on basaltic bedrock. The mare soils can be further subclassified as to high or low titanium content. Highlands soils are relatively enriched in aluminum and calcium, while mare soils are relatively enriched in iron, magnesium, and titanium. Average major element chemistry of these three types is given in Table 1.

Table 1. Major Element Chemical Composition of Lunar Soils and Soil Simulants

	Lunar Highlands Soils* (%)	Lunar Low Titanium Mare Soils[†] (%)	Lunar High Titanium Mare Soils[§] (%)	Hawaiian Basalt** (%)	High Titanium Mare Simulant[†] (%)
SiO$_2$	45.0	46.4	42.0	46.4	41.7
TiO$_2$	0.5	2.7	7.5	2.4	7.5
Al$_2$O$_3$	27.2	13.5	13.9	14.2	12.8
Fe$_2$O$_3$	–	–	–	4.1	3.7
FeO	5.2	15.5	15.7	8.9	12.8
MgO	5.7	9.7	7.9	9.5	8.5
CaO	15.7	10.5	12.0	10.3	9.2
Total	99.3	98.3	99.0	95.8	96.2

*Average composition of Apollo 16 soils compiled from *Handbook of Lunar Soils* (Morris, 1983).
[†]Average composition of Apollo 12 soils from Taylor (1975), p. 62.
[§]Average composition of Apollo 11 soils from Taylor (1975), p. 62.
**Composition of Hawaiian basalt HAW-11 from *Basaltic Volcanism on the Terrestrial Planets*, p. 166.
[†]Calculated composition from recipe in Table 2. Iron in ilmenite as FeO.

CHANGES IN SOIL FROM MOON TO LABORATORY

Soil cannot be removed from the surface of the Moon without altering at least some of the *in situ* characteristics such as bulk density and stratigraphy. The least physically disturbing way of sampling the lunar soil was with the large diameter core tubes used

Table 2. Changes in Soil from Moon to Lab

	Conditions	Changes
Moon	Impact–derived particle packing High vacuum	
Curatorial Facility	Dry nitrogen	Loss of original packing Adsorb water (minor)
Laboratory	Laboratory atmosphere	Adsorb water (major) Oxidation

on Apollo 15, 16, and 17 (Carrier *et al.*, 1971). Soil undergoes still further changes in the experimenter's laboratory (Table 2). On the lunar surface soil particles reside in a hard vacuum, free of water molecules and other atmospheric gases. The packing of particles is affected by continual meteorite bombardment. The dominant effect of this pounding is to pack the soil more tightly, although occasionally soil particles on the surface are ejected and then settle to a less dense configuration on crater rims (Carrier, 1973).

In the lunar sample curatorial facility, "pristine" samples are stored and handled only under dry nitrogen. Even so, small amounts of water and other gases are probably adsorbed on the highly reactive surfaces of lunar soil grains. The soil grains have lost their original packing during excavation, transit to Earth, and laboratory handling.

Furthermore, the ambient atmosphere of the experimenter's laboratory, with its relatively high water vapor and oxygen content, causes much more water to be adsorbed on the grain surfaces and some oxidation to occur. For example, the abundant metallic iron in lunar soil rusts easily.

SOME CRITICAL DIFFERENCES BETWEEN SIMULANTS AND LUNAR SOIL

Solar radiation and meteorite impacts, large and small, alter soil grains in ways that are difficult to duplicate on Earth. Also, lunar minerals are compositionally different, on a minor scale, due to the lack of volatile elements and reduced amounts of oxygen when the minerals were formed. Some of these unique lunar characteristics can be reproduced in very small quantities of simulant in experimental guns, charged particle beams, or furnaces. However, it is not practical to make usable quantities of simulants by these methods. Since simulants will probably be made using crushed, naturally–occurring minerals, they will be different from true lunar soil in several ways (Table 3).

Agglutinates, Iron Metal

Since the Moon has no atmosphere, very small meteorites impact the soil at high velocity, melting and shocking the rocky soil grains. Evidence of an impact on a 1 mm

Table 3. How Successful is a Simulant?

Can Simulate	Difficult to Simulate
Grain size distribution	Agglutinate glass with dispersed metal, grain shape
Gross mineral composition	Solar wind nuclei implantation
	Shock effects (grain strength)
General chemical composition	Mineral chemistry (reduced elements, no hydration)

diameter glass sphere taken from lunar soil is shown in Fig. 1. The splatters of glass from many repetitions of such micrometeorite impacts can glue tiny grains together in convoluted structures called agglutinates (Fig. 2). Iron metal blebs of 10 nm diameter are distributed throughout the agglutinic glass, making the glass magnetic. Agglutinates can make up over 50% of a mature lunar soil. This gluing together of smaller grains into larger ones is part of two competing processes, for impacts also break down soil grains into smaller ones.

Solar Wind

Because the Moon does not have a global magnetic field, high velocity nuclei from the solar wind impinge directly on small soil grains. These nuclei, of which hydrogen and helium are the most common, become implanted in the outer few angstroms of soil grains, creating an amorphous layer. In mature soils this solar wind hydrogen can exceed 100 ppm.

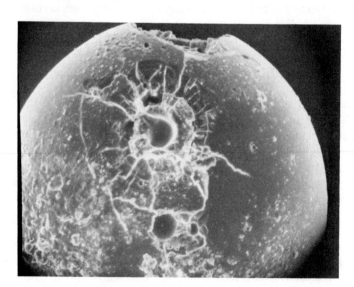

Figure 1. One millimeter diameter lunar glass sphere with micrometeorite impact pit. Photo courtesy of D. S. McKay (S-71-48106).

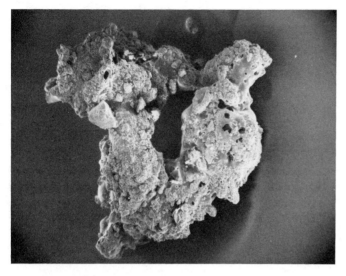

Figure 2. One millimeter diameter agglutinate. Photo courtesy of D. S. McKay (S-71-24575).

Shock Effects

The shock effects of meteorite impacts are commonly retained in lunar soil grains. Impacts fracture and weaken the mineral grains found in the lunar soil.

Mineral Chemistry

The major lunar minerals (anorthite, pyroxene, ilmenite, olivine) are similar in gross aspects to their terrestrial counterparts. However, the lunar minerals do not contain bound water in the crystal structure and have not been altered by hydration reactions on grain boundaries. Due to extremely low oxygen fugacities at the time of crystallization, several elements in lunar minerals are found in a more reduced state. Combined iron is almost totally ferrous iron, and iron metal may be found in interstitial phases and dispersed in glass. Titanium and chromium occur in the more reduced valence states of +3 and +2, respectively. Lunar ilmenite does not contain hematite as many terrestrial ilmenites do.

LUNAR FINES AS EXPERIMENTAL SAMPLES

Lunar samples are allocated very sparingly, and investigators are encouraged to work on the smallest possible samples. For example, scientific investigators typically determine major element chemistry from only 50 mg of material. Since engineering and industrial studies often require much larger sample size, experimenters must, when possible, scale down their experiments and make use of simulants.

Any lunar samples that may be available for technology studies will probably come from the residue of fines left in the Apollo collection bags. Early missions collected fewer, but larger, soil samples, On later missions, samples were smaller, more carefully chosen to sample different phenomena, and placed in individual bags.

Table 4. Grams of Sample Bag
Residues from Apollo Missions

Apollo 11	55 g
Apollo 12	–
Apollo 14	225 g
Apollo 15	335 g
Apollo 16	1808 g
Apollo 17	3525 g
Total	5948 g

The residue of fine material remaining in the rock and soil sample bags (about 5 kg total) could be pooled and homogenized for each mission except Apollo 12 (Table 4). This would result in a mixture of fines, representing an average chemical composition for each site of large enough size to serve as a standardized sample. However, these samples would not be representative of a true soil since rock dust would be admixed. Soil maturity (degree of exposure to micrometeorites and solar wind), as determined by fine-grained metallic iron content (Morris, 1978), would give a general indication of proportion of soil to rock dust. Investigators concerned with agglutinate, metal, and solar wind content could then make adjustments for under-representation of these components in the pooled fines.

As a standard sample, these pooled fines would be of known composition, grain size distribution, and maturity. This would be advantageous for comparisons among experiments. Use of these bag residues would be an efficient use of the Apollo collection, since their mixed origin makes them less valuable scientifically.

SIMULANTS FOR EXPERIMENTS

Since the properties to be simulated and degree of fidelity required are different for laboratory experimentation than for testing equipment and structures, simulants for these two uses are discussed separately. In general, simulants for laboratory experimentation require greater fidelity to chemical and mineral composition, in addition to grain size distribution. In creating simulants, costs must be weighed against benefits of increasing fidelity to lunar soil. The approach described below is a "middle-of-the-road" effort, when compared to the low cost extreme of using the nearest sand or crushed rock and the high cost extreme of creating micrometeorite impacts and solar wind implantations one by one in experimental guns and ion beams. Simulating the lunar soil for laboratory experimentation is approached from three aspects: soil grain size distribution, soil particle type distribution, and particle chemistry.

Grain Size Distribution

Grain size distribution curves have been compiled that encompass most Apollo soils (Carrier *et al.*, 1973). The grain size distribution of simulants should be created with the fewest sieve sizes that adequately characterize the grain size distribution curve and yet are practical to use. Thus, simulant composition should be defined as 90% finer than

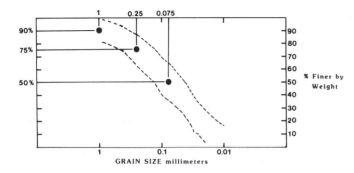

Figure 3. Grain size distribution curves encompassing most Apollo soils. Plot and data adapted from Carrier et al. (1973). Percent finer by weight at sieve sizes 1.0, 0.25, and 0.075 mm are used to define simulant characteristics.

1 mm, 75% finer than 0.250 mm, and 50% finer than 0.075 mm (Fig. 3). The distribution curve is not precisely simulated below 0.075 mm, because it is impractical to sieve large volumes of rock below this limit. Therefore, it is important to calibrate the pulverizing process in the small size range.

Particle Type Distribution

Nearly all particles comprising the lunar soil are lithic (chiefly breccia and poikilitic rocks in the highlands and breccia and basalt in the maria), mineral or glass fragments or agglutinates. The simulation is simplified by using crushed basalt or minerals to substitute for the lithic and mineral fragments and by using crushed glass to substitute for glass fragments and agglutinates. Although lunar particle type distribution varies with maturity of the soil, source rock type, and particle size, both mature highlands and mature mare soils can be approximated with a mineral or rock to glass ratio of 1:1 for sieve fraction <0.250 mm and a ratio of 3:1 for sieve fraction >0.250 mm. [These proportions were calculated from data for sample 60010 given in McKay et al. (1977) and from data for sample 71016 given in Heiken and McKay (1974).]

A Highlands Simulant

The target chemical composition for the highlands simulant is the average of Apollo 16 soils as given in Table 1. Normative calculations (Chayes and Metais, 1964), based only on Si, Al, Fe, Mg, and Ca indicate that a 3:1 weight ratio of anorthite ($CaAl_2Si_2O_8$) to pyroxenes (of mixed composition) would approximate this chemical composition. Adding pyroxenes raises both the iron and magnesium content.

Unaltered anorthite is not common on Earth. The least altered anorthite crystals can be found associated with frothy glass near some andesitic volcanoes, such as Miakejima near Tokyo. Anorthite also is found mixed with other minerals in andesitic areas and as anorthosite rock in layered intrusions, of which the Stillwater complex in Montana is an example.

The orthopyroxene bronzite is a close practical substitute for the norm-calculated pyroxene ratios of hypersthene (orthopyroxene) to diopside (clinopyroxene) of 6:1.

Glass of the highlands composition given in Table 1 can be made by Corning Glass Company by the dri-gauge method (Minkin et al., 1976).

In summary, a highlands simulant can be made by combining crushed anorthite, pyroxene, and synthetic glass in proportions based on grain size, lithic or glassy character, and chemistry. A sample recipe of this type is given in Table 5.

A High Titanium Mare Simulant

The target chemical composition for the high titanium mare simulant is the Apollo 11 soil average given in Table 1. The lithic and mineral component can be approximated by terrestrial basalts plus ilmenite. HAW-11 (Basaltic Volcanism Study Project, 1981), whose chemical composition is also given in Table 1, is an example of a suitable basalt. Combining this basalt with ilmenite ($FeTiO_3$) in a 9:1 proportion raises the titanium content of the mixture to that of the Apollo 11 soil. The resulting mixture also improves the calculated fit to Si, Fe, and Mg percentages of the Apollo soil (Table 1).

Glass of high titanium mare composition can also be made from melting oxides. Naturally occurring, basaltic-composition volcanic glass, such as is found in Hawaii could be used, but probably will not have a titanium concentration as great as the high titanium mare soils.

In summary, a high titanium mare simulant can be made by combining crushed Hawaiian basalt HAW-11, ilmenite, and synthetic glass in proportions based on grain size, lithic or glassy character, and chemistry. A sample recipe of this type is given in Table 5.

SIMULANTS FOR TESTING EQUIPMENT AND STRUCTURES

Important parameters to simulate for testing equipment and structures include bulk density and porosity. Grains of correct size distribution and specific gravity are needed,

Table 5. Recipes for Lunar Soil Simulants

	<0.075 mm	0.075 to 0.25 mm	0.25 to 1.0 mm	>1.0 mm
Sample Highlands Simulant: Anorthite to Pyroxene Ratio 3:1*				
Anorthite	18.8	9.4	8.4	5.6
Pyroxene	6.2	3.1	2.8	1.9
Glass	25.0	12.5	3.8	2.5
Sample High Titanium Mare Simulant: Basalt to Ilmenite Ratio 9:1*				
Basalt	22.5	11.3	10.1	6.8
Ilmenite	2.5	1.2	1.1	0.7
Glass	25.0	12.5	3.8	2.5

*To make 100 g of simulant, mix components by grams indicated in table.
The same grain size fractions and lithic to glass ratios were used for both simulants: >1 mm =0.10; 0.25-1 mm =0.15; 0.075-0.25 mm =0.25; <0.075 mm =0.50 (total = 1.00). The lithic to glass ratio for >0.25 mm =3:1 and for <0.25 mm =1:1.

so chemistry and mineralogy are less important. Also, since much larger quantities of simulant are needed (tons), crushing and grinding of a single component, usually basalt, on commercial size equipment would be used. Nearly 2500 kg of a basalt simulant was fabricated and characterized for testing the lunar rover (Mitchell and Houston, 1970; Green and Melzer, 1971). Crushed basalt also has been used for lunar resource utilization studies (Steurer, 1982).

The importance of packing the simulant properly after grinding is illustrated in the testing of the Apollo lunar surface drill by Martin Marietta. The simulant, used during design of the drill, was packed to a lesser density than was actually encountered on Apollo 15. The surprisingly dense soil at Hadley Rille made the drilling effort more difficult than expected. The density of the entire Apollo 15 drill sample was 1.75 g/cm^3, but the deepest section was 1.93 g/cm^3 (Carrier, 1974). Therefore, in preparation for subsequent drill testing, engineers recompacted the simulant to the densities encountered at Hadley Rille. The difficult task of achieving this high density for crushed vesicular glass and lithic particles was accomplished using electric tampers to compress each shallow layer (3–6 inches thick) as it was added to the test bed (Britton, personal communication, 1985).

CONCLUSIONS

When planning experiments for activities to take place on the Moon, investigators should remember the following.

1. Lunar soil in the laboratory does not accurately represent lunar *in situ* conditions. The Apollo soils have lost their original particle packing and have adsorbed volatiles.

2. Simulants can be made by ordinary means that reproduce specific properties of lunar soil such as grain size distribution, gross mineral composition, or general chemical composition.

3. Certain lunar soil characteristics are difficult to duplicate in simulants. These include agglutinates with their convoluted shapes and iron metal, implanted solar wind nuclei, impact shock effects on grains, and minerals with reduced elements.

4. A very small amount of lunar soil will be available for experimentation. Investigators should scale down their experiments and use simulant whenever possible.

Acknowledgments. *This paper was substantially improved through suggestions given by the following thoughtful reviewers: D. P. Blanchard, D. S. McKay, W. W. Mendell, and R. J. Williams. Their comments are much appreciated. This work was supported by contract NAS-9-15425.*

REFERENCES

Basaltic Volcanism Study Project (1981) *Basaltic Volcanism on the Terrestrial Planets.* Pergamon, New York. 1286 pp.

Carrier W. D. III (1973) The relative density of lunar soil. *Proc. Lunar Sci. Conf. 4th,* pp. 2403–2411.

Carrier W. D. III (1974) Apollo drill core depth relationships. *Moon, 10,* 183–194.

Carrier W. D. III, Johnson S. W., Werner R. A., and Schmidt R. (1971) Disturbance in Samples recovered with the Apollo core tubes. *Proc. Lunar Sci. Conf. 2nd,* pp. 1959–1972.

Carrier W. D. III, Mitchell J. K., and Mahmood A. (1973) The nature of lunar soil. *J. Soil Mech. Found. Div.*, American Society of Civil Engineers, 99, 813–832.

Chayes F. and Metais D. (1964) On the relationship between suites of CIPW and Barth-Niggli Norms. *Carnegie Inst. Washington, Year Book 63*, 193–195.

Green A. J. and Melzer K. J. (1971) *Performance of Boeing LRV Wheels in a Lunar Soil Simulant. Rep. 1, Effect of Wheel Design and Soil.* Tech. Rep. M-71-10, U.S. Army Engineer Waterways Experiment Station, Vicksburg. 83 pp.

Heiken G. (1975) Petrology of lunar soils. *Rev. Geophys. Space Phys., 13*, 567–587.

Heiken G. and McKay D. S. (1974) Petrography of Apollo 17 soils. *Proc. Lunar Sci. Conf. 5th*, pp. 843–860.

McKay D. S., Dungan M. A., Morris R. V., and Fruland R. M. (1977) Grain size, petrographic, and FMR studies of the double core 60009/10: A study of soil evolution. *Proc. Lunar Sci. Conf. 8th*, pp. 2929–2952.

Minkin J. A., Chao E. C. T., Christian R. P., Harris E. E., and Norton D. R. (1976) Three synthetic lunar glasses. *Meteoritics, 11*, 167–171.

Mitchell J. K. and Houston W. N. (1970) Mechanics and stabilization of lunar soils. In *Lunar Surface Engineering Properties Experiment Definition*, Final Report: Vol. 1 of 4, Space Sciences Laboratory Series 11, Issue 10. University of California, Berkeley. 166 pp.

Morris R. V. (1978) The surface exposure (maturity) of lunar soils: Some concepts and I_s/FeO compilation. *Proc. Lunar Planet. Sci. Conf. 9th*, pp. 2287–2297.

Morris R. V., Score R., Dardano C., Heiken G. (1983) *Handbook of Lunar Soils.* NASA Planetary Materials Branch Publication No. 67, NASA/Johnson Space Center, Houston. 913 pp.

Steurer W. H. (1982) *Extraterrestrial Materials Processing.* JPL Publication 82-41, pp. 27–39. Jet Propulsion Laboratory, Pasadena.

Taylor S. R. (1975) *Lunar Science: A Post Apollo View.* Pergamon, New York. 372 pp.

FRACTIONAL DISTILLATION IN A LUNAR ENVIRONMENT

Donald R. Pettit

Los Alamos National Laboratory, MS P952, Los Alamos, NM 87545

The establishment of a permanent lunar base will undoubtedly employ distillation operations as a routine practice. Reclamation of vital fluids along with products from chemical processes will lend itself to fractional distillation. The lunar environment, with reduced gravity and pressure, will dictate design modifications and offer some pleasant advantages. Column area will increase to maintain the same flow rates as Earth-based counterparts. Plate efficiencies can increase, allowing shorter columns. Thermal insulation will be facilitated by the lunar atmosphere, as well as low pressure "vacuum" distillation.

INTRODUCTION

With the development of a reliable space transport system, extraterrestrial engineering is becoming a respectable field of endeavor. The detailed engineering for maintaining a space station or lunar base, along with possible manufacturing processes, presents a challenge for scientists and engineers.

The establishment of a permanent lunar base will employ separation techniques as part of routine necessity. Recycling precious body fluids, in addition to solvents and products of chemical manufacture, could lend itself to fractional distillation. The lunar environment, with reduced gravity and pressure, will offer some unique possibilities for clever designs with a concurrent struggle to overcome the hardships.

Why use an age-old process like distillation when there are many "space age" separation techniques (such as membrane technology)? Distillation uses simple, hearty equipment that operates in a dependable manner, equipment that is not easily damaged if operation is in error. Many of the construction materials could ultimately be derived from lunar sources, saving the transportation costs of Earth-based goods. Most important, distillation uses heat energy as the main driving force for separation.

On the Moon, shaft and electrical energy will be at a premium. Whether from solar, combustion, or nuclear sources, heat energy will be more abundant and more efficiently obtained than shaft or electrical energy. In the allocation of such a valuable commodity, it makes good sense to employ processes that utilize heat directly, saving the shaft and electrical energy for those processes that cannot be driven any other way. Waste heat from ongoing processes may be of a quality suitable for driving distillation, thus realizing further economy.

The lunar environment will offer some unique advantages for distillation processes. Vacuum distillation will be possible due to the cryogenic temperatures available on the Moon. With vacuum distillation lower heat loads are realized with cleaner separations and with the possiblity of breaking azeotrope systems. With a radiation barrier the distillation columns will be essentially enclosed in a giant "thermos bottle," realizing very low heat losses, so the energy injected into the processes will be used efficiently for driving the separation. The constant nature of the lunar atmosphere will facilitate the process control, resulting in consistent product output and quality. In contrast, heat loss from Earth-based columns is of major concern, especially coupled with changing weather patterns that complicate the process control.

Fractional distillation is a mature engineering field backed with years of experimentation that resulted in practical design. The approach taken here is to utilize this existing knowledge, coupled with dimensional analysis and scaling arguments, to modify the design of Earth-based columns for lunar operation.

BACKGROUND IN DISTILLATION EQUIPMENT

The anatomy of a fractional distillation process is shown in Fig. 1. The heart of the unit consists of the fractionating column, basically a tall vertical pipe where the liquid and vapor experience intimate contact and where mass transfer between phases is effective, thus achieving component separation. At the bottom of the column is the reboiler, a vat of boiling liquid from which the vapors flow upward into the bottom of the column while the condensed liquid emanating from the column flows downward into the reboiler. The heat for driving the separation is injected into the reboiler.

At the top of the column is an overhead condenser that converts the enriched vapor effluent into a liquid, rejecting heat into the environment. A portion of this liquid is tapped off as head product, the balance being returned to the top of the column as reflux. The ratio of the amount of liquid returned to the amount tapped off is called the reflux ratio and is an important quantity in the design of a distillation process.

Most distillation processes are designed to operate on a continuous basis, unlike the familiar connotations of a moonshiner's batch still for making "white lightning." A continuous feed is introduced into the column at the point where the concentration of components in the feed matches that in the column. The enriched head product is continuously withdrawn from the top while depleted bottoms are continuously removed from the reboiler.

The design of what goes inside the column to achieve the intimate contact between vapor and liquid is somewhat of an art as well as a science. The column can be filled with plates, each having a standing pool of liquid that vapors bubble through, giving discrete or stage-wise contact. The column can be packed with irregular objects, providing continuous contact between the liquid trickling down and vapor percolating up. Presently, the most popular column design uses plates, with future trends leaning towards packed columns. This paper will deal with plate-type columns, with packed columns being the subject of another study.

There are many types of plate designs, with sieve tray plates being the most common. The sieve tray plate will be considered here initially, with the scaling arguments derived being general for most types of plate columns. Figure 2a shows a cross section of a

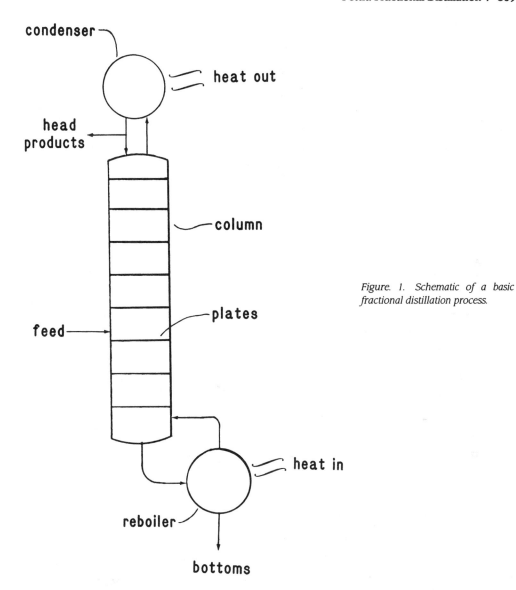

Figure. 1. Schematic of a basic fractional distillation process.

portion of a column containing a sieve tray plate, while Fig. 2*b* shows the top view. A pool of liquid, usually 10–20 cm deep, stands on top of a perforated plate with holes ranging from 4–15 mm in diameter. The liquid is kept from weeping through the holes by a steady stream of vapor pushing upward, emanating from the liquid on the plate immediately below. The vapor, with intimate contact, bubbles through the pool of liquid and thereby condenses. The heat released upon condensing vaporizes a corresponding amount of liquid, which pushes upwards as vapor to bubble through the next higher

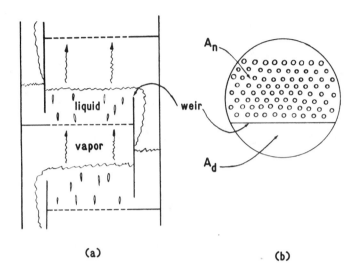

Figure 2. (a) Cross section of a column with sieve tray plates; (b) top view of a sieve tray plate.

plate. Insulation is critical in the operation, because the energy for vaporizing the liquid on a plate comes from the condensing vapors; any heat loss hinders this interplay.

The liquid level is maintained by a lip or weir, over which the liquid can splash and flow down a passage called a downer to the next lower plate. A liquid seal is provided so the vapors cannot flow "up the downer" and are forced to percolate through the sieve tray holes. The weir usually cuts some chord length of the column cross section, separating the downer area A_d, from the net area of the sieve tray, A_n (A_n is based on the sieve tray area, not the hole area), as seen in Fig. 2b.

To achieve a given head product purity, thermodynamics will dictate the ideal number of plates, assuming equilibrium is reached. A plate efficiency, E, is used to determine the number of real plates from this ideal case. Plate spacing is usually between 30–50 cm, so column height is then specified. Typical industrial columns may range from 1/2–2 m in diameter, 5–50 m high.

LUNAR MODIFICATIONS

Dimensional analysis and scaling arguments can be used to modify Earth-based columns for use on the Moon. The prime consideration is the reduction of gravity to one-sixth of that found on the Earth. Gravity-driven buoyant forces are responsible for moving the two-phase fluid system and will affect A_n, A_d, and E. Vapor-liquid thermodynamics will remain the same between the Earth and the Moon so that the number of ideal plates needed for a given separation will remain constant. Hydraulic similarity between Earth-based plates and Moon-based ones should be maintained through dimensional analysis so that the column operation will remain consistent with Earth-bound operations.

In the design of a distillation column, the feed rate and desired component separation are given; column pressure, temperature, number of ideal plates, and reflux ratio are then specified from a blend of thermodynamic and economic arguments. With these parameters fixed, the internal flow rates of vapor and liquid are also known. The column diameter is dictated by the A_n and A_d required to handle these internal flows, and the height is specified by the number of real plates calculated from the ideal number and E.

The net column area A_n is correlated to the internal volumetric gas (vapor) flow rate Q_G. The gas velocity V_F is defined as Q_G/A_n and is given by Treybal (1980) as

$$V_F = C_F \sqrt{\frac{\rho_L - \rho_G}{\rho_G}} \tag{1}$$

where ρ_L and ρ_G are the liquid and gas phase densities and C_F is the flooding coefficient, a constant determined from the details of plate geometry.

Equation (1) has no theoretical derivation; it is based on empirical correlation of experimental data for the prevention of droplet entrainment in the rising vapor. Realizing that such criterion is based on a balance of forces experienced by the droplets, it is recognized that the $(\rho_L - \rho_G)/\rho_G$ term in (1) is due to gravity-driven buoyant force between the liquid droplet and the gas. The density term must be multiplied by g, the acceleration due to gravity, in order to render the quantity into a proper buoyant force, which would certainly result if a theoretical derivation of (1) could be undertaken. Since (1) is developed empirically from Earth-based data, the acceleration due to gravity, which is not considered a separate parameter, would be buried in the flooding coefficient by the mechanics of the correlation process. It is not expected to find the gravitational acceleration anywhere in the equation. Therefore, the effect of gravity must enter in the flooding coefficient, resulting in V_F being proportional to $\sqrt{g}$.

The terminal velocity of a bubble in liquid (or a droplet in gas) has a well-known solution (Bird et al., 1960) and can be used to reinforce the arguments applied (1), realizing that such analysis is an oversimplification of the actual flooding process. Considering a sperical-shaped bubble

$$F_d = \frac{\pi \rho_L V_B^2 d^2 \theta}{8} \tag{2}$$

will be the drag force F_d, where V_B is the bubble velocity, d is the diameter, and θ is the drag coefficient. The buoyancy force F_b will be as follows

$$F_b = \frac{(\rho_L - \rho_G) g \pi d^3}{6} \tag{3}$$

Equating the drag force to buoyant force and solving for the bubble velocity gives

$$V_B = \sqrt{\frac{4\ dg}{3\theta}\left(\frac{\rho_L - \rho_G}{\rho_L}\right)} \tag{4}$$

For Reynold's numbers greater than 10, which applies to the flow regimes found in plate–type columns, the drag coefficient is constant.

This balance is essentially the same for a droplet falling in gas, except that the density term is $(\rho_L - \rho_G)/\rho_G$ because the drag force in (2) is based on the external flow of the medium around a sphere. When applied to a falling droplet, (4) is remarkably similar in form to (1). For the onset of flooding, the gas velocity V_F must be of the order of the droplet velocity, which yields V_F proportional to $\sqrt{g}$ the same as the result deduced from the empirical correlation in (1). Substituting Q_G/A_n for V_F and solving for A_n gives the proportion

$$A_n \propto \frac{Q_G}{\sqrt{g}} \tag{5}$$

For the specified feed, reflux ratio, and column pressure, Q_G will be essentially the same between the Earth and the Moon, so the net area ratio will scale as

$$\frac{A_n\big|_M}{A_n\big|_E} = \sqrt{\frac{g_E}{g_M}} \tag{6}$$

where the subscripts E and M differentiate between the Earth and the Moon.

The scaling of downer area A_d will be dictated by the effects of gravitational forces on liquid flowing downwards in a closed conduit. Considering laminar flow in a vertical pipe, the liquid flow rate Q_L can be expressed (Bird *et al.*, 1960) as

$$Q_L = \frac{(\pi R^2)^2 \, \rho_L g}{\pi 8 \mu_L} \tag{7}$$

where R is the radius and μ_L is the liquid viscosity. In general, for a closed conduit, Q_L will be proportional to gA_d^2. For a fixed volume of liquid flow, A_d will be as follows

$$A_d \propto \frac{1}{\sqrt{g}} \tag{8}$$

which gives the scaling ratio

$$\frac{A_d\big|_M}{A_d\big|_E} = \sqrt{\frac{g_E}{g_M}} \tag{9}$$

From (6) and (9), the lunar values of A_n and A_d increase by a factor of 2.45 in order to compensate for the one-sixth lunar gravity. For an earthly column 1 m in diameter, the corresponding lunar column would be 1.6 m.

The formation of bubbles with their corresponding interfacial surface area and rising velocity are the most important hydraulic concerns that affect plate efficiency. On a real plate, bubble–liquid interactions are complex. A simplified approach will be used to evaluate the major role of gravitational forces where single bubbles are rising in a body of liquid.

Assuming each plate is well mixed, the efficiency can be expressed as the proportion (Treybal, 1980)

$$E \propto 1 - e^{-k_L ah/V_B} \tag{10}$$

where E is called the Murphree plate efficiency, k_L is the bubble mass transfer coefficient base on the liquid phase, V_B is the bubble velocity, a is the total interfacial surface area, and h is the plate liquid depth. Equation (10) yields the ratio

$$\frac{\ln(1-E)\big|_M}{\ln(1-E)\big|_E} = \left(\frac{a_M}{a_E}\right)\left(\frac{h_M}{h_E}\right)\left(\frac{k_L\big|_M}{k_L\big|_E}\right)\left(\frac{V_B\big|_E}{V_B\big|_M}\right) \tag{11}$$

Based on penetration theory, k_L for a rising bubble is equal to (Treybal, 1980)

$$k_L = \left(\frac{D_{ab}}{\pi t}\right)^{1/2} \tag{12}$$

where D_{ab} is the diffusion coefficient and t is a fluid packet–bubble contact time. The contact time will be proportional to bubble diameter divided by bubble velocity, which gives

$$k_L \propto \left(\frac{V_B}{d}\right)^{1/2} \tag{13}$$

The diffusion coefficient is independent of gravity, thus being dropped as an argument. Equation (13) yields the mass transfer coefficient ratio

$$\frac{k_L\big|_M}{k_L\big|_E} = \left(\frac{d_E}{d_M}\right)^{1/2} \left(\frac{V_B\big|_M}{V_B\big|_E}\right)^{1/2} \tag{14}$$

which combined with (11) gives

$$\frac{\ell n\,(1-E)\big|_M}{\ell n\,(1-E)\big|_E} = \left(\frac{a_M}{a_E}\right)\left(\frac{h_M}{h_E}\right)\left(\frac{d_E}{d_M}\right)^{1/2}\left(\frac{V_B\big|_E}{V_B\big|_M}\right)^{1/2} \tag{15}$$

The rising velocity of a bubble has already been evaluated in (4) and is proportional to $\sqrt{dg}$.

The ratio of bubble velocity between the Earth and the Moon will be as follows

$$\frac{V_B\big|_M}{V_B\big|_E} = \sqrt{\left(\frac{g_M}{g_E}\right)\left(\frac{d_M}{d_E}\right)} \tag{16}$$

where the bubble diameter ratio has been included as a possible adjustable parameter.

Consider a bubble forming from gas percolating upwards through a plate hole into a body of liquid. It is important to determine the dependence of bubble mass (hence surface area) to hole diameter and the acceleration due to gravity coupled with the governing fluid properties. In the flow regime for bubble formation typically found on plates, surface tension has the dominating effect with the dependence of fluid viscosity being small. Using the Buckingham Pi method of dimensional analysis, the dimensionless pi group that arises is as follows

$$\pi = \frac{Mg}{\sigma D} \tag{17}$$

where M is the bubble mass, σ is the vapor-liquid surface tension, and D is the plate hole diameter. In order to assure hydraulic similarity, this dimensionless group is held constant between the Earth and the Moon, giving

$$\frac{Mg}{\sigma D}\bigg|_E = \frac{Mg}{\sigma D}\bigg|_M \tag{18}$$

There are several possibilities for juggling the parameters described in (15), (16), and (18) in order to scale the plates and determine their efficiencies.

Case I.   Constant Bubble Mass

The most likely choice is to maintain constant bubble mass between the Earth and the Moon, which will assure the same bubble diameter and interfacial surface area for mass transfer. The velocity ratio from (16) will then be the following:

$$\frac{V_B\big|_M}{V_B\big|_E} = \left(\frac{g_M}{g_E}\right)^{1/2} \tag{19}$$

and

$$\frac{\ln(1-E)\big|_M}{\ln(1-E)\big|_E} = \left(\frac{h_M}{h_E}\right)\left(\frac{g_E}{g_M}\right)^{1/4} \tag{20}$$

will be the plate efficiency.

The bubble velocity will be 41% less than that on the Earth due to the one–sixth gravity on the Moon. For the same liquid depth on the plates the contact time will be longer, thus increasing the efficiency and requiring fewer plates for a given separation and a corresponding reduction in column height. The liquid depth could be reduced on lunar plates to maintain the same efficiency, so the number of plates would remain unchanged with plate spacing being reduced. From consideration of plate maintenance and column operation the standard spacings are the most practical, so liquid depth should be kept the same, realizing a shorter column from the increased efficiency. For constant liquid depth, Table 1 shows typical Earth plate efficiencies and their corresponding lunar efficiencies given by (20), which are enhanced by an average of 25%.

Table 1.   Earth and Lunar Plate Efficiencies for Constant Bubble Mass and Liquid Depth

$E\big|_E$	$E\big|_M$
0.4	0.55
0.5	0.66
0.6	0.76
0.7	0.85
0.8	0.92

If bubble mass is to remain the same, (18) can be used to determine the hole diameter in the lunar plates, yielding

$$\frac{D_M}{D_E} = \frac{g_M}{g_E} \tag{21}$$

According to (21), the plate hole diameter for equal bubble masses will have to be six times smaller on the Moon; instead of holes 4–15 mm in diameter, the corresponding lunar perforations will be 0.67–2.5 mm. The pressure drop caused by the smaller holes could possibly increase, preventing the column from operating under vacuum conditions. From Treybal (1980), the pressure drop due to the perforated plate P, is proportional to

$$P \propto \frac{V_h^2}{g} \tag{22}$$

where V_h is the gas velocity through the holes. For holes placed on the corners of equilateral triangles and for equivalent hole diameter to pitch ratios, the increased value of A_n can increase the available hole area, decreasing the gas hole velocity. This can result in a pressure drop on the same order as Earth-based plates. In some instances it may be impossible to specify the smaller holes needed for constant bubble mass without dramatically increasing the pressure drop, making Case I an impractical approach.

Case II. Constant Plate Hole Diameter

For pressure drop consideration, the hole diameter in the plates will remain the same between the Earth and the Moon. From (18) the bubble mass ratio will then be as follows

$$\frac{M_M}{M_E} = \frac{g_E}{g_M} \tag{23}$$

which corresponds to lunar bubbles with six times the mass of earthly ones.

It follows from (23) that

$$\frac{d_M}{d_E} = \left(\frac{g_E}{g_M} \right)^{1/3} \tag{24}$$

is the bubble diameter ratio. The interfacial surface area can be approximated by the area of a single bubble times the number of bubbles. For a fixed Q_G, the number of bubbles, N, will scale inversely with the bubble mass

$$\frac{N_M}{N_E} = \frac{g_M}{g_E} \tag{25}$$

which gives an interfacial surface area ratio of

$$\frac{a_M}{a_E} = \left(\frac{d_M}{d_E}\right)^2 \left(\frac{N_M}{N_E}\right) = \left(\frac{g_M}{g_E}\right)^{1/3} \tag{26}$$

Using (16) and (24), it follows that

$$\frac{V_B\big|_M}{V_B\big|_E} = \left(\frac{g_M}{g_E}\right)^{1/3} \tag{27}$$

will be the bubble velocity ratio.

For constant hole diameter, the lunar plates will produce bubbles with 1.8× the diameter and 55% of the rising velocity and interfacial surface area. Combining (15), (24), (26), and (27) gives

$$\frac{\ln(1-E)\big|_M}{\ln(1-E)\big|_E} = \left(\frac{g_M}{g_E}\right)^{1/3} \tag{28}$$

for constant liquid depth. Table 2 shows the lunar plate efficiencies given by (28). The efficiencies decrease by an average of 34%, primarily the result of a significant decrease in the surface area due to the larger bubble diameter.

Table 2. Earth and Lunar Plate Efficiencies for Constant Plate Hole Diameter and Liquid Depth

$E\big|_E$	$E\big|_M$
0.4	0.25
0.5	0.32
0.6	0.40
0.7	0.48
0.8	0.59

Case III.   Constant Bubble Velocity

Another scaling criterion would be to maintain constant bubble velocity between the Earth and the Moon. From (16), the lunar bubble diameter would have to be 6× larger, which corresponds to a bubble with 216× the mass. Equation (18) dictates a lunar plate hole 36× larger, which does not yield a practical engineering design.

THE LUNAR ENVIRONMENT

The lunar atmosphere has a pressure of 10^{-12} torr, which for most considerations is a total vacuum. One would initially think the way to maintain a vacuum distillation process is to utilize the lunar atmosphere as a giant sink, but there are several reasons why this cannot be done. Purging materials into the lunar atmosphere would be a dreadful waste of resources; these materials, especially organics, will be too valuable to lose even a few percent. The void of the lunar atmosphere itself is also a valuable resource (it is noteworthy to point out that the absence of anything can be a resource). Many scientific investigations can capitalize on the combination of a gravitational setting with a vacuum environment, and the scientific value of a lunar base would significantly decrease if this atmosphere were to be contaminated.

Vacuum distillation will be maintained through the use of the cryogenic temperatures available on the Moon. Temperatures as low as 59 K can be obtained through radiation into space. The column pressure is specified by the lowest temperature available in the overhead condenser; in a lunar environment this will correspond to as low a column pressure as desired. Even "fixed gases" like oxygen, nitrogen, and carbon dioxide can be condensed, eliminating the need for vacuum pumps. The extra cost for vacuum distillation will be in the capital equipment needed to handle the radiation heat loads.

An example of a lunar distillation process would be the production of ethanol from fermentation of organic wastes. The column pressure would be maintained so the fermenter functions as a reboiler, where the alcohol is continuously boiled off at a temperature for optimum yeast growth. The azeotrope could be broken under the low pressure so absolute alcohol would be produced. A two-stage overhead condenser would first remove the condensable vapors, with a second-stage cryogenic condenser that condenses the carbon dioxide (as a solid) and any other fixed gases.

SUMMARY

The establishment of a permanent lunar base will offer some interesting possibilities for the design of distillation processes. The lunar environment will make possible convenient vacuum distillation and will facilitate column insulation. For a lunar column, the net plate and downer areas will increase by a factor of 2.45. The lunar plate efficiencies will either increase by about 25% (for constant bubble mass) or decrease by 34% (for constant hole diameter), the choice depending on the imposed engineering constraints.

REFERENCES

Bird R. B., Stewart W. E., and Lightfoot E. N. (1960) *Transport Phenomena*, Wiley, New York. 332 pp.
Treybal R. E. (1980) *Mass-transfer Operations*, pp. 139–210. McGraw-Hill, New York.

LUNAR MACHINING

William Lewis

Electrical and Computer Engineering Department, Clemson University, Clemson, SC 29634-0915

Chip-making machine tools have traditionally been used to shape iron. Regolith is several tenths percent iron nodules by mass, which could be melted by solar process heat, then cast. Chip-making machine tools may well be used to further shape these castings. Little serious consideration has been given to machine tool design within a lunar context, however. This is a first survey of the problems and opportunities of lunar machining. A conceptual framework for machine tool design is given, then applied assuming lunar operations. It is concluded that there will be a need for small machines operating in shirtsleeve environments, that larger machines will have to be outside in vacuum because they require large, rigid foundations that can sink vibration, and that productivity will be extremely important due to very high labor costs.

Serious consideration is being given to the establishment of a lunar base during the first half of the next century (von Puttkamer, 1976; Duke, 1984). Whether a scientific, industrial, or growth rationale is selected, machine tools will almost certainly be used at the base site. Machine tools may first be used in repair and maintenance of the base itself. Later, more machines might be brought in as support for construction of scientific instruments, construction of industrial plants, construction of more habitats, and manufacture of capital goods.

It is of some intellectual and practical interest to consider the design challenges that must be answered before machine tools can be used in a lunar environment. This short paper will content itself with identifying design challenges and opportunities. Ways of meeting them will be treated only briefly.

KINDS OF MACHINE TOOLS CONSIDERED

Many promising means have been proposed for production of industrially useful raw materials in the lunar environment. The basic physics of the environment are well described in *Advanced Automation for Space Missions* (Freitas and Gilbreath, 1982). Several specific processes have since been proposed, including vitrification of lunar soil (Meek et al., 1984), production of glass fiber cables (Steurer, 1984), production of metal powder (Criswell, 1983), and the carbonyl process (Steigerwald, 1984; Lewis and Meinel, 1983). Fabrication of these raw materials has received less attention, although high technology methods have been discussed, such as David Criswell's work concerning powder metallurgy technology (Criswell, 1983) and an intersecting beam method of mold fabrication coupled with powder metallurgy suggested by David Brin (personal communication, 1984; Schwerzel et al., 1984).

Lunar soil is several tenths of a percent iron spheroids that can be magnetically separated from regolith (Arnold, 1984). Iron may also be a by-product of lunar oxygen production (Cutler, 1984a,b). Furthermore, chip-making machine technology is reliable, simple, and mature. It was the critical manufacturing technology of the Industrial Revolution 200 years ago. It is thus a prime candidate for use under the primitive conditions that will characterize early lunar industry (Cutler, personal communication, 1984).

This paper will consider adaptation of chip-making machines to the lunar environment. It will emphasize adapting the lathe and the milling machine, which may encompass most significant challenges. Iron mining, extraction, and casting are beyond the scope of this paper.

PRINCIPAL DESIGN FACTORS OF MACHINE TOOLS

On Earth, the principal machine tool design factors may be thought of as precision, power, and economy (Doyle, 1961). These are apt to remain the principal factors on the lunar surface. Precision depends on avoidance of inaccuracy in construction, deflection under static or dynamic load, wear, and thermal expansion.

Inaccuracy in construction is avoided through careful control of factory environment and operations, use of precision machine tools, and careful inspection after each stage of manufacturing.

Deflection is minimized by rigidity and properly chosen natural frequency. Deflection can be caused by static loading, due to workpiece and frame weight, and also by dynamic loading, usually caused by eccentric rotation of a mass. Of the two, dynamic loading is usually the greater challenge, as it can cause large amplitude vibration if it is at the natural frequency of the machine tool's frame.

Static deflection is often minimized by massive iron alloy frames, which, for larger machine tools, are coupled to massive concrete foundations.

As a rule of thumb, dynamic deflection can be minimized by a combination of rigidity, low mass, and damping that gives the machine tool frame a natural frequency at least $\sqrt{2}$ (4 is better) times that of the highest exciting force frequency. Alternatively, a combination of rigidity, high mass, and damping can give the frame a natural frequency less than $1/\sqrt{2}$ (1/4 is better) that of the lowest frequency exciting force frequency (Tobias, 1965). However, the spectrum of the exciting force depends largely on the angular velocity of the eccentric mass, which varies during machining. Exciting force frequencies from 0–4000 Hz are commonly encountered during machine start-up and operations (the higher frequencies come from impact loads or gears), and an exciting force spectrum can have several peaks. The frame's natural frequency can also change with machine configuration.

On Earth, an industrial machine tool is rigidly attached to a massive, rigid foundation. If this is not possible, it is attached to a vibration isolation system. This drastically reduces the machine's natural frequency. Typically, one tries to put the natural frequency of the isolation system as far as practical from the nearest peak in the exciting force spectrum characteristic of normal operation and relies on damping to soak up noise and transient resonances during start-up. If the isolation system needs augmenting, dynamic deflection of the machine tool can be reduced by vibration dampers (Tobias, 1965) that dissipate vibrational energy most effectively over some narrow frequency range.

Wear of precision-located sliding way surfaces is reduced by hardened steel ways, by dirt shields, and sometimes by plastic inserts in which chips become embedded before they can damage the ways. Avoiding wear of precision-located shafts may involve a pressurized lubrication system or roller bearings. Should wear occur, it could make the precision surface curved, or could worsen fit and introduce motion hysteresis in the force vs. motion curve. It is difficult for a control system to compensate for either of these. Wear also promotes chatter and vibration with consequent degradation of workpiece surface finish and possible fracturing of the cutting insert. Should significant wear occur, the worn surfaces must either be restored or the machine must be discarded.

Thermal expansion is avoided by not exposing the machine tool to direct sunlight and keeping ambient air at a constant temperature. Of these two, shielding from sunlight is the more important, because local heating can distort the frame. Constant ambient temperature becomes important for precise work such as metering nozzle fabrication.

Power is provided by electric, hydraulic, or pneumatic drives. Conventional drives are bulky and frequently cannot be connected directly to the load. Mechanical energy must be conveyed from drive to load by mechanical elements. For rotary motion, gears and belts are commonly used. For linear motion, the pinion and rack, the screw and nut, and the crank are commonly used. All of these mechanical elements add flexibility, stick slip, and, frequently, force vs. motion hysteresis. Such effects are minimized only by maintenance of very small tolerances, which make these elements difficult to produce, hence expensive.

Economy of operation involves proper operating controls, provisions for safety, and facilities for changing jobs. Operating controls have tended to become increasingly automatic, so that for some systems of machine tools no direct labor is required, only monitoring. Safety features also have tended to involve automatic operation. Job changeover is a surprisingly important part of machine economy. The machine produced nothing during changeovers, so changeover time must be minimized. Contemporary practice favors use of pallets, downloading NC programs, and various shop floor control methods. General purpose machines tend to change over more rapidly than special purpose machines, accept a wider range of work, and are idle less. Conversely, they cannot be made as rigid as special purpose machines, and hence they remove metal more slowly.

Economy in construction and maintenance may involve the use of standardized component parts, such as the base, headstock, or saddle. These can be combined at times in novel ways to produce a special purpose machine tool.

This is summarized in Table 1, Primary Challenges.

ADDITIONAL DESIGN FACTORS FOR THE LUNAR SURFACE

Precision, power, and economy will be just as important on the lunar surface as they are on Earth but may be designed into the tools differently. Let us first consider the unique characteristics of the lunar operating environment.

Pressure Vessels

A large machine tool must generally be fixed to a massive foundation that provides rigidity and couples vibration to the soil. Such large tools cannot be simply bolted to

Table 1. Primary Challenges

Precision:
- Vibration (grounding, or isolation and damping)
- "Unsagging" in reduced gravity
- Wear (in vacuum operation)
- Thermal environment

Power:
- Heat dissipation
- Energy source (beyond scope of this paper)

Economy:
- Labor requirements
- Transportation costs from Earth

---

a pressure vessel. They would distort under load, lose accuracy, and shake the entire vessel. These effects are especially pronounced at low cutting speeds, which generate high cutting forces at low frequency. The cutting forces can be reduced by use of lubricating cutting fluid or a modified cutting tool geometry (Trent, 1977).

Smaller machine tools do not require rigid foundations. Resilient supports can be placed directly between the machine tool and the shop floor if the exciting forces are small in comparison with the machine tools' weight and if the machine tool frame is sufficiently rigid. In most cases, these supports need not even be fixed to the floor, since hardly any dynamic force is developed at the point of support (Makhult, 1977).

If the natural freqencies of the pressure vessel can be determined, vibrational dampers tuned to these frequencies can be added. Practical dampers frequently use rubber for the spring/dashpot. Effectiveness depends on many factors, including the natural .requencies of the machine tool, the spectrum of the exciting force, and the range of movement tolerable in the damper (Tobias, 1965).

Environment within the vessel is affected by debris from the machine tool. This includes vapor and droplets of cutting fluid, ozone and heat from electric motors, chips and chip fragments, lubricant vapors, and the many other things that make a machine shop a messy environment, wherever it may be. High accuracy operations require constant air temperature. This should be part of the base thermal control system's design criteria.

Vacuum

If a machine tool is operated in vacuum, heat dissipation by convection cannot occur. Heat dissipation by conduction and radiation is typically less efficient than convective cooling. Most of the waste heat from chip formation stays in the chip and thus will probably not be a serious problem (Trent, 1977). Heat dissipation from power and control units could be a challenge. Finally, exposure to direct sunlight could cause significant local heating and thermally warp the frame.

Vacuum operation could also create unoxidized free surfaces, particularly during chip formation. If the surface freshly uncovered during chip formation is not exposed to an oxidizing agent, chip–tool relations are changed. Edward Trent (1977) observed substantial increases in chip thickness and in cutting force when machining iron at atmospheric pressures below 0.001 mbar, apparently due to a substantially increased area of contact between chip and tool. Introduction of air, even at very low pressure, eliminated this effect. Trent suggested that unoxidized free surfaces seize against the cutting tool more strongly than do oxidized surfaces.

Wear at exposed bearing surfaces could lead to vacuum welding and rapid failure. Contamination by lunar dust is an additional environmental hazard. On the other hand, deliberate creation of precision unoxidized free surfaces in conjunction with locator pins/holes could permit use of vacuum welding in construction of structures.

Gravitational Field

Reduced weight in the weaker lunar gravitational field will reduce frame self loading and will reduce the resulting sag to 1/6 of its Earth value (assuming elastic deformation and superposition). If a complex geometry superimposes several strains, a significant loss of accuracy may result. Machining forces are fairly small [several tens or hundreds of Kgf or lbf, several thousand N (Trent, 1977)] compared to workpiece weight, which is at most comparable to frame weight. Wayne R. Moore, a machine tool designer and fabricator, has put together a very interesting book (Moore, 1970) on machine tool accuracy. He emphasizes the importance of inspecting and correcting high precision machines under expected conditions of use and the importance of appreciating droop caused by cantilevering. Reduction of frame self loading could cause precision linear surfaces machined into the top of cantilevered frame members to curve upward into a ski jump shape. This could be significant for large or very high precision machine tools.

Workpiece weight will also be reduced. The primary effect may simply be to ease loading and unloading, since fixturing and not workpiece weight holds the workpiece to the work table. However, an eccentrically mounted, massive, and rapidly rotating workpiece could conceivably throw itself and an unsecured machine tool off its foundation in 1/6 g.

Reduced acceleration of freely falling bodies in lunar gravity will affect dispersion patterns of debris leaving the machine tool. Chip and cutting fluid dispersion would be increased. More seriously, rotating parts released through latch failure would travel further or would hit overhead surfaces at greater velocity than on Earth.

Transportation Costs

Transportation costs to the lunar surface are thought to be in the range of $3,000 to $15,000 per kilogram (Duke, 1984). While not directly affecting precision or power, these costs will have a strong influence on machine tool design. One could design very light and compact machine tools, or design heavy machine tools and fabricate the most massive components from lunar materials on the lunar surface.

Transportation costs will determine labor costs. People must be transported to the lunar base, and until controlled ecological life support system (CELSS) technology is developed and applied, their food must be brought up from Earth as well. The labor pool will accordingly be small. An initial base complement of something less than 20, and perhaps as few as 2, seems plausible. Support facilities will be less developed than those on Earth, including both recreation and training facilities. Additionally, work in a vacuum will involve either pressure suits or teleoperators, both of which reduce effectiveness. Work inside pressure vessels could be hampered by limited area and volume. The net result could be a small, expensive, but not very productive work force. Automation, teleoperation from Earth, and attention to habitat design could increase work force fitness.

Transportation costs will also determine material supply costs. Spare parts from Earth will be expensive, as will consumables such as cutting inserts, cutting fluid, and lubricants. Importation of an adequate inventory may well be deferred indefinitely. In the worst case, one could combine highly expensive repairs with long waits for spare parts. Proper attention to inventory requirements during planning will be essential.

Lunar design challenges are described in Table 2.

Table 2.   Recommendations

Vibration:
  ● Operate smaller machines inside pressure vessel and isolate/damp vibration.
  ● Operate large machines in vacuum and sink vibration into massive foundation.
Unsagging:
  ● Correction if necessary at the lunar base.
Wear (in vacuum):
  ● Avoid vacuum welding.
  ● Avoid lunar grit contamination.
Thermal environment:
  ● Avoid sunlight in vacuum operation.
  ● Keep air temperature constant for high precision work in pressure vessels.
  ● Alternately, fabricate machine tools from zero coefficient of expansion composite material.
Power:
  ● Consider heat dissipation in base thermal budget and in machine design.
Economy:
  ● Automate or teleoperate material handling, setup, and operation to minimize labor requirements.
  ● Minimize transport cost by designing light machine tools *or* fabricating some machine tool elements at the lunar base.

DISCUSSION

The design challenges for a lunar base suggest that there will be two classes of machining: light duty machining inside and heavy duty machining outside. Light duty machining will be needed for base maintenance. Fortunately, inside, light duty machining requires only the design of a light duty, vibration isolated/damped, general purpose, but productive and accurate machine tool. Heavy duty machine tools pose more of a design challenge. They must operate in vacuum, require minimal oversight, and involve minimal haulage costs from Earth.

Technology Mix

The design challenges listed above are not those of Earth; consequently, we can expect a different technology mix. In the early days, the lunar machine tool will be entirely Earth-made and will combine low mass, reliability, versatility, and high productivity with high purchase price and a limited work volume. It may employ conventional technology within these limits, or may employ unusually high levels of automation, composite material, and unusually high power, high torque electric motors. The design goal would be to maximize productivity and minimize mass and shipping bulk.

From the second generation on, machines could be made partly on the Moon and partly on Earth. The economic issues here are more complex than may at first be apparent. Goldberg and Criswell (1981) have considered the general case. For the specific case of machine tools, the "make/buy" decision on the Moon will primarily involve the number of hours required to make an item, the marginal cost of these hours, the mass and geometry of the item to be made, the marginal cost of shipping this mass and geometry, and (as a secondary consideration) the salary of the people doing the work and the cost of the item if bought. Ordinarily, these discussions would be dominated by the primary considerations; one would choose the cheaper of marginal labor and marginal shipping. If assembly is feasible, one would expect massive parts to be made on the Moon [a 37 kW, 127-cm swing lathe masses about 21,400 kg (Doyle, 1961), most of which is frame] and precision parts, electronics, software, motors, and other goods whose manufacture requires extensive industrial plant, to be brought from Earth. The lunar machine tool may thus, in the medium range, be a mix of sophisticated, low mass components brought from Earth and very crude, but high mass components made at the lunar base.

Drivers

Over time, one can say that the technology mix will be driven first by labor costs, then by shipping costs from Earth, and eventually by material import costs. These different drivers will give rise to a mix of sophisticated and crude technologies that, however inappropriate they would be on Earth, will be quite functional on the Moon.

Side Effects

Lunar machine tools will generate seismic waves and will emit gases and vapors. These effects would not materially add to those of a moderate-sized (20-person) lunar base with supply rockets, construction equipment, and mining operations. A small, vibration isolated machine tool should have few or no significant side effects if kept in a pressure vessel.

SUMMARY OF RECOMMENDATIONS

The lunar environment poses unique, but partially predictable, challenges in machine tool design. The primary initial challenge appears to be the design of a light duty, small table, ultralight, ultrareliable, ultraversatile, vibration isolated, vibration damping, numerically controlled machine tool to be operated in a shirtsleeve environment for base maintenance. It might machine parts of up to 1 foot (30 cm) in longest dimension. This machine will be needed soon after the establishment of a permanently manned base.

A second challenge appears to be the design of a large, light, ultrareliable, numerically controlled lathe to be operated on a massive foundation in a vacuum. It might have a 30-inch (76-cm) throw, 7 feet (2 m) between centers and require 25 kW. Operation of this machine must be automated to the maximum extent possible. Consideration should be given to artificial intelligence, robotics, teleoperation from the lunar base, and teleoperation from Earth. If several are installed, attention should be given to material handling between machines.

The large lathe could machine massive frames for small rolling mills, supports for solar furnaces, and frames for unpressurized but radiation-shielded work areas. It could machine and polish large iron mirrors for a solar furnace. It could, in conjunction with the light duty machine tool, make most elements of crude, but effective, machine tools. Much of the mass in lunar industry would be in such elements.

Table 3 describes a starter kit that could address the first two challenges.

Table 3.   Starter Kit

- First, introduce a small, light duty, versatile machine tool for use inside base for maintenance and repair.
- After base is well established, introduce a large, heavy duty machine tool for use in vacuum, to machine large structural elements.
- Machine iron castings to make frames for mining equipment, machines, solar mirrors, machine tools, and other massive capital goods. This will require both the small and large machines above. Import fittings from Earth, assemble on Moon.
- Benefit is greatest for more massive frames and smaller assembly and machining times.

The final foreseeable challenge is expansion. Processes will become less crude and the range of available materials higher as the economy grows. This will mark the emergence of a self-sustaining lunar economy using lunar materials, a springboard to a true lunar industry and a true lunar society.

Acknowledgments.   *Many thanks are due to the NASA Summer Faculty Program and the California Space Institute, which were responsible for my introduction to this area of study and for several of the ideas presented above.*

REFERENCES

Arnold J. R. (1984) Lunar materials: domestic use and export (abstract). In *Papers Presented to the Symposium on Lunar Bases and Space Activities of the 21st Century*, p. 100. NASA/Johnson Space Center, Houston.

Criswell D. R. (1983) The roles of powder metallurgy in the development of space manufacturing. In *1982 National Powder Metallurgy Conference Proceedings* (J. G. Bewley, ed.), pp. 115–145. Metal Powder Industries Federation, Princeton.

Cutler A. H. (1984a) A carbothermal scheme for lunar oxygen production (abstract). In *Papers Presented to the Symposium on Lunar Bases and Space Activities of the 21st Century*, p. 22. NASA/Johnson Space Center, Houston.

Cutler A. H. (1984b) Plasma anode electrolysis of molten lunar materials (abstract). In *Papers Presented to the Symposium on Lunar Bases and Spaces Activities of the 21st Century*, p. 23. NASA/Johnson Space Center, Houston.

Doyle L. E. (1961) *Processes and Materials for Engineers.* Prentice Hall, New York. 797 pp.

Duke M. B. (1984) *Report of the Lunar Base Working Group.* LALP-84-43, Los Alamos National Laboratory, Los Alamos. 41 pp.

Freitas R. A. and Gilbreath W. P. (editors) (1980) *Advanced automation for space missions.* NASA CP-2255. NASA, Washington, D.C. 393 pp.

Goldberg A. H. and Criswell D. R. (1981) The economics of bootstrapping space industries—development of an analytic computer model. In *Space Manufacturing, 4,* pp. 137–149. AIAA Publication 81-22787. AIAA, New York.

Lewis J. S. and Meinel C. (1983) Asteroid mining and space bunkers. *Defense Science 2001+, 2,* 33–67.

Makhult M. (1977) *Machine Support Design Based on Vibration Calculus.* Akademiai Kiado, Budapest. 136 pp.

Meek T., Cocks F. H., Vaniman D. T., and Wright R. A. (1984) Microwave processing of lunar materials: potential applications (abstract). In *Papers Presented to the Symposium on Lunar Bases and Space Activities of the 21st Century,* p. 130. NASA/Johnson Space Center, Houston.

Moore W. R. (1970) *Foundations of Mechanical Accuracy.* Moore Special Tool Co., Bridgeport. 353 pp.

Schwerzel R. E., Wood V. E., McGinniss V. D., and Verber C. M. (1984) *Three-dimensional photochemical machining with lasers.* Batelle Colombus Laboratories, Colombus. 7 pp.

Steigerwald L. J. (1984) *Mold making by the vaporformed process.* Formative Products, Troy, MI. 5 pp.

Steurer W. H. (1984) Use of lunar materials in the construction and operation of a lunar base (abstract). In *Papers Presented to the Symposium on Lunar Bases and Space Activities of the 21st Century,* p. 78. NASA/Johnson Space Center, Houston.

Tobias S. A. (1965) *Machine Tool Vibration.* Wiley, New York. 351 pp.

Trent M. E. (1977) *Metal Cutting.* Butler and Tanner, London. 203 pp.

von Puttkamer J. (1976) Developing space occupancy: perspectives on NASA future space programme development. *J. Br. Interplanet. Soc., 29,* 147–173.

PRAWLINGS 85

8 / OXYGEN: PRELUDE TO LUNAR INDUSTRIALIZATION

P ROJECT APOLLO WAS INITIATED and then truncated by the political process in the United States. Very probably the next manned landings on the Moon will occur as the result of political decisions. As long as political motivations are the principal drivers for lunar activity, then realignments of national priorities also can terminate future programs.

In 1969, the Apollo flights were the only manned space activity. In the 21st Century, lunar surface activities will be an extension of a manned operations in near-Earth space, including routine flights between Earth and LEO and continuous presence in one or more LEO space stations. Although future flights to the Moon will not seem as exotic as they once did, lunar operations will not be secure until the surface of the Moon is viewed not as an outpost but as an integral part of the space infrastructure. To achieve this end, the lunar activity must return some benefit to the system.

For the foreseeable future, large scale operations in space will be hampered by the large energy required to launch from the surface of the Earth. Payloads can be launched from the Moon to LEO with an energy expenditure that is more than an order of magnitude less. Although hypothetical lunar products might have a lower transportation cost, the capital investment required for establishing and operating lunar manufacturing facilities might well make the concept infeasible. Two lunar products with potentially large markets yet requiring minimal processing are simple regolith for use as shielding mass and liquid oxygen for spacecraft propellant.

Production of oxygen propellant is attractive because it immediately relieves some of the burden on the transportation system for lunar operations. To place one kilogram on the lunar surface requires launching approximately seven kilograms to LEO, most of which is fuel for the trip. A small production plant capable of fueling a reusable vehicle shuttling between the base and lunar orbit halves the required launch mass to LEO. If the plant produces enough oxygen for the whole transportation system, then only approximately 1.2 kilograms are required from the Earth to send one kilogram on to the Moon.

A lunar production facility capable of supporting lunar operations and supplying excess propellant to LEO appears feasible. Simon addresses the key question whether the price of lunar oxygen in LEO can be less than that for propellant shipped from Earth. Although an accurate answer cannot be given at the present time, Simon's methodology explicitly identifies the relevant parameters and points to key technologies where research is needed.

An important element of the total system analysis is the process chosen to produce the oxygen. Several are possible, but the reduction of the mineral ilmenite by hydrogen has been studied somewhat more than others. Gibson and Knudsen discuss the process and potential reactor designs. Williams presents experimental data on ilmenite reduction. Cutler and Krag describe an alternative process that could be easily augmented to produce steel for use on the Moon or as an export for space construction.

A major irreducible cost element in production of lunar oxygen is the shipping cost of imported hydrogen. Whether hydrogen of solar wind origin can be recovered from the lunar regolith becomes an interesting technical issue. Carter constructs a simple model of the hydrogen rich layer on lunar grains. Tucker *et al.*, explore the use of microwave energy to free hydrogen in the soil thermally. White and Hirsch suggest the possibility of microbial processing of lunar regolith to extract absorbed volatile gases. Blanford *et al.*, have implanted ions into minerals to determine experimentally the behavior of gases embedded in the crystal lattice.

Friedlander suggests the delivery of hydrogen to the Moon in the form of methane or ammonia. Storage and handling of these compounds is much simpler than for liquid hydrogen, and the critical elements carbon and nitrogen can be made available for life support systems.

A PARAMETRIC ANALYSIS OF LUNAR OXYGEN PRODUCTION

Michael C. Simon

General Dynamics Space Systems Division, P.O. Box 85357, Mail Zone CI-9530, San Diego, CA 92138

A methodology has been developed for evaluating alternative approaches to utilization of space resources. The specific application discussed in this paper is a parametric cost analysis of liquid oxygen (LO_2) production on the Moon. A baseline scenario for production of LO_2 from lunar ore and its delivery to low Earth orbit (LEO) is defined, and the major systems required to support such an enterprise are identified. Fifteen principal variables influencing the capital emplacement and operations costs are defined and organized into a parametric cost model. A sensitivity analysis is performed to identify the impact of each of these variables on the costs of lunar LO_2 production. The cost model and sensitivity analysis are structured with emphasis on permitting evaluation of alternatives to the baseline scenario, rather than for developing conclusive cost estimates at the present time. Economic weighting factors are derived to provide a measure of the potential impact on cost from each of a number of key technologies. Using these factors, certain types of near-term technology development activities are recommended to maximize future cost-effective utilization of space resources.

INTRODUCTION

Utilization of space resources, *i.e.*, raw materials obtained from non-terrestrial sources, has often been cited as a prerequisite for large scale industrialization and habitation of space. While transportation of extremely large quantities of material from Earth would be costly and potentially destructive to our environment, vast quantities of usable resources might be derived from the Moon, the asteroids, and other celestial objects in a cost-effective and environmentally benign manner.

Of more immediate interest to space program planners is the economic feasibility of using space resources to support near-term space activities, such as scientific and commercial missions in the 2000–2010 time frame. The purpose of the analysis described in this paper is to identify the factors most likely to influence the economics of near-term space resource utilization, based on development of a baseline scenario for liquid oxygen production from lunar ore. Liquid oxygen as a propellant for space-based transportation systems appears to be an obtainable space resource with the firmest near-term requirement for quantities sufficient to be produced economically in an extraterrestrial setting.

ANALYSIS METHODOLOGY

The primary purpose of the parametric cost model developed as part of this study is to identify the factors that have the greatest influence on the economics of space resource utilization. In the near term, this information can be applied to the establishment of technology development strategies, so that space resource utilization capabilities are developed in a manner likely to achieve cost–effective results.

It is important to note that predicting the actual costs of particular space resource utilization scenarios is only a secondary objective of this analysis. Estimates are made and dollar values are assigned principally as a relative measure of alternative options. Since the technologies for space resource utilization are in an early stage of development, it is premature to state conclusively whether mining the Moon, asteroids, or other celestial bodies makes economic sense. For these reasons the parametric model is designed more for flexibility than for precision. Although preliminary estimates indicate that production of oxygen from lunar ore is one of the space resource projects most likely to yield an economic payback, this activity was selected as a "baseline scenario" primarily because the requirements for such an endeavor can be relatively well defined. Hence, identification of the major systems required to support the baseline scenario could be made without much difficulty as being (1) a processing and storage facility to manufacture liquid oxygen from lunar ore and store it on the Moon, (2) a system of habitable lunar base elements to support a small, full-time crew, (3) a power system to provide the energy required for processing and storing the LO_2 and to support other lunar operations, and (4) a transportation and logistics system to deliver and support lunar base elements and to transport the lunar LO_2 to LEO.

Once these major support systems were defined, 15 key variables were identified as influencing the costs of developing and operating these systems (Table 1). Cost variables were limited to general categories so that the parametric model could be adapted in the future to evaluation of alternative scenarios. Next, equations were developed to calculate capital and operations costs as functions of these variables. Utilizing the codes and units detailed in Table 1, these equations are as follows:

$$\text{Capital cost} = (p \times c_p) + (n_t \times c_n) + (n_m \times c_u) + c_f + c_t \times [(p \times m_p) + (n_m \times m_m) + m_f]$$

$$\text{Operations cost} = c_t \times [(n_r \times m_m) + (1 - d) \times (125{,}000)] + (n_b \times n_f \times \$100{,}000)$$

where the capital cost is defined as the total cost of developing, building, and installing the lunar base systems (including transportation costs), and the operations costs are the annual costs of manufacturing 1 million kg (1,000 metric tons) of LO_2 per year and delivering to LEO as much of this LO_2 as possible.

The first two constants that appear in the operations cost equation reflect the assumptions that a portion of the LO_2 produced on the Moon is utilized as propellant to deliver the remaining LO_2 to LEO, and that 1 kg of hydrogen must be delivered from Earth to the Moon for every 8 kg used as propellant en route from the Moon to LEO. A higher-than-usual mixture ratio of 8:1 was selected for the baseline case after initial analyses showed the resultant reduction in hydrogen requirements to offer substantial economic benefits. The third constant in the operations cost equation is the cost per

Table 1. Lunar Oxygen Production—Major Cost Variables

Variable	Code	Units of Evaluation
Power required	p	Megawatts of installed capacity
Cost of power	c_p	Non-recurring cost per megawatt of installed capacity
Number of types of lunar base modules	n_t	Number of units
Cost of modifying space station modules	c_n	Non-recurring cost for adapting each type of module ($)
Number of lunar base modules	n_m	Number of units
Unit cost of lunar base modules	c_u	Recurring cost of producing each lunar base module ($)
Processing/storage facility cost	c_f	Development and production cost ($)
Power system mass	m_p	Kilograms per megawatt of installed capacity
Earth-to-Moon transportation cost	c_t	Cost ($) per kilogram delivered from Earth to the Moon
Mass of lunar base modules	m_u	Kilograms
Mass of processing/storage facility	m_f	Kilograms
Number of lunar base resupply missions/year	n_r	Number
Net lunar oxygen delivered to LEO	d_o	Fraction of lunar LO_2 produced that is delivered to LEO
Ground support manpower		Number of full-time equivalent heads
Ground support overhead	n_f	Multiple of manpower needed for total

man-year of ground support manpower. The variable that provides this constant, n_f, is a ground support overhead factor that is multiplied by manpower costs to obtain total ground support cost.

After setting up these costs equations, baseline values were assigned to each cost variable, with the ground rule that lowest risk technologies would be utilized for each system. Lunar base modules, for example, were assumed to be modified versions of the habitat, laboratory, and logistics modules that will be developed for NASA's LEO space station.

Another key ground rule was that selection of the lunar processing site would be made on the basis of scientific data whose costs would not be included in this model. It was further assumed that an initial lunar base would be in place prior to the LO_2 production activity and that this facility would be scaled up to meet the LO_2 production requirements. The costs included in this model are only the marginal costs of expanding this initial facility to meet LO_2 production objectives.

Although some of these ground rules resulted in lower capital and operations cost estimates, the specification of lowest risk technology prevented the incorporation of many potentially cost-reducing technologies into the baseline scenario. The probable impact of these ground rules is analyzed later in this paper.

RESULTS OF THE ANALYSIS

Once baseline values were assigned to the cost variables, a simple calculation was made to obtain baseline capital and operations cost estimates. These costs were determined to be $3.1 billion for baseline capital cost and $885 million/per year for baseline operations costs.

An analysis of the performance of lunar OTVs indicated that 49.2% of the LO_2 produced would be delivered to LEO. Consequently, the unit cost of LO_2 delivered to LEO, assuming 10-year amortization of capital costs, was determined to be $2,430/kg ($1,100/lb). This cost is about one third less than the current cost of utilizaing the STS for propellant delivery to LEO and roughly equivalent to STS cost projections in the most optimistic, high flight-rate, low operations cost scenario. It should be reemphasized, however, that these costs are based on a specific set of assumptions and are for comparative purposes only.

The most important objectives of this analysis were the assignment of uncertainty ranges to each of the cost variables, the calculation of the sensitivity of LO_2 production costs to each of these variables, and the analysis of the technical and programmatic assumptions used to arrive at values for each variable. The data that were developed to support the sensitivity analysis of space resource utilization cost variables are summarized in Table 2. The baseline, best case, and worst case values assigned to each cost variable are shown, along with the impact of each variable's best case and worst case values on capital and operations costs. For example, as power requirements vary from a low value of 4 Mw to a high value of 12 Mw, with all other variables held at their baseline values, the capital cost for establishing the baseline capability ranges from $2.30 billion to $3.90 billion.

From this table it is evident that the principal driver of capital costs is the lunar base power requirement, while the Earth-to-Moon transporation cost is the most important operations cost driver. Since capital costs are amortized over a 10-year period, the Earth-to-Moon transportation cost has a much greater overall impact on the cost of lunar LO_2 in LEO. If this cost could be reduced from its baseline value of $10,000/kg delivered to the Moon to its best case value of $5,000/kg, capital costs would drop from $3.1 billion to $2.45 billion, operations costs would decline from $885 million/yr to $468 million/ yr, and the cost of lunar LO_2 would be reduced from $2,430/kg to $1,450/kg. Conversely, at its worst case value of $15,000/kg, the Earth-to-Moon transportation cost would drive capital costs up to $3.75 billion, operations costs to $1.3 billion/yr, and the cost of lunar LO_2 to $3,410/kg.

An alternative approach to showing the sensitivities of the cost variables is illustrated in Table 3, where percent changes in capital costs, operations costs, and the costs per kilogram of LO_2 produced (with 10-year amortization of capital costs) are shown. In this table the variables are ranked and listed in order of their impact on the LO_2 cost/ kg. The influence of each variable is calculated as an "impact factor" equal to the average of the best case and worst case percent changes in LO_2 cost/kg due to extreme values of the variable. From these impact factors it is clear that two of the cost factors are

Table 2. Capital and Operations Costs—Sensitivity to Key Variables

	Baseline Case Most Likely Value	Best Case Value	Best Case Result	Worst Case Value	Worst Case Result
Capital Costs					
1. Power required	8 Mw	4 Mw	$2.30 B	12 Mw	$3.90 B
2. Cost of power	$100 M/Mw	$50 M/Mw	$2.70 B	$200 M/Mw	$3.90 B
3. Number of types of lunar base modules	1	0	$2.80 B	2	$3.40 B
4. Cost of modifying space station modules	$300 M	$100 M	$2.90 B	$500 M	$3.30 B
5. Number of lunar base modules	1	1	$3.10 B	3	$3.90 B
6. Unit cost of lunar base modules	$200 M	$100 M	$3.00 B	$300 M	$3.20 B
7. Processing/storage facility cost	$500 M	$300 M	$2.90 B	$1.0 B	$3.60 B
8. Power system mass	10,000 kg/Mw	5,000 kg/Mw	$2.70 B	15,000 kg/Mw	$3.50 B
9. Earth-to-Moon transportation cost	10,000/kg	5,000/kg	$2.45 B	15,000/kg	$3.75 B
10. Mass of lunar base modules	20,000 kg	15,000 kg	$3.05 B	30,000 kg	$3.20 B
11. Mass of processing/storage facility	30,000 kg	15,000 kg	$2.95 B	50,000 kg	$3.30 B
Operations Costs					
1. Number of lunar base resupply missions/yr	1	1	$885 M/yr	3	$1.285 B/yr
2. Net lunar oxygen delivered to LEO	49.2%	70%	$625 M/yr	30%	$1.125 B/yr
3. Ground support manpower	20	10	$860 M/yr	50	$960 M/yr
4. Ground support overhead factor	25	5	$845 M/yr	50	$935 M/yr
5. Earth-to-Moon transportation cost	$10,000/kg	$5,000/kg	$468 M/yr	15,000/kg	$1.303 B/yr
6. Mass of lunar base modules	20,000 kg	15,000/kg	$835 M/yr	30,000 kg	$985 M/yr

Table 3. Sensitivity of Capital, Operations, and Oxygen Production Costs to Uncertainty Ranges of Key Variables

Variable	Sensitivity Ranking	Best Case		Worst Case		Impact Factor
		Change in Capital Cost	Change in LO$_2$ Cost/kg	Change in Capital Cost	Change in LO$_2$ Co st/kg	
Capital Costs						
Earth-to-Moon transportation cost	1	-21%	-40%*	+21%	+40%	40
Power required	2	-26%	- 7%	+26%	+ 7%	7
Mass of lunar base modules	3	- 2%	- 4%	+ 3%	+ 9%*	7
Cost of Power	4	-13%	- 3%	+26%	+ 7%	5
Number of lunar base modules	5	0%	0%	+26%	+ 7%	4
Power System mass	6	-13%	- 3%	+13%	+ 3%	3
Processing/storage facility cost	7	- 6%	- 2%	+16%	+ 4%	3
Number of types of lunar base modules	8	-10%	- 3%	+10%	+ 3%	3
Cost of modifying space station modules	9	- 6%	- 2%	+ 6%	+ 2%	2
Mass of processing/storage facility	10	- 5%	- 1%	+ 6%	+ 2%	2
Unit cost of lunar base modules	11	- 3%	- 1	+ 3%	+ 1%	1
Operations Costs						
Net lunar oxygen delivered to LEO	1	-29%	-45%	+27%	+97%	71
Earth-to-Moon transportation cost	2	-47%	-40%*	+47%	+40%*	40
Number of lunar base resupply missions/yr	3	0%	0%	+45%	+13%	7
Mass of lunar base modules	4	- 6%	- 4%*	+11%	+ 9%*	7
Ground support manpower	5	- 3%	- 3%	+8%	+ 6%	5
Ground support overhead factor	6	- 5%	- 3%	+6%	+ 4%	4

*Impact based on changes in both capital costs and operations costs.

by far more important than all the rest: Earth-to-Moon transportation costs and net delivery of LO_2 to LEO. The net percentage of LO_2 delivered to LEO is important because of its double impact. As the percentage of LO_2 delivered declines, LO_2 cost/kg increases not only because less LO_2 is delivered, but also because more hydrogen must be provided from the Earth to augment the LO_2 used as propellant from the Moon to LEO.

Table 3 also shows that costs associated with establishing crew accommodations on the Moon, *e.g.*, the number and cost of lunar base modules, have a relatively small impact on total cost. Lunar base resupply requirements, however, are a major driver, as is the mass of lunar base modules. Factors influencing the mass and cost of the power system and the processing/storage facility have a significant impact on capital costs but have a much smaller effect on LO_2 cost/kg. Factors affecting ground support costs have a moderate impact on operations costs and LO_2 cost/kg.

It is not surprising that the six operations cost variables are among the nine most important cost factors. Operations costs are significant because they are driven by the high cost of Earth-to-Moon transportation, while the impact of capital costs is reduced because they are amortized over a 10-year period. The relative significance of the operations cost leads to the important observation that LO_2 production costs may be reduced substantially by increasing capital expenditures on technologies that can reduce operations costs.

INTERPRETATION OF ANALYSIS RESULTS

During this study, a more detailed sensitivity analysis was performed to obtain a preliminary indication of which technologies need to be developed to maximize the probability that lunar LO_2 production will become a cost-effective enterprise. It became evident that the cost-effective performance of such an operation will depend primarily on the achievement of three major cost-reducing objectives: (1) reducing or eliminating the need to transport hydrogen from Earth to the Moon; (2) reducing space transportation costs, particularly the cost of Earth-to-Moon transportation; and (3) reducing lunar base resupply requirements.

If all three of these objectives were met to the greatest extent possible, *i.e.*, if hydrogen transportation requirements were eliminated, Earth-to-Moon transportation costs were reduced to its best-case value, and lunar base resupply requirements were eliminated, the cost/kg of lunar LO_2 delivered to LEO would be reduced from $2,340/kg to $600/kg, or about $270/lb. These figures assume no change in capital costs; but even if capital costs were doubled to achieve these capabilities, LO_2 cost/kg would be reduced to approximately $1,100/kg, less than half the baseline cost.

Twenty-five key technology issues influencing these and the other LO_2 production cost factors are presented in Table 4. In this table, the impact of each technology issue on the 15 cost variables is shown; a darkened square indicates a strong impact; a light square represents a moderate effect; no square indicates little or no impact. It should be emphasized that the selection and evaluation of these technology issues was based on the subjective judgment of a panel of experts convened to support this study, rather

Table 4. Impact of 25 Key Technology Issues on Space Resource Utilization Cost Factors

Legend:
- ■ heavy impact
- □ moderate impact
- ☐ little or no impact

Technology Issue	Unit cost lunar base modules	Cost modifying space station modules	No. types of lunar base modules	Mass of processing/storage facility	Power system mass	Processing/storage facility cost	Number of lunar base modules	Ground support overhead factor	Ground support manpower	Cost of power	Power required	Mass of lunar base modules	Number of lunar resupply missions/yr	Earth-to-Moon transportation cost	Net Lunar LO2 delivered to LEO
Lunar base power source (nuclear vs. solar)									■	■	■				
Scalability of small (<100 Kw) power systems				□				■	■						
Electrical vs. thermal energy									■	□					
Power consumption of processing technique(s)									□	■		□			
Complexity of power system installation									■	■					
Maintainability of power system										■				□	
Pressurized volume required for lunar operations			■			■					■		□		
Duration of lunar base crew shifts						□						■	■		
Degree of automation of lunar base operations			■	□		■			□	■	□	□	■		
Lunar base module commonality w/space station nodules	■	■	□		■	■									
Lunar base shielding requirements	□	□				□					■				
Space station interfaces		□													
Scalability of initial lunar research facilities									□	□					
Complexity of lunar factory processes			□	□	■	□		□	□	■	□	□			
Number of lunar factory processes			□		■	□		□	□	■	□	□			
Commonality of processing facility w/space station lab modules					□										
Commonality of LO2 storage unit w/OTV propellant depot					■										
Performance and cost of SDLV/HLLV (if available)														■	
Performance and cost of OTVs (if available)														■	
Size of lunar base crew						■	□		■	■	□	■			
Degree of closure of lunar base life support system	■	□				■	□		■	■	□	■			
Availability of aerobrake for LO2 delivery														■	
Availability of lunar hydrogen				□			■							■	
Self-sufficiency of lunar operations	■	□			■	■	■		■	■	■	■			
Ground support philosophy							□								

than a quantitative analysis. Based on the sensitivity analyses presented in this paper, the 15 cost factors are listed along the top of Table 4 in descending order of importance from left to right. Hence, a visual scan of Table 4 provides an immediate glimpse of the relative importance of the technologies based on three considerations: total number of squares, number of dark squares, and distribution of squares to the left of the chart (*i.e.*, toward the most important cost factors).

To quantify the impact of these 25 technology issues on the economics of the baseline space resource utilization scenario, a technology weighting factor of 3 was assigned to each dark square and 1 to each light square. These technology weighting factors then were multiplied by the impact factor (see Table 3) for each cost factor that the technology issue affects. A total economic weighting factor was calculated for each technology issue, based on the sum of the products across each row. For example, the lunar base power source has a heavy impact on power system mass and cost of power, for a total economic weighting factor of $(3 \times 5) + (3 \times 5) = 30$.

The 10 most important technology issues, based on their total economic weighting factors, are listed in Table 5. All 10 issues relate in some way to space transportation costs and/or lunar base resupply requirements, with performance and cost of space-based orbital transfer vehicles (OTVs) being the most critical technology issue. Developing a low cost OTV is a fundamental requirement for cost-effective space resource utilization because the OTV is the single most effective means of reducing Earth-to-Moon transportation costs.

The second most important issue is the availability of lunar hydrogen, since production of hydrogen on the Moon currently appears to be the most promising solution to the problem of costly hydrogen deliveries from Earth. The technology issue ranked third is availability of an OTV aerobrake; the capability for aeroassisted return to LEO is a requirement for achieving the efficiency in LO_2 delivered to LEO, thereby halving the effective LO_2 cost/kg. The fourth most important technology issue is the performance and cost of shuttle-derived launch vehicles (SDLVs) or heavy-lift launch vehicles (HLLVs), *i.e.*, the Earth-to-LEO transportation cost. However, it should be pointed out that cost reductions in this area will also reduce the cost of delivering LO_2 from Eath to LEO, making it more difficult for the lunar LO_2 production scheme to compete.

Table 5. Major Technology Issues

1. Performance and cost of OTVs (345)
2. Availability of lunar hydrogen (254)
3. Availability of aerobrake for LO_2 delivery (213)
4. Performance and cost of SDLV/HLLV (120)
5. Degree of automation of lunar base operations (119)
6. Self-sufficiency of lunar operations (97)
7. Size of lunar base crew (79)
8. Degree of closure of lunar base life-support system (71)
9. Complexity of lunar factory processes (51)
10 Number of lunar factory processes (48)

Numbers in parentheses are economic weighting factors.

The remaining six of the "top ten" technology issues all pertain to lunar base resupply requirements. The extent of lunar base automation and self-sufficiency are the most important of these issues, with the next two factors relating to the resupply needs of the lunar base crew. The ninth and tenth most significant technology issues relate to the complexity and number of lunar factory processes.

CONCLUSIONS AND RECOMMENDATIONS

On the basis of the analysis presented in this paper, it is not possible to establish conclusively whether production of liquid oxygen from lunar materials can be justified on economic grounds. Although the cost estimates for the baseline scenario are encouraging, a number of technologies with significant impact on lunar LO_2 production costs must be explored.

The principal conclusion is that the costs of any space resource utilization venture are likely to be dominated by transportation costs. This situation may change when the focus of human space activities moves sufficiently far from Earth (e.g., once large scale space industrialization is underway) but will probably remain valid as long as space resources are developed for use in near-Earth space. For this reason, it is recommended that high priority be given to development and evaluation of more cost-effective space transportation systems, particularly low-cost OTVs with an aeroassisted return-to-LEO capability.

Another key issue relating the baseline scenario is the cost of Earth-derived hydrogen. Production of lunar LO_2 would be far more cost-effective if a capability for the co-production of lunar hydrogen could be developed, even if capital costs were increased substantially. Although relatively large quantities of lunar ore would need to be processed, the additional costs of lunar hydrogen production could be offset by savings of over $600 million/ yr in transportation costs. Production of some alternative propellant constituent, such as aluminum, also might offer an opportunity for reducing or eliminating costly imports of fuels from Earth. However, this latter approach would require an entirely new development, the design of aluminum-burning space engines.

A third category that seems to have a substantial impact on the economics of lunar resource utilization are the technologies influencing lunar base resupply requirements. Increasing lunar base automation, closing the lunar base life support system, and other steps to reduce the frequency and scale of resupply missions appear to have a high likelihood of providing economic benefits and should be given particular emphasis in future study efforts.

Finally, it is important that parametric cost analyses be used in the assessment of a variety of space resource utilization scenarios. Use of lunar ore for production of construction materials is one such option, although this type of enterprise would probably require a dramatic increase in the level of space activity to be cost effective. Another option that merits careful consideration is the development of asteroidal resources. Both rocket propellants and construction materials could be derived from asteroids, and while the up-front costs of asteroid utilization would probably exceed the capital expenditures required for lunar development, operations costs could be substantially lower. Further analysis of all these opportunities needs to be conducted over the next several years before a commitment is made to any particular plan for space resource utilization.

As these and related technologies are developed, the reliability of space resource utilization cost estimates will improve. Eventually, it will become possible to generate cost estimates of sufficient fidelity to support detailed definition of space resource utilization objectives. An important step in this process will be the adaptation of this parametric model and similar techniques to the evaluation of a broad range of space resource development options.

Acknowledgments. The author wishes to acknowledge James Carter (University of Texas at Dallas), Andrew Cutler (University of California, San Diego), Rocco Fazzolare (University of Arizona), Joel Greenberg (Princeton Synergetics), and Robert Salkeld for their assistance in developing and analyzing the lunar resource utilization scenario described in this paper. Their valuable help in generating the study conclusions and supporting data is greatly appreciated.*

LUNAR OXYGEN PRODUCTION FROM ILMENITE

Michael A. Gibson and Christian W. Knudsen

Carbotek, Inc., 2916 West T. C. Jester, Houston, TX 77018

Lunar oxygen production from ilmenite appears to be the simplest means available. Fixed- or fluidized-bed, gas-solid reactor processing systems appear suitable for development; terrestrial analogs exist. Scale of operations for 1000 metric tonnes per year seems feasible. Severity of the lunar environment calls for particular design attention to fluid-solid flow, heat exchange, power supply, minimum maintenance, and modular replacement.

INTRODUCTION

In any future human colonization of the Moon, oxygen is clearly one of the most important materials to be supplied. It is required for both life support and propulsion. Incentives for lunar oxygen production from lunar raw materials as opposed to supplying it solely from Earth have been discussed by Criswell (1983) and Davis (1983). Both conclude that more efficient use of Earth-supplied hydrogen, total propellants, and available payload weights and volumes all result from lunar oxygen production.

Two types of lunar materials have been proposed as raw materials for oxygen production: ilmenite, $FeTiO_3$, and silicates such as anorthite, $CaAl_2Si_2O_8$ (Williams and Erstfeld, 1979; Kesterke, 1971; Steurer, 1982; Carroll, 1983). Both are lunar-surface-minable, occurring in soils, breccias, rocks, and basalts. The silicates are considerably more abundant than ilmenite, and this would suggest a preference for them as source materials. However, the silicates present more difficult process engineering problems because they must be reduced at temperatures of 1100°C or more by fluxed molten-phase electrolysis (Kesterke, 1971). Alternatively, silicates can be processed at still higher temperatures by plasma processes or, if flux is not used, in electrolysis processes (Steurer, 1982; Carroll, 1983). Under these conditions, degradation of electrode and/or container materials and difficult recovery of Earth-imported electrolyte fluxing agents present severe challenges that are, at present, difficult to overcome. Ilmenite, on the other hand, can be reduced at temperatures of 1000°C or below, and it can be separated and concentrated from a natural abundance of approximately 5 wt % to perhaps 90 wt % in the process feedstock (Williams *et al.*, 1979). Williams *et al.* (1979) have proposed electrostatic and magnetic schemes for beneficiation; Agosto (1984) has performed initial and encouraging experimental tests of one of these beneficiation schemes. These circumstances make ilmenite appear to be the preferable raw material.

The available processing techniques include hydrogen-reduction, chemical reductions with other reagents, and electrochemical reduction (Rao *et al.*, 1979). We have concluded that probably the best route to lunar oxygen is via hydrogen-reduction of ilmenite. This is best done with some kind of continuous or semi-continuous process cycle using recycled, pressurized hydrogen as the working fluid. This paper discusses the various reduction options along with the reasons for our design choices.

ILMENITE REDUCTION

Choice of Reductant

Hydrogen was chosen over several other potential reducing agents or methods, *e.g.*, CH_4, CO, and electrolysis, for the following reasons:

- Hydrogen will probably have to be imported for propulsion and life support anyway, so it is likely to be the easiest material to supply for make-up of processing losses caused by leakage from the recycle loop.
- With carbon-bearing reductants (CH_4 and CO), part or all of the oxygen product must be won by electrolyzing mixtures containing CO and CO_2. This step often leads to solid carbon deposits that are hard to remove continuously. With H_2, the electrolysis of liquid H_2O, at least, is straightforward, and this one step liberates the oxygen product and regenerates the reducing agent.
- As discussed below, use of H_2 may present some heat integration opportunities with large energy-saving consequences. These would not be possible with an electrochemical or CH_4-reduction technique, neither of which allow regenerating the reducing agent at reduction temperature.
- Electrochemical methods require solutions to the difficult container and electrode materials problems mentioned above.

Process Scheme

The basic reactions are shown below.

$$\text{FeTiO}_3 + H_2 \rightleftharpoons \text{Fe} + \text{TiO}_2 + H_2O \xrightarrow{\text{electrolysis}} H_2 + \tfrac{1}{2}O_2 \tag{1}$$

recycle

ilmenite feed — solid product — oxygen product

The process requires heating and contacting of gas and solids, and, to be fully continuous, a way of introducing solid ilmenite feed and removing spent, reduced ilmenite product must be provided. It would be very desirable to use countercurrent, gas-solids heat exchange to reduce overall energy requirements and to reject heat in spent solids at as low a temperature as possible. It would also be desirable to conduct the H_2O electrolysis at reaction temperature (vapor phase electrolysis), if possible, to avoid cooling and reheating the recycled H_2 to condense H_2O.

Figure 1 presents a conceptual design of a fluidized bed reactor/vapor phase electrolysis flow plan that at least on paper accomplishes the objectives above. The reactor

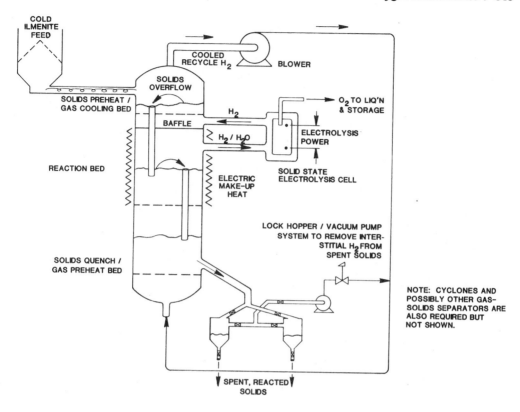

Figure 1. Continuous, fluid bed ilmenite reduction/O_2 production.

shell actually contains three staged fluidized beds stacked vertically with gas in upward flow and solids in countercurrent downward flow. Cold ilmenite feed from a lock hopper at reactor pressure enters the top, solids-preheat bed via a screw feeder. Here it exchanges heat with hot recycled H_2. This stage preheats the solids and cools the H_2 sufficiently for it to be circulated reliably by the blower. The solids overflow into the reaction bed where the H_2 reduction occurs. Electrical make-up heat must be supplied to this stage on temperature control to bring gas and solids to the desired reaction temperature and to supply the endothermic heat of reaction.

Finally, spent, reacted solids overflow into the solids' quench/gas-preheat bed where they exchange heat with cold, recycled H_2. The cooled solids are discharged through an alternating lock hopper/vacuum pump apparatus to minimize loss of interstitial H_2 gas to the lunar vacuum.

To extract product oxygen, the H_2/H_2O mixture leaving the reaction bed is sent to an electrolyzer using a solid-state, ceramic electrolyte and operating at reactor temperature. This device was suggested by the prototype studies of Weissbart and Ruka (1962). Gaseous oxygen is evolved on the product side of the solid electrolyte and is sent to liquefaction and storage. The H_2O is reconverted to H_2 on the feed side of the electrolyte, and the H_2 is recycled.

Staged, fluidized-bed reactor/heat exchangers such as in Fig. 1 are in commercial use for terrestrial applications such as limestone calcining and sulfide roasting (Perry and Chilton, 1973). Their use was also proposed for fluidized iron ore reduction (FIOR). This was a process developed and studied intensively in the 1960s and 1970s by Exxon Corporation and several steel companies. Iron ores were reduced in fluidized beds with H_2, natural gas, and other reducing gases as a possible replacement for conventional blast furnaces. Changing energy economics prevented exploitation of the process, but its strong analogies to H_2-reduction of ilmenite are evident.

As noted in Fig. 1, efficient cyclones and possibly other types of gas-solids separators will be required but, for clarity, are not shown. Clean, gas-solid separations will be especially important upstream of the blower and the solid-state electrolysis device. Proper sizing of feed solids and rejection of fines will also alleviate entrainment and elutriation problems.

In addition to heat integration, the scheme of Fig. 1, if workable, would have significant heat transfer equipment advantages. All heating of gases and solids, except for O_2 cooling and liquefaction, occurs in fluidized beds with very high transfer coefficients. In particular, if the solid-state, high-temperature electrolysis proved feasible, the poor-coefficient process of condensing dilute H_2O vapor from a "fixed" gas, H_2, would be avoided.

Thermodynamic Considerations

The ilmenite reduction step of (1) is mildly endothermic; $\Delta H = +9.7$ kcal/gm-mol at 900°C as calculated from Williams and Erstfeld's data (1979). Reduction is, however, reversible and strongly equilibrium-limited. Figure 2 shows equilibrium conversion of a pure H_2 feed as a function of temperature. At 900°C, the per-pass conversion of H_2 cannot exceed 7.4%. This is important because it means that required H_2 circulation rates are large compared to O_2 production rates since

$$\text{Required molar } H_2 \text{ circulation rate} = \frac{2 \cdot \text{Required molar } O_2 \text{ production rate}}{(H_2 \text{ conversion})} \quad (2)$$

Large circulation rates imply substantial premiums on devising energy-efficient ways to circulate the recycled H_2 with minimal temperature cycling. Both the countercurrent gas/solids heat exchange and the solid-state electrolyzer of Fig. 1 contribute to this type of improved energy efficiency.

Scale of Operations

Initial lunar base planning indicates the need for an oxygen production facility producing approximately 1000 metric tonnes of liquid O_2 product annually. Reasonable assumptions of reactor operating conditions are as follows: temperature = 900°C; per-pass hydrogen conversion = 5% (about 2/3 of the equilibrium value); superficial gas velocity in fluid

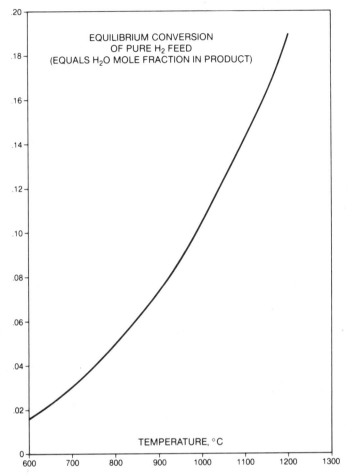

Figure 2. Equilibrium for the reaction $FeTiO_3 + H_2 \rightleftarrows Fe + TiO_2 + H_2O$.

beds = 1 ft/s; gas velocity in circulating line = 50 ft/s. These assumptions lead to an estimated hydrogen circulation rate of about 2200 standard cubic feet per minute , about the capacity of a 14- to 16-inch diameter ventilating fan. The reactor and circulation line diameters can be estimated as a function of operating pressure as shown in Table 1. Note that these assumptions fix only the reactor *diameter* and not the bed heights or volumes. Specifying heights or volumes requires reaction rate, heat transfer, and fluidized bed expansion data. We are presently obtaining these data in our experimental and engineering study.

These sizes appear to pose no problem, particularly at the higher pressures. Engineering evaluations are necessary to find the best balance between smaller diameter and thicker walls at high pressures versus larger diameters but thinner walls at low pressures.

Table 1. Reactor and Circulating Line Diameters

Operating Pressure, KPa	Reactor Diameter	H_2Circulation Line Diameter
100 (1 atm)	13.5'	12"
200 (2 atm)	9.5'	10"
500 (5 atm)	6.0'	6"
1000 (10 atm)	4.5'	4"

Other Design Considerations Specific to Lunar Operations

Tables 2–4 list lunar environmental features differing greatly from Earth and identify specific consequences of these features on the design. Tables 2 and 3 discuss the more familiar, broad implications applying to design of any lunar processing plant. Table 3, in particular, shows design responses required to minimize adverse effects on plant operability/service factors. Table 4 concentrates on reduced gravity effects and shows the quantitative effect of gravitational acceleration on several important fluidized bed parameters. These results were derived from general fluidization correlations collected in Davidson and Harrison (1963) and Kunii and Levenspiel (1969). The overall conclusion from Table 4 is that, compared to terrestrial fluid beds with similar gas rates, lunar fluidized

Table 2. Lunar Versus Terrestrial Environment—Effects on Processing

Environmental Feature	Comparison to Earth	Processing Consequences
Gravity	Moon: 1/6 g Earth: 1 g	Major effects on fluidized beds, gas–solids transport systems, gravity flow of liquid and particulate solids
Surface Temp. Range	Moon: About 290°C (-140°C–+150°C) Earth: 30°C	Widely fluctuating as-mined feed-solids temperature
Atmosphere/Coolants	Moon: Air/Water Absent Earth: Air/Water Abundant	Only closed-loop fluid systems usable; final heat rejection by radiation or heat pipe; unlimited hard vacuum available
Conventional Fuels	Moon: Absent Earth: Plentiful	Process heating by electricity or direct solar; power generation by nuclear or solar
Human Access	Moon: Difficult/Minimal Earth: Easy/Frequent	Extreme emphasis on mininum maintenance , modular replacement

Table 3.   Other Lunar Environment/Design Effects

Environmental Feature	Design Response
Fluctuating Surface (Feed Solids) Temperature	Provide agitated holding bins to average out Overdesign preheat capacity
Lack of Coolants/Conventional Fuels	Use heat integration to reduce energy demand, heat rejection duties
	High heater-to-process coefficients desirable for make-up heat supply Efficient, low-weight radiators desirable

Difficult Human Access	Redundancy/automated change-out for high-maintenance items:
	  Pumps and blowers
	  Solids feeders
	  Electric resistance heaters
	Overdesign/minimize use of high-wear items
	  Shaft seals
	  Rotating surfaces in dusty or gritty service

beds will need larger particles and/or lower fluidizing velocities and will expand substantially more. The larger particles may not have the adverse effect of larger bubbles/poorer contacting efficiency that they would have on Earth.

Table 4.   Reduced Gravity Effects on Equipment Design

Parameter	Approximate Dependence on g	Lunar vs. Terrestrial Design
Fluidized Bed Reactors, Solids Standpipes		
Minimum Fluidization Velocity, U_{mf}	$g^{1.0}$	Operable gas velocity range is from U_{mf} to U_t; must use larger particles or lower velocities
Particle Terminal Velocity, U_t	$g^{2/3}$-$g^{1.0}$	Larger particles→larger bubbles→poorer contacting efficiency
Bubble Diameter	$g^{0.4-1.0}$	Smaller bubbles mean better contacting efficiency; gravity effect counters particle size effect on bubble size
Bed Expansion	$1/(g^{0.7-1.0})$	Taller bed required for same inventory
Standpipe Throughput	$g^{0.5}$	Taller standpipes for same throughput
Fixed Bed Reactors		
No major effects		
Liquid Pumps		
Suction Head	$g^{1.0}$	Taller suction legs or low NPSH pumps required

REFERENCES

Agosto W. N. (1984) Electrostatic concentration of lunar soil ilmenite in vacuum ambient (abstract). In *Papers Presented to the Symposium on Lunar Bases and Space Activities of the 21st Century*, p. 24. NASA/Johnson Space Center, Houston.

Carroll W. T. (editor) (1983) Research on the use of space processes. *JPL Publication 83-36*, Jet Propulsion Laboratory, Pasadena. 200 pp.

Criswell D. R. (1983) A transportation and supply system between low earth orbit and the moon which utilizes lunar derived propellants (abstract). In *Lunar and Planetary Science XIV, Special Session Abstracts*, pp. 11–12. Lunar and Planetary Institute, Houston.

Davidson J. F. and Harrison D. (1963) *Fluidized Particles.* Cambridge University Press, Cambridge. 155 pp.

Davis H. R. (1983) Lunar oxygen impact upon STS effectiveness (abstract). In *Lunar and Planetary Science XIV, Special Session Abstracts*, pp. 13–14. Lunar and Planetary Institute, Houston.

Kesterke D. G. (1971) Electrowinning of oxygen from silicate rocks. *U. S. Bureau of Mines Report of Investigations 7587*, Reno, Nevada. 12 pp.

Kunii D. and Levenspiel O. (1969) *Fluidization Engineering.* Wiley, New York. 534 pp.

Perry R. H. and Chilton C. H. (editors) (1973) *Chemical Engineers' Handbook,* 5th Edition. McGraw-Hill, New York. 1958 pp.

Rao D. B., Choudary U. V., Erstfield T. E., Williams R. J., and Chang Y. A. (1979) Extraction processes for the production of aluminium, titanium, iron, magnesium, and oxygen from nonterrestrial sources. In *Space Resources and Space Settlements* (J. Billingham, W. Gilbreath, B. O'Leary, and B. Gosset, eds.), pp. 257–274, NASA SP-428, NASA, Washington.

Steurer W. H. (editor) (1983) *Extraterrestrial materials processing.* JPL Publication 82-41, Jet Propulsion Laboratory, Pasadena.

Weissbart J. and Ruka R. (1962) Solid electrolyte fuel cells. *J. Electrochem. Soc., 109*, 723–726.

Williams R. J. and Erstfeld T. E. (1979) *High Temperature Electrolyte Recovery of Oxygen from Gaseous Effluents from the Carbo-chlorination of Lunar Anorthite and the Hydrogenation of Ilmenite: A Theoretical Study.* NASA TM-58214, NASA, Washington. 51 pp.

Williams R. J., McKay D. S., Giles D., and Bunch T. E. (1979) Mining and beneficiation of lunar ores. In *Space Resources and Space Settlements* (J. Billingham, W. Gilbreath, B. O'Leary, and B. Gosset, eds.), pp 275–288. NASA SP-428, NASA, Washington.

OXYGEN EXTRACTION FROM LUNAR MATERIALS: AN EXPERIMENTAL TEST OF AN ILMENITE REDUCTION PROCESS

Richard J. Williams

NASA/Johnson Space Center, Code SN12, Houston, TX 77058

The reaction of ilmenite with hydrogen to produce water has been studied experimentally in order to evaluate the effectiveness of using a cold trap to improve yields in a continuous flow process. Yields were enhanced, but not to the degree observed in batch processing systems. The terrestrial simulant used in these studies contained traces of iron sulfide, which released H_2S during processing with a deleterious effect on several components of the test system. More sophisticated testing should be undertaken to obtain kinetic data and attention given to the removal of sulfides in the pre-process beneficiation.

INTRODUCTION

Several studies (*e.g.*, Driggers, 1976; Bock, 1979) have indicated that large quantities of propellant are required to support orbital transfer operations in an expanded space transportation system. A large energy penalty is paid for lifting propellant from the Earth's surface to orbits at which it is required. If a source for propellant were available in space, or at least from sources with a shallower gravity well than the Earth's, the efficiency of an orbital transfer system could be improved (Criswell, 1983; Davis, 1983). Other studies (*e.g.*, Rao *et al.*, 1979; Criswell, 1978, 1980) have proposed that the rocks and minerals found on the Moon are a source of oxygen for use as the oxidizer in H_2/O_2 propulsion systems. The reduced energy cost associated with lifting the oxygen into space from the Moon's relatively shallow gravity well, combined with the fact that oxygen makes up 86% by weight of propellant in LOX–H_2 systems, suggests that substantial net benefit could accrue to the overall transportation system if the production of LOX from lunar materials is feasible.

One of the most simple of the possible extraction processes involves the reaction of hydrogen with iron-oxide-bearing lunar minerals to produce water, from which hydrogen and oxygen are then extracted by electrolysis. Ilmenite ($FeTiO_3$) is an ideal lunar mineral on which to base an extraction process: it is the most abundant of the oxide minerals (up to 10%, modal, in some Apollo 11 soils); it can be readily separated from other non-reactive soil components (Agosto, 1983); it is relatively rich (58% by weight) in the reducible FeO component; and it is chemically quite pure, so that the possibility of side reactions is reduced. In addition, the production reaction, $H_2 + FeTiO_3 = Fe + TiO_2 + H_2O$, is very similar (chemically and thermodynamically) to those involving other iron oxides and iron silicates; thus, a study of it serves as a good general model. The major technical problem is that at moderate temperatures (700–1000°C) the per pass conversion of H_2 to H_2O is less than 5%.

Although the conversion can be enhanced by running the process at higher temperatures, major problems (e.g., degradation of materials, hydrogen loss by diffusion, and sintering of solid reactants and products) occur that limit the viability of this solution. Williams et al. (1979) suggest that the per pass yield could be substantially improved by buffering the H_2O pressure over the reaction at the water liquid–vapor equilibria. Essentially, a cold trap is included in the system; water is pumped from the reaction zone producing a greatly enhanced yield. Williams and Mullins (1983) reported preliminary experimental results that verified that the process modification by Williams worked in a batch mode. It is the purpose of this report to present the results of a study of a variation of the process suitable for operation in a continuous mode.

THEORY

If the solid phases are pure and if the pressures and temperatures are such that the gases behave ideally, the reaction $H_2 + FeTiO_3 = Fe + TiO_2 + H_2O$ is described by the equation

$$K = PH_2/PH_2O \tag{1}$$

in which K is function of temperature only and is computable from basic thermochemical data, and PH_2 and PH_2O are the equilibrium partial pressures of hydrogen and water. Defining $P_{l/v}$ to be the pressure of H_2O in equilibrium with liquid and constraining $PH_2O = P_{l/v}$, (1) becomes

$$K = PH_2/P_{l/v}$$

Since $P_{l/v}$ is primarily a function of the cold trap temperature, it will not be the same as PH_2O (equation 1) except fortuituously; thus, PH_2 must be different from that in (1). We will call it PH_2' and write

$$K = PH_2'/P_{l/v} \tag{2}$$

The per pass conversion of H_2 is defined as

$$C = 1 - (P - P^iH_2)/P^iH_2$$

where P is either the hydrogen pressure given by (1) or (2), and i denotes the initial hydrogen pressure. The calculated conversions are summarized in Table 1. Comparing the conversions, one sees that the incorporation of a cold trap theoretically can improve yields by as much as 23-fold. Of course, the calculations assume thermodynamic equilibrium. Williams and Mullins (1983) reported on experiments that demonstrated that in the batch mode (that is, the one in which overall equilibrium should be attained)

Table 1.   Theoretical Yields

T(°C)	C(W/O Cold Trap)	C(W/Cold Trap)
600	0.031	0.73
700	0.044	0.81
800	0.057	0.85

Notes: C (W/O Cold Trap) computed from Equation 1.
C (W/O Cold Trap) from Equation 2, using an initial hydrogen pressure
at 10 psi and cold trap temperature of 10°C.
Thermodynamic data from Robie *et al.* (1978).

enhancements in yield were obtained using "cold-trap technology" that were consistent
with theory. Here we discuss follow-on experiments in which recirculation of hydrogen
was simulated; in this type of process, equilibrium will only be *approached* due to flows
in the system. The major issue here is whether any enhancement in yield is produced
using a cold trap.

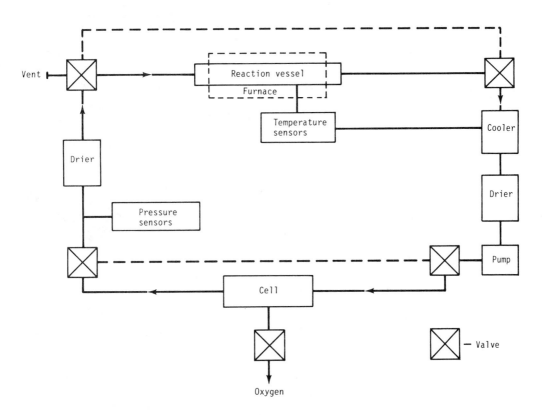

Figure 1.   *Schematic of Oxygen Production Test System. Arrows show direction of gas flow. Dashed lines are
bypass used during purge of system. Power, control, and data acquisition subsystems are not shown.*

EXPERIMENTAL APPARATUS

The apparatus (see Fig. 1) used for the test was constructed from commercial components with the exception of the reactor vessel that was designed and fabricated in-house and driers that were constructed from nylon rod. Stainless Swagelock or Cajon-type fittings were used throughout the system (which was constructed using stainless tubing). The reactor vessel was fabricated from 304 stainless steel and used a copper gasket seal; it was designed to ASME Boiler and Pressure Vessel Code specifications (Section III, division 2) for 1500°F and 100 psi. The vessel had an outside diameter of 1-1/2", a bore of 3/4", and was 11" long. The furnace was a commercial nichrome-wound tube furnace controlled by a thermocouple-driven proportional controller. The cooler was a spiral of copper tubing soldered to a bank of four thermoelectric coolers; an instrument cooling fan provided air circulation for removing heat from thermoelectric devices. The pump was a light duty circulating pump that provided about 2 psi pressure head. Finally, the cell was a General Electric hydrogen generator with the electrolysis membrane reversed so that hydrogen was released on the "dry" side of the system. The masses of the subsystems and the total system volume is given in Table 2.

The temperature, pressure, and the power consumption of the various active components were measured during testing. The small quantities of water produced during the test could not be recycled and measured simultaneously without recourse to extremely long experiments. Thus, yields were measured by absorbing the water produced and comparing the yields from experiments with and without a cooler.

TESTING AND DATA

Results from the following tests are the basis of this paper:

1. Hydrogen flow, cooler on, Al_2O_3 granules in the reactor
2. Hydrogen flow, cooler off, ilmenite in the reactor
3. Hydrogen flow, cooler on, ilmenite in the reactor

Table 2.   Component Masses

Components	Mass (Kg)
Pump	3.5
Furnace	2.7
Vessel (empty)	0.8
Electrolysis cell	2.0
Power supply for cell	3.0
Condenser	1.1
Power supply for condenser	3.9
Tubing, miscellaneous	1.5

Total volume of closed loop (including vessel and filters) 202 cm^3.

Table 3.   Chemical and Mineralogical Analyses

	Wt %
SiO_2	2.72
TiO_2	38.75
Fe_2O_3	54.21
MnO	0.45
MgO	2.02
CaO	0.39
SO_3	0.90

Mineralogical Analysis: X-ray analysis indicates that the sample is a mixture of ilmenite with calcic plagioclase, olivine, and hornblende. SEM analysis of the ilmenite indicates a composition of $Ilm_{97} Hm_3$. It also revealed the presence of a minor sulfide with the composition FeS.

Nominal test conditions were: P_{H2} = 10 psi; reactor temperature = 705°C; condenser temperature = 3°C. The system, set for bypass (see Fig. 1), was thoroughly dried and tested for leaks using helium flow between tests. The vessel contained approximately 90 g of ilmenite (black, massive variety from Quebec; see Table 3 for chemical and mineralogical analysis) that had been crushed, sieved to less than 150 microns, and maintained in a vacuum oven at 80°C until used. The ilmenite was not changed between Test 2 and 3. Water yield was measured by weighing the drier (loaded with "drierite") before and after the tests, each of which was nominally one hour in duration. The data from the three tests are shown in Table 4. Voltages and current measurements indicate that about 2400 watts were required to maintain the system under operational conditions (see Table 5 for breakdown by subsystem).

Table 4.   Test Results

Weight	Gain	Duration
Blank, cold trap	3 mg	49 min.
Ilmenite, no cold trap	15 mg	63 min.
Ilmenite, cold trap	65 mg	76 min.

Reactor temperature = 703 ± 5°C.
Cold trap = 3 ± 2°C.
Room temperatures = 22 ± 0.5C.

Note: ± are observed fluctuations; values are average of measured values.

Table 5. Power Consumption

Source	Watts
Furnace	1475
Electrolysis cell	677
Cooler	89
Pump	77
Sensor, *etc.*	86

Note: Values are averages of measured values.

ANALYSIS

Table 6 presents the yield of water (mg/min) and compares Test 2 (no cold trap) with Test 3 (cold trap). It is assumed that the yields are linear with time, and that the minor variations in parameters from experiment to experiment and within a given experiment as a function of time do not significantly affect the results. The data indicate that yields and the specific yeilds (yield/unit mass and yield/unit power) are all higher by a factor of 3–4 for the test with cooling versus the equivalent test without cooling. Thus, enhancements noted by Williams and Mullins (1983) for batch operation are evident in continuous mode, although they are not as large as the factor of 20 suggested by theory.

SUMMARY

The enhancement of water production from the hydrogen reduction of ilmenite by the use of cold traps has been demonstrated for a continuous flow process. However, the improvements were not as dramatic as those predicted from theory or those obtained in batch operation. The result is not particularly surprising, since the calculated enhancements will apply only to a system of overall equilibrium—that is, only static systems, like batch processing, will show the full benefit of cold-trapping. It is significant that some enhancement has been noted for the continuous flow process, since this suggests that an optimal continuous flow process may exist in which the benefits of enhanced yield can be combined with those of continuous processing. Studies to define this optimum would be useful.

Table 6. System Performance

Test No.	Yield (mg/min)	Power (W)	mg/KW-Hr	Mass (Kg)	mg/Kg-min
2	0.18	2315	6.17	13.5	0.013
3	0.80	2404	21.34	18.5	0.043

Notes: Yield is net from Table 2 minus baseline (0.06 mg/min). Mass is from Table 1; condenser and its power supply mass have been removed from mass budget for Test 2.

The current data are limited by the number of tests performed and duration of those tests. However, further testing would require significant redesign of the experiment in two ways:

1.   The relative size of the electrolysis cell and the reaction vessel need to be better matched. The cell used in these experiments is at least a factor of 10 too large; consequently, an indirect (water absorption) technique had to be used to measure production. A much smaller cell or a much larger vessel would have permitted direct observation of water levels during the experiment.

2. The terrestrial ilmenite used in these studies contains minor sulfide. The H_2S produced by the reaction of the sulfides with hydrogen attacked the electrolysis cell membrane; it was this that forced termination of testing. Lunar ilmenite is also associated with sulfides. Thus, any process will have to deal with these contaminants, using physical or chemical techniques to purify starting materials or efficient gases. Technologies (see, for example, King, 1974) exist that can handle such problems; however, these add complexities and costs to the system that have not been fully appreciated.

We believe that the test program described in this paper was successful in demonstrating the principles of operation, quantifying yields, and isolating design problems for future researchers. However, it must be emphasized that we were mainly dealing with feasibility and preliminary concept testing; thus, we could be sloppy to some degree—for example, the presence of ferric iron in our ilmenite is not critical. The next generation of tests will require more sophistication and a more complete simulant to obtain quantitatively relevant process data. When better simulants are available, further testing should proceed to explore the behavior over longer times and to optimize production. Hopefully, our experience is a useful introduction to such research.

Acknowledgments.   *This work was supported by Johnson Space Center Discretionary Funds. The technical support of O. Mullins, A Lanier, W. Carter, and R. Ybanez of LEMSCO was instrumental in designing and constructing the test system and conducting the tests. Kathie Ver Ploeg obtained the X-ray and SEM analysis for the study. Finally, I wish to thank L. A. Taylor for his efforts in reviewing and improving the paper.*

REFERENCES

Agosto W. N. (1983) Electrostatic separation of binary comminuted mineral mixtures. In *Space Manufacturing 1983*, Advances in Astronautical Sciences, Am. Astronaut. Soc. Rpt. 83-231, pp. 315-334. American Astronautical Society, Tarzana. CA.

Bock E. (1979) *Lunar Resource Utilization for Space Construction.* GDC-ASP-75-001, General Dynamics Convair Division, San Diego. 109 pp.

Criswell D. R. (1978) *Extraterrestrial Materials Processing and Construction, Final Report,* NASA CR-158870, NASA, Washington. 476 pp.

Criswell D. R. (1980) *Extraterrestrial Materials Processing and Construction, Final Report,* NASA CR-167756, NASA, Washington. 655 pp.

Davis H. R. (1983) Lunar oxygen impact on STS effectiveness. *Eagle Engineering Rpt. No. 8363,* Eagle Engineering, Inc., Houston. 44 pp.

Driggers G. W. (1976) Some potential impacts of lunar oxygen availability on near-Earth space transportation (abstract). In *Lunar Science VII, Special Session Abstracts*, pp. 26–34. Lunar Science Institute, Houston.

King W. R. (1974) Air pollution control. In *Riegel's Handbook of Industrial Chemistry* (James A. Kent, ed.), pp. 874–893. Van Nostrand, New York.

Rao D. B., Chowdery U. V., Erstfeld T. E., Williams R. J., and Chang Y. A. (1979) Extraction processes for the production of aluminum, titanium, iron, magnesium, and oxygen from nonterrestrial sources. In *Space Resources and Space Settlements* (J. Billingham, W. Gilbreath, B. Gossett, and B. O'Leary, eds.), NASA SP-428, pp. 257–274. NASA, Washington.

Robie R. A., Hemingway B. S., and Fisher J. R. (1978) Thermodynamic properties of minerals and related substances at 298.15 K and 1 Bar (10^5 Passals) pressure and at higher temperatures. *U.S. Geol. Surv. Bull. 1452.* U.S. Govt. Printing Office, Washington. 456 pp.

Waldron R. D. and Criswell D. R. (1984) Materials processing in space. In *Space Industrialization*, Vol. 1, pp. 97–130. CRC Press, Cleveland.

Williams R. J., McKay D. S., Giles D., Bunch T. E. (1979) Mining and beneficiation of lunar ores. In *Space Resources and Space Settlements* (J. Billingham, W. Gilbreath, B. Gossett, and B. O'Leary, eds.), NASA SP-428, pp. 275–288. NASA, Washington.

Williams R. J. and Mullins O. (1983) Enhanced production of water from ilmenite (abstract). In *Lunar and Planetary Science XIV, Special Session Abstracts*, pp. 34–35. Lunar and Planetary Institute, Houston.

A CARBOTHERMAL SCHEME FOR LUNAR OXYGEN PRODUCTION

Andrew Hall Cutler

California Space Institute, A-005, University of California at San Diego, La Jolla, CA 92093 and Energy Science Laboratories, Suite 113, 11404 Sorrento Valley Road, San Diego, CA 92121

Peter Krag

Department of Metallurgical Engineering, Colorado School of Mines, Golden, CO 80401

Carbothermal reduction may be the process of choice for winning oxygen from lunar materials. A wealth of terrestrial engineering experience can be applied to development of a lunar plant. Coproducts, such as slag and steel, can be utilized in lunar operations and may be useful lunar exports. Carbon is a more efficient reductant on a per mass basis than is hydrogen in the competing ilmenite reduction process. An accurate plant design awaits further research to address uncertainties in process kinetics, behavior, catalysts, and reactor materials.

INTRODUCTION

The most useful material to produce on the Moon is oxygen for propellant (Cole and Segal, 1964). During the Apollo flights, about 75% of the Saturn V's effort placed propellant in lunar orbit and on the lunar surface to enable return to Earth. A lunar source of oxygen could double or triple the net lunar base mass (and capability) for a given Earth to LEO mass throughput. It also can provide inexpensive gas for a number of uses in lunar exploration. Lunar oxygen may even be exported to LEO at costs arguably below Earth to LEO launch costs (Salkeld, 1966; Davis, 1983; Andrews and Snow, 1981; Cutler, 1984; Cutler and Hughes, 1985; Simon, 1985).

Transportation requirements and transportation economics for the utilization of lunar resources in Earth orbit are reasonably well understood (although market models are somewhat ill-defined). However, the actual production of basic commodities from available lunar minerals is poorly understood (Duke and Arnold, 1977; McKay and Nozette, 1985). The lunar environment is sufficiently different from the terrestrial environment that plant design and process chemistry will differ substantially from that used on Earth. Currently no accurate plant designs exist that can be used in systems studies or mission definition.

A candidate approach for winning oxygen from lunar minerals is the carbothermal process (Fig. 1), which yields steel as a necessary by-product. The process combines chemistry from steel-making and from coal synthesis gas reforming with electrolysis or thermolysis of water. Research was carried out on the carbothermal reduction process in the early 1960s and was reported in the technical literature and NASA technical reports (Rosenberg *et al.*, 1963a,b; Rosenberg *et al.*, 1964a,b,c,d,e; Rosenberg *et al.*, 1965a,b,c,d,e,f).

Every step in this process has been used extensively in engineering practice. The terrestrial data provide an excellent framework for research on and possible development of this process for the extraterrestrial environment.

The near-term benefits of lunar oxygen are recognized. Lunar steel may also be a near-term useful product. It could be utilized in many lunar base components such as roadbeds, landing pads, instrument foundations, walkways, stairways, pressure vessels, pipes, drill rigs, and cables. A source of lunar steel combined with lunar base propellant production could enable various large scale space projects. If we continue to transport all needed materials from Earth, space activities in the 21st Century will be as they are now—limited and expensive. We must learn to use space resources if we are to change our presence in space from tentative peeks to vigorous exploration and exploitation. The carbothermal processing plant is a good example of a technology that takes maximum advantage of terrestrially derived engineering knowledge to start using extraterrestrial resources.

PROCESS CONSIDERATIONS

The carbothermal process for producing lunar oxygen (Fig. 1) starts with mining regolith and separating out a desirable mineral fraction, probably ilmenite. About 100,000 tons of lunar regolith must be mined per year to produce 1000 tons of oxygen, based on a 10% usable ilmenite content. This corresponds to digging a pit 100-m × 100-m

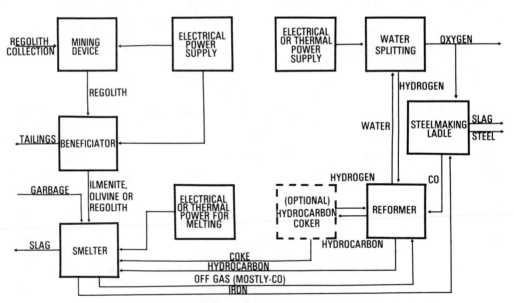

Figure 1. *Flow diagram of the carbothermal oxygen production plant. Regolith is mined; steel and oxygen are produced; and tailings, iron-making slag, and steel-making slag are discarded.*

× 5-m deep over the year, about the amount of material excavated in building 1.5 miles of interstate highway. A beneficiator removes oversize material and separates the remainder into tailings and feed material, about 90% ilmenite combined with 10% flux. The tailings are disposed of down a conveniently located slope, and the feed materials charged to the smelter.

In the proposed lunar system, the feed contains ilmenite, anorthite as a fluxing agent, and any carbon-bearing solids garnered from garbage or the off-gas stream. Phase diagrams of the system, $TiO_2 - SiO_2 - Al_2O_3 - CaO$ (Nurnberg, 1981), show that a 10% addition of anorthite is sufficient to lower the slag melting point below that of the iron product.

The overall process chemistry in the plant is:

$$FeTiO_3 \rightarrow Fe + TiO_2 + 1/2\,O_2 \qquad (1)$$

This is composed of the following steps (not showing partial reduction of TiO_2):

$$FeTiO_3 + (1 + x)C \rightarrow FeC_x + CO + TiO_2 \quad \text{(slag–metal bath reduction)} \qquad (2)$$

$$FeC_x + x/2\,O_2 \rightarrow Fe + x\,CO \quad \text{(iron decarburization)} \qquad (3)$$

$$yCO + (2y + 1)\,H_2 \rightarrow yH_2O + C_yH_{2y+2} \quad \text{(reforming)} \qquad (4)$$

$$C_yH_{2y+2} \rightarrow C_y + (y + 1)H_2 \quad \text{(hydrocarbon cracking)} \qquad (5)$$

$$H_2O \rightarrow H_2 + 1/2\,O_2 \quad \text{(electrolysis)} \qquad (6)$$

In the smelter, feed material is mixed with a carbonaceous reductant and heated until melting, reduction, and slag/metal separation take place. The reductant should be as hydrogen poor as possible to keep the volume of smelter off-gas down and to reduce heat loss in this stream. The reductant is recycled carbon, reformed from monoxide to waxes by Fischer-Tropsch synthesis and is perhaps cracked (4,5). Unfortunately, these waxes do not coke well, and research and development may be required to improve the yield of crackable products. The light ends can be reformed either by partial reoxidation or arc cracking to give materials more suitable for coking, at the cost of substantially increased power consumption.

Carbon is the refractory of choice for this reactor due to its high strength to weight ratio, its resistance to attack by the (titanium carbide saturated) iron product, and its availability from electrode wear, garbage, and the recycled off-gas. Carbon is used in terrestrial ilmenite smelters and, with the slag skull, is found to be semi-permanent in service (Noda, 1965). Some operating data are available (Knoerr, 1952; Noda, 1965; Grau and Poggi, 1978) from plants that use a thermal gradient stabilized slag skull to protect the refractory lining from chemical attack. Process power consumption is about 3.5 MWh/ T Fe (megawatt-hours per ton of iron), and arc heating electrode consumption is between

5 and 10 kg/T Fe, which translates to converting about 0.75% of the carbon flow in the total process into electrodes. Arc heating uses low voltage DC power like that generated in photovoltaic or nuclear thermionic processes.

Substantial research has been done on processes for iron production by introducing hydrocarbon, coal, or coke into a slag–metal bath (Eketorp *et al.*, 1981). Because reduction in the bath (2) is an endothermic process, it has not proven possible to get rapid reaction on the scale of typical terrestrial iron reduction furnaces (10,000 tons of hot metal produced per day). However, calculations indicate that heat transfer in the bath is more than adequate for the small scale of operation relevant to lunar iron reduction plants.

One advantage of smelting ilmenite is that under process conditions the titania will be partially reduced. This increases oxygen recovery per unit throughput and decreases the amount of iron that must be decarburized per unit oxygen produced. However, feed minerals for lunar iron production have low iron content compared to terrestrial ores, causing a high ratio of slag to iron and increasing the heat demands and reactor size per unit oxygen (or iron) production.

The pig iron product will contain carbon, and decarburization (3) is required to make a usable product and to obtain high carbon recycle efficiency. Decarburization could employ steam, oxygen, or wet hydrogen. Low lunar gravity is helpful to decarburization by allowing longer gas/liquid contact times. Steam decarburization is endothermic and thus would not lead to hot spots in the bath or the severe refractory wear associated with oxygen decarburization. Alternatively, oxygen decarburization is exothermic, so the vessel would need no heat source. The overall oxygen/steam ratio could be adjusted to bring the reactor into thermal balance. Although power availability and materials arguments imply that consumable electrode arc heating would be the most practical approach to heat introduction, induction heating, which also removes the need for hydrocarbon cracking and electrode production, is possible. Modern induction power converters have masses ranging from 2 metric tons at 1MW to 3.5 metric tons at 2MW. Conversion efficiency is typically 96%, adding 40–80 kW to the heat load that must be rejected at low temperatures.

Counter-current heat exchange between pelletized feed and reactor off-gas in a moving bed could decrease the electrical energy requirement. It would also cool the off-gas to suitable temperature for subsequent processing and would condense volatile impurities.

Both the decarburization and reduction off-gases will contain sulfur that could poison the Fischer-Tropsch catalysts. The best solution is to use resistant catalysts, or else to pass the off-gases through a water spray made basic by alkali oxide fumes. (This water spray is a small added inventory of water on its way to the electrolyzer.)

Water is a by-product of Fischer-Tropsch synthesis; its electrolysis (6) leads to the product oxygen, which is liquified and stored. Electrolysis has been extensively studied in a space context. Literature also exists on using thermal cycles for production of hydrogen and oxygen from water, often using solar heat. Thermal water-splitting, if available, would significantly reduce electrical power demands. Although processes making direct use of solar radiation are known (Raissi and Antal, 1985), further research and design studies are needed.

SYSTEM CONSIDERATIONS

A variety of considerations apply to any system involving the production of oxygen on the Moon and its delivery to Earth orbit. Tradeoffs must be made involving solar versus nuclear power supply and continuous versus batch processing. Systems design and economic criteria will be quite different if a lunar oxygen producing plant is installed before a lunar base or as part of a lunar base program.

A coexisting lunar base would have several beneficial effects on lunar factory operations and economics, since routine and non-routine repairs could be attended to quickly and would be a marginal cost (the marginal cost of extra person-hours) rather than a direct cost with the addition of transport charges. Propellant oxygen for transportation from the Moon is an economical product even under circumstances where lunar oxygen export to Earth orbit may prove uneconomical. Inexpensive plant by-products, such as cast slag bricks or formed steel parts, may have some utility and economic value at the lunar base.

Without a coexisting base, reliability would become very important and would have to be ensured even at a high cost in mass or initial factory cost. One simple means of enhancing product supply reliability is to install over-capacity, so that unexpected downtime does not deplete product stores. Another approach is to have redundant plant elements so that unscheduled maintenance does not stop all production. The costs and benefits of over-capacity are obvious. The change in capacity required for reasonable reliability (10–20%) is not enough to affect the plant systems design substantially. Redundancy may be a different story because the total capacity is shared by multiple small plants. Since plant masses (and capital costs) scale according to the 2/3 power of capacity, the ratios of costs for no redundancy, dual redundancy, and triple redundancy are 1:1.26:1.44.

For a carbothermal oxygen plant, the reduction in capacity of each process reactor with increased redundancy means that the smelting reactor will become more difficult to operate in a continuous mode. The costs of batch and continuous processes scale differently, with large scale favoring continuous processing. Continuous operation is very desirable on the Moon due to the ease of automation and automatic control, simplified design, lower thermal cycling, and facile use of gravity feed when compared to batch processing. If the capacity reduction needed to make redundant oxygen factories forced a transition from continuous to batch smelting, the capital cost penalties would increase substantially. Electrolysis, reforming, and cracking units are operated economically on Earth in continuous mode on a scale similar to that necessary for a lunar factory capable of producing 1000 tons of oxygen per year. Despite possible increased costs, redundancy is very effective in reducing the likelihood of unexpected supply interruptions if a plant must be operated at an unmanned lunar facility. Single versus redundant plants must be traded off in any systems study of lunar oxygen production at an unmanned facility.

Continuous process ilmenite smelters are operated on Earth at 10–20 times the capacity needed in a 1000 ton per year of oxygen lunar factory, but the thermal balance in these

Table 1. Energy and Power Requirements per Ton of Product Oxygen

Process Step	Energy (GJ)
Reduce 3.68 tons of iron (75% efficiency)	12.1
Heat to melt 3.68 tons of iron (500 kWh/ ton)	6.6
Heat content of 4.25 tons of slag (470 kWh/ ton)	7.2
Heat content of off gas (1350 oC effective heating)	8.0
Energy to electrolyze water (60% efficiency)	28.9
Energy to liquify oxygen	5.4
Total energy consumption, carbothermal process	68.2
Total power requirement for 1000 tons of O_2 per year (100% duty cycle)	2.16 MW
Nuclear power plant capacity (90% duty cycle)	2.40 MW
Solar power plant capacity (40% duty cycle)	5.40 MW

Methane reductant, no heat regeneration.
Data scaled from that presented by Rosenberg *et al.* (1965f) for a 12,000 lb/mon oxygen production facility without heat pumping. Power was assumed to scale linearly.

smelting units is delicate. The possibility of scaling a continuous smelting reactor down to the capacity needed for carbothermal production of lunar oxygen must be the subject of further study.

In contrast, all terrestrial steel-making is done in batch processes. Capacities up to several hundred tons of steel per hour are common. It seems unlikely that lunar steel-making can be done in a continuous mode.

Manufacturing plant power demand will be several megawatts (Table 1), well in excess of power demand for other functions of a lunar base. Nuclear energy is likely to be used to power a manned surface facility, due to the difficulty of storing solar derived energy for use during the lunar night. Continuous energy supply may not be a requirement for a lunar factory. Since factory power demands are much higher than and have a different time phasing and quality from other power demands, a total systems design will be required to select the optimum power system for the complex. In fact, the factory and base may have separate power systems.

There are two competing factors in the tradeoff between solar and nuclear energy for the lunar factory power supply: (a) the relative power to mass ratios of solar and nuclear power systems, and (b) the overall plant masses required to produce a given amount of product when the power supply and plant duty cycles are taken into account. Plant mass breakdowns for solar and nuclear powered plants are given in Table 2. The simple assumptions used in this table give solar and nuclear powered plant masses that are essentially equal. Accurate projections of solar and nuclear power system power to mass ratios as well as accurate scaling laws derived from careful design studies will be needed before any decision can be made about the selection of power sources.

The delivery of a lunar oxygen factory to the Moon's surface and the storage and transportation of lunar oxygen are separable from the lunar oxygen manufacturing plant itself in a systems sense. Transportation system characteristics are more likely to be

Table 2. Processing Plant Mass per Unit Output of Product

Product	100% Duty Cycle		90% Duty Cycle		40% Duty Cycle	
	Oxygen	Iron	Oxygen	Iron	Oxygen	Iron
Processing plant	30.4	8.69	32.6	9.32	56.0	16.0
Mining and beneficiation	10.8	3.09	12.0	3.43	27.0	7.7
Oxygen plant mass			44.6		83.0	
Power plant (Table 1):						
Nuclear (40 W/kg)			60			
Solar (190 W/kg)					28.5	
Total mass			104.6		111.5	

In tons per 1000 tons per year.
Data scaled from that presented by Rosenberg *et al.* (1965f) for a 12,000 lb/month oxygen production facility without heat pumping. A decarburization reactor (taken as equal in mass to the reduction reactor) was added. Decarburization is assumed to consume 5% of the ultimate O_2 production. Plant mass is assumed to scale as the 2/3 power of capacity.

determined by space station and lunar base needs than by lunar oxygen factory needs. Nevertheless, there exists a design interaction between maximum size and mass of factory components and the payload capacity of the transportation system.

DISCUSSION

The carbothermal reduction process has been discussed in an appropriate thermodynamic context, and plant mass and power estimates have been made (Rosenberg *et al.*, 1965f). Mass estimates (Table 2) are 32.6 tons of plant and power requirements equal to 60 tons of 1990 (projected) nuclear power supply and mining and beneficiation equipment of about 12 tons per 1000 tons of oxygen production per year. Mining equipment mass is taken from Gertsch (1983), and beneficiation is assumed to be similar to that reported by Agosto (1985). Assuming a mass ratio of iron to oxygen of 3.5:1 (as for stoichiometric oxide), these correspond to 9 tons of plant supply, 17 tons of power supply, and 3 tons of mineral handling equipment per ton of steel produced per day in an ilmenite smelting process.

The carbothermal process for lunar oxygen production and a process based on the hydrogen reduction of ilmenite bear the same relationship to each other as direct reduction and smelting do in terrestrial engineering. It is thus reasonable to expect them to be competitive technologies as they are on Earth.

The hydrogen reduction process reacts the mined ilmenite with hot hydrogen to produce water, iron, and rutile. The water is electrolyzed to produce oxygen and regenerate hydrogen. Thermodynamic equilibrium limits the per pass conversion of hydrogen to water to about 5%. Subsolidus hydrogen reduction of ilmenite produces an intimate mixture of iron and rutile and would require a subsequent thermal processing step to separate the iron.

There are several factors that seem to favor the carbothermal reduction based process. Makeup carbon is available from a variety of sources, such as garbage and scrap. Hydrocarbons and carbon are much easier to handle than hydrogen. This is particularly significant when some reagent inventory must be kept on hand for makeup between supply deliveries from Earth. As process temperature increases, significantly more oxygen can be recovered from a given weight of ilmenite. Reduction carried out under slagging conditions will only require 1/2 to 2/3 as much ilmenite per unit oxygen recovery as would subsolidus reduction at 700°C–900°C.

The major problem with hydrogen reduction of ilmenite appears to be the unfavorable equilibrium constant for the conversion of hydrogen to water. This ranges from 0.031 at 600°C (Williams, 1983) to 0.117 at 1300°C (Shomate *et al.*, 1946). If the hot, wet hydrogen is cooled down to condense out the water and then reheated, an excessive heat demand is imposed on the system. Developing a high temperature electrolysis cell using a ceramic oxide ion conductor to dewater the hydrogen without cooling it adds development costs and risks. In contrast, carbon reacts completely in a single pass. It also extracts more oxygen from ilmenite in a single pass than does hydrogen on a mass basis. (A mass unit of carbon extracts 1.33 to 1.45 mass units of oxygen per pass, while a mass of hydrogen extracts 0.25 to 0.94 mass units of oxygen per pass, depending on the equilibrium conversion limit.)

The following processes that have been proposed to extract oxygen from lunar material cannot be compared directly with the carbothermal process but are considered to have significant technical weakness:

1. Electrolysis of molten lunar minerals has been studied as an oxygen producing process. However, no anode material tried to date has clearly demonstrated adequate corrosion resistance (Haskin and Lindstrom, unpublished data, 1984), leaving some possibility that the process is impractical.

2. Vapor phase pyrolysis of lunar minerals involves vaporizing lunar soil or selected mineral separates and then rapidly quenching the vapor. The hot gas contains some metal atoms and oxygen molecules. Literature data (Borgianni *et al.*, 1969) make it seem unlikely that quenching of the hot gas can be made rapid enough to prevent loss of oxygen through reoxidation.

3. The hydrofluoric acid leach process as described by Waldron (1985) involves dissolving bulk lunar soil in a fluoride based acid, separating the resulting mixed salt solution into pure metal fluorides, reducing these metal fluorides with sodium and potassium, hydrolyzing the resulting alkali fluorides to regenerate hydrofluoric acid, and producing oxygen by electrolysis of molten alkali hydroxides. Design-based mass estimates are 77 tons of plant per ton of soil input per day, or 586 tons of plant per thousand tons of oxygen per year (assuming 40% oxygen by mass of lunar soil and a 90% duty cycle), much higher than for competing processes.

RESEARCH NEEDS

Conceivably, fatal flaws could exist in the carbothermal process in any of three areas where research is badly needed: (a) kinetics and phenomenology of (simulated) lunar ilmenite reduction, (b) carbide solubility in typical slags, and (c) iron decarburization. Experimental data on these key issues would narrow design uncertainties in the two least well understood parts of the system, the smelter and the steel-making reactor.

Once it has been verified that the basic process chemistry for carbothermal production of lunar oxygen is sound, the following secondary research projects would define system performance and design. A Fischer-Tropsch catalyst is needed that gives readily crackable product with a low H to C ratio and which is also economical in the lunar context. Appropriate hydrocarbon coking and electrode production techniques need development. Thermal methods of producing hydrogen and oxygen from water should be explored and traded off against electrolysis in system design studies. Various methods of heat rejection and thermal control should be studied and compared to each other at the system level. All of the above studies would have to be performed to determine whether plant development is warranted.

The choice between a nuclear or a solar powered system depends on the relative power-to-mass ratios and on the plant mass scaling law. The tradeoff seems close now and cannot be resolved without careful systems studies shortly before plant emplacement.

Separating water into hydrogen and oxygen by thermolysis is an interesting possibility for a carbothermal process plant and for a variety of other types of plants. Water thermolysis should be examined in a space context independently of other projects.

A lunar oxygen factory in the context of a lunar base program can be the source of inexpensive by-products (slag and steel) that would be very useful to the base. The properties of slag or steel manufactures cannot be predicted solely on the basis of compositional data. If these by-product materials were to be useful in lunar base activities, their properties would have to be determined experimentally using simulants of appropriate composition.

If development seems to be a possibility, the interface between the oxygen producing factory (after the oxygen liquifier) and the storage and transportation system will have to be studied. If the oxygen storage facilities on the lunar surface are considered part of the transportation system, parameters such as mission frequency have practically no effect on the oxygen producing plant, and few plant parameters have any effect on the transportation system.

CONCLUSIONS

Carbon is an economical reductant in the lunar context because it removes more than 1.25 times its mass of oxygen per pass. Hydrogen, though lighter per mole, only removes 0.4–0.9 times its own mass per pass due to thermodynamic limitations. Makeup carbon might be readily available from spent factory delivery stages, scrap, or lunar base garbage.

Thus the carbothermal route to lunar oxygen is economically and technically attractive. There is low technical and performance risk in plant design and development due to extensive Earth-based experience with each element of the process. The basic research needed to demonstrate the feasibility of this concept is straightforward and well defined. It consists of quantifying carbon loss in slag and steel and studying process behavior in the reduction reactor. Further definition should be pursued, both in the laboratory and through selected system studies.

REFERENCES

Agosto W. N. (1985) Electrostatic concentration of lunar soil ilmenite in vacuum ambient. This volume.

Andrews D. G. and Snow W. R. (1981) The supply of lunar oxygen to low earth orbit. *Space Manufacturing, 4*, 173–179.

Borgianni C., Capitelli M., Cramorossa F., Triolo L., and Molinari E. (1969) The behavior of metal oxides injected into an argon induction plasma. *Combustion and Flame, 13*, 181–194.

Carroll W. F. (editor) (1983) *Research on the Use of Space Resources*. JPL Publication 83-36, Pasadena, CA. 325 pp.

Cole D. M. and Segal R. (1964) Rocket propellants from the Moon. *Astronaut. Aeronaut., 2*, 56–63.

Cutler A. H. (1984) Transportation economics for lunar oxygen utilization (abstract). In *Abstracts of Papers, Third Cal-Space Investigator's Conference*, 55–59. California Space Institute, La Jolla.

Cutler A. H. and Hughes M. L. (1985) Transportation economics of extraterrestrial resource utilization. *Proc. 7th Princeton Space Manufacturing Conf.*, pp. 223–244. AIAA, New York.

Davis H. P. (1983) *Lunar Oxygen Impact Upon STS Effectiveness*. Eagle Engineering Report EEI-8363, Houston. 63 pp.

Duke M. B. and Arnold J. R. (editors) (1977) *Summer Workshop on Near Earth Resources*. NASA CP-2031, NASA, Washington, DC. 105 pp.

Eketorp S., Wijk O., and Fukagawa S. (1981) Direct use of coal for production of molten iron. In *Extraction Metallurgy '81* p. 184.

Gertsch R. E. (1983) A method for mining lunar soil. *Proc. 6th Princeton/AIAA Space Manufacturing Conf., 53*, pp. 337–346. Adv. Astronaut. Sci., Univelt, San Diego.

Grau A. and Poggi D. (1978) Physico-chemical properties of molten titania slags. *Can. Metall. Quart, 17*, 97–102.

Knoerr A. W. (1952) World's major titanium mine and smelter swing into full scale production. *Eng. Min. J., 153*, 72–79.

McKay D. S. and Nozette S. D. (1985) Technological springboard to the 21st Century. *Proc. 1984 NASA-ASEE Summer Study at La Jolla*. In press.

Noda T. (1965) Titanium from slag in Japan. *J. Metals., 7*, 25–32.

Nurnberg K. (editor) (1981) *Slag Atlas*. Verlag Stahleisien, Dusseldorf. 282 pp.

Raissi A. T. and Antal M. J. (1985) *Effects of Heating Rate on Solar Thermal Decomposition of Zinc Sulfate*. Proceedings A.I.Ch.E. Symposium, Series 81, #245, pp. 204–212. Denver.

Rosenberg S. D., Beegle R. L. Jr., Guter G. A., Miller F. E., and Rothenberg M. (1965f) The manufacture of propellants for the support of advanced lunar bases. In *Papers Presented to the National Aeronautic and Space Engineering and Manufacturing Meeting*, pp. 1–16. Society of Automotive Engineers, New York.

Rosenberg S. D., Guter G. A., and Miller F. E. (1963a) *Research on Processes for Utilization of Lunar Resources*. NASA CR-52318, NASA, Washington DC. 29 pp.

Rosenberg S. D., Guter G. A., Miller F. E., and Jameson G. R. (1963b) *Research on Processes for Utilization of Lunar Resources*. NASA CR-58956, NASA, Washington DC. 71 pp.

Rosenberg S. D., Guter G. A., and Miller F. E. (1964a) *Research Processes for Utilization of Lunar Resources*. NASA CR-56885, NASA, Washington DC. 35 pp.

Rosenberg S. D., Guter G. A., and Miller F. E. (1964b) *Research on Processes for Utilization of Lunar Resources*. NASA CR-59167, NASA, Washington DC. 72 pp.

Rosenberg S. D., Guter G. A., and Miller F. E. (1964c) *Research on Processes for Utilization of Lunar Resources*. NASA CR-59197, NASA, Washington DC. 73 pp.

Rosenberg S. D., Guter G. A., and Miller F. E. (1964d) *Research on Processes for Utilization of Lunar Resources*. NASA CR-59633, NASA, Washington DC. 25 pp.

Rosenberg S. D., Guter G. A., and Jameson G. R. (1964e) *Research on Processes for Utilization of Lunar Resources.* NASA CR-53749, NASA, Washington DC. 29 pp.

Rosenberg S. D., Guter G. A., Miller F. E., and Jameson G. R. (1964f) *Catalytic Reduction of Carbon Monoxide with Hydrogen.* NASA CR-57, NASA, Washington DC. 68 pp.

Rosenberg S. D., Guter G. A., Miller F. E., and Beegle R. L. Jr. (1965a) *Research on Processes for Utilization of Lunar Resources.* NASA CR-57484, NASA, Washington DC. 50 pp.

Rosenberg S. D., Guter G. A., Miller F. E., and Beegle R. L. Jr. (1965b) *Research on Processes for Utilization of Lunar Resources.* NASA CR-67821, NASA, Washington DC. 118 pp.

Rosenberg S. D., Guter G. A., Miller F. E., and Beegle R. L. Jr. (1965c) *Research on Processes for Utilization of Lunar Resources.* NASA CR-62857, NASA, Washington DC. 34 pp.

Rosenberg S. D., Guter G. A., and Miller F. E. (1965d) The utilization of lunar resources for propellant manufacture. *Adv. Astronaut. Sci., 20,* 665–680.

Rosenberg S. D., Guter G. A., and Miller F. E. (1965e) Manufacture of oxygen from lunar materials. *Annals of the New York Academy of Science 1106,* pp. 1106–1122.

Salkeld R. J. (1966) Economic implications of extracting propellants from the Moon. *J. Spacecraft and Rockets, 3,* 254–260.

Shomate C. H., Naylor B. F., and Boericke F. S. (1946) *Thermodynamic Properties of Ilmenite and Selective Reduction of Iron in Ilmenite.* U.S. Bureau of Mines Report of Investigation 3864. 19 pp.

Simon M. (1985) A parametric analysis of lunar oxygen production. This volume.

Steurer W. H. (editor) (1982) *Extraterrestrial Materials Processing.* JPL Publication 82-41, Pasadena, CA. 147 pp.

Waldron R. W. (1985) Total separation and refinement of lunar soils by the HF acid leach process. In *Proc. 7th Princeton Space Manufacturing Conf.,* pp. 132–149. AIAA, New York.

Williams R. J. (1983) Enhanced production of water from ilmenite: An experimental test of a concept for producing lunar oxygen (abstract). In *Lunar and Planetary Science XIV,* pp. 34–35. Lunar and Planetary Institute, Houston.

LUNAR REGOLITH FINES: A SOURCE OF HYDROGEN

James L. Carter

*University of Texas at Dallas, Programs in Geosciences, P.O. Box 830688,
Richardson, TX 75083-0688*

The theoretical evaluation of the lunar regolith fines as a primary source of hydrogen reveals that a minimum order of magnitude increase in hydrogen content is possible in beneficiated fines because both particle size and particle shape play a significant role in the relationship of volume percent of surface coating to grain size. The lunar regolith fines meet the basic requirement for beneficiation because a major portion (minimum two-thirds) of the hydrogen occurs in the less than 20-μm-size fraction, a relatively small part of the fines. Beneficiation should be accomplished by a combination of vibratory screening followed by cyclone and/or possibly electrostatic separation. Early exploitation of the lunar regolith fines for hydrogen probably will be limited to hydrogen obtained as a by-product or co-product from the mining and processing for other elements or materials because a minimum of about 13,600 tons to about 19,600 tons of 100 ppm hydrogen-bearing lunar regolith fines will have to be processed with about 3,100 tons to about 4,500 tons, respectively, of concentrate heated to supply 1 ton of hydrogen, yielding a recovery of about 74% to about 51%, respectively, of the hydrogen.

INTRODUCTION

The interest in establishing a permanent manned presence on the Moon makes it imperative to determine if indigenous lunar material can be mined and extracted economically to provide materials needed for a lunar base operation (McKay and Williams, 1979). Importing materials to the lunar surface from Earth initially will cost approximately $25,000 per pound in 1984 dollars—over five times the value of gold on Earth! This paper (a) explores the possibility of the lunar regolith fines as a primary source of hydrogen—a critical element from the viewpoint of consumables (*e.g.*, propellant, water, and hydrogen-bearing reagents) and (b) compares the known lunar data with the developed hypothesis.

The known concentration range of hydrogen in lunar regolith fines is shown in Table 1. Such values are often used as evidence that the Moon is devoid of water even though 100 ppm of hydrogen is equivalent to 0.09 wt % water. It would, however, take 10,000 tons of 100 ppm hydrogen-containing lunar regolith fines to produce 1 ton of hydrogen at 100% recovery. Therefore, if lunar regolith fines are to become an economical source for hydrogen, they must be beneficiated because at least 99.98% of this type of lunar material is "waste" (Table 1) and would consume energy during the extraction of hydrogen. Beneficiation of lunar regolith fines can only occur if a relatively small portion of the material in the fines contains a significant amount of the hydrogen and also if that material can be separated, which implies it must have unique physical or chemical properties. In addition, the process must be economical, that is, not be labor intensive or technically complex.

Table 1. Range of Hydrogen Content of Bulk Lunar Regolith Materials in ppm from Apollo Missions

Apollo Mission	Hydrogen (ppm)
11	38– 60
12	26– 80
14	61–106
15	15–120
16	10– 79
17	17–211

Data from Taylor (1975); Bustin *et al.* (1984).

PREVIOUS MODEL STUDIES

Implanted solar wind particles, of which hydrogen is the most abundant component (Cameron, 1973), are known to penetrate to depths greater than 1000 Å, but most of the particles are trapped within 200 Å of the surface (Leich *et al.*, 1973, 1974; DesMarais *et al.*, 1974). DesMarais *et al.* (1974) recognized both surface- and volume–correlated components of hydrogen in lunar fines by means of analyses of the distribution of hydrogen with respect to particle size assuming spherical shapes. Bogard (1977) developed the idea that solar wind implanted gases become increasingly volume–correlated by incorporation into agglutinates as a soil matures. Frick *et al.* (1975) developed the concept of spherical shells containing an equilibrated gas concentration to allow for a considerable penetration depth of an incident noble gas particle. However, the abstract by Carter (1984) is the first attempt to quantitatively evaluate the relationship of volume of surface coating to particle size or shape.

THEORETICAL CONSIDERATIONS

General

Consider a perfectly spherical soil grain of diameter D with a gas–enriched outer layer of thickness t. The volume of the enriched layer is related to the total volume of the grain by the formula: $V_{layer} = V_{grain} [6(t/D) - 12(t/D)^2 + 8(t/D)^3]$. If the thickness of the layer is small compared to the diameter of the grain, then the fraction of the volume that is enriched can be estimated adequately using only the first term of the expansion. For example, a 200-Å coating on a 20-μm grain would represent an enrichment of only 0.6% of the total grain volume. Since large grains would require very thick coatings to provide any significant enriched volume, all subsequent discussions will be confined generally to data for particles 20 μm or less in diameter.

Surface Coating Volume Relationships

The importance of both grain diameter and thickness of coating to volume percent of surface coating on spheres is shown in Fig. 1. A significant amount of coating volume

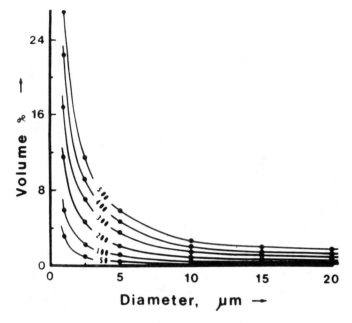

Figure 1. *Relationship of volume percent of surface coating versus diameter in microns for coating thicknesses from 50-500 Å on spheres.*

(greater than 1%) occurs (a) on particles 20 μm or less in diameter for all surface coating thicknesses greater than about 334 Å and (b) on particles with diameters less than about 3 μm for all surface coating thicknesses of at least 50 Å.

The volume percent coating on spheres versus surface coating thickness up to 200 Å for various particle diameters is shown in Fig. 2. The intersection of the dashed horizontal line with each labeled particle diameter line reveals the minimum coating thickness

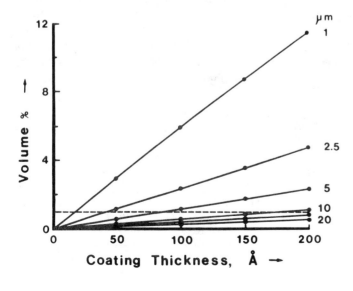

Figure 2. *Relationship of volume percent of surface coating on spheres versus coating thickness in angstroms for particle diameters of 1–20 μm. Area above dashed line represents volume percent of surface coatings greater than 1%.*

Table 2. Relationship of Percent of Surface Coating Volume Versus Maximum Particle Length in μm for 200-Å-thick Surface Coating on Platic and Prismatic Shapes, Where the Maximum Length is Twice the Shortest Length

Maximum Length, μm	% Volume	
	Plate*	Prism**
1.0	15.2128	18.7456
2.5	6.2728	7.7968
5.0	3.1681	3.9490
10.0	1.5920	1.9872
15.0	1.0631	1.3277
20.0	0.7980	0.9968

*% Volume $= 100[\{(A)^2 \times 0.5A\} - \{(A-0.04)^2 \times (0.5A-0.04)\}]/\{(A)^2 \times 0.5A\}$; where A = maximum particle length.
**% Volume $= 100[\{(A) \times (0.5A)^2\} - \{(A-0.04) \times (0.5A-0.04)^2\}]/\{(A) \times (0.5A)^2\}$; where A = maximum particle length.

necessary for a coating of 1 vol %. For example, for a 2.5-μm-size grain a minimum thickness of surface coating to give 1 vol % is 42 Å. Figure 2 also can be used to give the maximum grain size in micrometers for a given thickness and a given volume percent of surface coating. For example, for an average 200-Å-thick coating on spheres, the volume of coating is greate. than 1% for all grain sizes less than 12 μm in diameter.

Particle Shape Relationships

The theoretical effect of grain shape on volume of coating to grain diameter has not previously been systematically examined (Carter, 1984). The importance of grain shape on the volume relationships is demonstrated by Tables 2 and 3 and by Figs. 3 and 4. For example, for a 200-Å-thick coating the irregular prismatic shape has up to 4.8 times more volume percent of coating material than the spheric shape with 1-μm-size particles having up to 55.6 vol % (Table 3, Fig. 4). This, however, is not the maximum amount of possible enrichment. The upper limit is unknown but could approach 100% for the highly convoluted shapes that are common in lunar regolith materials. However, the spheric

Table 3. Relationship of Percent of Surface Coating Volume Versus Particle Width in μm for 200-Å-thick Surface Coating on 1-μm-length Platic and Prismatic Shapes

Width, μm	% Volume	
	Plate	Prism
1.000	11.5264	11.5264
0.750	12.7552	13.9669
0.500	15.2128	18.7456
0.250	22.5856	32.2624
0.125	37.3312	55.6096

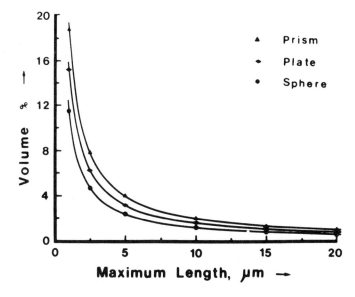

Figure 3. Relationship of volume percent of 200 Å surface coating versus maximum particle length in microns for particle shapes: sphere • , plate ✦, and prism ▲ .

shape is one end member of a family of curves and is the worst case (Fig. 3) because, since a sphere has the minimum surface area to volume ratio of any shape, it has the minimum volume percent of surface coating of any shape.

Percent Total Surface Coating Volume Relationships

The percentage of total surface coating volume is plotted versus grain diameter in Fig. 5, assuming a coating thickness of 200 Å. The calculation was done for a range of diameters from 1.0 μm to 1000 μm, although the plot is truncated at a diameter of 20 μm. The curve in effect depicts the relation of cumulative volume enrichment as a function of grain size for a purely fictitious soil that has equal weight percents distributed in each of the calculated size bins. The figure shows that 90.9% of the total volume of surface coating occurs on the less than 20-μm-size particles and that 82.1% occurs on the less than 10-μm-size particles. Since real lunar soils often have higher weight percentages in the finer size fractions, a real cumulative fraction could be even more favorable for beneficiation of the lunar fines. Calculations using other coating thicknesses show similar trends.

LUNAR REGOLITH FINES

Bustin *et al.* (1984) found that 63% of the hydrogen in Apollo 15 sample 15021, which has an average of 41 ppm hydrogen, was contained in the less than 20-μm-size fraction, which made up 23 wt % of the sample; similar results were found by DesMarais *et al.* (1974). The data from sample 15021 (Bustin *et al.*, 1984) were recalculated and are listed in Table 4 along with the theoretical calculations for 200-Å-thick coated spheric and prismatic shapes. A comparison of the lunar data with results of the theoretical

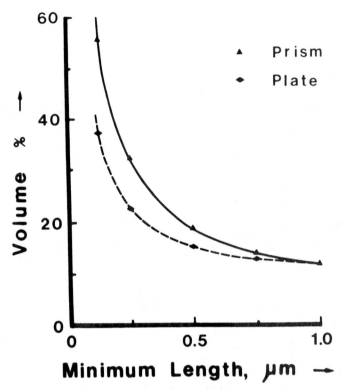

Figure 4. Relationship of volume percent of 200 Å surface coating on 1 μm maximum length particles versus shortest particle length in microns for particle shapes: plate • and prism ▲

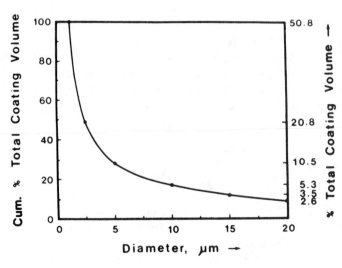

Figure 5. Relationship of cumulative percent of total coating volume versus diameter in microns for 200 Å surface coating on spheres.

Table 4. Relationship of Percent of Surface Coating Volume and Percent of Total Coating Volume to Particle Diameter and Particle Shape for 200-Å-thick Surface Coating

Average Particle Diameter, μm	Spheric % Vol.	% Coating	Prismatic* % Vol.	% Coating	% Hydrogen Content†
10.0	1.195	56.46	1.987	56.41	51.34
32.5	0.369	17.42	0.614	17.43	17.11
60.0	0.200	9.44	0.333	9.45	9.73
82.5	0.145	6.87	0.242	6.88	6.38
120.0	0.100	4.72	0.167	4.73	5.03
200.0	0.060	2.83	0.100	2.84	4.36
375.5	0.032	1.51	0.053	1.51	4.03
750.0	0.016	0.76	0.027	0.76	2.01
Total	2.117	100.01	3.523	100.00	99.99

*Maximum length is twice the width.
†Calculated from μg/g data by Bustin *et al.* (1984) for Apollo regolith fines sample 15021.

calculations (Table 4, Fig. 6) reveals excellent similarity, but there is a slight decrease of hydrogen overall in the less than 100-μm-size fractions and a significant enrichment of hydrogen in the greater than 82.5-μm-size fractions.

The presence of constructional particles (McKay *et al.*, 1971; Carter, 1971), namely agglutinates (DesMarais *et al.*, 1974) and other dust-welded particles, and other non-spherical particle shapes, would increase the hydrogen content for those grain sizes containing these types of particles. Constructional particles result from the welding together

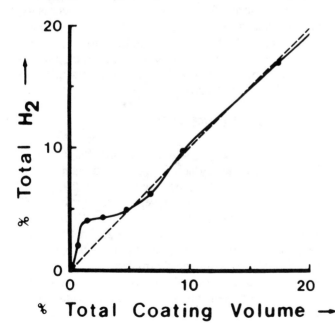

Figure 6. Relationship of percent of total hydrogen (μm g/g) in grain size separates from Apollo 15 sample 15021 (listed in Table 4) versus percent of total surface coating volume on spheres. Dashed line is 1:1 relationship.

of regolith particles with liquid silicates during meteorite bombardment (McKay *et al.*, 1971) and should contain hydrogen values greater than those predicted by overall grain size alone. On the other hand, because a certain amount of the implanted hydrogen was released during the formation of the silicate liquid, as attested to by the presence of metallic iron spherules of non-splashed nature (*e.g.*, Carter, 1973; Morris, 1980), an agglutinate or any other type of constructional particle, unless exposed to solar wind bombardment after formation, would have hydrogen contents lower than the theoretical predictions from total combined surface area of all the very fine size particles attached to or imbedded within its surface. However, unless there was loss of hydrogen during the formation of a constructional particle below the value predicted from overall grain size and there was no subsequent exposure of the particle to solar wind implantations, constructional particles would always contain excess hydrogen for a given grain size. This finding is supported by the research of DesMarais *et al.* (1974), who found agglutinates to be enriched in hydrogen relative to other particle types of similar grain size.

BENEFICIATION

It is clear from Tables 2–4 and from Figs. 1–5 that, theoretically, the less than 20-μm-size particles and certainly the less than 10-μm-size particles contain significant surface coating volumes and will have to be physically separated to increase the surface coating materials to acceptable levels for exploitation. One possible way to accomplish this for lunar materials is to remove the material coarser than 100 μm by vibratory screening and then separate the less than 20-μm-size fraction by direct gaseous classification using turboscreening (Perry and Chilton, 1973). Finer size fractions, if necessary, can be made with modified cyclone techniques or possibly by electrostatc separation (Perry and Chilton, 1973; Agosto, 1984). Similar techniques are routinely used on Earth to classify particles to less than 2 μm in diameter (Perry and Chilton, 1973). The optimum techniques can be determined best by experimentation with lunar simulants followed by testing of actual lunar materials under "lunar" conditions.

EXTRACTION OF HYDROGEN

One aspect of the question of economics is how difficult and expensive it is to extract the hydrogen from the beneficiated (classified) lunar regolith fines. The hydrogen release pattern for a sample of Apollo 11 regolith fines (10086) (Gibson and Johnson, 1971) was recalculated and plotted in Figure 7. Their recalculated data reveals that approximately 81% of the hydrogen is released below 600°C using a heating rate of 4 °C/min. This means that only a moderate amount of thermal energy should be required to extract a significant portion of the hydrogen, especially if advantage is taken of daytime temperature on the lunar surface. Similar results are predicted from the thermal release patterns of samples of Apollo 12, 14, and 15 regolith fines (Gibson and Johnson, 1971; Gibson and Moore, 1972). A microwave technique has also been proposed to extract water from water-bearing lunar materials (Meek *et al.*, 1984).

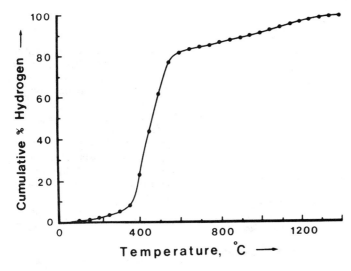

Figure 7. Relationship of cumulative percent of hydrogen versus temperature in degrees centigrade for Apollo 11 sample 10086,16. Data calculated from Gibson and Johnson (1971).

DISCUSSIONS AND CONCLUSIONS

The theoretical rationale of lunar regolith fines as a primary source of hydrogen has been formulated and tested (Tables 2–4, Figs. 1–6). Similar arguments can be made for other elements implanted by solar wind bombardment or as surface condensates associated with volcanic eruptions (*e.g.*, Chou *et al.*, 1975; Meyer *et al.*, 1975) or with meteorite impacts. The results listed in Tables 2–4 and displayed in Figs. 1–5 demonstrate that particles less than 20 μm in size and especially particles less than 10 μm in size have the theoretical capacity for significant concentration of hydrogen. Furthermore, the lunar regolith fines meet the basic requirement for beneficiation because a major portion of the hydrogen occurs in a relatively small part of the fines (Table 4). Particle shape is also a significant factor. For example, the prismatic shapes have up to 4.8 times more surface coating per maximum particle length than the spheric shape (Figs. 3 and 4).

The data shown in Figs. 1–6 result in mandatory size classification of the less than 20 μm fraction. This should be accomplishable by a combination of vibratory screening followed by cyclone and/or possibly electrostatic separation. These methods are routinely used on Earth (Perry and Chilton, 1973) and should be adaptable to the lunar environment. However, the 1/6 g of the lunar environment may result in lower separation efficiencies. Only moderate amounts of thermal energy should be required to separate the hydrogen from the lunar regolith fines because approximately 81% of the solar wind trapped hydrogen is released by heating lunar regolith fines below 600°C (Fig. 7) (Gibson and Johnson, 1971; Gibson and Moore, 1972), or possibly, the hydrogen may be extracted with microwaves (Meek *et al.*, 1984).

Finally, the amount of material that would have to be processed to supply one ton of hydrogen is significant even when all efforts have been made to enhance the production

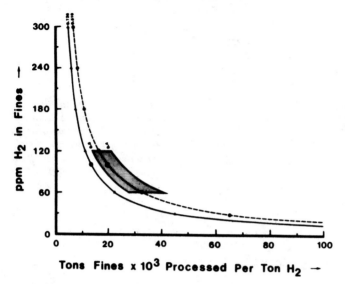

Figure 8. Relationship of number of tons of lunar regolith fines processed versus ppm hydrogen in the fines to yield one ton of hydrogen. (Number of tons of fines processed per ton of hydrogen = $10^6/A \times 10^2/B \times 10^2/C$; where A = ppm hydrogen in fines; B = % of hydrogen in fraction heated; and C = % of hydrogen liberated during heating. Fractions are total recovery values.)

(Fig. 8). For example, assuming (a) an average of 100 ppm hydrogen (Table 1), (b) 90.9% of the hydrogen in the less than 20-μm-size fraction (theoretical limit for spherical shapes), (c) 81% recovery of the hydrogen during heating of the concentrate (Fig. 7) and (d) 23 wt % of the lunar regolith fines in the less than 20-μm-size fraction (Bustin et al., 1984) a minimum of 13,582 tons of lunar regolith fines would have to be processed (Fig. 8) with 3,124 tons of concentrate heated yielding a theoretical recovery of 73.6% of the hydrogen. For 63% of the hydrogen in the less than 20-μm-size fraction of Apollo 15 sample 15021 (Bustin et al., 1984), a minimum of 19,596 tons of this type of lunar regolith fines would have to be processed (Fig. 8) with 4,507 tons of concentrate heated yielding a recovery of 51.0% of the hydrogen. The stippled area of Fig. 8 represents the most likely ranges for "typical" lunar fines. Although these tonnages are not high by terrestrial standards, they will probably limit early exploitation of the lunar regolith fines to hydrogen obtained as a by-product or co-product from the mining and processing for other elements or materials.

Acknowledgments. This paper is based in part on research supported by NASA-JSC grant NAG-9-99.

REFERENCES

Agosto W. N. (1984) Electrostatic separation and sizing of ilmenite in lunar soil simulants and samples (abstract). In Lunar and Planetary Science XV, pp. 1–2. Lunar and Planetary Institute, Houston.

Bogard D. D. (1977) Effects of soil maturation on grain size–dependence of tapped solar gases. Proc. Lunar Sci. Conf. 8th, 3705–3718.

Bustin R., Kotra R. K., Gibson E. K., Nace G. A., and McKay D. S. (1984) Hydrogen abundances in lunar soils (abstract). In Lunar and Planetary Science XV, pp. 112–113. Lunar and Planetary Institute, Houston.

Cameron A. G. W. (1973) Abundance of the elements in the solar system. Space Sci. Rev., 15, 121–146.

Carter J. L. (1971) Chemistry and surface morphology of fragments from Apollo 12 soil. *Proc. Lunar Sci. Conf. 2nd*, 873–892.

Carter J. L. (1973) Morphology and chemistry of probable VLS (Vapor-Liquid-Solid)-type of whisker structures and other features on the surface of breccia 15015,36. *Proc. Lunar Sci. Conf. 4th*, 413–421.

Carter J. L. (1984) Lunar regolith fines: A source of hydrogen (water) (abstract). In *Papers Presented to the Symposium on Lunar Bases and Space Activities of the 21st Century*, p. 27. NASA/JSC, Houston.

Chou C.-L., Boynton W. V., Sundberg L. L, and Wasson J. T. (1975) Volatiles on the surface of Apollo 15 green glass and trace-element distributions among Apollo 15 soils. *Proc. Lunar Sci. Conf. 6th*, 1701–1727.

DesMarais D. J., Hayes J. M., and Meinschien W. G. (1974) The distribution in lunar soil of hydrogen released by pyrolysis. *Proc. Lunar Sci. Conf. 5th*, 1811–1822.

Frick U., Baur H., Ducati H., Funk H., Phinney D., and Signer P. (1975) On the origin of helium, neon, and argon isotopes in sieved mineral separates from an Apollo 15 soil. *Proc. Lunar Sci. Conf. 6th*, 2097–2129.

Gibson E. K. Jr. and Johnson S. M. (1971) Thermal analysis-inorganic gas release studies of lunar samples. *Proc. Lunar Sci. Conf. 2nd*, 1351–1366.

Gibson E. K. Jr. and Moore G. W. (1972) Inorganic gas release and thermal analysis study of Apollo 14 and 15 soils. *Proc. Lunar Sci. Conf. 3rd*, 2029–2040.

Leich D. A., Tombrello T. A., and Burnett D. S. (1973) The depth distribution of hydrogen and fluorine in lunar samples. *Proc. Lunar Sci. Conf. 4th*, 1597–1612.

Leich D. A., Goldberg R. H., Burnett D. S., and Tombrello T. A. (1974) Hydrogen and fluorine in the surfaces of lunar samples. *Proc. Lunar Sci. Conf. 5th*, 1869–1884.

McKay D. S. and Williams R. J. (1979) A geologic assessment of potential lunar ores. In *Space Resources and Space Settlements*, (J. Billingham and W. Gilbreath, eds.) pp. 243–255. NASA SP-428. NASA/Johnson Space Center, Houston.

McKay D. S., Morrison D. A., Clanton U. S., Ladle G. H., and Lindsay J. F. (1971) Apollo 12 soil and breccia. *Proc. Lunar Sci. Conf. 2nd*, 755–772.

Meek T. T., Cocks, F. H., Vaniman D. T., and Wright R. A. (1984) Microwave processing of lunar materials: Potential applications (abstract). In *Papers Presented to the Symposium on Lunar Bases and Space Activities of the 21st Century*, p. 130. NASA/JSC, Houston.

Meyer C. Jr., McKay D. S., Anderson D. H., and Butler P. Jr. (1975) The source of sublimates on the Apollo 15 green and Apollo 17 orange glass samples. *Proc. Lunar Sci. Conf. 6th*, 1673–1699.

Morris R. V. (1980) Origins and size distribution of metallic iron particles in the lunar regolith. *Proc. Lunar Planet. Sci. Conf. 11th*, 1697–1712.

Perry R. H. and Chilton C. H. (editors) (1973) *Chemical Engineer's Handbook*. McGraw-Hill, New York. 106 pp.

Taylor S. R. (1975) Lunar Science: A Post-Apollo View. Pergamon Press, New York. 372 pp.

HYDROGEN RECOVERY FROM EXTRATERRESTRIAL MATERIALS USING MICROWAVE ENERGY

D. S. Tucker, D. T. Vaniman, J. L. Anderson, F. W. Clinard, Jr., R. C. Feber, Jr., H. M. Frost, T. T. Meek, and T. C. Wallace

Los Alamos National Laboratory, Los Alamos, NM 87545

The feasibility of recovering hydrogen from extraterrestrial materials (lunar and martian soils, asteroids) using microwave energy is presented. Reasons for harvesting, origins, and locations of hydrogen are reviewed. Problems of hydrogen recovery are discussed in terms of hydrogen release characteristics and microwave coupling to insulating materials. From results of studies of hydrogen diffusivities (oxides, glasses) and tritium release (oxides), as well as studies of microwave coupling to ilmenite, to basalt and to ceramic oxides, it is concluded that using microwave energy in hydrogen recovery from extraterrestrial materials could be the basis for a workable process of hydrogen recovery.

WHY HARVEST HYDROGEN FROM EXTRATERRESTRIAL MATERIALS?

As manned space efforts proceed from the low Earth orbit activities of the latter part of this century toward colonization of the Moon, and ultimately Mars, in the 21st Century, propulsion and life support will place the greatest demands on technology and resources. Since the most advanced chemical propulsion systems employ hydrogen and oxygen as propellants, the availability of water becomes a significant issue. Although water is prevalent on the Earth and on solar system objects farther from the sun, the Moon exhibits no substantial evidence for it in free or bound (crystalline) states. However, schemes have been proposed to extract bound oxygen from lunar minerals and import hydrogen from the Earth in condensed forms. An alternate solution is to directly harvest hydrogen or water from extraterrestrial materials (lunar and martian soils, asteroids, *etc.*).

Lunar surface soils do contain hydrogen implanted from the solar wind, but in small concentrations. Much interest has been directed toward the possible presence of water ice within permanently shadowed craters at the lunar poles (Arnold, 1979), but without direct evidence it is best not to rely on its existence for current scenarios of Moon-base operations. Hydrogen occurs both as water ice and in hydrous minerals of carbonaceous chondrites of the asteroid belt. The farther outward we travel from Earth's moon, the better our chances of finding usable sources of hydrogen or water. Since water is vital to human use of space, we may find a useful trade-off between Earth-export plus local harvesting for lunar development and complete reliance on local harvesting for exploitation of Mars and the asteroids.

This paper deals primarily with the technical feasibility of recovering hydrogen from lunar soil using microwave energy as the heating source. Since hydrogen is abundant on Mars and in certain types of asteroids, this technique could also be viable in hydrogen or water recovery from these sources.

ORIGINS OF HYDROGEN ON THE MOON, MARS, AND ASTEROIDS

Hydrogen occurs almost entirely as soil-wind implanted hydrogen within the soil components. The fraction of this hydrogen that can occur as water is miniscule (probably 10^{-3} to 10^{-4}; Blanford *et al.*, 1984). Thus the chances of obtaining primary water from lunar surface samples, as far as we know, are nil. Our knowledge of volatile-element systematics of lunar materials indicates that water is also internally deficient on the Moon, because the propellant gases for lunar volcanic gas-fountains were dominated by sulfur rather than by water and carbon-oxygen compounds as on Earth. In fact, the lowest hydrogen contents observed among lunar surface soils (0.2 ppm) occur in the gas-fountain orange glasses of Apollo 17 (Epstein and Taylor, 1973).

In contrast to Earth's moon, the hydrogen of Mars and of the carbonaceous chondrites appears to occur almost always as water ice or as water and structural OH within minerals. Moreover, the origins of this water are closely related to primordial formation or to atmospheric evolution (Mars). Primary water is directly available on Mars and on the carbonaceous chondrites.

LOCATIONS OF HYDROGEN ON THE MOON, MARS, AND ASTEROIDS

Although lunar hydrogen is rare, the hydrogen that we know about is concentrated in lunar surface soils. In a depth profile obtained by studying the Apollo 15 deep drill core, Des Marais *et al.* (1974) found hydrogen concentrations of 124–184 $\mu g/g$ (or ppm) above 150 cm depth, and concentrations of 84–104 $\mu g/g$ below that depth. Bustin *et al.* (1984) report a range of 17–106 $\mu g/g$ hydrogen in 17 bulk soils representing all Apollo missions. Hydrogen contents in some bulk soils range up to almost 200 $\mu g/g$ (Epstein and Taylor, 1973; Des Marais *et al.*, 1974). A useful average concentration is 50 $\mu g/g$, but this hydrogen is surface-correlated and is therefore most abundant in the finest soil fractions (Bustin *et al.*, 1984). However, our knowledge of how hydrogen is held by lunar soil particles is poor (Blanford *et al.*, 1984). For example, there is evidence that hydrogen abundance correlates with TiO_2 content of a soil regardless of grain size (McKay, 1984).

How much hydrogen could we get from lunar surface soils? If an average surface soil abundance of 50 $\mu g/g$ is assumed, then a totally efficient extraction system would obtain about 50 grams of hydrogen, which could be converted to 0.45 liters of water, from each metric ton of soil processes. Obviously the extraction methods will not be totally efficient, and much of the surface area around a lunar base would have to be harvested to obtain appreciable amounts of water. Cost effectiveness of lunar hydrogen extraction will have to be weighed against the costs of hydrogen export from Earth in alternate forms such as condensed methane or ammonia (Friedlander, 1984).

Mars and the carbonaceous chondrites, compared to Earth's moon, are oases in space. In addition to having water ice at its polar caps, Mars may have permafrost extending to lower latitudes; the martian soil apparently contains hydrous minerals. Suggestions have been made that the hydrous mineralogy of martian soils at the Viking Lander sites may be dominated either by clay (Banin *et al.*, 1981) or by palagonite formed as an alteration product from the abundant martian basalts (Allen *et al.*, 1981). In either case the water can be extracted easily from these minerals by low-temperature heating. Carbonaceous chondrites contain serpentine-like hydrous silicates and perhaps some surface clays; spectral reflectance studies suggest that about 80% of the main-belt asteroids are carbonaceous chondrites (McCord and Gaffey, 1977) with water contents that may range from 1–20% (O'Leary, 1977). For extended human presence on Mars or among the asteroids, exportation of hydrogen from Earth may be necessary during initial exploration but should not be required for the development of permanent or semi-permanent bases.

HYDROGEN RECOVERY

Hydrogen Release Characteristics

For any type of hydrogen recovery process a major concern is the efficacy of the release of hydrogen from extraterrestrial materials, particularly from the lunar soils in which the concentration is so small. Several processes that appear to influence the rate of hydrogen release from a heated solid include the following:

1. Hydrogen solubility in the solid
2. Hydrogen diffusivity in the solid
3. Thermodynamic equilibria (H_2/H_2O)
4. Desorption kinetics
5. Grain boundary migration
6. Percolation outward through a packed bed

Each of these processes will have to be considered in some detail before the recovery of hydrogen from extraterrestrial material can be evaluated properly. Considerable data on hydrogen diffusivity in ceramics and glasses exists, much of it accumulated as part of the technology program for magnetic fusion energy. In that context, the studies were concerned with recovery of tritium (^3H) bred in a lithium-containing blanket material from $^6Li(n, \alpha)t$ and $^7Li(n,n'\alpha)t$ reactions. If this tritium is bred in a solid material such as Li_2O, $LiAlO_2$, Li_4SiO_4, or $LiZrO_3$, it must be recovered with high efficiency. Studies of these recovery processes have generated much of the available tritium release data, which can readily be applied to hydrogen. Additional data are available on hydrogen and deuterium diffusivities in other ceramics and in quartz, pyrex, and borosilicate glass. A portion of these data is summarized in Table 1. In both pyrex and borosilicate glass the hydrogen species appear to migrate as an hydroxyl (OH) species. The hydrogen is released from these materials principally as water (Yamamoto, 1984).

Table 1.   Selected Deuterium and Tritium Diffusion Coefficients

Diffusing Species	Material	Do (cm^2/s)	Q, (kJ/mole)	Temp. Range C	Reference
D$_2$	Pyrex glass	5.4	43.5	100–300	Laska et al. (1969)
T$_2$	Quartz	1–100	175	20–1000	Matzke (1967)
T$_2$	Fused silica	1–100	218	20–1000	Matzke (1967)
T$_2$	TiO$_2$	1.8×10^{-2}	107	500–900	Cathcart et al. (1979)
T$_2$	Al$_2$O$_3$	6.4×10^{-5}	132	300–1000	Cathcart et al. (1979)
T$_2$	BeO	1.3×10^{-6}	129	300–1200	Fowler et al. (1977)

D = Do exp (–Q/RT)

Tritium extraction from irradiated metallic aluminum–lithium alloys has been a familiar process since the 1950s. The technique involves heating the irradiated targets in a retort to drive off the tritium. Studies have also been conducted on extraction of tritium from irradiated lithium oxide, lithium aluminate, and lithium silicate targets. In laboratory experiments, ~95% of all tritium is recovered when the pre-irradiated samples are held at 800°C for 4 hours (Johnson et al., 1976). Progressively greater recovery is obtained at 800°C with extended anneals at this temperature. Anneals of 10–15 minutes at 1300°C appear to result in almost complete tritium recovery from the aluminate and silicate targets.

The main fraction of the gases evolved from the target capsules during laboratory extractions is condensible (HTO and T$_2$O). Other experimenters have likewise observed that at temperatures above 600°C a major fraction of the released tritium is indeed in the condensible form. In addition to the silicate and aluminate, a limited number of tests were performed with petalite (Li$_2$O–Al$_2$O$_3$–8SiO$_2$). Tritium release characteristics appeared to be quite similar for both aluminate and petalite.

Consideration of the available data on hydrogen, deuterium, and tritium diffusivities and release characteristics in glasses and ceramics leads one to conclude that if extraterrestrial materials can be heated to 600°–1000°C, there is reasonable probability for release of the hydrogen from these materials. Whether the hydrogen is recovered as gaseous H$_2$ or as the oxide (H$_2$O) is uncertain. Some insight into the relative probabilities of these two alternatives is doubtless possible from thermochemical techniques using complex chemical equilibrium calculations. Such calculations would predict the equilibrium composition of gas released from lunar soils containing implanted hydrogen when heated.

Application of Microwave Heating in Hydrogen Recovery

The use of microwave energy would offer a new, very efficient method of heating lunar materials. Not only could extraterrestrial materials be heated with less energy than with conventional methods, but the heating would be accomplished in a controlled manner and in much less time than with a method such as solar heating. This is due in part to the ability to control field intensities, heating rates, and maximum heating temperature during microwave heating. Also, the thermal conductivity of the lunar soil surface is estimated by Langseth *et al.* (1972) to be 60 μW/cm-K, which is two to three orders of magnitude less than that for oxides such as alumina. Thus, the use of focused solar radiation to quickly heat the lunar-soil volumes appears impractical. The top centimeter or so of lunar soil would strongly impede the diffusion of heat, possibly fusing the soil and preventing release of subsurface hydrogen. In contrast, microwaves tend to heat a soil volume uniformly because their low attenuation results in a fairly uniform released-energy density.

As a specific example, microwave generators could be attached to a mobile vehicle, allowing *in-situ* heating of the upper regolith layer. If the hydrogen was released principally as H_2O, the water vapor could be condensed on a cold platter immediately above the surface. If only hydrogen is released, it will evolve from the lunar surface in molecular form, presenting a significant collection problem. Since such a system would eliminate the extensive mining and hauling associated with a retorting process, it will be important to study evolved lunar soil volatiles and to investigate innovative collection techniques for molecular hydrogen in the lunar environment. An important aspect of the recovery process is the coupling of microwave energy in insulating materials such as basalts, ilmenite, and feldspars, which are widely present on the lunar surface and likely to be found on Mars. Microwave energy at 2.45 GHz is known to couple to such minerals as well as activate dielectric loss mechanisms such as ion migration, ionic vibration, and impurities. As the temperature increases from approximately 25°C, thermally activated processes (such as thermal activation of electrons from valence to conduction bands) that affect conductivity will also impact overall loss characteristics.

Since the early works by Von Hippel (1954a,b), there has been interest in how microwave energy couples into oxide materials. Perhaps the best work done in this area has been by Tinga (1969, 1970), by Tinga and Edwards (1968), by Tinga and Nelson (1973), and by Tinga and Voss (1969). Table 2 lists some materials that have been thermally processed along with observed heating rates for three glasses using microwave energy. All work to date has been accomplished at the two allowed industrial frequency bands (915 MHz and 2.45 GHz).

Table 2. Some Materials Heated by Microwave Energy at Los Alamos National Laboratory

Material	Melting Temperature	Observed Heating Rate
OI – 1756 Glass	462°C	12,000 C/hr
OI – 0038 Glass	735°C	200,000 C/hr
OI – 1613 Glass	1450°C	33,000 C/hr
Alkali Basalt	1200°C	—
Al_2O_3	2059°C	—

High absorptivity of microwave energy is obviously desirable if processing by use of this energy source is to be feasible. Energy absorption of a plane wave in a dielectric material increases explicitly with frequency and also with the product of loss tangent and dielectric constant (Frost, 1985). Loss tangent shows much more variability with frequency than does the dielectric constant and is therefore the focus of most attention when considering dielectric losses. Impurities and defects play an important role in absorption and thus in the coupling of energy to the material in question. At a given temperature and frequency, there is an "intrinsic" loss characteristic of the chemical formulation and crystal structure of the material. As a rule, loss tangent is higher than this intrinsic value due to the presence of chemical and structural defects. Whether energy absorption is due to oscillation of a charged species (*e.g.*, ion or electron) or to scattering from grain or interphase boundaries, higher frequencies such as those mentioned earlier may be expected to couple more effectively into loss processes associated with defects of quite small scales such as might exist in lunar soils.

Mobile defects and ions that enhance the conductivity of oxide materials are known to couple to microwave energy at the 2.45 GHz and 915 MHz frequencies, increasing energy absorption. For example, beta alumina, which contains 11% sodium by weight, can be heated from room temperature to its sintering temperature (1850°C) in just a few seconds when placed in a 2.45 GHz microwave field of 400 watts power. Materials such as $Cu_{2x}O$, ZnO, and ZrO_2 also should couple efficiently because they are defect-controlled semiconductors. In order to accomplish heating of traditional oxide materials such as alpha alumina, we incorporate materials that do couple at 2.45 GHz radiation such as glycerol or sodium nitrate. These materials cause the oxide to heat to a few hundred degrees Celsius, after which the oxide itself will then couple because the loss tangent will increase sufficiently with temperature to allow the material to absorb electromagnetic energy efficiently.

Initial work with terrestrial alkali basalt has shown that in order to heat this material, a coupling agent is needed. Recently we have shown the ability to heat an ilmenite-rich rock to its melting temperature using 2.45 GHz microwave energy without a coupling agent. Since ilmenite is present on the lunar surface in abundant quantities, it could act as a coupling agent to allow the initial heating of those lunar materials that may not couple at ambient temperature.

Also, it is known that most lunar soil, at least down to a depth of one foot, contains about 10^6 cosmic or solar fossil particle tracks per cc. The defects induced into this soil over millions of years of exposure to solar wind and cosmic ray events should enhance the loss tangent of this material and perhaps allow lunar materials to be heated without the use of coupling agents in a microwave field.

One variable that has not been investigated but that has much control over whether a material couples or not, is the field intensity (E,H). On the Moon, heating will be in a vacuum (high dielectric breakdown strength) where very high fields can be used. Consequently, materials that would not couple on Earth may heat very easily and quickly on the Moon.

CONCLUSIONS AND RECOMMENDATIONS

This paper has discussed hydrogen recovery from extraterrestrial materials in general and in particular from lunar soils, using microwave energy. Results have been presented that are relevant to the prospects for microwave processing. One can conclude that if enough effort is invested in further study, this process could produce techniques in which energy can be used in a viable hydrogen recovery process.

Much work remains to fully characterize some of the phenomena observed to date with microwave-heated oxide and composite materials. For example, characterization of the species in which hydrogen is released from oxides will require chemical equilibrium calculations, diffusion modeling, study of reaction and sintering kinetics, and characterization of microwave coupling processes. Also, in order to be certain that lunar materials can be heated in a microwave environment, some small amount of well characterized lunar material should be thermally processed in such an environment and fully restudied after processing.

REFERENCES

Allen C. C., Gooding J. L., Jercinovic M. J., and Keil K. (1981) The surface soil of Mars (abstract). In *Third International Colloquium on Mars*, pp. 1–3. Lunar and Planetary Institute, Houston.

Arnold J. R. (1979) Ice in the lunar polar regions. *J. Geophys. Res., 84,* 5659–5668.

Banin A., Rishpon J., and Margulies L. (1981) Composition and properties of the Martian soil as inferred from Viking biology data and simulation experiments with smectite clays (abstract). In *Third-International Colloquium on Mars*, pp.16–18. Lunar and Planetary Institute, Houston.

Blanford G., Børgesen P., Maurette M., and Meller W. (1984) "On-line" simulation of hydrogen and water desorption in lunar conditions (abstract). In *Papers Presented to the Symposium on Lunar Bases and Space Activities of the 21st Century*, p. 124. NASA/Johnson Space Center, Houston.

Bustin R., Kotra R. K., Gibson E. K., and Nace G. A. (1984) Hydrogen abundances in lunar soils (abstract). In *Lunar and Planetary Science XV*, pp. 112–113. Lunar and Planetary Institute, Houston.

Cathcart J. V., Perkins R. A., Bates J. B., and Manley L. C. (1979) Tritium diffusion in rutile (TiO_2). *J. Appl. Phys., 50,* 4110–4119.

DesMarais D. J., Hayes J. M., and Meinschein W. G. (1974) The distribution in lunar soil of hydrogen released by pyrolysis. *Proc. Lunar Sci. Conf. 5th*, pp. 1811–1822.

Epstein S. and Taylor H. P. Jr. (1973) The isotopic composition and concentration of water, hydrogen, and carbon in some Apollo 15 and 16 soils and in the Apollo 17 orange soil. *Proc. Lunar Sci. Conf. 4th*, pp. 1559–1575.

Fowler J. D., Chandra D., Elleman T. S., Payne A. W., and Verghese K. (1977) Tritium diffusion in Al_2O_3 and BeO. *J. Amer. Cer. Soc., 60,* 155–161.

Friedlander H. N. (1984) An anlysis of alternate hydrogen sources for lunar manufacture (abstract). In *Papers Presented to the Symposium on Lunar Bases and Space Activities of the 21st Century*, p. 28. NASA/Johnson Space Center, Houston.

Frost H. M. (1985) Facility for measuring microwave properties of radiation damaged ceramics for Rf windows and other fusion applications. In *Seventh Annual Progress Report on Special Purpose Materials for Magnetically-Confined Fusion Reactors*, Report DOE/ER 0113/4. In press.

Johnson A. B. Sr., Kabele T. J., and Gurwell W. E. (1976) *Tritium production from ceramic targets: A summary of the Hanford Co-Product Program.* BNWL-2097, Battelle Pacific Northwest Laboratory, Richland. 142 pp.

Langseth M. G. Jr., Clark S. P. Jr., Chute J. L., Jr., Keihm S. J., and Wechsler A. E. (1972) Heat-flow experiment. In *Apollo 15 Preliminary Science Report.* NASA SP-289, pp. 11-1 to 11-20. NASA/Johnson Space Center, Houston.

Laska H. M., Doremus R. H., and Jorgensen P. J. (1969) Permeation, diffusion, and solubility of deuterium in pyrex glass. *J. Chem. Phys., 50,* 135-137.

Matzke H. J. (1967) Diffusion of tritium in quartz and quartz glass. *Z. Naturforsch. A, 22,* 965-969.

McCord T. B. and Gaffey M. J. (1977) Remote determinations of the compositions of Earth approaching asteroids. In *New Moons: Special Session of the 8th Lunar Science Conference on Towing Asteroids into Earth Orbits for Exploration and Exploitation,* pp. 3.1-3.15. Lunar and Planetary Institute, Houston.

McKay D. S. (1984) Evaluation of lunar resources (abstract). In *Papers Presented to the Symposium on Lunar Bases and Space Activities of the 21st Century,* p. 121. NASA/Johnson Space Center, Houston.

O'Leary B. J. (1977) General overview of the development, deployment and cost of a mass driver tug and retrieval of an Earth approaching asteroid. In *New Moons: Special Session of the 8th Lunar Science Conference on Towing Asteroids into Earth Orbits for Exploration and Exploitation,* pp. 1.1-1.19. Lunar and Planetary Institute, Houston.

Tinga W. R. (1969) *Multiphase Dielectric Theory—Applied to Cellulose Mixtures,* pp. 62-65. Ph.D. Thesis, University of Alberta.

Tinga W. R. (1970) Interactions of microwaves with materials. *Proc. IMPI,* pp. 19-29.

Tinga W. R. and Edwards E. M. (1968) Dielectric measurement using swept frequency techniques. *J. Microwave Power, 3,* 144-175.

Tinga W. R. and Nelson S. O. (1973) Dielectric properties of materials for microwave processing-tabulated. *J. Microwave Power, 8,* 23-66.

Tinga W. R. and Voss W. A. G. (1968) Materials evaluation and measurement technique. In *Microwave Power Engineering* (E. Okress, ed.), pp.189-194. Academic, New York.

Von Hippel A. R. (1954a) *Dielectrics and Waves.* Wiley and Sons, New York. 284 pp.

Von Hippel A. R. (1954b) *Dielectric Materials and Applications.* Wiley, New York. 438 pp.

Yamamoto Y. (1984) Diffusion analysis on the thermal release of tritium from a neutron absorbing material. *J. Nucl. Mat., 120,* 161-165.

MICROBIAL EXTRACTION OF HYDROGEN FROM LUNAR DUST

David C. White

Department of Biological Science, 310 Nuclear Research Building, Florida State University, Tallahassee, FL 32306-3043

Peter Hirsch

Institut für Allgemeine Mikrobiologie, Universität Kiel, Biozentrum, Olshausenstrasse 40/60, D-2300 KIEL, West Germany

If molecular hydrogen in lunar dust can be made available to the hydrogenases of bacteria, then several microbial pathways exist for the potential liberation of hydrogen, carbon dioxide, and methane using relatively simple apparatus. Intermediate products include microbial biomass and short chain organic acids such as acetate. The hydrogen could be harvested, and carbon dioxide, phosphate, nitrogen, and trade nutrients could be recycled. All reactions suggested in this paper, or similar ones, have been demonstrated on Earth, with the exception of the initial utilization of hydrogen from lunar fines by bacterial hydrogenases. However, it is possible to test these reactions on extant samples of lunar fines. If potentially toxic elements in lunar soils present problems for such processes, bacterial tolerance can be induced by plasmid transfer or by selection among cells subjected to increasing levels of these elements.

INTRODUCTION

A future permanent human colony on the Moon should be as independent from Earth as possible. Rocket fuel, for instance, should be produced from lunar resources to reduce the high costs of transportation from Earth (Duke *et al.*, 1985). Detailed analyses of samples brought back during the Apollo missions have shown that lunar regolith fines contain, mostly in their 20 μm size fraction, substantial amounts of extractable hydrogen (Büstin *et al.*, 1984). The H_2 concentrations varied from 50–220 ppm; average values given were 100–150 ppm, or 100–150 μg/g regolith fines. This hydrogen can be released by heat. The economic viability of the lunar base could be greatly improved if the hydrogen held in the lunar regolith fines could be harvested by bacteria, which would be needed anyway to aid in the recycling of human organic wastes.

Bacteria offer particularly versatile mechanisms of harvesting lunar hydrogen as they contain self-correcting features such as multiple homeostatic abilities to regulate their microenvironment. They show genetic versatility that allows them to compensate for a wide variety of conditions. Bacteria can be thought of as self-replicating catalysts with the power to adapt to their environment.

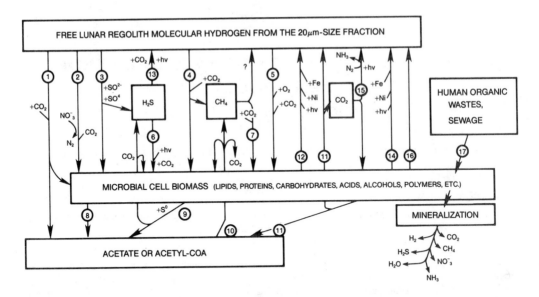

Figure 1. Possible reactions by which molecular hydrogen could be harvested from lunar dust with microbial activities. 1-Acetobacterium woodii, A. carbinolicum, Methanobacterium thermoautotrophicum, Clostridium thermoaceticum, Sporomusa sphaeroides. 2-Paracoccus denitrificans. 3-Desulfovibrio vulgaris; Campylobacter *sp.* 4-Methanobacterium *spp.* 5-see Table 2. 6-Chromatiaceae, chlorobacteria. 7-Rhodopseudomonas gelatinosa. 8-Acetobacterium carbinolicum, Ilyobacter polytrophus. 9-Thermococcus celer, Desulfuroccus mucosus. 10-Methanosarcina barkeri. 11-Acetomicrobium flavidum, Acidaminobacter hydrogenoformans. 12-Purple bacteria. 13-Chromatiaceae. 14-Rhodospirillaceae. 15-Cyanobacteria. 16-Escherichia coli. 17-Anaerobic sewage fermentations; aerobic treatment.

We propose that especially anaerobic bacteria could serve as a mechanism for harvesting the lunar hydrogen. The microbial biomass and organic compounds into which the hydrogen would be fixed initially could then be converted to molecular hydrogen by employing special bacterial strains or consortia and special conditions (Fig. 1). With the exception of the initial reaction whereby hydrogen is released from lunar dust, and whereby the bacteria are subjected to possibly toxic regolith mineral components, all subsequent steps are well established in microbial systems existing on Earth.

In a first step, anaerobic bacteria utilizing the hydrogen could create organic acids, CO_2, H_2S, or CH_4, as well as biomass. As in terrestrial sewage plants, anaerobic digesters could produce more methane from this biomass. Immobilized anaerobic and phototrophic bacteria in illuminated reaction chambers could generate CO_2 and H_2. After harvesting the hydrogen, the CO_2, phosphate, nitrogen, and trace nutrients could be recycled.

At a later stage, with oxygen available from iron–oxide–containing minerals, aerobic processes could generate water from the bacterial oxidation of hydrogen. Also, aerobic methane–oxidizing bacteria would convert CH_4 and CO_2 into water and biomass. The water could be dissociated into oxygen and hydrogen. Aerobic hydrogen release by N$_2$-fixing cyanobacteria (blue–green algae) would be another possibility that could be explored.

ENVIRONMENTAL CONDITIONS NEEDED FOR BACTERIAL CULTIVATION

The conditions required are relatively simple to engineer. The necessary bacterial cultures can be brought from Earth in the form of either lyophilyzed or deeply-frozen strains. Lyophilization (freeze-drying) is performed by exposing well-grown cultures to very high vacuum in the cold. The dried cells are sealed in glass vials while attached to the vacuum line. Lyophilized cells are extremely resistant to noxia such as temperature changes, *etc.* As they do not grow in this dry state, they cannot express mutations occurring occasionally due to radiation. Transportation from Earth to the Moon in a spaceship would, therefore, be possible. No measurable losses of viability were noted with *Hydrogenomonas eutropha* Z1 and *Escherichia coli* K 12 when flown aboard the Russian satellite Cosmos 368 or aboard the automatic lunar station Zond 8 (Taylor, 1974).

The pressure would be optimized by engineering studies and would most likely not exceed several Earth atmospheres. Fermenters for all of these reactions could be opaque, but in the case of photodissociation of organics by purple or green bacteria, some protection from high light intensities and from short wavelength ultraviolet light would be needed. Nitrogen, sulfur, and possibly phosphorus, as well as most trace minerals necessary for the functioning of specific enzymes, would be available from lunar regolith fines. The relative concentrations of these might have to be altered.

In the first stage of an optimal system, the lunar fines would be added to anaerobic fermenters and the hydrogen harvested in the form of bacterial biomass and fermentation products. The second stage would utilize reactors in which phototrophic anaerobes (purple or green bacteria) could be immobilized on special surfaces and generate H$_2$ in the light. The reactor fluid and other gases containing CO$_2$, as well as phosphate, sulfur, nitrogen, trace minerals, vitamins, and water would be recycled, partly by microbial processes ("mineralization"), partly chemically.

All organic wastes and sewage generated by the lunar population would be collected and fermented much as in anaerobic sewage treatment facilities on Earth, to provide additional products for the photoreactors.

In the third stage, with oxygen available from iron-oxide-rich lunar minerals, aerobic autotrophic bacteria or unicellular green algae would be employed as well. Especially selected or engineered aerobic strains with high hydrogenase activity would be able to tolerate stress conditions of the lunar environment. These aerobic organisms could extract H$_2$ from the regolith in explosion-proof fermenters; their Knallgas reaction would form water and biomass.

Some recently discovered archaebacteria (*Sulfolobus ambivalens*) grow either anaerobically with H$_2$, S^0, and CO$_2$ to form H$_2$S and cell biomass or aerobically with

S^0, O_2, and CO_2 to form sulfate (Zillig *et al.*, 1985). To work with bacterial strains of great metabolic versatility would certainly be advantageous.

HYDROGEN UTILIZATION

The initial reaction, which is critical to the whole operation, requires the release of hydrogen from the lunar soil. Although pretreatments may be required, the essential interaction is between the enzyme hydrogenase in the bacteria and the lunar hydrogen. Well-known hydrogenases for the initial anaerobic step occur in two candidate microbes: the sulfate-reducing and the methanogenic bacteria. Sulfate-reducing bacteria form biomass and H_2S from SO_4^{2-}, H_2, and CO_2 (plus trace minerals and possibly some vitamins). The Michaelis-Menten constant K_m, for growth with hydrogen-limiting conditions for a number of SO_4 reducers, is $2 - 4 \times 10^{-6}$ Molar (Robinson and Tiedje, 1984). (The K_m is the concentration of H_2 at which growth is at the half maximal rate.) Methanogenic bacteria that form CH_4 and biomass from H_2 and CO_2 have K_m values of $6 - 7 \times 10^{-6}$ Molar hydrogen. Sulfate reducers thus can outcompete the methanogens for hydrogen. The H_2S generated from sulfate-reducing bacteria may pose corrosion problems in the reactors, as it does on Earth, but, of the organisms that have been well studied, these have the highest known affinity for hydrogen.

Table 1. Anaerobic Bacterial Hydrogen Consumption Yielding Various Reduced Compounds and Cell Biomass (CBM)

Organisms	Reactions	References
Paracoccus denitrificans	$5H_2 + 2H^+ + 2NO_3^- \longrightarrow N_2 + 6H_2O + CBM$	Schlegel and Schneider (1978)
Campylobacter sp.	$H_2 + S^0$ (or $S_2O_3^{2-}$, or SO_3^-)$\longrightarrow HS^- + H + CBM$	Laanbroek *et al.* 1978
Desulfovibrio vulgaris (strain Marburg)	$4H_2 + H^+ + SO_4^{2-} \longrightarrow HS^- + 4H_2O + CBM$	Schlegel and Schneider (1978)
Methanobacterium sp.	$4H_2 + CO_2 \longrightarrow CH_4 + 2H_2O + CBM$	Mah and Smith (1981)
Methanosphaera stadtmaniae	$H_2 + CH_3OH \longrightarrow CH_4 + H_2O + CBM$	Miller and Wolin (1985)
Vibrio succinogenes	$H_2 + fumarate^{2-} + \longrightarrow succinate^{2-} + CBM$	Schlegel and Schneider (1978)
Acetobacterium carbinolicum	$4H_2 + 2HCO_3^- + H^+ \longrightarrow acetate + 4H_2O + CBM$	Eichler and Schink (1984)
Acetobacterium woodii	$H_2 + CO_2 \longrightarrow acetate + CBM$	Ragsdale and Ljungdahl (1984)
Methanobacterium thermoautotrophicum	$8H^+ + 2CO_2 \longrightarrow acetyl\text{-}CoA$ (acetate) $+ CBM$	Stupperich and Fuchs (1984)
Sporomusa sphaeroides	$28H^+ + 7CO_2 \longrightarrow acetate + CBM$	Möller *et al.* (1984)
Clostridium thermoaceticum	$H_2 + CO_2 + CoASH + methyl\text{-}THF \longrightarrow acetyl\text{-}CoA + THF + CBM$	Pezachka and Wood (1984)

Since the reaction of the lunar hydrogen with these bacteria is the untested portion of our harvesting scheme, we propose that a large number of different bacterial types be tested. Suggested anaerobic bacteria are listed in Table 1. Products of hydrogen fixation are water, sulfide, methane, succinate or acetate, and cell biomass.

It is also possible to create microbial biomass and water from hydrogen aerobically, once oxygen becomes available either as an import or as a locally produced commodity. The process requires digesters with vigorous aeration, the engineering of which is well known. A selected list of aerobic bacterial candidates is given in Table 2. The list contains representatives of very different physiological groups. Some aerobic H$_2$ bacteria excrete large amounts (up to 5g/l) of organic compounds, especially when NH$_4^+$ and oxygen are limiting (Vollbrecht et al., 1978).

Table 2. Aerobic Bacteria with the Ability for Hydrogen Consumption and the Formation of Water and Cell Biomass (CBM)

Organisms	References	Organisms	References
Pseudomonas saccharophilia	Schlegel and Schneider (1978)	Corynebacterium sp.	Jochens and Hirsch, (personal communication (1985)
Alcaligenes eutropha	Schlegel and Schneider (1978)	Azospirillum lipoferum	Malik and Schlegel (1981)
Seliberia sp.	Schlegel and Schneider (1978)	Derxia gummosa	Malik and Schlegel (1981)
Comamonas sp.	Schlegel and Schneider (1978)	Rhizobium japonicum	Malik and Schlegel (1981)
Hydrogenobacter sp.	Schlegel and Schneider (1978)	Microcyclus aquaticus	Malik and Schlegel (1981)
Xanthomonas autotrophicus	Schlegel and Schneider (1978)	Microcyclus eburneus	Malik and Schlegel (1981)
Rhizobium leguminosarum	Schlegel and Schneider (1978)	Renobacter vacuolatum	Malik and Schlegel (1981)
Mycobacterium phlei	Hirsch (1961)	Rhodopseudomonas capsulata (dark)	Gogotov (1984)
Nocardia opaca	Hirsch (1961)	Rhodospirillum rubrum (dark-grown)	Gogotov (1984)
Nocardia hydrocarbonoxides	Hirsch (1961)	Rhodopseudomonas acidophila (dark)	Schlegel and Schneider (1978)
N. petroleophila	Hirsch (1961)	Gloeobacter sp. 7421	Howarth and Codd (1985)
N. autotrophica	Hirsch (1961)	Synechococcus sp. 6307	Howarth and Codd (1985)
Nocardia sp. from Antarctic rocks	Hirsch (personal communication, 1985)		

METHANE UTILIZATION

Organic wastes from human activities and the combined cellular biomasses produced anaerobically or aerobically would be fermented in anaerobic digesters resulting in the production of methane, carbon dioxide, and possibly some short chain organic acids, using technology already well studied on Earth.

Numerous species of aerobic bacteria are known to oxidize methane (Whittenbury *et al.*, 1970): *Methylomonas, Methylycoccus, Methylosinus* spp. From the anaerobic fermenters they could harvest methane, a particularly attractive product as it is easily handled. These bacteria form water and cell biomass in the presence of oxygen. They can also fix nitrogen if the level of oxygen is low. The water produced would be harvested and electrochemically dissociated into hydrogen and oxygen. It could also be dissociated biologically by using immobilized cyanobacteria (Kayano, *et al.*, 1981).

BIOLOGICAL RELEASE OF HYDROGEN

The terminal stage of the process involves the production of H_2. Numerous phototrophic bacteria can be grown under conditions where hydrogen is released in the light. We have summarized some of these reactions in Table 3; they are examples of a large number

Table 3. Microbial Hydrogen Evolution from Various Substrates

Organisms	Reactions	References
Rhodopseudomonas sp., *Thiocapsa* sp., *Chromatium* sp., *Thiocystic* sp.	lactate, acetate, ethanol, S^0, or S^{2-} $\xrightarrow[\text{anaerobic}]{\text{light}}$ $\longrightarrow H_2$ + oxidized substrate + CBM	Gogotov (1984); Matheron and Baulaigne (1983)
Rhodopseudomonas sp. (strain MPBE 2271)	malate $\xrightarrow[\text{anaerobic}]{\text{light}}$ H_2 + CO_2 + CBM	Mitsui *et al.* (1983)
Escherichia coli	sugars, pyruvate, or formate $\longrightarrow X + 2H^+ + CBM + \longrightarrow H_2$	Adams *et al.* (1981)
Acetomicrobium flavidum	hexose + $2H_2O$ $\xrightarrow{58°C}$ $4H_2 + 2CO_2$ + 2 acetate + CBM	Soutschek *et al.* (1984)
Clostridium sp.	pyruvate or molasses $\xrightarrow{\text{ferredoxin}}$ H_2 + CBM	Schlegel and Schneider (1978); Suzuki *et al.* (1983)
Rhodopseudomonas gelatinosa	$CH_4 + CO_2$ $\xrightarrow[\text{anaerobic}]{\text{light}}$ CBM + H_2?	Wertlieb and Vishniac (1967)
Nostoc sp. *Gloeobacter* sp. 7421	$N_2 + CO_2$ $\xrightarrow[\text{air}]{\text{light}}$ $H_2 + H_2O + O_2 + NH_3$ + CBM	Kerfin and Böger (1982), Howarth and Codd (1985)
Synechococcus sp.	$N_2 + CO_2$ $\xrightarrow[\text{air}]{\text{light}}$ $H_2 + H_2O + O_2 + NH_3$ + CBM	Mitsui *et al.* (1983)
Acetaminobacter hydrogenoformans	glutamate $\xrightarrow[\text{anaerobic}]{}$ H_2 + acetate + CO_2 + propionate or formate	Stams and Hansen (1984)

of similar reactions known to occur in bacteria and even in green algae. In the latter case, strains of *Chlorella pyrenoidosa, C. vulgaris,* or *Selenastrum gracile* actively produce hydrogenase approximately four hours after onset of anaerobiosis (Kessler and Maifahrth, 1960). The hydrogen production of immobilized cells of *Clostridium butyricum* has been studied quantitatively by Suzuki *et al.* (1983). Three kilograms of wet cells produced, from molasses, 400–800 ml/min^{-1} H$_2$ continuously over a 2-month period. An engineering analysis of photosynthetic hydrogen production systems was carried out by Herlevich *et al.* (1983), who calculated the cost of production.

Possibly a reisolation of the *Rhodopseudomonas* observed by Wertlieb and Vishniac (1967) would be attempted. This bacterium oxidized CH$_4$ with CO$_2$ in light, anaerobically, with the formation of cell biomass. Under nitrogen-limiting conditions, it might be possible to induce the organism to release hydrogen in the light. Of course, it is also possible to develop an abiotic process to photochemically release hydrogen from methane.

The metabolic versatility of non-sulfur purple bacteria (*Rhodospirillaceae*) could be advantageous. These organisms readily utilize hydrogen as an electron donor for photoautotrophic growth (Stanier *et al.*, 1976). The mechanisms used by these anaerobes resemble those of aerobic hydrogen bacteria (Schneider and Schlegel, 1977), but aerobic hydrogen consumption requires the absence of light. In the presence of light, under anaerobic conditions, nitrogen limitation induces N$_2$-fixation, and their nitrogenase releases H$_2$.

INTERMEDIATE REACTIONS YIELDING ACETATE

Conversion of various organic excretion products to acetate may become necessary. Acetate is a possible precursor for methane (Mah *et al.*, 1978) or photosynthetic hydrogen evolution (Table 3). *Ilyobacter polytrophus* ferments 3-hydroxybutyrate, crotonate, pyruvate, citrate, glucose, fructose, malate, or fumarate, to form acetate as well as some formate, propionate, and butyrate (Stieb and Schink, 1984). Another acetogenic bacterium is *Acetobacterium carbinolicum*, which ferments formate, methanol, aliphatic alcohols, 1,2-diols, 2,3-butanediol, acetoin, glycerol, lactate, pyruvate, or hexoses (Eichler and Schink, 1984). Glutamate is converted to acetate, CO$_2$, formate, propionate, and H$_2$ by *Acidaminobacter hydrogenoformans* (Stams and Hansen, 1984). Such versatile bacteria would be preferred to fermentation specialists that could utilize only few compounds.

FORMATION OF REDUCED SULFUR COMPOUNDS

Hydrogen fixation by sulfate-reducing bacteria has already been mentioned (Table 1). Several newly discovered anaerobic bacteria utilize organic compounds as hydrogen donors and S^0 as the hydrogen acceptor. *Desulfurococcus mucosus* and *D. mobilis* grow with yeast extract or casein as carbon and hydrogen source and need S^0 as the acceptor, producing sulfide and CO$_2$ (Zillig *et al.*, 1982). These organisms are extremely thermophilic, requiring 85°C for optimal growth. Similar reactions with peptides are carried out by *Thermococcus celer*, a moderately halophilic thermophile that requires 88°C and 40 g/l of NaCl for growth (Zillig *et al.*, 1983). Finally, sulfide production from S^0 and acetate has been described for *Desulfuromonas acetoxidans*, an anaerobic bacterium that lives syntrophically with green phototrophic sulfur bacteria of the genus *Chlorobium* (Pfennig and Biebl, 1976):

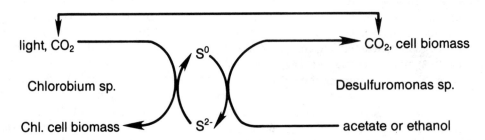

The sulfide–hydrogen that accumulates in such reactions would be transferred into purple bacterial biomass (*Chromatiaceae*) through their anaerobic photosynthesis. Under aerobic conditions, the sulfide-reducing equivalents could be harvested by sulfide-oxidizing bacteria such as *Beggiatoa* spp., *Thiothrix* spp., *Thiobacillus* spp., etc.

POSSIBLE TOXICITY OF LUNAR REGOLITH COMPONENTS

In the first reactions proposed here, the H_2-utilizing bacteria and their enzymes will be in direct contact with the lunar regolith fines. Analyses of Apollo soil samples taken from mare, highland, or basin ejecta show relatively high concentrations of some potentially toxic elements (Table 4). Although some of the elements occur in higher concentrations

Table 4. Concentration Ranges of Selected Lunar Soil Elements with Potential Toxicity to Microorganisms

Element	Concentration Range in Lunar Soil (μg/g soil)
As	0.01–0.41
Ba	85.7–767.5
Be	1.2–5.5
Cd	0.03–0.8
Cr	700.0–3600.0
Cu	6.4–31.0
F	37.0–278.0
Mn	500.0–1900.0
Ni	131.0–345.0
Pb	0.8–10.0
Se	0.03–0.39
Sr	104.2–234.0
U	0.26–3.48
Zn	6.3–49.0

From Freitas and Gilbreath (1982)

on Earth, their combined presence could be detrimental to normal terrestrial bacteria. Lunar samples containing high chromium and manganese may have to be tested for such toxic effects.

Microorganisms can develop extreme tolerances to heavy metals and some rare elements (Tyler, 1981). The detoxification can be accomplished by several different methods. Cell walls of bacilli and filamentous fungi can sequester heavy metals. Chitin or chitosan of *Rhizopus arrhizus* can sequester uranium or thorium at amounts of up to 20% of the organism's dry weight (Volesky *et al.*, 1983). Cadmium resistance in *Staphylococcus aureus* is, for example, plasmid-bound, due to an increased Cd–efflux system [2 Cd^{2+}/ 2H$^+$ antiporter (see Foster, 1983)]. Arsenate resistance is also plasmid-bound and linked to arsenite and antimony resistance (Foster, 1983). In *Pseudomonas putida*, cadmium resistance is due to synthesis of three novel proteins rich in cysteine [MW 4000–7000 (see Higham *et al.*, 1985)].

Information on the physiological effects of most of these potentially toxic regolith components on microorganisms is scanty. Although some of the elements listed in Table 4 are probably needed as "trace elements," the concentration ranges offered to the bacteria are extremely critical. Therefore, experiments will have to be conducted in which a large number of bacterial test strains are exposed to various lunar soil samples under simulated lunar reactor conditions.

SOME FINAL THOUGHTS

At the present time, it does not seem to be possible to send experimental cultures of bacteria to the Moon to study their longevity and activities under lunar conditions. However, the "Biostack" experiments of the Apollo 16 and 17 missions have shown that living microorganisms can indeed be carried safely to the lunar surface and back (Taylor, 1974). *Bacillus subtilis* spores embedded in polyvinyl alcohol sheets were subjected to high-energy, multicharged ("HZE") particles, as indicated by tracks registered in special dosimeters. After this treatment, the *Bacillus subtilis* spores germinated, grew, and elongated just as well as the untreated ground controls did.

The quantitative aspects of the hydrogen-harvesting processes are difficult to assess. We assume an average bacterial cell composition as described in Table 5. We can further assume for the sake of calculation that respiration or other processes would not result

Table 5.   Composition of an Average, Growing Bacterium

Water	80 %	
Dry matter	20 %	
carbon		45–55 %
oxygen		20 %
nitrogen		10–15 %
hydrogen		10 %
phosphorus		2–6 %
other elements		5 %

From Schlegel, 1984.

in loss of cell substance during the synthesis of 1 g of microbial biomass. Living bacterial substance should then be composed of water (800 mg/g) and dry matter (200 mg/ g). This dry biomass would then consist of approximately 50% (or 100 mg) carbon, 20% (or 40 mg) oxygen, 10% (or 20 mg) nitrogen, 10% (or 20 mg) hydrogen, 5% (or 10 mg) phosphorus, and 5% (or 10 mg) of other elements.

To extract 20 mg hydrogen from an average lunar soil (size fraction < 20 μm, containing 100 ppm hydrogen), we would need 200 g regolith fines. Since energy is indeed required to construct a living organism, one might have to spend perhaps 1–2 kg of lunar dust for the production of 1 g of living bacteria. With respect to carbon, and assuming that 5–10 g of soil contain 1 mg C, the construction of 1 g living bacteria would require at least 0.5–1.0 kg of soil. The availability of carbon for the bacteria has not been considered in the calculation, but elemental carbon has not been known to be utilized by bacteria. For the nitrogen, 200 g of soil would probably be sufficient. In total, 3–5 kg of lunar soil would allow the construction of 1 g living bacteria—always assuming that the elements can be offered in a "palatable" form.

The availability of versatile bacteria with many different metabolic possibilities for the utilization of lunar regolith hydrogen has been demonstrated. Next should be an assessment of actual yields of these reactions under (simulated) lunar conditions. Combinations of different organisms to form active consortia or syntrophic mixtures would have to be tested for optimal metabolic activity and longevity. Genetic engineering might be necessary to "train" the selected bacteria for tolerances to lunar stress conditions. Thus, artificial microcosms (ecosystems) will have to be created, and their homeostasis (ability to return to original activity after a perturbation) will have to be optimized, possibly by combining a large number of different microorganisms to perform a given reaction. This latter activity represents a challenge to present-day microbiologists who plan future lunar colonies.

REFERENCES

Adams M. W. W., Mortenson L. E., and Chen J. S. (1981) Hydrogenase. *Biochim. Biophys. Acta, 594*, 105–176.

Bustin R., Kotra R. K., Gibson E. K., Nace G. A., and McKay D. S. (1984) Hydrogen abundances in lunar soils (abstract). In *Lunar and Planetary Science Conference XV*, pp. 113–114. Lunar and Planetary Institute, Houston.

Duke M. B., Mendell W. W., and Roberts B. B. (1985) Towards a lunar base programme. *Space Policy, 1*, 49–61.

Eichler B. and Schink B. (1984) Oxidation of primary aliphatic alcohols by *Acetobacterium carbinolicum* spp. nov., a homoacetogenic anaerobe. *Arch. Microbiol., 140*, 147–152.

Foster T. J. (1983) Plasmid-determined resistance to antimicrobial drugs and toxic metal ions in bacteria. *Microbiol. Rev., 9*, 361–397.

Freitas R. A. Jr. and Gilbreath W. P. (1982) Nonterrestrial utilization of materials: Automated space manufacturing facility. In *Advanced Automation for Space Missions*, pp. 77–188. NASA CP-2255, NASA, Washington, DC.

Gogotov J. (1984) Hydrogenase of purple bacteria: Properties and regulation of synthesis. *Arch. Microbiol., 140*, 86–90.

Herlevich A., Karpuk M., and Lindsey H. (1983) Engineering analysis of potential photosynthetic bacterial hydrogen production systems. *Ann. N.Y. Acad. Sci., 413*, 531–534.

Higham D. P., Sadler P. J., and Scawen M. D. (1985) Cadmium-resistant Pseudomonas putida synthesizes novel cadmium proteins. *Science, 225*, 1043–1045.

Hirsch P. (1961) Wasserstoffaktivierung und Chemoautotrophie bei Actinomyceten. *Arch. Mikrobiol.*, *39*, 360–373.

Howarth D. C. and Codd G. A (1985) The uptake and production of molecular hydrogen by unicellular cyanobacteria. *J. Gen. Microbiol.*, *131*, 1561–1569.

Kayano H., Karube I., Matsunaga T., Suzuki S., and Nakayama O. (1981) A photochemical fuel cell system using *Anabaena* N-7363. *Europ. J. Appl. Microbiol. Biotechnol.*, *12*, 1–5.

Kerfin W. and Böger P.(1982) Light induced hydrogen evolution by blue-green algae (cyano-bacteria). *Physiol. Plant. Copenhagen*, *54*, 93–98.

Kessler E. and Maifahrth H. (1960) Vorkommen und Leistungsfähigkeit von Hydrogenase bei einigen Grünalgen. *Arch. Mikrobiol.*, *37*, 215–225.

Laanbroek H. J., Stal L. H., and Veldkamp H. (1978) Utilization of hydrogen and formate by *Campylobacter* spp. under aerobic and anaerobic conditions. *Arch. Microbiol.*, *119*, 99–102.

Mah R. A. and Smith M. R. (1981) The methanogenic bacteria. In *The Prokaryotes. A Handbook on Habitats, Isolation and Identification of Bacteria* (M. Starr *et al.*, eds.) pp. 956–977. Springer-Verlag, Berlin.

Mah R. A., Smith M. R., and Baresi L. (1978) Studies on an acetate-fermenting strain of *Methanosarcina*. *Appl. Environm. Microbiol.*, *35*, 1174–1184.

Malik K. A. and Schlegel H. G. (1981) Chemolithoautotrophic growth of bacteria able to grow under N$_2$-fixing conditions. *FEMS-Microbiol. Letters*, *11*, 63–67.

Matheron R., and Baulaigue R. (1983) Photoproduction d'hydrogène sur soufre et sulfure par des Chromatiaceae (French with English Summary). *Arch. Microbiol.*, *135*, 211–214.

Miller T. L. and Wolin M. J. (1985) *Methanosphaera stadtmaniae* gen. nov., spp. nov.: A species that forms methane by reducing methanol with hydrogen. *Arch. Microbiol.*, *141*, 116–122.

Mitsui A., Philips J. E., Kumazawa S., Reddy K. J., Ramachandran S., Matsunaga T., Haynes L., and Ikemoto,H. (1983) Progress in research toward outdoor biological hydrogen production, using solar energy, seawater and marine photosynthetic synthetic microorganisms. *Ann. N.Y. Acad. Sci.*, *413*, 514–530.

Möller B., Ossmer R., Howard B. H., Gottschalk G., and Hippe H. (1984) *Sporomusa*, a new genus of gram-negative anaerobic bacteria, including *Sporomusa sphaeroides* sp. nov. and *Sporomusa ovata* sp. nov. *Arch. Microbiol.*, *139*, 388–396.

Pezacka E. and Wood H. G. (1984) The synthesis of acetyl-CoA by *Clostridium thermoaceticum* from carbon dioxide, hydrogen, coenzyme A, and methyltetrahydrofolate. *Arch. Microbiol.*, *137*, 63–69.

Pfennig N. and Biebl H. (1976) *Desufuromonas acetoxidans* gen. nov. and sp. nov., a new anaerobic sulfur-reducing, acetate-oxidizing bacterium. *Arch. Microbiol.*, *110*, 3–12.

Ragsdale S. W. and Ljunqdahl L. (1984) Hydrogenase from *Acetobacterium woodii*. *Arch. Microbiol.*, *139*, 361–365.

Robinson J. A. and Tiedje J. M. (1984) Competition between sulfate-reducing and methanogenic bacteria for H$_2$ under resting and growing conditions. *Arch. Microbiol.*, *137*, 26–32.

Schlegel H. G. (1984) *Allgemeine Mikrobiologie*. G. Thieme-Herlag, Stuttgart. 571 pp.

Schlegel H.G. and Schneider K. (1978) Distribution and physiological role of hydrogenases in micro-organisms. In *Hydrogenases: Their Catalytic Activity, Structure and Function* (H. G. Schlegel and K. Schneider, eds.) E. Goltze K. G., Göttingen.

Schneider K. and Schlegel H. G. (1977) Localization and stability of hydrogenases in aerobic hydrogen bacteria. *Arch. Microbiol.*, *122*, 229–238.

Silverman M. P., Munoz E. F., and Oyama V. I. (1971) Effect of Apollo 11 lunar samples on terrestrial microorganisms. *Nature*, *230*, 168–169.

Soutschek E., Winter J., Schindler F., and Kandler O. (1984) *Acetomicrobium flavidum* gen. nov., sp. nov., a thermophilic anaerobic bacterium from sewage sludge forming acetate, CO$_2$, and H$_2$ from glucose. *Syst. Appl. Microbiol.*, *5*, 377–390.

Stams A. J. M. and Hansen T. A. (1984) Fermentation of glutamate and other compounds by *Acetaminobacter hydrogenoformans* gen. nov., sp. nov., an obligate anaerobe isolated from black mud. Studies with pure cultures and mixed cultures with sulfate-reducing and methanogenic bacteria. *Arch. Microbiol.*, *137*, 329–337.

Stanier R. Y., Adelberg E. A., and Ingraham J. (1976) The Microbial World. Prentice-Hall, Englewood Cliffs, NJ. 871 pp.

Stieb M., and Schink B. (1984) A new 3-hydroxybutyrate fermenting anaerobe, Ilyobacter polytrophus, gen. nov., sp. nov., possessing various fermentation pathways. *Arch. Microbiol., 140*, 139–146.

Stupperich E. and Fuchs G. (1984) Autotrophic synthesis of activated acetic acid from two CO_2 in *Methanobacterium thermoautotrophicum*. I. Properties of the *in vitro* system. *Arch. Microbiol., 139*, 8–13.

Suzuki S., Karube J., Matsuoka H., Ueyama S., Kawakubo H., Isoda S., and Murahashi T. (1983) Biochemical energy conversion by immobilized whole cells. *Ann. N.Y. Acad. Sciences, 413*, 133–143.

Taylor G. R. (1974) Space Microbiology. *Ann. Rev. Microbiol., 28*, 121–137.

Tyler G. (1981) Heavy metals in soil biology and biochemistry. In *Soil Biochemistry*. (E. A. Paul and I. N. Ladd, eds.) pp. 371–414. Marcel Dekker, New York.

Volesky B., Sears M., Neufeld R. J., and Tsezos M. (1983) Recovery of strategic elements by biosorption. *Ann. N.Y. Acad. Sciences, 413*, 310–312.

Vollbrecht D., El Nawawy M. A., and Schlegel H. G. (1978) Excretion of metabolites by hydrogen bacteria. I. Autotrophic and heterotrophic fermentations. *Europ. J. Appl. Microbiol. Biotechnol., 6*, 145–155.

Waldrop M. M. (1984) Asking for the moon. *Science, 226*, 948–949.

Wertlieb D. and Vishniac W. (1967) Methane oxidation by a strain of *Rhodopseudomonas gelatinosa*. *J. Bacteriol., 93*, 1722–1724.

Whittenbury R. K., Phillips K. C., and Wilkinson J. F. (1970) Enrichment, isolation and some properties of methane-utilizing bacteria. *J. Gen. Microbiol., 61*, 205–218.

Zillig W., Geast S., and I. Holz (1985) Plasmid-related anaerobic autotrophy of the novel archaebacterium *Sulfolobus ambivalens*. *Nature, 313*, 789–790.

Zillig W., Holz I., Janekovic D., Schäfer W., and Reiter W.(1983) The archaebacterium *Thermococcus celer* represents a novel genus within the thermophilic branch of the archaebacteria. *Syst. Appl. Microbiol., 4*, 88–94.

Zillig W., Stetter K. O., Prangishvilli D., Schäfer W., Wunderl S., Janekovic D., Holz I., and Palm P. (1982) *Desulfurococcaceae*, the second family of the extremely thermophilic, anaerobic, sulfur-respiring *Thermoproteales*. *Zentralbl. Bakt. Hyg., Abt. 1. Orig., C3, 304–317*.

HYDROGEN AND WATER DESORPTION
ON THE MOON:
APPROXIMATE, ON-LINE SIMULATIONS

G. E. Blanford[1], P. Børgesen[2], M. Maurette[3], W. Möller[2], and B. Monart[3]

[1]*University of Houston-Clear Lake, Houston, Texas 77058 U. S. A.*
[2]*Max-Planck-Institut für Plasmaphysik, D-8046 Garching, W. Germany*
[3]*Laboratoire René Bernas, F-91406 Orsay, France*

To help assess possible water reserves on the Moon derived from the solar wind (which could supply a lunar base) silicon, sapphire, and oligoclase (feldspar) have been irradiated with a beam of 2.5 keV/amu deuterium ions to simulate solar wind proton bombardment of lunar materials. For silicon and sapphire the areal density of deuterium $D(\phi)$ (atoms/cm^2) increases with the incident ion fluence ϕ until a critical re-emission fluence is reached ($\phi_{re} \simeq 3 \times 10^{17}$ D/cm^2). At this point deuterium begins to be lost from both targets until $D(\phi)$ reaches a saturation value $D_s \simeq 3 \times 10^{17}$ D/cm^2 at $\phi_s \simeq 10^{18}$ D/cm^2. Oligoclase begins losing deuterium when the beam is turned on, and $D(\phi)$ never reached a saturation value up to the limit of our experiment ($\phi \simeq 3 \times 10^{18}$ D/cm^2). There was also a small yield of $\sim 10^{-4}$ D$_2$O$^+$/D$_{incident}$ during the implantation of oligoclase. The similar saturation values for the dissimilar materials silicon and sapphire and the lack of a saturation value for oligoclase are unexplainable by models previously used to describe the build-up of solar wind species in lunar samples. Also, the thermal release patterns for our targets have three generally lower temperature peaks when compared to those of returned lunar soils. These differences may be the result of imperfect simulation conditions or the hydration of the returned samples.

INTRODUCTION

It has been argued that solar wind (SW) hydrogen implanted into lunar dust grains could result either in the storage of hydrogen by OH bonds in oxygen-rich grains (Zeller *et al.*, 1966) or in the release of small molecules such as H$_2$ and H$_2$O during pyrolysis of lunar soil samples (Gibson and Moore, 1972; Desmarais *et al.*, 1974; Bustin *et al.*, 1984). The latter authors found that ~100 ppm water is released from lunar soil samples. At this level of abundance, water will have to be found at the polar regions of the Moon (Watson *et al.*, 1961; Arnold, 1979) or hydrogen will have to be brought from Earth for maintaining life at a lunar base. We believe that it is useful to study the uptake and release of SW hydrogen in lunar analog materials to assess whether there are components of the lunar soil containing a much higher concentration of SW hydrogen that may be economically extractable and also to provide data to better estimate the SW contribution to possible lunar ice reservoirs.

The first step in this assessment is to question the validity of the basic models that have been used over the last 15 years for describing the accumulation and release of

SW-related species at lunar conditions. These models have been thoroughly reviewed by Pillinger (1979). In the most frequently quoted model, it has been assumed that the surface concentration C_s of an implanted species reflects the value of the implant fluence required to sputter away a thickness of target material corresponding to the projected range of the incident ions, that is, $C_s \propto Y_s^{-1}$, where Y_s is the sputtering rate of the incident ions (Carter and Colligon, 1968; Carter et al., 1972). Another model requires that the areal density D_s of the implanted species be somewhat scaled by the value of the amorphizing fluence ϕ_a, which forms an amorphous layer on most lunar silicate minerals (Bibring et al., 1974b). In this paper we present recent results concerning the validity of these models, which were obtained by using a reaction chamber developed in fusion reactor research. Our results show that the models for accumulation and release of SW species do not appear to be totally valid and suggest that mature lunar surface soils may act as efficient converters for transforming the flux of energetic SW protons into a flux of low energy hydrogen and water molecules continuously injected into the lunar atmosphere.

EXPERIMENTAL DETAILS

Reaction Chamber

The reaction chamber used in this work was developed a few years ago for investigating the retention in the near-surface range, the diffusion into the bulk material, and the release of hydrogen isotopes from metals (Scherzer et al., 1983). This chamber is equipped with a quadrupole mass spectrometer on-line with both a 30 kV accelerator for ion implantation and a 2.5 MV Van de Graaff accelerator for analysis. The mass spectrometer can monitor the simultaneous release of four distinct masses in the mass range 1–100 during ion implantation and during thermal desorption runs when the target can be heated up to ~900° C. To minimize mass interferences associated with terrestrial contaminants, which yield strong backgrounds at masses 1, 2, 3, 18, and 20, we implanted with ^{2}D instead of ^{1}H at an energy of 5 keV/atom. We simulated the effects of He-damage, which create H trapping sites, with ^{3}He pre-implants. We monitored the release of D_2 and D_2O molecules. The residual concentration as well as the depth concentration profile of deuterium in the targets were measured using the $D(^3He, \alpha)H$ nuclear reaction with a ~1 mm^2 beam of 750 keV ^{3}He. It is worth noting for our subsequent discussion of water generation at lunar conditions that the vacuum and the residual H_2O pressure in the chamber were ~10^{-8} and ~10^{-10} mbar, respectively.

Choice of Targets

Except for the work of Arnold and Doyle (1982) on fused silica, only materials relevant to fusion reactor technology, such as metals (Möller, 1983) and some "poor" conductors such as graphite (Sone and McCracken, 1982), TiC (Doyle et al., 1981), and TiB$_2$ (Doyle and Vook, 1979), have been investigated using sophisticated "on-line" techniques. For our investigation we have chosen terresterial analogs of lunar materials and silicon and sapphire. In this paper we report our early results for silicon, sapphire, and oligoclase.

Most of the lunar surface is covered with amorphous and crystalline silicates that belong to the specific family of insulators known as nuclear track detectors. These solids register tiny regions of radiation damage along the path of both high ($E \simeq 1$ MeV/amu) and low energy ($E \simeq 1$ keV/amu) ions. At a particular critical implant dose ϕ_a, they develop an amorphous coating of radiation damage showing a drastically increased chemical reactivity. The hydration ability of this coating can be several orders of magnitude greater than that of undamaged material (Dran *et al.*, 1984). Because planetary scientists have assumed that radiation damage or sputtering rates play a dominant role in the accumulation of SW species in the lunar regolith, we have used targets of silicon and sapphire, which, as well as the metals and poor conductors, are different from the solids classed as nuclear track detectors. Their stability against radiation damage spans >1000 fold range (Naguib and Kelly, 1975), and their sputtering rates span >10 fold range compared to those of the lunar nuclear track detectors (Carter and Colligon, 1968).

PRELIMINARY RESULTS AND DISCUSSION

Build-up of Saturation Concentrations: Results

During the 2.5 keV/amu deuterium implantation at room temperature, we continuously monitored the variation of the areal density $D(\phi)$ (atom/cm^2) of deuterium retained in the targets as a function of the incident ion fluence ϕ. A simple and similar trend is observed for both silicon and sapphire. At first the areal density of D in the target linearly increases with ϕ, with no D_2 re-emission being detected with the on-line mass spectrometer. At a critical re-emission fluence $\phi_{re} \simeq 3 \times 10^{17}$ D/cm^2, deuterium begins to be lost from the targets. Finally $D(\phi)$ reaches a saturation value $D_s \simeq 3 \times 10^{17}$ D/cm^2 at ϕ_s $\simeq 10^{18}$ D/cm^2. The values of D_s and ϕ_s noted for silicon and sapphire are similar. It is remarkable that this behavior corresponds to that of metals at liquid nitrogen temperature (Möller, 1983) and of fused silica at room temperature (Arnold and Doyle, 1982) with approximately the same saturation fluence $\phi_s \simeq 10^{18}$ D/cm^2. On the other hand, oligoclase drastically departs from this simple behavior and loses deuterium at a low fluence ($\sim 10^{15}$ D/cm^2), that is, when the beam is turned on. The areal density $D(\phi)$ continuously increases but never reaches a saturation value up to the limit of our experiment at $\phi \simeq 3 \times 10^{18}$ D/cm^2. The continuous loss of deuterium from oligoclase is possibly related to the "percolation" of deuterium through radiation damage islands produced by ion implantation. This mechanism has been proposed to explain the greatly enhanced hydration ability of ion-implanted silicates when exposed to water or water vapor (Borg *et al.*, 1982). They found that the increased hydration ability begins with extremely low fluences of implanted hydrogen ($\sim 10^{-2}\phi_a$).

Build-up of Saturation Concentrations: Inadequacy of Previous Models

According to proposed models, the saturation concentration C_s should vary inversely with the sputtering rate Y_s (Carter and Colligon, 1968; Carter *et al.*, 1972), or the areal density D_s of the implanted species should vary with the amorphizing fluence ϕ_a (Bibring *et al.*, 1974b). But the targets used in our experiments clearly have similar values of

C_s and ϕ_a despite their drastic differences in sputtering rates and stability to radiation damage. Consequently, models claiming a simple relationship between C_s and Y_s or D_s and ϕ_a require revision. Our depth profile measurements indicate that the bulk concentration of hydrogen is limited to a fixed value that depends on the material and the re-emission of excess atoms. In this work we find that this maximum concentration is ~0.3 D atom/ matrix atom.

Thermal Desorption of Hydrogen Isotopes in Terrestrial Targets

The thermal release patterns of silicon, sapphire, and oligoclase were obtained by continuously monitoring the release of D_2^+ ions with the mass spectrometer while the sample was being heated (Fig. 1). These patterns show at least three distinct peaks and are more complex than the single peak structure observed for metals and fused silica. This multi-peak structure, which signifies complex D-traps in the material, is unique for each solid as opposed to the approximately constant values of ϕ_s. Furthermore, heating the material well below melting temperature totally extracts the trapped deuterium. After the highest temperature release peaks, observed at ~690°C, ~830°C, and ~430°C for silicon, sapphire, and oligoclase, respectively, no deuterium can be detected in the samples with the ^{3}He analysis beam. This type of behavior has also been observed for metals (Möller, 1983), graphite (Sone and McCracken, 1982), TiC (Doyle et al., 1981), and TiB$_2$ (Doyle and Vook, 1979) implanted at low temperatures and fused silica implanted at room temperature (Arnold and Doyle, 1982).

Thermal Desorption of Hydrogen Isotopes in Lunar Soil Samples

The thermal release patterns observed in our experiments are markedly different than those observed for lunar soil samples (Fig. 1). The thermal release pattern shows two distinct peaks at ~600°C and ~1200°C (i.e., at melting temperature) for most solar wind related species (including hydrogen) (Gibson and Moore, 1972; Desmarais et al., 1974; Bustin et al., 1984). Clearly it is necessary to investigate whether integrated temperature release peaks observed in our preliminary work can at least model the low temperature peak found in lunar soils and also whether our low temperature peaks can be shifted to higher temperatures by additional effects. Some of these effects include lunar conditions that have not been simulated such as implantation at the effective day temperature of the Moon (~130°C) (Taylor, 1975), predamage associated with other SW implanted species, especially helium, the chemisorption of hydrogen on SW He bubbles tightly bound to structural defects, and the multiple implants of H, N, C, and other SW ions. An earlier work indicates that the pyrolysis of feldspar targets subjected to multiple implants with ^{2}D, ^{15}N, and ^{13}C ions better reproduces the two peak release pattern observed for lunar soils (Bibring et al., 1974a). Another consideration is the increased hydration ability of hydrogen implanted silicates (Dran et al., 1984). The best preserved lunar soil samples have been kept in "dry" nitrogen that still contains ~200 ppm residual water. Because Dran et al. (1984) found that hydrogen implanted silicates have increased hydration abilities, this water may quickly permeate the entire depth of the amorphous coatings formed on lunar silicates and mix with the SW trapped species (Leich et al., 1974). The

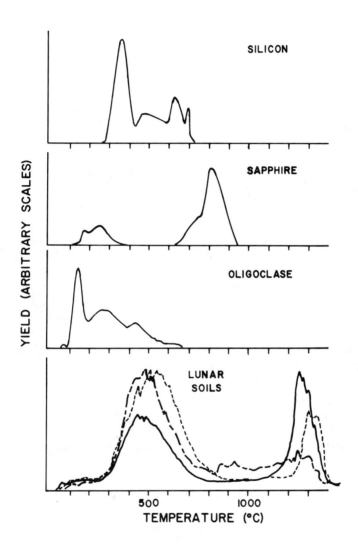

Figure. 1. *Thermal desorption of implanted D_2 from silicon, sapphire, and oligoclase and of solar wind H_2 from lunar soils [solid line = 14163,178, dotted line = 15021,21, dot-dash line = 15601,31; from Gibson and Moore (1972)].*

resulting hydrated layer may form new families of high temperature traps that would behave differently than those in truly pristine lunar samples and our targets. It has also been suggested that the high temperature peak may result from SW hydrogen redistributing deeply into agglutinates during their formation (D. S. McKay, personal communication, 1984). In previous thermal release experiments on lunar soil samples it has been assumed that terrestrial water would be weakly adsorbed on the grain surfaces and would result in the lowest temperature release peak, which would be easily distinguishable from the high temperature release peaks expected for SW hydrogen. Because hydration ability has been found to be increased by several orders of magnitude by radiation damage (Dran et al., 1984), we believe that there is reason for questioning this assumption. This would imply that the value of ~100 ppm SW hydrogen published for lunar soil samples needs to be reassessed.

The Lunar Surface: A Solid State Converter for the Continuous Generation of H_2 and H_2O Molecules in the Lunar Atmosphere

We have reported on the re-emission of the D_2 molecules from all targets during either ion implantation at room temperature or during thermal desorption runs at T > 150°C. We have also detected a small yield of ~10^{-4} $D_2O^+/D_{incident}$ during the implantation of oligoclase starting at a low implantation fluence. To be sure that we were not measuring the "memory" of the chamber walls to previous experiments, we implanted ^{3}He in a similar target pre-implanted with a high dose of D. This target released a much lower yield of D_2O from the D implanted, oxygen-rich targets. If lunar silicates behave like oligoclase, the re-emission of H_2 and H_2O would be initiated at low SW ion fluences. Consequently, during the lunar day, areas of mature regolith on the lunar surface, which are estimated to be about 1/3 of the total surface (Langevin, 1978), may act as efficient "converters" for transforming the flux of SW protons (~3 x 10^8 H.cm^{-2}.s^{-1}) into an outward flux of low energy H_2 and H_2O molecules. This continuous outgassing of the lunar regolith may contribute significant amounts of water to permanently shadowed regions near the lunar poles (Watson et al., 1961; Arnold, 1979) and may initiate a weak hydration of the lunar regolith materials. Such a water cycle may have implications for other astrophysical sites as well.

Acknowledgment.   *This work has been supported in part by NASA grant NSG 9043.*

REFERENCES

Arnold G. W. and Doyle B. L. (1982) Trapping and release of implanted D/H ions in fused silica. *Nucl. Instrum. Methods, 194,* 491–495.

Arnold J. R. (1979) Ice in the lunar polar regions. *J. Geophys. Res., 84,* 5659–5668.

Bibring J. P., Burlingame A. L., Chaumont J., Langevin Y., Maurette M., and Wszolek P. C. (1974a) Simulation of lunar carbon chemistry: I. Solar wind contribution. *Proc. Lunar Sci. Conf. 5th,* pp. 1747–1762.

Bibring J. P., Langevin Y., Maurette M., Meunier R., Jouffrey B., and Jouret C.(1974b) Ion implantation effects in "cosmic" dust grains. *Earth Planet. Sci. Lett., 22,* 205–214.

Borg J., Dran J. C., Langevin Y., Maurette M., Petit J. C., and Vassent B. (1982) The planar track model and the prediction of alpha-recoil aging in radwaste materials. *Radiat. Eff.*, *65*, 173–181.

Bustin R., Kotra R. K., Gibson E. K., Nace G. A., and McKay D. S. (1984) Hydrogen abundances in lunar soils (abstract). In *Lunar and Planetary Science XV*, pp. 112–113. Lunar and Planetary Institute, Houston.

Carter G. and Colligon J. S. (1968) *Ion Bombardment of Solids*, Elsevier, New York. 446 pp.

Carter G., Baruah J. N., Grant W. A., and Whitton J. L. (1972) Ion retention and saturation effects during implantation of semiconductors. *Radiat. Eff.*, *16*, 108–114.

DesMarais D. J., Hayes J. M., and Meinschein W. G. (1974) The distribution in lunar soil of hydrogen released by pyrolysis. *Proc. Lunar Sci. Conf. 5th*, pp. 1811–1122.

Doyle B. L. and Vook F. L. (1979) Hydrogen trapping and re-emission in TiB2 coatings for tokamaks upon thermal, pulsed electron, and laser annealing. *J. Nucl. Mater.*, *85/86*, 1019–1023.

Doyle B. L., Wampler W. R., and Brice D. K. (1981) Temperature dependence of H saturation and isotope exchange. *J. Nucl. Mater.*, *103/104*, 513–518.

Dran J. C., Langevin Y., Petit J. C., Chaumont J., and Vassent B. (1984) Thermal annealing behavior of ion implanted muscovite mica: Implications for its defect structure. *Nucl. Instrm. Methods, Sect. B*, *229*, 402–408.

Gibson E. K. and Moore G. W. (1972) Inorganic gas release and thermal analysis study of Apollo 14 and 15 soil. *Proc. Lunar Sci. Conf. 3rd*, 2029–2040.

Langevin Y. (1978) *L'étude de l'évolution des surfaces des petits corps de la système solaire*. Dissertation Doctorat ès Science, Université de Paris Sud, Paris.

Leich D. A., Goldberg R. H., Burnett D. S., and Tombrello T. A. (1974) Hydrogen and fluorine in the surfaces of lunar samples. *Proc. Lunar Sci. Conf. 5th*, pp. 1869–1884.

Möller W. (1983) The behaviour of hydrogen atoms implanted into metals. *Nucl. Instrum. Methods*, *209/210*, 773–790.

Naguib H. M. and Kelly R. (1975) Criteria for bombardment induced structural changes in non-metallic solids. *Radiat. Eff.*, *25*, 1–12.

Pillinger C. T. (1979) Solar wind exposure effects in the lunar soil. *Rep. Prog. Phys.*, *42*, 897–961.

Scherzer B. M. U., Ehrenberg J., and Behrisch R. (1983) High fluence He implantation in Ni: Trapping, re-emission and surface modification. *Radiat. Eff.*, *78*, 417–426.

Sone K. and McCracken G. M. (1982) The effect of radiation damage on deuteron re-emission and trapping in carbon. *J. Nucl. Mater.*, *111/112*, 607–615.

Taylor S. R. (1975) *The Moon: A Post Apollo View*, Pergamon, New York. 327 pp.

Watson K., Murray B. C., and Brown H. (1961) The behavior of volatiles on the lunar surface. *J. Geophys. Res.*, *66*, 3033–3045.

Zeller E. J., Ronca L. B., and Levy P. W. (1966) Proton-induced hydroxyl formation. *J. Geophys. Res.*, *71*, 4855–4860.

AN ANALYSIS OF ALTERNATE HYDROGEN SOURCES FOR LUNAR MANUFACTURE

Herbert N. Friedlander

Monsanto Research Corporation/Mound[1], P. O. Box 32, Miamisburg, OH 45342
Author's correspondence address: 5940 Pelican Bay Plaza, Gulfport, FL 33707
[1]Mound is operated by Monsanto Research Corporation, a subsidiary of Monsanto Company, for the U.S. Department of Energy under Contract No. DE-AC04-76-DP00053.

Earth's relatively large mass places a high energy cost on the use of terrestrial materials in space. The availability of minerals on the Moon at one-sixth Earth gravity, coupled with ample solar energy, ensures that some materials required for permanent lunar bases and space exploration will be manufactured there. Terrestrial hydrogen must be imported in some form to supply the critical need for water and oxygen for life support until lunar vacuum pyrolysis is established and thereafter, as well, if the supply of lunar hydrogen is inadequate. The importation of liquid hydrogen is generally considered, but liquid methane and liquid ammonia should also be evaluated. A weight penalty is associated with importing hydrogen as methane or ammonia; however, higher boiling points reduce boil-off during transport and storage, offsetting this penalty. In addition, methane and ammonia can be chemically converted into other compounds by industrial processes for which effective catalysts have been developed, and they can be used for reduction of ilmenite.

INTRODUCTION

A limitation to the exploration and use of space for scientific and industrial enterprise is the high cost of energy needed to overcome the Earth's gravitational field in order to lift terrestrial materials into space. The reusable space shuttle reduces the capital cost of ferrying such materials to near-Earth orbit but has little impact on the energy cost. The material requirements of servicing satellites in geosynchronous orbit and of exploring and colonizing the Moon and nearby planets will be even more energy-expensive if all that material must be of terrestrial origin. One cost-reducing alternative is to use materials already available in near-Earth space, that is, on the Moon and on asteroids with Earth-approaching orbits (Steurer, 1982).

The availability of minerals on the Moon, coupled with ample solar energy, ensures that some materials required for permanent lunar bases will be manufactured there. The lower energy cost of material in space, resulting from the one-sixth lower gravity of the Moon compared with the Earth, assures the use of lunar materials for space exploration as well.

Lunar material has been sampled by astronauts from the U.S. Apollo missions and by automatic remote samplers of the Soviet Union. From extensive chemical analysis of lunar rocks brought back to Earth, considerable information is available on the chemical

Table 1. Lunar Rock Composition

Major elements, > 1%	Minor elements, > 0.1%	Trace elements
O	Na	C
Si	Cr	N
Al	K	F
Fe	Mn	Ni
Ca	P	Ba
Mg	S	Sr
Ti		Zr
		H
		He
		Cl
		Co
		all others

composition of lunar surface rocks and soil of both mare and highland areas (Arnold and Duke, 1977; Criswell, 1983a).

Table 1 lists the elements found in lunar material in order of abundance (Criswell, 1983b; Phinney et al., 1977). Oxygen is the most abundant, followed by silicon, aluminum, and iron. Carbon, nitrogen, and hydrogen are found only in trace amounts in the upper layers of soil. Their presence is attributed to the deposition of ions from the solar wind and possibly from meteorites. All of the other elements found on Earth are also present in varying but limited amounts.

The most desirable materials for early lunar manufacture are those required for life support, shelter, construction, and rocket propulsion. Using available lunar elements to meet these needs suggests that the first materials to be manufactured on the Moon will be water, oxygen, iron, titanium, silicon, aluminum, and ceramic heat and radiation shields (Phinney et al., 1977). Water and oxygen are essential for life support. In addition, oxygen is an important component for chemical rocket propulsion. It accounts for the major mass needed for space transportation in the vicinity of Earth. Looking ahead to a permanent lunar colony suggests that carbon and nitrogen compounds will be needed for plant and animal life support.

To make water and important carbon and nitrogen compounds requires hydrogen. In addition, hydrogen is an effective reductant for lunar minerals for production of oxygen and metals. Therefore, this analysis of alternate hydrogen sources for lunar manufacture should be of value in planning for future missions for manned exploration and settlement of the Moon.

LUNAR PRODUCTION OPTIONS

Water is needed to support life. Only traces of water have been found on the Moon (Criswell, 1983a). Free water vapor is precluded by the near-vacuum of the lunar atmosphere with a pressure of less than 10^{-12} torr in the daytime and less than 10^{-14} at night (Criswell,

1983a). Soil at the lunar surface holds between 50 and 100 ppm of water originating from reaction of solar wind protons with lunar oxide minerals (Criswell, 1983a). The water can be freed by vacuum pyrolysis at 200°-600°C. There is less than 10 ppm of bound water in the lunar rocks, which requires temperatures above 1000°C for release (Criswell, 1983a).

Vacuum pyrolysis yields other volatile components besides water (Gibson and Hubbard, 1972). Any free hydrogen in the volatile components can be oxidized to water by reduction of suitable lunar mineral oxides such as ilmenite. However, vacuum pyrolysis would require mining and handling more than four metric tons of lunar soil for each gallon of water produced.

A detailed analysis has been made by Arnold (1979) of the possibility that water may accumulate in cold traps at the lunar poles. Areas in perpetual darkness in polar craters and rock crevices are estimated never to exceed about 100 K. Some of the water vapor released on sunlit lunar surfaces at above 400 K, because of the long mean free path for gases in the high vacuum lunar atmosphere, would quickly diffuse to the traps and be held there for many millenia. Confirmation of the speculation on trapped ice awaits exploration of the lunar poles.

Obviously, water used on the Moon, whether found, produced, or imported, would be carefully conserved, purified, and recycled. During the early stages of lunar base development and if the lunar sources prove inadequate, water will have to be imported. To have water on the Moon, only the hydrogen portion need be imported since oxygen is available there. Alternate options for hydrogen import to the Moon will be evaluated.

Because both carbon and nitrogen are critical to animal and plant life, adequate supplies must be available to maintain permanent lunar colonies. The solar wind deposits both in lunar surface soil. Carbon has been found in lunar rocks at 20–200 ppm and freed by vacuum pyrolysis at temperatures up to 600°C as carbon dioxide and methane and above 1000°C as carbon dioxide and carbon monoxide (Criswell, 1983a). Similarly, 50–150 ppm of nitrogen has been found and isolated as molecular nitrogen by vacuum pyrolysis at above 1000°C (Criswell, 1983a). If vacuum pyrolysis proves inadequate, terrestrial C and N will have to be imported. The simplest form for their import is combined with hydrogen as methane and ammonia.

There is ample oxygen on the Moon in the form of oxidic minerals (Arnold and Duke, 1977; Criswell, 1983a). Because oxygen is so critical for life support and is a major component by weight for rocket propulsion, its production has received considerable attention. Oxygen probably will be the first lunar resource to be exploited for space exploration.

Oxygen can be isolated by direct pyrolytic or electrochemical treatment of lunar silicates (Criswell, 1978; Jarrett et al., 1980). Other methods, which require importing reagents from Earth, include carbo–chlorination and HF, acid, or alkaline decomposition of silicates followed by electrochemical separation of components (Criswell, 1978; Jarrett et al., 1980). More desirable methods are based on the reduction of lunar minerals, such as ilmenite, followed by electrolysis of the water produced to yield oxygen along with hydrogen, which can be recycled to the reduction process (Williams, 1983).

Hydrogen compounds are required for reduction of ilmenite. Prior to lunar application of vacuum pyrolysis, or if the lunar hydrogen supply is inadequate, some hydrogen will have to be imported from the Earth. The importation of liquid hydrogen is generally favored, but alternate hydrogen sources such as liquid methane and liquid ammonia should also be evaluated.

LUNAR HYDROGEN OPTIONS

A source of hydrogen will be needed on the Moon by explorers or colonizers to ensure an ample supply of water. With water available, a supply of oxygen can also be ensured. The decomposition of water into its elements, hydrogen and oxygen, is shown as:

$$2 \, H_2O \rightleftharpoons O_2 + 2 \, H_2 \tag{1}$$

The forward reaction can readily be carried out electrochemically to produce oxygen along with hydrogen for recycle. Membranes, catalysts, electrodes, and other equipment for electrolysis are commercially available because the power generating industry produces hydrogen for use as a gaseous lubricant for high tolerance electric generator bearings in this way (Nuttall, 1980).

The reverse reaction to produce water can be carried out by reacting hydrogen with oxygen in a suitable burner. However, it is better to use the hydrogen for direct reduction of lunar oxide minerals to produce water and metal from the mineral. This reaction is illustrated below in (2) for the reduction of ilmenite, an iron titanate mineral found in ample supply in the lunar mare regions (Arnold and Duke, 1977).

$$FeTiO_3 + H_2 \rightarrow Fe + TiO_2 + H_2O \tag{2}$$

Hydrogen reduction of ilmenite has been studied (Williams, 1983; El–Guindy and Davenport, 1970). High temperature is required to drive the reaction to the right. Further development is likely to yield catalysts that accelerate the reaction at lower temperatures. The products of this reaction, all useful on the Moon, are metallic iron, titania, which has a high albedo, and water for direct use or for electrolysis to oxygen and hydrogen. The hydrogen can then be recycled for further ilmenite reduction.

Clearly, lunar hydrogen and water will require considerable mining and material handling for vacuum pyrolysis of lunar surface particles to yield the small percentage they contain. Also, until further exploration is undertaken, the existence of trapped water at the lunar poles must be considered speculative. If these lunar sources prove to be inadequate, or too costly to develop, hydrogen will have to be imported from Earth.

The most convenient form for shipment of terrestrial hydrogen to the Moon is as liquified gases such as molecular hydrogen, methane, and ammonia. Relevant properties of these liquids are in Table 2.

Hydrogen can be obtained from methane with carbon monoxide and carbon dioxide as by-products. The petrochemical industry applies the water/gas shift reactions to reform

Table 2. Hydrogen Imports

	Hydrogen	Methane	Ammonia
Molecular weight	2	16	17
Hydrogen % by weight	100	25	17.6
Boiling point, °C	-252.8	-161.5	-33.35
Liquid density at bp, g/ml	0.070	0.424	0.68
Hydrogen weight at bp, g/ml	0.070	0.106	0.120
Heat of vaporization at bp, J/g	452.2	577.8	1369

methane for production of hydrogen, carbon monoxide, and carbon dioxide from natural gas (Richardson, 1973; Czuppon and Buividas, 1980). Most of the industrial hydrogen used for hydrogenation and to make ammonia is produced by this chemistry.

The conditions of temperature, pressure, concentration, contact time, and catalyst variations to allow industrial interconversion of these compounds are well known (Bridger and Chinchen, 1970). This chemistry is illustrated in (3), (4), and (5).

$$CH_4 + H_2O \rightleftharpoons CO + 3 H_2 \quad \Delta H_{25°C} = 49.26 \text{ kcal/mol} \tag{3}$$

$$CO + H_2O \rightleftharpoons CO_2 + H_2 \quad \Delta H_{25°C} = -9.84 \text{ kcal/mol} \tag{4}$$

$$CH_4 + 2 H_2O \rightleftharpoons CO_2 + 4 H_2 \quad \Delta H_{25°C} = 39.42 \text{ kcal/mol} \tag{5}$$

Chemical engineers have developed proven industrial processes and plant designs for carrying out these interconversions (Moe, 1962). Adapting this terrestrial chemical technology to lunar conditions will require extensive further development.

Chang (1959) has shown that ilmenite can be reduced to iron in 80% yield in one hour at 1100°C with carbon monoxide, hydrogen, or their mixtures. Russian scientists, because of the large deposits of ilmenite and natural gas found in Siberia, have studied the reduction of ilmenite with methane (Reznichenko et al., 1983). They obtained 85–90% reduction of ilmenite as 0.25–0.5 mm particles with natural gas in a fluidized bed at 1000°–1030°C in 5–7 minutes.

Although, methane may be undergoing the reforming reactions of (3), (4), and (5), and the ilmenite reduction is actually effected by carbon monoxide and/or hydrogen, the overall reduction of lunar oxidic minerals such as ilmenite by methane may be summarized by (6):

$$4 \text{ FeTiO}_3 + CH_4 \rightarrow 4 \text{ Fe} + 4 \text{ TiO}_2 + 2 H_2O + CO_2 \tag{6}$$

Reaction (6) does yield products useful on the Moon, namely iron, titania, water, and carbon dioxide. Further extensive development will be required to adapt the Russian process to lunar conditions.

Hydrogen can also be readily freed from ammonia by reversal of the ammonia synthesis reaction, (7).

$$2\ NH_3 \rightleftharpoons N_2 + 3\ H_2 \quad \Delta H_{500°C} = 26\ kcal/mol \tag{7}$$

Synthesis is favored by low temperature and high pressure and decomposition by high temperature and low pressure (Bridger and Snowdon, 1970). Iron catalyzes both the ammonia synthesis and reverse reactions (Czuppon and Buividas, 1980). The small amount of particulate metallic iron, which can be separated magnetically from the lunar surface soil, may well be an effective catalyst (Steurer, 1982). The nitrogen produced can be used to dilute oxygen for breathing or oxidized to nitric oxides electrochemically. The product of these nitric oxides and additional ammonia is ammonium nitrate, an effective fertilizer for plants and an explosive.

However, it may prove possible to find the proper conditions of temperature and pressure to control the ammonia–hydrogen equilibrium in order to reduce ilmenite directly with ammonia as illustrated in (8).

$$3\ FeTiO_3 + NH_3 \rightarrow 3\ Fe + 3\ TiO_2 + 3\ H_2O + N_2 \tag{8}$$

Otherwise, the hydrogen from the ammonia decomposition can reduce the ilmenite directly according to (2) with the nitrogen only acting as a diluent. Again, development of the necessary reaction conditions and engineering for the lunar environment will require further study.

CRITERIA FOR CHOICE OF HYDROGEN IMPORT

A weight penalty is associated with importing hydrogen as methane or ammonia compared with molecular hydrogen (see Table 2). For every ton of hydrogen imported as methane, three tons of carbon must also be imported; for every ton of hydrogen imported as ammonia, 4.7 tons of nitrogen must be imported. If the carbon or nitrogen would have to be imported anyway, bringing hydrogen along as methane or ammonia is reasonable, since other factors offset part of the weight penalty. For example, based on liquid density at the boiling point, liquid hydrogen requires a vessel 9.7 times larger than that for an equal weight of liquid ammonia, or 6 times larger than one for liquid methane.

Based on hydrogen content alone, the vessel to transport one ton of hydrogen as liquid hydrogen would be 1.5 times larger than one to transport a ton of hydrogen as liquid methane and 1.7 times larger than one to transport a ton of hydrogen as liquid ammonia. Thus, for the same hydrogen payload, the additional weight of the carbon or nitrogen is partly offset by the lighter weight of a smaller vessel with thinner walls.

In addition, the higher boiling points of methane and ammonia compared with hydrogen helps offset the weight penalty through reduced boil-off during transport and storage and simpler containment, insulation, and refrigeration systems. A simpler container and a lower insulation requirement translates into less container weight for equal boil-off.

This effect is enhanced by the three-fold higher heat of vaporization in joules per gram of ammonia at its boiling point compared with hydrogen at its boiling point. The heat of vaporization of methane is 1.3 times that for hydrogen.

The temperature on the surface of the Moon varies from 400 K during the lunar day to 100 K during the lunar night (Criswell, 1983a; Williams and Jadwick, 1980). At a depth of 150 cm below the surface, the temperature is estimated to be nearly constant at about 253 K (Langseth, 1976). The boiling point of liquid ammonia is just 13 K lower, so that a buried liquid ammonia pressure vessel with no boil-off can be considered. The boiling point of liquid methane is 141 K and of liquid hydrogen is 233 K lower than this subsurface temperature. The larger the differential, the faster the boil-off, or the more insulation required.

CONCLUSION

For sustained lunar colonization, carbon and nitrogen will be needed for plant and animal growth and for production of useful synthetic organic materials. Lunar minerals contain little carbon and nitrogen. Since they may have to be imported, a useful way to import them could be as methane and ammonia. Methane can be converted into carbon dioxide, carbon monoxide, and hydrogen by well known industrial processes for which effective catalysts have been developed, and it can also be used directly for reduction of ilmenite. Ammonia can be converted to nitrogen and hydrogen. Carbon dioxide fed directly to plants along with ammonia and its by-products form the essential chemicals required for sustained plant life in a self-sufficient lunar habitat.

The well developed industrial chemical processes for utilizing these gases on Earth need to be adapted for lunar operation. For example, carbon monoxide and hydrogen can produce a wide array of hydrocarbons via such synthetic gasoline processes as Fischer-Tropsch (Dry, 1976; Vannice, 1976). Thereby, broad vistas of well known organic chemical processes needed for life on the Moon are opened.

Based on known chemistry, the complex task of engineering a process chemical industry for automated operation in the lunar environment is essential to colonization of the Moon.

REFERENCES

Arnold J. R. and Duke M. B. (1977) *Summer Workshop on Near Earth Resources.* NASA CP-2031, NASA, Washington. 95 pp.

Arnold J. R. (1979) Ice in the lunar polar regions. *J. Geophys. Res.,* 84, 5659–5668.

Bridger G. W. and Chinchen G. C. (1970) Hydrocarbon-reforming catalysts. In *Catalyst Handbook* (Industrial Chemicals Industries, Ltd., Agricultural Division, eds.), pp. 64–96. Springer-Verlag, New York.

Bridger G. W. and Snowden C. B. (1970) Ammonia synthesis catalysts. In *Catalyst Handbook* (Industrial Chemicals Industries, Ltd., Agricultural Division, eds.) pp. 126–146. Springer-Verlag, New York.

Chang M. C. S. (1959) U. S. Patent 2912320. U. S. Patent Office, Washington.

Criswell D. R. (1978) *Extraterrestrial Materials Processing and Construction.* NASA CR-158870, NASA, Washington. 476 pp.

Criswell D. R. (1983a) Lunar utilization. In *Extraterrestrial Materials Processing and Construction*, p. 2.2, 2.16–2.19, 4.9, 4.12–4.19, 4.4. NASA CR-167756, NASA, Washington.

Criswell D. R. (1983b) Roles of powder metallurgy in the development of space manufacturing. *Prog. Powder Metall.*, 38, 115–146.

Czuppon T. A. and Buividas L. J. (1980) Hydrogen for ammonia production. In *Hydrogen: Production and Marketing* (W. N. Smith and J. G. Santagelo, eds.), p. 47–66. American Chemical Society, Washington.

Dry M. E. (1976) Advances in Fischer-Tropsch chemistry. *Ind. Eng. Chem. Prod. Res. Dev.*, 15, 282–286.

El-Guindy M. I. and Davenport W. G. (1970) Kinetics and mechanism of ilmenite reduction with graphite. *Met. Trans.*, 1, 1729–1734.

Gibson E. K. Jr. and Hubbard N. J. (1972) Thermal volatilization studies on lunar samples. *Proc. Lunar Sci. Conf. 3rd*, 2003–2014.

Jarrett N., Das S. K., and Haupin W. E. (1980) Extraction of oxygen and metals from lunar ores. *Space Solar Power Review*, 1, 281–287.

Langseth M. G., Keihm S. J., and Peters K. (1976) Revised lunar heat-flow values. *Lunar Sci. Conf. 7th*, 3143–3171.

Moe J. M. (1962) Design of water-gas shift reactors. *Chem. Eng. Prog.*, 58, 33–36.

Nuttall L. J. (1980) Production and application of electrolytic hydrogen. In *Hydrogen: Production and Marketing* (W. N. Smith and J. G. Santangelo, eds.), p. 191–212. American Chemical Society, Washington.

Phinney W. C., Criswell D. R., Drexler E., and Garmirian J. (1977) Lunar resources and their utilization. In *Space-Based Manufacturing for Nonterrestrial Materials* (G. K. O'Neill, ed.), p. 97–123. *Prog. Astronaut. Aeronaut.*, 57, American Institute of Aeronautics and Astronautics, New York.

Reznichenko V. A., Galushko Yu. S., Karyazin I. A., and Morozov A. A. (1983) Mechanism of fluidization, heating, and reduction of iron-titanium concentrates in a fluidized bed by gases containing methane. *Izv. Akad. Nauk. S.S.S.R., Mettally.*, No. 6. 9–14.

Richardson J. T. (1973) SNG (Substitute Natural Gas) catalyst technology. *Hydrocarbon Process.*, 52, 91–95.

Steurer W. H. (1982) *Extraterrestrial Materials Processing.* NASA CR-169268, NASA, Washington. 147 pp.

Vannice M. A. (1976) Catalytic synthesis of hydrocarbons from hydrogen/carbon monoxide mixtures over the Group VIII metals. IV. The kinetic behavior of carbon monoxide hydrogenation over nickel catalysts. *J. Catalysis*, 44, 152–162.

Williams R. J. and Jadwick J. J. (1980) *Handbook of Lunar Materials.* NASA RP-1057, NASA, Washington. 133 pp.

Williams R. J. (1983) Enhanced production of water from ilmenite: An experimental test of a concept for producing lunar oxygen (abstract). In *Lunar and Planetary Science XIV, Special Session Abstracts*, pp. 34–35. Lunar and Planetary Institute, Houston.

AG-UNIT #3

9 / LIFE SUPPORT AND HEALTH MAINTENANCE

I N ITS EARLIEST MANIFESTATIONS, the lunar base simply will be a space station. Crews can be ferried there and returned home. Consumables can be replenished periodically, and wastes removed. Of course, the transportation system will be more elaborate than the Earth-to-orbit system needed to support the LEO space station. However, any space platform beyond LEO (e.g., at geosynchronous Earth orbit, or GEO) would have similar transport requirements. Consequently, no new major technical issues are raised, in general, by a lunar base program.

As mentioned earlier, a distinguishing quality of a planetary base is the availability of resources for expansion or sustenance of the operation. If the technology and the resource base is sufficient to make the facility self-sustaining, then the space station becomes transformed into something very much more—a colony.

True self-sufficiency implies closure of the life support system, which does represent new technology. For short duration missions of the space shuttle, food, water, and oxygen are supplied; wastes are collected and returned. Some chemical regeneration of water and air will take place in the space station, but food must be imported and wastes exported.

A closed system that supplies food must have a biological component. NASA has sponsored research on such systems under the acronym CELSS for Controlled Ecological Life Support System. MacElroy et al., review the CELSS concept and discuss the evolution of complexity of life support systems. Salisbury and Bugbee present data on an extraordinarily energy efficient plant that could form the basis of a lunar agriculture. Sauer cites the metabolic needs of human crews which must be met by the system. Sedej describes a new engineering concept for a water recycling subsystem.

A second aspect of self-sufficiency not explicitly addressed by the papers in this section concerns expanding the life support system to support a growing population. Increasing food production can only be accomplished by an increase in the inventory of the critical biological elements in the system. Yet, the Moon is entirely lacking in water and quite deficient in all volatile elements, as far as we know. Carbon, nitrogen, and hydrogen seem to exist only as solar wind implanted gases in small grains of the lunar regolith. Although the total quantity spread over the entire Moon may be quite large, the concentrations are very small, making the resource difficult to exploit. Some scientists speculate that primordial indigenous lunar gases or volatile residues from comet impacts are cryogenically trapped in the bottoms of eternally shadowed polar craters. Unfortunately, this suggestion cannot be evaluated until a polar orbiting survey satellite can take measurements. Therefore, in early development stages the critical volatile elements must be imported and carefully husbanded. On a longer term, tourism or business trips to the Moon may prove to be a source of valuable biological waste.

All manned activities beyond the Van Allen belts will expose astronauts to radiation doses from the galactic cosmic ray flux. On the Moon, regolith can be used for radiation shielding of habitats and work spaces. Silberberg *et al.*, present important data on the nature of the radiation hazard and secondary nuclear interactions that occur in the shielding material.

Another complex element of life support is health maintenance. The planning of medical facilities is difficult for any long duration, isolated activities, whether at sea, at the South Pole, in space, or on the Moon. In some ways, lunar gravity eases the handling of medical emergencies compared to a space station environment. No paper was submitted that discussed the general topic of health maintenance philosophy, but several authors from the Los Alamos National Laboratory present some research on the effects of low gravity and on advanced monitoring systems. Lehnert *et al.*, examine the possibility of pathogenic effects from aerosols. Atchley *et al.*, report progress on a new and sensitive technique for monitoring radiation damage to human cells. Jett *et al.*, point out the advantages gained by development of a space-qualified flow cytometer.

THE EVOLUTION OF CELSS FOR LUNAR BASES

R. D. MacElroy and Harold P. Klein

NASA/Ames Research Center, Moffett Field, CA 94035

M. M. Averner

Complex Systems Research Center, University of New Hampshire, Durham, NH 03824

INTRODUCTION

The prospect of returning to the Moon to establish a growing lunar base is an exciting one for many reasons. It would mean that, as a nation, we are again looking beyond our immediate problems, and it would mean that a significant commitment has been made to scientific investigation of the Earth–Moon system.

While the excitement about a return to the Moon can run high, and we can speculate about the kinds of scientific questions that can be probed with the help of lunar materials, the question of more direct interest to us here is: How can we *stay* on the Moon long enough to begin serious exploration? How can a lunar base evolve from a small, occasionally occupied outpost, to a continuously inhabited base, to a self-sufficient habitat? In this context, our primary interest is life support. The issues we have been most recently addressing are associated with problems of supporting in space at first a few, and then increasing numbers of people for short, medium, and long periods of time.

HUMAN LIFE SUPPORT REQUIREMENTS

The major human life support requirements are well known (Fig. 1). A person's requirement for food is not just caloric, because food provides the body with construction materials, as well as with energy. Additionally, there are more subtle requirements for human metabolism that were suspected even by ancient peoples, but which only began to be discovered within the last 200 years. Examples of such needs are ascorbic acid, along with other vitamins, and iodine. It is unlikely that all of these requirements are known even now, but, for the most part, a deficit of any of them is unlikely to appear except after many years of deprivation.

CURRENT TECHNOLOGY

Presently, life support needs in space are met by taking all of the needed materials along: food, water, and oxygen. Waste materials are collected in various ways and stored.

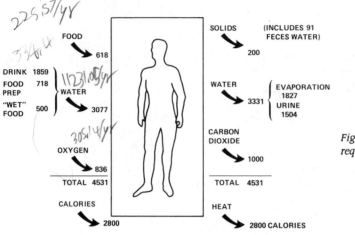

Figure 1. Human metabolic requirements (grams/man-day).

FROM: GUSTAN, E. AND VINOPAL, T., CONTROLLED ECOLOGICAL LIFE SUPPORT SYSTEM: A TRANSPORTATION ANALYSIS. NASA-CR-166420. NASA-Ames Research Center MOFFETT FIELD, CA (1982)

These methods of life support are appropriate only for moderate-sized crews who are in space for relatively short periods of time.

When the crew size increases, and/or the duration of the mission increases, two possible methods are available to meet crew life support requirements (Fig. 2): either the mission can be resupplied with the materials needed, or the necessary materials can be regenerated. The cost of resupplying food, air, and water can rapidly become prohibitive (Gustan and Vinopol, 1982); and materials intended for life support resupply will compete in the flight manifest, in weight (Fig. 3) and in volume, with other things such as equipment. It has long been of interest, therefore, to explore various ways of regenerating life support materials.

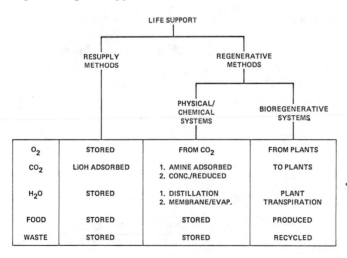

Figure 2. Comparison of crew life support options.

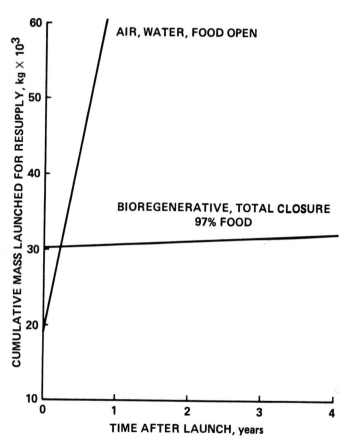

**ESTIMATED BREAKEVEN TIME
MISSION: LEO (4 PERSON)**

Figure 3. Cumulative mass launched vs. mission time for life support options.

REGENERATIVE LIFE SUPPORT

The next phase of life support technology uses physical and chemical techniques to regenerate oxygen and water. The carbon dioxide produced by the crew can be concentrated and processed to release the oxygen it contains. Similarly, used water in urine and wash water can be reclaimed by removing materials dissolved or suspended in it. The equipment necessary for these processes has been developed under programs operating through NASA/Ames and NASA/Johnson Space Centers and by several private companies (Schubert *et al.*, 1984; Quattrone, 1984).

Regeneration of part of the water and part of the oxygen needed for life support goes a long way to decrease the size and frequency of resupply missions. However, because the recycling of materials is incomplete, and because food is not regenerated, resupply is required at rates that increase as the size of the crew increases.

BIOREGENERATIVE LIFE SUPPORT

Although chemical and physical methods of regenerating all materials, including food, are imaginable, current technology is not available to handle some of the finer points, such as building food polymers that provide complete nutritional requirements for humans or removing undigestible enantiomers of common organic compounds. Methods exist to do such syntheses and separations compactly and with very little energy input. The "techniques" were invented during the course of evolution of organisms, and are those used by the photosynthetic organisms that are the fundamental suppliers of all of the food we eat.

Photosynthesis has the advantage that it directly uses the major human metabolic waste product, carbon dioxide, and combines it with water to create organic material that is food, as well as the essential gas oxygen. In addition, since water is the transporter of materials in plants and is rapidly passed from the plant to the atmosphere, higher plants can act to regenerate pure water.

An engineered life support system conceivably could be based upon the same processes that support life on Earth. However, it is important to make a distinction between the way in which the Earth's life support system works, and the way in which an engineered one would work. The difference is primarily one of complexity. Each component or living organism in the natural system is connected with many others through a large number of interfaces.

BIOREGENERATIVE LIFE SUPPORT AND ECOLOGY

An engineered bioregenerative life support system in space will require many of the same physical structures and processes as the terrestrial life support system, but to a significant extent these processes will have to be stringently controlled. For example, the terrestrial ecological system depends upon the existence of enormous buffers of gases and water (the atmosphere and the oceans). It will be necessary to engineer the functions of these buffers into a bioregenerative life support system by significantly reducing the size of the reservoirs and by changing the rates of the processes involved in buffering activity. Similarly, the variety of organisms that constitute the web of life on Earth is unacceptable for a bioregenerative life support system; there is no need for a Noah's ark when the intention is exclusively human life support.

CONTROLLED ECOLOGICAL LIFE SUPPORT SYSTEMS (CELSS)

The product of bioregenerative life support research is expected to be a Controlled Ecological Life Support System (CELSS). At the present time much of the activity associated with NASA's CELSS program is focused on possible use of the system as a part of a future

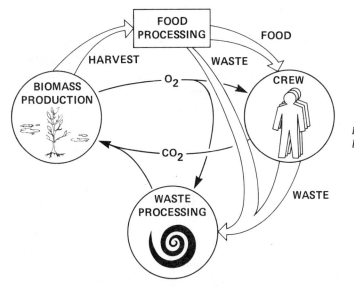

Figure 4.   Material cycling within a bioregenerative life support system.

space platform. Even when partially regenerative physical and chemical methods are used to recycle water and oxygen, and when resupply is possible every 90 days, the use of a bioregenerative system can result in a major reduction in the cost of resupply of food.

The concept of a CELSS is illustrated in Fig. 4. The crew is supplied with water, food, and oxygen by a module that is capable of growing plants or algae, or a combination of both. The organisms consume carbon dioxide and water (metabolic products of the crew) and use energy to convert these materials into food. The issues that are addressed are therefore associated with weight, volume, and cost of operation.

CELSS COMPONENTS

A number of mismatches will occur in the rates at which products are supplied and used, especially between the bioregenerative and the crew modules. For example, some of the solid and liquid wastes of the crew cannot be directly used by the photosynthetic organisms without processing; nor will all parts of the plant biomass produced by the bioregenerative system be useful as food to the crew. As a consequence, a waste processor must be introduced to break down plant and human waste. In the terrestrial ecosystem, waste processing is the function of microbes that take in organic material and produce water and carbon dioxide. It is likely that a biological waste treatment system analogous to those on Earth would be too large for use in space; microbes work too slowly, and a CELSS will be too small. Physical–chemical systems are more attractive (Moore *et al.*, 1982); however, biological systems that partially degrade plant wastes and allow the recovery of usable materials from traditionally non–edible plant fractions will be beneficial to a CELSS.

Another mismatch occurs between the gas demand and production of the crew and bioregenerative modules. The crew's demand for oxygen and its production of carbon dioxide are both variable, depending upon physical activity. The state and the stage of growth of the photosynthetic organisms affects the rates at which they use and produce oxygen and carbon dioxide. A solution to this problem in a small volume, such as that available in a space station, will require an air processing device and associated reservoirs to hold gases that are temporarily in excess and to supply gases that are temporarily in demand.

CONTROL OF BIOREGENERATIVE LIFE SUPPORT SYSTEMS

The minute size of the storage reservoirs and the limited capabilities of the buffers in a CELSS, compared to the immense size of those that make up the terrestrial ecosystem, will mean that the system cannot rely on the responses of the organisms to maintain a constant environment. Rather, it will be necessary to control the movement of materials and to speed up such movements greatly. This will require that information be available on all flows and compositions and that adjustments be made to maintain crew health and safety as well as the stability of the system as a whole. Cybernetic control is demanded by a CELSS, and in its absence the system will go out of control and fail very rapidly (Auslander *et al.*, 1984).

CELSS CONSTRAINTS AND DRIVERS

Research done in the CELSS program has explored the possible use of various photosynthetic organisms and has determined that both unicellular algae and higher plants appear to be worthwhile for consideration in a CELSS. The productivity of some higher plants, such as wheat, for example, is such that surface area of about 20 m^2, cultivated under moderate light intensity with hydroponic nutrient delivery, is sufficient to suuply the caloric requirements of one person continuously.

Through the use of special environmental conditions, such as high carbon dioxide concentrations, higher temperatures, and special attention to nutrient delivery, it will be possible to increase higher plant productivity and thus significantly decrease the area necessary for growth. Some simple processing and extraction procedures can significantly increase the amount of food material produced. A shift in the ratio of edible to non-edible biomass can be expected, for example, from dwarf wheat varieties. Another major improvement in productivity can result from genetic engineering techniques applied to increase the rate and extent to which plants partition the products of photosynthesis into food materials. It is possible that the area required for plant growth can be reduced to less than 10 m^2.

Along with such a reduction in area comes a decrease in the amount of light energy needed for plant illumination. Also, lamp design that more efficiently produces the wavelengths needed by plants is expected to decrease the power requirement by as much as 30%. Various modifications in other lighting techniques, including the use of fibre optics to deliver natural sunlight, can further reduce power requirements. It is thus foreseeable that the obvious improvements in technology and in cultivation techniques can make the use of bioregenerative systems competitive with other life support approaches. Practically speaking, the current level of technology permits the development of a CELSS, even though the effects on productivity of real technological breakthroughs cannot be easily evaluated in such areas as genetic engineering, lamp design, solar power collection, or waste heat removal.

EARLY LUNAR BIOREGENERATIVE LIFE SUPPORT

Even the first lunar outpost could have a module capable of bioregenerative life support. The current CELSS program plan calls for the addition of a small experimental CELSS module to the space station in the year 2000. It would have the capability of supporting the equivalent of two people with food, oxygen, carbon dioxide removal, and water. Although experimental in the sense that it would function to explore the entire process of bioregeneration in micro-gravity and space radiation environments, the module would also operate to supplement the standard life support system and would be connected into that system.

A space station module containing CELSS technology on the lunar surface could be placed into full operation within about 60 days after landing. The environment of the Moon will be, in the initial stages of lunar base development, more similar to the space station environment than to the Earth. The lunar surface has 1/6 of the gravity of the Earth, a characterisitic that may be advantageous to the growth of plants. The system still will be required to be closed and to be limited in weight and volume. Because the system is closed, and because it is intended to recycle approximately 97% of the mass that it contains, it will be advantageous to early lunar development. There will be no appeciable accumulation of waste material, and the system can be modified or expanded relatively easily.

POWER REQUIREMENTS

The most worrisome aspects of utilizing CELSS technology for life support on the lunar surface are its energy demands. We estimate that, in the 2000–2005 time frame, the power generating requirements for life support on the lunar surface, assuming incorporation of the most obvious technological and cultivational advances, and exclusive of the costs of waste heat rejection, will be approximately 5.5 KW/person, during the lunar day. The scenario assumes continuous 28-day illumination of photosynthetic organisms, direct fibre optics-delivered natural light for 14 days, and stored, fuel cell-generated power for 14 days. Plants will require 3 KW/person electric illumination during the lunar night and regeneration of fuel cell capabilities.

It is possible that certain tactics of cultivation and organism selection can be employed to reduce the power requirement by 50% or more, but at the expense of increasing the cultivated area per person. For example, it appears possible to alter, or subvert, the plants' "season sensing" that causes triggering of different growth phases, and then to alter the environment (*i.e.*, temperature, carbon dioxide concentration, and lighting) to place the organisms into a kind of suspended animation for the duration of the 14–day lunar night. When daylight again arrives, the normal growth cycle could be resumed. If the base were sited at one of the poles these problems would be eliminated because sunlight would be available constantly.

A LUNAR CELSS

The lunar surface rock is rich in oxygen, an essential component of a life support system; however, hydrogen, carbon, and nitrogen abundances are very low. These materials will have to be brought from Earth, and they will have to be recycled. All organic "waste" materials must be conserved from the initiation of lunar base buildup because they will be valuable assets once a bioregenerative system is introduced to the lunar environment. It is also worthwhile to consider the possibility that building and construction materials that are brought from Earth could be composed of biologically relevant materials.

The evolution of a lunar base to structures capable of housing and supporting hundreds and even thousands of people understandably will require a bioregenerative life support system. At some point, when the crew complement passes a certain size, perhaps as few as 20 people, pressures will arise to shift the life support system from "vegetable" to "animal–vegetable." The reasons for the shift will be numerous but will be heavily influenced by human food preferences. Photosynthetic organisms can supply all of the known nutrient requirements of humans, except for small amounts of vitamins and co-factors that can be easily added to food supplies. However, the human animal has evolved as an omnivore, and a large fraction of the extant human race includes meat in its diet. While the tastes and textures of a variety of meats can be simulated using plant protein, it is possible that deprivation of meat for long periods of time will result in a psychologically-based preference demand.

The conversion of vegetable material to meat is an expensive process. The current American method of feeding grain to cattle is incredibly energy consuming and could not be supported in the spartan economies of the initial bioregenerative life support systems. However, animals that can thrive on plant materials that humans cannot digest are common, and small animals, such as insects, mollusks, and fishes are possible candidates for early inclusion in a CELSS. Small mammals (*e.g.*, rabbits) and birds (*e.g.*, the Japanese quail) are likely to be slightly lower on a priority list. Larger mammals, such as goats and sheep, are probably lower still. However, the data on energy requirements and sources necessary for these animals has not been thoroughly examined yet, and no hard data comparisons have been made.

LIGHT, HEAT, AND POWER

It is likely that the area required for bioregenerative life support and the power required to support photosynthesis will be substantial. Efficiencies of photosynthesis vary between

1%, which is the usual figure for most crops, to 3%, the value generally assumed for careful agriculture, to about 10%, obtained in CELSS experiments, to 16%, obtained with carefully controlled algal growth. These values refer to the energy of photons arriving at the organisms' energy-capturing apparatus compared to the energy that is actually captured into chemical compounds.

Some of the photons in the solar spectrum are harmful. Wavelengths shorter than about 400 nm are considered deleterious, and those shorter than 3000 nm are definitely injurious. At wavelengths longer than 700 nm photons are marginally useful, and those beyond 800 nm are ineffective. Ideally, the radiation to be used for photosynthesis should be most intense at the wavelengths corresponding to the absorbtion peaks of the two major chlorophyll components of plants or algae, and injurious or non-effective radiation should be excluded.

Recent advances in commercially available light technology might be used in a lunar base (Mori, personal communication, 1983). In the new system, a Fresnel lens is carefully focused on a fibre optics collector. By adjustment of the geometry of the lens and collector, only certain wavelengths are transmitted into the fibre optics system, allowing the elimination of injurious ultraviolet light and photosynthetically ineffective infrared. The light is diffusively radiated at the termination of the fibre optics system, providing plants or algae with sufficient light for normal growth. The system may also be used as part of a solar concentrator.

While it is possible, and even likely, that the photosynthetic efficiency of higher plants can be increased to match that of algae, 80–90% of all of the light brought into the system for photosynthesis will be transformed into heat. Some of this substantial heat load may be useful for heating other parts of the lunar base, but the major part will have to be radiated or otherwise directed out of the system. At the moment we do not have sufficient data to estimate the heat dissipation load.

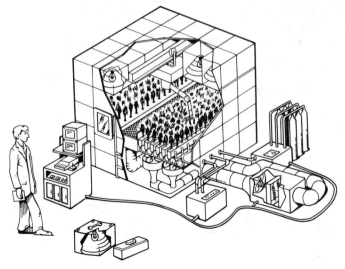

Figure 5. Artist conception of a sealed plant growth chamber with robotics.

CELSS AND HUMAN LABOR

The concept of a lunar base includes sealed chambers (Fig. 5) devoted to the cultivation of photosynthetic organisms and animals necessary for life support. Algal cultivation can be automated easily, but human activities necessary for cultivation of plants in intensive controlled environment agriculture could demand a considerable labor investment.

The CELSS program in NASA is initiating studies of the use of robotics devices to accomplish many of the routine tasks required for higher plant cultivation. Examples include spacing of plants for optimal light use, sampling plants for determination of growth stage and for disease detection, harvesting crops that do not produce fruit uniformly (e.g., tomatoes), harvesting uniformly ripening crops (e.g., wheat), processing crops to remove non-edible materials, and converting crops into food (e.g., milling wheat).

It may be advantageous to cultivate crops entirely without human intervention. This would eliminate a major mechanism for the introduction of unwanted organisms, such as viruses and bacteria, that may be deleterious to growth. It would also allow the maintenance of an environment for plant growth that might be difficult for humans to tolerate, such as high concentrations of carbon dioxide, low concentrations of oxygen, high temperatures, etc.

CONCLUSIONS

Our current picture of a bioregenerative life support system for a lunar base is one of a small but highly automated system capable of very intensive agriculture, maintained in a stable state by computer control, and responding in every way as a typical manufacturing plant. The input and output materials will be known, both in composition and in rate, and the system will be directed to respond to the requirements of the crew, which will be its *raison d'etre*. The system will be capable of expanding to meet the demands of additional crew and of being subdivided to ensure system safety. It will be designed to address the fundamental requirements of a functioning, economically independent lunar base: total separation from the external environment and complete recycling of all of the expensive materials that originated on Earth.

REFERENCES

Auslander D., Spear R., Babcock P., and Nadel M. (1984) *Control and Modelling of a CELSS (Controlled Ecological Life Support System)*. NASA CR-177324. Berkeley College of Engineering, Berkeley. 85 pp.

Barney G. O. (1980) *The Global 2000 Report to the President*. Doc. No. 0-256-752. U.S. Govt. Printing Office, Washington. 766 pp.

Gustan E. and Vinopal T. (1982) *Controlled Ecological Life Support System: A Tansportation Analysis*. NASA CR-166420. Boeing Aerospace, Seattle. 126 pp.

Moore B., Wharton R. A. Jr., and MacElroy R. D. (1982) *Controlled Ecological Life Support System: First Principal Investigators Meeting*. NASA CP-2247. NASA/Ames Research Center, Moffett Field. 90 pp.

Quattrone P. D. (1984) Extended mission life support systems. *Advances in the Astronautical Sciences*, Sci. Technol. Ser., 57, 131–162.

Schubert F. H., Wynveen R. A., and Quattrone P. D. (1984) Advanced regenerative environmental control and life support systems: Air and water regneration. In *Advances in Space Research: Life Sciences and Space Research XXI* (2) (H. Oser, J. Oro, R. D. MacElroy, H. P. Klein, D. L. De Vincenzi, and R. S. Young, eds.), pp. 279–289. Pergamon, Oxford.

WHEAT FARMING IN A LUNAR BASE

Frank B. Salisbury and Bruce G. Bugbee

Plant Science Department, Utah State University, Logan, UT 84322-4820

Green plants in a lunar base could remove CO_2 from and add O_2 to the atmosphere, produce food, recycle most waste products, and contribute to a water purification system. We have studied wheat in the context of a bioregenerative, life-support system because of its suitability as food, its vertical leaf orientation, its excellent growth under continuous light, and available background information. Theoretical photosynthetic efficiencies suggest that yields could never exceed 195 g m^{-2} day^{-1} of dry matter when plants are irradiated with 1000 μmol of photons s^{-1} m^{-2}, an irradiance that can easily be achieved with high-pressure sodium lamps. In practice, yields are limited by incomplete light absorption, percentage of edible biomass (harvest index), digestibility of biomass, and efficiency of lamps. Considering these factors, minimum figures per person might be about 6 m^2 of growing area and 3.55 kW of electrical energy. Based on yields currently achieved, which greatly exceed the best field yields (in terms of primary biomass production), minimum figures are 24 m^2 and 13.4 kW person^{-1}. If these numbers are doubled to provide a margin of safety, a lunar farm could support 100 people in an area of about 5000 m^2, the size of an American football field. Our harvest index is presently low (20–25%) because of poor seed set. Future yields might be increased by manipulating temperature, humidity, nutrients, CO_2, and the radiation environment, especially if the harvest index is improved and if early canopy development is promoted. Selecting and breeding suitable cultivars appears especially promising. Some problems of constructing and operating a lunar farm are noted.

INTRODUCTION

Growing green plants in the closed environment of a lunar base would accomplish some of the same important functions that are performed by green plants in the closed (with respect to matter) system of planet Earth. Carbon dioxide from the atmosphere is used in the photosynthetic production of organic matter, and oxygen is released as a byproduct of the reduction of water. Only relatively small amounts of water are involved in the synthesis of carbohydrate, but much larger quantities are transpired from plant leaves and other surfaces. On Earth, this evaporated water eventually condenses as rain or snow, a purification process that would also be used as part of the water purification system in a lunar base (condensation on cooled surfaces). The green plants would also be a part of the waste recycling system, utilizing mineral elements from partially or completely oxidized organic waste products. With some plants, urine could be used directly or after some dilution. Based on current technology, such an agricultural system could become a functional part of the bioregenerative life-support system in a lunar base, but future research on both biological and engineering problems is necessary to improve efficiency.

For the past four years, we have been supported by the NASA controlled-environment, life-support-system (CELSS) program. Although a variety of crop plants would be grown in a CELSS, we have studied wheat for the following reasons:

1. The vertical leaf orientation allows wheat to efficiently absorb high levels of solar radiation and convert this energy into a high food yield per unit area. Crop plants with horizontal leaves grow well at low light levels but generally cannot achieve such high productivities per unit area.

2. Wheat can be processed into a variety of food products that can supply a major portion of dietary carbohydrate and protein. Lettuce, for example, can supply only a small portion of dietary calories before its vitamin A becomes toxic.

3. Wheat, rice, and maize are the major food crops of the world. Much is known about wheat physiology, and this knowledge can be rapidly adapted to a new environment. At the beginning of our project, much wheat expertise was already available at Utah State University.

4. Much is also known about wheat genetics, so it is possible to quickly select and breed new cultivars for a new environment.

5. Wheat forms flowers in response to long days (*i.e.*, it is a long-day plant) and responds well to continuous light, which results in a maximum use efficiency per unit mass of the lighting system. Short-day crop plants such as rice have an obligate requirement for a dark period (about 8–12 hours, depending on species and cultivar) before they initiate seed production. Tomatoes cannot grow under continuous light, which causes their leaves to become yellow (chlorotic) and eventually die. The physiological mechanisms underlying these responses are not yet completely understood, but crop plants that are efficient food producers in a range of photoperiods are highly desirable in a CELSS.

THE COSTS OF A BIOREGENERATIVE SYSTEM

The feasibility problems for a CELSS in an orbiting spacecraft or in a lunar base are similar—although resolution of the problems seems much more straightforward with a lunar base. In either case, one must first reckon the costs, calculating what must be transported or, as in the case of the Moon, be constructed from local materials. Plants require relatively large quantities of water (which, of course, can be recycled) and relatively small quantities of mineral nutrients. In addition, they require carbon dioxide for photosynthesis. So far, in our musing about a CELSS in a spacecraft, we have tacitly assumed that the carbon dioxide would be produced by the respiration of astronauts. One can visualize a lunar base as being much larger, however, so it might well be necessary to transport carbon to the lunar base to be sure that ample CO_2 would be present in the atmosphere, especially at the beginning before any recycling had occurred. Fairly sophisticated equipment is required to grow plants under completely controlled conditions, and in either case this would have to be manufactured on Earth and transported to the spacecraft or lunar base—until manufacturing capabilities at the lunar base had advanced to a relatively high level of technology. Once that has occurred, it might be possible to supply water and mineral nutrients from lunar materials. We can visualize the construction of basic growing facilities from metals and perhaps glass produced from lunar materials, but it would probably be some time before many of the necessarily advanced environmental control systems could be manufactured on the Moon.

A second consideration is the time required of astronauts or inhabitants of a lunar city to maintain the functioning agricultural system. This is an important continuing part of the cost. A third consideration is the energy requirement. Significant energy is needed to operate the farming system, and much more energy might be required to provide artificial light during the two-week lunar night.

In many ways, a CELSS on the Moon has several advantages over one in an orbiting space station. It would be much simpler to construct growth units for the lunar surface than it would be for an orbiting space station. The presence of even one-sixth of the Earth's gravitational field would greatly simplify the construction or assembly of facilities and would also reduce many of the biological and engineering problems related to the growing of plants. It is still not known how well plants will respond to microgravity; some evidences suggest that plants can be grown efficiently in such an environment, but since plants on Earth normally respond to gravity (they grow upright as controlled by a delicate gravity-sensing system) in various subtle as well as obvious ways, it is not surprising that some features are abnormal when plants are grown in microgravity (*e.g,* Conrad, 1968). Evidence also shows that some of the symptoms plants exhibit in microgravity might also be exhibited, albeit to a lesser extent, at one-sixth g (Salisbury and Ross, 1969).

One problem presents features that are somewhat common to the spacecraft and the lunar surface: The light-dark cycle differs markedly from the 24-hour cycle experienced by plants on Earth. A low-Earth-orbit space station would experience 60 minutes of sunlight and 30 minutes of darkness. This cycle could be quite deleterious to the growth and development of many plants, and we are currently investigating its effect on wheat. The 29.5-day light/dark cycle on the Moon is clearly a problem. No crop plant could remain productive after 15 days of darkness, so light would have to be provided during the dark intervals, although the light could be at irradiance levels well below sunlight.

THE SIZE OF A LUNAR FARM

The immediate goal of our research effort is to determine the controlled-environment food-production efficiency of wheat per unit area, per unit time, and per unit energy input.

The Theoretical Minimum Size of a Lunar Farm

At the CO_2 concentrations present in the Earth's atmosphere, species with C_4 photosynthesis (*e.g.,* maize, sugar cane) are often more efficient than species with C_3 photosynthesis, which includes wheat and most crops (summary in Salisbury and Ross, 1985). At elevated CO_2 levels, however, C_3 plants are significantly more efficient than C_4 plants. Therefore, C_3 plants are a good choice for a CELSS or a lunar station, where CO_2 levels are expected to be elevated. From the stoichiometry of electron transport in photosynthesis and a proton requirement of three for ATP synthesis (Handgarter and Good, 1982), a theoretical minimum of 9 mol of photons are required to fix 1 mol of CO_2 into carbohydrates. In addition, some energy is required for nitrate reduction, some is lost to fluorescence, and some is absorbed and reradiated as heat by non–photosynthetic pigments, so the best conversion efficiency that has been achieved in single leaves of higher plants is 12 mol of photons per mole of carbohydrates (Ehleringer and Pearcy, 1983; Osborne and Garrett, 1983). This is close to the conversion efficiencies achieved with algae.

With a 12–photon requirement and assuming a continuous flux of 1000 μmol s^{-1} m^{-2} of visible radiation (about one–half full sunlight at the Earth's surface) and 10% loss for root respiration, we could theoretically produce 195 g m^{-2} d^{-1} of dry biomass. If all of this biomass were edible, if the human body could metabolically obtain 4 kcal from each gram, and if 3100 kcal were consumed per person per day, then each person could be fed from the production on only 4 m^2. This is the highest possible efficiency that could be achieved by any plant species.

Theoretical Energy Requirements

McCree (1972) calculates that 5 μmols of photons s^{-1} produced by high–pressure sodium lamps in the photosynthetic part of the spectrum (400–700 nm) represent almost exactly one watt of energy. Thus, if high–pressure sodium lamps can be made 40% efficient at producing photosynthetic energy (efficiency of 37.6% is noted below), an input of 500 W m^{-2} could produce 1000 μmol s^{-1} m^{-2}. If 4 m^2 were required per person, the energy input could be as low as 2 kW per person, using only artificial light.

Potentially Achievable Size and Energy Requirements Using Higher Plants

Four factors reduce the achievable productivity of plants below theoretical: light absorption, harvest index, digestibility, and energy conversion.

1. *Light absorption.* Plant leaves never absorb all the incident radiation. Our measurements suggest that, under ideal conditions, 5% of the radiation is reflected, and 1% is transmitted, even by a dense canopy with vertical leaves. It is unlikely that absorbed energy will ever exceed 95% of incident energy.

A more significant absorption problem occurs during the early stages of plant growth when small plants do not intercept all the incident irradiation. Wheat is grown at densities up to 1500 plants m^{-2} (6.7 cm^2 plant^{-1}, 2.6 cm between plants). This is 3–6 times normal planting densities in the field, but plant leaves absorb only 50% of the irradiance when they are 14 days old and 90% when 18 days old. After day 18, light interception continues to be excellent until harvest at day 60. The germinating seeds do not require light until emergence on day 3, but absorption efficiency is low from day 3 to about day 18. In our current system, this loss is about 20% of the total area and energy required to grow the crop. A mechanical system to alter plant spacing during early growth (so plants are moved apart as they mature) could eliminate some of this loss. Such systems are being used in commercial controlled–environment food production.

2. *Harvest Index.* The most significant limitation to food production is that not all the biomass produced by the plants is edible. The edible divided by total biomass (both dry) is called the harvest index. A lettuce crop has about 80% edible leaves and 20% inedible stem and roots. Potatoes can have a harvest index of edible tubers of 80% of

the total biomass, and wheat can reach 60% edible grain on a dry-mass basis. Under the best conditions, there is a 20–40% loss from inedible plant materials. These could be consumed by animals (chickens, pigs, rabbits, *etc.*) to produce edible protein for humans, although this would introduce some complications.

Many authors have suggested crops with edible roots, leaves, and reproductive structures; sweet potatoes and sugar beets are examples. In most cases, however, only the *young* leaves are edible, although it is the mature tubers, roots, fruits, or seeds of such plants that are normally harvested. Unusual food crops should be considered for a CELSS, but claims of high productivity and high harvest index often cannot be substantiated.

3. *Digestible energy per unit edible biomass.* When the energy content of oven dry wheat is determined by combustion in a bomb calorimeter, values as high as 3.94 kcal per gram are obtained, but the digestible energy is only about 3.7 kcal. This relationship also holds for other food commodities.

4. *Energy conversion.* High-pressure sodium lamps produce 376 W of energy between 400 and 700 nm per 1000 W input power. This makes them 37.6% efficient (Chris Mpelkas*, personal communication, 1985). Their output, however, must be reflected down onto the plants. The best reflectors are about 90% efficient. This makes the overall efficiency of the system 33.8%. Efficiencies of 26% have been achieved on a commercial scale. The Phytofarm in Dekalb, Illinois, has an energy input of 304 W m^{-2} from high-pressure sodium lamps and a photon output of 400 μmol s^{-1} m^{-2} or 80 W m^{-2} of photosynthetic irradiance (Maynard Bates, personal communication, 1984).

Considering these four factors, the potential size and energy requirements that can be achieved in a lunar farm can be calculated as follows:

Theoretical (with 1000 μmol s^{-1} m^{-2} light input)	195 g m^{-2} d^{-1}
90% light absorption over life cycle	175 g m^{-2} d^{-1}
80% harvest index	140 g m^{-2} d^{-1}
Multiplied by 3.7 kcal g^{-1} (92.5% digestible)	518 kcal m^{-2}
Assume 3000 kcal per person per day:	
3000 kcal person^{-1} d^{-1} divided by 518 kcal m^{-2} d^{-1} = 5.79 m^2 person^{-1}	
Round to:	6 m^2 person^{-1}
Energy requirement:	
1000 μmol s^{-1} m^{-2} =	200 W m^{-2}
200 W m^{-2} divided by 0.338 efficiency =	592 W m^{-2}
592 W m^{-2} $\times$ 6 m^2 = 3552 W person^{-1}	3.55 kW person^{-1}

These theoretical efficiencies would be very difficult to achieve with a crop plant (such as strawberries) that is chosen for its aesthetic qualities and flavor rather than

---

*at Sylvania Test and Measurements Laboratory, Danvers, MA.

for its productivity. Nonetheless, research will need to be done on all species grown in a CELSS to optimize their edible productivity.

Currently Achievable Productivities with Wheat

During the past year, after spending much time on designing and building research chambers to create optimum environmental conditions for studies on wheat productivity, reproducible production data have been obtained that can be used to estimate the size of a lunar farm that could be built today. So far, we have been highly successful in converting photosynthetic irradiance into biomass but much less successful in converting total biomass into edible yield.

We measure short-term rates of carbon fixation in wheat canopies with a gas exchange system that includes a pressurized growth chamber (Salisbury, 1984). A canopy of 0.8 m^2 is grown in this chamber with the roots in a sealed, recirculating, hydroponic system (roots fed with nutrient solutions). A light input of 1000 μmol s^{-1} m^{-2} in an atmosphere enriched to 1700 ppm CO^2 has resulted in photosynthetic rates as high as 58 μmol s^{-1} m^{-2} of carbon dioxide absorbed by the leaves. If we subtract for root respiration and multiply by the photoperiod each day, this figure can be converted into a daily growth rate. Root biomass in our hydroponic systems is typically only 10% of the total (20–30% in the field). Subtracting this estimated 10% respiratory loss and assuming continuous light, this photosynthetic rate should result in a growth rate of 135 g m^{-2} d^{-1}. This compares well with the theoretically achievable growth rate noted above of 175 g m^{-2} d^{-1} (at 90% light absorption).

We measure actual growth rates at weekly intervals by removing a 0.2 m^2 section of plants (about 200 plants in a rigid support), blotting the roots dry, weighing the section, and returning it to the hydroponic solution. A few plants are destructively harvested and dried to determine percent dry-mass, from which dry-mass growth rates can be calculated. We have measured growth rates of 875 g m^{-2} week^{-1} or 125 g m^{-2} d^{-1}. This growth rate is close enough to that predicted from the gas exchange measurements to serve as a validation of the short-term photosynthesis measurements. Unfortunately, it takes about 22 days for a group of plants to reach this growth rate, and the rate gradually decreases as the plants mature. These factors combine to make an average growth rate of 90 g m^{-2} d^{-1} over a 60–day life cycle.

The production of 90 g m^{-2} d^{-1} total biomass is truly remarkable by conventional agricultural standards. Typical field productivities are less than 10 g m^{-2} d^{-1}, and 20 g m^{-2} d^{-1} is exceptional. Wheat is obviously stressed even in the best field conditions. The stress factors could be low carbon dioxide and/or low light, neither of which would be economical to change in the field.

These high growth rates are the good news. The bad news is that we have not yet been able to cause wheat growing at high rates to partition a normal percentage (40 to 50%) of its total biomass into edible grain. A crop producing 90 g m^{-2} d^{-1} should have a grain yield of 35 to 45 g m^{-2} d^{-1}; our crops have produced only 20 to 25 g m^{-2} d^{-1}. We expect to solve this problem, but at the moment, the reasons for this low harvest index remain unclear. A comparison of our yield components with field production data offers some clues (Table 1).

Table 1. A Comparison of Controlled Environment and Field Productivities

	Life cycle (days)	Seed yield (g m^{-2})	Harvest index (%)	Heads per m^2	Seeds per head	Mass per seed (mg)
Controlled environment	60	1300	25%	3000	15	29
High yield from field	100	800	45%	800	30	33

Continuous light and a constant high temperature (27°C) are principal factors responsible for shortening the life cycle from 100 to 60 days. These same two factors may also be partly responsible for the low seed number per head, which is associated with our low harvest index. Low seed number per head is the result of few spikelets formed on the head (spike) during the floral induction phase (days 15–22) and/or poor seed set during and following the pollination period (called *anthesis*: days 30–37). There is published evidence that the shortening of the growth period associated with long photoperiods results in the production of fewer spikelets per spike during floral induction (Rawson, 1970; Lucas, 1972).

The main problem appears to be poor seed set in existing florets. Wheat is self-pollinated, and the anthers (male flower parts) do not appear to shed pollen normally in our conditions. This inhibits fertilization and thus seed set. We are just beginning to study the problem.

Our reproducible seed yields of 20–25 g m^{-2} d^{-1} give a harvest index of about 20%, and it is reasonable to expect that, based on a better understanding of floral initiation and pollination, we can double this to about 40% (40–50 g m^{-2} d^{-1}) without any additional energy inputs.

With a harvest index of 40% (instead of 80%) the above size and energy figures for the lunar farm must be doubled: 12 m^2 and 6.68 kW person^{-1}. With a harvest index of 20%, the figures are multiplied by four: 24 m^2 and 13.4 kW person^{-1}. Even if these figures were doubled again to provide a large margin of safety plus room for working aisles between groups of plants and for other work areas, they would not be discouraging from the standpoint of a lunar base. A medium-sized classroom has on the order of 50 m^2, and 100 humans could be supported by a lunar farm of 5000 m^2 at most. Based on the most optimistic figures given in the above sections, this could be reduced to about 600 m^2. (A standard American football field, including end zones, has an area of 5364 m^2.) We estimate that the lunar farm designed to feed 100 people might be operated by a staff of 2–10 lunar farmers.

This is not to suggest that a lunar farm would be inexpensive and easy to construct. It might have to consist of relatively small, self-contained modules, all initially brought from Earth. If sunlight were used directly, the farm would need a transparent or translucent covering strong enough to withstand internal atmospheric pressure (probably reduced from that on Earth), and, more importantly, micrometeorite bombardment. The initial quantities of equipment, water, carbon dioxide, and minerals that had to be brought from

Earth would be formidable—but quantities of food required to support 100 inhabitants of a lunar city would provide an even more formidable continuing transport problem.

The energy required to operate such a farm completely with artificial light from high-pressure sodium lamps would also be very large (334–1340 kW, based on the above figures). If solar cells are used to collect energy, it must be realized that only about 5% of sunlight will eventually be converted to light from the lamps, assuming highly efficient solar cells and lamps. With these ideas in mind, it is important to consider direct use of sunlight during the lunar day. The slowly changing position of the Sun in the lunar sky might be a problem, but that could probably be solved with reflectors, translucent and diffusing glass, or even bundles of fiber optics. It is claimed that fiber optics can transmit as much as 50–68% of the light, but even if light is first greatly concentrated by fresnel lenses, the size of the required bundles of fiber optics to irradiate a lunar farm is a bit staggering. If the lunar station is powered by a fairly large nuclear plant, as is often proposed, power for irradiating the plants might not be a serious problem. It is important, however, for engineers to be aware of the high light levels required by plants for optimum yields. Adequate illumination for an office environment is clearly not sufficient light for growing plants in a lunar farm.

INCREASING THE YIELDS

We calculated above that a theoretical maximum production of dry matter when 90% of 1000 μmol s^{-1} m^{-2} of light was absorbed was about 175 g m^{-2} d^{-1}. Our figure of 90 g m^{-2} d^{-1} represents an efficiency of about 51%. The challenge is to close the gap between 51 and 100%. There are many parameters to manipulate. Consider a few.

Temperature. So far we have not really studied temperature. We use values (27°C) reported to be optimal for wheat with CO_2 enrichment. Higher temperatures could shorten the life cycle but might decrease yield per day. With a few plants, varying temperature on a 24-hour cycle increases yield, but this does not seem likely for wheat. There could be surprises.

Humidity. There are two possible adverse effects if humidity is too high. First, because transpiration is reduced under such conditions, leaf cooling is less, and leaf temperatures may increase above optimal levels when irradiation is as high as it is in our chambers. Second, because transpiration is reduced, mineral uptake may be reduced. Evidence from recent experiments in our laboratory suggest that this is an important effect when CO_2 levels are elevated, which causes partial stomatal closure. (Stomates are the pores on a leaf surface through which water evaporates and CO_2 enters.) It is easier for us to maintain optimal nutrient conditions within plant tissue when humidities are lowered. This does not appear to be the case when plants are growing under less ideal conditions.

Nutrients. Plants are grown with their roots in aerated, circulated, nutrient solutions. We find that mineral nutrient concentrations in these solutions can be very critical, and responses to nutrients can change as other parameters change—as just noted for humidity. We have expended much time developing adequate nutrient solutions and techniques to provide them, but we have not yet solved all the problems. Our youngest plants sometimes show deficiency symptoms that disappear as the plants mature (*i.e.*, when they reach about two weeks of age). Iron, manganese, phosphate, and other nutrients can be problems, especially as the pH increases rapidly as nutrients are absorbed. We have been able to control pH within fairly narrow limits by providing a balance of ammonium and nitrate ions and by using an automated system to add acid when needed. Ammonium ions are exchanged for hydrogen ions produced in the plant roots, decreasing pH, and nitrate ions are exchanged for bicarbonate ions from the roots.

Carbon dioxide concentration. CO_2 is typically limiting at ambient levels (320 ppm = 15 mmol m^{-3} at sea level). Yields are greatly increased when CO_2 levels are raised around the plant leaves. We elevate to 1700 ppm (60 mmol m^{-3} at our elevation). Stomates tend to close completely when CO_2 levels are elevated too high, but we are not yet sure of the upper limits. It would be possible to manipulate other gases, and lowering oxygen levels would also increase rates of photosynthesis, probably without stomatal closure. So far we have not invested the time and money required for such a study.

The radiation environment. There are several aspects of the light environment that must be studied:

1. Light level (irradiance)—Increasing irradiance would not help in the above example to raise efficiency; indeed, it might lower the efficiency of photosynthesis if the process had already reached light saturation. If saturation had not been achieved, however, an increase in light level might raise absolute yields expressed as g m^{-2} d^{-1}. With today's technology, it is difficult to get and expensive to maintain light levels much above the 1000 μmol s^{-1} m^{-2} that we have used, although we have now outfitted one growth chamber so that we obtain 2000 μmol s^{-1} m^{-2}, equivalent to sunlight at noon. Preliminary results show that photosynthesis increases considerably compared with half of sunlight, but photosynthesis was not quite doubled.

2. Light quality (spectrum)—The balance of wavelengths can be modified in an almost infinite variety of ways, so there is much room for experimentation. One of these approaches has been taken with rather interesting results. Healthy wheat plants have been grown that produce normal grain under low-pressure sodium lamps. The energy from these lamps is nearly all confined to one line in the spectrum at 589 nm. The lamps are efficient at producing light energy, so they might be of use in the lunar farm during lunar night. Furthermore, it was found (Guerra *et al.*, 1985) that secondary metabolites (specifically lignin) are more dilute in tissues, and activities of two key enzymes (PAL and TAL) in the synthesis of secondary metabolites are greatly reduced in plants grown under these lamps. This could mean that primary metabolites (starch, protein, fat) could be higher in the plants, although this has not yet been shown to be true.

In general, light quality has many important photomorphogenetic effects. For example, light quality might influence the partitioning of assimilates in such a way that harvest index is increased.

3. Light cycling—Daylength (photoperiod) has profound effects on many plant responses including flowering, seed filling, tillering (formation of axillary stems in grass plants), and dormancy. Hence, photoperiod can and does influence the duration of the life cycle as well as the harvest index. We are manipulating the photoperiod to see if we can increase seed set.

4. Canopy development—Our results so far show that canopy development and harvest index are probably the most important factors limiting our grain yields. The above calculation assumes that 90% of the incoming radiation is absorbed, but this is far from true when the plants are small. The leaf-area index (LAI) expresses the number of layers of leaf tissue through which a given ray of light must pass (on average) before it strikes the substrate. For maximum absorbance of incoming radiation, a high LAI is essential. Wheat reaches an LAI of 6–8 in the field and has reached 14 in our controlled conditions. At that point, light at the bottom of the canopy is only about 10–20 μmol m^{-2} s^{-1}. Absorption is extremely high, and photosynthesis reaches about 77% of the theoretical maximum. This suggests that we cannot improve photosynthetic efficiency more than perhaps 10–20% by manipulating parameters as suggested above—although it remains worthwhile to attempt to do so.

THE MOST PROMISING FUTURE WORK

Yield per unit area per unit time is much more than photosynthetic efficiency of a mature canopy. There are two especially important characteristics of a wheat farm that strongly influence yield and can still be manipulated: the time to canopy closure and the harvest index. Both can be influenced by manipulation of the environment and of the plant's genetics.

As noted, plants may be moved apart as they mature, but such techniques could be more trouble than they are worth. It is also possible that environmental manipulations could produce a leafy plant quicker, and/or at a savings of energy input. Since wheat is a facultative long-day plant, plants could be started under short days, perhaps with somewhat reduced irradiance levels. This would save electrical energy if it were done during the 14-day lunar night, and the retarded flowering might allow the development of a more leafy plant before its energy resources were directed toward flower production and seed filling.

Cultivars are being selected with a high genetic potential for rapid canopy closure. (Two members of our team, Rulon Albrechtsen and Wade Dewey, are wheat breeders.) Particularly promising for rapid canopy closure are uniculm cultivars that produce only one or a few tillers per plant. Normal wheat plants close the canopy by sending out as many as 3–15 tillers, each of which produces a head of wheat. This takes time, and plants cannot be too closely spaced at the beginning or they become overcrowded in the field. A cultivar that produced only one to a few tillers could be planted in a dense pattern to begin with, so that the canopy was rapidly closed. Five segregating generations, incorporating some desirable agronomic characteristics into a uniculm cultivar have been completed. We have also selected six generations of dwarf wheat (35 cm tall) under CO_2-enriched, continuous-light environments. In essence, a special wheat plant for a CELSS is being designed.

There is certainly much potential for increasing yields by increasing harvest index (by producing more wheat seeds per plant). Again, the most promising approaches are to manipulate environmental factors and to select suitable cultivars. As noted, there is good reason to believe that manipulation of photoperiod will increase seed set. We have now tested about 600 cultivars and find great differences in their growth, yield of grain, and harvest index. Harvest index is usually highest in dwarf (30 cm) cultivars, but so far their overall yields are relatively low.

SOME CONCLUSIONS

It should be quite feasible and probably profitable (depending on the permanence of the lunar base) to establish a wheat farm on the Moon. There are serious problems (*e.g.*, the long lunar night), but solutions are presently available, and future research could provide even better solutions. Based upon current data, it appears that about 6–25 m^2 should be sufficient to provide food for an active adult. A lunar farmer would not be at the mercies of unpredictable weather, as earthly farmers are; rather, he or she would be at the mercies of the inherent tendencies for mechanical equipment to falter and of his or her own propensities to make human errors. This being the case, an inhabitant of a lunar city might feel more at ease if there is ample area of a lunar farm dedicated to producing food for his or her survival.

REFERENCES

Conrad H. M. (1968) Biochemical changes in the developing wheat seedling in the weightless state. *BioSci.*, *18*, 645–652.

Ehleringer J. and Pearcy R. W. (1983) Variation in quantum yield for CO_2 uptake among C_3 and C_4 plants. *Plant Physiol., 73*, 555–559.

Guerra D., Anderson A. J., and Salisbury F. B. (1985) Reduced phenylalanine ammonia-lyase and tyrosine ammonia-lyase activities and lignin synthesis in wheat grown under low pressure sodium lamps. *Plant Physiol., 78*, 126–130.

Handgartner R. P. and Good N. E. (1982) Energy thresholds for ATP synthesis in chloroplasts. *Biochem. Biophys. Acta, 681*, 397–404.

Lucas D. (1972) The effect of daylength on primordia production of the wheat apex. *Aust. J. Biol. Sci., 25*, 649–656.

McCree K. J. (1972) Test of current definitions of photosynthetically active radiation against leaf photosynthesis data. *Agri. Meteor., 10*, 443–453.

Osborne B. A. and Garrett M. K. (1983) Quantum yields for CO_2 uptake in some diploid and tetraploid plant species. *Plant, Cell, Environ., 6*, 135–144.

Rawson H. M. (1970) Spikelet number, its control and relation to yield per ear in wheat. *Aust. J. Biol. Sci., 23*, 1–15.

Salisbury F. B. (1984) Achieving maximum plant yield in a weightless, bioregenerative system for a space craft. *Physiologist, 27*, S-31–34.

Salisbury F. B. and Ross C. W. (1969) *Plant Physiology*, pp. 492–493. Wadsworth, Belmont, CA.

Salisbury F. B. and Ross C. W. (1985) *Plant Physiology.* Third edition, pp. 195–228. Wadsworth, Belmont, CA.

METABOLIC SUPPORT FOR A LUNAR BASE

R. L. Sauer

NASA/Johnson Space Center, Houston, TX 77058

A review of the metabolic support systems used and the metabolic support requirements provided on past and current spaceflight programs is presented. This review will provide familiarization with (1) unique constraints of space flight and (2) technology as it relates to inflight metabolic support of astronauts. This information, along with a general review of the NASA effort to develop a Controlled Ecological Life Support System (CELSS) will define the general scenario of metabolic support for a lunar base. A phased program of metabolic support for a lunar base will be elucidated. Included will be discussion of the CELSS water reclamation and food recycling technology as it now exists and how it could be expected to be progressively incorporated into the lunar base. This transition would be from a relatively open system in the initial development period, when mechanical phase change water reclamation and minimal plant growth are incorporated, to the final period when practically total closure of the life support system will be proved through physicochemical and biological processes. Finally, a review of the estimated metabolic intake requirements for the occupants of a lunar base will be presented.

INTRODUCTION

The objective of this paper is to impart a general understanding of the metabolic support requirements of a manned lunar base and how these needs might be provided. In the context of this paper, metabolic support includes the oxygen, water, and nutrition intake and waste output (feces, urine, insensible water, and carbon dioxide output) of man in space.

ANALYSIS AND DISCUSSION

Figure 1 shows the average input requirements in grams per day of oxygen, water, and food for supporting a human in a space environment. While these requirements

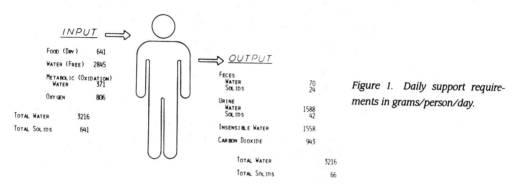

Figure 1. Daily support requirements in grams/person/day.

reflect those currently used to support manned spaceflight, it is projected that they will not change substantially in order to support man on the lunar surface (Thornton and Ord, 1977; Waligora and Sauer, unpublished data). For a six-person contingent and for a period of one year, these daily input requirements translate to the totals shown in Table 1. Using the estimated cost of $13,000 (Duke *et al.*, 1984) to launch from Earth and deliver a kilogram of material to a lunar base, these quantities would cost over $130 million simply for transportation (Duke *et al.*, 1984; Waligora and Sauer, unpublished data). Figure 1 also depicts the average metabolic output in grams per day of a human in space.

These economic considerations quickly lead to the need to consider the recycling of metabolic materials, *i.e.*, water, oxygen, and food, in order to reduce resupply costs. Table 2 shows how these materials have been provided in previous U.S. missions and are projected to be provided in future missions. In some respects, recycling has already been practiced in spaceflight. For example, in Apollo and Shuttle, water has been produced as a byproduct of fuel cell operation. The primary purpose of the fuel cells was to produce electricity through the combining of oxygen and hydrogen. In all missions, a form of atmospheric recycling has been accomplished through the adsorption of CO_2 on lithium hydroxide and trace contaminants on charcoal.

While these reclamation processes have been effective in spaceflight missions, they will not be adequate for longer term stays such as the space station. For these missions,

Table 1. Metabolic Requirements (6 Persons for 1 Year)

Item	Weight (kg)	Cost*($\times 10^6$)
Food	1404	18.2
Water	7043	91.6
Oxygen	1765	22.8
Total	10212	132.6

*Estimate: $13,000/kg

Table 2. Space Program Environmental Control Life Support Systems (ECLSS)

Program	Food	Water	Oxygen
Mercury	Stored	Stored	Stored: CO_2–LiOH
Gemini	Stored	Stored	Stored: CO_2–LiOH
Apollo	Stored	Fuel Cell	Stored: CO_2–LiOH
Skylab	Stored	Stored	Stored: CO_2–MolSieve
Shuttle	Stored	Fuel Cell	Stored: CO_2–LiOH
Space Station	Stored	Reclaimed	Produced
Lunar Base	CELSS and stored	CELSS	CELSS

it is projected, as shown in Table 3, that water will initially be reclaimed from humidity condensate and wash water and eventually from urine through physico-mechanical/chemical means. Oxygen will be reclaimed from expired carbon dioxide through reduction of carbon dioxide to methane and water (Sabatier process) followed by water electrolysis or reduction of carbon dioxide to carbon and oxygen (Bosch reactor). Inflight food production will be limited to the growing of specialty items such as sprouts and salad materials.

The lunar base will initially utilize the space station developed recycling technology. Eventually, however, the lunar base will require a higher degree of life support system closure than provided for the space station. This is because of the more permanent nature of the lunar base and the greater distance from Earth and, therefore, increased transportation costs of providing expendables from Earth. As shown in Table 4, metabolic support systems are envisioned to be transitional through the anticipated different phases of the lunar base maturation process. The initial phase will use systems similar to those used in the space station, while the advanced phase will utilize Controlled Ecological Life Support System (CELSS) technology that will permit essentially autonomous operation free from Earth support.

Table 3.   Space Station ECLSS Candidate Systems

Water Reclamation
    Vacuum compression – distillation : urine
    Multi-filtration : humidity condensate
    Thermally integrated membrane evaporation : urine
    Chemical precipitation filtration : wash water

Oxygen Reclamation
    Bosch reactor: $CO_2 \rightarrow C + O_2$
    Sabatier: $CO_2 + H_2 \rightarrow CH_4 + H_2O$
    Water electrolysis: $H_2O \rightarrow H_2 + O_2$

Food
    Sprout culture
    Limited vegetable culture

Table 4.   Phases of Lunar Base Metabolic Support

	Initial	Intermediate	Advanced
Oxygen	Physico/chemical	Physico/chemical, bioregenerative	Bioregenerative, physico/chemical
Water	Physico-mechanical/chemical	Physico-mechanical/chemical, bioregenerative	Bioregenerative, physico-mechanical
Food	Resupply; limited plant growth	Resupply; bioregenerative	Bioregenerative; resupply

The food nutrient requirements as currently established for spaceflight are shown in Tables 5 and 6 (Sauer and Rapp, 1983; Stadler *et al.*, 1982). It is projected that these requirements will not change significantly for future space missions, including the lunar base.

Table 5. Daily Macro Nutrient Needs

	Weight (g)	Kcal	Percent Daily Requirements
Normal Range			
Carbohydrate (CHO)			52–60%
Protein			15%
Fat			25–33%
Shuttle Menu (3000 kcal)			
Carbohydrate	420.6	1682	56%
Protein	126.2	505	17%
Fat	92.7	834	28%
Total	639.5	3021	100%

Table 6. Minimum Daily Nutritional Levels

Nutrient	Amount
Kilocalories	3,000
Protein	56 g
Vitamin A	5,000 IU
Vitamin D	400 IU
Vitamin E	15 IU
Ascorbic Acid	45 mg
Folacin	400 μg
Niacin	18 mg
Riboflavin	1.6 mg
Thiamin	1.4 mg
Vitamin Bμ	2.0 mg
Vitamin B$_{12}$	3.0 μg
Calcium	800 mg
Phosphorus	800 mg
Iodine	130 ^{TM}g
Iron	18 mg
Magnesium	350 mg
Zinc	15 mg
Potassium	70 mEq
Sodium	150 mEq

SUMMARY

The metabolic support requirements for a lunar base will not vary significantly from those currently provided for manned space missions. The degree of closure of the life support cycles, however, will increase and will evolve from the relatively open-loop cycle of the systems used today to the controlled ecological life support systems envisioned in the CELSS program.

REFERENCES

Duke M. B., Mendell W. W., and Keaton P. W. (1984) *Report of the Lunar Base Working Group.* LALP 84.43, Los Alamos National Laboratory, Los Alamos. 16 pp.

Leach C. L., Leonard J. L., Rambaut P. C., and Johnson P. C. (1978) Evaporative water loss in man in a gravity-free environment. *J. Appl. Physiol.: Respir. Environ. Exercise Physiol., 45,* 430–436.

Sauer R. L. and Rapp R. M. (1983) *Shuttle OFT Medical Report.* NASA TM-58252. NASA/Johnson Space Center, Houston. 102 pp.

Stadler C. R., Bourland C. T., Rapp R. M., and Sauer R. L. (1982) Food system for Space Shuttle Columbia. *J. Am. Diet. Assoc., 80,* 108–114.

Thorton W. E. and Ord J. (1977) Physiological mass measurements in Skylab. In *Biomedical Results from Skylab,* pp. 175–181, NASA SP-377. NASA/Johnson Space Center, Houston.

IMPLEMENTING SUPERCRITICAL WATER OXIDATION TECHNOLOGY IN A LUNAR BASE ENVIRONMENTAL CONTROL/LIFE SUPPORT SYSTEM

Melaine Meyer Sedej

NASA/Johnson Space Center, Mail Code EC2, Houston, TX 77058

A supercritical water oxidation system (SCWOS) offers several advantages for a lunar base environmental control/life support system (ECLSS) compared to an ECLSS based on space station technology. In supercritically heated water (630 K, 250 atm) organic materials mix freely with oxygen and undergo complete combustion. Inorganic salts lose solubility and precipitate out. Implementation of SCWOS can make an ECLSS more efficient and reliable by elimination of several subsystems and by reduction in potential losses of life support consumables. More complete closure of the total system reduces resupply requirements from the Earth, a crucial cost item in maintaining a lunar base.

INTRODUCTION

Living and working on the Moon is no longer a far-fetched dream. Today the technology is available to establish a primitive habitat/workstation on the Moon (Duke *et al.*, written communication, 1984; Roberts and Duke *et al.*, personal communication, 1984), and NASA is studying the feasibility of establishing a lunar base as a follow-up or parallel venture to the space station. For these studies, commonality of lunar base systems with their counterparts on the space station is generally taken as a baseline planning assumption for saving development costs. However, this policy should not exclude the consideration of different systems that may offer substantial advantages for lunar base application. This paper addresses conceptual designs for a lunar base environmental control/life support system (ECLSS) based on a new, potentially cost-saving technology: a supercritical water oxidation subsystem (SCWOS).

BASIC LIVING REQUIREMENTS FOR A LUNAR BASE HABITAT

The basic living requirements (*i.e.*, maximum partial pressure of CO_2, minimum partial pressure of O_2) (Lin and Meyer, 1983) for a lunar base are the same as those for the space station. However, the ground rules and methods for meeting those requirements may be different for life on the Moon because of the different properties of the lunar environment and the different objectives of lunar missions. Such factors as radiation level, thermal extremes, the 30-day diurnal cycle, gravity, terrain, and lunar soil composition will impact designs (Smith and West, 1983). Also, mission objectives such as studying the lunar atmosphere or taking sensitive optical measurements may preclude the use of subsystems that would alter the natural environment. Dumping or venting may be prohibited, at least in the vicinity of cameras, telescopes, or other sensitive instruments.

The lunar terrain may provide structural support for pressure vessels or entire subsystems considered too large, too noisy, or otherwise unsuitable to be operated next to the crew's living quarters. Although oxygen can be recovered from the lunar soil and might prove to be a local resource for the ECLSS of future habitats, an oxygen plant is unlikely to be built before the first lunar habitat.

The lunar gravity (one–sixth Earth gravity) will facilitate many operations on the lunar base, such as two–phase separation and convection in the pressurized modules. Therefore, several ECLSS subsystem designs, not the least of which are the shower and toilet, can be simplified compared to their counterpart low Earth orbit (LEO) subsystems.

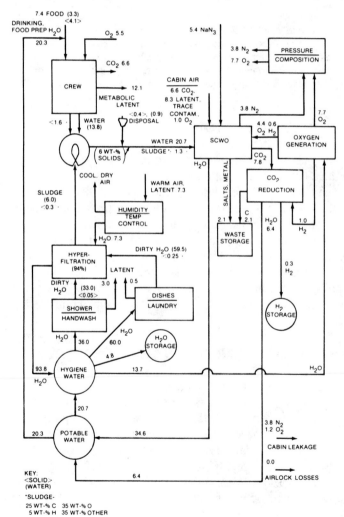

Figure 1. SCWOS-ECLSS daily mass balance for a three-member crew.

The problem of inaccessibility to civilization for resupplies is much more profound for Moon missions than for space station LEO missions. The astronauts will have to take everything they need (for themselves and for the base) to survive 90 days or longer. Therefore, the less dependent on terrestrial resupplies the base is, the more flexible the mission can be and the less time wasted on housekeeping.

CANDIDATE ECLSS CONCEPTS

Many elements are common to the lunar base ECLSS proposed below (Fig. 1) and a space station type ECLSS (Fig. 2). Both have the same atmospheric pressure/composition control subsystem, O_2 generation subsystem, CO_2 reduction subsystem, and hygiene facilities. In fact, most of the hardware in the two ECLSS's is the same, but the few differences make the two concepts quite dissimilar.

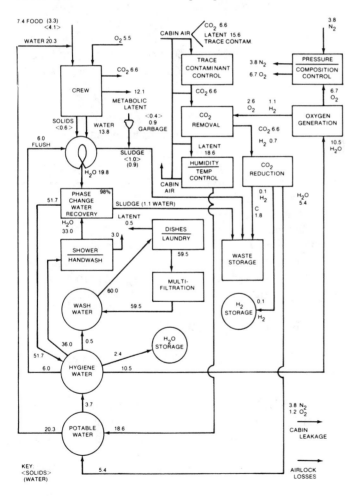

Figure 2. Space station-type ECLSS for a lunar base; daily mass balance for a three-member crew.

SCWOS-ECLSS

The SCWOS technology is based on the physics and chemistry of water molecules (H_2O) at conditions above their supercritical pressure and temperature (at 25.3 MN/m^2 [250 atm] and 627.59 K [670 °F] (Bratt, written communication, 1982; Josephson, 1982; Timberlake *et al.*, 1982; Modell *et al.*, 1982; Swallow, personal communication, 1984). Under these conditions the dielectric constant of H_2O weakens, causing two important phenomena to occur: hydrocarbons and other normally immiscible organics become miscible in the water medium, and normally dissolved inorganic salts precipitate out of the solution. At the high temperature, instantaneous complete combustion of the organics results if sufficient oxygen is present yielding H_2O, carbon dioxide (CO_2), and nitrogen (N_2). The precipitated solid salts can be separated from the process stream in the same solids separator that removes any metal particles found in solution.

One way to achieve and sustain the high temperature for the supercritical combustion would be to preheat the feed electrically. However, the effect of the extreme conditions on the heat exchanger would make corrosion and structural problems difficult to control (Modell, personal communication, 1984). These problems could be avoided by introducing oxygen (O_2) and hydrogen (H_2) to the feed mixture for their "heat of reaction" value ($O_2 + 2H_2 \rightarrow 2H_2O + heat$) (Modell, personal communication, 1984). Maintaining the temperature is a matter of "superinsulating" the sytem, possibly by utilizing the lunar vacuum. The heat of combustion of the reactants ensures that the temperature during reaction would not fall below the lower limit for rendering complete combustion. Reaching and maintaining the desired pressure is also achievable using current technology.

Food, water post-treatment supplies, and a nitrogen–containing solid that is discussed later are the required resupplies for the ECLSS designed around an SCWOS. The by-products would be salts, minerals, dense carbon, and excess water and hydrogen, all of which could be used elsewhere on the lunar base. A mass balance and a functional schematic for the SCWOS-ECLSS is shown in Fig. 1.

Five subsystems would make up the air management group of the SCWOS-ECLSS: the atmospheric pressure/composition control subsystem, the O_2 generation subsystem, and SCWOS (for CO_2 removal, trace contaminant control, N_2 makeup, and partial humidity control), the CO_2 reduction subsystem, and the humidity/temperature control subsystem. The multifunctional SCWOS also would be part of the waste management group and the water management group, which are discussed later in more detail.

The atmospheric pressure/composition subsystem would be similar to that of the space shuttle, but the sources of the gases (O_2 and N_2) would be different. Oxygen would be generated by water electrolysis ($2H_2O + electral \ power \rightarrow O_2 + 2H_2$). Nitrogen would be derived from the SCWOS.

Since not enough N_2 for ECLSS needs could be generated by the normal wastes fed to the SCWOS (urine, feces, garbage, dirty water, and trace contaminants) (Marrero, 1983), the SCWOS feed could be supplemented with a nitrogen–containing solid or liquid compound supplied from Earth. There are several compounds to choose from (Table 1) that would benefit the lunar base in ways beyond N_2 generation. For example, several

Table 1. Nitrogen-Containing Compounds That Are Candidate Reactants for Nitrogen Generation (From Sax, 1965; Weast, 1977)

Compound and Descriptions	Molecular Formula	Molecular Weight, g/g-mole	Density or Specific Gravity*	Melting Point	Boiling Point	Heat of Formation kcal/g-mole	Comments†
Ammonium hydroxide Colorless liquid	NH_4OH	35.05	—	-77°C	—	-87.64 (aqueous)	—
Hydrazine Colorless fuming liquid, white crystals	N_2H_4 (NH_2-NH_2)	32.05	1.011^{15} (liquid)	1.4°C	113.5°C	+12.05	Flash point is 126°F (open container) Auto-ignition occurs at 518°F
Hydrazine azide White powder	N_2H4-HN_3	75.07	—	75.4°C	—	—	—
Hydrazoic acid (Azoimide) Colorless liquid	HN_3	43.03	1.09^{25}_4	80°C	37°C	70.3	May be used to sustain SCWOS reaction temperature
Sodium amide White crystalline powder	$NaNH_2$	39.02	—	210°C	400°C	-28.04	Yields heat with moisture
Sodium azide Colorless hexagonal crystals	NaN_3	65.01	1.846^{20}	—	—	—	Decomposes in a vacuum
Sodium nitride Dark gray crystals	Na_3N	82.98	—	300°§	—	—	To be considered only if sodium ions are highly desirable
Sodium nitrite Slightly yellowish or white crystals	$NaNO_2$	69.00	2.168^0	271°C	320C§	-85.9	The oxygen elements may fuel the SCWOS combustion reaction
Sodium nitrate Colorless crystals	$NaNO_3$	84.99	2.261	306.8°C	380°C§	-101.54 (crystalline) 106.65 (aqueous)	The oxygen elements may fuel the SCWOS combustion reaction

*Superscripts and subscripts are temperatures in degrees centigrade

†Many of the compounds are dangerous, some explosive; however, if there are ways to minimize the danger, they have been retained for comparison

§Decomposition point

nitrogen–containing compounds in the table could be decomposed to produce sodium hydroxide (NaOH), utilized in some of the chemical processes for claiming O_2 from the lunar soil (Duke, written communication, 1984). As discussed later in the comparison between ECLSS's, the most important consequence of the use of this N_2 generation concept is that the compound could be resupplied as a solid or liquid. If this N_2 generation scheme is rendered undesirable, the SCWOS–ECLSS could revert to the same N_2 suply subsystem to be used for the space station (probably cryogenic storage).

The CO_2 reduction subsystem in the air management group would receive all the H_2 from the O_2 generation subsystem except that used in the SCWOS. All the CO_2 that entered the SCWOS with the process air and that formed by combustion inside the reactor would leave the SCWOS in a concentrated stream. The CO_2 reduction subsystem would receive the CO_2 stream and convert the CO_2 and H_2 into water and dense carbon. The excess H_2 would be stored for other lunar base needs.

The SCWOS–ECLSS water management group would consist of two water loops: the potable water loop and the hygiene water loop. Normally, to save energy and expendables, the hygiene water would not be made potable. "Prepotable" water would come from urine, water vapor in the air (e.g., metabolic latent), SCWOS combustion product water, and CO_2 reduction product water. Potable water would be derived from two intense sterilizing processes that operate at temperatures above 533 K (500°F): the CO_2 reduction subsystem and the SCWOS. The sterile water would be chemically enhanced for flavor and for bacterial growth prevention to yield potable water. Once the potable water tanks were full, the processed water would be redirected to the hygiene water supply. In fact, the mass balance (Fig. 1) shows that there would be enough of this redirected water to be used for taking showers or for rinsing in the dishwasher and laundry machine.

Ordinarily, hygiene water would be used for laundering, dishwashing, showering, and handwashing. Surplus hygiene water could be stored for other lunar base operations. Dirty hygiene water and whatever humidity condensate was not processed by the SCWOS would be cleaned by reverse osmosis, a selective regenerable filtering process. After post–treatment, the clean water would be returned to hygiene water storage.

SPACE STATION TYPE ECLSS

One space station ECLSS (SS–ECLSS) concept (Johnson Space Center, 1983; Lin, personal communication, 1984) being considered for a lunar base habitation is depicted in Fig. 2. This ECLSS concept is closed and resupply-free except for water filters, post–treatment chemicals, N_2 makeup, and food. Feces, garbage, hygiene sludge, and carbon (C) are returned to Earth in storage containers at each resupply interval. Excess clean water has many other uses outside the ECLSS. The system needs little scheduled maintenance except for frequent water filter changes. The biggest time demand occurs at resupply/return changeout, when transferring storage tanks is the major task for ECLSS recharge.

Seven subsystems make up the air management group: the atmospheric presure/composition control subsystem, the N_2 supply subsystem, the O_2 generation subsystem, the CO_2 removal subsystem, the CO_2 reduction subsystem, the trace contaminant control system, and the humidity/temperature control subsystem. The water management group has three reclamation subsystems: one for producing drinking water, one for hygiene water, and one for wash water (laundry and dishwashing). Having these three water groups minimizes energy consumption and expendables.

COMPARISON OF THE LUNAR BASE ECLSS CONCEPTS

The differences between the two ECLSS concepts go beyond what appears on the schematics. In particular, several SCWOS–ECLSS processes save significant resupply weight and volume. Since transportation costs have been found to dominate cost estimates of lunar base development, recycling of life support consumables is a crucial design consideration.

The handling of wastes (trace contaminants, feces, trash, and garbage) by the SCWOS–ECLSS saves significant resupply weight and volume in terms of filters, bactericides, and waste containers. The wastes (solid, liquid, and gaseous) would be broken down into harmless combustion products. Bacteria would be destroyed, so concern about masking or filtering odors, resupplying bactericides, or venting and dumping wastes would be greatly reduced. In fact, the materials derived from the SCWOS–ECLSS waste reduction would be incorporated back into the ECLSS to help futher close the system: CO_2 would go to CO_2 reduction, H_2O would go to potable water storage, and N_2 would go to the atmospheric pressure/composition control subsystem.

The N_2 supply concept of the SCWOS–ECLSS may be preferable to cryogenic storage. Cryogenic N_2 flow cannot be turned off indefinitely, like a water faucet, because of possible over-pressurization with heat. Venting of N_2 from cryogenic storage vessels is potentially more wasteful than using the SCWOS N_2 production method. In addition, cryogenic tanks are generally spherical and therefore complicate compact packaging efforts. High pressure gaseous storage is costly in terms of volume and weight. These problems may be eliminated in the SCWOS–ECLSS where a powder, a grindable solid, or a liquid that is rich in elemental nitrogen (N) can be reduced to N_2. Such a compound, being solid or liquid, could be packaged in any desired shape for resupply. Several of the candidate compounds (Table 1) would break down into wastes that might reduce resupply weight elsewhere. Carrying nitrogen in solid or liquid form would greatly simplify logistics.

The air management group of the SCWOS–ECLSS is simpler than that of the SS—ECLSS. In one package, the SCWOS would remove the CO_2, the trace contaminants, and more than half of the water vapor from the air. Essentially two and one-half SS—ECLSS air management subsystems would be replaced by the SCWOS. Having fewer unique subsystems would reduce the crew's training load and cut down on the spare parts inventory, not to mention increasing the reliability and decreasing the maintenance of the ECLSS.

The water management group of the SCWOS–ECLSS is also simpler than that of the SS—ECLSS. The former has two water loops; the latter, three. The mass balances mentioned earlier indicate that the potable water supply level of the SCWOS-ECLSS compared to that of the SS-ECLSS is less critically dependent on subsystem production and consumption rates. In the SCWOS–ECLSS, twice as much potable water is produced daily as is used for drinking and food preparation. In the SS–ECLSS, the ratio of "produced" to "humanly ingested" potable water of nearly one to one signifies greater dependence on timing among the ECLSS entities. The relative abundance of potable water in the SCWOS–ECLSS opens up new integration possibilities, such as potable water showers and potable water rinse cycles for the laundry machine and dishwasher. These luxuries cannot be afforded as easily in the SS–ECLSS.

STATUS OF SCWOS DEVELOPMENT

Substantial work has been done to understand the chemistry of supercritical water oxidation and ways to develop the technology. Many sludges and solutions have been successfully converted to the products of complete combustion (CO_2, H_2O, and N_2) in a breadboard reactor. However, there are still chemical and mechanical difficulties associated with processing some ECLSS wastes. For instance, although a recent discovery led to the complete combustion of urea (a major component of urine) at a lower than expected temperature (Swallow, personal communication, 1984), the processing of urine has not been successful to date. (The as yet uncontrolled precipitation of urinary salts has clogged the reactor.) Furthermore, the preparation of trash and garbage for processing has not been successful. (Very little work has been done in this area, however.) Much development work lies ahead in readuction optimization, design optimization, and automation to make this technology useful for a lunar base ECLSS.

For comparison and other candidate ECLSS waste management subsystems, the estimated SCWOS power level for processing the wash water, urine, and feces of an eight-person crew is 300–400 W, continuous (Thomason, personal communication, 1985). This power level excludes the energy for producing supplementary oxygen but does not take credit for the partial carbon dioxide removal, the trace contaminant control, and the humidity reduction that results from the waste processing. (The process uses cabin air for the combustion oxygen supply.) Since the SCWOS process can be compared favorably with other candidate waste management subsystems, an ECLSS designed around an SCWOS should certainly do well in comparison with the more conventional partially closed ECLSS designs being considered for space station use.

CONCLUSIONS AND RECOMMENDATIONS

The following are some of the many reasons given in this paper for supporting the candidacy (and development) of the SCWOS–ECLSS for operation in the lunar base habitat.

1. Trace contaminants would be controlled without the consumption of expendables.
2. Waste management would be simplified and would require less storage room and maintenance.

3. The SCWOS would make useful by-products out of trace contaminants and wastes.
4. Air management would be simplified.
5. Nitrogen logistics would be more manageable.
6. Resupply would be reduced and facilitated in many ways.
7. Water management would be simplified.
8. Luxuries such as bathing in potable water or having potable water rinse cycles for the laundry machine and dishwasher could be afforded.
9. The SCWOS-ECLSS would allow more mission flexibility.

Hopefully these qualitative advantages will stimulate interest in learning more about the quantitative differences (power consumption, heat rejection, weight, and volume differences) between the SCWOS-ECLSS and other candidate lunar habitat ECLSS's. To perform the quantitative analyses, a preliminary set of ground rules for lunar base habitation is required. Such specific criteria as resupply period, crew size, frequency and duration of outdoor activity, plans for lunar resource utilization, and environmental policies must be established or assumed. Establishing the criteria will initiate the next phase of the trade-off study to choose a lunar base ECLSS, the quantitative study phase.

The discovery of supercritical water oxidation could be a key development for space exploration. Although this technology, as discussed here, would enhance long-duration lunar missions, quantitative analyses are needed to gain a better appreciation for the advantages and disadvantages of competing ECLSS concepts. Preliminary calculations encourage optimism toward the use of the SCWOS-ECLSS as a step toward self-sufficiency for a lunar settlement.

REFERENCES

Johnson Space Center (1983) *Conceptual Design and Evaluation of Selected Space Station Concepts.* JSC 19521, NASA/Johnson Space Center, Houston. 64 pp.

Duke M. B., Mendell W. W., and Roberts B. B. (1984) Toward a lunar base programme. *Space Policy, 1,* 49–61.

Josephson J. (1982) Supercritical fluids. *Environ. Sci. Technol., Amer. Chem. Soc., 16,* 548A–551A.

Lin C. H. and Meyer M. S. (1983) Systems engineering aspects of a preliminary conceptual design of the space station environmental control/life support systems. Thirteenth Intersociety Conference on Environmental Systems, Society of Automotive Engineers, San Francisco, CA. 17 pp.

Modell M., Gaudet G. G., Simon M., Houg G. T., and Biemann, K. (1982) Destruction of hazardous waste using supercritical water. *Solid Waste Manage., 55,* 202–212.

Sax N. I. (1965) *Dangerous Properties of Industrial Materials.* Reinhold, New York. 1343 pp.

Smith R. E. and West G. S. (Compilers) (1983) *Space and Planetary Environment Criteria Guidelines for Use in Space Vehicle Development,* Vol. 1. NASA TM-82478 NASA, Washington, D.C. 246 pp.

Timberlake S. H., Houg G. T., Simon M., and Modell M. (1982) *Supercritical Water Oxidation for Wastewater Treatment: Preliminary Study of Urea Destruction.* SAE 82-0872. Society of Automotive Engineers, Warrendale, PA. 10 pp.

Weast R. C. (editor) (1977) *Handbook of Chemistry and Physics,* CRC Press, Cleveland.

RADIATION TRANSPORT OF COSMIC RAY NUCLEI IN LUNAR MATERIAL AND RADIATION DOSES

R. Silberberg, C. H. Tsao, and J. H. Adams, Jr.

E. O. Hulburt Center for Space Research, Naval Research Laboratory, Washington, D. C. 20375-5000

John R. Letaw

Severn Communications Corporation, Severna Park, MD 21146

The radiation environment on the lunar surface is inhospitable. The permanent settlers may work ten hours per 24-hour interval for the two-week-long lunar day on the lunar surface, or 20% of the total time. At moderate depths below the lunar surface (<200 g/cm^2) the flux of secondary neutrons exceeds considerably that in the upper atmosphere of the Earth, due to cosmic-ray interactions with lunar material. The annual dose equivalent due to neutrons is about 20 or 25 rem within the upper meter of the lunar surface. The dose equivalent due to gamma rays generated by nuclear interactions near the lunar surface is only on the order of 1% of that due to neutrons. However, gamma-ray line emission from excited nuclei and nuclear spallation products generated by cosmic rays near the lunar surface is of considerable interest: these lines permit the partial determination of lunar composition by gamma spectroscopy.

INTRODUCTION

The cosmic-ray environment on the lunar surface is inhospitable for permanent settlement. There is no radiation-absorbing atmosphere and no overall magnetic field that deflects charged particles. The annual dose equivalent due to cosmic rays at times of solar minimum is about 30 rem. Also, the lunar surface is not protected from solar flare particles; at energies above 30 MeV, the dose equivalent over the 11-year solar cycle is about 1000 rem. Most of those particles arrive in one or two gigantic flares, each lasting only about two days. These doses greatly exceed the permissible annual dose—0.5 rem for the general public and 5 rem for radiation workers. This difficulty can be overcome, however, by adequate shielding. For permanent lunar residents, it is necessary to construct shelters several meters below the lunar surface. In this paper we estimate the thickness of lunar regolith that must be used for shielding of habitats using the results of our radiation transport calculations.

The primary cosmic-ray nuclei (discussed by Adams and Shapiro, 1985) propagate in the lunar soil and undergo nuclear transformations. Our radiation transport calculations include nuclear interactions, ionization losses, and solar modulation for the stable as well as unstable cosmic-ray nuclides from H to Ni. Also, the production of neutrons and neutron-generated nuclear recoils are taken into account. For radiobiological analysis, the cosmic-ray energy spectra are converted into LET (Linear Energy Transfer or ionization energy deposition) spectra. These are then converted into absorbed doses and dose equivalents as a function of depth of lunar soil and compared with the permissible dose limit of 5 rem/year for radiation workers. The quality factors used are those of Silberberg *et al.* (1984), which are based on and are practically identical to those of the RBE committee to ICRP and ICRU of 1963. The quality factor for neutrons is that of Armstrong *et al.* (1969), *i.e.*, 10 for $0.2 \le E$ (MeV) ≤ 3, and somewhat smaller outside this energy interval.

THE PROPAGATION EQUATION

Cosmic-ray nuclei fragment in collision with the atomic nuclei in the lunar soil. The fundamental equation for cosmic-ray propagation that includes the effects of nuclear transformations and energy losses based in Ginzburg and Syrovatskii (1964), is

$$\frac{\partial J_i}{\partial x} = \frac{-J_i}{\lambda_i} + \sum_{j>i} \frac{J_i}{\lambda_{ij}} + \frac{\partial}{\partial E}\left[J_i \left(\frac{dE}{dx} \right)_i \right] \tag{1}$$

Here J_i is the differential flux of cosmic-ray particles of isotopes of type i; x is the path length in units of g/cm², dE/dx is the (positive) stopping power; λ_i is the fragmentation mean free path of a nucleus of isotope i; and λ_{ij} is the mean free path of a nucleus of type j yielding one of type i. The cross sections used are those of Silberberg and Tsao (1973), Letaw (1983), and Letaw *et al.* (1983). For a composite material, λ_i and λ_{ij} are weighted over the nuclei of a mixture, with N decomposed so as to represent the individual number of atoms/cm³ of a given type in the lunar material. For our calculation we adopted the composition given by Reedy (1978), with the relative abundances of nuclei as shown in Table 1. The cosmic-ray fluxes and energy spectra used in our calculations are those of Adams and Shapiro (1985).

CALCULATION OF DOSE

The output of the propagation program yields the energy spectra $J_i(E)$ of all nuclear species at various depths of a given material. For the calculation of the dose, the energy spectra are converted into rate of ionization energy loss (or LET) spectra. Using the abbreviated notation $\frac{dJ_i(S)}{dS} = J_i'(S)$, where S is the stopping power or dE/dx, the integral LET spectrum $N_i(S_o)$ is given by

$$N_i(S_o) = \int_{S_o}^{\infty} J_i'(S)\, dS \tag{2}$$

The absorbed dose rate from nuclides of type i, with stopping power $S > S_o$ is given by

$$\dot{D}_i(S_o) = \int_{S_o}^{\infty} J_i'(S) S\, dS \tag{3}$$

Table 1. Relative abundances of the nuclei of the more
common elements in lunar soil

Element	Abundance (%)
O	61
Mg	4
Al	9
Si	16
Ca	6
Fe	4

If x is in units of g/cm^2, J in units of particles/cm^2 s, and E is in units of 100 ergs, then $\dot{D}$ is given in units of rad/s. For the dose equivalent, the integral of (3) contains the quality factor Q(S), defined in terms of LET intervals and approximated as in Silberberg et al. (1984). The dose equivalent rate is given in units of rem/s. The doses have been calculated for energy deposition in water, i.e., for biological tissue–like material.

DOSE DUE TO COSMIC RAYS

Figure 1 shows the LET spectra and the integral absorbed dose rates as a function of shielding in lunar material, from 1–200 g/cm^2. The total absorbed dose rate in units of rad/y can be read at the left hand side. (Example: If we want to determine the annual dose at values of LET exceeding 30 MeV/g/cm^2, at a depth of 50 g/cm^2 of shielding, we locate the point on the curve 50, vertically above a LET of 30 MeV/g/cm^2, and read the dose of 0.1 rad/y on the axis, horizontally from the above point.)

Figure 2 shows the corresponding LET spectra with the quality factor included in the integration of (3), i.e., the integral dose equivalent rate, from 1–200 g/cm^2. The units

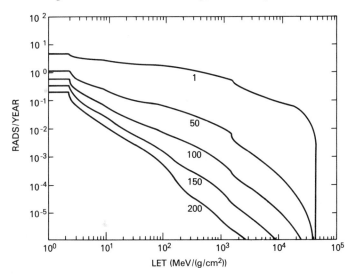

Figure 1. The integral absorbed rates in units of rads/y, as a function of the LET distribution, with shielding from 1 to 200 g/cm^2 as a variable parameter.

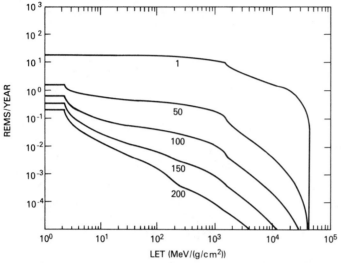

Figure 2. The integral dose equivalent rates in units of rems/y, as a function of the LET distribution, with shielding from 1 to 200 g/cm^2 as a variable parameter.

are rem/y. In both Fig. 1 and Fig. 2, the shoulder above approximately 1500 MeV/(g/cm^2) results from the contribution of the highly ionizing iron nuclei. The large reduction of the dose at high values of LET is due to the depletion of cosmic–ray iron with shielding, both because of its large spallation cross section and high rate of ionization loss, as a result of which slower iron nuclei stop in the shielding.

Figure 3 shows the attenuation of the annual integral absorbed dose and dose equivalent due to cosmic–ray nuclei. After about 100 g/cm^2, the dose equivalent due to nuclei is similar to that of the absorbed dose, because of the breakup of heavy nuclei. However, as we show later, when neutron generated nuclear recoils are considered, the difference between the absorbed dose and the dose equivalent persists.

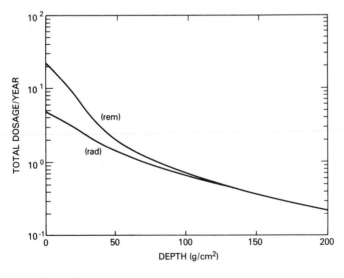

Figure 3. The attenuation of the annual dose due to cosmic-ray nuclei with shielding. The upper and lower curves show the dose equivalent and absorbed dose rates, respectively.

Table 2. The annual dose equivalent due to cosmic-ray
generated neutrons

Depth (g/cm^2)	Annual Dose (rem/y)
0	1.5
10	3
20	5
100	13
200	12
300	8
400	5
500	2

DOSE DUE TO NEUTRONS

The dose rate due to neutrons is calculated using, first, the neutron depth profile in lunar material measured by Woolum and Burnett (1974) and the calculations of Lingenfelter *et al.* (1972); second, the energy spectrum of neutrons in lunar soil, calculated by Reedy and Arnold (1972); and third, the relationship between the neutron flux and the absorbed dose and dose equivalent, as a function of energy, given by Armstrong *et al.* (1969). The quality factor for neutrons thus is that of Armstrong *et al.* (1969). Table 2 gives the annual dose equivalent of the neutron dose in lunar material, as a function of depth.

PERMISSIBLE DOSE AND SHIELDING REQUIREMENTS

We note from Table 2 that only for >400 g/cm^2 does the annual dose equivalent become smaller than 5 rem, the permissible annual dose for radiation workers. At the time of a giant flare, like that of February, 1956, the dose over the two–day duration of the flare exceeds the annual dose of Table 2 by an order of magnitude. At the time of such a flare, a shield of 700 g/cm^2 is required to reduce the dose to the level permissible for radiation workers.

For a few astronaut–volunteers over 30 years of age, the Radiobiological Advisory Panel (Langham, 1970) has permitted higher dosages: an annual dose of 38 rem and a lifetime limit of 200 rem. The latter limit is reached in about ten years on the lunar surface even in the absence of solar flares.

Figure 4 shows a comparison of the annual dose equivalent due to cosmic-ray nuclei and neutrons, as a function of depth in lunar material, down to 500 g/cm^2. It can be seen that for a shielding >400 g/cm^2, the annual dose is brought down to the level permissible for radiation workers. Even with such shielding, one receives a dose of 200 rem in 40 years, the permissible lifetime dose for a few astronaut-volunteers. On rare occasions, a few days per 11–year solar cycle, additional shielding is needed at the time of giant solar flares.

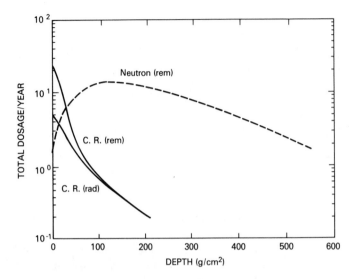

Figure 4. A comparison of the annual dose equivalent due to secondary neutrons and cosmic-ray nuclei, as a function of shielding. Also, the absorbed dose rate due to cosmic-ray nuclei is shown.

The introduction of materials that have large neutron cross sections (Li, Gd) would help to some extent; however, the cross section is large below 0.3 MeV, while a large fraction of neutrons have higher energies and thus are not absorbed; the neutron energy spectrum is given by Reedy and Arnold (1972).

GAMMA-RAY LINES

The biological effects of gamma rays induced by cosmic ray and solar flare particle interactions in the lunar soil are relatively minor. On the other hand, the gamma–ray lines are likely to be useful for mineral prospecting on the Moon. Concentrations of elements like U, Th, Ti, and K can be located as well as the more common elements shown in Table 1. The emission rates of gamma–ray lines on the lunar surface have been explored by Reedy (1978).

CONCLUSIONS

Permanent residents on the Moon can spend about 20% of the time (or 40% of the two–week daylight time) without significant shielding. Most of the time should be spent in shelters of >400 g/cm^2 or about two meters of densely packed lunar soil, either below the surface or at the surface beneath a shielding mound. At the time of rare gigantic flares, shelters >700 g/cm^2 are needed; such a protection is particularly important for radiation–sensitive fetuses.

REFERENCES

Adams J. H., Jr. and Shapiro M. M. (1985) Irradiation of the Moon by galactic cosmic rays and other energetic particles. This volume.

Armstrong T. W., Alsmiller R. G., Jr., and Barish J. (1969) Calculation of the radiation hazard at supersonic aircraft altitudes produced by an energetic solar flare. *Nucl. Sci. Eng., 37,* 337–342.

Ginzburg V. L. and Syrovatskii S. I. (1964) *The Origin of Cosmic Rays.* Pergamon, New York. 426 pp.

Langham W. H. (editor) (1970) *Radiation Protection Guides for Space-Mission and Vehicle-Design Studies Involving Nuclear Systems.* Radiobiological Advisory Panel, Committee on Space Medicine, Space Science Board, NAS/NRC, Washington, D.C.

Letaw J. R. (1983) Proton production cross sections in proton–nucleus collisions. *Phys. Rev. C, 28,* 2178–2179.

Letaw J. R., Silberberg R., and Tsao C. H. (1983) Proton–nucleus total inelastic cross sections: An empirical formula for E $>$ 10 MeV. *Ap. J. Supp., 51,* 271–274.

Lingenfelter R. E., Canfield E. H., and Hampel V. E. (1972) The lunar neutron flux revisited. *Earth Planet. Sci. Lett., 16,* 355–369.

Reedy R. C. (1978) Planetary gamma ray spectroscopy. In *Gamma Ray Spectroscopy in Astrophysics* (T. L. Cline and R. Ramaty, eds.), pp. 98–148. NASA TM-79619, NASA, Washington, D.C.

Reedy R. C. and Arnold J. R. (1972) Interaction of solar and galactic cosmic ray particles with the Moon. *J. Geophys. Res., 77,* 537–555.

Silberberg R. and Tsao C. H. (1973) Partial cross sections in high-energy nuclear reactions, and astrophysical applications. *Ap. J. Suppl., 25,* 315–333.

Silberberg R., Tsao C. H., Adams J. H., Jr., and Letaw J. R. (1984) Radiation doses and LET distributions of cosmic rays. *Rad. Res., 98,* 209–226.

Woolum D. S. and Burnett D. S. (1974) In-situ measurement of the rate of 235-U fission induced by lunar neutrons. *Earth Planet. Sci. Lett., 21,* 153–163.

AEROSOL DEPOSITION ALONG THE RESPIRATORY TRACT AT ZERO GRAVITY: A THEORETICAL STUDY

B. E. Lehnert, D. M. Smith, L. M. Holland, M. I. Tillery, and R. G. Thomas[1]

Life Sciences Division and Health, Safety, and Environment Division, Los Alamos National Laboratory, Los Alamos, NM 87545
[1]*Also at the Office of Health and Environmental Research, U. S. Department of Energy, Washington, DC 20545*

Significant fractions of airborne particles composing inhaled aerosols can deposit along the respiratory tract during breathing. Depending on the environmental condition, some particles that enter the body via the respiratory route can pose health hazards. On Earth, three general rate mechanisms are active in this deposition process: (1) inertial impaction, (2) diffusion, and (3) gravity-dependent sedimentation. Spacecraft, stations, and bases represent unique settings where potentially pathogenic aerosols may be encountered under the condition of zero or reduced gravity. The present study was undertaken in order to predict how particle deposition in the human respiratory tract at zero gravity may differ from that on Earth. We employed the aerosol deposition model of the Task Group on Lung Dynamics to assess the regional deposition of particles ranging from 0.01-10 μm diameter at two particulate densities, 1 and 4, during simulated tidal breathing and breathing during moderate-heavy exercise. Our results suggest that the gas exchange regions of the lungs of space travelers and residents are afforded some protection, relative to their earth-bound counterparts, against the deposition of particles due to the absence of gravity; an approximately two- to tenfold reduction in the efficiency of collection of particles >0.5 μm in diameter occurred in the pulmonary region during resting conditions and exercise. Deposition along the tracheobronchial tree, however, is not markedly altered in the absence of gravity, indicating airway sites contributing to this structure remain susceptible to insults by inhaled aerosols.

INTRODUCTION

According to their size and level of physical activity, adult humans breathe 1-2 × 10^4 L of air daily. Contained in this air are contaminating particles, the mass concentration of which varies from one environmental condition to another. Depending on their geometric dimensions, densities, and physiohemical characteristics, particles composing breathed aerosols can represent a significant health hazard upon deposition in the lungs. How aerosols deposit in the lungs of humans under the special condition of reduced gravity could be of major importance to the well-being of astronauts and mission specialists who reside for long periods of time in a setting that favors the stability of aerosols arising from a variety of real and potential sources, *e.g.*, dusts, microorganisms, dander, flatus, electrical fires, etc.

The respiratory tract, beginning with the nose and nasal cavities, acts like a filter in that a significant fraction of an aerosol present in inspired air may be removed during its movement into and out of the lungs during breathing. The deposition of aerosol particles in the lung occurs primarily by three mechanisms: (1) inertial impaction, (2) gravitational settling or sedimentation, and (3) Brownian diffusion.

Inertial impaction occurs when a particle leaves the airstream and collides with a stationary component of the airway, as the stream follows a bend in its path. The probability for a particle to deposit by impaction (I) is proportional to the airstream velocity ($\dot{V}$) in which it is contained, the aerodynamic diameter of the particle (d^2):a descriptor of the particle's aerodynamic behavior that takes into account the particle's size, mass, and shape, and airway branching angle (ϕ), and is inversely proportional to airway radius (R):$I \alpha \dot{U} d^2 \sin \phi/R$. Large particles, particles with high densities, high airstream velocities such as those that occur during labored breathing during exercise, large branching angles in the airways, and small airway radii all favor deposition by this process. Impaction occurs predominantly in the upper regions of the respiratory tract, including the nasopharyngeal region and upper portions of the tracheobronchial tree, and is involved in the deposition of particles ranging from 2–20 μm in diameter.

The deposition of particles in the 0.05–20 μm aerodynamic range is governed by gravitational settling. Deposition by sedimentation is determined by the particle's terminal settling velocity (V_t), which is that velocity due to the force of gravity opposed by the viscous resistance of the air through which the particle is falling: $V_t = (\rho - \sigma) \, gd^2/18\gamma$, where γ = the viscosity of air, ρ and σ = the densities of the particle and air, respectively, d = particle diameter, and g = acceleration of the particle due to gravity. Overall, the rate of particle settling is proportional to the square of the particle's diameter, and the distance traversed along a horizontal axis is, in turn, proportional to V_t and $\dot{V}$ in an airway. The predominant sites for particle deposition by gravitational settling include the mid- and peripheral airways and the lung parenchyma.

Brownian diffusion becomes the most prevalent deposition mechanism for particles <0.5 μm in diameter. Particles of this size that enter the airways are displaced by the random bombardment of gas molecules and thereby may subsequently collide with airway walls. Particle displacement (δ) is inversely proportional to the viscosity of air (γ) and the diameter of the particle and is directly proportional to the residence time (t) of the particle in a given air space (Stuart, 1984): $\delta = [(RT/N) \, (C_t/3\pi\gamma d)]^{1/2}$, where R, T, and N are the ideal gas constant, the absolute temperature, and Avogadros' number, respectively. C is a slip correction factor that takes into account that the spaces between molecules of air are no longer negligibly small relative to particle size (Cunningham, 1910). Accordingly, the probability of deposition by diffusion (D) increases as the displacement motion is increased relative to the size of the confining space (Landahl, 1950). The major sites for particle deposition by diffusion are the nasopharyngeal region, small airways, and the gas exchange regions of the lung. Deposition by this process can be considerable. It has been shown experimentally that the total respiratory deposition of particles of approximately 0.02 μm diameter is $\approx$40–60% (Hursh and Mercer, 1970; George and Breslin, 1967).

From an experimental perspective, total deposition resulting from the above mechanisms can be determined by inhale–exhale studies by evaluating the difference between the inspiratory and expiratory concentrations of a breathed aerosol and the total volume of air breathed, *vis.* [Inspiration (I)] – [Expiration (E)] = mass 1_I^{-L} – mass 1_E^{-L} = mass 1^{-L} deposited and $\dot{V} \cdot t = L \, min^{-1}$ = volume breathed; the product of the two, mass

$1^{-L}_{Retained} \cdot L_{Breathed}$ = total mass deposition. Presently, however, it is not possible to achieve a detailed measurement on the distribution of the deposited aerosol at various depths in the lung. Such information can only be obtained by theoretical methods or inferential treatment of available lung deposition and retention data. With regard to the latter, for example, the fractional mass clearance of deposited, insoluble particles from the lung that occurs up to 24 h after its deposition is often interpreted to index tracheobronchial deposition, although it is likely that some percentage of this cleared mass is of alveolar origin. Nevertheless, theoretical treatments of total respiratory tract deposition of aerosols that employ idealized models of human lung architecture, impose ventilatory cycles and patterns, and incorporate the three prominent deposition mechanisms have provided results for total deposition that closely approach those obtained experimentally. Such theoretical modeling of deposition, as well, have provided predictive data comparable to experimental interpretations of aerosol deposition patterns in the nasopharyngeal, tracheobronchial, and pulmonary regions of the lung, Figs. 1 and 2 (see Stuart, 1984).

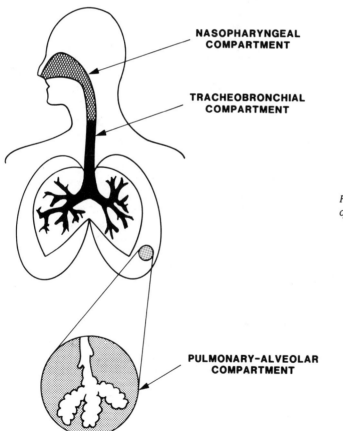

NASOPHARYNGEAL COMPARTMENT

TRACHEOBRONCHIAL COMPARTMENT

PULMONARY-ALVEOLAR COMPARTMENT

Figure 1. Regional compartments of the respiratory tract.

The present study, which employs predictive modeling of aerosol deposition, was undertaken in order to assess how particle deposition in the respiratory tracts of humans residing in reduced gravity conditions, *i.e.*, zero gravity, might compare with deposition patterns that occur on Earth. Such information is useful for assessing the potential risk to individuals that breathe real and potential aerosols aboard shuttlecraft and the future space station, as well as to provide some insight as to how aerosol deposition may be affected in individuals occupying a lunar base at $\approx 1/6$ gravity.

METHODS

The lung deposition model employed in the present study is similar to that of the Task Group on Lung Dynamics (1966), which incorporates the aforementioned deposition processes as well as idealized airway anatomy for an adult's lung. The theoretical subject was a 20-year-old male, nose-breathing tidal volumes of 770 or 1490 ml at breathing frequencies of 12 and 15 breaths/min, respectively. Tidal volumes of 770 and 1490 ml were used in order to model deposition during resting tidal breathing and while ventilating during moderate–heavy exercise. A square-wave breathing pattern was imposed; inspiratory, expiratory, and end-expiratory pause times represented 43.5%, 51.5%, and 5% of the total breathing cycle times, respectively. The breathing cycle times were 5 s at 12 breaths/min and 4 s at 15 breaths/min. The theoretical subject was "exposed" to particles ranging in size from 0.01–10 μm (geometric diameter) at particulate densities (ρ) of 1 and 4 under one gravity ($1\times$g) and zero gravity ($0\times$g) conditions. The slip correction factor used was that of Davies (1945). The collection or deposition efficiency of any given site in the respiratory tract was assumed to be the same during inhalation and exhalation, and the particles were assumed not to experience any hygroscopic growth changes while

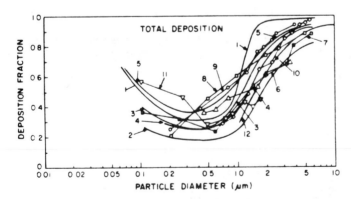

Figure 2. Predictive models and experimental values for total deposition of inhaled particles: [1] predictive, Findeisen (1935), 200 ml· s⁻¹, 14 respirations·min⁻¹; [2] predictive, Landahl (1950), 300 ml· s⁻¹, 5 respirations·min⁻¹, tidal volume 450 ml; [3] predictive, Landahl (1950), 300 ml.s⁻¹, 7.5 respirations·min⁻¹, tidal volume 900 ml; [4] predictive, Landahl (1950), 1000 ml.s⁻¹, 15 respirations·min⁻¹, tidal volume 1500 ml; [5] theoretical, Beeckmans (1965), 5 respirations. min⁻¹; [6] experimental Wilson and LaMer (1948), 5.5 respirations·min⁻¹, [7] experimental, Landahl et al., (1951), 7.5 respirations·min⁻¹, tidal volume 900 ml; [8] experimental, Gessner et al., (1949), 15 respirations·min⁻¹, tidal volume 700 ml; [9] experimental, Van Wijk and Patterson (1940), 19 respirations·min⁻¹, tidal volume 700 ml; [10] experimental, Dennis (1961) 13.3 respirations·min⁻¹, tidal volume 720 ml; [11] experimental, Dautrebande and Walkenhorst (1966), 10 respirations·min⁻¹, tidal volume 990 ml; [12] experimental, Davies (1972), 15 respirations·min⁻¹, tidal volume ml.

in the respiratory tract that could affect their deposition pattern. Deposition was generalized to three lung compartments (Fig. 1). The nasopharyngeal compartment extends from the anterior nares to the larynx; the tracheobronchial compartment includes the trachea down to the terminal bronchioles, and the pulmonary-alveolar compartment, or the gas exchange region, extends from the respiratory bronchioles down to, and includes, the alveoli or air sacs of the lung. Depositions in these compartments are expressed as a percentage of the total inhaled aerosol.

RESULTS

Estimates of aerosol deposition in the various compartments of the respiratory tract during normal breathing at rest, and breathing during moderate-heavy exercise at $1 \times g$ and $0 \times g$ are shown in Tables 1 and 2. At resting tidal volume, particles ranging in size from 0.01–0.2 μm in diameter deposit in the three respiratory tract compartments essentially identically in the presence or absence of gravity, at both densities studied. On the other hand, the percentages of deposition of particles ranging from 0.5–5.0 μm in the pulmonary or gas exchange compartment markedly decreased approximately two-to tenfold under $0 \times g$ conditions with the greatest effect seen with the particles having the higher density. The deposition pattern of the large particles (10 μm), which normally deposit by inertial impaction, was unaffected by gravity. During moderate exercise where the tidal volume and air flow velocities are increased, the pulmonary-alveolar compartment deposition of the 0.10–0.2 μm diameter particles was, as expected for particles whose deposition is governed primarily by diffusion, independent of the influence of gravity. At this higher tidal volume, however, deposition by impaction for particles in the 0.5–2.0 μm range markedly increased in the nasopharyngeal compartment secondary to increased air flow velocities, and alveolar deposition was less at both $1 \times g$ and $0 \times g$. Tracheobronchial deposition of the particles was largely unaffected by gravity during normal and high tidal volume breathing.

Table 1. Deposition* for Adult Breathing at a Resting Tidal Volume = 770 ml, 12 breaths·min^{-1}

Geometric Diameter (μm)	Nasopharyngeal Compartment (%)				Tracheobronchial Compartment (%)				Pulmonary-Alveolar Compartment (%)			
	1 g		0 g		1 g		0 g		1 g		0 g	
	$\rho = 1$	$\rho = 4$	$\rho = 1$	$\rho = 4$	$\rho = 1$	$\rho = 4$	$\rho = 1$	$\rho = 4$	$\rho = 1$	$\rho = 4$	$\rho = 1$	$\rho = 4$
0.01	0.3	0.3	0.3	0.3	24.8	24.8	24.8	24.8	59.7	59.7	59.7	59.7
0.10	0.0	0.0	0.0	0.0	1.6	1.8	1.6	1.8	45.5	46.1	45.2	45.1
0.20	0.0	0.0	0.0	0.0	0.9	1.3	0.9	1.2	30.5	33.8	29.1	29.0
0.50	0.0	15.3	0.0	15.2	1.0	3.3	1.3	2.7	23.0	31.8	15.1	12.6
1.00	12.5	41.1	12.5	41.1	2.8	7.1	2.3	5.8	28.8	30.2	8.1	5.3
2.00	39.6	68.2	39.5	68.1	6.8	10.5	5.6	8.5	30.4	16.3	3.4	1.6
5.00	76.4	100.0	76.3	100.0	10.2	0.0	8.4	0.0	10.8	0.0	0.7	0.0
10.00	100.0	100.0	100.0	100.0	0.0	0.0	0.0	0.0	0.0	0.0	0.0	0.0

*Respiratory tract deposition as a function of particle size, particle density (ρ), and tidal volumes for $1 \times g$ and gravity-free environments

Table 2. Deposition for Adult Breathing at a Tidal Volume = 1490 ml, 15 breaths•min^{-1}

Geometric Diameter (μm)	Nasopharyngeal Compartment (%)				Tracheobronchial Compartment (%)				Pulmonary-Alveolar Compartment (%)			
	1 g		0 g		1 g		0 g		1 g		0 g	
	$\rho = 1$	$\rho = 4$	$\rho = 1$	$\rho = 4$	$\rho = 1$	$\rho = 4$	$\rho = 1$	$\rho = 4$	$\rho = 1$	$\rho = 4$	$\rho = 1$	$\rho = 4$
0.01	0.1	0.2	0.1	0.3	13.9	17.4	13.9	24.8	78.9	73.1	78.9	73.1
0.10	0.0	0.0	0.0	0.0	1.0	1.4	1.0	1.4	41.8	46.5	41.3	45.1
0.20	0.0	0.0	0.0	0.0	0.7	1.4	0.7	1.3	24.7	32.4	23.1	26.4
0.50	8.4	28.8	8.4	28.8	1.6	3.9	1.6	3.6	17.0	27.4	10.0	9.0
1.00	34.2	54.6	34.2	54.6	4.0	7.1	3.8	6.6	19.6	25.4	4.3	3.3
2.00	61.2	81.7	61.2	81.6	7.3	7.9	7.1	7.4	20.0	8.8	1.5	0.7
5.00	98.0	100.0	98.0	100.0	1.3	0.0	1.3	0.0	0.6	0.0	0.0	0.0
10.00	100.0	100.0	100.0	100.0	0.0	0.0	0.0	0.0	0.0	0.0	0.0	0.0

DISCUSSION

On Earth, particle sedimentation due to gravity is a predominant mechanism for the deposition of particles in the lung. The results of our theoretical study of particle deposition at 1×g and 0×g indicate that the deposition of particles that normally deposit in the gas exchange regions of the lung by this process on Earth is substantially diminished in the absence of gravity. These findings, which are directionally similar to those of Knight et al., (1970), who modeled particle deposition at 1/6 × g, suggest that space travelers are afforded some protection, relative to their earth-bound counterparts, from the potential pathogenic (disease-producing) effects of inhaled aerosols, including airborne bacteria whose deposition can bring about pulmonary infection. Reductions of mass deposition in space may be of particular importance aboard spacecraft, where particles that would tend to settle on Earth under the influence of gravity remain suspended in the air in the absence of gravity, especially in isolated areas of a spacecraft where the air is not readily recirculated and filtered.

On the other hand, our results indicate that the tracheobronchial tree generally is not protected from the deposition of aerosols at 0×g, and, accordingly, remains a target site for potential insults by hazardous aerosols, including microorganisms, as does the nasopharyngeal region. For relatively insoluble particles, at least, the rapid rate of clearance from the tracheobronchial region of the respiratory tract will limit their residency time in the lung and thereby decrease the likelihood of particle-tissue interactions that otherwise could bring about deleterious effects, assuming clearance is unaffected by reduced gravity.

We emphasize that the above results are derived from computer modeling, and, at best, provide directional information on particle deposition at 0×g. Actual aerosol deposition studies conducted in space are required, which may employ an animal model and a concurrent, parallel Earth-based study, in order to provide both direct information on the actual importance of the sedimentation component of particle deposition at 1×g, and to experimentally identify target sites along the respiratory tract for shuttle, space station, and lunar base associated aerosols.

Acknowledgments. *This work was supported by the U. S. Department of Energy Office of Health and Environmental Research. The authors gratefully acknowledge the managing editors of Environmental Health Perspectives for allowing us to reproduce the figure shown as Figure 2 of this report, and we acknowledge Dr. B. O. Stuart as the originator of this figure and collector of the data compiled therein (Stuart, 1984).*

REFERENCES

Beeckmans J. M. (1965) The deposition of aerosols in the respiratory tract. (I) Mathematical analysis and comparison with experimental data. *Can J. Physiol. Pharmacol., 43*, 157–172.

Cunningham E. (1910) On the velocity of steady fall of spherical particles through a fluid medium. *Proc. Roy. Soc. London., A-83*, 357–365.

Dautrebande L. and Walkenhorst W. (1966) New studies on aerosols. IV. *Arch. Int. Pharmacodyn., 162*, 194–203.

Davies C. N. (1945) Definitive equations for the fluid resistance of spheres. *Proc. Phys. Soc. London, 57*, 259–268.

Davies C. N. (1972) An algebraic model for the deposition of aerosols in the human respiratory tract during steady breathing. *J. Aerosol Sci., 3*, 297–309.

Dennis W. L. (1961) A formalized anatomy of the human respiratory tract. In *Inhaled Particles and Vapours* (C. N. Davies, ed.), pp. 88–94. Pergamon Press, London.

Findeisen W. (1935) Über das Absetzen kleiner, in der Luft suspendrerten Teilchen in der menschlichen Lunge bei der Atmung. *Pflugers Arch. Ges. Physiol., 236*, 367–379.

George A. C. and Breslin A. J. (1967) Deposition of natural radon daughters in human subjects. *Health Phys., 13*, 375–384.

Gessnes H., Ruttner J. R., and Buhler H. (1949) Zur Bestimmung des Korngrossenbenerches von silikogenem Staub. *Schweiz. Med. Wochenschr., 79*, 1258–1262.

Hursh J. B. and Mercer T. T. (1970) Measurement of ^{212}Pb loss rate from human lungs. *J. Appl. Physiol., 28*, 268–274.

Knight V., Couch R. B., and Landahl H. D. (1970) Effect of lack of gravity on airborne infection during space flight. *JAMA, 214*, 513–518.

Landahl H. D. (1950) On the removal of airborne droplets by the human respiratory tract. I. The lung. *Bull Math. Biophys., 12*, 43–56.

Landahl H. D., Tracewell T. N., and Lassen W. H. (1951) On the retention of airborne particulates in the human lungs. *Arch. Ind. Hyg. Occup. Med., 3*, 356–366.

Stuart B. O. (1984) Deposition and clearance of inhaled particles. *Environ. Health Perspect., 55*, 369–390.

Task Group on Lung Dynamics (1966) Deposition and retention models for internal dosimetry of the human respiratory tract. *Health Phys., 12*, 173–207.

Van Wijk A. M. and Patterson H. S. (1940) The percentage of particles of different sizes removed from the dust-laden air by breathing. *J. Ind. Hyg. Toxicol., 22*, 31–35.

Wilson I. B. and LaMer V. K. (1948) The retention of aerosol particles in the human respiratory tract as a function of particle radius. *J. Ind. Hyg. Toxicol., 30*, 265–280.

TOWARD THE DEVELOPMENT OF A RECOMBINANT DNA ASSAY SYSTEM FOR THE DETECTION OF GENETIC CHANGE IN ASTRONAUTS' CELLS

Susan V. Atchley[1,2], David J.-C. Chen[2], Gary F. Strniste[2],

Ronald A. Walters[3], and Robert K. Moyzis[2]

[1]*Cell Biology Department, University of New Mexico, Albuquerque, NM 87131*
[2]*Los Alamos National Laboratory, Life Sciences Division, Genetics Group, Los Alamos, NM 87545*
[3]*ADCEL, Los Alamos National Laboratory, Los Alamos, NM 87545*

We are developing a new recombinant DNA system for the detection and measurement of genetic change in humans caused by exposure to low level ionizing radiation. A unique feature of this method is the use of cloned repetitive DNA probes to assay human DNA for structural changes during or after irradiation. Repetitive sequences exist in different families. Collectively they constitute over 25% of the DNA in a human cell. Repeat families have between 10 and 500,000 members. We have constructed repetitive DNA sequence libraries using recombinant DNA techniques. Repeats used in our assay system exist in tandem arrays in the genome. Perturbation of these sequences in a cell, followed by detection with a repeat probe, produces a new, multimeric "ladder" pattern on an autoradiogram. The repeat probe used in our initial study is complementary to 1% of human DNA. The X-ray doses used in this system are well within the range of doses received by astronauts during spaceflight missions. Due to its small material requirements, this technique could easily be adapted for use in space.

INTRODUCTION

The space radiation environment poses a continual threat to the genetic integrity of spaceflight personnel. Prolonged manned spaceflights expose astronauts to galactic radiation, a mixture of low- and high-LET (linear energy transfer) particles that may deliver exposures of 100 mrem per day or greater to the human body (Benton *et al.*, 1984). Much in-flight variation exists due to the influences of altitude, orbital inclination, and geomagnetic shielding; nevertheless, the intensity and energy spectra of particles composing galactic radiation, along with their penetrating nature, gives them the potential to do serious harm to human cells. The biological effects of low- and high-LET radiation are poorly understood, especially with regard to the long-term somatic and genetic consequences of exposure. Cancers are among the principal late somatic effects known to be produced in humans by acute, high doses of ionizing radiation (Kato and Schull, 1982). Others include depressed fertility (Ash, 1980), altered immune responsiveness (Gofman, 1981), and accelerated aging (Altman and Gerber, 1983). Doses in the kilorad range also cause non-regenerating mammalian cell types such as kidney (Yang *et al.*, 1977) and brain cells (Forssberg, 1964) to become metabolically inactive. Since only aberrations induced in mature germ cells or their precursors may be transmitted to subsequent generations, no heritable human defects directly attributable to the effects of low level ionizing radiation have yet been observed.

How dangerous are the relatively low doses of radiation that will be commonly encountered during space travel and colonization? Living organisms have always had to contend with the genotoxic effects of terrestrial radiation and have evolved cellular responses to deal with damage to their genomes. The space radiation environment, however, is more powerful than that of Earth. Data from previous manned spaceflight missions show that doses of several rems are routinely delivered to the body surfaces of crew members (Benton *et al.*, 1984). As mission duration increases so will the radiation doses received. The scientific basis for generating risk estimates for low doses of ionizing radiation is inadequate. Figures are highly dependent on which of many possible projection models are employed to extrapolate from high-dose data, and they usually consist of inferences drawn from laboratory animal experimentation. Current radiobiological knowledge is not extensive enough to have confidence in the theoretical bases for extrapolation to low-dose irradiations of humans as the groundwork for establishing exposure limits in space.

With these considerations in mind we undertook experiments with human cells exposed to small doses of X-rays. The so-called "low-dose question" (National Research Council, 1980) seemed to us to be particularly amenable to attack using molecular biological techniques, since the genetic material (DNA, or deoxyribonucleic acid) is the critical cellular target of radiation damage. We are currently developing a new molecular system for assaying changes in DNA that has many distinct advantages over existing methods.

Over 25% of the DNA in each human cell consists of a class of moderately to highly reiterated sequences, known collectively as "repetitive DNA." Individual members of repeat families are present in from ten to five hundred thousand copies in each cell. We have constructed repetitive DNA sequence libraries using recombinant DNA techniques and have isolated and characterized individual repeating elements that together comprise between 75-90% of the mass of human repeats. Approximately one-third of the repeats in our libraries are organized in human DNA in tandem arrays. Since all repeats arranged this way have restriction enzyme sites in the same places, a Southern blotting experiment, in which one of the elements is used as a probe, produces a series of discrete bands on an autoradiogram. Qualitative and quantitative changes in the particular repetitive sequence under analysis result in changes from the normal band pattern. By far the greatest strength of our technique is its sensitivity. Using DNA harvested from as little as one milliliter of human blood, the thousands of human repetitive DNA sequences in a given family can be assayed for changes using a single cloned repeat probe. If different probes are employed, up to 10% of the human genome can be assayed at one time, providing an increase of several orders of magnitude in detection power over existing systems.

Here, we discuss the effect of acute, low-dose X-irradiation on DNA from normal human fibroblasts (skin cells). The purpose of this study was to qualitatively ascertain the effects of low doses of ionizing radiation on DNA in human cells. Preliminary evidence indicates that damage to the genome, perhaps in the form of interstrand crosslink formation, is more extensive than we had anticipated for the dose employed, but that after one day, the autoradiographic pattern approaches the normal one of DNA from unirradiated controls.

MATERIALS AND METHODS

Cell Culture

Normal human diploid fibroblasts (strain designation HSF-22) cultured in α-MEM medium supplemented with 10% fetal bovine serum were passaged in plastic tissue culture flasks (T-75's, Falcon).

Exposure of HSF-22 Cells to X-Rays

To damage cells with X-irradiation, each flask was placed on a tray of ice, to inhibit repair processes. Cells were exposed aerobically to radiation emitted by a 250-KVP, 30-Ma, General Electric Maxitron. Following irradiation, half of the cells (5 flasks) were lysed immediately and their DNA processed for digestion. The other half were placed in an incubator (37°C, 5% CO_2) and allowed to "recover" for one day. After twenty-four hours, their DNA was also isolated.

DNA Isolation and Purification

DNA was isolated from HSF-22 cells by pipetting approximately 5.0 ml of lysis buffer (10 mM Tris, pH 8.0; 100 mM EDTA; 10 mM NaCl) containing 0.5% SDS into each flask. Cell lysis caused the solution to become viscous within one minute. The solution containing DNA, RNA, and cell debris was removed and brought to 500 μg per ml in proteinase K (Beckman Instruments, CA). Proteolysis was allowed to occur overnight at 37°C. Following repeated extractions with phenol:chloroform:isoamyl alcohol (25:24:1), the aqueous phase was dialyzed against several changes of sterile 0.1 × sodium chloride, sodium citrate buffer (0.015 M NaCl, 0.0015 M Na citrate, pH 7.0). Solid CsCl (Gallard-Schlesinger Corp., NY) was added to each sample to produce a final density of 1.700 g/cm³. The solution was pipetted into sterile Oak Ridge tubes and centrifuged 72 hours at 42,000 rpm and 20°C. Following gradient collection all samples were dialyzed extensively against sterile 1.0 M NaCl, 0.1 M EDTA, pH 8.0, and, finally, against 0.1 × SSC.

Restriction Enzyme Digestions and Agarose Gel Electrophoresis

Two micrograms of each purified DNA were digested with restriction enzyme EcoRI (New England Biolabs, MA) in a total reaction volume of not more than 50 μl. The high salt buffer in which the restrictions took place consisted of 100 mM NaCl, 50 mM Tris-Cl, pH 7.5, 10 mM $MgCl_2$, and 1 mM dithiothreitol. Digestions were incubated for two hours at 37°C. We observed that to digest DNA from irradiated cells, a vast excess of enzyme was required. For a discussion of this point, see Results.

Digested DNA fragments were electrophoresed in a 0.5% agarose gel made with TA buffer (0.04 M Tris, 0.02 M Na acetate, and 0.002 M EDTA, pH 7.8). Separation was carried out for 16–20 hours using a submerged format at 22 V.

Southern Blotting Hybridization and Autoradiography

Southern blot hybridizations were performed as described by Southern (1975). For a schematic depiction of the procedure, see Fig. 1. Prehybridization and hybridization of the nitrocellulose filters were carried out at 68°C in a plastic bag containing 30 mM Tris–Cl pH 8.0, 300 mM NaCl, and 3.0 mM EDTA, pH 8.0; 10X Denhardt's solution (0.2% bovine serum albumin, 0.2% polyvinylpyrolidone, and 0.2% Ficoll); 0.5M NaCl; 0.05% SDS, and 100 μg/ml sonicated *E. coli* DNA. Reaction volumes were 20 ml. The only difference between the prehybridization and hybridization reactions was that in place of some water volume, the hybridization reaction mixture contained a radioactively labeled repetitive DNA probe [$\alpha(^{32}P)$-deoxynucleotide triphosphates, New England Nuclear, MA]. Prehybridization reactions were carried out for 2–5 hours. Hybridizations were incubated for approximately 16 hours. Following the hybridization reaction, nitrocellulose filters were washed at 68°C for two one-hour periods in prehybridization solution containing 0.2% sodium pyrophosphate and 0.1% SDS. Three final washes, also at 68°C, consisted of 20 mM Tris–Cl, pH 8.0, 300 mM NaCl, 3.0 mM EDTA, pH 8.0, 1X Denhardt's solution, 0.2% sodium pyrophosphate, and 0.1% SDS.

Air-dried filters were placed under XAR-5 X-ray film in an X-Omatic Cassette with intensifying screens (Eastman Kodak Co., NY). Autoradiographic exposure was at -70°C for 4 hours.

RESULTS

The experimental design employed in the present study comprises a mixture of techniques. Figure 1 illustrates the study scheme sequence. We set about looking for radiation-induced genetic change by isolating DNA from cells that had been irradiated and cells that had been irradiated and allowed to recover for one day. The basis for recognizing changes in these molecules is the manner in which restriction enzymes cut them. Restriction endonucleases are proteins that cut DNA at specific nucleotide sequences usually composed of four or six base pairs (Nathans and Smith, 1975). Every time the specific base sequence appears, the enzyme will hydrolyze both strands of the DNA molecule at that point. The double helix is cut into a random series of fragments that are displayed according to their length by electrophoresis on agarose gels. Reproducible migration of each fragment relative to the others is an inverse function of its molecular weight. One then takes the agarose gel and makes a replica by transferring the restriction fragments to a piece of nitrocellulose paper. This procedure is called Southern blotting in honor of its originator, E. M. Southern (Southern, 1975). The paper sheet is a sturdy template. Each restriction fragment is located on it in the same relative position it was in the gel.

The process of finding a pattern or a pattern change is called hybridization. Complementary single strands of DNA reform double helices during this reaction. One partner of the potential pair is tightly bound to the paper. The other is a radioactive probe added to a bag containing the paper and hybridization solution. The radioactive DNA is used to localize specific nucleic acid sequences in restriction fragment mixtures.

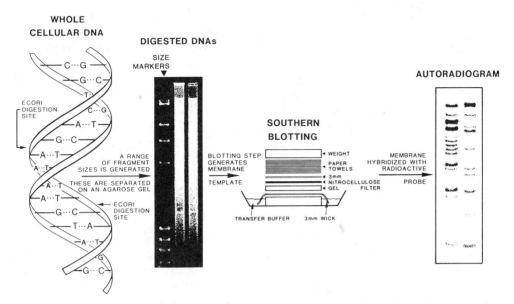

Figure 1. The experimental design. Left to right: cellular DNA is digested with a restriction enzyme. Digested DNA is fractionated by electrophoresis on an agarose gel. The Southern blotting step creates a nitrocellulose template. Finally, a highly radioactive DNA probe of known sequence is hybridized to the filter to detect complementary fragments. Patterns of hybridization are revealed by autoradiography.

The locations of the fragments that react with the probe are identified by exposing the hybridized paper to X-ray film. A dark band on the autoradiogram reveals the location of the cellular DNA fragment that contains a sequence complementary to the probe DNA. If there has been no change due to radiation exposure, the fragments that contain the same sequence as that of the probe will have migrated in the gel to the same locations in both the irradiated and control DNAs. If a perturbation has occurred, the band pattern will be different.

Preliminary experiments have utilized a repeat named pHur-22 (plasmid *Hu*man *r*epeat), which is reiterated sixty-thousand times in a human cell. Fibroblasts were irradiated with doses of from 5R to 1000R of X-rays. Half of the cells were lysed immediately, and their DNA was harvested. The other half were lysed twenty-four hours after treatment. Under all experimental conditions, multimeric bands were produced in the DNA of the treated cultures. After one day, the pattern had begun to approach the normal one produced by DNA from unirradiated control cells, except in cells treated with 1000R. In this case, only very high molecular weight bands were observed. Changes in band patterns can be produced in several ways. Restriction enzyme sites can be gained internal to some of the fragments to generate new fragments and to cause the disappearance of an original one. Such a process has not occurred in this case, because smaller, not larger, fragments would have arisen, and none of the control bands have vanished or diminished in intensity. Conversely, a site could have been lost, with the concomitant generation of a higher molecular weight band on the autoradiogram. Destruction of a restriction enzyme site in a tandem array of repeating elements will produce two bands from the original one. Further removal of sites will produce a ladder of bands corresponding to the dimeric, trimeric, and higher multimeric units of the array. Partial digestion of a tandem block, where the enzyme is prevented from cutting the DNA at all possible recognition sites, even though the sites physically still exist, will produce the same autoradiographic ladder.

Our interpretation of these data is that the "new" bands we observe have been generated by restriction enzyme inhibition. This kinetic argument is justified on several grounds. DNA isolated from unirradiated controls is cut to completion using the correct number of units of enzyme: one unit cuts one microgram of DNA in thirty minutes at 37°C. DNA from irradiated cells is not cut at all with this amount of enzyme. To generate even the partial digests observed, a thirty-fold excess of enzyme is necessary. In other words, sixty units of EcoRI were required to digest two micrograms of this DNA. DNA from cells given a day to recuperate began to resemble the normal band pattern.

Since we were using one enzyme, called EcoRI (recognition sequence: GAATTC) to sample for DNA changes, we could not detect all of the perturbations that arose as a result of radiation exposure. To expand the study and improve the detection capabilities of the assay, we are currently utilizing other enzymes and a battery of different repeat probes. In this initial experiment, owing to the use of the pHuR-22 element probe, which is complementary to 1% of the DNA in a human cell, we are measuring change in orders-of-magnitude more DNA than has ever been done before. This is especially important in the context of developing such methodology to detect potential damage to the human genome as a result of the exposure of spaceflight personnel to ionizing radiation or chemicals.

DISCUSSION

The experimental approach described in this report has significant potential application in a lunar base or on board a space station. It has minimal material requirements in terms of chemical reagents and equipment. In terms of DNA source, astronauts could donate one ml of blood at intervals throughout a mission, and samples could be frozen in shielded liquid nitrogen containers, or samples could be drawn before and after a mission for a less detailed comparison as an indication of accumulated damage. With mechanical modification to account for weightlessness, and the use of fluorescent rather than radioisotopic detection systems, there is no reason that the entire procedure could not be done in space.

It remains to be shown that interstrand DNA–DNA crosslinks really are the radiation-induced events that lead to the kind of changes observed in our study. Whether or not we have correctly identified the damage mechanism at this stage, the kinetics of enzyme inhibition is an index of the amount of damage initially incurred by the cell. We are in the process of administering different doses of X-rays to fibroblasts to attempt to locate a threshold dose above which changes become permanent.

As far as some highly speculative aspects of biological manipulation in space are concerned, consider this. It is known that radio-protective chemicals, administered prior to irradiation, can reduce radiation death in mammals three-fold (Yuhas and Storer, 1969; Maisin, 1983). The intriguing possibility exists that prior exposure of astronauts to diminutive radiation doses could induce the expression of repair enzymes in most of their cells, providing a temporary but effective shield against the increase in particle bombardment. We could use repetitive sequences to measure the relative effectiveness (or lack thereof) of such future protocols.

It may be that the abrogation of the initial effect we observe is due to the induction of repair enzymes in the fibroblast, a terminally differentiated cell type that may not ordinarily need to express those protein genes. Once they become completely activated (presumably in a matter of hours), damage in the DNA is expeditiously removed. It has been proposed, however, that entire regions of the genome exist that are physically inaccessible to repair enzymes (Wheeler and Wierowski, 1981; Wheeler *et al.*, 1983). If this is the case, our observation may be due to a simple dilution effect brought about by continued cell division during the "repair" phase, or a restructuring of chromatin domains in response to radiation. The radio-sensibilities of various differentiated and undifferentiated human cell types should be examined in order to support or refute this interpretation. DNA-DNA cross-linked complexes may be a prerequisite intermediate in allowing damaged regions to be searched out for repair by enzymes. Indeed, the processing of damaged molecules into structures refractory to restriction may be an indication of the primary stages of repair. Cross-linked structures could be part of a homeostatic mechanism that serves to alter DNA's susceptibility to degradation, thereby preserving the molecules for repair.

Our detection system will lead to a better understanding of the mechanisms of genetic change caused by exposure to radiation and/or chemicals. An estimation of risk to individuals (or populations) from such exposure is currently impossible to obtain directly, either from animal models, predictive extrapolation from microbial assays, or from human epidemiology. The development of such analytical systems as the one we are using as a direct indicator of genotoxic exposure is, from both individual and societal perspectives, imperative.

Acknowledgments. *This work was supported by the U.S. Department of Energy and the Los Alamos National Laboratory. S.V.A. is the recipient of an Associated Western Universities Laboratory Graduate Fellowship.*

REFERENCES

Altman K. I. and Gerber G. B.(1983) The effect of ionizing radiations on connective tissue. *Adv. Radiat. Biol., 10*, 237–304.

Ash P. (1980) The influence of radiation on fertility in man. *Br. J. Radiol., 53*, 271–278.

Benton E. V., Almasi J., Cassou R., Frank A., Henke R. P., Rowe V., Parnell T. A., and Schopper E. (1984) Radiation measurements aboard Spacelab 1. *Science, 225*, 224–226.

Forssberg A. (1964) Effects of small doses of ionizing radiations. *Adv. Radiat. Biol., 1*, 117–156.

Gofman J. W. (1981) *Radiation and Human Health.* Sierra Club Books, San Francisco pp. 660–699.

Kato H. and Schull W. J. (1982) Life span study report 9, part 1. Cancer mortality among atomic bomb survivors, 1950–1978. *Radiation Effects Research Foundation Technical Report 12-80.* Radiation Effects Research Foundation. 39 pp.

Maisin J. R. (1983) Chemical protection against the long-term effects of ionizing radiation. In *Radiation Research: Somatic and Genetic Effects* (J. J. Broerse, G. W. Barendsen, H. B. Kal, and A. J. Van der Kogel, eds.), pp. C2–21. Martin Nijhoff Publishers, The Netherlands.

Nathans D. and Smith H. O. (1975) Restriction endonucleases in the analysis and restructuring of DNA molecules. *Ann. Rev. Biochem., 44,* 273–293.

National Research Council (Committee on the Biological Effects of Ionizing Radiations) (1980) *The Effects on Populations of Exposure to Low Levels of Ionizing Radiation.* National Academy of Science, Washington, DC. 523 pp.

Southern E. M. (1975) Detection of specified sequences among DNA fragments separated by gel electrophoresis. *J. Mol. Biol., 98,* 503–527.

Wheeler K. T. and Wierowski J. V. (1983) DNA repair kinetics in irradiated undifferentiated and terminally differentiated cells. *Radiat. Environ. Biophys., 22,* 3–19.

Wheeler K. T., Wierowski J. V., and Ritter P. (1981) Are inducible components involved in the repair of irradiated neuronal and brain tumor DNA? *Int. J. Radiat. Biol., 40,* 293–296.

Yang T. C. H., Blakely E., Chatterjee A., Welch G., and Tobias C. A. (1977) Response of cultured mammalian cells to accelerated Krypton particles. In *Life Sciences and Space Research: Proceedings of the Open Meeting of the Working Group on Space Biology of the Nineteenth Plenary Meeting of COSPAR.* (R. Holmquist and A. C. Strickland, eds.), pp. 169–174. Pergamon Press, New York.

Yuhas J. M. and Storer J. B. (1969). Chemoprotection against three modes of radiation death. *Int. J. Radiat. Biol., 15,* 233–238.

FLOW CYTOMETRY FOR HEALTH MONITORING IN SPACE

James H. Jett, John C. Martin, George C. Saunders, and Carleton C. Stewart

Experimental Pathology Group, Los Alamos National Laboratory, Mail Stop M-888, Los Alamos, NM 87545

Monitoring the health of space station or lunar base residents will be necessary to provide knowledge of the physiological status of astronauts. Flow cytometric techniques are uniquely capable of providing cellular, chromosome, hormone level, and enzyme level information. The use of dyes provides the basis for fluorescently labeling specific cellular components. Laser-induced fluorescence from stained cells is quantitated in a flow cytometer to measure cellular components such as DNA, RNA, and protein. One major application of a flow cytometer will be to perform a complete blood count including hematocrit, hemoglobin content, and numbers of platelets, erythrocytes, granulocytes, lymphocytes, and monocytes. A newly developed flow cytometry based fluoro-immunoassay will be able to measure levels of serum enzymes and hormones. It will also be possible to quantitate radiation exposure and some forms of chromosome damage with flow cytometric measurements. With relatively simple modifications to existing technology, it will be possible to construct a flight-rated cytometer.

INTRODUCTION

The health maintenance facility of the space station, lunar base, or any extended voyage to Mars or beyond must be capable of routine hematological, immunological, and blood chemistry measurements. Even with an expert in these areas available on such missions, routine measurements and assays would require several conventional clinical instruments. Flow cytometry, a technology under development for the past twenty years (Steinkamp, 1984; Braylan, 1983; Melamed *et al.*, 1979), is uniquely suited to perform the types of measurements required in a health maintenance facility. In addition, the capabilities of the flow system described below will be a valuable source of information in space adaptation research.

The ability of a flow cytometer to make rapid measurements on cells and subcellular components requires the localization of the particles to a small probe volume (10^{-12} l) and rapid transport of the particles through that probe volume. By employing the principle of hydrodynamic focusing, first used by Crossland-Taylor to count blood cells (Crossland-Taylor, 1953) (see Fig. 1), the cells in the flowing sample stream are confined to a central core on the order of 10 μm in diameter as the cells pass through a focused laser beam. The laser light interacts with the cell in a number of ways. Measurement of the light scattered by a cell gives information about cell size in the forward direction and about cell shape and surface morphology when measured at 90° to the laser beam propagation

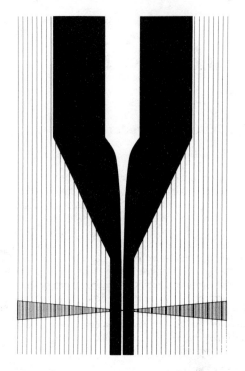

Figure 1. Schematic drawing depicting hydrodynamic focusing. The sample (white) is surrounded and focused by the sheath fluid (black) as it flows downward. The quartz walls of the flow cell are striped. The probe volume at the intersection of the focused laser beam and sample stream is typically 10^{-12} liters. The linear velocity of the cells through the laser beam is on the order of 10 m/s. The resulting pulse widths are one microsecond when the laser beam is focused to a height of 10 μm.

direction. In some systems, it is also possible to determine cell size by electrical resistance measurements via the Coulter principle.

The laser beam will also excite fluorescent dyes used to label specific cellular constituents. Photodetectors measure the amount of fluorescence emitted by the dye molecules as a cell passes through the laser beam, providing a quantitative measurement of the stained cellular components. Table 1 lists a number of the cellular constituents

Table 1. Cellular Properties or Constituents Measured with Fluorescent Dyes by Flow Cytometers in Which the Intensity of the Signal is Proportional to the Measured Quantity

Cellular Property	Reference
DNA	Kruth, 1982
RNA	Traganos et al.,1979
Protein	Crissman and Steinkamp, 1982
Enzyme levels	Dolbare, 1983
Cell surface antigens	Loken and Herzenberg, 1975
Membrane potential	Shapiro, 1981
Cellular pH	Visser et al., 1979
Mitochondria	James and Bohman, 1981
Cell viability	Hamori et al., 1980

Table 2. Examples of Cellular Properties not Determined by
Fluorescence Intensity Measurements with Flow Cytometers

Cellular Property	Reference
Membrane fluidity by fluores- cence polarization	Jovin, 1979
Molecular proximity by energy transfer	Jovin, 1979
Cell size by pulse widths	Steinkamp and Crissman, 1974
Cell size by pulse risetimes	Leary *et al.*, 1979
Cell size by axial light loss	Steinkamp, 1983
Cell size by light scatter	Salzman, 1982

that have been quantitated by fluorescence intensity measurements and references to publications describing the measurements.

Measurements other than fluorescence intensity that have been made with flow cytometers are listed in Table 2. In addition, the time at which a cell passes through a flow cytometer can be recorded, thus correlating any flow measurement with time to provide a history of a changing parameter such as a fluorescence (Martin and Swartzendruber, 1980). This type of kinetic measurement can be used to study the turnover rate of fluorogenic substrates to give a measure of enzyme activity or can be used for dye-binding studies.

The efforts of several groups have contributed to the development of flow cytometry to make the technology an important tool in a number of fields of biomedical research. Blood cell volume distributions, one of the first measurements made with a flow system (Coulter, 1956), are now routinely made in most hospitals. The first charged droplet deflection cell sorting was based on electronic cell volume measurements (Fulwyler, 1965).

Measurements of fluorescence from cells first quantitated the amount of DNA per cell by measuring the amount of fluorescence emitted by a dye stoichiometrically bound to the cellular DNA as cells, illuminated by a filtered mercury arc lamp, passed through a flow cell (Van Dilla *et al.*, 1969). Since 1969 the technology and its applications have expanded rapidly. The Stanford group was the first to use fluorescence as a decision basis for cell sorting (Hulett, 1969). The addition of multiple measurements on each cell began in the 1970s. Now it is routine to measure up to 8 parameters on a cell-by-cell basis. The history of the development of flow cytometry and cell sorting is documented in Melamed *et al.* (1979) and Steinkamp (1984).

As an example of the type of data that can be obtained from a flow cytometer, Fig. 2 shows a typical DNA histogram for an exponentially growing cell population. The histogram, which contains data from approximately 75,000 cells, was obtained in less than two minutes. In a flow system specially designed by the Lawrence Livermore National Laboratory group (Peters *et al.*, 1985), cells and chromosomes can now be analyzed at rates up to 20,000 per second.

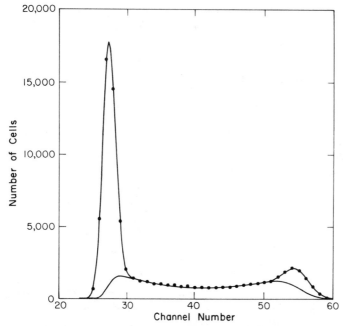

Figure 2. Histogram of cellular DNA content. This histogram is the result of measuring the fluorescence emitted by a DNA-specific dye as 75,000 cells pass through a flow cytometer. The peak at approximately channel 27 is due to cells in G_1. The peak at channel 55 is due to cells in G_2+M. The cells synthesizing DNA fall inbetween the two peaks. The upper solid line represents the results of a computer analysis of the data to partition the total area under the histogram into the contributions due to G_1, S, and $G_{2+}M$ cells. The area under the lower solid line is the S-phase cell contribution.

MEASUREMENTS RELEVANT TO SPACE MEDICINE

There are several classes of flow cytometric measurements that provide information important in health maintenance. They include routine hematological assays, immunological assays including immune cellular function as well as determination of serum enzyme and hormone levels, quantitation of radiation exposure, and quantitation of chromosome damage.

Hematological Measurements

One major application of a flow cytometer is the complete blood count, using only microliters of blood. Such a blood count would include the hematocrit and hemoglobin content as well as the numbers of platelets, erythrocytes, granulocytes, lymphocytes, and monocytes.

Red Blood Cells. The number of red blood cells per cubic millimeter, their volume, and hematocrit distribution can be determined by a flow cytometer. The number density is measured by adding a small volume of plastic microspheres at a known density to the sample. The size of the microspheres is selected such that they do not interfere with the blood cell volume measurements. By counting both the number of microspheres and the number of cells, the volume of the sample analyzed is determined, from which the density of cells can be calculated. With an electronic cell volume measurement station in the flow cytometer, the volume distribution of the red blood cells can be measured. From this information, the hematocrit of the erythrocytes can be determined. Since hemoglobin absorbs strongly at 420 nm, it will also be possible to measure the hemoglobin content on a per cell basis.

White Blood Cells. The number of white blood cells per cubic millimeter can be determined by measuring samples with microspheres at known concentrations as described above. As shown in Fig. 3, forward and 90° light scattering measurements can be used to subdivide the white blood cells into granulocyte, lympocyte, and monocyte subpopulations. Using fluorescently labeled monoclonal antibodies (Hoffman *et al.*, 1980) specific for these subsets of leukocytes, it will also be possible to apply and refine currently available methodologies to resolve the granulocyte subsets: neutrophils, eosinophils, and basophils and the lymphocyte subsets: B cells, T helper cells, and total T cells (as shown in Fig. 4). The methods can be made simple and packaged in the form of a prepared kit to which a drop of blood from the finger would be added and then analyzed. After measurement, the data would be either analyzed on board or transmitted to Earth for further analysis and interpretation.

Functional Activity of Leukocytes

The leukocyte measurements cited above provide information about the immunological status of an individual. In addition to just enumerating the different cell types, functional assays can also be performed. For example, the phagocytic activity of monocytes and granulocytes can be measured by incubating the cells with small (1-

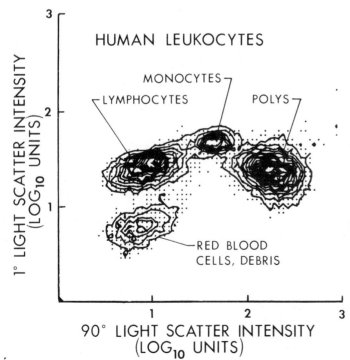

Figure 3. *Contour representation of a two-dimensional histogram of forward and right angle light scatter from whole blood. The labeled islands are identified as being due to lymphocytes, monocytes, polymorphonuclear leukocytes, and red blood cells. This figure demonstrates that a number of cell types can be identified on the basis of light scatter alone.*

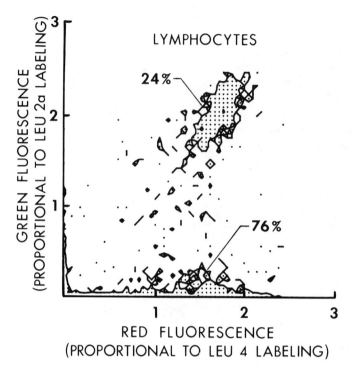

Figure 4. *Contour representation of a two-dimensional histogram of log green fluorescence (proportional to monoclonal antibody Leu-2a labeling) and log red fluorescence (proportional to monoclonal antibody Leu-4 labeling). Of the cells that are Leu-4 positive, indicating that they are T cells, 24% are Leu-2a positive, indicating that they are T suppressor cells.*

2 μm) fluorescent microspheres for a short time and then analyzing the sample in a flow system to determine the amount of microsphere fluorescence that is associated with each cell (Steinkamp *et al.*, 1982). It is also possible to determine the number of particles, up to approximately 20, ingested by each cell. There are a number of techniques for determining cell viability with fluorescent dyes. Thus, cytotoxicity assays can be performed with a flow system (Horan and Kappler, 1977).

In addition to phagocytosis and cell viability, it is also possible to assess the microbicidal activity of granulocytes and monocytes. The major biochemical pathway for bacterial killing by these leukocytes is the hexose monophosphate shunt, which generates superoxide anion and hydrogen peroxide. The activity of the shunt can be quickly assessed on a cell-by-cell basis by measuring the autofluorescence above 400 nm after excitation at 350 nm. Thus, both phagocytosis, the major ingestive pathway, and microbicidal activity can be quickly measured using a flow cytometer.

Serum Hormone and Enzyme Levels

The levels of a variety of hormones and enzymes in blood serum are indicators of a number of aspects of an individual's state of health. For example, elevated levels of creatine kinase are indicative of a recent heart attack.

A method of measuring the concentration of serum proteins and enzymes of importance using a flow cytometer has been developed (Saunders *et al.*, 1985). The basis

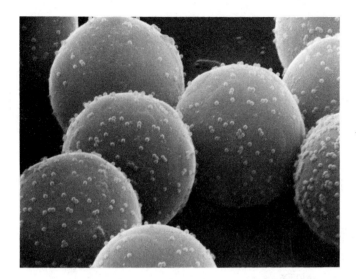

Figure 5. Scanning electron micrograph of small (0.25 μm) fluorescent antigen-coated microspheres bound to large (10 μm) antibody-coated microspheres by antigen-antibody interactions.

of this new assay is similar to a competitive binding radio–immunoassay except that small fluorescent microspheres are displaced instead of radioactively labeled molecules. The assay consists of coating large (10 μm diameter) non-fluorescent microspheres with antibodies to the molecule of interest. These large antibody-coated spheres are then incubated in the liquid sample being assayed. The free antigen molecules in the sample bind to the large spheres via the antibody–antigen interaction. The number of remaining free antibody sites on the large spheres is inversely proportional to the amount of free antigen in the sample. At this point, small (0.1 μm diameter) fluorescent antigen–coated microspheres are added to the sample for another incubation period. These small fluorescent spheres bind to the remaining free antibody sites on the large spheres (Fig. 5).

Without separation of bound from free fluorescent spheres, the whole sample is analyzed by a flow cytometer. There is a very small DC fluorescent level present, due to the few free small spheres in the probe volume, that is ignored electronically. The data acquisition system is set to record the fluorescence only when a 10-μm sphere is detected by light scatter. As the free antigen concentration increases and binds more antibody sites on the large sphere, fewer small fluorescent microspheres can bind, and there is less fluorescence associated with the large spheres. This loss of fluorescence is proportional to the antigen concentration (Fig. 6).

With this assay, a detection limit of 10^{-12} molar has been achieved for horseradish peroxidase. In a sandwich assay based on similar principles, a detection limit of 10^{-14} molar has been reached for the same antigen. In theory, this type of assay could be developed for any molecule that is immunogenic.

One of the advantages of this type of assay is that it is homogeneous, which alleviates the necessity of separating bound from free label chemically—a procedure that both takes time and results in decreased precision because the equilibrium is disturbed. Further

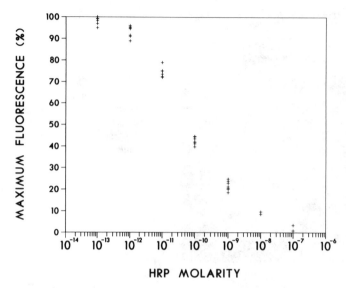

Figure 6. Immunofluorescence displacement curve. The percent of the maximum fluorescence measured with no antigen present is plotted versus the concentration of antigen present. For this set of experiments, run on different days, the detection limit for horseradish peroxidase was 10^{-12} molar. The tight grouping of the data points is an indication of the interday reproducibility of the assay.

advantages include the fact that no radio–labeled compounds are used, and the sensitivity of the system is better than many radio–immunoassays.

Determination of Radiation Exposure

It has been recently shown (Nusse and Kramer, 1984) using a tissue culture cell system that the dose of radiation received by cells can be determined by a flow cytometric assay. The assay consists of determining the number of micronuclei formed in the cultured cells after irradiation. The cell suspension is treated with a detergent, stained with a DNA–specific dye, and analyzed on a flow cytometer. The micronuclei appear in the distribution as objects with low DNA content. By plotting the ratio of the number of micronuclei detected to the total number of cells analyzed as a function of dose, a dose response curve can be generated. A flow cytometer with reasonable sensitivity is needed, since the DNA content of the micronuclei is on the order of that of the smallest human chromosomes, which is approximately 1/100 of the human genome.

Chromosome Damage

A relatively recent development in flow cytometry is flow karyotype analysis (Carrano et al., 1979). The measurement basis for a flow karyotype is quantitation of chromosomal DNA content. Some techniques use two lasers to excite two dyes with different base pair specificities. Present-day techniques resolve the human karyotype into 21 groups with relative ease based on two–color fluorescence measurements. Since, at the present time, chromosome banding patterns cannot be resolved by flow measurements, only DNA content per chromosome can be measured. Thus, not all types of subtle chromosome damage can be detected by flow cytometric measurements. However, gross deletions, breaks, and insertions can be detected.

SYSTEM CONSIDERATIONS FOR THE SPACE ENVIRONMENT

There are no perceived incompatibilities between zero or low gravity environments and flow technology. However, some design changes will be necessary. Since lasers are bulky, very inefficient at producing light, and have high power requirements, another illumination source will be necessary. Several flow systems have been built with mercury arc lamps for light sources. Another possibility is to use a sun tracking system to provide illumination.

Air pressurized fluid tanks will have to be replaced with pulsation-free fluid pumps to avoid problems in a microgravity environment. Analog and digital electronic components can be miniaturized to meet requirements of size, power consumption, and cooling. All data are recorded in digital form and are compatible with data links for transmission back to Earth. A system can be designed with automated alignment procedures controlled by a preprogrammed microprocessor. With additional development, it will be possible to record images of cells as they pass through the light beam. Images recorded in this manner can be analyzed for morphological features to determine cell types, abnormal cytology, and other parameters of interest. In addition to the instrumentation development, sample preparation protocols will have to be modified and/or developed with appropriate packaging of reagents for prolonged flights. We do not envision any difficulty in the development of one- or two-step kits for any of the desired analytical end points.

CONCLUSION

A space flow cytometer will provide the means of obtaining a large variety of information necessary for providing the health of astronauts on missions to the space station, the Moon, or beyond. Current advances in flow cytometry are primarily in the area of new probes for measuring new attributes of cells, chromosomes, and molecules. Hence, in addition to the measurements briefly described here, new capabilities are being continuously added to the list of measurements routinely made by flow cytometers.

Acknowledgments. *The authors wish to thank S. Nachtwey and G. R. Taylor of NASA/Johnson Space Center, and M. D. Enger of Los Alamos National Laboratory for helpful discussions.*

REFERENCES

Braylan R. C. (1983) Flow cytometry. *Arch. Pathol. Lab. Med., 107,* 1–6.

Carrano A. V., Gray J. W., Langlois R. G., Burkhart-Schultz K. J., and Van Dilla M. A. (1979) Measurement and purification of human chromosomes by flow cytometry and sorting. *Proc. Natl. Acad. Sci. USA, 76,* 1382–1384.

Coulter W. H., (1956) High speed automated blood cell counter and cell sizer. *Proc. Natl. Electronics Conf. 12,* pp. 1034–1040.

Crissman H. A. and Steinkamp J. C. (1982) Rapid, one-step staining procedures for analysis of cellular DNA and protein by single and dual laser flow cytometry. *Cytometry, 3,* 84–90.

Crossland-Taylor P. J. (1953) A device for counting small particle suspended in a fluid through a tube. *Nature, 171,* 37–38.

Dolbare F. (1983) Flow cytoenzymology—an update. In *Oncodevelopmental markers: Biologic, Diagnostic, and Monitoring Aspects* (W. H. Fishman, ed.), pp. 207–217. Academic Press, New York.

Fulwyler M. J. (1965) Electronic separation of biological cells by volume. *Science, 150,* 910–911.

Hamori E., Arndt-Jovin D. J., Grimwade B. G., and Jovin T. M. (1980) Selection of viable cells with known DNA content. *Cytometry, 1,* 132–135.

Hoffman R. A., King P. C., Hansen W. P., and Goldstein G. (1980) Simple and rapid measurement of human T lymphocytes and their subclasses in peripheral blood. *Proc. Natl. Acad. Sci. USA, 77,* 4914–4917.

Horan P. K. and Kappler J. W. (1977) Automated fluorescent analysis for cytotoxicity assays. *J. Immunol. Methods, 18,* 309–316.

Hullett H. R., Bonner W. A., Barrett J., Herzenberg L. A. (1969) Automated separation of mammalian cells as a function of intracellular fluorescence. *Science, 166,* 747–749.

James T. W. and Bohman R. (1981) Proliferation of mitochondria during the cell cycle of human cell line (HL-60). *J. Cell Biol., 89,* 256–260.

Jovin T. J. (1979) Fluorescence polarization and energy transfer: Theory and application. In *Flow Cytometry and Sorting* (M. R. Melamed, P. F. Mullaney, and M. L. Mendelsohn, eds.) pp. 137–165. Wiley and Sons, New York.

Kruth H. S. (1982) Flow cytometry: Rapid biochemical analysis of single cells. *Ana. Biochem., 125,* 225–242.

Leary J. F., Todd P., Wood J. C. S., and Jett J. H. (1979) Laser flow cytometric light scatter and fluorescence pulse width and pulse rise-time sizing of mammalian cells. *J. Histochem. Cytochem., 27,* 315–320.

Loken M. R. and Herzenberg L. A. (1975) Analysis of cell populations with a fluorescence-activated cell sorter. *Ann. NY Acad. Sci., 254,* 163–171.

Martin J. C. and Swartzendruber D. E. (1980) Time: A new parameter for kinetic measurements in flow cytometry. *Science, 207,* 199–201.

Melamed M. R., Mullaney P. F., and Mendelsohn M. L. (editors) (1979) *Flow Cytometry and Sorting.* Wiley and Sons, New York. 716 pp.

Nusse M. and Kramer J. (1984) Flow cytometric analysis of micronuclei found in cells after irradiation. *Cytometry, 5,* 20–25.

Peters D., Branscomb E., Dean P., Merill T., Pinkel D., Van Dilla M., and Gray J. W. (1985) The LLNL high-speed sorter: Design features, operational characteristics, and biological utility. *Cytometry, 6,* 290–301.

Salzman G. C. (1982) Light scattering analysis of single cells. In *Cell Analysis* (N. Catsimpoolas, ed.), pp. 111–143. Plenum, New York.

Sanders G. C., Jett J. H., and Martin J. C. (1985) Amplified flow cytometric separation free fluorescence immunoassays. *Clinical Chem.* In press.

Shapiro H. M. (1981) Flow cytometric probes of early events in cell activation. *Cytometry, 1,* 301–312.

Steinkamp J. A. (1983) A differential amplifier circuit for reducing noise in axial light loss measurements. *Cytometry, 4,* 83–87.

Steinkamp J. A. (1984) Flow cytometry. *Rev. Sci. Instrum., 55,* 1375–1400.

Steinkamp J. A. and Crissman H. A. (1974) Automated analysis of deoxyribonucleic acid, protein, and nuclear to cytoplasmic relationships in tumor cells and gynecological specimens. *J. Histochem. Cytochem., 22,* 616–621.

Steinkamp J. A., Wilson J. S., Saunders G. C., and Stewart C. C. (1982) Phagocytosis: Flow cytometric quantitation with fluorescent microspheres. *Science, 215,* 64–66.

Traganos F., Darzynkiewicz Z., Sharpless T., and Melamed M. R. (1977) Simultaneous staining of ribonucleic and deoxyribonucleic acids in unfixed cells using acridine orange in a flow cytofluorometric system. *J. Histochem. Cytochem., 25,* 46–56.

Visser J. W. M., Jongeling A. A. M., and Tanke H. J. (1979) Intracellular pH-determination by fluorescence measurements. *J. Histochem. Cytochem., 27,* 32–35.

10 / SOCIETAL ISSUES

E STABLISHING A PERMANENTLY MANNED BASE on the Moon is, of necessity, a large and visible exercise of engineering and technology. Some will see no more than that, but such a reductionist viewpoint misses the whole that is greater than the parts. A lunar settlement continues humanity's movement to accessible frontiers. It may start as a statement of national resolve or as a monument to international cooperation. It could be an heroic enterprise of epic dimensions or the stimulus for democratization of space through economic growth. One thing is clear—the Moon sits on the lip of the confining terrestrial gravity well and thus is the stepping stone to the solar system.

The space program blends a curious mixture of romance and pragmatism. Goals are set by dreamers and implemented by realists. Many of yesterday's visions have been realized, but the awe-inspiring accomplishments rest on carefully executed, often mundane contributions from tens of thousands of people. The inspirational and the commonplace are both aspects of the human condition to be reflected in the space activities of the 21st Century, and both are discussed in the contributions here.

Any grand achievement by society must begin as an expression of the ordinary processes of decision making. Logsdon, an experienced observer of the space program, looks at the initiation of past endeavors for clues to the key elements of consensus. The debate over allocation of public resources revolves around the worth and expense of any program. Sellers and Keaton analyze historical space expenditures in terms of the gross national product of the U.S. and predict the availability of funding for major ventures over the next two decades.

One school of thought on the worth of the space program sees space technology as a surrogate for military technology. On the theory that military expenditures are supported by the self-interest of a technology intensive industrial complex, competition in space is viewed as a peaceful replacement for competition in armaments. Alternatively, a large space endeavor can be a focus for international cooperation. The exploration of the solar system constitutes a politically neutral enterprise wherein new mechanisms for dialogue, exchange, and cooperation can be fostered. Smith draws on extensive experience with international relations in Antarctica to evaluate the prospects for cooperation on the Moon. Oberg, a longtime observer of Soviet space activities, comments on the possibility of an independent, competitive lunar project complementing a U.S. initiative.

Contemporary space law asserts that space, including celestial bodies, is not subject to national appropriation. Whether mining the Moon for use of its resources in space constitutes "appropriation" is not entirely clear. The "Moon Treaty," which has been ratified to date only by a few non-spacefaring nations, is much more restrictive in its provisions. Moore looks at implications of space law for proposed lunar initiatives. Joyner and Schmitt go one step further and propose new principles that could serve to extend existing legal mechanisms to allow investments in planetary development.

As technical capability increases and as more people work in space, management functions and eventual control of social structures must shift to the space facility from the Earth. Finney looks at living in space through the eyes of an anthropologist and urges the study of evolving social systems there as part of ongoing research. Lawler and Jones both present historical analogies where resource-poor colonies with long supply lines sometimes succeeded and sometimes failed.

To date, space travel has been the privilege of the few but the dream of many. Knox surveys expression of the soaring human spirit as it accompanies those first explorers to the frontier of space.

DREAMS AND REALITIES: THE FUTURE IN SPACE

John Logsdon

Graduate Program in Science, Technology, and Public Policy, School of Public and International Affairs, George Washington University, Washington, DC 20052

What factors must converge to create a societal commitment to creating a permanent human base on the Moon? There have been three major decisions to start manned space programs in the past 25 years—those which began Apollo, the space shuttle, and the space station. An examination of these decisions suggests that no one particular situation facilitates major commitments. Rather, a commitment results from the convergence of the political context; the goals of political leaders, particularly the President; the needs of various space institutions, particularly NASA; the success of earlier programs; and the options available at a particular decision point. While many of these factors are beyond the control of advocates of a lunar base, there are a variety of steps that can be taken over the next few years to improve the chances of program approval at some future time. However, advocates of a lunar base should recognize that it is unlikely that the political leadership will be willing to support another major space program until the space station is nearing operational status sometime in the 1990s.

INTRODUCTION

It is argued that "when the requisite technology exists, the U.S. political process inevitably will include lunar surface activities" (Duke *et al.*, 1985, p. 50). To a student of the political process, such an assertion of inevitability must be viewed with some skepticism, especially when it seems linked primarily to the existence of "requisite technology." It is important for those who want to see a "return to the Moon" goal accepted as an important aspect of future U.S. space policy to recognize that, while developing the requisite technology is a necessary condition for a lunar base program, it is far from a sufficient one.

What factors are essential for creating a political commitment to a lunar base? It is impossible to forecast, at least in specific terms. The United States government *has*, in the past three decades, initiated three major manned space programs—Apollo, space shuttle, and space station; thus it certainly is not impossible to organize and sustain a political commitment to a multi-year, multi-billion dollar enterprise in space, even though such politically supported undertakings are the exception rather than the rule in the United States. Initially, this paper explores these three decisions in order to identify their major characteristics; then, these characteristics will be compared in order to make some useful general comments on the conditions that might lead to the hoped-for commitment to a lunar base program at some future date.

Apollo as a Crisis Decision

In times of crisis—situations that allow only a short time for response without extensive prior planning and where the issues at stake are of great importance—many of the barriers to rational, "top–down" decision-making disappear. Such a situation occurred in April, 1961 (Logsdon, 1970). The self-image and international standing of the United States had been stung in rapid succession by the success of the Soviet Union in orbiting the first man, and by the failure of the United States to follow through in support of the invasion of Cuba by U.S.-trained forces at the Bay of Pigs. In a memorandum dated April 20, 1961, President John F. Kennedy asked:

> Do we have a chance of beating the Soviets by putting a laboratory in space, or by a trip around the moon or by a rocket to land on the moon, or by a rocket to go to the moon and back with a man? Is there any other space program which promises dramatic results in which we could win?

What Kennedy hoped was to demonstrate to the world, through space achievements, that the United States remained the leading nation in technological and social vitality. Almost equally important, though not as clearly articulated, Kennedy saw such achievements as a means of restoring American pride and self-confidence. The Soviet Union's surprising demonstration of technological and strategic strength through its series of space firsts leading to Yuri Gagarin's flight had shaken our image of the United States as the unchallenged technological leader of the world.

After two weeks of assessing alternative answers to these questions, Kennedy's advisers, led by Vice-President Lyndon Johnson, agreed that the United States had at least a fifty-fifty chance of winning a competition to complete a successful manned lunar expedition and that no other alternative provided a better combination of achievement, risk, and cost. Kennedy accepted this assessment, and the Apollo program was born.

The memorandum to the President recommending the lunar landing effort was signed by Johnson, NASA Adminstrator James Webb, and Secretary of Defense Robert McNamara. It stated clearly the rationale underpinning the enterprise:

> It is man, not merely machines, in space that captures the imagination of the world. All large-scale projects require the mobilization of resources on a national scale. They require the development and successful application of the most advanced technologies. Dramatic achievements in space therefore symbolize the technological power and organizing capacity of a nation. It is for reasons such as these that major achievements in space contribute to national prestige.

Space achievements developed prestige, they asserted, in the power struggle between the United States and the Soviet Union; the United States should thus undertake a manned mission to the Moon, even if the scientific or military grounds were lacking:

> Major successes, such as orbiting a man as the Soviets have just done, lend national prestige even though the scientific, commercial, or military value of the undertaking may by ordinary standards be marginal or economically unjustified. . . . Our attainments are a major element in the international competition between the Soviet system and our own. The non-military, non-commercial, non-scientific but "civilian" projects such as lunar and planetary exploration are, in this sense, part of the battle along the fluid front of the cold war.

Kennedy accepted these arguments. In announcing this decision on May 25, 1961, he told the Congress and the nation:

> If we are to win the battle that is going on around the world between friends and tyranny, if we are to win the battle for men's minds, the dramatic achievements in space which occurred in recent weeks should have made clear to us all, as did the Sputnik in 1957, the impact of this adventure on the minds of men everywhere who are attempting to make a determination of which road they should take. . . . We go into space because whatever mankind must undertake, free men must fully share.

Apollo emerged from a crisis atmosphere, and stands as a powerful example of the fact that government can make and can keep a commitment to multibillion-dollar, long-term programs when they serve broad national purposes and are begun with adequate political support. The existence of a crisis situation made the Apollo commitment possible; it did not make it inevitable. Other circumstances had to converge to make Apollo happen. They include:

1. Enough prior research to assure decisionmakers that the proposed undertaking was technologically feasible; a manned lunar mission had been under serious examination for several years prior to the Kennedy decision, and no technological obstacles had been identified. NASA had selected a lunar landing as the appropriate long-term objective of its manned flight program over a year before May, 1961.

2. The undertaking was the subject of enough political debate that groups interested in it and opposed to it were identified and their positions and relative strengths were evaluated, and potential sources of support had time to develop. Both Lyndon Johnson and James Webb had effective working relationships with the leaders of Congress, and obtained pledges of support for an accelerated space program. The President and Vice-President also consulted non-governmental leaders to test their reaction to a vigorous U.S. response to the Soviet challenge in space.

3. In the political system, there were individuals in leadership positions whose personalities and visions supported the initiation of large-scale government activities aimed at long-term payoffs and who had the political skill to choose the situation in which such activities could begin with a good chance of success.

When Kennedy announced his decision to go to the Moon in May, 1961, there were no significant negative reactions, and the funds required to accelerate NASA's program passed Congress quickly and with little opposition. The program was well underway before such opposition first developed in 1963.

The Shuttle: a Bad Bargain?

Apollo, as a crisis decision, was an exception to how policy choices are usually made in the United States. The normal process of policy-making involves a wide variety of participants; it is characterized by bargaining among players positioned within various government organizations. Individuals and groups outside government participate in this process and can be very influential, but their power lies primarily in influencing those within the government who control the resources required to undertake a new course of action. Decisions are almost always made, not by one central decisionmaker, but by a process of interaction among various government organizations and individual political actors. The process leading to the 1972 decision to begin the space shuttle program is an example of the normal policy process in operation (Logsdon, 1979a,b).

In the shuttle decision, major participants were:

- NASA, both as an engineering organization eager to take on a new and challenging technology development job after Apollo and as a government agency interested in maintaining its budget, institutional base, and status;

- Department of Defense, attracted by the potentials of the proposed shuttle for various national security missions in space;

- the aerospace industry, interested in undertaking another major effort along the lines of Apollo;

- the Congress, still supportive of space but unwilling to approve another Apollo-like project aimed at, for example, manned planetary exploration;

- scientists skeptical of the value of or need for another major manned program to follow Apollo;

- analysts who, for the first time, examined a major space initiative in terms of its cost effectiveness;

- the Office of Management and Budget, protective of the budget and unconvinced that the shuttle was a good investment of public funds;

- the President's Science Advisor and his supporting staff and advisory committees, who believed that some sort of shuttle program was an appropriate post-Apollo space initiative, but who were skeptical of the NASA-defined shuttle as being the best approach to lowering the cost and increasing the ease of access to space;

- President Nixon and his policy advisers, skeptical of the future political payoffs from major space programs but unwilling to take the United States out of manned space flight and concerned about the employment impacts of programs such as the shuttle in key electoral areas like California.

The shuttle decision was the end product of a high-pressure, broadly-based, sometimes confusing debate that extended from early 1969 to early 1972, reaching a peak in the second half of 1971. The shuttle that President Nixon finally approved for development was dramatically different in both design and estimated cost from that which NASA had originally hoped to develop. NASA's planned shuttle had been part of a grand design for the post-Apollo space program aimed ultimately at a manned mission to Mars, with a space station and a lunar base as intermediate goals. The final shuttle design emerged from a process of negotiation, compromise, and conflict; it had the rationale, technical characteristics, and cost implications required to gain the support of the President and his advisers, the Department of Defense, and a majority of Congress, while still meeting most of the needs of NASA and its contractors. This coalition was able to overcome continuing opposition from the scientific community and the President's budget office, and thus provided enough support for the program to gain approval.

It was barely enough support, however, and the compromises made to make the decision politically acceptable made program success difficult to attain. NASA agreed to tightly constrained annual and total budget ceilings for the shuttle program, with little flexibility to accommodate technical problems that might arise. Some aspects of shuttle design may have been underexamined in the rush to make a decision, and NASA may have been overly optimistic in assessing the risks and technological readiness of various elements of the shuttle program, particularly the main engines and thermal protection system.

Further, what political support the program had (beyond NASA and its contractors) was not very intense. Only a few in the Department of Defense were involved in the decision process; the bulk of the Air Force was unenthusiastic. Presidential support was neither active nor strong, as had been the case with Apollo. Neither the President nor Congress had accepted, at the time of the decision, a vision of the nation's objectives in space that gave purpose and priority to the shuttle program. Not until the shuttle was threatened with cancellation in 1979 did the top leaders of the country decide it was critical to the verification of the proposed strategic arms limitation agreement (SALT II) and thus deserving of the support required to make it successful.

Selling the Space Station

NASA was forced in 1972 to accept a scaled-down shuttle program as all that it could "sell" to the nation, given the political context of the time. As shuttle development neared completion, the incoming NASA leadership in mid-1981 identified a permanently-manned space station as the agency's top choice for its post-shuttle program. Two and a half years later, after an intensive coalition-building effort, NASA was able to obtain approval to begin station development. Thus the decision to build the space station was another product of normal, non-crisis policy-making, but this time the President *was* active and supportive, and in the end that support proved decisive in allowing NASA to proceed with its top priority program (Waldrop, 1984).

Major participants in the space station decision were:

- NASA, needing another major development program to keep its technical capabilities fully occupied;

- the aerospace industry, hoping to continue to receive major NASA contracts but also beneficiaries of a major defense buildup;

- the Congress, which had been pushing for several years for a statement of long-range goals in space;

- the scientific community, determined to oppose any large new NASA program that would compete with space science missions for resources;

- the Office of Management and Budget, more convinced than ever that major manned space programs were an unneeded drain on the federal budget;

- the Department of Defense and other elements of the national security community, which opposed the space station both because it was not essential to any military need and because it might compete with higher priority DOD programs for funding;

- an emerging community of potential space station users and organizations committed to developing commercial applications of space technology;

- President Reagan and his policy advisers, who saw space leadership as important both symbolically and economically and who accepted NASA's argument that the space station was the logical next step in maintaining that leadership.

NASA had been studying various types of space stations for two decades prior to 1981; these study efforts were coalesced into an agency-wide task force in May, 1982. This task force identified mission requirements, assessed technological requirements, and defined a space station architecture; thus various technological uncertainties were being reduced as the decision process proceeded. NASA also asked the National Academies of Sciences and Engineering to assess the station's potential.

Thus when the station decision came before the President, the technical, policy, and budgetary aspects of the undertaking had been fully articulated, and all interested parties had had an opportunity to make their views known. The President could apply his judgment in order to resolve the conflicts between NASA's proposal and the views of other agencies. He did so in a way that linked the space station to broad national objectives such as national pride, international leadership, and economic growth. Even in the face of growing budget deficits, Ronald Reagan was willing to use his Presidential prestige in support of the space station.

Whether such strong Presidential support has created a political base for the station program solid enough to withstand criticisms and attacks is yet to be seen, although the first year budget for the station program was approved essentially unchanged. Just as it had taken several years to develop the support that led to a Presidential go-ahead in January, 1984, it may take several years to assess the lasting power of that support.

Some Observations

Perhaps the most basic comment to be made regarding these brief case studies is that they demonstrate how a major space commitment can emerge from three very different situations. Of the ingredients for program approval (at least in a form facilitating program success), only one appears essential: strong support from the President. It is basically impossible to begin and complete a large-scale, long-term government program without a lasting bankable commitment from the White House.

The word "bankable" is important here. While it is conceivable that President Reagan, acting upon the recommendations of the National Commission on Space, could announce before 1989 a commitment in principle to a long-range plan for space exploration that includes establishing a lunar base, that commitment would have limited significance until it is translated into the resources required to implement the program.

The fact that there was approximately a decade between the Apollo and shuttle and between the shuttle and station commitments suggests that the President and the rest of the policy system are likely to be willing to provide substantial funding for only one major manned space project at any particular time. While the priority assigned to the multi-billion dollar space program among various government programs has been both high and low, it is hard to imagine the President ever according the civilian space program enough priority to accommodate two or more simultaneous large development efforts. If this conclusion is valid, then lunar base advocates are likely to have to wait until the 1992–1995 period, when station funding and personnel requirements decrease and when the success of the station program is evident, before they have much hope of receiving the kind of Presidential support that commits substantial resources to their favored program.

An earlier Presidential commitment in principle to a lunar base program would, of course, significantly increase the odds of a second, more meaningful commitment later. But it does not guarantee such a commitment. Decisions to begin a large-scale program are very much a product of the favorable convergence, at a particular time, of a number of factors, including:

- the specific political context;

- the visions, values, and styles of individuals in key leadership positions, particularly the President;

- the ambitions and needs of the organization that would carry out a proposed program, particularly as interpreted by the leaders of that organization;

- the ambitions and needs of other organizations that view themselves in competition for the same share of limited national resources required to carry out the program under consideration;

- the outcome of earlier programs of the same character; program success not only in technical terms but also in political terms is essential to approval of any "logical next step;"

- the program choices available, their technical, budgetary, and political characteristics, and their potential payoffs.

The preceding historical review suggests how the interplay among these factors in 1961, 1969–72, and 1981–84 led to decisions to allocate substantial resources to major new undertakings in space. In retrospect, it is clear that many of the factors that made those decisions possible were well beyond the control of those advocating a major new start in space, and so it is likely to be at the time when a lunar base proposal appears on the White House and Congressional agendas.

To say that advocates of a lunar base (or any other large scale program requiring government funding) cannot control the policy process determining the program's fate is not to say that they have no influence on policy-making. There are two general categories of actions that can have such influence: (1) providing a sound technical basis for decision-makers; and (2) developing and honing a convincing program rationale and attempting to broaden the base of those who accept that rationale and are willing to advocate it.

Effective studies and preliminary research and development activities can combine both "technical" and "advocacy" components. A major role of conceptual studies and exploratory research is to reduce technical uncertainties about the character and consequences of proposed courses of action. All participants in policy-making want to understand the payoffs, the cost, and the risks associated with proposed actions, and technical studies can reduce uncertainties about such outcomes. Studies can cast light on the technical, economic, organizational, legal, and perhaps even the political consequences of a particular program, and thereby help policy-makers understand the stakes involved in their actions.

Another way technical studies can make a general contribution to policy choice is by providing the basis for an extremely persuasive argument in support of a particular course of action. If one participant in policy-making has an articulate case in support of his point of view (note that this is different from having an objectively conclusive analysis), he has a powerful asset in the policy-making process. Policy-making is not only a competition among powerful groups; some ideas and concepts also confer power on those who put them forward. In making policy in technology-intensive areas, arguments cloaked in the garb of technical analysis have particular potency.

These comments are intended to suggest an agenda for those who are convinced that a lunar base program is in the national, indeed the world's, interest. While they wait with, hopefully, controlled impatience for the time, some years in the future, when a Presidential go-ahead on such a program is at least potentially attainable, program advocates should be attempting to convince those who control the relevant year-by-year budgets to provide enough support to carry out the studies and exploratory research needed to reduce those unknowns that can be explored without a major investment of resources. They should use their technical work as the basis for building a case for a lunar base that, for the time being, does not claim more than can be demonstrated rather conclusively, i.e., their advocacy should not outrun their data. They should continue to communicate their ideas to a broader audience, but not attempt to mobilize broad-scale support until it can count in policy-making.

This is a recommendation for moderation in advocacy, and is not likely to be palatable to those who want to move ahead as quickly as possible. It derives from several decades of careful observation of how the policy process works. As long as government funding continues to be absolutely required for enterprises like the lunar base, then persons interested in seeing those enterprises come into being must accept the reality of government decision-making. Wishing away the normal policy process won't work, at least in the absence of some significant action-forcing stimulus—a crisis.

In describing the decision to go to the Moon, I suggested that "the politics of the moment had become linked to the dreams of centuries and the aspirations of the nation" (Logsdon, 1970, p. 130). There is no way to *make* this happen, but it seems to be the necessary condition for making the dream of a permanent lunar base become a reality.

REFERENCES

Duke M. B., Mendell W. W., and Roberts B. B. (1985) Towards a lunar base programme. *Space Policy, 1*, 49–61.

Logsdon J. M. (1970) *The Decision to Go to the Moon: Project Apollo and the National Interest.* MIT Press, Cambridge. 187 pp.

Logsdon J. M. (1979a) The policy process and large-scale space efforts. In *Space Humanization Series*, pp. 65–79.

Logsdon J. M. (1979b) The Space Shuttle decision: Technology and political choice. *J. Contemp. Bus., 21*, 13–30.

Waldrop M. W. (1984) The selling of space station. *Science, 223*, 793–794.

THE BUDGETARY FEASIBILITY OF A LUNAR BASE

Wallace O. Sellers

Merrill Lynch & Co., 165 Broadway, 46th Floor, New York, NY 10080

Paul W. Keaton

Los Alamos National Laboratory, MS D434, Los Alamos, NM 87545

At a time when there is almost constant concern and discussion about fiscal deficits and the need for budgetary stringency, the financial feasibility of a lunar base is certainly a legitimate issue. Conservative cost estimates have indicated that a permanent lunar facility with significant scientific and/or industrial capability can be established over twenty years for well under $100 billion in current dollars. The Apollo project was carried out in half that time, for a cost of approximately $80 billion in current dollars, during a period when the actual GNP was only half of what it is today. Projecting the GNP to the end of this century and assuming historically consistent outlays on space activities, we conclude that a lunar base program can be carried out in an evolving space program without extraordinary commitments such as occurred in 1961.

The establishment of a permanent lunar base will require large sums of money over long periods of time, and this program is under discussion at a time when there is great concern over the size of the federal deficit. The question addressed here is whether a program of this sort is affordable at the present time. The conclusion is that, in fact, a permanent lunar base can be financed without increasing NASA's historical budgetary allocation.

In our analysis, it is assumed that the basic costs of space infrastructure will have to be borne by the federal government, and what part private sector funding might contribute to this project is not considered. Eventually, however, there may well be opportunities for corporate activities in space that lessen the pressure on the public sector for the finance of the space program. Neither have we taken into account the possibility of other nations participating in a return to the Moon and sharing the costs. This certainly would be desirable, but we suspect that even if there should be other participants in the program, the major burden of the cost would be borne by the United States.

A NASA/Johnson Space Center team (Roberts, 1984) has made various estimates of the cost of a lunar base. That team studied three separate scenarios with distinct emphasis on resource utilization, colonization, or science. In each case, a space transportation system capable of delivering payloads to lunar orbit was assumed to exist in the latter part of this century. Transportation elements developed specifically for the exploration of the Moon and for establishment of the base were charged to the program. Other cost items included surface facilities and transportation costs such as fuel and operations expense. The derived costs range from a low of $79 billion for a program with an emphasis on utilizing lunar resources to a high of $90 billion for one with an

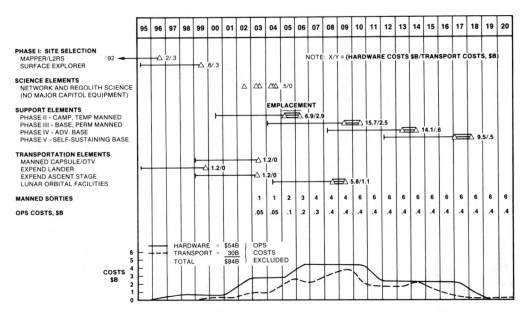

Figure 1.   Cost estimates of lunar base colonization.

emphasis on self-sufficiency and colonization. Figure 1 shows the most expensive scenario. To be conservative, we can say that with the establishment of a proper transportation system, it will cost under $100 billion over 25 years to put a base on the Moon. For comparison purposes, the Apollo costs to place man on the Moon were about $80 billion over eleven years. All of these costs are quoted in 1984 dollars.

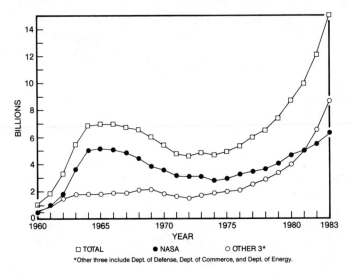

Figure 2.   Dollars authorized by the Federal Government for ●-NASA, ○-Departments of Defense, Commerce, and Energy, and □-the total federal outlay.

Timing of expenditures is almost as important as totals in a program of this sort. The ·NASA team estimated maximum yearly expenditures for hardware and transport to be on the order of $6–9 billion in 1984 dollars, depending upon the activity emphasized by the base. The period of 2006–2010 or even later would require the heaviest annual funding as a percentage of federal outlays.

Figure 2 shows dollars authorized by the federal government for the space program, and Fig. 3 shows total expenditures for the space program, and the NASA budget as a percentage of federal outlays.

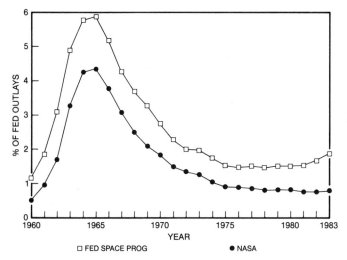

Figure 3.   Percent of federal outlays for o-the total space program and ●-the NASA budget.

An examination of the relationship between total federal government outlays, total expenditures for the space program, and the NASA budget shows that our nation's largest space effort was made during the 1960s when we made a determined national commitment to place a man on the Moon (Fig. 2). This effort was initiated by President Kennedy but remained a real national goal under succeeding administrations with full support from the Congress. During this period, total outlays for the space program rose to almost 6% of federal outlays, and the NASA budget was over 4½% of federal outlays. Looking back to this 1960s period, we see how extraordinary this effort was. It required the spur to national pride of Sputnik's success, the full commitment of an eloquent President, and a period of relative prosperity to bring forth this sort of effort.

During the early 1970s, we had a steady decline in the share of federal outlays dedicated to NASA and the total space program. By 1975, expenditures for the overall space effort leveled out at about 1½% of Federal outlays, and they have leveled out at 0.8% of federal outlays since 1978. The stability of NASA's share of outlays in recent years would seem to indicate that there is a base of support in Congress for at least this level of expenditure, reflecting the fact that American public opinion toward the nation's space program is more favorable than at any time in our history. For our projections, we assume that

Table 1. Federal Outlays in Relation to Gross National Product

Year	GNP Calendar Year (billions)	Outlays Fiscal Year (billions)	Outlays as % of GNP
1960	$ 507	$ 92.2	18%
1965	691	118.4	17%
1970	993	195.7	19%
1975	1,549	324.2	20%
1980	2,632	576.7	21%
1985	3,951*	925.4	23%

*1985 Merrill Lynch Estimates

NASA will be accorded at least 0.8% of federal outlays as its share in the next 35 years, the period over which a permanent lunar base could be established.

The next problem is to determine the likely size of federal outlays in the period 1995 to 2020, a difficult problem, especially in view of present efforts to reduce the federal budget.

The federal outlay percentage of GNP has increased from 18% in 1960 to 23% in 1985 (Table 1). It seems unrealistic to expect that outlays will be reduced to the 17–19% level of the 1960s because of the many new domestic programs that have been added, the increasing use of indexation, and the higher level of defense spending. We use an estimate of 20% as a base case for federal outlays as a percentage of GNP, although we also show the figures for federal outlays at 22½% of GNP.

The U.S. GNP over the relevant time period remains to be estimated. Economic forecasting of GNP for a specific year in the future has not developed a particularly positive image in recent years. On the other hand, over the very long run, growth in GNP in the U.S. has been quite consistent. If one looks at the historical record over the past 100 years (Table 2), the GNP has shown a compound growth rate of slightly over 3%. In the 1960–82 period, the compound rate of growth of real GNP has been 3.6%, although

Table 2. Growth Rate of Gross National Product in the United States

Initial Year	Terminal Year	Growth Rate
1874*	1970	3.6%
1884*	1970	3.3%
1890	1970	3.3%
1900	1970	3.2%
1910	1970	3.0%
1920	1970	3.3%
1930	1970	3.5%
1940	1970	3.9%
1950	1970	3.6%
1960	1970	4.0%

*Average for decade

Table 3.   Available Funding

GNP Growth	3%		3¼%		3½%	
Federal Outlays as % of GNP	20%	22½%	20%	22½%	20%	22½%
Available to NASA at 80% of Federal Outlays 1995–2010 ($ Billions)	$306	$344	$325	$366	$346	$389

the range for individual years went from a high of 10.8% from 1971–1972, to a low of –1.9% from 1981–1982 (U.S. Bureau of Census, 1983). Some idea of the consistency of the growth rate of Gross National Product in the United States (U.S. Bureau of Economic Analysis, 1973) can be seen in Table 2.

These projections estimate growth in GNP at 3%, fractionally below the history of the last 100 years. Fiscal 1984 GNP is estimated at $3,581 billion, an average of the fourth quarter of 1983 and the first three quarters of 1984.

Table 3 shows the monies available for NASA, assuming this 3% GNP growth, plus federal outlays at 20% (below the experience of recent years but high historically), and a NASA budget of 0.8% of federal outlays, the lowest level since 1961. In spite of these relatively conservative assumptions, the anticipated budgetary allocation to NASA of over $300 billion will exceed the full cost of a lunar base by more than threefold. Budgetary constraints should not be a significant hurdle to a permanent lunar base. To put it quite simply, if we want it, we can afford it.

Table 3 also includes what might be available to NASA assuming somewhat higher growth rates in GNP and somewhat higher federal outlays. It demonstrates the obvious: the higher the growth in GNP, the more easily programs can be financed.

The present problem of the federal budget deficit amounts to about 22% of all federal outlays. We expect that this problem will be addressed by a modification of the entitlement program, defense spending, tax revenue rates, or growth in the economy, and most likely all four. We certainly do not expect that the 22% deficit problem would be addressed by, for example, cutting NASA's share of federal outlays to 0.4% from 0.8%. Even if that were done, NASA could still afford to adopt a lunar base as a primary long range goal providing no other major civilian space initiatives were made.

One factor that should be kept in mind is that the commercial benefits that will be derived from space over the long run are likely to be substantial. The two key contributions to be made by the federal government are (1) a reduction in space transportation costs and (2) simply providing experience for human activities in space. With assistance from the government in these two areas over the next decade or so, it is likely that commercial exploration of space will lead to substantial financial benefits for the private sector and, eventually, for the public sector.

The decision to go forward with a lunar base program is quite obviously a political decision that is uniquely difficult because the program is such a very long one that cannot be funded in fits and starts. It is essential to understand that subject to the appropriation process of Congress and the performance evaluation of the Executive branch, these expenditures would continue into the next century. Once initiated, we are truly committed.

The commitment should be made now; in the long run, it is probably inevitable. The sooner it is made, the more intelligently and economically the lunar base can be planned and implemented. We believe that the program is affordable.

REFERENCES

Roberts B. B. (1984) The initial lunar base and its growth potential (abstract). In *Papers Presented to the Symposium on Lunar Bases and Space Activities of the 21st Century,* p 132. NASA/Johnson Space Center, Houston.

U. S. Bureau of the Census (1983) *Statistical Abstract of the United States.* U. S. Government Printing Office, Washington, DC.

U. S. Bureau of Economic Analysis (1973) *Long Term Growth 1860–1970.* U. S. Government Printing Office, Washington, DC.

LUNAR STATIONS: PROSPECTS FOR INTERNATIONAL COOPERATION*

Philip M. Smith

Executive Officer, National Academy of Sciences and National Research Council, Washington, DC 20418

In the 1960s, two years before the first Apollo landing, it was my good fortune to take the leaders of the nation's space program to Antarctica. Bob Gilruth, Wernher von Braun, Max Faget, Ernst Stuhlinger, several of the Apollo astronauts, and others participated. The purpose was to look at all dimensions of the Antarctic program—the scientific program and its planning, the logistics necessary for its support, search and rescue procedures, the effects of long stay-times in isolation and other human factors, and international cooperation among the nations working on that continent. These space leaders felt Antarctica was the best Earth model in actual operation on a scale of projected space activity from which to draw ideas about space stations, and particularly lunar station planning. That expedition stimulated my own interest in lunar stations, one that has continued to this day. Thus, when offered the opportunity to attend and participate in this conference, I immediately responded in the affirmative.

My subject is the prospect for international cooperation in lunar station planning. In considering policies for a U.S. program, and as other nations consider their policies for future lunar station planning, it is inevitable that international cooperation will be a dimension carefully reviewed and discussed. Without commenting as an advocate or a critic on the desirability of a lunar station or a series of them, the time of their construction, or the desirability of international cooperation, it is possible to project several modes of scientific, logistics, private-public sector, and other cooperation that could take place on the Moon. My objective is to lay out a few ideas concerning different modes of cooperation in the hope that these thoughts will contribute to and perhaps stimulate some of the discussion over the next few years.

ASSUMPTIONS

I begin by stating that my crystal ball gets quite cloudy beyond the first three decades of the 21st century, so my remarks are really focused on the next fifty years. My further planning assumptions are five in number.

First, a return to the Moon for extended stay-times will take place within the next fifty years.

---

* Views of the author, not those of the National Academy of Sciences or the National Research Council.

Second, at this point in the discussion of future lunar station activity, it is unwise to advocate any one model of national or international activity. Nations should understand various modes of cooperation, but, as nations, should keep options open until renewed lunar activity comes into clearer view.

Third, while transnational enterprise is a worthy goal, national enterprise, that is, the activity of nation-states, will continue to be the important and dominant factor in the development of space for quite some time to come. The first three decades of space have been dominated by national objectives although limited private sector economic cooperation has begun. National activity will probably be the continuing condition in the decades ahead. On the other hand, dominance of space activity by national enterprise has not ruled out international cooperation in space missions, nor would it in lunar base development.

Fourth, in early years of extended stay-time on the Moon, the principal rationale will be scientific observation and the logistic support of those observations by humans or automated facilities. Another early thrust may be the use of the Moon as a logistic or resource platform for other space activities. True lunar industrialization or commercial activity in the economic sense, at least in my view, lies somewhere beyond the next fifty-year period. Thus it is somewhat beyond the framework of my discussion. If for some reason economic or commercial development becomes attractive at an earlier point in time, it will inevitably change the character of international cooperation that may ensue.

Fifth, I assume that the Moon will not become a military platform and will remain a body of peaceful enterprise where a nation's national security interest is maintained and expressed through ability to participate in scientific and other lunar ventures.

NATIONAL ENTERPRISE

The exploration of the Earth over the last several centuries has been dominated by nationalism—the mounting of national exploratory expeditions, generally nationalistic and oftentimes economic in purpose. The national enterprise has been centered in exploration of new territory, its settlement, and its potential economic exploitation. This, of course, is the obvious fact of history. However, it is interesting to recall that in this framework of nationalistic activity there has long been an associated observational or scientific activity. And increasingly in this century, scientific cooperation has been a principal means of maintaining national visibility and has been accepted as a premise in international relations and in development of some international law.

In the 1800s some of the great national enterprise expeditions were essentially geographical and scientific. The Lewis and Clark expedition was an exploring and scientific expedition, sent out under the auspices of a President and Congress. In the late 1700s and early 1800s, Great Britain, the United States, France, and Russia all mounted major Pacific Ocean and Southern Ocean exploratory expeditions. In their day they were akin to a lunar or martian expedition: one left with the expectation of being away from home for several years, possibly not returning at all.

A number of national polar expeditions were mounted in 1898 as a part of an International Polar Year. A second Polar Year occurred in the early 1930s, again with nations contributing their scientific observations to a shared good. But at the same time, in Antarctica especially, these national enterprises furthered national objectives. The first expeditions of the late Admiral Richard Byrd took place in that period. In Antarctica, in fact in this century, there was a period of fifty years of relatively intense nationalism, in exploration and science, before the new period of international activity that was ushered in by the International Geophysical Year.

The Antarctic Treaty that has governed that continent's research and other activities was signed in 1959. It became a model, or at least a vision, of the way space activity could proceed. The Antarctic Treaty resolved or set aside some vexing legal issues, but activity there has continued primarily as national activity. It is interesting to note that national pride, expressed through the fulfillment of scientific objectives, continues to play an evolutionary role in Antarctica. There are a number of new entrants into Antarctic activity—West Germany and more recently India and Brazil. Within the last few weeks, the People's Republic of China has announced that it will send an expedition of some 500 individuals, counting those individuals on board two ships, to Antarctica for a 1984-1985 summer expedition, a forerunner of a year-round station established for the purpose of undertaking scientific research, but also fulfilling national objectives comparable to the activities of other large nations working in Antarctica.

Today it may be easier in the United States for us to think about international activity on the Moon than it is in some other nations, because we have already been there. We were the first to arrive and plant a flag, a source of pride for all people, but a special source of joy for Americans. There may be, among some Americans, a more generous attitude concerning international cooperation than may exist in other nations. We should remind ourselves of our own history to appreciate how others may see us in relation to their own national enterprise objectives in space. During the exploration of the American West, the spirit of "manifest destiny" was rampant. It was a time in which one would have heard little about international cooperation for any frontier enterprise whatsoever—scientific or otherwise. Thus it is conceivable that other nations may wish to achieve their own "manifest destiny" through expeditions to the Moon.

In our time, since the great success of the International Geophysical Year, there have been many international cooperative ventures in science. They provide concepts of several ways we might proceed on the Moon. In science, aspirations for international cooperation, peaceful development, and common purpose have now become one of the cardinal features in a century that is truly the century of science. The vision of space and of the Moon as an opportunity for a successful new international or transglobal beginning is attractive. However, if history is any guide, perhaps we should think of an evolution that progresses toward that point out of continued national enterprise.

Thus, I suggest, as one projects ahead to the end of the 20th century and early into the 21st century, that national enterprise might play a very large role in lunar activity. The United States, the Soviet Union, and other major players in the space arena will undoubtedly continue to express national pride by way of national scientific and technological prowess in space. Japan, a nation striving very hard to be totally competitive with the United States and other industrial powers, could well find its own national interest served in a manned expedition to the Moon.

INTERNATIONAL ACTIVITY IN THE CONTEXT OF NATIONAL ENTERPRISE

What are the implications for cooperation in lunar activity if national enterprise continues to be a strong and dominant feature of space exploration? There are a number of possible scenarios.

It would be possible for the United States to define and finance most of its own plans for lunar station activity but to offer participation to other nations and to their scientists. This form of international cooperation would be somewhat along the lines of the arrangements worked out for the development of the space lab by the European Space Organization.

There are a number of Earth models of this kind of cooperation. The International Program of Ocean Drilling, a highly successful program in which five countries participated, was in its initial years the Deep Sea Drilling Project of the United States, funded through the National Science Foundation. The program had an interesting evolution. After its initial eight years as a U.S. enterprise, but with scientists from other nations involved, the United States extended fuller membership in financing and planning to the U.S.S.R., Federal Republic of Germany, Japan, France, and the United Kingdom. Budgetary pressure forced the international phase of ocean drilling beginning in 1975. After political developments surrounding the Soviet entry into Afghanistan in 1979, that country withdrew. We or another nation could proceed with lunar development along the lines of this model. The lunar settlement for scientific work could be initiated with some guest scientists and eventually fuller operational partnership offered to other nations.

There is another scenario. If other nations are only partly convinced to work within the program of a dominant space power such as the United States because of their own desires to prove fuller partners or competitors, it is conceivable that there could be more than one national expedition to the Moon by the early part of the next century. There could in fact be independent expeditions by several nations. In such a case, it would be possible to construct a framework for international cooperation such as those established through the International Council of Scientific Unions. By common agreement through ICSU, observations in space from the Earth or unmanned space platforms, cooperation in the atmospheric sciences, oceanography, glaciology, and other disciplines was established in the IGY and has proceeded since. The Committee on Space Research (COSPAR) has played this planning role in space science. There are a number of major activities in Antarctica that go forward this way planned under SCAR, the Scientific Committee on Antarctic Research. The International Antarctic Glaciological Project, for example, in which France, the Soviet Union, the United States, Australia, and Great Britain have participated, is a decadal survey of the East Antarctic ice sheet. The project was planned and is being coordinated through SCAR. Independent national activity coordinated through an ICSU-like arrangement would in itself be a successful method of international cooperation on the lunar surface.

A variation on this model is a cooperative mode buttressed by both an international planning organization and a non-governmental international scientific union. In the highly successful Global Atmospheric Research Program (GARP), individual nations pool their scientific and logistic efforts. GARP is planned scientifically through ICSU and nationally and logistically through the member states of the World Meteorological Organization (WMO). While WMO is under the overall aegis of the United Nations, I do not stress that aspect. Indeed, the fact that the WMO has been uncharacteristically free of the political agenda of the developing nations is the ingredient that has made the WMO-ICSU arrangement for GARP successful. For over a decade, a series of observational programs culminating in the Global Weather Experiment has been carried out. One could imagine the formation of a specialized international governmental organization like WMO and an ICSU scientific union that would together coordinate a lunar observational program of several decades duration where national expeditions of varying length, unmanned observations, and permanent stations on the Moon could be established. The participating nations would budget and mount their own lunar efforts, but under these newly established international organizations. Their efforts would be coordinated to achieve more effective strategies than would be obtained by uncoordinated effort.

Another direction lunar cooperation might take would be the development of bilateral arrangements between two countries electing to pool their resources. In the post-World War II period there have been many bilateral agreements for cooperation in research. Many are negotiated at the level of heads of government. Many are at an agency level; some are arranged by other organizations within nations. Bilateral agreements arranged at the level of heads of government have often been very stable over successions of governmental leadership. Most of these bilateral agreements have been prompted by a combination of scientific or technical interests and political objectives that could be served by cooperation. As one surveys programs of the industrialized nations in research and development today, one finds an increasing effort to utilize bilateral activities for the conduct of expensive longer term research programs. For example, there are international bilateral activities to promote fusion research.

The most famous bilateral activity in space, of course, is the Apollo-Soyuz linkup a decade ago. If, by the turn of the century, the United States and the Soviet Union remain the principal spacefaring nations they could decide to pool resources. With impoverished communications between the nations it is hard to think about it, but one could imagine the United States and Soviet Union arranging—several decades from now— a bilateral agreement for a joint lunar effort. In this case there would develop a lunar program that was cost-shared and planned according to agreed-upon principles of joint funding and management. The two countries, as principal partners, could extend cooperation to other nations.

It is also conceivable that two or three nations could team up in a bilateral or trilateral effort to pool their energies to match or parallel a strong unilateral effort of the United States. For example, the emerging industrial Pacific Rim high technology nations such as Japan, the People's Republic of China, Korea, and others could among them create a bilateral or trilateral effort for lunar activity.

TRANSNATIONALISM

There has been much discussion about transnationalism in which there would be, in effect, total international cooperation amongst a consortium of nations and their scientific and engineering communities for the planning of lunar stations. Transnationalism seems to work better in theory than in practice. Nonetheless, if one looks at Western European cooperation today, as compared to European cooperation of two to three decades ago, there is much evidence of growing transnationalism in scientific and economic enterprise. One must conclude that true transnationalism in lunar base planning is a possibility, even though I have suggested earlier that I believe it will be slower in developing on the Moon than some other modes of cooperation. A discussion of the economic and political conditions necessary for true transnational cooperation in lunar station development would require a separate, extended discussion. But what if there were a real sea change in the affairs of nations? Transnationalism in lunar scientific or observational cooperation could be possible.

As a model, there is the highly successful transnational enterprise in high energy particle physics. The European Laboratory for Particle Physics—its French acronym is CERN (Council of Europe for Research Nuclear)—was organized in the early 1950s and started with eleven member nations that pooled their resources for the construction of the several accelerators located at the edge of Lake Geneva. Thirteen nations now contribute. Collectively, CERN nations compete with the U.S. program in high energy physics and those of other nations. The United States helped organize CERN and contributes as a limited partner to its costs through the Department of Energy. U.S. scientists are CERN experimenters. There have been many problems over the years. Today, for example, Great Britain is suggesting that it may withdraw. These difficulties have been resolved successfully for two decades. CERN is a truly representative model of a kind of transnational science activity that could go forward on the Moon.

PRIVATE SECTOR ENTERPRISE

To complete this survey of prospective modes of international cooperation, let me mention still another possibility, although it may presume more actual economic potential in lunar exploration in the next fifty or so years than could take place. This would be the development of a non-nation state international consortium of private interests that would, in fact, plan lunar station activities.

The International Telecommunications Satellite Consortium (INTELSAT) is such a model. On Earth, there are many energy, mining, and construction consortia that are of a similar character. Were such a lunar venture to emerge, one might suspect, however, that national interests of the several nations interested in lunar and space exploration might come into play in a regulatory sense or because of other considerations. These discussions seem to arise for several reasons, including the need for continued and substantial subsidization of the private enterprise by public funds and the national interests that must be protected. So, one could anticipate that if a non-nation state consortium for international cooperation on the Moon were to develop, it might be heavily regulated by the several governments of the venturing partners. Or, the consortium could get swept up in north-south political questions through the United Nations unless the newly industrializing nations have by the early part of the next century sufficient economic self-interests in lunar efforts that this model would appeal.

Unlike Antarctica, where exploration of mineral resources and hydrocarbons is still seen as economically distant after 160 years of exploration, an extractive resource industry on the Moon may certainly develop comparatively early in the history of lunar occupancy. The extractive resource activity could well be operated by commercial, private sector companies. These companies, within the time frame of my assumptions, would doubtless be under contract to a government or governments.

A more likely variant involving commercial or private economic interests could be the utilization, by nations, of private sector enterprises to operate lunar research facilities. Again, there are successful models to be examined. Consortia of universities in the United States operate astronomy observatories for the National Science Foundation. Since World War II, the Energy Department and its predecessor agencies have operated, under service contract agreement with companies, facilities such as the Oak Ridge National Laboratory. NASA has contracted with companies for the operation of its facilities. In the lunar case, such service and support ventures could be international in scope but private sector enterprises.

CONCLUSIONS

I have not stressed truly transglobal enterprise on the Moon, an international effort by all nations on the Earth carried out under the United Nations or some subsidiary. It would, on the basis of past experience, require a collective spirit of enterprise that is difficult to imagine in the fifty-year time span I have been addressing. At every level of human endeavor such transglobal enterprises run into barriers. Even mountain climbers who have attempted multinational assaults on Mount Everest and other mountains have had mixed success. The scaling up from that comparatively simple level of effort to any other presents insurmountable difficulty today because of the debate between the industrialized and the developing nations. Perhaps it will be different by 2020.

In the absence of a transglobal United Nations-like effort, there are a variety of ways in which international cooperation in lunar station planning might proceed. There may be no preferred strategy, nor a model from past or current experience that will apply early in the next century. A range of opportunities for cooperation will exist for the United States and other nations. National and international interests will have to be balanced and weighed. Since scientific cooperation may offer the only early rationale for extended lunar activity, the traditions of international cooperation in science forged in our time will certainly be examined and re-examined by nations planning lunar stations some decades from now. Indeed, our successes in international scientific cooperation are a great legacy to give to those who will plan the return to the Moon.

SOVIET LUNAR EXPLORATION: PAST AND FUTURE

James E. Oberg

Soaring Hawk Ranch, Route 2, Box 350, Dickinson, TX 77539

After a hiatus of more than a decade, Soviet lunar exploration is now expected to resume in 1989 with the launch of an unmanned polar orbiting survey probe. Now is thus a good point to take stock of Soviet space capabilities applicable to advanced lunar missions and to make estimates of future options based on these capabilities. To do this, a brief but full chronicle of past Soviet lunar exploration is required.

The successful launching of Luna-1 on January 2, 1959—the first space probe to escape from Earth—had been preceded by at least three launch failures starting the previous September (according to recently released government documents at the LBJ Presidential Library in Texas). It was followed by another failure in June and then two successes in the fall, a lunar impact and a lunar farside photographic fly-by. Two additional farside photographic missions the following spring failed. So of the first series of Soviet lunar probes, there were three successes in nine attempts, a success rate not unlike that of the early U.S. Pioneer program.

Beginning in January 1963, Soviet hard-landing spacecraft were launched toward the Moon. Success was finally achieved with the Soviet Luna-9, four years and several failures later. This feat was followed within months by the five (of seven attempted) NASA Surveyor soft landings; one later Soviet hard-lander, Luna-13, was also successful.

The distinction between "soft-lander" (<1 m/s touchdown speed) and "hard-lander" (5–10 m/s touchdown) has generally been forgotten, since the only American hard-lander program was the unsuccessful early Ranger/Block II series in 1962. Surveyor, Viking, and the Apollo lunar module were all soft-landers, as were the large Soviet Luna vehicles of 1970–1976. The early Soviet hard-lander probes (1963–1966) relied on collapsible structure that allowed their high-shock-resistant electronics to survive. Such a design is simpler to build but severely restricts the range of research apparatus capable of being carried safely to the surface.

The characteristic string of flight failures that almost always inaugurated the appearance of every new class of Soviet deep space probes should be interpreted less as evidence for engineering incompetence than as the result of a conscious—and not at all unreasonable—development philosophy. Soviet space engineers evidently concluded that it could be cheaper to perform shakedown flights in orbit than to undergo a long series of expensive ground tests and computer analyses; American space engineers were unwilling to face the political costs of a deliberate approach that almost guaranteed a string of publicized and embarrassing failures. Soviet spacecraft were thus committed to flight earlier in their development than corresponding American vehicles, but they would take a longer flight program to fully mature.

In the summer of 1966, three Soviet probes were successfully put into lunar orbit for science and photographic surveys. At least one of the payloads failed without sending any data.

MAN-TO-THE-MOON RACE

A separate Soviet lunar program eventually appeared under the cover name "Zond" (or "probe"), which had earlier been used for a series of interplanetary missions. This lunar Zond series involved the launching of modified "Soyuz" man-rated spacecraft on a translunar loop and its return to Earth. The first attempts were made early in 1967, with a successful mission coming eighteen months and four or five failures later. (Reportedly, one capsule came back far off course and was recovered by the Chinese, who for many years exhibited it in the Red Army Museum in Beijing.)

The Zond program was evidently part of the secret Soviet man-to-the-Moon race, in competition with Apollo. But American astronauts orbited the Moon first (in December, 1968) and seven months later were walking on its surface.

If there is one major disappointment in the way in which Apollo has passed into our history books, it is related to the purpose of John Kennedy's original 1961 vision. He wanted the project to demonstrate American technological superiority to the world, and particularly to the Soviet Union. At the time of his famous speech, the world believed that the United States was years behind in the "Space Race". Hindsight shows the Soviets were orchestrating "space spectaculars" with little more than a year's head start on a moderately bigger booster rocket, but at the time it seemed to require a miracle for the U.S. to catch up.

Unfortunately, when the miracle was ultimately delivered, the heart and soul of it—the victory in the Moon Race—was torn out of it by a clever Soviet propaganda ploy. In a special Apollo 11 fifth anniversary program in 1974, Walter Cronkite summed up the consensus of his media colleagues (but NOT of the experts) that there had never been a Moon Race after all.

Recent contributions have been made to the mountain of evidence showing that the Soviets in the 1960s were trying their utmost to build spacecraft and boosters to fly cosmonauts around the Moon and, ultimately, to land them on the surface. This new evidence includes testimony of Soviet emigrants as well as uncoordinated public statements by official Soviet spokesmen who have disclosed some revealing new details.

In the early 1980s, amid worldwide concern over falling Soviet satellites with nuclear reactors aboard, it was learned that Kosmos-434 was about to reenter. When launched during the era of the Moon Race, it had been thought by Western observers to possibly be an unmanned flight test of a man-rated lunar spacecraft. This judgment was based on the fact that it maneuvered through its orbits in a pattern analogous to that a lunar spacecraft would have to follow near the Moon; also, telemetry signals characteristic only of man-related spacecraft were picked up from it. But at the time it was launched, the Soviets officially referred to it (and three other flights like it) as routine "scientific satellites for research in outer space." Then, after many years of gradually slipping closer and closer to the atmosphere, the satellite finally burned up over Australia.

To allay fears "down under," an official spokesman of the Soviet Foreign Ministry in Moscow assured Canberra that there was no radioactive material aboard Kosmos-434—the satellite was merely "an experimental lunar module!" This, of course, is what Western observers had suspected all along.

"Lunar modules" (in Russian, "lunnaya kabina") is the same term used for the Apollo lunar module and indicates craft that are built and tested to be flown by men, not by robots. The Soviet space secrecy curtain had parted, and another small piece of the true history of the Moon Race was unveiled.

The summer of 1984 was the fifteenth anniversary of the first Apollo landing. It was also the fifteenth anniversary of another lunar flight-related event, which significantly shaped the space exploration paths of Earth's two major spacefaring nations. In 1969, on a launch pad north of Tyuratam in Soviet Central Asia, a giant booster rocket physically twice the size of America's Saturn-V (the "SL-X" or "G-class") was destroyed in a spectacular explosion. Two later launch attempts also failed, sounding the death knell for any Soviet attempt to match or surpass Apollo in the 1970s.

With a little change of luck, the history of manned lunar exploration could have turned out entirely different. This is another wrong impression given by the history books, that things happened the way they did because that was the way they had to happen. Far from it—it was often by the narrowest of margins and by the extremes of cleverness that success was separated from setback and catastrophe.

UNMANNED SOVIET LUNAR PROBES OF THE 1970s

The Soviet lunar program pushed on with a large unmanned soft-lander program, which also began in 1969. By late 1970, success was achieved. Luna-16 landed, picked up soil samples, and returned to Earth. A month later, Luna-17 deposited a "Lunokhod" robot lunar buggy onto the surface for long, slow traverses.

The "scooper" mission was successfully repeated twice more, while at least three missions failed. A second "lunokhod" was landed in 1972, after which the program was terminated. Two spare "lunokhod" modules were modified to act as lunar orbiters. But by 1976, this entire series of three types of spacecraft had ceased.

Why? One possible answer is that funding for Soviet lunar exploration was cut off when the need to "show the flag" vis-a-vis American lunar exploration vanished.

There may have also been serious questions about the additional scientific value of more samples. American visitors to the Vernadskiy Institute in Moscow in the mid 1970s noticed that American lunar samples delivered in exchange for Soviet samples had not even been opened, long after the Soviets had obtained them. This suggested a serious bottleneck in the delicate process of extraterrestrial materials analysis, a bottleneck that could have led Kremlin budget cutters to decide that if their scientists could not handle the samples already on hand, there was no use in spending hundreds of millions of rubles to obtain more samples. In addition, more advanced systems were feasible for sample return, and these may have looked more cost-effective.

Photographs of Soviet Venus probes being prepared for launching in the late 1970s show at least one lunar "scooper" vehicle in the background of the assembly hall. This was years after the last known Soviet lunar probe, and it suggests that additional launchings were considered possible but still uncertain. Indeed, one or more launchings may have been made but encountered booster failure.

THE NEXT SOVIET LUNAR PROBE

Soviet lunar scientists regularly attend the annual Lunar and Planetary Science Conference held in Houston every March and throughout the late 1970s kept telling their Western colleagues of their own attempts to obtain funding for advanced lunar missions such as farside landers/scoopers and a polar orbit survey probe. In mid 1984 such approval came, and the following March, at the Sixteenth such conference, Dr. Valeriy Barsukov described what was now on the official Soviet space schedule.

The next project is to be the Lunar Polar Orbiter in 1989–1990. If tradition is followed, it will be Luna-25. Barsukov publicly gave no hint whether it would be a single spacecraft or a series, as is more traditional. Orbital altitude is to be about 100 km. Its aim is to gather information on the chemical and mineralogical composition of the entire surface in order to give a basis for the scientific selection of sites for the return of new samples.

As presented by Barsukov in Houston in March, 1985, the roster of instruments included: a TV camera; gamma spectrometer ("super-pure germanium"); X-ray spectrometer; neutron detectors ("to test the hypothesis of Jim Arnold on polar water ice, and also to map KREEP basalt locations"); spectrophotometer; IR-spectrometer; altimeter; spectrometer of charged particles; a plasma study complex; micrometeorite analysis apparatus; magnetometer; spectrometer of reflected electrons; scintillation counter of reflected electrons; scintillation gamma spectrometer.

No data was given on spacecraft weight, power, or data rate, but available boosters can place up to several tons in the prescribed orbit. The last "Luna" in 1976 weighed 5300 kg pounds on its translunar trajectory. For comparison, a proposed U.S. lunar polar orbiter in the mid 1970s had an injection mass of approximately 500 kg.

But what kind of spacecraft will it be? A major Soviet design theme is commonality and economy, with the same basic bus often appearing for many years in different variations. Since the orbiters of the early 1970s were modified Lunokhod chassis modules now long out of production, a new spacecraft will be needed. One obvious candidate is available: the highly successful Venera-class bus.

Analysis by spacecraft expert and sovietologist David Woods shows that the standard Venera vehicle does not have enough tankage to carry the fuel required for braking into the planned low lunar orbit. However, the addition of jettisonable tanks (a technique used by the Soviet lunar orbiters of the mid 1960s) would solve this problem. In addition, the Venera power system (which depends on full sunlight) would have to be augmented with battery packs for the periodic excursions into the lunar shadow.

Such a program offers the opportunity for cooperation. Western instruments could be included, and Western tracking facilities could be utilized for data communications. The degree of influence of the renewed American interest in lunar exploration on the Soviet decision to fund this new project is uncertain but is justifiably subject to speculation.

Soviet planners could reasonably have assumed that a small U.S. lunar polar orbiter might appear in the early 1990s, so this decision would be consistent with the traditional Soviet desire (which has its American counterpart) to score another scientific first.

POSSIBLE NEAR-TERM FOLLOW-UP LUNAR MISSIONS

Barsukov did not want to discuss follow-up lunar projects, and obviously none have been approved. He did not disclose any specific desires or recommendations from his own office on this subject. However, it should be reasonable to expect that his old desire for farside sampler missions is still alive. Whether such probes would just be rebuilt samplers of the types launched in 1970, or entirely new models, is not known. The 1970-era model was limited to a narrow equatorial region near the Moon's trailing limb, due to severe constraints on the return stage's guidance and propulsion capability. Any new system would have to overcome these constraints.

Another leading possibility for a follow-up lunar mission is a combined operation of rovers (with sample collection capability), coordinated with a return capsule that can be loaded with the cannister filled by the rover. The rover would be landed at sites located by the geochemical surveys from the polar orbiter and could operate for several months before the return module is sent. The final rendezvous on the lunar surface might be difficult but correspondingly valuable.

A third type of mission might involve very high-resolution photography. Soviet military reconnaissance satellites in low Earth orbit weigh about six or seven tonnes and have recently been upgraded to allow direct imaging data transmission back to ground sites. A modified version of this payload might serve well.

SOVIET TRANSLUNAR TRANSPORTATION CAPABILITIES

The feasibility of such near-future options depends on the Soviet transportation capability to lunar orbit. The current standard deep space carrier is the "Proton" (SL-12 in Pentagon parlance, "D"-class in the Western civilian catalogs), while possible new vehicles include the SL-X-16 and the SL-W Saturn-V-class "super booster."

The SL-12 "Proton" (also known as the "D-1") has been operational for more than twenty years. It can carry 42,000 pounds into low Earth orbit, and (in the "D-1-e" variant, which includes a fourth stage) it can inject upwards of 12,000 pounds on a translunar trajectory. More than half of this can be placed into low lunar orbit; at least two tons of payload can be landed softly on the lunar surface.

Although some sources have suggested that the Proton may be phased out in the coming years in favor of the slightly smaller (and still untested) "SL-X-16," other data items suggest otherwise. For one, the annual launch rate has in the last five years climbed from the 6-8 typical of the 1970s to an impressive 11-13, suggesting that additional production facilities (reflecting significant capital investment) have recently come on line. Also, the Soviets have made some attempts to offer the Proton commercially on the world geosynchronous traffic market, an unlikely step if the booster were about to be scrapped.

Late in 1984, the Soviets made a space test flight (called "Kosmos-1603"), which analysts at the time interpreted as demonstrating a new high-energy upper stage, probably fueled with liquid hydrogen. The test was repeated in mid 1985. If Soviet engineers were to proceed with development of J-2-class and Centaur-class engines and were to build new upper stages for the Proton's three-million-pound thrust first stage, the booster could attain impressive levels of performance: up to 70,000 pounds in low Earth orbit and almost 30,000 pounds on a translunar trajectory. This appears to be a very logical line of development. Such an improved Proton could be operational by 1990, even though it probably would not be needed for the announced lunar polar orbiter.

Vague allusions to nuclear-powered upper stages also appear in the Soviet press from time to time. A Nerva-class "space tug" is another viable future option for 1990s Soviet space operations and would promise additional performance improvements based on the tried-and-true Proton first stage.

The purported "SL-X-16" booster, reportedly undergoing final preparations for flight tests in 1985-1986, is supposed to have a payload capacity of about 30,000 pounds in low Earth orbit. There are no suggestions that it has any deep-space applications; analysis of the launch pad structures allegedly indicates that it is designed to be a quick reaction military payload carrier. Since even when operational it will not have anywhere near the performance of the already existing Proton booster, its immediate relevance to future Soviet lunar exploration appears minimal. However, it may have profound long-term relevance, since it is supposedly a component of an even larger Soviet space booster, the "SL-W." According to published Pentagon analyses (backed up by independent European sources), this "SL-W" is to have a payload capacity of up to 400,000 pounds in low Earth orbit. This translates to at least 150,000 pounds on a translunar trajectory.

However, this booster may not become operational until the next decade and may be devoted to more near-Earth military-related or space station-related missions. Still, no matter what the original motivation for its development, its availability in the 1990s could be a major temptation for applying it to a greatly accelerated Soviet manned lunar program.

FUTURE SOVIET MANNED LUNAR ACTIVITY

In the 1970s, Soviet space officials talked freely about possible future manned lunar activity. In 1979, Georgiy Narimanov told a reporter: "I think that stations designed for lunar studies will figure prominently in future space exploration. Using such stations put into lunar orbit, it will be feasible to periodically take cosmonauts to the lunar surface aboard small expeditionary ships. Such stations will be assembled in Earth orbit and then sent to the Moon." Boris Petrov, then head of the "Intercosmos Council," wrote in much the same vein: "In the future there will be a need for a lunar orbital station, which could be assembled in near-Earth orbit and then towed into lunar orbit." Yet these comments probably reflect more the desires than the concrete plans of the officials involved.

Any consideration regarding Soviet manned lunar flight must face the issues of cost versus justification. Chief spacecraft designer Konstantin Feoktistov addressed this question in a newspaper article in late 1984, when he wrote: "One frequently asks when will the Soviet Union send its cosmonauts to the Moon? I answer that question with a question: why do in space what has already been done by others when there is an enormous number of other unresolved problems? If we do this, then it should be at a new, significantly higher level. If we talk about the Moon, this means that it makes no sense to send brief expeditions there and with the same radius of action on the lunar surface. Sufficiently practical and significant goals—scientific and national economic—are needed. And these goals are not yet evident in development of the Moon. Even more so with respect to the expenditures that would be required. Thus, no one plans any longer to go to the Moon."

CONCLUSIONS

Against these conflicting opinions, several observations can be made about the range of future possibilities for Soviet lunar exploration.

1. The Proton booster and reasonable upgradings of it will be available for at least the next ten years and will be capable of carrying between six and twelve tons per launch toward the Moon. This payload weight is more than adequate for advanced samplers, advanced landed rovers, and possible even both on the same mission.

2. The renewed commitment to Soviet lunar exploration, as indicated by the polar orbiter(s), speaks well for the probable approval of future probes to make use of the lunar surface surveys to be produced. Again, sample return associated with some surface mobility is the most reasonable and likely development.

3. While no manned lunar capability appears to be a primary target today, other programs are requiring the development of hardware (such as the Salyut manned module or the SL-W giant booster), which in the 1990s could well be readily, rapidly, and cheaply turned to manned lunar missions. A single Proton launch could carry a team of cosmonauts into lunar orbit and back; an SL-W launch could place a well-stocked Salyut module in low lunar orbit (or the Earth-Moon L1 point, a mission described in recent Soviet technical journals). Modified landers could emplace supplies, shelters, and man-carrying rovers on the lunar surface. Reasonably modified landers of the 1970–1976 generation could serve as stripped-down "space jeeps" to transport cosmonauts between lunar orbit and the surface, perhaps even in a bare-bones EVA mode with the space-suited crewmembers sitting in open space for the several hours required for the transfer.

Insofar as such exploration serves scientific tasks, it benefits the whole world and should thus be encouraged. Since no reasonable military utility for manned lunar flight has been proposed, sharing of American Apollo-era experiences and technology cannot adversely affect the international military balance. But sharing such projects can—and, in my opinion, should—be allowed to affect positively the international diplomatic balance.

Past Soviet Lunar Probes

Luna (first series): 1958–1960

1	1958 Jun 22	none	launch failure
2	1958 Sep 24	none	launch faiure
3	1958 Dec 04	none	launch failure
4	1959 Jan 02	Luna-1	passed moon
5	1959 Jun 18	none	launch failure
6	1959 Sep 12	Luna-2	impact
7	1959 Oct 04	Luna-3	photo fly-by
8	1960 Apr 15	none	launch failure
9	1960 2nd qtr	none	launch failure

Luna (second series): 1963–1968

1	1963 Jan 30	none	stuck in low orbit
2	1963 Feb 03	none	launch failure
3	1963 Mar 02	none	launch failure
4	1963 Apr 02	Luna-4	payload failure
5	1963 Jul 03	none	launch failure
6	1964 Mar 21	none	launch failure
7	1964 Apr 20	none	launch failure
8	1964 Jun 04	none	launch failure
9	1965 Mar 12	Kosmos-60	stuck in orbit
10	1965 May 09	Luna-5	destroyed on impact
11	1965 Jun 08	Luna-6	payload failure
12	1965 Oct 04	Luna-7	destroyed on impact
13	1965 Dec 03	Luna-8	destroyed on impact
14	1966 Jan 31	Luna-9	sucessful hard landing
15	1966 Mar 31	Luna-10	lunar orbiter
16	1966 Apr 30	none	launch failure
17	1966 Aug 24	Luna-11	payload failure
18	1966 Sep 25	none	launch failure
19	1966 Oct 22	Luna-12	lunar orbiter
20	1966 Dec 21	Luna-13	successful hard landing
21	1968 Feb 07	none	launch failure
22	1968 Apr 07	Luna-14	pre-Zond nav, comm tests

Lunar Zond: 1967–1970

1	1967 Mar 10	Kosmos-146	stuck in low orbit
2	1967 Apr 08	Kosmos-154	stuck in low orbit
3	1967 Nov 22	none	launch failure
4	1968 Mar 02	Zond-4	phantom moon fly-by
5	1968 Apr 22	none	launch failure
6	1968 Sep 14	Zond-5	flyby & splashdown
7	1968 Nov 10	Zond-6	flyby & USSR landing
8	1968 Dec 06?	none	cancelled manned launch
9	1969 Aug 07	Zond-7	flyby & USSR landing
10	1970 Oct 20	Zond-8	flyby & splashdown

Luna (third series): 1969–1976

1	1969 Jan 20	none	launch failure
2	1969 Apr 22	none	launch failure
3	1969 Jun 05	none	launch failure

4	1969 Jul 13	Luna-15	destroyed on impact
5	1969 Sep 23	Kosmos-300	stuck in low orbit
6	1969 Oct 22	Kosmos-305	stuck in low orbit
7	1970 Feb 06	none	launch failure
8	1970 Aug 08	none	launch failure
9	1970 Sep 12	Luna-16	success-sampler
10	1970 Nov 10	Luna-17	success-lunokhod
11	1971 Sep 02	Luna-18	destroyed on impact
12	1971 Sep 28	Luna-19	success-orbiter
13	1972 Feb 14	Luna-20	success-sampler
14	1973 Jan 08	Luna-21	success-lunokhod
15	1974 May 29	Luna-22	success-orbiter
16	1974 Oct 28	Luna-23	payload failure (sampler)
17	1975 Oct 15	none	launch failure
18	1976 Aug 09	Luna-24	success-drill sampler
19	1978 ?	none	cancelled? (sampler)

SLX-X ("G-class") Blow-ups: 1969–1971

1	1969 Jul 04?
2	1971 Jun 24
3	1972 Nov 22

Lunar Module (LEO tests): 1969–1971

1	1969 Nov 28	none	D-1 launch failure
2	1970 Nov 24	Kosmos-379	A-2
3	1970 Dec 02	Kosmos-382	D-1
4	1971 Feb 26	Kosmos-398	A-2
5	1971 Aug 12	Kosmos-434	A-2

The launch failure dates are reconstructed from declassified U.S. Government documents and from analyses by the British Interplanetary Society; the author considers them reliable. Soviet spokesmen have never admitted any launch failures and in fact explicitly deny them.

LEGAL RESPONSES FOR LUNAR BASES AND SPACE ACTIVITIES IN THE 21ST CENTURY

Amanda Lee Moore

Aerospace Law Consultant, 7 Brookside Circle, Bronxville, NY 10708

The Lunar Bases and Space Activities of the 21st Century Symposium from which this volume resulted made a very persuasive case for a lunar base as the next logical step in the exploration and use of outer space in the next century. The feasibility of this goal—a lunar base and human inhabitation of the planets—was presented in great detail from the science, technology, and policy viewpoints. One of the great unknowns, however, is the legal response to a lunar base initiative. Building from the general principles of the 1967 Outer Space Treaty, a whole range of legal regimes is possible.

The range of legal regimes is a distinct advantage as manned space activities in the 21st Century will be like no others. The problems must be defined and models to solve them chosen. These actions should reveal more clearly the necessary legal underpinnings. Successful completion of the U.S. space station will provide considerable experience at working in the unique environment of space. It will also test the legal relationships between the U.S. and its partners (U.S. nongovernmental users, foreign governments, and foreign commercial participants) in terms of their efficacy in promoting and safeguarding respective rights and interests. Any legal regime touching on lunar activities must meet the realities of that environment, if the feasibility of exploiting lunar resources is to be proved and the benefits made available to the international community.

CONCERNS

In all the rhetoric about lunar activities and legal regimes, two major concerns emerge: (1) How can lunar resources be used on the scale necessary to support lunar and other space activities in the face of the explicit prohibition of national appropriation under international space law? and (2) How can the role and rights of private enterprise be guaranteed?

The national appropriation prohibition is most simply stated in the 1967 Treaty on Principles Governing the Activities of States in the Exploration and Use of Outer Space, including the Moon and Other Celestial Bodies (Outer Space Treaty). Specifically, Article 2 states that "outer space, including the Moon and other celestial bodies, is not subject to national appropriation by claim of sovereignty, by means of use or occupation, or by any other means." This document is in effect the "Magna Carta" for the exploration and use of outer space, the Moon, and celestial bodies. It has been ratified by more States than any other general space law treaty (Lee and Jasentulyana, 1979). Most importantly, its general principles have been accepted by over 75 States, including the major space powers, in particular the U.S., U.S.S.R., France and the other member States of the European Space Agency, Japan, and China.

An identical prohibition is contained in Article 11 of the Agreement Governing the Activities of States on the Moon and Other Celestial Bodies (Moon Agreement), except that "any" has been added before "claim" (U.S. Congress, Senate Committee on Commerce,

Science, and Transportation, 1980). The Moon Agreement was open for signature on December 18, 1979, and entered into force on July 11, 1984. However, its ratifying States to date have not been those nations with extensive space programs. By November, 1985, the Moon Agreement had been ratified by Austria, Chile, the Netherlands, the Philippines, and Uruguay. Domestic opposition has been sufficiently strong to prevent U.S. ratification of the treaty. The U.S.S.R., in turn, usually becomes party to a space treaty only when the U.S. does. Nevertheless, the nonappropriation principle as stated in the Outer Space Treaty is sufficiently normative in character so as to be considered a valid principle of international law in both treaty and custom (Joyner and Schmitt, 1984).

Clearly it is permissible to use lunar materials for scientific research when results of that research are disseminated to the public and international scientific community. This has already been done with lunar rocks brought back to Earth by the U.S. and U.S.S.R. Similarly lunar resources may be used in experimental and pilot projects to prove a technology or the feasibility of a new development. A claim to a portion of the Moon or its resources, however, is not lawful under existing space law. Proponents of private enterprise argue that without such an internationally recognized national claim to lunar resources (in effect a valid title to the resources) commercial development will not be possible (Dula and Lingl, 1984).

The roles and right of private enterprise will always be derived from national law (Finch and Moore, 1985). International space law has already recognized that space activities may be conducted by organizations other than governmental agencies. The Outer Space Treaty, in Article 6, makes States internationally responsible for national activities in space whether conducted by governmental agencies or nongovernmental entities. These nongovernmental entities, in turn, must have authorization and continuing supervision by the appropriate State. The amount of that authorization and supervision is left to the State alone. Even with an international organization conducting activities in space, treaty compliance must be ensured by both the international organization and the member States who are party to the Outer Space Treaty. International space law does not prohibit commercial activities in space. Instead it recognizes that such activities will take place, but always with a State responsible for those actions. Therefore, whatever role private enterprise will have on the Moon, it must look to national governments to promote, support, and protect it.

LUNAR ENTERPRISES

A lunar base in the 21st century is a goal with a timeframe stretching over 50 years from now. A lot will happen in that time, principally in science and technology but also in economics and international affairs. A series of breakthroughs, or even just one, could change the whole equation for achieving the goal. The best way is not yet written in stone. Therefore, it will always be necessary to study and keep options open.

National activities in space will continue to dominate the field for the next dozen years to come. Similarly, the present rationale for a lunar base is strictly scientific and as a logistic base for further planetary exploration, e.g., to undertake a mission to Mars, activities accomplished almost exclusively by government support. Commercial development is not expected until the last half of the 21st century.

As all the possible scientific uses of the Moon can only be projected, so the full range of commercial opportunities can only be imagined (Finch and Moore, 1985). Most often it is proposed that lunar and asteroid resources be processed for the raw materials to build large structures in space. These structures in turn would (1) support the collection of solar energy for relay to Earth for conversion into electricity, (2) provide habitats for the necessary personnel who may be in space for a limited or permanent duration, and (3) include facilities to manufacture products better made in space or simply unavailable on Earth. On the Moon itself, a profitable enterprise might be to simply transport and house tourists who wish to journey to the Moon. The most profitable uses of the Moon are most likely ones not yet envisioned.

MODELS

The implications of such enterprises are that each specific activity will have its own legal requirements to succeed. With a specific activity in mind, a model method of operation could be chosen and the most appropriate legal rules determined. Over the years, numerous uses have been made of space and its resources without national appropriation. To date these resources have been mainly intangible, *e.g.*, the radio frequency spectrum and slots on the geostationary orbit for communication satellites. Various models of operations have evolved and most contain international elements. As such, the legal arrangements are not "national" appropriation, but the international ordering of resources for use on a nondiscriminatory basis.

Following are just a few models for a lunar base operation:

1. A U.S. enterprise with participation and funding from other participants, as planned for the U.S. space station
2. Independent activities and expeditions coordinated through an established international organization
3. Bilateral endeavors
4. User-based international organization along the line of Intelsat and Inmarsat
5. A truly transnational enterprise, such as the one for building a pipeline between Europe and the U.S.S.R.
6. A consortium of private business concerns to conduct the particular space activity, as U.S. companies built and operate the Alaskan pipeline
7. An international authority that franchises States or private enterprises to exploit resources under specified conditions, such as the International Seabed Authority for exploitation of minerals under the high seas.

Certainly with any space enterprise involving the U.S., private enterprise is already a partner as the contracted supplier of goods and services. Already private enterprise competes for the contracts to supply the ground and space segments of Intelsat. In this particular model, the participation of private enterprise in the international organization itself is left to the State, which may designate any entity it wishes as its representative. Access to the service of the organization provided through the designated entity is similarly left up to the State (Finch and Moore, 1985).

MECHANISMS

Based on the desired model, the necessity of and mechanisms for legal rules may be determined. Where coordination is through an existing organization, procedures in place or a specifically convened meeting of the members should suffice. Bilateral efforts or national endeavors with multilateral participation can be conducted through normal channels of international discourse. A consortium of private businesses or a specially formed transnational enterprise would probably reach agreement among themselves, and then publicly seek any needed permissions, financing, or goods and services. The influence of the funding sources should not be underestimated at either the national or private level.

A user-based international organization like Intelsat would entail specifically convened negotiations among the interested States to establish the desired entity. Arrangements would most likely include decisions by weighted voting based upon investment in and use of the enterprise. As an example, Joyner and Schmitt suggest an organization called "INTERLUNE" for the provision of "lunar base facilities, services and access of high functional potential, quality, safety, and reliability to be available on an open and nondiscriminatory basis to all peaceful users and investors (Joyner and Schmitt, 1984). In their view, the organizational concept would be tailored to provide cooperative management of a lunar base to the benefit of its members, users, and investors. INTERLUNE would provide such management through "sharing international opportunities, rather than through unilateral control by any one nation or set of competing nations."

In Article 11, the Moon Agreement envisions negotiations among its States Parties under United Nations auspices to establish an international regime, including appropriate procedures, to govern the exploitation of the natural resources of the Moon "as such exploitation is about to become feasible." Such a conference would give form and substance to the concept of the common heritage of mankind, rather than philosophical speculation (U.S. Congress, Senate Committee on Commerce, Science, and Transportation, 1980). A review of the Agreement itself can take place in 1994 (ten years after entry into force), or as soon as 1989 if one-third of the States Parties so request and a majority concur, as provided by Article 18.

IN CONCLUSION

Conceivably, a lunar facility could be established, lunar resources extracted for their component materials on a profit-making basis, and other activities commenced using the Moon as a base, with no further legal notice on an international scale. Informal agreements among the interested States could remove any points of contention and the individual countries would see to the interests of their individual entities, be they governmental or private. This is an optimal scenario, but not a likely one. Too many questions have been raised in both the public and private sectors for any State to proceed with a lunar base initiative without taking into account the interests of other potential users and the international community.

It must be stressed again that private or commercial entities must look to their national governments and their national laws for the most effective promotion of their interests. If what is wanted is a sure set of rules, then private enterprise must make their views known to the respective governments at every stage of a space initiative from planning to operation. The most effective input is that given during the planning stage. However, flexibility must be included in any final arrangements as circumstances change in the actual operation of a space station or lunar base. In the U.S., for example, such input may be made by membership in advisory groups and delegations, testimony before Congress, the conduct of seminars and meetings of interested government and nongovernment personnel, and even the release of private studies.

All these actions facilitate the communication of private industry's needs for effective participation in space activities, be they near-Earth, in geostationary orbit, on the Moon or beyond. International space law requires little beyond peaceful uses in accordance with international law on a nondiscriminatory basis, with government authorization and supervision and the dissemination of results. There is no rule of law precluding profit or private enterprise from space.

REFERENCES

Dula A. and Lingl H. (1984) A debate concerning the merits of competing international legal frameworks governing the utilization of lunar resources (abstract). In *Papers Presented to the Symposium on Lunar Bases and Space Activities of the 21st Century*, p. 43. NASA/Johnson Space Center, Houston.

Finch E. R. and Moore A. L. (1985) *Astrobusiness: A Guide to the Commerce and Law of Outer Space*, pp. 55–70. Praeger, New York.

Joyner C. C. and Schmitt H. H. (1984) Lunar bases and extraterrestrial Law: General legal principles and a particular regime proposal (abstract). In *Papers Presented to the Symposium on Lunar Bases and Space Activities of the 21st Century*, p. 40. NASA/Johnson Space Center, Houston.

Lee R. S. and Jasentulyana N. (editors) (1979) *Manual of Space Law*, Vols. I–IV. Oceana, Dobbs Ferry, NY.

U.S. Congress, Senate Committee on Commerce, Science, and Transportation (1980) *Agreement Governing the Activities of States on the Moon and Other Celestial Bodies*, Committee Print, 96th Congress, 2nd session.

EXTRATERRESTRIAL LAW AND LUNAR BASES: GENERAL LEGAL PRINCIPLES AND A PARTICULAR REGIME PROPOSAL (INTERLUNE)

Christopher C. Joyner

Department of Political Science, George Washington University, Washington, DC 20052

Harrison H. Schmitt

P.O. Box 8261, Albuquerque, NM 87198

Contemporary international law recognizes the validity and relevance of extraterrestrial legal principles to outer space activities, including the conceivable establishment of permanently manned bases on the Moon. By applying the legal principles of free access to outer space, non-appropriation of celestial bodies, use of space for peaceful purposes only, and development of outer space for mankind's benefit, this paper suggests a possible regime—INTERLUNE—for governing international activities on the Moon. Modeled on the INTELSAT system, INTERLUNE would be comprised of an Assembly of Parties, a Board of Governors, a Board of Users and Investors, and a Director General. INTERLUNE represents a model organization tailored to facilitate cooperative international management of a lunar base to benefit its member states, users, and investors. INTERLUNE could provide such management through a sharing of both sovereignty and opportunity. Importantly, such a regime would enhance international cooperation by precluding the possible usurpation of control over the Moon and its natural resources by any one state or group of competing states.

INTRODUCTION

Serious contemplation of returning to the Moon necessarily invites critical consideration of assessing and evaluating the international legal implications of performing that mission. Relatedly, there is also the fundamental priority of considering what type of regime should be formulated to carry out activities on the Moon peacefully for the benefit of all peoples.

This paper therefore aims to achieve two broad purposes. First, it identifies and sets out in clearer relief the relevant extraterrestrial legal principles that currently are deemed applicable to establishing and operating lunar basing facilities. Second, it seeks to propose a prospective regime, "INTERLUNE," for overseeing international activities on the Moon peacefully, efficiently, and in accord with the recognized principles of extraterrestrial law. While obviously not intended to be either a panacea or a definitive legal schema for managing man's affairs on the Moon, the "INTERLUNE" proposal hopefully will stimulate serious discussion and reasoned debate about the nature of a lunar regime, as well as the appropriate political, economic, and legal prerequisites for its promulgation.

PRINCIPLES OF EXTRATERRESTRIAL LAW

During the past 25 years, extraterrestrial law, or space law, has become a recognized and indeed vital branch of international law (Christol, 1982; Lachs, 1972; Kolosov, 1974; Williams, 1981). Though admittedly specialized and directly pertinent only to the conduct of a relative paucity of states, at least four major multilateral treaties have been negotiated and are now legally in force: (1) the Treaty on Principles Governing the Activities of States in the Exploration and Use of Outer Space, including the Moon and Other Celestial Bodies; (2) the Agreement on the Rescue and Return of Astronauts, and the Return of Objects Launched into Outer Space; (3) the Convention on International Liability for Damage Caused by Space Objects; and (4) the Convention on Registration of Objects Launched into Outer Space. All these agreements today are active facets of international law, to which both the United States and the Soviet Union, as well as other major space-resource states, have legally obligated themselves as parties. Consequently, these four relevant treaties compose the contemporary international legal framework for space law and, coincidentally, for implementing a lunar base program.

Of these agreements, the 1967 Outer Space Treaty furnishes the most far-reaching contributions for fashioning extraterrestrial law. Today 86 states are parties to this agreement. Moreover, at least five fundamental principles of space law can be distilled and crystallized from the Outer Space Treaty's operative provisions. Significantly, these cardinal principles have been accepted in the practice of states as constituting legally salient norms for regulating the conduct of space-related activities. Put succinctly, these principles assert that: (1) space, including the Moon and other celestial bodies, is the province of mankind and should be developed for its benefit (Article I); (2) space, including the Moon and other celestial bodies, should be free for exploration and use by all states. Equality in and free access to all areas shall be available to all states, and freedom of scientific investigation shall be ensured to any interested party (Article I); (3) space, including the Moon and other celestial bodies, is "not subject to national appropriation by claim of sovereignty, by means of use or occupation, or by any other means" (Article II); (4) space, including the Moon and other celestial bodies, shall be used exclusively for peaceful purposes. Accordingly, "[t]he establishment of military bases, installations and fortifications, the testing of any type of weapons and the conduct of military maneuvers on celestial bodies shall be forbidden." Of especial import, "States parties to the [Outer Space] treaty undertake not to place in orbit around the Earth any objects carrying nuclear weapons or any other kinds of weapons of mass destruction, install such weapons on celestial bodies, or station such weapons in outer space in any other manner" (Article IV); and, (5) international law as formulated here on Earth does extend extraterrestrially to space and celestial bodies. Hence, general principles of law in the Outer Space Treaty embody expectations of the parties, and also articulate the paramount recognition that all mankind possesses a common interest in the progressive exploration and use of outer space for peaceful purposes (Article III).

These five fundamental principles undergird the corpus of extraterrestrial law and have developed only since 1957. Of special note, they have been derived from the traditional sources of international law, namely, international treaties and conventions, customary state practice, so-called "general principles of law," and the writings and opinions of legal scholars and jurists. These principles were initially expressed in "codified legal form" in the 1967 Outer Space Treaty but have been subsequently integrated as cardinal provisions into the later multilateral space-related conventions. In essence they form the foundation upon which extraterrestrial law rests, and thus concomitantly become the critical legal considerations that must be accounted for in the actual establishment of future lunar basing facilities (Adams, 1968; Kopal, 1973; Lachs, 1972; Christol, 1982; Menter, 1979; Kulebyakin, 1971).

Inasmuch as these fundamental principles of extraterrestrial law are all expressed in the Outer Space Treaty, this agreement may be properly regarded as a covenant of outer space law. Respective to the Moon, however, one principle stands out as being more significant legally and politically than the others, namely, the non-appropriation provision in Article II. At present, this principle asserts that no portion of the Moon or other celestial bodies is susceptible to any state's sovereign claim, national title, or territorial jurisdiction. In short, there can be no private or state ownership of the Moon (or for that matter, of any other celestial body). This point requires some clarification. Utilization of the Moon's resources—be it removal of Moon rocks and taking other samples or the implementation of techniques to extract oxygen from the lunar soil—is not in question; such scientific and life-support activities would be permissible under the legal provisions of the Outer Space Treaty. It is the national appropriation of the Moon as sovereign territory belonging to some polity on Earth that is unacceptable. As the direct consequence, the legal status of the Moon was transformed from a condition of *terra nullius* (*i.e.*, vacant land that belonged to no one and therefore was available to claim by anyone) to that of *res extra commercium*, the status of legally not being susceptible to any possibility of national appropriation. For the present, therefore, effectuation of national sovereignty on the Moon perforce has been eliminated from legitimate consideration (Adams, 1968; Bhatt, 1968; Christol, 1982; Williams, 1981; DeKanozov, 1975).

THE MOON AND "THE COMMON HERITAGE OF MANKIND"

The most recent international effort aimed at defining more precisely the Moon's legal status culminated in the negotiation in 1979 of the Agreement Governing the Activities of States on the Moon and Other Celestial Bodies, more commonly known as the "Moon Treaty." In general, the Moon Treaty, which only recently came into force, merely reaffirms and only slightly extends the existent corpus of extraterrestrial law. However, the glaring exception to these progressive developments is found in Article XI of the Moon Treaty, which: (1) asserts that the Moon and other celestial bodies are the "common heritage of mankind"; (2) reiterates that the Moon is not subject to national appropriation; (3) stipulates anew that the surface and subsurface, inclusive of *in situ* resources, may not become the property of any state, international organization, non-governmental entity, or natural person; (4) posits that parties would enjoy non-discriminatory access to the Moon; and (5) asserts that at the time when exploitation of the Moon becomes feasible, states party to the the treaty will establish an international regime to "govern exploitation of the natural resources." In this respect, the regime purportedly would operate to ensure orderly development and rational management of lunar resources. Less clear and more polemical, however, is the regime's attendant purpose of facilitating an equitable sharing of benefits derived from those resources for parties, whereby "special consideration" would be given to "the interests and needs of the developing countries as well as the efforts of those countries which have contributed either directly or indirectly to the exploration of the Moon" (Christol, 1980; Cocca, 1974; Dekanozov, 1978; Smith, 1980; Larschan and Brennan, 1983).

It is primarily because of this "common heritage of mankind" (CHM) provision that the Moon Treaty only recently entered into force in late 1984 by securing ratification by the requisite number of five states—Austria, Chile, Netherlands, the Philippines and Uruguay—none of which is a significant space-faring power. Furthermore, it is the CHM notion that has fostered confusion about the precise status of the Moon under contemporary international law.

Respective to the Moon, the most that can be posited about the status of CHM as a legal concept is that it *may* inculcate an emergent principle of international law. At present, CHM is a notion containing more latent than apparent actual value, with more nascent and potential legal ramifications than manifest implications or binding obligatory force. As a consequence, CHM remains less than a bona fide principle of international law, posing only inferential relevance to the Moon's legal status (Gorove, 1972; Kosolov, 1979).

The upshot of these observations suggests that certain general principles of extraterrestrial law have been established, are recognized in the practice of states, and are currently applicable to the Moon (Galloway, 1981; Matte, 1978). Nonetheless, space law remains in a process of continuous evolution, with much still left open to diverse national interpretation. Notwithstanding this caveat, broad international consensus maintains that outer space, the Moon, and other celestial bodies should be open for exploration and use for the benefit of mankind, that principles of international law are applicable to these opportunities, and that the Moon and other celestial bodies should be used exclusively for peaceful purposes. To these ends, the Moon and other celestial bodies are today regarded as being legally immune from sovereign claim or territorial acquisition (Dolman, 1981).

Yet, extraterrestrial law is not static. It inculcates an evolutionary process that is capable of growing to meet new challenges and adapting to formulate new norms. Very likely, the international law pertaining to outer space will continue to evolve as new technologies are developed and further exploration is undertaken (Dula, 1979; Menter, 1983; Smith *et al.*, 1983; Vassilovskaya, 1974). Certainly in this regard, establishment of a lunar basing facility will engender widespread international concern about the type of regime necessary for managing various activities on the Moon. The next section of this paper turns to address this critical legal concern. It formulates a regime of international law that is consistent with both accepted principles of space law and the philosophy of CHM but avoids the practical difficulties of the more extreme proposals for the implementation of CHM in law.

THE INTERLUNE CONCEPT: USER-CONTROLLED INTERNATIONAL MANAGEMENT SYSTEM FOR A LUNAR BASE

The existing regimes for space discussed above clearly place significant legal constraints on governments interested in the establishment of a permanent lunar base. The obvious practical difficulties that the world is experiencing with one-nation, one-vote international organizations also suggest that notable pragmatic constraints exist for states desiring to participate in founding such a base or settlement. Fortunately, international experience has provided a successful model of a high-technology management system that meets the legal, operational, and self-interest constraints attendant to international operations in space. That model is INTELSAT, a user-based management system that works to coordinate operation of international communications satellites (Smith, 1980).

In this paper, a version of the INTELSAT model is suggested as being especially appropriate for the international management of a lunar base. It is submitted here that "INTERLUNE," as the organization is termed, will aptly satisfy the aforementioned legal constraints, as well as hold consistent with the principles of free enterprise that are shared by the world's democracies. More importantly, INTERLUNE would bring into the management of a lunar base those states and other interests that evince the greatest motivations for ensuring successful implementation of that managerial system. INTERLUNE does not require that sovereignty be given up in space; it does not require that free-enterprise opportunities be abandoned in space; it merely requires that sovereignty and opportunity be shared.

Technological advancements have produced a trend toward recognizing a "common heritage of mankind" in certain international resources. This trend is most apparent in negotiations regarding the resources of the sea and outer space. It indicates a general realization that states possess common interests in sharing benefits from the exploitation and environmentally sound use of those resources.

The Moon can become a common heritage resource for mankind. However, without a feasible administrative system and a peaceful management environment, lunar opportunities will remain unavailable and moribund. An institutional arrangement should be possible that would vest operation and control of lunar bases in an organization comprised of states that actively participate in creating such bases, in association with users of the bases or investors in their operations. Such states and related entities would be united by a common bond of policy and purposes focused on the technical and financial success of the enterprise.

INTERLUNE'S MANAGEMENT STRUCTURE

The conceptual advantages of a regional organization such as INTERLUNE could be realized only if the actual institutional structure were designed to provide an equitable system for various interests to exert influence and control, as well as to furnish efficient and proper management of the base.

The main functioning bodies INTERLUNE is comprised of are the Assembly of Parties, the Board of Governors, the Board of Users and Investors, and the Director's Office. The

Assembly of Parties would exert policy authority over the Board of Governors, which would in turn exert functional authority over the Director General. The Board of Users and Investors, operating within the policy framework set down by the Assembly of Parties, would develop recommendations regarding operational issues affecting the Board's interests. These recommendations would then be presented to the Board of Governors through the Board of Users and Investors' formal representatives on that Board.

The basic national membership of INTERLUNE would be constituted in the Assembly of Parties. Each Assembly member's interest in INTERLUNE would be proportional to its investment of equivalent capital in the creation and initial operation of the lunar base. A member's proportional interest at the beginning of any operating year would set the number of votes to which that member would be entitled during deliberations of the Assembly of Parties.

The principal function of the Assembly of Parties would be to establish policy within the legal parameters of an INTERLUNE charter previously negotiated and agreed to by its member states. Under its charter, INTERLUNE might express as its prime goal "the provision of lunar base facilities, services and access of high functional potential, quality, safety, and reliability to be available on an open and non-discriminatory basis to all peaceful users and investors." As ancillary goals, INTERLUNE would be expected: (1) to seek a return on investment in its assets and operation, not exceeding a stipulated annual percentage, while remaining consistent with meeting its primary goal; and (2) to ensure the neutrality and security of activities under its jurisdiction. In addition, the Assembly of Parties would supply a mechanism and forum for the peaceful settlement of disputes relating to provisions in the INTERLUNE Charter or any resultant policy derived therefrom.

Several specific mandates for the Assembly of Parties would be incorporated in the INTERLUNE Charter, including the following: (1) to provide general policy guidelines and specified long-term objectives to meet the primary goal of INTERLUNE; (2) to establish general rules concerning rates of charge for use of INTERLUNE's facilities and services on a non-discriminatory basis; (3) to consider and adjudicate complaints submitted to it by states, competent international organizations, users and investors; (4) to maintain a body of laws, rules, procedures, and instructions for dealing with normal operations and dispute settlement; and (5) to establish general guidelines for the financial participation of potential investors in INTERLUNE.

The Assembly of Parties would be comprised of one representative from each member state. Decisions on all matters would be taken by three-quarters majority vote.

The Board of Governors would have the responsibility for the operation and maintenance of INTERLUNE's facilities and services, as well as for the design, development, construction, improvement, upkeep, and general operation of INTERLUNE. As conceivably defined, some specific duties of the Board of Governors would include: (1) to adopt policies, plans, and programs aimed at enhancing and sustaining the environmentally balanced operation of INTERLUNE; (2) to create and implement annual budgets; (3) to establish periodically rates of charge for utilizing INTERLUNE's facilities and services in accordance with the general rules set by the Assemby of Parties; (4) to solicit capital; (5) to appoint the Director General and to approve senior staff appointments; and (6) to arrange contracts with a state, organization, or institution relating to the performance, functioning, and operation of INTERLUNE's facilities and services.

The Board of Governors would be comprised of up to 15 members. One governor each would be drawn from those states, or groups of states, who have made major space investments in support of INTERLUNE (e.g., the United States, the Soviet Union, and the European Space Agency); two governors would be selected to represent the Board of Users and Investors; and the remainder would represent those states, or voluntary pairs of participating states, that would qualify according to a formula based on actual commitment of resources in INTERLUNE's behalf.

Voting participation by the Board of Governors would be defined by the Assembly of Parties and the relationship set for Assembly deliberation. The governors should endeavor to make all decisions by consensus. However, if consensus is not possible, governors would each participate in the deliberation process commensurate with the voting proportion of their respective states. A three-quarters majority of the total voting participation would be necessary for substantive decisions, while a simple majority would be necessary for procedural decisions. The governors representing the Board of Users and Investors would have voting particpation proportional to the cost paid to, or capital invested in INTERLUNE.

INTERLUNE's Board of Users and Investors would have the responsibility to advise INTERLUNE on all matters of policy and operations that affect the use and financial viability of INTERLUNE's facilities, services, management efficiency, and future explansion. Initially, all committed users and investors would be invited by the Assembly of Parties to a Charter Conference to establish the organizational structure of the Board of Users and Investors. Upon the Charter's acceptance by three-quarters majority of the users and investors and ratified by the Assembly of Parties, the Board of Users and Investors would receive staff and financial support from INTERLUNE, and as aforementioned, would be granted two representatives on the Board of Governors.

The executive body, or staff component, of INTERLUNE would be headed by the Director General. Among his specific duties would be (1) to serve as the legal representative of INTERLUNE and be responsible for all administrative and personnel functions; (2) to contract out to various competent entities technical and maintenance functions associated with INTERLUNE's operation, with due regard to cost and consistency *vis-a-vis* competence, effectiveness, and efficiency (as provided for in the Charter agreement, such entities would be comprised of various nationalities, or could be an international corporation owned and controlled by INTERLUNE); and, (3) to serve as the principal negotiator on behalf of INTERLUNE.

Ultimately INTERLUNE would require the establishment of a dispute settlement system. A first level of this system might be arbitration under the auspices of the Assembly of Parties. A second level could be a judicial tribunal, created by the Assembly of Parties, which would serve as a final court of appeals for unresolved disputes, as well as for criminal or civil violations under INTERLUNE's jurisdiction. Importantly in this regard, adoption of a code of criminal and civil law for INTERLUNE would of necessity be agreed to in an addendum to its basic charter, being subject to modification of amendment only by the Assembly of Parties voting through their secondary vote procedure.

THE IMPLEMENTATION OF INTERLUNE

The legal initiation and viable implementation of any international idea or organization never comes easily or simply. INTERLUNE will be no exception. Nonetheless, activating INTERLUNE clearly will remain possible, so long as the major space-faring powers—particularly the United States—sustain unequivocal commitment to the establishment of a lunar base, with the attendant political will to search out and secure international participation in such an endeavor. To do otherwise would seem regrettable in that a great opportunity for increased legal cooperation and international trust among traditionally competing states would be lost. Thus, viable implementation of the INTERLUNE regime in all likelihood would hinge upon substantial participation by the United States.

Presuming commitment to a lunar base by the United States, the next logical step toward implementation ostensibly would be the convening of an international conference to consider and negotiate a draft INTERLUNE Charter. All states would be invited to send representative delegates or observers, and potential users or investor entities should be invited as observers, as well as encouraged to participate as members of official delegations.

A critical consideration lies in the respective roles that the Soviet Union and the developing states should play in the creation and implementation of the INTERLUNE regime. To facilitate their participation in INTERLUNE, the opportunity should be made so attractive that they cannot refuse. Such an offer is inherent in, first, an unequivocal commitment by the United States, Western Europe, and Japan in the project; second, a manifest willingness by all parties to share sovereignty, opportunity, and technology; and third, a clear articulation of the direct economic, legal, and political benefits accruing to all participant states. Once an established and successful reality, INTERLUNE would surely attract additional states that at first may have been reluctant to participate. Though purportedly conceived as an international self-regulating monopoly, INTERLUNE should always remain open to new members and investors. In this manner, the regime's humanistic goals, as well as its specific economic and technical purposes, could be most fully achieved in the interest of all mankind.

CONCLUSION

Existent extraterrestrial law, as well as the fundamental interests of space-faring states, are consistent with the inception of a user-based international organization for managing a lunar regime. Through an Assembly of Parties, a Board of Governors, a Board of Users and Investors, and a Director General, INTERLUNE would meet its primary goal of providing open access to and available facilities and services for a lunar base founded on principles of non-discrimination and peaceful purposes only. The internal structure and guiding philosophy of INTERLUNE allows for all participants to share representation in decisions affecting its activities; additionally, moreover, INTERLUNE would assure effective and responsive management of the lunar facility and the Moon's natural environment.

The INTERLUNE proposal is a model organization concept tailored to provide cooperative management of a lunar base in order to benefit its members, users, investors, and, indeed, all mankind. Significantly, INTERLUNE would provide regime management through sharing both sovereignty and opportunity, rather than through unilateral control by any single state or set of competing states. Surely this is the extraterrestrial legal precedent that we wish to establish for mankind at the now not-so-distant shores of the new ocean of space.

REFERENCES

Adams T. R. (1968) The outer space treaty: An interpretation in light of the no-sovereignty provision. *Harvard J. Int. L., 9*, 140–157.

Bhatt S. (1968) Legal controls of explorations and use of the moon and celestial bodies. *Indian J. Int. L., 8*, 33–48.

Christol C. Q. (1980) The common heritage of mankind provision in the 1979 agreement governing the activities of states on the moon and other celestial bodies. *Int. Lawyer, 14*, 429–465.

Christol C. Q. (1982) *The Modern International Law of Outer Space.* Pergamon, New York. 945 pp.

Cocca A. A. (1974) The principle of the "common heritage of all mankind" as applied to natural resources from outer space and celestial bodies. *Proc. 16th Colloq. L. Outer Space*, pp. 174–176.

Dekanozov R. V. (1975) Juridical nature of outer space including the Moon and other planets. *Proc. 17th Colloq. L. Outer Space*, pp. 200–207.

Dekanozov R. V. (1978) Draft treaty relating to the moon and the legal status of its natural resources. *Proc. 20th Colloq. L. Outer Space*, pp. 198–203.

Dula A. (1979) Free enterprise and the proposed moon treaty. *Houston J. Int. L., 3*, 3–35.

Dolman A. (1981) *Resources, Regimes, World Order.* Pergamon, New York. 425 pp.

Galloway E. (1981) Issues in implementing the agreement governing the activities of states on the moon and other celestial bodies. *Proc. 23rd Colloq. L. Outer Space*, pp. 19–24.

Gorove S. (1972) The concept of "common heritage of mankind": A political, moral or legal innovation. *San Diego L. Rev., 9*, 390–404.

Kolosov Y. M. (1979) Legal and political aspects of space exploration. *International Affairs (Moscow)*, March 1979, pp. 86–92.

Kolosov Y. M. (1974) Interrelation between rules and principles of international space law and general rules and principles of international law. *Proc. 16th Colloq. L. Outer Space*, pp. 45–48.

Kopal V. (1973) The development of legal arrangements for the peaceful uses of the moon. *Proc. 15th Colloq. L. Outer Space*, pp. 149–164.

Kulebyakin V. (1971) The moon and international law. *International Affairs (Moscow), 9*, 54–57.

Lachs M. (1972) *The Law of Outer Space.* Sijthoff and Noordhoff, Rockville. 212 pp.

Larschan B. and Brennan B. (1983) The common heritage of mankind principle in international law. *Columbia J. Int. L., 21*, 305–331.

Matte N. M. (1978) The draft treaty on the Moon, eight years later. *Ann. Air Space L., 3*, 511–544.

Menter M. (1979) Commercial space activities under the Moon Treaty. *Syracuse J. Int. L. and Comm., 7*, 213–238.

Menter M. (1983) Peaceful uses of outer space and national security. *Int. Lawyer, 17*, 581–595.

Smith D. D., Paptkieviez S., and Rothblatt M. (1983) Legal implications of a permanent manned presence in space. *W. Va. L. Rev., 85*, 857–872.

Smith D. D. (1980) The Moon Treaty and private enterprise. *Astronaut. and Aeronaut., 18*, 62–63.

Vassilovskaya E. G. (1974) Legal problems of the exploration of the moon and other planets. *Proc. 16th Colloq. L. Outer Space*, pp. 168–171.

Williams S. (1981) International law before and after the Moon agreement. *Int. Relations, 7*, 1168–1193.

LUNAR BASE: LEARNING TO LIVE IN SPACE

Ben Finney

Department of Anthropology, University of Hawaii, Honolulu, HI 96822

Going back to the Moon to establish a permanent base presents a social as well as a technical challenge. The experience of small, isolated groups in highly stressed environments points to the need for appropriate systems of social organization to enable groups to work safely and productively in space. Based on a survey of the literature and on experience working with small groups as an anthropologist, suggestions are put forward as to the need to: (1) make social research and planning part of the lunar base program; (2) make learning how to live in space an iterative effort, starting now with the space station and carrying on beyond the lunar base; (3) simulate space communities in realistic mock-ups on Earth before testing them out in space; and (4) make self-design of space communities by those who will actually live in space a basic element of planning.

We need to pay attention to the nature of the social systems to be created with the first lunar bases. The composition, organization, and governance of those first lunar communities will be vital to their success and, ultimately, to the goal of learning to live in space. We should start now on a research and development program directed toward developing social systems designed so that people can safely and productively live and work on the Moon.

To highlight this need, let me cite an example drawn from the history of Antarctic exploration. In the Antarctic summer of 1911 two teams raced for the then unconquered South Pole: the British under the command of Robert Falcon Scott of the Royal Navy, and the Norwegians, led by the veteran explorer Roald Amundsen. Both made it to the Pole, but only Amundsen and his men, who got there first, made it back alive. Scott and his men perished on the terrifying march back (Huntford, 1984).

Amundsen, an accomplished Arctic explorer who had been the first to force the Northwest Passage, did everything right. Combining elements of western technology with Eskimo-style clothing and dog sleds, and with Nordic skiing techniques, his race to the Pole and back went like clockwork. His men were well chosen and trained, and their individual roles and the organization of the team were meticulously planned and rehearsed. In contrast, Scott, although he had already led one Antarctic expedition, paid little heed to the merits of Eskimo and Nordic Arctic technology, experimented fruitlessly with ponies and tractors, and finally settled on the killing strategy of "man-hauling" sledges across the glaciers. What is more, many of Scott's men were ill-chosen and inadequately trained, and their duties as well as the structure of the team were not made clear. According to a recent biographer (Huntford, 1984), Scott followed a tenet then popular in the Royal Navy: "an officer does not worry about details but stands ready to improvise."

This race to the South Pole illustrates how crucial appropriate planning and preparation in both hardware and human relations are to the success of hazardous exploratory expeditions. NASA, of course, passed that test in the Apollo program. The question now is, how will it do for the return to the Moon?

No one seriously advocates going back to the Moon the way it was done 15 years ago. We now want to develop permanent bases there. This new conception requires technological developments beyond those used in Apollo—reusable space transport, closed or nearly closed ecological life support systems, and techniques for mining and processing lunar ores, to name several of the most obvious. Just as hardware appropriate to staying on the Moon has to be developed, so, too, do we need to develop an appropriate sociology for living and working on the Moon.

To be sure, the astronauts and cosmonauts have done a superb job in pioneering manned spaceflight. But, as representatives from both groups admit (Bluth, 1981; Carr, 1981), the problems of living and working together in space for prolonged periods are far from solved. For example, Gerald Carr, commander of the 84–day Skylab mission, has gone on record that he expects that "the sociological problems will prove to be more difficult to solve than the technological ones." Gone are the days when space operations involved just a few male astronauts, alike in background and training, going out for short-duration spaceflights. On the Moon, life will get much more complicated.

First, a lunar base, or at least a mature one, will involve far more people than have hitherto been together in space for long periods. Second, lunar base inhabitants are likely to be heterogeneous. For example, a community might include scientists, technicians, and medical personnel as well as more traditional astronauts; it might have a number of women as well as men; people from several nations are likely to be represented; and, private sector employees may work alongside civil servants and military personnel. Third, all these people will have to stay together for many months, perhaps a year or longer, in confined quarters located in an unearthly environment separated from Earthside family, friends and familiar sights, sounds and smells. Judging from partly analogous situations in the Antarctic and elsewhere, all this will add up to significant stresses and strains on every individual's psyche and on the social fabric of the group (Bluth, 1984; Connors et al., 1985; Helmreich et al., 1980). A social organization designed to minimize such stresses and strains, or at least one adapted to dealing with them, would help ensure that a lunar base would be a successful and productive community.

During the 1960s NASA did commission a considerable number of studies of naval and Antarctic analogues to prolonged space missions, and sponsored still other studies of how men adapted to living in cramped and hazardous undersea habitats. These were directed toward identifying the social-psychological problems that might be encountered on prolonged space missions and suggesting strategies for dealing with these. Yet, by the late 1970s enthusiasm for such studies fell to a low point, and relatively few new ones were being commissioned (Mutschlechner, 1979). Furthermore, Robert Helmreich (1983, p. 447), a psychologist who participated in many of these studies, has stated that "there is no available evidence. . .that these research programs have had any influence on the conduct of past operations or the organization and planning of future missions."

Social scientists have been tempted to explain this state of affairs by saying that engineers do not understand them, or that astronauts feel threatened by social scientists because of their experience with psychologists and the latter's role in selection for and de-selection from the program. However, those on the technical and operational side can just as easily accuse the social scientists of being trivial in their research and incomprehensible in their publications. While not denying that there may be real problems here of communication among sub-cultures, the time may now be ripe for a renewal of social science research on space living. During the 1970s NASA was forced to shelve its expansive plans and restrict its horizons. Social scientists wanting to talk about long duration missions had no audience. Now, however, the agency is being encouraged to develop a space station and to think ahead about returning to the Moon, as well as other visionary projects. My plea is that research and development on appropriate social systems be part of that forward-looking effort.

However, if given the green light to go ahead, don't expect the social science community to immediately bring forth a unified and empirically validated program. Just as the prospect of mining and processing lunar materials has resulted in a wide variety of proposals, so, too, would the prospect of social science research on lunar living elicit a wealth of ideas, models, and methodologies. My own viewpoint on the subject is derived from experience as an anthropologist working with small groups, including those involved in maritime exploration. Let me briefly outline four points that would be important to a social science research and development program.

First, don't separate social science from everything else. As Miller (1984) points out, we are dealing with living systems that are at once biological and social. And, of course, they are technological as well for they will not exist on the Moon without all the hardware and procedures for getting people there, housing them, and keeping them alive. Social scientists must work closely with biologists, human factors specialists, architects, and ultimately, the engineers and managers who conceive, design, and operate the whole system.

Second, make the planning of an appropriate lunar social system part of a larger, iterative program for learning how to live in space, whether in orbit, on the surface of the Moon, or on some other body. This program should build upon previous experiences— in space and in analogous situations on Earth. It should focus intensively now on the space station, then apply the lessons from the space station to the lunar base, then learn from the first lunar communities, and so on.

Third, conduct realistic simulations of space social systems before they are put into operation. While it may be far too early to start simulating lunar communities, soon we should have enough design information to start space station simulations.

Utilizing realistic mock-ups of a space station, experiments could be conducted to investigate various hypotheses on crew composition and structure. For example, do one simulation with a crew organized along hierarchial lines with a commander in complete control, as a captain on a ship, and then do another simulation in which authority is shared according to specified roles and responsibilities. Test various personnel combinations—female/male ratios, proportion of scientists to traditional astronauts, and so forth. Investigate optimum crew size and rotation systems by actually trying them out. From such simulation experiments and from other research and experience, an appropriate space station social organization could be designed, then tested in space and modified according to experience.

By the time the space station is operational, plans for a lunar base may have advanced to the point whereby a lunar community could be simulated. Using findings from the space station simulations, and then actual operations, it would be possible to refine models for long-duration space living, tailor them for the special conditions of lunar living, and then test them with simulations.

Simulation experiments should use realistic mock-ups for the space station, lunar base, or whatever the relevant system under investigation. They should be conducted for long periods equal to, or at least approaching, the length of the projected missions. And the participants should be given real scientific, materials-processing and maintenance tasks to perform throughout the duration of the simulation.

Such an ambitious project would be expensive and difficult to undertake. But if it could isolate factors critical for space living and thereby help lead to the design of social systems for space that would enhance safety and productivity, such a simulation program would repay the investment.

Fourth, include self-design by those who will actually live in space. Make them active participants in the research and development of the social systems in which they will live, instead of passive and perhaps alienated subjects of experiments and plans developed solely by others. This last point raises the issue of lunar community autonomy. Although a lunar base might start out being totally, or almost totally, dependent on Earth for materials and supplies of fuel, food, oxygen and other vital items, the ideal would be to develop progressively greater degrees of self-sufficiency. Greater material self-sufficiency implies increasing social self-sufficiency ranging, for example, from local initiative in research to self-governance of the community itself. Such close supervision from Earth as is involved in day-to-day scheduling by mission control, or in the step-by-step direction of experiments by ground-based principal investigators will have to give way to some measure of lunar base autonomy—if the benefits of adapting to this new environment are truly to be realized.

This trend toward autonomy should be anticipated, not ignored or resisted. If we really want to learn how to live in space, the locus of creativity must someday shift from Earth to space. As lunar communities grow in size and competence, so they should be encouraged to develop their own solutions to the problems of living in space. In so doing they would be developing the first of many space-based cultures that could enhance humanity's future.

REFERENCES

Bluth B. J. (1981) Soviet space stress. *Science 81, 2,* 30–35.

Bluth B. J. (1984) Sociology and Space Development. In *Social Sciences and Space Exploration* (S. Cheston, C. M. Chafer, and S. B. Chafer, eds.), pp. 72–78, National Aeronautics and Space Administration, Washington, D.C.

Carr G. (1981) Comments from a Skylab veteran. *The Futurist, 15,* 38.

Connors M. M., Harrison A. A., and Akins F. A. (1985) *Living Aloft: Human Requirements for Extended Space Missions.* NASA, Washington. In press.

Helmreich R. L. (1983) Applying psychology in outer space. *Amer. Psychol., 38,* 445–450.

Helmreich R. L., Wilhelm J. A., and Runge T. E. (1980) Psychological considerations in future space missions. In *Human Factors of Outer Space Production* (T. S. Cheston and D. L. Winter, eds.), Westview, Boulder, CO. 206 pp.

Huntford R. (1984) *Scott and Amundsen: The Race to the South Pole*. Atheneum, New York. 576 pp.

Miller J. G. (1984) A living systems analysis of space habitats (abstract). In *Papers Presented to the Symposium on Lunar Bases and Space Activities of the 21st Century*, p. 36. NASA Johnson Space Center, Houston.

Mutsschlechner M. (1979) Living in space stations, an interview with J. P. Kerwin. *Spaceflight, 21*, 271–274.

LESSONS FROM THE PAST:
TOWARD A LONG-TERM SPACE POLICY

Andrew Lawler

Contributing Space Editor, The Futurist Magazine, 4416 St. Elmo Ave., Bethesda, MD 20814

Space is a new environment, but there are examples of past human migrations that can provide valuable lessons to settlement planners. The colonization of the isolated, resource-poor, and previously uninhabited island of Barbados in the early 17th century reveals the weaknesses of corporate-sponsored settlement. The Antarctican experience of the late 1950s, on the other hand, underscores problems with government-organized projects. The unique requirements of space exploration and development in the long term call for a hybrid of these institutions—corporations and government. Combining the best of both may be a wise space policy objective.

INTRODUCTION

The exploration and development of space promises both economic benefits and advances in scientific knowledge. Now with the lure of potential new products and services, industry is beginning to speculate about new medicines, alloys, and crystals that could be produced in orbit. Space scientists, engineers, and even sociologists recognize that a whole array of new products and new fields of knowledge await discovery.

The central space policy question today is what system will direct this growth. Until recently, space exploration and development has been a government operation. Now, with the lure of potential new products and services, industry is taking a more direct role in space enterprises. The legal, financial, and political barriers to commercial space development will not disappear in the near future, but they do not appear to be insurmountable hurdles. Yet it is unlikely that the organizations that spawned space exploration, national governments, will simply fade out of the picture. It is not yet certain what system—governmental, corporate, or some new combination—will prevail in space in the coming decades.

Pertinent lessons from past migrations can help us clarify the uncertain future of space development. Historical analogies obviously have their limits; nevertheless, they are our only guides to the future. By examining past colonization attempts, we can make more informed decisions on what system may, and should, work best in space.

Only governments and corporations can afford the enormous capital investments required for extensive space exploration and development. This is the starting point in our search for relevant historical analogies. Other key factors include the absence of native peoples in orbit or on the Moon, the radically different environment, and the need to supply even bare essentials, especially in the early years, in order for human settlers to survive. Our search, then, is for settlement attempts that developed under similar conditions, but under different institutions.

BARBADOS

*Planting countries is like the planting of woods; for you must
make account to leese almost twenty years profit, and expect
your recompense in the end. For the principal thing that hath
been the destruction of most plantations, hath been the base
and hasty drawings of profit in the first years. It is true, speedy
profit is not to be neglected, as far as may stand with the food
of the plantation, but no further.*
—Francis Bacon, 1597 (Miller, 1983, p. 81)

The colonization of the West Indian island of Barbados by a group of merchants
is a fascinating story of corporate colonization. This resource-poor, uninhabited, and
subtropical island rivaled early Virginia in population and wealth, though endowed with
only a fraction of the land. The boom was short-lived. Within two decades of its founding,
Barbados was locked into economic specialization and dependence that still cripples
it today. Assuming that space development might also hinge largely on corporate
involvement and the potential for great economic profit, 17th-century Barbados could
provide the late 20th Century with some valuable lessons.

The attempt to settle this small and isolated island began in London in the early
seventeenth century. The survival of the Virginia colony, organized and financed by London
merchants, was in serious doubt until settlers began to earn a profit from tobacco. The
market for the leaf grew rapidly and attracted other London entrepreneurs. One wealthy
merchant, William Courteen (originally a Dutchman), organized a syndicate that he hoped
would reap profits from the new product. He raised £10,000 and secured a royal patent
to the island of Barbados.

The Courteen Syndicate sent a ship in 1627 to settle and develop the island, which
is isolated from neighboring isles by the prevailing easterlies. Today we know that
Amerindians once inhabited Barbados prior to the European settlement, but the leader
of the expedition concluded that there were "noe people on the island untill they [the
English] came" (Harlow, 1926, p. 40). "Wild pigs ran free, but there were no staples such
as corn, and no cleared land" (Bridenbaugh and Bridenbaugh, 1972, p. 63). There was
not even a large supply of fresh water. One early settler complained that on Barbados,
"water is to be prized above any thing else" (Bridenbaugh and Bridenbaugh, 1972, p.
63).

The expedition was forced to divert to the Spanish Main to trade with natives "for
all things to be gotten for the planting of this island Barbadoes" (Harlow, 1925, p. 40).
A small number of Arawaks, the indigenous people, asked to be taken along by the
British, and they were granted a plot of land on Barbados. Within a matter of a few
years, however, all were enslaved and soon died, and their important knowledge of the
environment was lost.

The island was out of the range of the aggressive Carib (from whom we get our
modern word cannibal). The Carib were a threat to Arawak and European peoples well

into the 1700s, but Barbados was the sole island in the Caribbean that was secure from their raids (Bridenbaugh and Bridenbaugh, 1972, p. 63). Unlike contemporary Virginia and New England, the Amerindians played a minimal role on Barbados.

The Courteen Syndicate, though founded in a capitalistic spirit, was not an innovative organization. The colonists were hired hands expected to grow tobacco in order to pay off the original investment of the merchants. As the historian Harlow reports, the settlers were more serfs than pioneers. "Neither the land nor the stock," he writes, "were their own; they were merely his [Courteen's] tenants at will, working the demesne of the lord of the manner" (Harlow, 1926, p. 7). The London merchants wanted to pay back their investments and make a profit; they did not link monetary gain with a healthy, well-adjusted settlement.

Most of the settlers were young, single men who came not to settle but to earn a wage and perhaps have some adventure. They cared little for the ecological health of the land, living facilities, a nutritious diet, sanitation, or their neighbors. Early on, writes one historian, "feud and faction became the order of the day in a little community faced by perils enough" (Harlow, 1926, p. 12).

Introduction of private land ownership in 1630 did not greatly alter the settlers' way of life. The economy remained totally focused on tobacco. But the tobacco grown on Barbados proved "very ill-conditioned, fowle, full of stalkes, and evil-coloured" (Bridenbaugh, 1968, p. 53). The product could not compete with the superior Virginia leaf. There was little profit to be made, but the colonists continued to shun subsistence foods, remaining dependent on imported food and drink and the island's wild pigs. One visitor predicted in 1631 that "this plentiful world of theirs is now passed" (Harlow, 1926, p. 24). The starving time of 1632–1633 followed.

Tobacco was gradually abandoned, and a new crop, cotton, became Barbadian gold. But it was only with the introduction of sugar in the 1640s that substantial wealth began to flow into Barbados. "Curiously enough," notes one historian, "this access to prosperity, advantageous as it was from an economic point of view, eventually proved to be the main cause of the island's decline" (Harlow, 1926, p. 43). This is not difficult to explain. Anyone with enough land or capital became, in the words of one contemporary, "so intent upon planting sugar that they had rather buy foode at very dear rates than produce it by labour, so infinite is the profitt of the sugar workes once accomplished" (Dunn, 1972, p. 53). Initial investments for sugar works were too high for the vast majority of settlers, and the larger landowners began to buy up the small farms. Slaves were brought from Africa to work these vast estates. Barbados became locked into an economic specialization that, however profitable in the short-term, was disastrous for the overall health of the island. It has never fully recovered.

A lack of control by the island government, the transient population, and the incredible concentration of people (in the 1640s Barbados had one of the densest populations in the European world) resulted in appalling living conditions, even for the period. Open sewers and polluted drinking water spread epidemics; mortality in 1647 reached such horrible proportions that in the main village of Bridgetown, they threw the dead directly into the river "so that many died in a few hours, poisoned by the drinking water" (Southey, 1827, p. 315).

Alcohol was available but expensive. Drinking was nevertheless the favorite pastime for the young men, who were thousands of miles and several months away from home. "If ye would but bridle ye excesse of drinkinge," wrote one despairing visitor to the islanders, "together with ye quarrelsom conditions," then, he believed, there might be some semblance of stability (Harlow, 1926, p. 66).

Early Barbados presents us with the disturbing picture of a chaotic settlement with a weak government, a vulnerable economy, and colonists who refused to adapt to new conditions. This last point is particularly important. Spanish settlers appear to have more readily adapted their agriculture, architecture, and clothing to the New World. On Barbados, by contrast, one visitor found wealthy plantation owners living in "timber houses, with low roofs, so low for the most part of them that I could not stand upright with my hat on" (Bridenbaugh, 1968, p. 39). When the visitor proposed to design an innovative house that would take account of the trade winds, pests, and the late afternoon sun, even the most educated "did not or would not understand" (Bridenbaugh, 1968, p. 34). The English simply could not adapt quickly enough to an environment far more alien to them than to settlers from Mediterranean regions.

The "gold rush" or "every-man-for-himself" economy undermined attempts to deal with problems in a systematic way. Unlike their Spanish contemporaries in the New World, the English did not plan towns, build public sanitation systems, or enforce strict laws. The cooperation needed to accomplish these things simply did not exist among settlers who, as a visiting Frenchman wrote, "all came in order to make money" (Bridenbaugh and Bridenbaugh, 1972, p. 35).

Based on the experience of Barbados, space development should avoid the pitfalls of economic specialization. A colony self-sufficient in basic needs with a diverse export product line would be more stable than a settlement largely dependent on the outside world, where forces are often beyond its control. In addition, a diverse population—including men, women, and children—would add permanence to a settlement.

The success (in both human and monetary terms) of a colony is dependent on strong coordination among the settlers, cultural flexibility, and the subordination of short-term gains for long-term returns. This latter may prove difficult for an industry eager to pay back nervous investors. A purely corporate development of space could follow some of the paths of the Barbadian settlement. Companies might be tempted to skimp on infrastructure (such as expensive redundant systems). Such cost-cutting could have catastrophic effects. Industry could also tie the economy to a single profitable service or good. If this occurs, price fluctuations on the international market could doom the colony's long-term growth.

Should the path of space development therefore be blazed by government? Let us examine an attempt by a national government to settle an uninhabited, remote, and resource-poor area.

ANTARCTICA

It is easier to escape the omnipresence of people in New York than in Antarctica.
—Ernst Stuhlinger (1969, p. 6)

The early years of settlement in the Antarctic provide an example of a government-sponsored project clearly analogous to the first steps we are making in space. The continent has no indigenous peoples, the production of the most basic essentials (such as water) requires great effort, and it is unlike any other environment on Earth.

Antarctica has been permanently settled only since 1954. During the International Geophysical Year (IGY) 1957–1958, a dozen nations joined together to build scientific stations on the continent. Planning the logistics of supplies, transportation, and coordination of the IGY was an enormous task. "Men and equipment had to be shipped in a scale hitherto undreamed of" (King, 1969, p. 241). The American project was the most ambitious, and the Navy was placed in charge of logistics. Five scientific bases were to be ready by early 1957, complete with two years' worth of supplies. Everything had to be imported, and the nearest airfields were thousands of kilometers away in New Zealand.

Erection of the South Pole station was particularly difficult. The pole is 1300 kilometers from the main base at McMurdo Sound. Located high on the Antarctic Plateau, the windiest, coldest, and driest desert in the world, the sole natives of the pole are a few colonies of microbes.

All parts of the base were specifically designed to fit the cargo bay of C–124s. The sections of the structure could then be dropped from the air. By the onset of the winter, Seabees (the Navy's construction battalion) finished the cluster of Jamesway huts. Connected by short passageways and equipped with a galley, radio room, bunkrooms, and ample storage space, the new base was luxurious compared to the past dwellings on the continent. But Paul Siple, a close friend of Admiral Byrd and one of the last of the original Antarctic explorers, watched in sadness as "these young titans of modern Antarctica threw together a series of Jamesway huts surrounded by a ring of litter that made the glamorous pole look like the corner of a city dump" (Lewis, 1965, p. 22).

Siple was appalled by the waste and poor work done by those "accustomed to the opulence of the military service" with an attitude of "get the job done and to hell with conserving supplies" (Lewis, 1965, p. 72). Siple understood that this attitude had no place in an environment that demanded efficiency, careful planning, and respect.

Inside the base there were no private rooms; the latrine was distant, and the temperature varied from 0°C on the floor to 40°C at the ceiling. The man with the bottom bunk froze while the man on the top sweated. Many work sites had no connecting passageways and could only be reached through the outside. Food was good and plentiful, but the diesel generators made an earsplitting racket that never ceased.

Water was particularly hard to supply. The pole station reported during one winter "twice a week over five tons of snow must be dug by hand for the snow melter in temperatures of -90°F and winds 20 mph—all at an altitude of 10,000 feet above sea level" (Lewis, 1965, p. 276). Despite this effort, the water continually tasted of the diesel fuel used to melt it.

The design, though revolutionary, lacked any cohesive plan. No private architects were consulted, no psychiatrists were asked for suggestions, and no construction firms offered alternative designs. Time constraints and the bureaucratic structure of the Navy precluded a more flexible approach. Within a few seasons the entire base at the South Pole was covered by blowing snow.

The permanent bases were a major departure from past exploration. After wintering over in 1958-1959, C. S. Mullin believed that "danger, hardship, or the direct effects of the cold did not represent important stresses" (Mullin, 1960, p. 322). The psychosocial environment assumed importance very quickly on the bases. During the first winter, one of the 18 men in the party developed "a frank and florid psychosis" (Nardini *et al.*, 1972, p. 97). There was, of course, no way to evacuate him, and there were "no provisions for adequately separating such a patient from the remainder of the group" (Nardini *et al.*, 1972, p. 97).

Selection procedures were subsequently tightened and proved fairly successful. But a deeper problem revolved around the organizational structure of the smaller Antarctican bases. In one case, overworked support staff began to question the right of scientists to avoid housekeeping duties. No open hostilities broke out, but the problem was "a cause of serious dissension and disruption in the group" and, more ominously, "the situation was never resolved" (Lewis, 1965, p. 273).

A winter base in Antarctica is a unique world, where the cook often has greater prestige than the officer-in-charge and the radio operator can have more influence than an accomplished scientist. The traditional hierarchical structure of the military, and of government as a whole, breaks down among a small group of people isolated from others for months at a time. This was a controversial and embarrassing realization for the Navy. Flexible authority and sharing of tasks among everyone are vital for the well-being of a small, isolated group. This can run against the grain of highly specialized scientists and career military officers. The absence of women was also a factor. Navy traditions excluded females from the continent, and this increased tensions.

Government sponsorship of operations in Antarctica during the late 1950s did not result in any attempts to develop self-sufficiency. The Navy made no moves to develop the native resources to lower the enormous costs of transportation. Windpower was just one of many alternative energy sources that were never tapped, and greenhouses to produce fresh vegetables might have had additional psychological benefits (the Soviets grew plants at an early stage in Antarctica, and now aboard the Salyut space station as well, B. J. Bluth, personal communication, 1984). A huge armada of ships and planes continues to supply the bases in Antarctica every year with energy and food.

Some lessons have been learned. With great reluctance, the Navy eventually allowed women on the continent, and a new base on the South Pole includes a geodesic dome that prevents snow build-up. A more flexible organizational structure is tolerated, and private enterprise is now providing some services and personnel. But the early years of inefficiency, poor design and construction, and inappropriate organization could be repeated if government plans long-term space development on a large scale.

The Antarctican experience reminds us that the danger of mutiny or psychosis in a space station or colony are as real as the threat of meteors or solar flares. The importance of the psychosocial environment in space development must not be underestimated. Not only crew selection, but station design and the organizational structure play a large role in the stability of a small group. Task generalization can defuse tensions; flexible authority is vital; and mixing sexes appears desirable. The past tendency of high-tech planners to ignore psychosocial considerations must be curbed as we enter space. These considerations can often be accommodated without undue difficulty. Self-sufficiency in basic needs, for example, can bring down costs (economic benefits) and also provide a hobby for settlers (psychological benefits).

Basing a space policy purely on scientific research, as is the case in Antarctica, is probably not an optimal choice. The rise and fall of NASA's budget over the past two decades underscores the difficulty that the scientific community has in controlling its own destiny. In times of fiscal restraint, those projects with the least public support are often the first cut, and research is often a prime target. An emphasis on pure research in space heavily ties policy to the political winds on Earth and slows the chances for steady growth.

Handing the reins of space development to the military might be just as unwise. Budgets would still depend on political considerations on Earth. The military of both superpowers already are changing their emphasis in space from surveillance to weapons. A scenario of space containing sophisticated weapons systems controlled by bureaucratic organizations and produced by classified research is not a very appealing vision. Industry would, of course, gain short-term profits. But a healthy and viable economy would be very unlikely to ever develop in a largely military scenario.

An essentially profit-oriented approach to space policy might, however, have similar disadvantages. A corporation, eager to pay back initial investments, could insist that settlements concentrate on the production of one or a few profitable products for export. Should a cheaper substitute or process be discovered that could produce the goods or services on Earth, space development could come to a halt. The space inhabitants might find themselves as dependent on the economic winds of Earth as their counterparts in the government-controlled scenario are tied to politics.

CONCLUSIONS

As we have seen, both a largely corporate and a predominantly governmental approach to space development have their respective dangers. Industry might be unable to wait patiently while settlements struggle through their early, most vulnerable, years. Government, on the other hand, could delay development through costly and inefficient management techniques.

A vision of large, permanently inhabited colonies in space should include highly self-sufficient settlements that produce a wide variety of goods and services. This future scenario should also include a tolerant organizational superstructure that would require an economic return from the colonies, yet be cautious not to demand too much too soon.

A successful long-term space policy cannot be built upon cries for free enterprise, nor can it rest on traditional central planning. A hybrid of institutions and ideologies could best overcome the deficiencies inherent in both systems. Both COMSAT and INTELSAT are beginnings. And Ariane is already influencing the policies of traditionally bureaucratic NASA, as well as the small private launch companies.

Government/corporate cooperation in space development is not just a passing accommodation. The unique physical, economic, and political conditions of space require a fresh look at our current institutions and ideologies. We owe it to future generations to choose a long-term policy that will ensure survival, health, and prosperity in space.

REFERENCES

Bridenbaugh C. (1968) *Vexed and Troubled Englishmen.* Oxford University Press, New York. 487 pp.

Bridenbaugh C. and Bridenbaugh R. (1972) *No Peace Beyond the Line.* Oxford University Press, New York. 440 pp.

Dunn R. (1972) *Sugar and Slaves.* University of North Carolina Press, Chapel Hill. 359 pp.

Harlow V. T. (1925) *Colonising Expeditions to the West Indies and Guiana 1623–1667.* Bedford Press, London. 262 pp.

Harlow V. T. (1926) *A History of Barbados 1625–1685.* Clarendon Press, Oxford. 347 pp.

King H. G. R. (1969) *The Antarctic.* Blandford Press, London. 276 pp.

Lewis R. S. (1965) *A Continent for Science.* Viking Press, New York. 300 pp.

Miller H. H. (1983) *Passage to America.* Division of Archives and History, North Carolina University (Department of Cultural Resources), Raleigh, North Carolina. 84 pp.

Mullin C. S. (1960) Some psychological aspects of isolated Antarctic living. *Amer. J. Psychiatry, 117,* 322–325.

Nardini J. E., Herrmann R. S., and Rasmussen J. E. (1962) Navy psychiatric assessment in the Antarctic. *Amer. J. Psychiatry, 119,* 97–105.

Stuhlinger E. (1969) Antarctic research: A prelude to space research. *Antarct. J., 1–7, J–F.*

Southey T. (1827) *Chronological History of the West Indies,* Vol. 1. Longman, Rees, Orne, Brown, and Green, London. 350 pp.

HISTORICAL PERSPECTIVES ON THE MOON BASE— COOK AND AUSTRALIA

Eric M. Jones

Earth and Space Science Division, Los Alamos National Laboratory, Los Alamos, NM 87545

Ben R. Finney

Department of Anthropology, University of Hawaii, Honolulu, HI 96822

Among the many historical episodes that have relevance to the establishment of a lunar base, the voyages of Captain Cook and the founding of Britain's Botany Bay colony in Australia seem particularly appropriate. The process resulting in the selection of Cook rewards study, as do his relations with the Admiralty, with the scientific establishment, and with the scientists who accompanied him. Britain's tight control of the Botany Bay settlement and its unwillingness to promote early self-sufficiency may have delayed the time when Australia became self-supporting. Structuring the lunar base to offer opportunities for private initiatives may hasten the day when it becomes a self-supporting settlement rather than an externally supported scientific base on an Antarctic model.

Learning to live and work in space is going to require some adaptation. However, we should remember that, in the words of historian Alfred Crosby (1985), "We have done all this launching out into space before." During the five million or so years that the hominid line has been on this planet, our ancestors have constantly probed the limits of human capabilities, learning through the development and use of culture and technology to live in environments for which they were not physically adapted.

The coming era of space development has many parallels in the past: the spread of hunting and gathering peoples across the face of the planet, the oceanic exploits of the Polynesians and the Vikings, the flowering of Greek culture around the Mediterranean during the classical period, and the stillborn Chinese maritime initiative of the early Ming Dynasty. Each of these, along with the later explosion of European mariners into the World Ocean, have lessons to teach us, providing both inspiration and caution. This essay highlights two related but distinct episodes from the British experience. These examples illustrate the interplay between individuals and institutions, between goals and means.

Captain Cook

In 1768, Britain was the superpower of the day. The long war with France in Canada was drawing to a close following the fall of Quebec. Tensions were beginning to build with the American colonies and would soon lead to revolution. Meanwhile, another sort of revolution, this one in industry, was altering the fabric of British society. Britain was beginning to dominate world commerce and the seas that were making global trade possible. However, in those middle decades of the eighteenth century, despite the fact that Europeans had been sailing the Pacific for nearly two centuries, much of that vast ocean was still uncharted owing to the fact that navigation still relied on dead reckoning and latitude sailing. The problem of determining longitude was unsolved, and, as a consequence, much of the map of the world was blank or filled with lands that were more fantasy than reality, particularly in the southern Pacific. However, in his three voyages James Cook would replace the mapmakers' fantasies with a known ocean and would make global voyages safer and more certain (Beaglehole, 1974).

A primary motivation for Cook's first voyage was scientific. One of the central problems of the day was that of astronomical distances. In 1767, the Royal Society urged His Majesty's government to send a party into the Pacific to observe the Transit of Venus. The government was persuaded, but the Lords of the Admiralty were looking beyond the narrow scientific question and seeing an opportunity to address more practical problems. Was there a southern, temperate continent in the south Pacific where Britain might profit? What resources and shelter were there in the Pacific to support British interests? And would the newly published Nautical Almanac bring longitudinal calculations within the reach of non-scientists? Consequently, the Admiralty wanted their own man as leader of the expedition; they chose not from the scientific establishment but from naval ranks.

They chose for skill rather than rank, plucking James Cook from North American service where he had been charting the coast of Newfoundland. Cook succeeded beyond all expectations, removing the mythical southern continent from the maps, adding the east coast of Australia, accurately plotting the positions of the far-flung islands of Polynesia, and correctly surmising the existence of Antarctica. He also proved the worth of the almanac and, during the second voyage, of the new timepieces with which the problem of determining longitude was solved. In the bargain he defeated scurvy and gave Europe detailed knowledge of the peoples and products of the Pacific. Perhaps the greatest tribute came from Benjamin Franklin, who in 1778, despite the fact that Britain and America were then at war, urged that ships in American service "treat the said Captain Cook and his people with all Civility and Kindness, affording them, as common Friends to Mankind, all Assistance in your Power."

Yet, however much the Cook saga may provide inspiration for those who would explore the ocean of space, it should also provide some cautions. Obviously, a man of Cook's talents was the right choice to command the voyages, rather than an astronomer or a certain geographer the scientific establishment had proposed. A practical seaman, an accomplished navigator, and a leader of men was needed for the three-year circumnavigation—not a scientist.

But Cook did make mistakes, particularly on the third voyage. By then he had become tired and was probably chronically ill as well. What is more, this bright star of the Royal Navy, who had already done so much to advance knowledge, had become so alienated from scientists, particularly as a result of episodes connected with the second voyage, that he would have none aboard his ships. So an exhausted Cook, without equals to advise or perhaps restrain him, sailed to his doom; he was killed in Hawaii, the victim of one of his own rare lapses in judgment.

In any undertaking of this kind there is conflict between the need for autonomy on the one hand and advice and supervision on the other. In Cook's time autonomy was a necessity; he was literally out of reach for years at a time. The Admiralty chose a man who had inspired the confidence of Hugh Palliser and other high-ranking officers with whom he had served in Canada, gave him broad instructions, and hoped for the best. Their trust was rewarded, although it is tempting to second guess and wish that Cook had had ranking advisors on the fatal voyage.

Autonomy on the scale that Cook experienced is no longer possible, necessary, or even desirable. The definite advantage to rapid communication is that experts can be called on to supplement the skills and judgment of the flight crew. However, as both we and the Russians have relearned in recent years, flight crews sometimes resent what they consider to be too-frequent interference from the ground. But, we do seem to be learning; at the close of a recent shuttle mission, ground personnel were complimented on their helpfulness and their willingness to accept the flight crew's judgment.

These recent examples and the Cook experience suggest that the best course of action is to choose the most qualified people and, as long as they are getting the job done, to offer advice and instructions only when absolutely necessary or when requested. On occasion Cook could have used a senior advisor, as he had during the first voyage in the person of Joseph Banks (later President of the Royal Society). But Cook's great success probably came because he had freedom to exercise his considerable talent and judgment. There is a middle ground, but it is probably best to err on the side of autonomy. We need to be reminded of that from time to time.

Australia

Cook's discovery of Australia's east coast led to the establishment in 1788 of a British colony (Blainey, 1968; Shaw, 1972). Botany Bay started out as a penal settlement. The American Revolution had created a crisis for the British penal system; convicts from the slums of the growing cities were no longer welcome in the Americas. In 1784 James Matra, who had sailed with Cook, promoted the idea of an Australian settlement, although as a haven for American loyalists and as a theatre for new commercial ventures. The government was not much interested in the commercial possibilities but was willing to entertain the idea, soon championed by Banks and by the Home Secretary, Lord Sydney, of a penal colony. Access to flax and timbers from tiny Norfolk Island may have been the principal attraction (Blainey, 1968).

As originally conceived, the settlement was soon to become self-sufficient, thanks both to the rich soil Banks described and to the toil of convicts. Unfortunately, it did not work out that way. When Captain Phillip arrived at Botany Bay in 1788 with a party of over 1000 people, he discovered that the soil was not at all suitable and that few of the over 700 convicts had any useful skills. By 1790 Phillip was writing to London: "The sending out of the disordered and helpless clears the gaols and may ease the parishes from which they are sent; but, Sir, it is obvious that this settlement, instead of being a Colony which is to support itself, will, if the practise is continued, remain a burden to the mother country." What Phillip desperately needed were people with the appropriate skill and with an interest in the future of the settlement. London balked, however, at the idea of granting land to convicts who had finished their sentences. The thought did not square with the perceived need for punishment. Would not news of their success incite further waves of crime in the cities? To add to their problem, Norfolk Island proved to make poor sail cloth, while the lack of good landing places on the island made it almost impossible to off-load the tall pine trunks needed for ships' masts.

Despite these setbacks, support for the colony continued. London realized that Australia's geographic position along the routes to Asia was of strategic importance. Not surprisingly, London was preoccupied with the coast, and it was decades before local interests began to turn inland.

Local interests did develop. Because there were few free settlers in the early years, the initial impetus came from the private interests of the military. Officers in the New South Wales Corps were, like their counterparts throughout the British Empire, often younger sons blocked from family wealth and titles by elder brothers. They had come to Australia to make their fortunes. A group of them, including the notorious John MacArthur, gained virtual control of the economy, buying grain cheaply in years of abundance and selling it dearly in the frequent years of devastating drought. In 1806, Governor Bligh tried to undermine the speculators, buying wheat at a fair price in a year of plenty and distributing meat from government supplies to farmers devastated by floods. MacArthur and his friends were not to be denied, however; they rebelled and eventually, through the aide of powerful friends in London, had Bligh recalled.

MacArthur and people like him had positive roles to play as well. They introduced sheep and, with their accumulated wealth, acquired the large tracts of land needed to support flocks in the poor climate. Wool was the first step toward self-sufficiency, the first exportable product of the settlement. Only wool was valuable enough that it could bear the cost of internal transportation in a land lacking navigable rivers, as well as the cost of shipment to Europe, and still yield a profit.

But who was to get land and at what price? Would convicts continue to work for private individuals and, in a kind of work-release program, be able to work for themselves part of the time? How would free settlers be encouraged to come? And how much control would the colonists be permitted to take of their own political destinies?

It would be decades before these issues were even partially sorted out. Transportation of convicts to the built-up areas of New South Wales ended in the 1840s. The discovery of gold brought a flood of new settlers in the 1850s, who then had to be absorbed into the mainstream of the settlement as the gold ran out. Despite the lure of the great expanses of open land, Australians learned slowly that much of their continent was not made for farmers; the lack of rivers and the natural link between farmer and market made it cheaper to feed Australia with Indian grain before the coming of the railroads. For this reason Australia has been an urban nation from the very beginning, largely confined to a narrow strip of fertile coast in the south and east.

Most of the problems of the early decades arose because Australia was a very different place than many had believed in the beginning. Although there are some parallels with the American experience, Australia presented some unique problems. Australia was very far from England, and, as the first parties discovered, a continent less blessed with Nature's bounty than supporters of the settlement had led the government to believe. It was a dry land, subject to frequent drought, and British crops did poorly. So, too, did the convicts who were cast into this new land without basic survival skills. And, for the first few decades, there were few free settlers; potential immigrants were kept away by distance, a lack of opportunity, and official policy that discouraged their coming before around 1820. But free settlers did come. They and the convicts who had served out their terms found ways to make livings and gradually to build a self-supporting community. All that happened in less than a lifetime, and by 1840 Australia was, for practical purposes, standing on its own feet.

What does the Australian experience have to teach us, as we contemplate a return to the Moon? Australia is not the Moon; it is a terrestrial environment far more benign than our airless satellite, but there are similarities. In broad brush a few pertinent features stand out. Living on the Moon is going to be a new and initially difficult experience. The lunar base will start small and be very dependent on Earth. The first "settlers" will have to be technically trained: astronauts, engineers, and scientists. Bureaucracies will oversee the operation from afar; administrative control will be and must be tight. The costs and difficulties of transportation will in large measure determine the pace of settlement and the products that can be profitably produced.

If the lunar settlement is to grow and eventually become self-supporting, some helpful features can be built into this social experiment from the beginning.

First, the stated purpose of the lunar base must be the eventual establishment of a self-supporting lunar settlement. If that purpose is clear from the beginning, then the inevitable transition from a tightly administered scientific base to a more open community may come more easily. If, on the other hand, we say that we are going for purely technical reasons, then the interests of entrepreneurs and private settlers (on whom ultimate success may well depend) might not receive proper attention.

Second, reducing the cost of transportation must be a main concern, although we anticipate that these costs will remain high for a period of time. During that initial stage subsidies of public and private ventures will be required.

Third, there should be emphasis on the development and encouragement of profitable enterprises through technology transfer, tax incentives, and the development and maintenance of basic services.

Fourth, there should be mechanisms established for the orderly transfer of control of lunar operations to the settlers as the population grows—mechanisms analogous to the Ordinance of 1787 in the United States and those used by Britain to create independent nations such as Australia, New Zealand, and Canada.

There will be conflicting interests, shifting purposes, short-sightedness, greed, and mistakes. The lunar base is going to be expensive and will remain so for longer than some of us would like. If we give careful thought to the design of the experiment now, however, considering the human as well as the technical aspects, and keep our eyes open both for the pitfalls and the opportunities that will come along later, perhaps self-sufficiency of the lunar settlement will come quickly.

Finally, we should remember that Britain supported the Australian settlement long enough for it to succeed. Let us hope that we can do the same for the lunar enterprise.

REFERENCES

Beaglehole J. C. (1974) *The Life of Captain James Cook.* Stanford Univ. Press, Stanford. 760 pp.

Blainey G. (1968) *The Tyranny of Distance.* St. Martin's Press, New York. 365 pp.

Crosby A. W. (1985) Life (with all its problems) in space. In *Interstellar Migration and the Human Experience* (B. R. Finney and E. M. Jones, eds.), pp. 210–219. Univ. California Press, Berkeley.

Shaw A. G. L. (1972) *The Story of Australia.* Faber and Faber, London. 336 pp.

SPACE POEMS: CLOSE ENCOUNTERS BETWEEN THE LYRIC IMAGINATION AND 25 YEARS OF NASA SPACE EXPLORATION

Helene Knox

Muhlenberg College, Allentown, PA 18104

The National Aeronautics and Space Administration was created in 1958, after the uproar in 1957 over Sputnik, Russia's (and the world's) first Earth-orbiting satellite. NASA is a federal agency charged with carrying out government policies in specific areas for research and development in science, engineering, and technology. Unlike many government efforts, NASA has generated activities and images with wide appeal to millions of people, not only in America, but around the world. The space program and people's responses to it are phenomena unique to our time; living in our time necessarily includes coming to terms with the fact that the human species is actually moving off the planet. From the beginning of the program, many people longed for an eloquent communication from spacefarers, something more than "Everything is A-OK—the view is really great up here!" Among the grumblings, the idea surfaced early that NASA ought to send a poet into space. The agency had other priorities, although it now has a valuable collection of paintings and drawings by artists who were invited to monitor its activities. An "Arts in Space" program is now in the long-range planning stage to send poets, painters, sculptors, composers, dancers, and other creative artists to the proposed Space Station to work for several weeks at their respective expressive arts. Meanwhile, American poets have been on Earth; how did they respond to more than 25 years of space exploration? This question can be answered by a survey of the patterns of space imagery the poets selected for literary representation in their work, and their attitudes toward it.

In the anthology *Inside Outer Space* (Vas Dias, 1970, p. 141), we find poet Robert Kelly saying, "What we have needed / is a language / always needed a tongue / to caress our technology." This means being more creative with the language we have inherited, learning to adjust both it and ourselves to the new world revealed to us by science and transformed before our very eyes by technology. Only thus can we live in the present and be late 20th Century poets. This challenge is no different in kind from that explained by Hart Crane (1966, pp. 261–263) in his essay "Modern Poetry" in 1929.

> *The function of poetry in a Machine Age is identical to its function in any other age; and its capacities for presenting the most complete synthesis of human values remain essentially immune from any of the so-called inroads of science . . .For unless poetry can absorb the machine, i.e., acclimatize it as naturally and casually*

> *as trees, cattle, galleons, castles and all other human associations of the past,*
> *then poetry has failed of its full contemporary function . . .Contrary to general*
> *prejudice, the wonderment experienced in watching nose dives is of less immediate*
> *creative promise to poetry than the familiar gesture of a motorist in the modest*
> *act of shifting gears. I mean to say that mere romantic speculation on the power*
> *and beauty of machinery keeps it at a continual remove; it cannot act creatively*
> *in our lives until, like the unconscious nervous response of our bodies, its*
> *connotations emanate from within—forming as spontaneous a terminology of*
> *poetic reference as the bucolic world of pasture, plow and barn.*

By 1985, in an America over 90% urbanized, many more Americans have seen computer terminals and digital watches, ultimately brought to you by your local space program, than have seen plows. Nevertheless, for space imagery to carry the convincing power of spontaneous expression arising unconsciously from the deepest self, the poet must be able to acknowledge that space is a place where human beings have lived and are going to live, and he or she must be able to imagine living there in full intensity of sensory perception, emotional response, and cognition and language. Some poets who have chosen to write using space imagery are still earthbound, and mightily resist the transition to interplanetary civilization. Some use space imagery only to talk about themselves; this is as legitimate as using anything else in the universe to do that, but these poems are not primarily concerned with the social process of adjusting to the migration into space, which is my concern in this paper.

This social process, seen in its literary representations, encompasses a wide range of emotions. Among them is a wry and cynical humor, as in John Ciardi's couplet "Dawn of the Space Age," where he quips: "First a monkey, then a man, / Just the way the world began" (Ciardi, 1962, p. 30). This alludes to the test flights of trained chimpanzees like Able, Baker, Enos, and Ham, who were blasted by rockets to unheard-of heights in cramped capsules jammed with electronics; they responded to the flashing signals almost as well as the astronauts did shortly thereafter. William Stafford (Vas Dias, 1970, p. 304), however, seems to have misgivings and anxiety about the evolution into space. In "Dog Asleep," he imagines that a sleeping dog, twitching its feet, is dreaming about Laika, the Russian dog who died on her space mission. He says we try to console ourselves with the words, "It's for the best, / and Laika volunteered—she wagged her tail." But then we twitch our feet. The dangers were real enough, and many of them were unknown; radiation was known. In "23rd Light Poem" (Vas Dias, 1970, p. 172), Jackson MacLow advises:

> *Let them carry freight in those ships,*
> *moon minerals scooped by machines*
> *resistant to the ruthless rays,*
> *not men or other sentient beings*
> *dear to the fathomless Buddha.*

Sean O'Meary (Marcus, 1975, p. 81) imagines the worst: "All that remains of Astronaut / Orbits the morbid moon / Satellite and skeleton / Together as one." However, when the first man actually did go into space and orbit the Earth, William Carlos Williams (Vas Dias, 1970, p. 355) reports that

> Gagarin says, in ecstasy,
> he could have
> gone on forever
>
> he floated
> ate and sang,

and that when he returned

> to take his place
> among the rest of us . . .
> he felt
> as if he had
> been dancing.

Around the time of the flight of Apollo 7, before we landed on the Moon or even orbited it, James Dickey (1968, p. 26) prophetically predicted the future of manned space exploration in his poem "A Poet Witnesses a Bold Mission":

> In a sense they are all poets, expanders of
> consciousness beyond its known limits. Because of them,
> the death-cold and blazing craters of the moon will
> think with us, and the waterless oceans of Mars;
> the glowing fogs of Venus will say what they are.
>   And those places will change us also. We have not
> lived them yet, and perhaps have no language
> adequate to them. But these men will find that, too,
> as they plunge with their fragile and full humanity,
> with their wives and children, with their gardens
> and grocery lists and head colds and ideas for poems . . .

With the first lunar orbiter, our cameras—those extensions of our power of vision— got far enough away from Earth to give us our first real picture of it in its true context: infinite blackness. The new perspective was both physical and mental, and was brilliantly described in a metaphor of Richard Peck (1971, p. 86): "He turned toward earth no longer at his feet: agate in onyx." May Swenson (1978, p. 94), in "Orbiter 5 Shows How Earth Looks from the Moon," perceives the Indian Ocean as a woman in 3/4 profile, sitting on her heels with her bare feet tucked beneath the tip of Africa. The woman has a holy jug in her right hand;

> Asia is
> light swirling up out of her vessel . . .
> Her tail of long hair is
> the Arabian Peninsula.
> A woman in the earth.
> A man in the moon.

Ernest Sandeen, in "Views of our Sphere" (Vas Dias, 1970, p. 272), writes, "We deserved that earth-shot from the / moon's asbestos-gray horizon: a / family portrait on the old homestead . . ." He adds, "what can we hope for but smaller and smaller snapshots of this place?"

As for our only natural satellite itself, Arthur C. Clark (1969, p. 295) has cogently observed, "Although the Moon has inspired more verse than all the rest of the heavens put together, few poets have ever thought of it as a planet rather than a conveniently discreet source of illumination for their serenading." Among those poets who willfully refused to change their cultural bias, and who would have to be dragged kicking and screaming into the 21st Century, was the illustrious W. H. Auden. His poem "Moon Landing" (Phillips, 1974, pp. 162–163) reveals some of the worst "Bah! Humbug!" attitudes my research has yet encountered. "Unsmudged, thank God, my Moon still queens the Heavens," he writes, "as She ebbs and fulls, a Presence to glop at." Archibald MacLeish, in contrast, did not consider it "glopping" to write, in his poem "Voyage to the Moon" (Phillips, 1974, pp. 141–142), published on the front page of the *New York Times* on the day of the Moon landing:

> Presence among us,
> wanderer in our skies,
> dazzle of silver in our leaves and on our
> waters silver,
> O silver evasion in our farthest thought—
> "the visiting moon" . . ."the glimpses of the moon" . . .
> and we have touched you!

He says of the Moon, "You were a wonder to us, unattainable," and he describes the three-day journey of the Apollo astronauts, who "steered by the farthest stars" through risk of death and "unfathomable emptiness" until they "set foot at dawn upon your beaches, sifted between our fingers your cold sand." Then comes the transformation: Earth replaces Luna as the celestial object overhead:

> We stand here in the dusk, the cold, the silence . . .
> and here, as at the first of time, we lift our heads.

Over us, more beautiful than the moon, a
moon, a wonder to us, unattainable . . .
Over us on these silent beaches the bright
earth,
     presence among us.

May Swenson (1978, p. 73) can imagine the same perspective: "All night it was day, you could say, / with cloud-cuddled earth in the zenith, / a ghost moon that swiveled."

Auden uses his poem to insult the rocket genius Wehrner von Braun, who was the software and the propulsive force behind the American space program, and also to assert the (alas! shortsighted) opinion that dying in the effort to colonize space, as several Americans and many Russians have already done, would have no meaning:

Worth going to see? I can well believe it.
Worth seeing? Mneh! I once rode through a desert
    and was not charmed: give me a watered
    lively garden, remote from blatherers

about the New, the von Brauns and their ilk, where
on August mornings I can count the morning
    glories, where to die has a meaning,
    and no engine can shift my perspective.

Auden's immovable perspective is deeply grounded in literary culture, myth, and legend. Lisel Mueller, too, understands that the Moon can never be the same again: "Goodbye crooked little man / huntress who sleeps alone / dear pastor, shepherd of stars . . .we trade you in as we traded / the evil eye for the virus . . .Scarface hello we've got you covered . . ." (Phillips, 1974, pp. 147-148). Mueller's sarcastic tone reveals that she is on Auden's side in this controversy. In his poem "Apollo: For the First Manned Moon Orbit," James Dickey (Vas Dias, 1970, pp. 63-64) also acknowledges that the Moon is a "smashed crust / of uncanny rock ash-glowing   alchemicalizing the sun / With peace: with the peace of a country / Bombed out by the universe." His poem recreates the astronaut's urgent longing

to complete the curve   to come back
singing with procedure . . .
And behold
The blue planet steeped in its dream
Of reality, its calculated vision shaking with
The only love.

The tone of Dickey's poem is 180 degrees from sarcasm. It recounts an ultimate adventure to a land wondrous strange: one without any human associations. The Apollo 8 astronauts are the first human beings to see the Moon as it really is: they come, says Dickey, "in the name of a new life." This tone of astonishment, of the wonder and joy of discovering "the magic ground of the dead new world," continues in a companion poem, "Apollo: The Moon Ground" (Vas Dias, 1970, pp. 65–66). The tough American voice is unmistakable. One Apollo 11 astronaut says to the other:

> *Buddy,*
> *We have brought the gods. We know what it is to shine*
> *Far off, with earth. We alone*
> *Of all men, could take off*
> *Our shoes and fly.*

Their mission is to collect Moon rocks for scientific research into the true 4.5-billion-year history of the Sun-Earth-Moon system. He says,

> *The ground looms, and the secret*
> *Of time is lying*
> *Within amazing reach . . .*
> *We leap slowly along it . . .*
> *The Human Planet trembles in its black*
> *Sky with what we do . . .*
> *We are this world: we are*
> *The only men . . .*
> *We laugh,*
> *with the beautiful craze*
> *Of static. We bend, we pick up stones.*

The astronauts have humanized the Moon merely by being on it. Culturally, they "are this world." Robert Kelly (Vas Dias, 1970, p. 141) also understood that what matters in this unprecedented situation is

> *that human*
> *breath will shape*
> *utterance on the unconscious moon*
> *wake it*
> *& us*
> *from the bitter long dream of silence*
> *by breath*
> *of a man's body*
> *by the weight of his weight*
> *breath*
> *breathed into the moon.*

Al Purdy (Vas Dias, 1970, p. 242–243) advises the astronauts to complete the range of human emotion while they are there, to "let a handful of moondust run thru your hand /

and escape back to itself / for those others / the ghosts of grief and loss / walking beyond the Sea of Serenity."

The mental jolt required to permit a new planet to swim into our ken can be considerable. In the poem "July 20, 1969," Robert Vas Dias (1970, p. 338) describes the sense of disorientation he experienced on the day of the Moon landing:

> All day commentators have been talking
> of eras, generations, voyages, all
> the extra-terrestrial wandering
> of minds hooked
> on distances, historic precedents
> and the mechanism of escape
> velocities / my ears float in
> waves of coded engineering
> data, lists of steps to program
> lunar orbital insertion, and I see
> them swim in the deadly anti-
> atmosphere, laying out
> experiments like picnic tables.
> I am exhausted by the matter-of-fact
> recital of the incredible
> and I begin to doubt I
> exist outside relay circuits
> to moon and back. . .

David Ignatow (Vas Dias, 1970, p. 118) describes a similar sensation:

> At dawn, as you look up from the pavement
> to the sky, feeling without foothold,
> a starry wall is moving steadily back.
> What do you say to that, you whining rockets?

The threat, real or imagined, perceived by many writers seems to be that, in losing a sense of being solidly grounded on this Earth (presumably where we belong), we will lose a sure sense of our own identity—who we are. Lois Van Houton (unpublished work, 1985) asks

> And what will become of us, suffixed
> in space
> like the hanged man of the Tarot deck?
> Will we be lost to ourselves   how
> will we compensate this grandeur of
> earth   with most foreign cold?

Rosemary Joseph asks, if men ever walk on Mars and Venus, "will they still be men in the / image of Adam?" (Joseph, 1966). Peter Viereck (Pater, 1981, p. 419), in "Space-Wanderer's Homecoming," imagines one of our remote descendants, "after eight thousand years among the stars," feeling a sudden nostalgia for August and Earth's version of nature. He returns, looking for his horse and harp, and says, "Oh my people . . .My name is—. / Forgot it, I forgot it, the name 'man'."

It may be this sense of disorientation that makes some writers fiercely resist the change to the space age. Maxine Kumin (Phillips, 1974, p. 156), for instance, reacted to a very early lunar probe ("They had meant / merely to prick," she says) by imagining that "the moon was undone: had blown out, sky high." The tides cease, "dogs freeze in mid-howl, women wind their clocks," and lunatics suddenly become sane. She had prophesied this in a dream, and because it came true, she has not dreamed since. May Swenson (1978, p. 214) ends her poem, "Landing on the Moon" with the question, "Dare we land upon a dream?" In her essay "The Experience of Poetry in a Scientific Age" (Nemerov, 1966, pp. 150–151), however, she clarifies that "My moon is not in the sky, but within my psyche." She does say she thinks man will "eventually infiltrate the solar system, and go beyond," but it may cost an evolutionary transformation into "homo mechanicus." Swenson in fact was fascinated by the space program until we stopped sending up manned missions in the early 1970s. Few poets seem to understand the beauty and excitement of the unmanned missions—robot spacecraft like the Pioneers to Venus, Jupiter, and Saturn; Mariner and Viking to Mars; and the sophisticated Voyager missions, which have shown us Jupiter and Saturn in breathtaking detail. Voyager 2 is en route to Uranus and Neptune, and I can hardly wait until 1986 to see the photos. In the poem that I wrote about this, humanity as a species has already become "homo technologicus":

> We were ready for this
> before we believed it could be
> possible—homo technologicus
> sending his eyes and ears, by remote
> control, over a billion
> miles away, to the outer
> planets, and all their planet-sized
> moons. To see these worlds
> for the first time.
>
> Funding problems: "Hey Voyager,
> we're gonna hafta
> pull your plug."
> Voyager: "Are you guys crazy?
> I just sent you fabulous
> pictures of Saturn! And I'm on course

for Uranus and Neptune, and you're gonna
NOT LISTEN?!?"

The data stream. Bit by bit,
the binary numbers transform
themselves into visions cold and
beautiful, or, like Io, bursting
with volcanoes. The sun but a bright
star, and fading all the time.
Radiation, survived
and measured—Jupiter's
magnetic tail a paper dragon
back in the lab.

I am one of very few poets to write about these extensions of human consciousness. Another among the few is Diane Ackerman, who wrote a book called *The Planets: A Cosmic Pastoral.* In her poem "Cape Canaveral" (Ackerman, 1976, p. 59) she saw the tremendous significance of the 1976 Viking probes, sent to look for life on Mars:

how we'd gathered
on these Floridian bogs
to affirm the sanctity of Life
(no matter how or where
it happens), and be drawn, like the obelisk we launch,
that much nearer the infinite.

Ben Belitt agrees: "It is time to reinvent life, / we say, smelling ammonia from Mars / in a photograph" (Taylor, 1974, p. 268). Robert Fitzgerald says, "I regret I shall not be around / to stand on Mars" (Wallace-Crabbe, 1980, p. 39). Ray Bradbury blesses our new home: "Old Mars, then be a hearth to us . . ." (Bradbury, 1982, p. 212). It will happen, perhaps within 20 or 30 years. The first step in this evolution was the first step on the Moon, which I regard as a greater evolutionary breakthrough than the first slither of whatever sea creature first finned its way onto terra firma, because that, at least, was on the same planet! When Neil Armstrong touched the lunar regolith, he said, "That's one small step for a man, one giant leap for mankind," and he was right. I have it on good authority from Thomas Paine, Administrator of NASA at the time, that Armstrong thought this up by himself and kept it a secret until the appropriate moment—just like a poet! In my poem "Apollo 11," I imagine Armstrong feeling, if not articulating, the following thoughts:

The powdery moondust has
just settled—disturbed for the first
time in 4½ billion years, by our

retrorocket. Only with these
electronic shells, this new carapace
we ourselves have made, can we
swim through the vacuum waves
to reach and touch this shore. The
waterplanet's salt surging through my
nerves, my hands tremble on the sturdy
ladder, like electrons excited in
their shells. Dripping with sweat
inside my custom-made spacesuit, I
shift my weight, so strangely
light, down onto my right foot, and touch
the moon. One small step,
launched from the shoulders of
all those who wanted to know, to
take that next step. My reptile
brain dimly recalls the feel of the first
claw that grasped the slimy ocean-edge, to
drag its living body, with its seed,
on through countless bodies, generations
of swimmers, runners, and
flyers, as far as
we can go.

A surprising number of poets are ready to step on the next spaceship, as Donald Hall cheerfully admitted to me that he would. Walter Lowenfels has a poem whose speaker sees "The Impossibilists" announcing, "Today is obsolete!"; they twist strands of DNA into new living creatures and predict Moon gardens (Vas Dias, 1970, pp. 168–169). X. J. Kennedy predicts what appear to be lunar mass drivers like those now being developed by Dr Gerard K. O'Neill's Space Studies Institute at Princeton: "Engines of slag careening from their track / into the unending dark, end over slow end," and he says, "It may well be that when I rev my car / and let it overtake and pass my thinking, / It's space I crave (Vas Dias, 1970, pp. 142–143). Stanley Kunitz's speaker in "The Flight of Apollo" (Vas Dias, 1970, p. 153) says from the Moon that he is

restless for the leap towards island universes pulsing
beyond where the constellations set. Infinite
space overwhelms the human heart, but in the middle
of nowhere life inexorably calls to life. Forward
my mail to Mars. What news from the Great Spiral
Nebula in Andromeda and the Magellanic Clouds?

And Richard Hugo (Vas Dias, 1970, pp. 115–116) welcomes our fusion into the future universe and the new forms of humanity into which we will evolve:

> Beyond Van Allen rings, the stars
> don't glitter, arrogant as moons.
> When did we start? Light-years ago.
> Why did we come? No matter. We
> are not returning to that world
> of ditch and strain, the research terms:
> Cryogenic fuels, free radicals,
> plasma jets, coordinated fusion.
> Only the last, in all this void, applies.
> A universe is fusing in our eyes.

> Why return to air and land, when
> free from weight and the weight
> of hope, we float toward that blue
> that kisses man forever out of form.
> Forget the earth, those images and lies.
> They said there'd be no wind out here,
> but something blows from star to star
> to clean our eyes and touch our hair.

Contemplating this vision and others like it, poets should not forget to acclimatize themselves to absorbing the machine that is making it all possible: the space shuttle. Diane Ackerman (1983, p. 32), in her poem on this latest hybrid between technology and humanity, one which allows us to transform the space beyond the atmosphere into more of the biosphere for ourselves, says of the shuttle astronauts:

> In zero gravity, their hearts will be light,
> not three pounds of blood, dream and gristle.
> When they were young men, the sky was a tree
> whose cool branches they climbed,
> sweaty in August, and now they are the sky
> young boys imagine as invisible limbs.

To summarize the 25 (by 1985, actually 27) years of NASA history and to predict its future, I offer my poem "The House that Jack Built."

In May, 1961, President John F. Kennedy told
Congress: "I believe that this nation should
commit itself to achieving the goal, before
this decade is out, of landing a man on the
moon and returning him safely to the earth."
On July 20, 1969, flowers appeared on Jack
Kennedy's grave, with the note, "Mr. President,
the Eagle has landed."

This is the house that Jack built.

This is the rocket that flew to the moon
and back to the house that Jack built.

These are computers all over the world
that monitor spacemen who walk on the moon,
and splash down to the house that Jack built.

These are the men and women alert
at the Deep Space Network, spanning Earth,
who navigate spacecraft among Saturn's moons—
a new wing for the house that Jack built.

This is an astronaut floating safe and free
because thousands in Houston hold the other end
of the invisible lifeline to Mission Control,
where they keep the Orbiters flying like moons
high through the house that Jack built.

This is New Jersey, a garden at last,
and West Virginia, healed of strip mines past,
with industry in orbit. Solar sails riding sunbeams
float perfect crystals from asteroids to Earth, Earth
rising, swirling crescent in the moon's black sky.
This is why: the house that Jack built.

This is the spaceship Enterprise, the first to fly,
then Columbia, Challenger, Discovery, and then
Atlantis, for all the lost worlds we will find.
We command the fleet that sails by fire
to spin domed cities off of this Earth,
and make new earths out of planets and moons.
This is the house that Jack built.

REFERENCES

Ackerman D. (1976) *The Planets: A Cosmic Pastoral.* Morrow Quill, New York. 159 pp.

Ackerman D. (1983) *Lady Faustus.* Morrow, New York. 94 pp.

Bradbury R. (1982) *The Complete Poems of Ray Bradbury.* Ballantine, New York. 288 pp.

Ciardi J. (1962) *In Fact.* Rutgers University Press, New Brunswick. 68 pp.

Clark A. C. (1969) *The Coming of the Space Age.* Meredith, New York. 301 pp.

Crane H. (1966) *The Complete Poems and Selected Letters and Prose of Hart Crane* (B. Weber, ed.). Liveright, New York. 302 pp.

Dickey J. (1968) "A Poet Witnesses a Bold Mission," *Life,* November 1, 1968, p. 26.

Joseph R. (1966) "These stars distract." *The New York Times,* July 12, 1966, op-ed page.

Marcus D. (editor) (1975) *Irish Poets, 1924–1974.* Pan Books, Berkeley. 203 pp.

Nemerov H. (1966) *Poets on Poetry.* Basic Books, New York. 250 pp.

Pater A. (1981) *Anthology of Magazine Verse, 1980.* Monitor, Beverly Hills. 640 pp.

Peck R. (1971) *Mindscapes: Poems for the Real World.* Delacorte, New York. 165 pp.

Phillips R. (editor) (1974) *Moonstruck: An Anthology of Lunar Poetry.* Vanguard, New York. 181 pp.

Swenson M. (1978) *New and Selected Things Taking Place.* Little, Brown, Boston. 301 pp.

Taylor H. (editor) (1974) *Poetry: Points of Departure.* Winthrop, Cambridge. 345 pp.

Vas Dias R. (editor) (1970) *Inside Outer Space.* Doubleday Anchor, New York. 398 pp.

Wallace-Crabbe C. (editor) (1980) *The Golden Apples of the Sun: Twentieth Century Australian Poetry.* Melbourne University Press, Melbourne. 243 pp.

11 / MARS

OVER THE PAST FOUR YEARS, I have spoken before many audiences on the capabilities of the Space Transportation System and the implications for space development. My talks generally follow the themes in this book and consequently emphasize exploration and utilization of the Moon. I find that almost all who have considered this topic view Mars as the long-term goal for human settlement. Schmitt capitalizes on the martian mystique to propose a national project as a symbolic theme for the technology and spirit of the new century. Many believe the Red Planet should be the next manned goal, bypassing the Moon because "we have already been there." King echoes these sentiments and lays out some arguments for that viewpoint based on scientific knowledge to be gained and the resources available for supporting exploration.

All of us have experienced traveling on the Earth and intuitively expect long journeys to take more energy and fuel than short ones. For destinations in the solar system, the bulk of the fuel is expended in the gravity field of the Earth. Therefore, a spacecraft capable of the round trip to GEO has the propulsive capacity to take payloads to lunar orbit or martian orbit. Payloads delivered to the martian system are smaller because propulsive departure from the Earth's orbit and propulsive insertion at the orbit of Mars demand additional fuel. Landing on and ascending from the surfaces of these planets requires extra capability. As O'Leary points out, the Space Transportation System of the next decades will open a variety of options for exploration, including the moons of Mars and the Earth-approaching asteroids. Since the energy calculations are based on favorable positions of the Earth and the destinations, launch windows for distant bodies are infrequent and travel times are usually long.

As awareness of the versatility of the Space Transportation System increases, consideration is being given to the role of regular missions to the martian system within that infrastructure. The deficiency of the Moon in volatile elements and water makes those materials very expensive outside the Earth's gravity well. Cordell suggests that the moons of Mars may be economically viable sources of water in support of activities in space. Squyres reviews a body of recent scientific thought on the role of water in the geologic evolution of the planet Mars. Obviously, martian water will be utilized by surface explorers, but current space development models do not include planetary surface water as a space resource.

A MILLENNIUM PROJECT—MARS 2000

Harrison H. Schmitt

P.O. Box 8261, Albuquerque, NM 87198

The establishment of a permanent martian base by 2010 would be the first firm step toward human settlement of the solar system away from Earth. Such a goal should be the foundation of an international Millennium Project that will mobilize the energies and imaginations of young people who are already looking beyond Earth orbit and the Moon. Although a Mars 2000 project should be the ultimate aim of space policy for the remainder of the century, it can benefit greatly from successful efforts to create an Earth Orbital Civilization of space stations and permanent lunar bases.

INTRODUCTION

The frontier of space—this new ocean of exploration, commerce, and human achievement—has produced a level of excitement and motivation among the young generations of the world that has not been seen for nearly a century. History clearly shows us that nothing motivates the young in spirit like a frontier. The exploration and settlement of the space frontier is going to occupy the creative thoughts and energies of major portions of human generations for the indefinite future. I find totally unacceptable and totally unrealistic the view that it will be 50 to 100 years before we reach major new milestones in space. People are not going to wait that long. The only principal historical issues in doubt are the roles that will be played by free men and women and how those roles will relate to the problems of the human condition here on Earth.

The return of Americans and their partners to deep space must be viewed in the context of the nation's overall perspective of its role in the future of humankind. That role in space has not been fully formulated into a national consensus. It seems safe to say, however, that for generations now alive we will be the free world's principal agent and advocate in interweaving the ongoing Age of Information, the soon-to-be Earth Orbital Civilization and the world's Millennium Project: Mars 2000.

First of all, we must recognize the obvious. Space provides a major part of the foundation for the Information Age into which we have entered. Satellites provide the essential links in the gathering and transmission of information on a world-wide basis. Reinforced by the advent of modern automatic data processing, weather forecasting, resource identification and monitoring, information systems and general telecommunication systems, satellites and transoceanic cables have created an expanding central nervous system for the planet Earth. This system not only provides a means for dramatically increasing the well-being of all people—particularly those in the less developed countries—but also for keeping the peace.

Second, the unique resources of near-Earth space provide the basis for the creation of an Earth Orbital Civilization. The weightlessness, the high vacuum at high pumping rates, and the unique view of the Earth, Sun, and stars can be utilized for industrial, public service, educational, research, and peacekeeping purposes. These new "resources" are not only accessible to Americans, but to all the free peoples of the world if we are willing to act as their agent.

Finally, the Moon and planets offer both challenges and opportunities to excite existing and future generations of free men and women just as the New World offered both challenge and opportunities to past generations. The extension of our civilization of freedom to the planetary shores of the new ocean of space should be our basic rationale for the world's Millennium Project: Mars 2000. With the successful completion of this project, namely, the establishment of a permanent martian base by 2010, we should see the first firm steps toward permanent human settlements away from Earth.

Our consideration of a return to deep space should take into account the requirements and realities of each of these phases of our national and international role in space. In short, there is as much need for a "Chronicles Plan" now as there was in 1978 when it was first proposed to the Carter Administration and the Congress (Schmitt, 1978).

THE INFORMATION AGE

The Soviet Union shocked the world in 1957 when it simultaneously launched the first artificial satellite of the Earth and the Age of Information. I was an exchange student in Norway at the time and observed firsthand the profound impact this event had on the international student community then in residence at the University of Oslo. There was wonder at this first reach by mankind into the new ocean of space; there was fear that the oppressive Soviet civilization would dominate humankind through use of its new technological prowess.

The collection and distribution of information on a worldwide basis via satellite has provided a distinct change in the course of human history. The most graphic demonstration of this change came when, on Christmas Eve, 1968, hundreds of millions of human beings throughout the world simultaneously had new thoughts about a familiar object in the night sky—the Moon. The men of Apollo 8 were there, and the Moon would never be the same for anyone. Now we realize that the world will never be the same. The Information Age can provide solutions to those age-old problems of the human condition on Earth if we are wise enough to reach out and grasp them.

Information systems technology, in the broadest sense, makes it possible to rationally imagine the gradual elimination of hunger, disease, poverty, and ignorance in underdeveloped portions of the world. These four horsemen of disaster are rushing down on humankind and freedom at unparalleled speeds. However, for the first time in human history, we can consider technically realistic means of stopping their onslaught.

Through information technology and know-how we can and should help underdeveloped nations create agricultural, health, resource, and educational systems that permit their entry into the economic twentieth century. As an astronaut traveling in Africa, Asia, and Latin America, I heard one message from those who do not want dictatorships of either the right or the left. "Send us know-how, not dollars. Dollars just go into the pockets of our leaders; know-how will go into our minds." This is what the Age of Information is all about.

The race to the Moon and the goal of putting men there and returning them safely to Earth provided the intellectual and technological stimulus for the ongoing information revolution. The establishment and support of deep space bases and settlements will provide continued incentives for the improvement of technologies related to computers, automation, data processing, systems longevity, and telecommunications. All such technological improvements, as well as the international collaboration involved in creating them, will increase the leverage we have to remove the underlying causes of much of the world's unrest that threatens the lives and well-being of its inhabitants.

In the crucible of modern history there are several significant but far too few foreign policy experiments showing that efforts by the United States to transfer know-how can be both successful and well received. For example, there was the early Peace Corps, before well-meaning but inexperienced college students replaced the highly motivated and knowledgeable American professionals from all walks of life that gave so much of themselves. There is the training in telecommunications provided by U.S. industry on behalf of INTELSAT and revived today by the privately sponsored United States Telecommunications Training Institute. There is educational cooperation such as that between New Mexico State University and the technical colleges of Mexico. There is the provision of national technical services and training through contracts between U.S. industry and Third World countries.

Such activities should be the wave of the future rather than the exceptions of the past. With their expansion, there finally can be hope that moderate political forces for freedom can replace the totalitarian forces for oppression.

EARTH ORBITAL CIVILIZATION

The tangible beginnings of the creation of an Earth orbital civilization came with the launch of Skylab in 1974. The Skylab missions, followed by those of Salyut, Spacelab, and the Space Shuttle, began the examination of many of the potential uses of the resources of near-Earth space. Skylab gave direction to our imagination about the use of these resources. The Salyut, Spacelab, and the Space Shuttle now give license to our imagination. Space stations will give reality to this imagination.

The resources of a near-Earth space civilization are basically three in number: (1) instantaneous and continuous view of the Earth, the Sun, and deep space; (2) an infinite quantity of clean, ultra-high vacuum at high-pumping rates; and (3) a weightless environment, that is, an environment free of most gravitational stress.

An instantaneous and continuous view of the Earth and its total environment makes possible a wide spectrum of space activities. Observatories become possible from which research and services in meteorology, oceanography, geology, and ecology can be conducted and from which broad-scale explorations for new terrestrial resources can be carried out. Man's application of his eyes, his mind, and his imagination to the direction of remote-sensing systems presently has unlimited potential to benefit humankind.

The Sun remains the major contributor to the stability and the changes of our magnetic, atmospheric, oceanic, and biological environments. However, our knowledge of how the Sun's energy and magnetism influence changes in our environment is extremely limited. A continuous view of the Sun outside the shielding effects of the Earth's atmosphere will first give us understanding of processes taking place on and in the Sun and of the nature of their interactions with the Earth. Then, eventually, this view will form the foundation for forecasts that will protect against, or take advantage of, the effects of interactions between Sun and Earth.

One of our greatest sources of intellectual strength and scientific vitality is in continued observation of the universe and attempts to understand the stars and interstellar space. The greatest discoveries of the future probably lie in investigations of stellar and interstellar phenomena. The nature of gravity; the origin of plants; the limits on our ability to manipulate matter, energy, and time; and our future as explorers of the universe are all issues at stake.

Nearer to home, major limitations to experimentation and manufacturing in many areas of physics and chemistry exist on Earth because of the difficulty in maintaining clean, ultra-high vacuums in large volumes. The existence and accessibility of such a vacuum resource in near-Earth space open new dimensions in the study and use of physical and chemical theory.

The continuous absence of gravitational stress, that is, weightlessness, provides a unique experimental and practical environment heretofore unavailable to man. Extensive experimentation and manufacturing has never before been possible where convection does not exist, where containers for fluids are not required, and where gravitationally unconstrained crystal or biological growth can occur. The absence of gravitational stress means that no containers are required to hold the experimental materials. Thus, new investigations of the dynamics and properties of fluids and materials formed from fluids and emulsions are now feasible. These investigations, along with other basic research in physics, chemistry, and biology, are leading to commercial applications in space manufacturing facilities.

Further, in medical laboratories in space we now can look continuously and in great detail at cell, animal, and human growth and function in the absence of gravitational stress. We can also look at the applications of such an environment in medical treatment and recovery as well as in the production of medicine. This is probably the appropriate place to point out that those men and women who are handicapped by gravity on Earth will have no significant handicap in the weightless environment of space. They can be just as productive living and working in this environment—where movement is accomplished merely by the push of a finger—as can those who are not gravitationally handicapped.

Finally, let us consider the possibility of a space education facility where one can conceive of students of all ages participating, limited only by interest and minor physical criteria. One now can conceive of spaceflight for students of all disciplines from the nuclear physics major to the medical student to the poet or novelist. One also can conceive of students in space who represent all nationalities and who will continue the great traditions of space that bring people and nations together.

The stimulation triggered in young minds by a week or summer in space defies the imagination. The fact that we can now consider education in space as a rational possibility is a measure of what transpired in the two decades of Apollo, Skylab, Salyut, Spacelab, and Shuttle. A change of course has been made; each generation, ours and those that follow, must determine what exact course to chart, but proceed we must; history does not allow us to stand still.

A return to deep space and the establishment of lunar and planetary settlements can play an extraordinarily important role in an Earth Orbital Civilization. The potential resources that can be derived from the surface materials on the Moon may well sustain both the transportation and manufacturing economics of that civilization.

The management (governance) of a lunar settlement that supports the activities of the stations of an Earth Orbital Civilization, as well as supporting its own existence, must conform not only to existing space law and precedents, but also to the cooperative urge that space endeavors have generated in the human psyche. The only practical way that this management system can come into being appears to be to follow the precedents set by the Intelsat and Inmarsat organizations. These are user-based and profit-making management systems rather than one-nation/one-vote or national systems. Such user-based systems for lunar base management, such as the Interlune proposal (Joyner and Schmitt, 1985) draw their strength from the economic self-interest each member nation has in the success of the enterprise. The integration of our own free-enterprise system with such organizations has been shown to work very well. Ultimately, we should expect a lunar settlement to become independent of Earth-based institutions as it develops diverse trading relationships and becomes largely environmentally self-sufficient.

THE MILLENNIUM PROJECT: MARS 2000

No matter what other justifications may be given, the ultimate rationale for today's generations to return to deep space and to establish a permanent presence there is to create the technical and institutional basis for the settlement of Mars. This will be the first great adventure for humankind in the third millennium after the birth of Christ.

Steadily increasing philosophical and psychological momentum for this adventure is building among the young people of the Earth. There is great privilege and enjoyment in spending hundreds of hours in the schools of America and the world talking with these future space travelers about Apollo's space experiences and listening to their view of the future. Let us examine that view more closely.

College and high school students seem to have their impressions of space flight and "what was it like" fairly well formed. Unlike most adults, however, the students of elementary school age ask questions purely and simply because they are curious. After a few encounters with adult audiences, most of the questions of succeeding audiences can be anticipated. However, one never can anticipate all or even most of the questions from students in elementary school who want to go to space.

In order to test awareness of young students, and before showing colored slides from my trip to the Moon, I ask a few questions.

To be sure of the general knowledge of the group, I ask, "What is gravity?" Lots of hands go up and the typical answer is, "It is what holds you down." Excellent answer; in fact, few adults know much more than that unless they are capable of grasping or debating Einstein's theory of general relativity.

The next question I ask is, "How many of you would like to go to the Moon someday?" About 75% of the hands of fifth-grade and younger students go up (older students are more reluctant to raise their hands if Dick and Jane aren't going to).

"Now, how many of you would like to go to Mars?" About 85% of the hands go up. Why 10% more? I asked some of those who raised their hands for Mars and not the Moon, "What is wrong with the Moon?" The answer: "You've already been to the Moon!"

Ten percent are never satisfied. Rather than the Moon, many have their eyes on Mars. They are the ones who will go to Mars. They are the ones, like most of our ancestors before them, who will never be satisfied with either the comforts or the restrictions of home and Earth. These are the parents of the first martians.

The importance to the parents of the first martians of a self-sustaining settlement on the Moon—trading directly with our Earth Orbital Civilization of permanent space stations—is that it gives us the technical and institutional basis to go to Mars with the purpose of establishing a permanent base on the first expedition. This expedition could be on its way by the end of the first decade of the third millennium. A permanent settlement will take a little longer, but a permanently occupied base—resupplied by regular interplanetary space stations—clearly will be possible as well as desirable soon after the establishment of a permanent lunar settlement and an Earth orbit space station.

Why the hurry? Why a Millennium Project that stretches our reach to the limit? The answer is in the minds of young people who will carry us into the third millennium. It is in the generations now in school, now playing around our homes, now driving us to distraction as they struggle toward adulthood. They will settle the Moon and then Mars. They will do this because they want to do this. They want to "be there." Our role is merely one of staying out of their way while we preserve and expand their opportunities.

The answer to "Why the hurry?" is also clear in the determination of the Soviet Union to establish its sovereignty in deep space and on Mars before the forces of freedom do so. Very long duration Earth-orbit flights by the cosmonauts, heavy-lift launch vehicle development, and their public emphasis on manned Mars exploration all tell us what the Soviets expect to do before the end of the twentieth century. An attempt to put Soviet cosmonauts in the vicinity of Mars by October 1992, the 75th Anniversary of the Bolshevik Revolution, is not only possible, it is highly probable. How sad if this adverse trend of political history is established in the 500th year after the discovery of America by Columbus.

Mars 2000 will be for the children of the free world what space stations were to their parents and what Apollo was to their grandparents: the total embodiment of the best in the human spirit. Maybe most importantly, if our determination is unequivocal, astronauts and cosmonauts may be able to join hands in this great adventure.

There is little technical distance between us today and the realization of all that I have suggested. Certainly, there is little to be done compared to the task that faced us when we began the race to the Moon. Whatever technical options may turn out to be appropriate, now is the time to create those options so that the next generation may proceed when they are ready.

The first martian base will probably be established using inflatable shelters in one of the deep equatorial valleys near areas that show strong photographic evidence of being underlain by water-rich permafrost. The Valles Marineris is such a valley. The deep valleys also will provide somewhat higher temperatures and atmospheric pressure than other possible sites such as the polar regions where water-ice is clearly present. However, one of the advantages of using a large interplanetary space station for the first trip to Mars is that the time necessary to examine the martian moons for resources and to select a proper site for the first base can be spent in orbit about the planet before committing to the first landing. If necessary, reconnaissance of several sites can be carried out before committing to a site for the base.

The most inconvenient aspect of living and working on Mars will be the dust. Unlike the Moon, which has no atmosphere, dust on Mars blows around in great global storms and settles very slowly. It may be that these storms will require the temporary confinement of the explorers to shelters much like in winter at Antarctic bases.

One of the major uncertainties that Mars 2000 will have to deal with is the chemical and agricultural nature of the martian soil. It is highly probable that, like the lunar soil and new volcanic ash, the martian soil is very fertile. In fact, it may be rich in particles of clay as well as impact and volcanic glass. However, the existence of a martian atmosphere, its oxidizing nature (the "red" planet), and the possible presence of sulfur in the soil crusts may mean that some treatment of the soil will be required before it can be farmed properly. An alternative for growing food may be hydroponics, depending on the availability of sufficient quantities of water. Again, the interplanetary space station used for transport should be large enough to outfit for either of these options as well as for large quantities of "imported" food.

The confidence we can have in discussing the establishment of a permanent base on Mars comes from two directions: the confidence and knowledge gained from the Apollo expeditions to the Moon and the spectacular and detailed data returned by the Viking landers and orbiters of Mars. We know nearly as much about Mars as a planet as we did about the Moon before Armstrong and Aldrin landed there in 1969.

However, space activities will be sustained less by technology and knowledge than by emotions: the emotions of young Americans and young people the world over. Indeed, even young Soviets, East Europeans, Chinese, and Cubans also must look to space as the Earth's frontier. As with our ancestors, their freedom lies across a new ocean, the new ocean of space.

The Millennium Project: Mars 2000 is their hope as well as our mission.

REFERENCES

Joyner C. C. and Schmitt H. H. (1985) Lunar bases and extraterrestrial law: General legal principles and a particular regime proposal (abstract). In *Papers Presented to the Symposium on Lunar Bases and Spaces Activities of the 21st Century*, p. 40. NASA/Johnson Space Center, Houston.

Schmitt H. H. (1978) National Space and Aeronautic Act of 1978. *Congressional Record*, Vol. 124, No. 167, Part IV.

MARS: THE NEXT MAJOR GOAL?

Elbert A. King

Department of Geosciences, University of Houston, Houston, TX 77004

The next major thrust into space may be strongly tied to the occurrence and availability of extraterrestrial resources. An attempt to exploit material resources at any extraterrestrial location will require a major commitment over a period of many years. The magnitude of this commitment probably will not allow simultaneous vigorous pursuit of other major projects. Thus, it behooves us to consider our mission options carefully and to design a program that will produce access to the most needed resources, such as hydrogen, oxygen, water, carbon dioxide, and other volatiles, at locations that will help us to achieve our most important goals. The exploration and exploitation of Mars and its moons offers an attractive alternative to possible lunar options and would ultimately lead to colonization of the planet. Serious planning must be initiated soon for manned missions to Mars and its moons if we are to make informed decisions about the next few decades of human activity in space.

INTRODUCTION

Virtually everyone believes that our civilization will send manned missions to Mars and that it is desirable to do so, and authors of NASA publications predict or assume that manned exploration of Mars will occur (*e.g.*, French, 1977; Glasstone, 1968). Mars has fascinated some high-level NASA officials who have published extensively on manned missions to Mars (*e.g.*,von Braun, 1962; Ley and von Braun, 1956), but while colonization of the "Red Planet" is fondly mentioned in some studies, the actual event is usually assigned to some indefinite date in the distant future. This need not be the case! The United States space program is now committed to the construction of a space station during the early 1990s. A properly designed, low Earth orbit space station, with capability for on-orbit assembly and servicing of stages and modules, would complete the infrastructure that is needed to support manned missions to Mars.

WHY GO TO MARS?

Stated very simply, Mars is the most "user friendly" planet (for humans) in the solar system next to Earth. Because of its surface environment, scientific importance, accessibility, and abundant resources (particularly volatiles), it is a logical place for a permanent base and by far the most attractive planetary surface locality for a colony. For details of Mars' atmosphere and surface environment see Scientific Results of the Viking Project, *J. Geophys. Res.*, 82, 3959–4681 (1977); *J. Geophys. Res.*, 84, 7909–8544 (1979); *J. Geophys. Res.*, 87, 9715–10,306 (1982). Let us enumerate some of the most attractive features of Mars.

Atmosphere

The general abundance of volatile elements on Mars and its size result in the presence of a thin atmosphere comprised chiefly of carbon dioxide with small amounts of water vapor, nitrogen, argon, and other gases. The total atmospheric pressure depends on the local elevation but averages approximately 6 mbar. This includes approximately 30 micrometers of precipitable water. The atmosphere is available everywhere on the martian surface and constitutes a resource that can be used for the production of life-supporting oxygen, potable water, and rocket fuel. The ready availability of carbon dioxide and water greatly simplifies possible agricultural activities. The presence of an atmosphere also provides the opportunity for aerobraking of spacecraft at Mars, thereby conserving fuel and total launch weight for missions outbound from Earth. In addition, the atmosphere provides complete protection from micrometeorites and, in fact, offers protection from meteorites up to several hundred grams in size! This atmosphere has provided Mars with a weathered surface that probably contains hydrated minerals and may contain other surface or near surface ore deposits such as weathering products, lag, placers, and evaporites. Because of the atmosphere, Mars has a certain amount of weather. Although occasionally adverse, weather conditions may serve to provide psychologically satisfying changes for the expedition members.

Size and Rotation

Mars' size, between that of the Earth and Moon, resulted in a geologically more active planet than the Moon. Such familiar features as many large volcanic constructs are clearly seen and mapped on the martian surface. How late this activity continued into Mars' history is not known, but the implied high level of volcanism probably provided greater opportunity for the occurrence of certain types of ore deposits, *e.g.*, hydrothermal ores, than on the Moon. The mass of Mars provides surface gravity of about 0.4 g, a comfortable and easy working environment. Mars' rather rapid rotation not only causes some weather, but results in a day/night cycle similar to that of the Earth that humans are likely to find familiar and easy for living adaptation.

Polar Caps and Permafrost

We know that there are large accumulations of volatiles, particularly water and carbon dioxide, in the martian polar caps. These volumes are so large as to constitute a virtually inexhaustible supply. Although the extents of the polar caps vary seasonally, they still contain at maximum recession sufficient volatiles for virtually any conceivable use. There is much visual evidence of volatiles, probably permafrost, in the shallow martian subsurface as shown by certain crater morphologies, patterned ground, and large effluent channels.

Surface and Shallow Subsurface Rocks

Photogeology from orbital imagery (Mutch *et al.*, 1976; Carr, 1981; also see the previously noted general information on Mars) reveals that many of the surface rocks of Mars are volcanic. Some of the volcanoes strongly resemble terrestrial basaltic volcanoes, and it is possible that more silica-rich differentiates are present. In addition, there are many wind-deposited sediments on the surface and, by inference, sedimentary rocks. The Viking landers provided us with chemical analyses of these sediments at two localities. Also clearly visible on Mars are impact craters of a wide range of sizes, which ensures the presence of rock glasses and fragmental debris commonly associated with such craters.

Arguments have recently been made that some meteorites, *i.e.*, SNC meteorites of basaltic affinities, originated as impact ejecta from Mars (Nyquist, 1983). Thus, we may have samples of some martian rocks already available for study. In any case, it appears that the range of surface compositions and, hence, variety of surface resources, is at least as great as the Moon and probably much greater.

Scientific Interests

While the scientific cream has already been skimmed off the top of lunar studies by the Apollo and Luna Programs, Mars still awaits detailed scientific exploration. Important questions remain about Mars relative to biological sciences. There is still the possibility of present or past martian life forms, although none were detected by our Viking landers and the prospects for a positive result do not appear encouraging. Planetary scientists are eager to establish martian geological and climatological history and to make comparisons with the Earth and Moon. Detailed investigations of the atmosphere, polar caps, and the results and dynamics of their interactions with the surface and subsurface await us on Mars. The list of physical sciences measurements desired is too long to be enumerated here, but it is much longer than that for the Moon because Mars is a much more complex planet (Mutch *et al.*, 1976; Carr, 1981; and previous note).

Phobos and Deimos

Mars has two moons that are important objects for scientific exploration in their own right. Most planetary scientists believe them to be captured asteroids. Their own albedos and densities indicate that they probably are composed of primitive, water-bearing carbonaceous chondrite material. Thus, they not only provide possibly important mission options for reconnaissance of the martian surface but offer the possibility of providing another source of mission expendables such as water, oxygen, and rocket fuel (O'Leary, 1984). Because of the low total energy required for a round trip from low earth orbit, as well as scientific and exploration considerations, Singer (1984) has proposed an early manned mission to Phobos and Deimos.

Public Interest

Mars is well known to the public. It has been popularized by science and science fiction alike for decades, *e.g.*, Percival Lowell, Carl Sagan, Buck Rogers, Edgar Rice Burroughs, Orson Welles, *etc.* Nonetheless, Mars is new! Men have not set foot there. The pure excitement of exploration will capture the public imagination as it did with Apollo. A return to the Moon would not stimulate such interest, but humans on Mars will.

Even with all of its attractive features, the low temperatures of the martian surface, lack of free oxygen in the atmosphere, and radiation background render Mars an extremely harsh environment for human habitation. Even so, it is a far better environment than the Moon!

HOW ACCESSIBLE IS MARS?

Mars is closer than you believe. In terms of propulsion energy, a round-trip mission to a moon of Mars from low Earth orbit requires less propulsion energy for an equivalent mass spacecraft (assuming aerocapture at Mars and Earth) than does a round-trip mission to the surface of the Moon (assuming aerocapture at Earth). With the same aerocapture assumptions, a round-trip mission to the surface of Mars requires a delta V of only approximately 1400 m/s more than a round-trip mission to the surface of the Moon. A detailed summary of the energies required for various mission options is given by O'Leary (1984). The difference is in travel times and frequency of launch opportunities. A one-way trip to Mars on a minimum energy trajectory requires approximately 270 days, while a one-way trip to the Moon requires only approximately 3 days. Launch opportunities to Mars from Earth are far less frequent than for the Moon, as are return opportunities. These and other considerations result in a general requirement for larger and more massive spacecraft for missions to Mars than to the Moon.

For any long duration, manned space mission we will require development of reliable long-term life support systems, whether going to the Moon or to Mars. The place to test such systems is in low Earth orbit as the Russians have done in recent years. Also, biomedical research must identify exercise, diet, and other factors that will prevent significant physiological deterioration during long periods of reduced gravity, and/or spacecraft must be designed to include artificial gravity systems. Adequate radiation protection must be provided for long duration flight crew members. This might be provided by covering most of the exterior of the crew compartments with bagged rock material collected at Phobos or Deimos and returned to low Earth orbit by an unmanned mission. Although manned Mars missions are possible with conventional hydrogen/oxygen propulsion, consideration should be given to the development of higher specific impulse propulsion systems, which would greatly enhance missions to deep space by shortening trip times and expanding launch opportunities. It appears that these developments are solvable engineering and medical problems, not greatly more complex than those that have been solved in many previous space flight programs. If the space station were to evolve into the assembly and servicing platform that we need for mission operations, we would then have all of the required infrastructure to support the manned exploration of Mars—except for the most important ingredient of all—a bold political decision!

NEW SPACE RACE TO MARS?

A new "space race" to Mars has already begun! The Russians are well along the way to a capability to send humans to Mars. A similar opinion has been expressed by Paine (1984). Their long duration Earth orbit missions have qualified their life support systems for deep space missions. The Soviet interest in Mars is long standing, as witnessed by their "flotilla" of four spacecraft launched to Mars in 1973 and their earlier efforts. The Russians have announced their intention to send an unmanned mission to Phobos/ Deimos in 1988. Furthermore, their development of large lift rockets will provide them with the ability to deliver large payloads to low Earth orbit and beyond.

The United States is not completely unprepared for this race. We have explored Mars with eight unmanned probes including three flybys, three orbiters, and two landers. We know more now about Mars than we knew about the Moon when the Project Apollo decision was made. We should not view our present second place in this race with undue alarm. After all, the winner of every peaceful space race is the human race, and there is always the possibility of international cooperation in space projects. Also, the race to Mars will not be won with a single manned landing as was the race to the Moon. The race to Mars will be won by the nation whose human expedition arrives at Mars with the means and the will to stay!

CONCLUSIONS

Most importantly, we must initiate serious planning for human missions to Mars. These plans must include mission profiles, engineering concepts, technology assessments, and cost analyses. These studies should constitute an important part of the basis for an informed political decision in the near future that will identify our national goals and future programs in space.

The design of the planned low Earth orbit space station should allow the flexibility to add facilities for the assembly and servicing of large interplanetary manned spacecraft.

We must continue to expand our Mars research and resource evaluation programs. This data base will continue to be of great value for mission planning.

Without the benefit of the actions above, we run the serious risk of making a decision that, in effect, will cede Mars to the Soviets for decades!

REFERENCES

Carr M. H. (1981) *The Surface of Mars.* Yale Univ. Press, New Haven. 232 pp.

French B. M. (1977) *Mars: The Viking discoveries.* NASA EP-146. NASA, Washington. 36 pp.

Glasstone S. (1968) *The Book of Mars.* NASA SP-179. NASA, Washington. 315 pp.

Ley W. and von Braun W. (1956) *The Exploration of Mars.* Viking Press, New York. 176 pp.

Mutch T. A., Arvidson, R. E., Head J. W. III, Jones, K. L and R. S. Saunders (1976) *The Geology of Mars.* Princeton Univ. Press, Princeton. 400 pp.

Nyquist L. E. (1983) Do oblique impacts produce martian meteorites? *Proc. Lunar Planet. Sci. Conf. 13th,* in *J. Geophys. Res., 88,* A785–A798.

O'Leary B. (1984) Phobos and Deimos as resource and exploration centers. In *The Case for Mars,* pp. 39–65. American Astronomical Society and Technology Series, v. 57. Univelt, San Diego.

Paine T. (1984) A timeline for martian pioneers. In *The Case for Mars II,* Univelt, San Diego. In press.

Singer S. F. (1984) The Ph-D proposal: A manned mission to Phobos and Deimos. *AAS, Sci. Technol., 57,* 39–65.

von Braun W. (1962) *The Mars Project.* Univ. Illinois Press, Urbana. 91 pp.

RATIONALES FOR EARLY HUMAN MISSIONS TO PHOBOS AND DEIMOS

Brian O'Leary

Science Applications International Corporation, 2615 Pacific Coast Highway, Hermosa Beach, California 90254

S. F. Singer proposed that Phobos and Deimos be manned bases for unmanned Mars exploration. A second rationale, described herein, concerns using the resources of the martian moons to lower the cost of the missions and to accelerate the onset of space industrialization utilizing non-terrestrial materials. This paper presents estimates of the ballistic velocity increments required to rendezvous with the two satellites, depart, and return to the Earth. When aerobraking at both Mars and Earth are considered, Phobos and Deimos are considerably more accessible in Δv than are the lunar surface, the martian surface, and most known Earth-approaching asteroids. Launch windows from Earth are usually more frequent than for the most accessible asteroids. Because of their low albedoes and mean densities, both satellites are likely to be carbonaceous, with large quantities of easily accessible water. Phobos and Deimos are ideal early resource targets and Mars exploration bases. These considerations suggest that the martian moons be early destinations that could dovetail two important goals in future space activity: the exploration of Mars and materials recovery.

S. F. Singer (1984) presented a scientific rationale and plan for the manned exploration of Phobos and Deimos (PhD) and subsequent unmanned exploration of the martian surface. The PhD mission, wrote Singer, would provide a number of advantages over either sending unmanned rovers or people directly to Mars first. One, telemetered control of rovers from the Mars-facing sides of the satellites serving as stable platforms will shorten the response time from fifty minutes to less than 0.2 seconds. Two, surface and sub-surface martian samples can be recovered at disparate sites by remote control and examined in a PhD laboratory rather than having to resort to a sophisticated laboratory on Mars or to run the risk of back-contamination by direct sample return to the Earth. Three, PhD-controlled surface rovers allow for sequential exploration of Mars at several locations rather than a single-mission, one- or two-location direct shot that may miss the most interesting spots (Viking and Apollo suffered from this problem). Four, unmanned landings are easier, less costly, safer, and could be done much sooner. And fifth, we could directly sample these mysterious moons to determine their history and fate, gaining valuable insight on the origin and evolution of small bodies in the solar system.

Adelman and Adelman (1984) have presented advantages of using Phobos as a base for gravity wave astronomy, radio astronomy, astrometry, and the study of asteroids and comets. Also, a Mars-pointing telescope on Phobos will permit resolutions an order of magnitude or more greater than that of the Viking orbiters (Sagan, personal communication, 1984).

O'Leary (1981, 1983, and 1984) has investigated the resource potential of the martian moons and their accessibilities. Both scientific and resource rationales are summarized in this paper. The Soviets are planning an unmanned reconnaissance mission to Phobos in 1988 (Covault, 1985). Clearly, the incentive to explore and utilize the satellites of Mars is increasing.

The very low mean densities of both moons (~2 g/cm^3) and their low reflectivities are strong indicators of a chemical content similar to that of the volatile-rich carbonaceous chondrite meteorites (Veverka et al., 1978). If this is true, easily extractable water is likely to be found on both moons.

Previous studies have examined the rationale (O'Leary, 1977, 1983, and 1984) and mission scenarios (O'Leary et al., 1979; O'Leary, 1983) for setting up resource recovery operations on selected Earth–approaching asteroids during special launch opportunities. The martian moons can also be considered as Earth–approaching asteroids with mining potential and the added features of more accessibility of volatiles and the opportunity to explore Mars. The most accessible known asteroid, 1982 DB, offers slightly better opportunities at decade intervals, but missions in the intervening years deteriorate rapidly (Hulkower, written communication, 1982).

Earlier studies (O'Leary et al., 1979; Staehle, 1983) have also pointed out that, in terms of the energy (and cost) of setting up an oxygen–extracting plant on a non-terrestrial body, the presence of water and other volatiles is far more preferable than the chemical processing of metal oxides that predominate in the lunar soil and many meteorite classes. Carbonaceous materials also make available hydrogen, carbon, and nitrogen—elements that will become useful for refueling and life support.

In the long run, the absence of these materials on the Moon or asteroid would necessitate expensive resupply from the deep gravity well of the Earth. Cordell (1984) has estimated that the energy cost of delivery of water to lunar base from Phobos and Deimos would be approximately 3 times less than that delivered from the Earth's surface.

The ballistic velocity increments (Δv) required to rendezvous with Phobos, depart, and return to the Earth have been estimated (Von Herzen, written communication, 1979). Table 1 shows the results for average launch windows from Earth every two years, assuming

Table 1. Mission Opportunities to the Most Accessible Known Extraterrestrial Resources

Target	Launch Date (or Frequency)	Round-Trip Travel Time	Velocity Intervals ΔV (km/s)			
			Escape	Rendevous (Land)	Depart	Total*
Lunar Surface	Frequent	$\geq$ 7 Days	3.2	2.7	2.4	8.3
Martian Surface (Atmospheric Braking)	Every 2 Years	$\geq$ 2 Years	3.6	$\geq$1.0	$\geq$5.6	$\geq$10.2
Phobos and Deimos	Every 2 Years	$\geq$ 2 Years	3.6	1.9	1.8	7.3
Phobos and Deimos (Mars Aerobrake)	Every 2 Years	$\geq$ 2 Years	3.6	$\geq$0.5	$\geq$1.9	$\geq$6.0
Asteroid 1982 DB	Sept. 2001	4 Months	3.2	5.8	4.3	13.3
Asteroid 1982 DB	Jan. 2001	2.0 Years	4.3	0.8	0.5	5.6
Asteroid 1982 DB	Dec. 2001	2.1 Years	4.5	0.5	0.7	5.7

*These figures assume aerobraking at Earth.

Hohmann transfer ellipses between circular, co-planar orbits. Primarily because of the eccentricity of Mars' orbit, the figures for each launch window will vary up and down from the values reported here, but they are representative values for mean opportunities, as indicated by cross-checking with analyses of specific mission opportunities (Staehle, 1983; Hoffman and Soldner, 1984).

Table 1 shows that the total mission velocity increment for Phobos, Deimos, and other targets can be divided into three principal maneuvers: escape from low Earth orbit onto the transfer ellipse, rendezvous with Phobos/Deimos either by an impulsive maneuver near Mars or by a circularization after one or more aerobraking maneuvers at the top of the martian atmosphere and return on a transfer ellipse to the Earth with either aerobraking reentry, aerobraking injection into low Earth orbit, or an impulsive braking at low Earth orbit. Alternatively, a Venus gravity assist inbound or outbound would shorten the total mission time to $\leq$ 700 days with similar total Δv's (O'Leary, 1985).

A large benefit comes from aerobraking at Mars. Once aerocapture is achieved, either through one pass or successive passes through the martian atmosphere, the Δv required to circularize at Phobos is 590 m/s and at Deimos, 667 m/s. The alternative is to break impulsively in the vicinity of Mars to eliminate the excess hyperbolic velocity, change planes, and then rendezvous with the satellites. These maneuvers consume a Δv to 1.5–2.0 km/s.

Aerobraking would also be used upon returning to the Earth or Earth orbit. Current studies of orbital transfer vehicle (OTV) systems to be used in the post-1996 time frame show an aerobrake capability (Hoffman and Soldner, 1984; Walberg, 1983). Because the velocities of atmospheric encounters projected for the Earth are greater than those at Mars, no significant design requirements need be added to a spacecraft with round trip capability to Mars. Although the heating is different and the spacecraft mass is greater at Mars than the Earth, heat shields can be distributed among spacecraft components to minimize the mass.

Table 1 shows a result that may be surprising at first glance: that even without aeroassist at Mars, the moons of Mars are more accessible to the Earth at biennial opportunities than is the Moon of the Earth. The chief difference is in the requirement to soft-land payloads on the lunar surface. With meager escape velocities of 11 and 6 m/s for Phobos and Deimos, the impulse required to "land" on these moons is negligible. Required maneuvers more resemble rendezvous and docking with a spacecraft rather than impulsive blasting into an out of a gravity well. As in the case of the asteroids, the PhD missions permit low impulse propulsion for the entire trip, opening the possibility of using solar electric engines, mass-drivers, tethers, and solar sails as the sources of propulsion.

The only advantages the Moon seems to offer are its proximity and launch window frequency of days versus months to years. However, once it has been established that humans can survive in space over long periods of time and once mining and exploratory operations begin and a pipeline of extraterrestrial materials starts to flow, it will probably not make much difference how close the source of materials is to the Earth. More significant will be the energy required to process and transport them, and Phobos and Deimos clearly provide an advantage. If the moons of Mars were moons of the Earth, we would be there by now.

Two Apollo circumlunar flights were attempted before a lunar landing. Likewise, safety factors and economics suggest going to Phobos and Deimos before attempting to land humans on the surface of Mars. The 1984 Office of Technology Assessment report to Congress on the U.S. space program refers to missions "to the vicinity of Mars" rather than Mars itself. Former Apollo astronaut and U.S. Senator Harrison Schmitt (1984) foresees a possible Soviet human mission to Phobos by 1992.

Phobos and Deimos offer more than enough materials to industrialize space. Within their volumes, they contain the equivalent of 10 million 100-m-diameter (1 million ton) objects! The milligravity environment of the martian moons will permit certain industrial operations that may be more easily carried out than either in weightlessness or under the influence of a significant gravitational field. Dust, equipment, and people will not float away, yet structures need not be built to withstand a planetary or lunar gravity.

Tables 1 and 2 summarize the relative accessibilities of Phobos, Deimos, selected Earth-approaching asteroids, the lunar surface, and the martian surface. From a resource retrieval perspective, Table 2 expresses the approximate ratio of returned to invested mass from these objects given assumptions that may apply to resource retrieval missions during the early 21st Century. A total outbound Δv of 5 km/s is assumed in each case; this reflects the velocity increment required to escape from low earth orbit and subsequently land on or rendezvous with the target object. While individual opportunities

Table 2. Allowable Returned Mass of Processed Metals as a Function of Ballistic Return Velocity Increment (Δv)*

Return ΔV (km/s)	Tons of returned mass per ton of invested mass	Examples	Launch Date (or Frequency)	Round-Trip Travel Time
0.1	100	1982 DB, at special	2000–01	2 Years
0.2	49	times	(Every two decades)	
0.5	19.2	1982 DB, 1943	Occasionally (For a	2–3 Years
1.0	8.9	Anteros, 433 Eros, and 1982 XB	given object, every decade or more)	
1.9	4.0	Phobos and Deimos	Every 2 years	2 Years
2.4	3.0	Lunar Surface	Frequently	~7 Days
~5	1.0	Martian Surface 1982 DB Several Near-Earth Asteroids	Every 2 Years Sept. 2001 Frequently	~2 Years 4 Months ~1–5 Years

*See text for assumptions.

will vary, this number is probably within 1 km/s of actual cases, yielding comparable mass fractions for outbound journeys.

It is further assumed that the same cryogenic rocket with exhaust velocity 4 km/s is used for the return trip as for the trip out and that the rocket is refueled with liquid oxygen (and possibly liquid hydrogen) obtained at the target object. The final assumption is that any excess hyperbolic velocity of the incoming resources to the vicinity of the Earth can be eliminated either by planetary/lunar gravity assists (O'Leary, 1979) or by aerobraking at Earth using the raw materials retrieved from the target object [Gaffey and McCord (1977) have proposed the vacuum foaming of asteroidal metals as a reentry heat shield that could be subsequently used as resources on the Earth or in Earth orbit]. The resultant mass fractions therefore reflect only the impulse required to depart from the target object on an Earthward trajectory; this supposition remains to be tested by more detailed spacecraft design and mission concepts.

It is clear from inspecting Tables 1 and 2 and from the preceding paragraphs that Phobos and Deimos offer a unique blend of exploration, resource processing opportunities, accessibility, and frequency of launches. These factors combine to create a powerful incentive to conduct a detailed analysis of mission scenarios (O'Leary, 1985).

As for which moon to target, Singer (1984) has proposed Deimos to be the first manned destination, primarily because the outer moon's orbital period is nearly synchronous with Mars' rotation, allowing for more continuous telemetered operation of surface rovers. Phobos, on the other hand, provides some advantages. In terms of Δv, it is more accessible to the martian surface. Its position deeper within the martian gravity well grants it slightly more accessibility for circularizing after an aerobraking encounter but normally slightly less favorable return-to-Earth circumstances (unless periapsis maneuvers are performed for Mars escape). Phobos also provides a higher resolution view of the martian surface. By the time a manned mission is launched, we are likely to have more knowledge about the composition of Phobos than that of Deimos. A confirmation of water in the forthcoming Soviet mission to Phobos might be the deciding factor.

In reality, it will be desirable to visit both moons to assess both bodies scientifically. From an operational point of view, redundancy is desirable. In the long run, the two satellites could serve complementary roles: Deimos for astronomy, planetary coverage, reconnaissance, and control of Mars rovers; Phobos as a site for volatile mining if water is found there, sorties to and from the martian surface, and closeup surveillance of interesting sites. It may also be convenient for Soviet bloc nations to focus on one moon and western countries on the other in a cooperative effort. The Δv between the two moons is a convenient 747 m/s for two burn Hofmann transfers.

The next logical step is to define mission scenarios, list scientific objectives, and perform cost and performance trades with other options, such as the direct manned mission from Earth to the martian surface. O'Leary (1985) constructed a reference scenario of an early manned mission to Phobos and Deimos. Features of the scenario would include lower overall cost and safer operations than a manned Mars landing, more quality martian science by sequential exploration using unmanned rovers at disparate sites telemetered from PhD, collection of Mars samples by unmanned vehicles and analyzing them in a PhD laboratory, sampling the martian moons themselves, and creating the infra-structure for processing fuels from PhD volatiles or oxides with rapid expansion thereafter. The only apparent missing feature of this first mission is the manned exploration of the martian surface. If, for political reasons, a manned Mars landing is warranted, the PhD scenario could include a sortie to the martian surface in a small vehicle at modest additional cost.

Many of the same ground rules for analyzing other manned missions—lunar bases, Mars landing mission, near-Earth asteroid missions—apply to the PhD case. Two studies carried out principally by the Schaumberg, Illinois group of Science Applications International Corporation (SAIC) provide the basic logistical requirements and costing assumptions for an initial lunar base (SAIC, 1984a), lunar base reconnaissance mission (SAIC, 1984b), a near-Earth asteroid mission (SAIC, 1984b), and a first manned Mars landing (Hoffman and Soldner, 1984; SAIC, 1984b).

In the last case, the investigators assumed a dual launch in 2003, aerocapture at both Mars and Earth, a 4-man crew, 30-day stay time, and minimal new technology. Their overall cost estimate of $39 billion (1984 dollars) could be reduced by a factor of 2 or more by eliminating the need for developing a man-rated Mars lander, including the fuel needed for an ascent stage to get out of Mars' gravity well (more shuttle flights, more OTVs). The manned Mars landing mission would more resemble a one-shot Apollo flight than the scenario of a sustained operation at a small base on PhD that would be manned initially and visited periodically for expanding the materials processing capability and upgrading the exploration program.

Because of increasing emphasis on the permanent presence of man in space and the developing technology to make that possible, the lunar base scenario may be more appropriate to the PhD missions than an Apollo scenario. One PhD workshop (O'Leary, 1984) considered an initial Phobos base consisting of two or three space station sized modules covered over with dust for radiation shielding (see artist's concept of Phobos base, Fig. 1). This modest facility could expand rapidly as we begin to industrialize space utilizing non-terrestrial materials. The economics of establishing and enlarging a PhD base appears to be more immediately feasible than those required to start and sustain a base on the martian surface deep within its gravity well. Table 2 shows that the ratio of returned mass to invested mass for delivery in free space is about 4 for Phobos and Deimos and only 1 for the surface of Mars. Phobos and Deimos will provide among the first of our non-terrestrial material resources.

The case for Phobos and Deimos enhances the case for Mars. We appear to have a situation of serendipity in science and economics. The PhD scenario requires new legitimacy previously not given it because of an inherent familiarity with doing things at the bottom of deep gravity wells. But the laws of physics and practical economics (e.g., O'Neill, 1977) dictate we take a second look at our future priorities in space exploration.

Mission planners should investigate the PhD options in as much detail as is now being given to manned Mars landings and lunar bases. These tasks should not be difficult because methods for determining most of the assumptions, mission parameters, and logistical requirements are common to all scenarios. The results will lead not to an either/or proposition about destinations involving irrevocable dead-ends, but a synergistic blend of missions that will grow rapidly. Phobos and Deimos appear to be at the focus of the initial step from which everything else will logically follow. The essence of martian exploration and space industrialization using non-terrestrial resources is embodied in the mysterious moons of Mars.

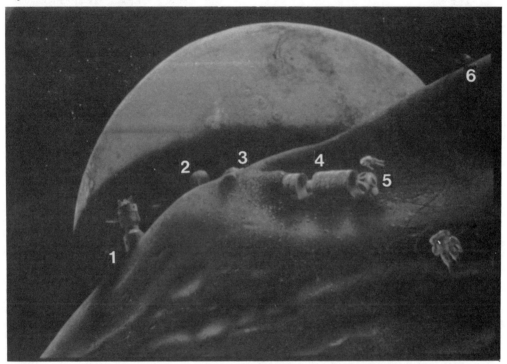

Figure 1.   1—Docking adapter for spacecraft; 2—Mars surface communications antenna; 3—airlock/access to buried modules; 4—modules based on space station technology; 5—multiple adaptor for future expansion (access port); 6—solar furnace for processing Phobos surface materials. Copy of painting by Michael W. Carroll.

REFERENCES

Adelman S. J. and Adelman B. (1984) The case for Phobos. In *The Case for Mars II.* Univelt, San Diego. In press.

Cordell B. M. (1984) The moons of Mars: Source of water for lunar bases and LEO (abstract). In *Papers Presented to the Symposium on Lunar Bases and Space Activities of the 21st Century,* p. 66. NASA/Johnson Space Center, Houston.

Covault C. (1985) Soviets in Houston reveal new lunar, Mars, asteroid flights. *Aviat. Week Space Technol., 122,* 18.

Gaffey M. J. and McCord T. B. (1977) An extraterrestrial source of natural resources. *Technol. Rev., 79,* 50.

Hoffman S. J. and Soldner J. K. (1984) Concepts for the early realization of manned missions to Mars. In *The Case for Mars II*. Univelt, San Diego. In press.

O'Leary B. (1977) Mining the Apollo and Amor asteroids. *Science, 197*, 363–366.

O'Leary B. (1979) Asteroid prospecting and retrieval. In *Space Manufacturing Facilities 3*, pp. 17–24. Proceedings of the 4th Princeton/AIAA Conference, AIAA, New York.

O'Leary B. (1981) The Fertile Stars, pp. 38–41. Everest House, New York.

O'Leary B. (1983) Mining the Earth-approaching asteroids for their precious and strategic metals. In *Space Manufacturing 1983*, volume 53, p. 386. Advances in the Astronautical Sciences, American Astronautical Society, New York.

O'Leary B. (1984) Phobos and Deimos as resource and exploration centers. In *The Case for Mars II*. Univelt, San Deigo. In press.

O'Leary B. (1985) Phobos and Deimos (PhD): Concept for an early human mission for resources and science. Princeton/AIAA Conference on Space Manufacturing. In press.

O'Leary B., Gaffey M. J., Ross D. G., and Salkeld R. (1979) Retrieval of asteroidal materials. In *Space Resources and Space Settlements* (J. Billingham, W. Gilbreath, B. O'Leary, and B. Gosset, eds.), pp. 173–189. NASA SP-428, NASA, Washington D.C.

O'Neill G. K. (1977) *The High Frontier*. Wm. Morrow and Co., New York. 288 pp.

Schmitt H. (1984) A millenium project: Mars 2000 (abstract) In *Papers Presented to the Symposium on Lunar Bases and Space Activities of the 21st Century*, p. 4. NASA/Johnson Space Center, Houston.

Science Applications International Corp. (SAIC) (1984a) *A Manned Lunar Science Base: An Alternative to Space Station Science?* Report No. SAIC-84/1502, SAIC, Los Alamos. 31 pp.

Science Applications International Corp. (SAIC) (1984b) *Manned lunar, asteroid, and Mars missions*. Report No. SAIC-84/1448. SAIC, Los Alamos. 82 pp.

Singer S. F. (1984) The PhD proposal: A manned mission to Phobos and Deimos. In *The Case for Mars*, pp. 39–65. Science and Techology Series, American Astronautical Society. Univelt, San Diego.

Staehle R. (1983) Finding "Paydirt" on the Moon and asteroids. *Astronaut. Aeronaut., 21*, 44.

Veverka J., Thomas P., and Duxbury T. (1978) The puzzling moons of Mars. *Sky & Telescope, 56*, 86.

Walberg G. D. (1983) Aeroassisted orbit transfer window opens on missions. *Astronaut. Aeronaut., 21*, 36.

THE MOONS OF MARS:
A SOURCE OF WATER FOR LUNAR BASES AND LEO

Bruce M. Cordell

General Dynamics/Space Systems Division, Advanced Space Programs,
C1-9530, P. O. Box 85990, San Diego, CA 92138

While oxygen is plentiful on the lunar surface, water and hydrogen are very scarce. The obvious source for these substances is Earth, although severe delta-V penalties are inevitable in this approach. Phobos and Deimos appear to be carbonaceous, volatile-rich moons. Up to 20% of these bodies may be loosely bound water; the total Phobos/Deimos water reservoir may be 10^{18} grams. The potential for Phobos and Deimos to contribute volatiles to the Moon and LEO is assessed. A concept is suggested for a mission originating at the lunar base that delivers large quantities of martian moon waters to the Moon. A Mars transfer craft aerobrakes at Mars before rendezvous with its satellites. A manned permanent station on Phobos and/or Deimos processes moon material to extract water. During Earth approach, aerobraking places the spacecraft into lunar trajectory. The advantages of this Moon–Phobos/Deimos loop include: (1) low delta-V's and cost compared to closed Earth-Moon circuits; (2) both LEO and the Moon are supplied with hydrogen and oxygen; and (3) this Moon–Phobos/Deimos loop provides an economic stimulus to further explore the martian moons and Mars itself.

INTRODUCTION

One of the major uses of the Moon will be as a source of raw materials. Life support systems, lunar base industrial processes, and spacecrafts and stations operating near the Moon and Earth will find many uses for lunar oxygen, metals, and silicates. However, water is unknown on the Moon and hydrogen is scarce; only traces (<50 ppm) of solar wind hydrogen are trapped within the regolith.

The possibility that ancient frozen cometary waters eternally shielded from the sun might have survived to the present at the lunar poles cannot be ruled out (Arnold, 1979). However, the total absence of moisture in the lunar samples suggests that possible polar waters should be viewed only as a potential bonus for lunar bases. Confirmation of lunar polar ice awaits the resumption of lunar exploration; thus, it is currently prudent to consider other possible sources of hydrogen to supplement the meager lunar supply.

An obvious source of hydrogen is Earth, due to its nearness to the Moon and abundance of hydrogen. Unfortunately, transportation costs resulting from Earth's well-known sizable gravitational well are substantial (characteristic velocities for Earth to LEO are 32,000 ft/s). These facts give impetus to the search for extraterrestrial and extralunar hydrogen reservoirs.

One attractive potential source of volatiles for cislunar operations is the population of Earth-approaching asteroids (O'Leary, 1977). Reflectance spectra for these objects and other evidence (*e.g.*, Gaffey *et al.*, 1977) suggest that some of these bodies are similar to Type I carbonaceous chondrites. Up to 20% of these tiny worlds may be loosely bound water in various mineral assemblages. A few of them are excellent space targets because of their accessibility and negligible surface gravities.

It is suggested here that two asteroid-like objects, which happen to be in orbit around Mars, may become the hydrogen/water source that will stimulate the initiation of lunar resource utilization and the establishment of manned bases on the Moon's surface. The moons of Mars combine very high scientific interest and resource potential with extraordinary accessibility at a level unmatched by any other known extraterrestrial object [e.g., see O'Leary (1984) and Cordell (1985)]. The total delta-V for an Earth-to-Phobos trip is less than that for a similar voyage to the lunar surface. Launch windows occur about every two years and one-way chemical propulsion times are several months. Phobos/Deimos round trip delta-V's compare favorably with those of any known Earth-approaching asteroids with 2-year launch windows (O'Leary, 1984). Additionally, a manned base on Phobos is an ideal platform from which to remotely study Mars itself (avoiding biological contamination and Earth-Mars radio time delay problems). Indeed, Phobos is an excellent staging area for manned missions to the martian surface.

Phobos and Deimos have low albedos, low densities, ancient surfaces, non-spherical figures, and reflection spectra that suggest they are relatively unmodified, volatile-rich, carbonaceous objects (e.g., Veverka and Burns, 1980). Water may compose up to 20% of these bodies (e.g., Hartmann, 1983) and could be retrieved for use in LEO/GEO or on the Moon. Propellant production facilities on Phobos/Deimos will make manned landings on Mars independent of terrestrial fuel supplies.

Furthermore, Phobos and Deimos are scientifically intriguing in their own right. Some cosmochemical theories claim that carbonaceous objects (e.g., Phobos and Deimos) could not have formed at Mars' solar distance. Thus, the martian moons might be asteroids and would have required capture by Mars. Phobos and Deimos exploration will provide an opportunity to test these ideas and possibly anticipate the scientific and resource bonanza awaiting us in the asteroid belt.

Mars itself constitutes a potential volatile source for lunar bases and LEO. Remote sensing and in situ investigations of Mars have indicated that Mars probably possesses Earth-like volatile abundances (e.g., Clifford, 1984). The potential for economically important mineral concentrations and precious ore bodies also appears to be very good—perhaps comparable with Earth (Cordell, 1984). Although Mars has probably not experienced plate tectonics (which is commonly associated with many mineral deposits on Earth) the existence of crustal swells, rifting, volcanism, impact cratering, and abundant water on Mars suggests that some ore-forming processes may have occurred during martian geological history. Terrestrial hydrothermal, dry-magma, and sedimentary mineral concentration processes have been identified that may have operated on Mars. In particular, tectonic similarities between mineral-rich Africa and portions of Mars suggest that the potential for mineral wealth on Mars is impressive (Cordell, 1984).

Despite the intriguing nature of the martian surface from scientific, resource, and adventure standpoints, Mars' gravitational well is deep enough to make direct transport of water or other substances to the Moon rather expensive.

THE FIRST MANNED MISSIONS TO PHOBOS AND DEIMOS

Initial Phobos/Deimos explorations are assumed to utilize advanced transportation systems (*e.g.*, Orbital Transfer Vehicles, OTVs) that are expected to be operational in the 1990s. This section presents a concept for the first manned mission to Phobos and Deimos utilizing trajectory data for the year 2001. Assuming reasonable funding levels, this is a conservative (*i.e.*, late) date. OTVs should be operational from the mid-1990s, and the LEO Space Station should achieve its growth configuration with full capability to support manned planetary missions. International cooperation and/or competition in space should certainly cause the initial manned Mars' moons expedition to edge toward the present (from 2001) rather than to recede into the future.

The only significantly new technology assumed is that associated with OTVs (*e.g.*, aerobraking capability) and human life support systems for long duration space missions. Although advanced techniques such as solar sails, tethers, and more powerful propulsion systems may indeed be available after the turn of the century, these initial manned Mars' moons missions have been conceived deliberately to be as simple and straightforward as possible.

Since this is a first mission, the general philosophy is to be safe, quick, and yet accomplish the crucial tasks. For this first manned visit to Phobos/Deimos, it is proposed to use a small crew of three, including one mission specialist astronaut trained as a geological scientist. The other two astronauts would be the commander/pilot and a physician/pilot. Both would actively support the geologic reconnaissance of Phobos and Deimos when in the vicinity of Mars and at other times would be engaged in monitoring spacecraft systems, crew condition, and cruise science.

For the 2001 opportunity, the astronauts would spend 60 days thoroughly mapping, exploring, and sampling the moons to ascertain their compositions and internal structures and to assess their resource potential. Complete high resolution imaging is essential for thorough topographic, structural, and albedo (later geologic) mapping. Remote sensing and imaging utilizes visual, infrared, radar, x-ray, and gamma-ray techniques. Internal structures are probed using geophysical packages (*e.g.*, active seismics) emplaced by the crew on the moons' surfaces. The acquisition of samples and cores of important type regions on the moons are of utmost importance for resource utilization plans.

A manned reconnaissance is preferred because exploration of Phobos and Deimos is essentially a geological task, *i.e.*, an activity best performed by astronauts actually on the moons. Highly trained, versatile astronauts can also be justified on an engineering basis (*e.g.*, equipment repair or modification). Indeed, the presence of men and women will make early manned Mars missions of great interest to the public—both in the U.S. and around the world. It is clear that the expansion of human activities to Mars, including the manned exploration of the martian satellite system and the actual first manned landing on Mars, will be among the most significant events in the entire history of our civilization.

It is assumed that the mission is to be assembled, launched, and later terminated at the LEO Space Station where inspection, testing, and analysis is performed on the samples and data from Phobos and Deimos. The delta-V's for the 2001 mission are shown in Table 1. Major propulsive burns are required only for trans–Mars and trans–Earth injections, because aerobraking at Mars and Earth is used to reduce spacecraft velocity relative to each planet. Rendezvous with the low–mass moons is trivial compared with the descent/ascent requirement associated with landing on Mars itself.

Table 1. ΔVs for the 2001 Phobos/Deimos Mission

Maneuver	ΔV		Wp(Klb)	OTV
	Km/s	fps		
Earth escape (from LEO)	3.57	11,688	225	1
			64.3	2
Outbound midcourse	0.2	655	10.3	
Mars capture (with aerobraking)	0.0	0.0	–	
Phobos/Deimos rendezvous/docking maneuvers	0.6	1,964	28.6	
Mars escape (from Phobos)	3.24	10,608	105.1	
Inbound midcourse	0.2	655	4.4	
Earth capture (with aerobraking)	0.1	327	2.1	
Total	7.91	25,897		

For simplicity, a stacked OTV, multiple perigee burn strategy (e.g., Friedlander et al., 1984) is utilized for injection into Mars transfer and for Mars escape. All propulsive burns (except launch from Earth's surface to LEO) are accomplished by OTVs. The OTV assumed here is based on the configurations developed in a recent General Dynamics/Convair report (Ketchum et al., 1984) for transport from LEO to GEO and the Moon. For the calculations, Isp = 485s for this cryogenic LH2–LO2 vehicle.

Trans–Mars injection (TMI) is accomplished using multiple perigee burns (e.g., Friedlander, 1984) to minimize gravity losses and yet achieve the relatively large delta-V (see Table 1). Two OTVs (with propellant weight = 225,000 lbs.) are required for completion of the mission. The first OTV and approximately 30% of the second OTV's propellants are required for TMI.

The trip to Mars consumes 186 days. Upon arrival at Mars, aerobraking in Mars' atmosphere slows the craft and places it in an elliptical transfer orbit to either Phobos or Deimos. Circularization burns required to rendezvous with either moon are relatively minor (600 m/s).

About one OTV is budgeted for the return to Earth. The 360–day trip home involves a swing by Venus that lengthens the trip time but reduces Earth encounter velocities and moderates the demands on the OTV aerobrake. Aerobraking at Earth and rendezvous with the space station require minimal propulsive burns. If aerobraking technology becomes very reliable prior to the time of this first manned Phobos/Deimos mission, a direct return to Earth, with shorter flight time, may be preferable.

The relatively short 606–day total mission time minimizes space–related hazards for the crew while still providing adequate Phobos/Deimos exploration opportunities. To avoid deleterious biological effects on the crew, the habitability modules will be spun to simulate low (1/3) gravity. Phobos/Deimos surface gravities are biologically negligible

(1 cm/s2). Unfortunately, high solar activity near 2001 could make any manned interplanetary travel more hazardous.

This first Phobos/Deimos sortie could return about 500 lbs. of moon material to the space station (and maybe Earth) for analysis. Presumably, this will confirm and enlarge our understanding of these objects as volatile-rich carbonaceous bodies (e.g., Veverka and Burns, 1980) with a number of relatively easily extractable substances, including water.

Prior to the launch of the first manned Phobos mission, promising technologies for extracting volatiles, particularly water, from supposed moon materials should be pursued intensely. This research and development phase will culminate in the return to Earth of the first samples from the moons of Mars. Specific methods will be empirically tested on the actual moon samples.

SUBSEQUENT MISSIONS TO PHOBOS AND DEIMOS

Should the demands for a source of water for activities on the Moon and in LEO become compelling and the desire to explore, utilize, and settle the martian environment become widespread, both goals could be accomplished by launching another more ambitious manned mission to Phobos and Deimos as early as 2005 (or two launch windows after the initial mission). This time, with full knowledge of the resource potential of these small but pivotal objects, astronauts would return with the following important objectives: (1) establish a permanent, human presence or base on these worlds; (2) construct and operate a water extraction plant (probably utilizing solar energy) as soon as possible; (3) in collaboration with research and development teams on Earth, experiment with possible technologies for producing propellants for spacecrafts from indigenous moon resources (e.g., from water); (4) embark upon the first close-range, continuous, manned reconnaissance of the planet Mars ever attempted; and (5) aggressively pursue Phobos/ Deimos science.

The specific technologies that will be utilized to extract water and process propellants on Phobos and Deimos will be dependent on the composition of the martian moons, available power sources, and weight constraints. The establishment of a water extraction system has not been considered in detail here and deserves study; this problem is likely to be particularly challenging because of the milli-g environment.

The scientific and strategic implications of Phobos/Deimos explorations, particularly with respect to the colonization of Mars, have been well elaborated upon by Singer (1984) and Adelman and Adelman (1984). Space precludes their repetition here, except to say that water on Phobos/Deimos can be the key to the exploration of Mars as well as a pivotal incentive to initiate major utilization of the Moon. Martian moon moisture can be used for life support, industrial processes, and rocket fuel (for interplanetary vehicles or descent/ascent to Mars itself) to support operations near Mars, Earth, and/or the Moon. Proper utilization of these martian resources will require the establishment of a permanent, relatively self-sufficient base on Phobos and Deimos. A small human presence near (and eventually on) Mars will facilitate the exploration, utilization, and settlement of Mars.

IRRIGATION OF THE MOON WITH MARTIAN WATERS

Once the Moon–Phobos loop is operating it should be possible to retrieve appreciable amounts of martian moon waters for use on the Moon and in LEO. Again, the intent here is to describe a simple set of operations, using primarily OTV propulsion and aerobraking technology, without resorting to more exotic transportation systems.

A typical water tanker mission could originate in low lunar orbit (delta-V's for this scenario are shown in Table 2). The event sequence would include the following: (1) multiple perigee burns by an OTV to insert the empty spacecraft tanker into Mars transfer orbit, (2) aerobraking, utilizing the martian atmosphere, (3) rendezvous with Phobos or Deimos requiring small impulses, (4) arrival at the manned moon bases where supplies are delivered, the OTV is fueled, and the water cargo is obtained, (5) separation from Phobos and insertion into Earth transfer orbit, (6) aerobraking at Earth and insertion into an elliptical lunar transfer orbit, (7) separation of a small tanker craft loaded with water that will aerobrake and rendezvous with the LEO space station (optional), and (8) propulsive burn to insert the spacecraft into a circular lunar orbit and rendezvous with the lunar space station.

Inspection of the total delta-V (Table 2) for the Phobos/Deimos–Moon loop versus the Earth–Moon round trip indicates that considerable advantages are obvious for Phobos/Deimos resource retrievals. The Phobos–Moon loops avoid descending into Earth's or Mars' gravitational wells and thus are not penalized in the way that Earth–Moon loops inevitably are. Additional advantages are evident in that at no time is the Phobos–Moon spacecraft ever immersed in a planetary atmosphere (except during aerobraking), thus

Table 2. ΔVs for Water Tanker Missions*, Phobos/Deimos–Moon vs. Earth–Moon

Earth–Moon Loop	ΔV		Phobos/Deimos –Moon Loop	ΔV	
Trajectory Leg	Km/s	fps	Trajectory Leg	Km/s	fps
Earth to LEO	9.7	31,758	Moon to lunar orbit	1.91	6,253
TLI from LEO	3.14	10,280	TMI from lunar orbit	1.6	5,238
Midcourse correction	0.10	327	Midcourse correction	0.2	655
Into lunar orbit	0.82	2,685	Mars moon rendezvous	0.6	1,964
Descent to Moon	2.10	6,875	TEI from Deimos	1.5–2.0	4,911–6,548
Ascent from Moon	1.91	6,253	Midcourse correction	0.2	655
TEI	0.82	2,685	Into lunar orbit	0.9	2,947
Capture into LEO	0.1	327	Descent to Moon	2.1	6,875
Descent to Earth	0.1	327	Ph/D to lunar surface and return	9.0–9.5	29,466–31,103
Earth to lunar surface and return	18.8	61,551	Ph/D to lunar surface and return	5.0–5.5	16,370–18,007
Earth to lunar station and return	14.8	48,455	Ph/D to LEO and return	7.9	25,865
Earth to LEO and return	9.8	32,085			

Note: Energy requirements to the lunar surface from Phobos/Deimos are one-fourth those from Earth
*Typical numbers for Hohmann trajectories.

conferring numerous design advantages on the tanker. Finally, it is seen that an enormous quantity of water can feasibly be delivered to the Moon from Mars' moons. To minimize delta-V costs, Phobos/Deimos-Moon water tanker missions should utilize conjunction-class trajectories.

It is assumed that advanced water loops involving the martian moons would utilize an unmanned single-stage OTV and a tanker with a fairly large capacity (e.g., 1/10 that of an external tank or 450,000 lbs. of water). It is also assumed here that spacecraft propellants can be produced on Phobos/Deimos so that OTVs refuel only near Mars. Thus, the spacecraft assembly launched from lunar orbit consists of an empty tanker and a partially fueled OTV; only 30,000 lbs. of propellants are required to rendezvous with Phobos/Diemos bases. Upon arrival at Phobos, the OTV is refueled to capacity, and the tanker is filled with water. The OTV has a propellant capacity of 400,000 lbs (inert weight of 26.2 klbs.) and requires about 355,000 lbs. for insertion of the "martian iceberg" into low lunar orbit.

From lunar orbit a smaller descent craft may shuttle small amounts of water from the tanker, which would remain in lunar orbit until it ran dry and began the cycle again. Launch windows occur every two years, and thus one tanker would supply the Moon with over 600 lbs. of water per (Earth) day. Of course, in a very advanced loop, more or larger tankers could be used elevating this amount to whatever is considered optimum. Phobos and Deimos are not likely to run dry soon since their total water reservoir (assuming C1 carbonaceous chondrite composition) exceeds 400 million external tank capacities.

It might be useful to return only hydrogen (instead of water) from the martian moons since the same tanker could retrieve much more payload and only hydrogen is needed on the Moon, not oxygen. The main disadvantages of this are the following: first, problems with leaks of cryogenics during the flight from Mars, and second, LEO will require both hydrogen and oxygen to sustain its activities. Problems with insulation of cryogenics over long time periods will probably be solved fairly easily within 20 years if the demands for martian resources become obvious. On the other hand, if the shuttle (or its evolved derivative) were utilized to transport large amounts of water into LEO, the water would be electrolyzed in LEO; only hydrogen would go to the Moon. Since other volatiles are also depleted on the Moon and are probably available on Phobos/Deimos, it might be more efficient to return methane or ammonia if simple techniques can be developed to produce these substances from Phobos/Deimos materials.

The main disadvantage of Phobos-Moon loops is that launch windows occur only every two years, while Earth-Moon loops are very frequent. Travel times from Mars are also typically several months, while Earth-to-Moon times are a few days.

Nevertheless, the delta-V penalties of Earth-Moon loops and apparent abundance of volatiles on Phobos/Deimos make water retrieval missions to Phobos and Deimos appear surprisingly attractive. The added scientific, economic, psychological, and political bonanzas inherent in manned Mars explorations and settlements (e.g., Singer, 1984; McKay and Stoker, 1984) are most impressive.

Simple water retrieval missions to Phobos and Deimos in support of space activities in LEO or at lunar bases may provide the economic incentive to begin the settlement of Mars as well as the industrialization of the Moon.

Acknowledgments. *I would like to thank Bob Drowns and Bill Ketchum of General Dynamics for their interest in and support of this work. Wendell Mendell and Hubert Davis made several comments that improved this paper.*

REFERENCES

Adelman S. J. and Adelman B. (1984) The case for Phobos. In *The Case for Mars II*, Amer. Astronaut. Soc., Sci. Technol. Ser. Vol. 62, pp. 245–252.

Arnold J. R. (1979) Ice in the lunar polar regions. *J. Geophys. Res. 84*, 5659–5668.

Clifford S. (1984) *Workshop on Water on Mars.* LPI Tech. Rpt. 85-03. Lunar and Planetary Institute, Houston, 98 pp.

Cordell B. M. (1984) A preliminary assessment of Martian natural resource potential. In *The Case for Mars II*, Amer. Astronaut. Soc., Sci. Technol. Ser. Vol. 62, pp. 627–639.

Cordell B. M. (1985) *Manned exploration of Phobos and Deimos utilizing OTVs.* Operations Research Memo 154-85-001, General Dynamics, Convair Division, San Diego, 17 pp.

Friedlander A. (1984) *Manned lunar, asteroid, and Mars missions.* Science Applications International Corp., Schaumburg, 82 pp.

Gaffey M. J., Helin E. F., and O'Leary B. (1977) An assessment of near-Earth asteroid resources. In *Space Resources and Space Settlements*, pp. 191–204. NASA SP-428, NASA, Washington, D.C.

Hartmann W. K. (1983) *Moons and Planets*, 2nd ed., Wadsworth Publishing Co., Belmont, 169 pp.

Ketchum B. (1984) *Orbital transfer vehicle concept definition and systems analysis study.* General Dynamics, Convair Division, San Diego, 517 pp.

McKay C. P. and Stoker C. R. (1984) Why Mars? In *The Case for Mars*, Amer. Astronaut. Soc., Sci. Technol. Ser. Vol. 57, pp. 19–27.

O'Leary B. (1977) Mining the Apollo and Amor asteroids. *Science, 197,* 363.

O'Leary B. (1984) Phobos and Deimos as resource and exploration centers. In *The Case for Mars II*, Amer. Astronaut. Soc., Sci. Technol. Ser. Vol. 62, pp. 225–244.

Singer S. F. (1984) The Ph-D proposal: A manned mission to Phobos and Deimos. In *The Case for Mars*, Am. Astronaut. Soc. Sci. Technol. Ser. Vol. 57, pp. 39–65.

Veverka J. and Burns J. A. (1980) The moons of Mars. *Annu. Rev. Earth Planet. Sci., 8,* 527–558.

THE PROBLEM OF WATER ON MARS

Steven W. Squyres

Theoretical Studies Branch, NASA/Ames Research Center, Moffett Field, CA 94035

The availability of water on Mars will be an important issue in planning the manned exploration of that planet. Water has played an important role in shaping the morphology of the martian surface. The present distribution of water on Mars is not completely understood. Important reservoirs include the regolith of the cratered highlands poleward of ±30° latitude, the polar layered deposits, and geochemically bound water in silicates. Other small reservoirs that may be simpler to exploit include atmospheric water and the polar perennial ice. Study of the water resource on Mars should be an important part of any precursors to manned exploration of the planet.

Ever since Schiaparelli's description of the now infamous "canals" on Mars, the subject of water on that planet has been of considerable interest, both within and outside the scientific community. While the canals of Schiaparelli were optical illusions, exploration of Mars to date has shown conclusively that H_2O, often in the liquid state, has played an important role in shaping the appearance of the planet's surface. With the renewed interest in manned exploration of Mars, the topic of water on Mars takes on considerable importance. Any manned mission to Mars will be much less expensive and more capable if water can be obtained on the planet's surface. While there is ample evidence for the action of water in the past, it is less clear where the water is at the present time. This paper briefly reviews the evidence for the former presence of large amounts of water on Mars and discusses the major sinks and present distribution of water on Mars.

One of the most startling and important discoveries of the Mariner 9 mission to Mars was the widespread evidence for modification of the martian surface by the action of liquid water (McCauley *et al.*, 1972; Milton, 1973). This evidence comes in the form of channels, which are often remarkably similar to terrestrial stream, river, and flood features.

The most striking channels are the outflow channels. These are most common in the equatorial regions of Mars and are concentrated along the northern lowland/southern highland boundary. They generally arise from the highlands and debouch onto the lowland plains. The source regions usually show very complex topography that earns them the name chaotic terrain. The appearance of the chaotic terrain strongly suggests removal of subsurface material and widespread collapse of topography. The channels arise fully formed from these chaotic regions and may extend for many hundreds of kilometers.

Although clearly the result of fluid flow, outflow channels bear only superficial similarity to terrestrial rivers. They are much more similar to the types of features formed by catastrophic floods on Earth (Milton, 1973; Baker and Milton, 1974; Baker and Kochel, 1979). The magnitude of the floods implied by the martian features is enormous. For example, Carr (1979) has estimated a peak discharge of as much as 5×10^8 m^3 s^{-1} for a flood originating in Juventae Chasma and extending across Lunae Planum. A reasonable mechanism for triggering flooding of this magnitude may be geothermal warming of ground ice (*e.g.*, McCauley *et al.*, 1972; Sharp and Malin, 1975). It may be that the water released in the floods was originally contained in a confined aquifer capped by an impermeable lid of permafrost (Carr, 1979). With tectonic warping of the aquifer (as in formation of the Tharsis bulge), a hydrostatic head sufficient to cause breakout and flooding may have been developed. Once released, the floods were of sufficient size that they could have proceeded for enormous distances across the martian surface even under the present climatic conditions. The density of impact craters superimposed on the outflow channels indicates, however, that they date from fairly early in martian history (Malin, 1976; Masursky *et al.*, 1977).

A second type of channel apparently caused by flow of liquid water is the valley network. These are more similar to terrestrial drainage systems, consisting of narrow, often sinuous valleys with tributary systems. They are dissimilar to terrestrial stream systems in a number of ways, however, and are more similar to terrestrial drainage systems formed by sapping (Pieri, 1980). While formation by precipitation cannot be ruled out in a few cases, most valley networks apparently had a sapping origin.

All of the valley systems on Mars are found in the ancient cratered highlands. The density of superimposed impact craters indicates that formation of valley systems was concentrated in the earliest part of martian history, probably more than 4 b.y. ago. Because the fluid discharges implied by the valley systems are quite modest, it is unlikely that they could have formed under the present climatic conditions. They therefore provide strong evidence that the pressure and temperature of the atmosphere very early in Mars' history were significantly higher than they are today. This clement era did not extend past the earliest part of martian history.

A number of types of features on Mars appear to owe their formation to removal of subsurface ice. One is chaotic terrain, and another is fretted terrain, first described by Sharp (1973). It consists of smooth, flat lowlands separated from older uplands by a complex pattern of escarpments. Fretted terrain is found primarily along the northern lowland/southern highland boundary. It probably formed by escarpment recession due to removal of ground ice. This may have taken place either by direct sublimation of ice or by emergence of groundwater. Other smaller areas are found where more limited collapse has occurred, forming tablelands with scalloped edges, and small, closed depressions. These features are similar to thermokarst features formed by melting of ground ice at high latitudes on Earth.

Many impact craters on Mars possess unusual ejecta deposits consisting of overlapping lobes of debris apparently fluidized at impact. These have commonly been referred to as rampart craters (McCauley, 1973; Carr *et al.*, 1977). Somewhat inconclusive morphologic evidence suggests that the flow was not gas supported, but instead involved lubrication by interstitial liquid, probably water. The putative liquid may have been generated from ground ice by impact heating, or may have been present as liquid in the subsurface material prior to impact.

A number of features on Mars resemble patterned ground common at high latitudes on Earth. Patterned ground on Earth forms as a result of repeated diurnal and seasonal freezing and thawing of ice-rich soil, causing movement of material and segregation by ice content or sorting by particle size. Resultant patterns include circles, stripes, and networks of polygons. Networks of polygonal fractures are the most common form of patterned ground on Mars. They could perhaps be frost related, but their scale is one to two orders of magnitude larger than that of terrestrial patterns. The larger size may be an indicator of longer timescale temperature cycles (Helfenstein, 1980) or may simply indicate that the polygons formed by some tectonic process not related to ground ice.

All of the landforms discussed thus far only provide evidence for the former presence of ground ice. A variety of other landforms apparently owe their morphology to the present existence of ground ice in sufficient quantities to alter the rheology of the surface materials. With large enough amounts of ice present in a matrix of silicate particles, creep deformation of the ice can cause the entire mass of material to undergo quasi-viscous flow. In order to learn more about the present distribution of ground ice on Mars, Squyres and Carr (1984) have recently mapped the global distribution of such features using high resolution Viking Orbiter images.

Three types of features were mapped: lobate debris aprons, concentric crater fill, and terrain softening. Lobate debris aprons are thick accumulations of debris at the bases of escarpments. They have distinct convex-upward topographic profiles indicating creep deformation throughout the entire thickness of the material (Squyres, 1978). They commonly show surface lineations parallel to flow (probably caused by inhomogeneities in the source region) and compressional ridges where the flow is obstructed. Their morphology is very similar to that of terrestrial rock glaciers. Concentric crater fill is apparently the same material confined within impact craters. Inward flow of the material gives rise to radial compressive stresses that produce crater fill with a pattern of concentric ridges. Terrain softening is a distinctive style of landform degradation apparent only in high resolution orbital images. It is revealed by extreme rounding of features that are elsewhere sharp (*e.g.*, crater rims) and marked convexity of slopes that are elsewhere straight or concave (*e.g.*, crater walls, erosional scarps). Lobate debris aprons and concentric crater fill probably require substantial amounts of interstitial ice. Terrain softening, because it preserves the large-scale components of the original topography, probably requires less ice.

Where these deformational landforms are observed, they are inferred to indicate recent or present existence of large amounts of ground ice. This inference is based on the following reasoning. (1) Ice may be present in sufficient quantities that its removal would have caused collapse, which is generally not observed (ice contents upwards of ~30% by volume are typically required for creep deformation in rock glaciers on Earth). (2) Removal of the ice would have caused a "stiffening" of the material, so that subsequent mass wasting and impact erosion would not have allowed the rounded morphology to persist. (3) Many of the lobate debris aprons and deposits of concentric crater fill are devoid of impact craters, indicating that flow sufficient to disrupt the surface morphology has taken place recently. Mapping of these features may therefore provide some of the most unambiguous evidence available for the presence of ground ice deep in the martian regolith.

The most striking characteristic of the distribution found is the nearly complete absence of all three classes of features from the equatorial latitudes. Virtually no examples of lobate debris aprons, concentric crater fill, or terrain softening are found equatorward of 30° latitude in either hemisphere. In the northern hemisphere, lobate debris aprons are most common in Tempe Fossae, Mareotis Fossae, the Phlegra Montes, and particularly in the fretted terrain between longitudes 280° and 0°. Concentric crater fill is also found in these areas, but is most common in Utopia Planitia. Terrain softening is most common in the portion of the cratered highlands lying between the fretted terrain and 30° north latitude. In the southern hemisphere, lobate debris aprons are common in the massifs surrounding the Argyre and Hellas basins. Concentric crater fill is observed primarily in the area east of Hellas. Terrain softening is observed in virtually all the high resolution images south of –30° latitude.

To summarize, all observed regions of old, heavily cratered terrain lying at latitudes poleward of ±30° have undergone terrain softening. This distribution may mean that the deep regolith equatorward of ±30° has been largely devolatilized, while that at higher latitudes still retains most of its original complement of outgassed H_2O. One possible explanation for the devolatilization is simply that ground ice at low latitudes is not in equilibrium with the atmosphere and that the regolith is too coarse grained to provide an effective diffusion barrier that would prevent its escape. The scale of topographic deformation observed suggests that this high ice content may extend to depths of 1 km or more. The amount of ice present in this reservoir is difficult to estimate, but it may form the largest present reservoir of H_2O on Mars.

The polar deposits are another large reservoir of water on Mars. The martian polar deposits exhibit a complex stratigraphy, but from the simplest standpoint can be considered to consist of three units. From base to top, these are the layered deposits, the perennial ice, and the seasonal frost cap. The layered deposits were first recognized in Mariner 9 images of Mars (Murray *et al.*, 1972; Soderblom *et al.*, 1973; Cutts, 1973). They are found at both poles and extend equatorward to 85°–80°. Individual layers are typically 10–50 m thick and extend laterally for hundreds of kilometers. The total thickness of the deposits is difficult to estimate accurately but may be 1–2 km in the south and 4–6 km in the north (Dzurisin and Blasius, 1975). The layered deposits are overlain by the perennial ice. At the north pole, the perennial ice reaches almost to the perimeter of the layered deposits, while in the south it covers a smaller area. Typical surface temperatures at the north pole in summer are 205 K (Kieffer *et al.*, 1976). This is substantially higher than the saturation temperature of CO_2 at the martian surface pressure and is clear evidence that the northern perennial ice is H_2O. The situation at the south pole is more complex (Kieffer, 1979), and the evidence there may indicate a thin layer of CO_2 frost underlain by H_2O ice. The thickness of the perennial ice is not well determined. The inferred thermal inertia implies a thickness of at least 1 m (Davies *et al.*, 1977), and the lack of observable topography suggests a maximum thickness of a few tens of meters. The seasonal frost cap consists of CO_2 and is deposited at each pole during the winter. It typically extends from the pole to 45°–40° in each hemisphere at its maximum extent.

The perennial ice is the most obvious reservoir of H_2O in the polar regions. The layered deposits may contain much more H_2O, however. The deposition of the layered deposits was first described by Cutts (1973). He noted that a substantial fraction of the atmospheric CO_2 is displaced into the seasonal frost cap each winter. He suggested that this poleward flux of CO_2 could entrain dust particles put into suspension by planet-wide dust storms. Entrained particles would then become embedded in the seasonal cap and be left behind when the cap evaporated in summer. Later models (Howard, 1978; Cutts et al., 1979; Squyres, 1979; Cutts and Lewis, 1982) have recognized the importance of the perennial ice in the process. The role of the perennial ice is indicated by the limited extent of the layered deposits. Seasonal dust deposition should take place at all latitudes covered by the seasonal caps, and indeed there is evidence in the Viking orbiter images for dust mantling down to the mid-latitudes. The layered deposits, however, are limited to within 5°–10° of the poles, which, at least in the north, coincides with the present limit of the perennial ice. It seems, then, that only dust that is deposited onto the perennial ice is incorporated into the layered deposits. The layered deposits are interpreted by most workers to be composed of a mixture of dust and H_2O ice. The layering may result from periodic climatic changes that vary the extent of the perennial ice and the dust-carrying capacity of the atmosphere (Murray et al., 1973). The ice content of the layered deposits is not known, but values as high as 85% have been suggested (Toon et al., 1980). They may therefore be another very important reservoir of H_2O on Mars.

There are several other ways in which water can be stored or lost on Mars. First, there is a small amount of resident H_2O in the martian atmosphere. Compared to other sinks, however, the water content of the atmosphere is quite small. Planet-wide, the largest concentrations occur near the north polar regions just after summer solstice, but even the largest column abundance measurement is only about 100 precipitable microns (Jakosky and Farmer, 1982). The maximum amount of water observed in the entire martian atmosphere at any time during the year is equivalent to just 1.3 km^3 of ice. Some water is adsorbed onto the surfaces of regolith grains. This water is available for exchange with the atmosphere, but its volume is probably also small. A large amount of water may be locked up geochemically as water of hydration in a variety of minerals. Finally, a significant amount of water has been lost to space from the atmosphere. The loss has taken place primarily by photodissociation of atmospheric H_2O and escape of H and O.

It is clear that water has been important in shaping the morphology of the martian surface, and it is clear that there is still a substantial amount of water on Mars today. It is less apparent how best to obtain the water that is present. In terms of total volume, the largest reservoirs of water on Mars are probably ground ice in the regolith poleward of ±30° latitude, the polar layered deposits, and geochemically bound water in silicates. In terms of accessibility, the best reservoirs are probably the atmosphere and the polar perennial ice, although the latter suffers from a climate that is extremely harsh, even by martian standards. When the time comes to make plans for manned martian exploration, it will be necessary to evaluate the costs and tradeoffs involved in extraction of water from each reservoir. For example, extraction of water from the atmosphere is attractive in principle, but requires turbine systems with high efficiency and throughput. Water may be available in the mid-latitude regolith in large quantities, but extraction would require that substantial effort be put into excavation. Water extraction from the polar regions would require operation in a very inhospitable environment. The most hospitable regions on Mars, near the equator, are the least attractive from the standpoint of water availability.

Before any plans are made for exploitation of the martian water resource, it will be necessary to study the present distribution and transport of water on Mars much more thoroughly. Problems of particular importance include the details of exchange of H_2O between the atmosphere and the regolith, determination of the ice content of the polar layered deposits, and verification of inferences made about the ice content of the regolith. The first steps toward these objectives will be made by the upcoming Mars Observer mission. Study of the water resource should also be an important part of any post-Mars Observer precursors to manned exploration.

Acknowledgments. *I am grateful to Steve Clifford for a helpful review. This work was supported by the NASA Planetary Geology Program.*

REFERENCES

Baker V. R. and Milton D. J. (1974) Erosion by catastrophic floods on Mars and Earth. *Icarus, 1823,* 27–41.
Baker V. R. and Kochel R. C. (1979) Martian channel morphology: Maja and Kasei Valles. *J. Geophys. Res., 1884,* 7985–7993.
Carr M. H. (1979) Formation of martian flood features by release of water from confined aquifers. *J. Geophys. Res., 1884,* 2995–3007.
Carr M. H., Crumpler L. S., Cutts J. A., Greeley R., Guest J. E., and Masursky H. (1977) Martian impact craters and emplacement of ejecta by surface flow. *J. Geophys. Res., 1882,* 4055–4065.
Cutts J. A. (1973) Nature and origin of layered deposits in the martian polar regions. *J. Geophys. Res., 1878,* 4231–4249.
Cutts J. A. and Lewis B. H. (1982) Models of climatic cycles recorded in martian polar layered deposits. *Icarus, 1850,* 216–244.
Cutts J. A., Blasius K. R., and Roberts W. J. (1979) Evolution of martian polar landscapes: Interplay of long term variations in perennial ice cover and dust storm intensity. *J. Geophys. Res., 1884,* 2975–2994.
Davies D. W., Farmer C. B., and LaPorte D. D. (1977) Behavior of volatiles in Mars' polar areas: A model incorporating new experimental data. *J. Geophys. Res., 1884,* 3815–3822.
Dzurisin D. and Blasius K. R. (1975) Topography of the polar layered deposits of Mars. *J. Geophys. Res., 1880,* 3286–3306.
Helfenstein P. (1980) Martian fractured terrain: Possible consequences of ice heaving. In *Reports of Planetary Geology Program, 1980* (H. E. Holt and E. C. Kosters, comp.), pp. 373–374. NASA TM-82385.
Howard A. D. (1978) Origin of the stepped topography of the martian poles. *Icarus, 1834,* 581–599.
Jakosky B. M. and Farmer C. B. (1982) The seasonal and global behavior of water vapor in the Mars atmosphere: Complete global results of the Viking atmospheric water detector experiment. *J. Geophys. Res., 1887,* 2999–3019.
Kieffer H. H. (1979) Mars south polar spring and summer temperatures: A residual CO_2 frost. *J. Geophys. Res., 1884,* 8263–8288.
Kieffer H. H., Chase S. C., Martin T. Z., Miner E. D., and Palluconi F. D. (1976) Martian north pole summer temperatures: Dirty water ice. *Science, 18194,* 1341–1344.

Malin M. C. (1976) Age of martian channels. *J. Geophys. Res., 1881*, 4825–4845.

Masursky H., Boyce J., Dial A. L., Schaber G. G., and Strobell M. E. (1977) Classification and time of formation of martian channels based on Viking data. *J. Geophys. Res., 1882*, 4016–4038.

McCauley J. F. (1973) Mariner 9 evidence for wind erosion in the equatorial and mid-latitude regions of Mars. *J. Geophys. Res., 1878*, 4123–4137.

McCauley J. F., Carr M. H., Cutts J. A., Hartmann W. K., Masursky H., Milton D. J., Sharp R. P., and Wilhelms D. E. (1972) Preliminary Mariner 9 report on the geology of Mars. *Icarus, 1817*, 289–327.

Milton D. J. (1973) Water and processes of degradation in the martian landscape. *J. Geophys. Res., 1878*, 4037–4047.

Murray B. C., Soderblom L. A., Cutts J. A., Sharp R. P., Milton D. J., and Leighton R. B. (1972) Geological framework of the south polar region of Mars. *Icarus, 1817*, 328–345.

Murray B. C., Ward W. R., and Yeung S. C. (1973) Periodic insolation variations on Mars. *Science, 18180*, 638–640.

Pieri D. (1980) Martian valleys: Morphology, distribution, age, and origin. *Science, 18210*, 895–897.

Sharp R. P. (1973) Mars: Fretted and chaotic terrain. *J. Geophys. Res., 1878*, 4073–4083.

Soderblom L. A. and Malin M. C. (1975) Channels on Mars. *Geol. Soc. Amer. Bull., 1886*, 593–609.

Squyres S. W. (1978) Martian fretted terrain: Flow of erosional debris. *Icarus, 1834*, 600–613.

Squyres S. W. (1979) The evolution of dust deposts in the martian north polar region. *Icarus, 1840*, 244–261.

Squyres S. W. and Carr M. H. (1984) The distribution of ground ice features on Mars (abstract). *EOS Trans. AGU, 1865*, 979.

Toon O. B., Pollack J. B., Ward W., Burns J. A., and Bilski K. (1980). The astronomical theory of climate change on Mars. *Icarus, 1844*, 522–607.

12 / A VISION OF LUNAR SETTLEMENT

MOST OF THE PAPERS in this book have addressed some limited aspect of establishing a manned base on the lunar surface. A few have taken an overview of the system required to emplace and/or sustain a surface facility. If one imposes a long range goal of true permanency, then certain technological and economic issues begin to emerge as key elements in the programmatic strategy. The high cost of access to the Moon must either be reduced by developing more efficient launch vehicles or else be circumvented by utilizing the lunar environment or resource base to augment the transportation system. Production of fuel on the Moon or installation of electromagnetic launchers for delivery of cargo to lunar orbit are examples of the latter. We have seen that achievement of closure in the life support system is a genuine technical challenge. Characteristics such as permanency and growth carry the implication of increasing organizational complexity, evolving management autonomy, and the eventual emergence of a culture. A number of scientists, engineers, planners, or scholars can claim to have pondered some one of these topics; very few have considered them in concert. Krafft Ehricke is one of the few.

Krafft Ehricke obtained a degree in aeronautical engineering at the University of Berlin where he also studied celestial mechanics and nuclear physics. As a teenager he founded a rocket society; two prewar patents of his dealing with rockets led to a transfer from the Russian front in 1942 to Peenemunde, where he participated in the development of the V2. After the war, he came to the U.S. and played a leading role in the design of the Atlas rocket. He is widely regarded as the father of the Centaur, advocating liquid hydrogen as a fuel when many thought it impractical. In 1965–1970 he began a series of studies entitled "The Extraterrestrial Imperative," introducing the concept of space industrialization. He published a three-part series under that title in the Journal of the British Interplanetary Society in 1977 and 1978. For the past 15 years he had been studying and writing on the subject of lunar industrialization and settlement, culminating in a book, *The Seventh Continent: Industrialization and Settlement of the Moon*, which is being edited for publication.

When we invited Dr. Ehricke to be a featured evening speaker at the Lunar Base Symposium in October, 1984, we were unaware of his serious illness. He eagerly agreed to appear and, despite our subsequent concern, made special arrangements for the difficult cross-country trip. I had never heard him speak and, like so many others in the audience, was enthralled by the panoply and the energy of his rich conceptualizations. Afterwards, I met him briefly and said that I felt like a student, working long and hard on a homework problem, finally to be shown the answer in the back of the book. Krafft Ehricke passed away at his home in La Jolla on December 11, 1984, a loss to us all.

Dr. Ehricke was working on the following paper at the time of his death, and it has been edited into final form by his long-time associate, Elizabeth Miller. The presentation of his ideas is at best a summary discussion, given the limitations imposed by our format. His forthcoming book will provide a more complete rendering. The paintings accompanying the paper are original art by Dr. Ehricke.

The facing piece for the section depicts Selenospheric civilization in 2062, during a return of Halley's Comet. Mare Orientale can be seen in the sunlit hemisphere of a waxing Moon. The network of lights on the earthward side represents Selenopolis, a fully developed lunar settlement. With a population of several million, it spans two million square kilometers, about one-fourth the area of the United States. Near the sunrise terminator are the lights of Cynthia, center of the Lunar Industrial Zone supporting mining, manufacturing, and power generation. The bright spots radiating outward from Cynthia are feeder stations, supplying raw materials from distant provinces. This section of Selenopolis, including slide landing strips for arriving traffic, comprises the lunar Technosphere. Expanding away to the horizon are the light of the lunar Sociosphere, a growing network of large multi-climatic and suburban housing for the Selenian population. On the horizon, at Sinus Medii, can be seen the bright lights of Novaterra, the large agricultural zone. During the lunar night, sections of Novaterra can be illuminated at an intensity of 0.6 solar constants by Soletta, an Oberth mirror swarm orbiting the Moon-Earth L1 Lagrangian point. The bright light near to and above Novaterra is an Androcell under construction in lunar orbit. This fusion-powered outer solar system research ship can carry a crew of thousands. The development of Selenopolis, and eventually the solar system, is open-ended.

LUNAR INDUSTRIALIZATION AND SETTLEMENT—BIRTH OF POLYGLOBAL CIVILIZATION

Krafft A. Ehricke

Space Global, 845 Lamplight Drive, La Jolla, CA 92037

This paper summarizes the major aspects of lunar industrialization and settlement. It identifies scientific and evolutionary facts leading to a definitive justification for why man must industrialize space, changing our present closed world into an open world. It defines three interlocking phases of open-world development: exo-industrialization, exo-urbanization, and extraterrestrialization. The Moon's environment and relative proximity render it the logical first location for this process to begin. Of all alternatives, the Moon offers the earliest, highest benefit relevance to terrestrial humanity and to the development of cislunar space. It is an invaluable proving ground for subsequent industrial developments and settlements in more distant, more expensive regions. Large orbiting space habitats should follow, not precede lunar development. A lunar development strategy consisting of five logical development stages (DS) is proposed. Their key accomplishments are summarized. DS-1 involves further synoptic prospecting for mineralogical provinces and candidate base site. DS-2 establishes a circumlunar space station, Moon Ferry, and automated laboratories and pilot facilities including an oxygen extraction plant on the surface. DS-3 establishes a first-generation nuclear-powered Central Lunar Processing Complex, and the first large-scale industrial production begins. DS-4 diversifies productivity by adding Feeder Stations in more distant metallogenic provinces. DS-5 establishes Selenopolis, a self-sustaining lunar civilization, founded on a powerful fusion energy base. Novel technological approaches are summarized, including the Lunar Slide Lander, Partially Enclosed Launch Track, the Selenopolis habitat, Lunetta, Soletta, and techniques for both mining, and helium and hydrogen production. Energy, transportation, products, markets, economic strategy and socio-anthropological factors are discussed. The central conclusion is the mandate for low-cost transport and energy, early self-sufficiency, early investment returns, industrial flexibility, and a systematic approach to creating a positive Earth-Moon balance of trade.

THE EXTRATERRESTRIAL IMPERATIVE AND LUNAR DEVELOPMENT

This paper sets forth key aspects of development of Earth's sister planet—Moon—as the first extraterrestrial world in this solar system. Its genesis is my book, *The Seventh Continent: The Industrialization and Settlement of the Moon*, which was written during the past ten years. It also considers a philosophy that I began developing in the early 1960s called The Extraterrestrial Imperative.

In the 60s, we heard statements like, "One square inch on Earth is more important than the entire surfaces of the Moon and Mars." But those who understood the long-term meaning of space not as a Roman circus or an ideological race between the United States and the Soviet Union realized that we will go into space precisely because that one square inch on Earth is so important. The space engineer is the environmentalist's greatest ally. Our work in space will change Earth's present closed-world environment into an open one with access to vast space resources and other critically needed benefits that will greatly improve the lives of all people, and preserve Earth at its best—as man's home and garden for the maximum human future.

The philosophy of The Extraterrestrial Imperative is a definitive justification for a long-term future based on man's ability to transcend the limits of one small planet. This philosophy is definitive because it is supported by scientific and evolutionary fact. The facts show that a closed-world crisis is imminent but surmountable, that such a crisis occurred once before on Earth, and that the solutions came through technological growth.

After its formative period, the young Earth, like a giant flower, soaked up solar radiation in the form of chemical energy (C–H compounds) until the first primitive life stirred. Life's technological level (enzymes and metabolic processes) limited its interaction with the environment to processing fossil C–H energy and food sources by various primitive chemosynthetic procedures, mostly fermentation processes. When the previously generated organic resources ran out, the life forms began consuming each other. This had to lead to a survival crisis that was as certain as the fact that our dependence on fossil energy sources must eventually come to an end.

This was life's First Great Crisis on Earth. But what seemed to be an absolute limit to growth was no limit to growth. It was the beginning of a series of evolutionary technological advances that led to more growth. Metabolic advances in the photoautotroph led to the chlorophyll molecule and photosynthesis. These inaugurated the first industrial revolution on this planet. It was not possible then, any more than it is for us today, to be totally planetogenic. So life began processing solar radiation energy to chemical energy and thus for the first time used an extraterrestrial component. This made the interactive environment no longer planetogenic and chemical, but astrogenic and umbilical to the vast resources of the primordial inorganic environment—primarily water, CO_2, and solar radiation. Life had gained control over the production of its basic staples of food and energy. This changed the First Earth to the Second Earth, where, with the aid of the parasitic oxygen metabolism to reconstitute the comparatively limited supply of CO_2, a growing and expanding network of biological niches eventually culminated in the magnificent negentropic creation of the biosphere. Out of its womb arose, after a certain gestation period, the human being as the seed of the next higher metabolic capability: *information metabolism*. This includes the special abilities of information collection, abstraction, and reconstitution as knowledge and ability, which are storable in the human brain or devices developed by it.

Unlike oxygen metabolism, which is parasitic because its survival depends on plant/ animal resources, information metabolism is umbilical—as is its forerunner, photosynthesis. The umbilical metabolism can interact between the negentropic sphere and the antropic wilderness on the outside. It need not be wholly dependent on organic fossil materials from one planet. Information metabolism can interact with the primordial environment and use inorganic matter, which is so abundant in the atom.

We all know what information metabolism has accomplished so far. It is the foundation of human civilization, including the industrial revolution that replaced slaves by machines. It is also the bridge to the cosmos. It can now enlarge our environmental base—not only for resources, but for growth in technology and, thereby, human existence.

The facts show that life now faces its Second Great Crisis on this planet. But life has, through advancing metabolic technologies, created open worlds. It will do so in the future because, for physical reasons, it can survive only in an energetically and materially open world. In an open world, there are no limits to growth. By capability and design, information metabolism can resolve the conflict that every umbilical metabolism has with the old environment. It can transcend the confines of one globe and become polyglobal. It has absolutely everything it needs to create a new and larger sphere of integration. I call this the *androsphere*.

While the biosphere is integrated by the chlorophyll molecule and photosynthesis, the androsphere is integrated by the human brain and information metabolism. The crucial difference between these spheres is this: the biosphere is integrated but non-modularized, permitting substances to move through everything—water, land, air—as is necessary for biotechnology to work. Biospheric technology uses some 20–25 elements and recirculates them because it is the only way to keep a materially open world. This non-modularized sphere is incompatible with information-metabolic technology—for example, nuclear technology. But the androsphere will be a modularized polyglobal structure without circulation through all parts. It is consistent with the evolutionary advancement of metabolic negentropism, and it is exactly what information metabolic technology and industry need to keep a materially open world. What is done in extraterrestrial environments does not necessarily effect Earth. By going beyond Earth, which is highly biophil, into environments that are largely technophil, we adapt the new environment to the requirements of information-metabolic androspheric technology. Building a selenosphere on the Moon is an example of androspheric modularization. Large habitats will be separated by vacuum from nuclear power plants, industrial plants, *etc.*, and what happens in these plants does not automatically drift into the habitats.

The technospheric evolution of life is clear in the main steps described. The sociospheric evolution has two analogues. In the biosphere, there are the species. In the androsphere, there are civilizations. Civilizations differ in different approaches to different things. But civilization is more. Civilization is ascendence beyond brutality, beyond the recognition of plurality. Civilizations will continue to be formed because they are dependent on infrastructural elements such as location, conditions, *etc.* Civilizations on the Moon and Mars will indeed be different from ours.

Taken together, these factors assure continuing growth—as long as we have an open world, and do not shy away from overcoming our problems by technology and *by our own development*. I must emphasize that technology is not the solution to our shortcomings. The solution is that we must grow and mature. But technology can make that easier. By contrast, a no-growth philosophy, which asks humans to live with less of everything, can regress us to the Middle Ages because a dog-eat-dog fight is bound to break out under such conditions. We have come too far. We have to go on. Life shows us that technological advancement is the road to take. But based on these technological advances, our species and civilization must advance also. Then we can proceed.

Based on the Extraterrestrial Imperative's epistomological infrastructure, I concluded at the end of the 1960s that three evolutionary phases of our expansion into space, following initial exploration, can be substantiated:

Space Industrialization —the capability of productive existence in the new environment

Space Urbanization —the capability of establishing large-scale settlements and extraterrestrial civilization to the extent to which it can be underwritten by industrial and biotechnical productivity

Extraterrestrialization —a prolonged process of socio-psychological development and anthropological divergence, based on the integration and further evolution of the first two phases, manifesting itself in the physiological, anatomical, immunological, esthetic, and general cultural sectors.

Accordingly, I have urged for the past 20 years that we concentrate on the industrial development of space in the post-Apollo era, first in near-Earth orbits with commercial transportation and small space facilities, then on the Moon.

THE LOGIC OF LUNAR DEVELOPMENT

It has been said, "If God wanted man to fly, He would have given man wings." Today we can say, "If God wanted man to become a spacefaring species, He would have given man a moon." We are fortunate to be part of a double planet system, with a nearby sister planet whose surface conditions, despite all their differences from ours, still bear closer resemblence to Earth than do any other accessible surfaces in this solar system.

The Moon is the logical proving ground for subsequent industrial developments and settlements elsewhere. Only 2–3 flight days away, it allows us to develop at our very doorstep the experience we need to operate successfully and cost-effectively in more distant regions. No other celestial body and no orbiting space station can more effectively permit development of the habitats, material extraction and processing methods, and in essence, all the science, technology, and sociology required for a responsible approach to extraterrestrial operations.

Moon *vs.* Mars

The alternative of choosing Mars over the Moon as the first location for extraterrestrial development would take more time and money. We would first have to develop a new propulsion system, because it makes no sense to go to Mars by the skin of your teeth— barely making it after maybe 150 days, staying there 8 days, and rushing off again. The development of adequate heliocentric transportation would be an enormous investment, requiring additional acceptance of the high costs and operational risks associated with more widespread launch windows and longer transit times that still cannot compete with those of lunar flight. Moreover, we already have most of the technology required to get to the Moon and back. We don't need much more than oxygen/hydrogen propulsion, especially since we will be exploiting lunar oxygen.

Lunar Base *vs.* Space Colonies

The alternative of beginning with giant orbiting space "colonies" is not only unnecessary, but it is the wrong agenda. The reasons touch on a variety of facts of life about space operations, ranging from concerns regarding the human to concerns regarding engineering, technological, economical, and other factors including investment returns. A brief evaluation of key differences between the orbiting space colony and the lunar base follows.

A lunar industrial facility will process lunar material *in situ*, precluding the need to transport large masses of raw materials for processing in orbit.

The weatherless lunar environment permits open storage of elements not immediately needed. This is preferable to the problems of mass distribution, orbital control, *etc.*, that stored surplusses would present to an orbiting facility.

The high vacuum and low gravity offered by the lunar surface are two key advantages for extraterrestrial industries. The high vacuum facilitates the generation and maintenance of cryogenic temperatures. One-sixth gravity (g) provides benefits for construction, transportation on and beyond the Moon, and operations. For example, the energy required to deliver cargo from the Moon to geosynchronous orbit and return to the Moon is 7.2% of that required for the same mission from Earth. Where 0 g is required, easily accessible circumlunar factories can be used. The energy to deliver cargo from lunar surface to orbit is less than 1/20 of that required for transport from Earth.

Apollo physiological experience shows that 1/6 g is a considerable, sufficient improvement over 0 g. One g is not needed and would actually be a hindrance. But higher g values can readily be generated centrifugally on the lunar surface. The lunar vacuum facilitates construction of very large habitat centrifuges equipped to house personnel for months at a time. The habitat centrifuge can run on magnetic cushions and be propelled by linear motors. It can provide habitats at 1 g, 0.4 g (in preparation for Mars missions), and at arbitrary levels from Earth-g to the g-levels on all other accessible surfaces in the solar system. At a radius of 1000 m, 1 g is generated at only 99 m/s = 356.6 km/h circumferential speed, or 0.95 rpm. It is, therefore, incorrect to claim that only one low-gravity level is available on the Moon.

The technophil lunar environment is favorable for powder metallurgical technology, electron beam and laser beam evaporation welding, and other advanced, efficient manufacturing processes. It is also favorable for the application of nuclear power, which is needed for the productive process to continue throughout the lunar night as a precondition for the economic viability of lunar operations. Lunar nights are for a major part illuminated brightly—up to 100 times the full Moon in the terrestrial sky—and are at least as suitable for mining, construction, and other outside work as are lunar days. But work at night requires an energy source that works at night, and the only one is nuclear

The need for protection from cosmic elements is another consideration. Orbiting facilities, especially if outside the radiation belt, are exposed to cosmic radiation, micrometeoroids, and solar proton wind from all directions (space angle 4π). But on the Moon, protection is offered from a space angle of at least 2π by the lunar mass. The use of topographical formations can increase the protective angle. Loose lunar material to provide protective shielding is abundantly available *in situ*. Lunar sand and breccia constitute suitable construction material. Breccia, sand, and dust are also the prime source of industrially valuable elements, and this fact greatly facilitates the mining of lunar crude.

Since raw materials prospecting, mining, and processing on a large scale are among the major objectives of space industrialization, permanent surface settlements are of course preferable to space colonies. The concept of a Lunar Industrial Zone (LIZ) on the lunar surface is considerably simpler and more economical than the process of supplying a space facility in circumferential orbit from the Moon's various mineralogical provinces.

With a surface base, the establishment of large hydroponic agricultural facilities is also possible, based on encouraging results from preliminary tests of lunar material with plant growth. These facilities would use lunar bulk material enriched by nutrients imported from Earth.

Finally, the following additional comparisons between the lunar surface settlement and the orbiting space colony are of interest. These comparisons consider the mandate for a cost-effective, flexible, stepwise approach to the evolving complexities of integrating the three fundamental subspheres of an extraterrestrial settlement: technosphere, biosphere, and sociosphere. The technosphere includes all of the technological and scientific development. The biosphere includes plants, animals, people and everything they need such as food production, environmental controls, vehicles, and various machines, *etc*. The proper integration of these spheres will require advanced concepts in modularization that permit experimentation, change, and improvement as the expanding body of extraterrestrial experience dictates.

A large facility, especially a non-modular orbiting space facility, must be fully laid out and frozen without the benefit of adequate extraterrestrial experience by its population. This inflexibility involves many risks, and costly corrections, if possible, later. The internal architecture of large future extraterrestrial settlements cannot be frozen today because of the progressive nature of extraterrestrialization, its concomitant social changes, and advancing technologies. Moreover, it is not even clear that it will be necessary to simulate terrestrial conditions virtually to the last detail, including 1 g artificial gravity.

But the lunar surface is forgiving. The environment and the initial development of smaller habitat systems facilitates the learning process in construction and in the social development of relatively smaller populations of hundreds to a few thousands. This permits rational progression to larger settlements and populations of various races parallel to the growth of industrial, technological, and economic capacity, until the state-of-the-art is reached that permits the layout and construction of giant bio-niches, which I call *Selenopolis*.

In far-flung, modularized Selenopolis, with its variety of bio-environments and climates, new Selenian social structures and pluralities can evolve. They are not enclosed in a shell floating in a vacuum. Outside is a vast, challenging, magnificent wilderness— mountains, plains, horizons bathed in the light of day and Earth-lit night.

In a cylindrical space colony there are no horizons. From rising circular slopes, people watch people. If they step out, there is only a weightless, horizonless, unchanging sunburned void in which their little world is imprisoned. Their inside and outside environments psychologically promote ingrowth and involution.

The lunar environment promotes growth and evolution, since Selenians live in a world that supplies them with resources, but also is open to the universe. We might call them Cosmopolynesians, for they live at the shores of space. Space will lure them always, like the vast Pacific lured the Polynesians whose cultural main bases were on island worlds, not in tiny cocoons.

The most important aspect of lunar development lies in the human sector. It bears repeating that technological progress and environmental expansion are no substitutes for human growth and maturity, but they can help the human reach higher maturity and wisdom.

Human growth is contingent not only on the absence of war, or overcoming hunger, poverty, and social injustice—but also on the presence of overarching, elevating goals, and their associated perspectives. Expanding into space needs to be understood and approached as world development, as a positive, peaceful, growth-oriented, macrosociological project whose goal is to ultimately release humanity from its present parasitic, embryonic bondage in the biospheric womb of one planet. This will demand immense human creativity, courage, and maturity.

But the human is the violent product of violent evolution on a violent planet in a violent universe. As he enters space, he carries the tensions of our time and a heritage of endless wars involving hundreds of thousands of years of ideological, sociological, religious, territorial, and political hostilities. This heritage is frozen into humans like the solar magnetic field is frozen in the solar wind as it leaves the sun.

If the human is to outgrow this terrible heritage and grow with his expanding world, a few limited activities in near-Earth orbit and limited scientific expeditions are not enough. Really great tasks that penetrate man's social structure, infuse a broad evolutionary perspective, tax his creativity, and challenge his mind are needed. The industrial development of the lunar world toward the first polyglobal civilization is one such task—indeed the first logical goal on the agenda of the Extraterrestrial Imperative.

In summary, it can be correctly concluded that:

1. The Moon is much more than a quarry for orbiting facilities. It is a logical target for open-world development with a significant technological, productive, socio-economic, *etc.*, consequence potential.

2. Open-world development will involve a difficult trial-and-error process that must progress in steps. Compared to an orbital (open-space) environment, especially outside the protection of the geomagnetic field, the lunar environment is more protective, more conducive to stepwise development and experimentation, and more forgiving of failures.

3. The Moon is the logical first target for open-world development on which we can cut our teeth relatively conveniently and economically. Nothing can be done in a space "colony" that cannot be done sooner, more efficiently, and more economically on the Moon, if a few basic conditions are met.

4. Therefore, large space habitats, which I call *androcells*, follow, rather than precede lunar development. Their purpose is different, serving primarily as links in the development of other solar system worlds.

PRINCIPLES OF SELENECONOMICS

Lunar development must be rooted in a viable lunar economy, which I call *seleneconomy*. A seleneconomy can be viable only if its sustained cost-yield prospects are favorable.

Lunar development strategy should be dedicated to steps that maximize investment returns, minimize return times, and keep investment sizes manageable, so that venture capital and private investment can be attracted as rapidly as possible. The usual method of raising a demand for billions of taxpayers' dollars and decades of time before any alleged returns can be realized is economically not meaningful.

The investment strategy—if it is rational and disciplined—is in turn a major steering factor for the technological development that must be subjected to controlled-sized investment quanta, must assure minimum non-productive freezing of investment capital, and must contribute to reduced operating costs, especially logistics from Earth.

This renders "independence"-oriented investments and technologies particularly important, along with production- and productivity-oriented investments. Independence-oriented investments fall essentially into three categories (which in turn cut through various technological disciplines and affect the other two investment categories mentioned before). The three independence-oriented investments will involve:

● Use of nuclear energy, primarily for power supply rather than for transportation. Without nuclear power plants and possibly other techniques for all forms of industrial and agricultural production, a viable seleneconomy does not appear attainable.

● Reduction of terrestrial propellant supply through transporation energy management, lunar oxygen (LULOX) production, and LULOX depots on the Moon and in relevant orbits, especially geosynchronous and near-Earth orbits. Thus the terrestrial supply equivalent specific impulse is raised to over 3000 seconds, a value not attainable by modestly advanced nuclear drives.

● Extensive use of lunar materials for construction, for shielding, for growth of food plants and other purposes. (One proposed application is harenodynamic heat rejection—a novel technique for using lunar sand for heat rejection. On the Moon, this technique can eliminate the need for large, heavy, expensive radiators. It also broadens the choice of reactor types in the direction of lower operating temperatures, since the fourth-power law loses its overriding influence.)

In summary, maximum value generation capability and flexibility should be achieved with minimum initial expenditures and lead time. Only on this basis can lunar industry be developed early, effectively, and in a financially responsible manner. And only rising productivity and sustained economic growth can sustain an ever-increasing lunar population and the development of high Selenian living standards.

LUNAR PRODUCTS

The development of the Moon's resources will result in a lunar industrial output whose ultimate magnitude is impossible to fully anticipate. But it is clear that it will include raw stock from mining and refining, as well as a vast number of semi-finished and finished products.

Products will include sheet metal and trusses of aluminum, magnesium, titanium, iron, or alloys; castings, bars, wires, powders of pure or alloyed materials; glasses; glass wool; ceramics; refractories; fibrous and powdered ceramics; insulation; conductors; anodized metals; coatings, including almost perfectly reflective sodium coating (since sodium can be freely used on the Moon and in orbits, whereas on Earth it reacts with water and is dulled by oxidation); thin film materials; silicon chips; solar cells; entire structures of various metals and alloys for lunar and orbital installations (they do not have to be made weather resistant); compound and fibrous materials; heat shields and insulation materials, as well as radiation shielding materials for space stations; propellant containers; entire orbiting facilities, such as space station and factory modules and liquid lunar oxygen depots; large portions of cislunar and interplanetary spacecraft; and so on.

Where 0 g is required for manufacturing, easily reached facilities in circumlunar orbit (CLO) can make crystal bole, fibers, solar cells and other special materials and products. Parts, components, subassemblies, and full assemblies can be integrated in CLO before being shipped to geosynchronous or other distant circumterrestrial orbits via electric freighters (which will eventually use lunar sodium as propellant).

PRIMARY MARKETS

Of all exo-industrial activities in geolunar space, lunar operations represent most closely a *total* industrial system. They are supported by carefully minimized, and eventually declining, non-selenogenic imports from Earth. In turn, they support orbiting and terrestrial industries with oxygen, helium, materials, products, and services. The four primary markets for lunar goods and services are described below.

1. Lunar surface. This market includes domestic demands for lunar industrial and habitational development; science and technology experiments for terrestrial customers; new forms of entertainment for terrestrial television viewers (low gravity, moonscape, and vacuum will permit "natural" special effects and later, new sports and cultural arts programs); and eventually tourism and retirement environments for Terrestrians.

2. Geosynchronous orbit (GSO). By 2000, GSO may hold well over 1000 service satellites that will be virtually indispensible socioeconomically. This market will demand spacecraft servicing; replacement parts; new components; partial or entirely new satellites; salvaging and recycling of inoperative systems, sections, components, and elements. In addition, LULOX filling stations may be established to reduce the cost of manned access or supply deliveries needed from Earth.

3. Near-Earth orbit (NEO). Orbiting manufacturing facilities will be likely buyers of lunar raw materials, capital equipment, entire production facilities, oxygen (not only for air but also for water—only hydrogen need then be imported from Earth, which means a large cost reduction), and eventually even some basic dehydrated foodstuffs.

4. Earth. This will be a major market for lunar raw materials, semi-finished products, and space-made components in larger quantities and involving larger masses than could be handled economically if the raw material should first have to be supplied from Earth to NEO manufacturing facilities. These lunar imports will sustain industries and create new job markets on Earth.

A LUNAR DEVELOPMENT STRATEGY

Lunar industry should be viewed as an organism that, over time, evolves to progressively more complex capabilities and generates sufficiently strong foundations for expansion. Lunar industry must be broad-based and diverse if it is to last. The need for economic feasibility and early returns will require a skillful interplay between market/customer-oriented products and services, and infrastructural investments such as transportation, energy, and surface/space installations that expand food production and diversify industrial productivity. Based on these considerations, on the seleneconomic principles mentioned earlier, and on basic evolutionary logic, the guiding principles of a lunar development strategy can be formulated:

Low-cost access to the lunar surface = low overhead and enhanced capability to provide services in geolunar space, due to low transportation and personnel costs.

Ample and low-cost energy assurance = high, cost-effective, and versatile value generation capacity.

Early self-sufficiency = low import costs, hence low operating costs and enhanced survival capability of lunar personnel.

Industrial flexibility = cost-effective means for increasing the capacity for diversification and adaptability to changes in market demands.

These principles have guided my studies since I called for the broad industrial use of extraterrestrial materials, specifically lunar materials, before NASA in 1971 and publicly, beginning in the late 1960s. They are designed to ensure steady progress; early economic viability through on-going productivity; and supply crisis resistance. (The latter ensures that lunar personnel do not have to return to Earth because they cannot sustain their lunar existence without basic inputs from Earth or do not have the "credit worthiness" to receive loans on the basis of a reasonably early payback capability—in principle, the situation that choked off the Apollo program.)

The Five Stages of Lunar Development

The lunar industrial establishment is likely to consist of two basic components: the Lunar Industrial Zone (LIZ) on the surface and an orbiting component, the Lunar Industrial Space Installation (LISI), in low CLO. Together they offer in one relatively small cosmic space the before-mentioned wide range of gravities, natural and artificial.

My studies have resulted in a number of novel technological approaches to lunar industrialization, as well as a consistent lunar development logic whose basic framework is briefly described here. There are five development stages (DS). There is an associated economic and industrial rationale for the transition from one DS to the next.

The achievements of each DS belong to one of three main sectors: *technosphere* (research, technology, industry); *biosphere* (plant/animal life, food production, general plant growth, selenobiosphere); and *sociosphere* (habitats, living and working spaces, society, economy, politics, and culture).

Stage One

DS-1 could be accomplished in this century. It involves synoptic prospecting to further detect metallogenic or mineralogenic provinces and obtain other advanced information for industrial site selection. Simplified Surveyor-type landers and at least one lunar polar orbiter can be used. Emphasis should be placed on sites in the western and far-western part of the near-Earth side north and south of, as well as at, the equator. This area holds promise as a base site not only due to its potential for resources, but for a variety of other reasons. These include its ready, low-cost Earth communications and transportation accessibility, and a flat area between 50° and 60° western longitude at Oceanus Procellarum. This area appears particularly suitable for what I call a slide landing strip, which would in turn influence the location of the first Central Lunar Processing Complex in DS-3.

DS-1 would also include establishment of a Lunetta reflector orbiter to illuminate the perpetually shadowed places at high latitudes and the polar regions to permit photography, cartography, and the possible identification of polar ice deposits, if any.

Stage Two

DS-2 is an important, cost-saving and indispensible preparation phase for those who will become "first generation" ground personnel. It involves further work toward surface base site selection and operations training prior to lunar base build-up.

A Circumlunar Space Station (CLSS) is established in ~100 km equatorial orbit and uses a Moon Ferry (MF) for limited manned surface missions. The CLSS serves as habitat, operations and training center, and laboratory for engineering, biological, and medical purposes. It will support experimentation with much larger quantities of lunar materials than could be economically delivered from the Moon to Earth. Most lunar materials will be brought from the surface to the CLSS by automated returners. The results should lead, about halfway through DS-2, to selection of the appropriate surface base site.

Experience in this DS will also provide medical and behavioral profiles of personnel who will spend time in the two worlds of lunar surface and space station and who will experience prolonged stay in orbit about another body, remote from Earth.

DS-2 continues with establishment of sophisticated, automated laboratories and pilot facilities on the surface. Modules are mailed from Earth to the selected base site. Using the MF, CLSS personnel descend to set up, start, and maintain the systems.

One of these systems can be the first small-scale automated oxygen extraction plant, providing gaseous O_2 for MF and CLSS. My version works on the basis of induction heating a mixture of lunar dust/sand and hydrogen, and on the formation of water and its electrolytic separation to oxygen and hydrogen. The latter is recycled. The former is stored in chilled, moderately compressed form. By omitting the liquefaction facility, the system is simplified. It consists of H_2 dust reaction chambers, H_2O condenser and storage container, electrolysis module and storage containers for H_2 and O_2. Powered by a 100 kWe M_{OU2} space reactor, the system produces about 2.8 t water in 22 (24-hr) days out of 11.9 t lunar dust, and generates electrolytically about 2.5 t O_2 in the remaining 8 days of a 30-day month.

A 10-person lunar crew consumes about 300 kg O_2 per month, so at least 2000 kg/ month could become available for propulsive purposes.

Other modules placed on the Moon during DS-2 can permit personnel to stay on the surface for weeks at a time. Figure 1 shows two laboratory/habitat modules and, in the background, a nuclear power station. Surface operations made possible by the existence of these and other modules will build a wealth of crew experience for work that will begin in DS-2 and further evolve during DS-3.

While existing transportation capabilities will be used to the maximum extent in this no-frills lunar development approach, DS-2 must be accompanied by major strides in research, development, testing, and engineering of special, highly cost-effective transportation systems for use in later DS.

Figure 1. First-generation lunar surface crews get started in laboratory/habitat modules mailed from Earth. The inverse converging shape of the modules maximizes shielding against corpuscular radiation, optimizes temperature control for placement in equatorial regions, and serves as an umbrella to provide shade in the module's immediate vicinity.

Stage Three

The transition from DS-2 to DS-3 initiates the operation of lunar industry. The payoff on initial investments begins here, at first modestly. This payoff is a prerequisite for expansion into larger-scale industrial production in DS-4, whereby lunar profitability will assist with the larger investments for the next stage, and will demonstrate credit-worthiness for attracting major terrestrial investment capital.

A first-generation nuclear-powered Central Lunar Processing Complex (CLPC) called "Cynthia" is established where there are favorable conditions for a slide-landing strip and for finding valuable raw materials nearby. As indicated earlier, a promising location for Cynthia is in the far western Oceanus Procellarum at the equator, just east of the south rim of Hevelius. It is not likely that this location contains higher concentrations of all desired minerals, if such a place exists at all. As Cynthia's production diversifies, minerals will have to be exploited in other places using Feeder Stations (FS) to be established in DS-4, since it would be uneconomical to place CLPCs at several locations.

Cynthia will first produce oxygen, then other materials. As a result of experience gained during DS-2, lunar personnel will by now have developed expertise in handling tools, equipment, and construction work. The latter will include cold-welding cut lunar rocks; producing lunar bricks and possibly cement with lunar sulfur as binder (instead of water); and compacting lunar fines (powdery to coarse sandy material) into building blocks for lunar igloos, or better, ligloos. These ligloos will have sprayed airtight inner liners and airlocks—providing shirtsleeve shelters, workshops, and "greenhouses" for growing food plants.

Crews will also manipulate teleoperators and supervise robots under various conditions, day and night. They will perform cold welding, laser and electron beam welding, surface mining and drilling. They will monitor and maintain a wide variety of equipment and agricultural modules.

Thus trained, the first-generation lunar crews can schedule Cynthia for production runs of oxygen, silicon, aluminum, iron, glasses, and other materials. From these raw materials, Cynthia can progress to powder metallurgy, vapor phase metallurgy, production of solar cells, computer parts, and eventually space habitat structures, communication platform stuctures, antennae, service satellite parts, reflector structures, and much more.

DS-3 will also involve the introduction of novel transportation arrangements that radically lower the costs of getting to and from the lunar surface and across cislunar space. The CLSS of DS-2 grows into a staging base, training second-generation selenauts and expanding into a 0 g factory.

Stage Four

DS-4 marks the expansion of industrial production and services. Diversification grows beyond extraction and semi-finished products to finished products and assemblies. Strategic economic positions must be attained for supplying orbital and terrestrial markets, yielding a high gross lunar product that not only builds a positive balance of trade, but also builds the infrastructure and establishes credit-worthiness for continuing expansion.

To broaden market response capability, Cynthia is augmented by FS in valuable, sometimes distant provinces identified during DS-1 and 2 as having an abundance of certain raw materials.

The FS are highly automated and basically simple. Most are unoccupied or intermittantly occupied. They are remotely controlled and operated from CPLC control by laser communication link via communication relays. The relays float on the supporting pressure of microwave beams, as would a ping-pong ball on a water jet. They are placed at appropriate altitudes, depending on FS distance. This communication concept precludes the need for either a satellite, which is interrupted by precession, or a tower, which would have to be 20 or 30 km high to maintain the connection. (The concept is also extremely practical for communications anywhere on the Moon, including control of electromagnetic trains, *etc.*)

Materials collected at the FS can be transported to Cynthia by various methods, depending on distance. Relatively close FS, say 120 miles away, can send cargo by electric cars. The most important FS can eventually deliver via high-speed electromagnetic trains. Meanwhile, distant FS can hurl cargo ballistically to receiver craters near Cynthia with great accuracy, thanks to low lunar gravity and high vacuum.

My design studies have considered a large number of ballistic systems for cargo transport. I am most favorably impressed by the suitability of a centrifugal launcher with curved launch tubes. It runs on magnetic cushions propelled by electromagnetic bilinear motors.

DS-4 also includes installation of fusion power plants and initial build-up of a solar reflector swarm, "Soletta," in L-1. Eventually, Soletta will reach the size of 120,000 km^2, illuminating a 200,000 km^2 area at lunar night around Sinus Medii for agricultural and biospheric purposes (Novaterra). Biospheric Novaterra and technospheric Cynthia become the pillars on which sociospheric Selenopolis and lunar civilization rest.

Stage Five

DS-5 establishes Selenopolis and the selenosphere—a fully developed lunar world with a large population underwritten by industry. This stage is contingent upon a strong economic foundation, a very high degree of self sufficiency, particularly in food production, and a powerful fusion energy base. Initially, it will require more massive imports from Earth. But its expansion should be commensurate with economic growth and the ability to sustain corresponding population increases, thereby financing its evolution to a high degree with lunar capital.

With this premise, we move into the twilight zone between economics and politics. Analysis of pertinent factors—social, economic, and the onset of the social and cultural extraterrestrialization process—suggests the inevitability of this development. Thus in DS-4, the issue will arise concerning the extent, in DS-5, of financing and controls by terrestrial institutional power.

As defined here, DS-5 is a state in which trade relations with Earth are based on rough commercial equality. This means mutually complementary value generation, where lunar civilization is not in a receiver position *vis-a-vis* Earth. The resulting high level of fiscal and economic self-determination, and the attraction of terrestrial investment capital, cannot help but encourage political implementation. Therefore, much will depend on lunar political status and prospects by the end of DS-4: will this be a colony of Earth, part of the common heritage of terrestrial mankind? Or will it be an independent political entity with Selenians in control of their own world? On a foundation of fusion power, the vast potential of the lunar economy renders the latter alternative possible and hence likely.

DS-5 progresses in a series of steps involving increasing advancements in industrial, agricultural, and energy production, in transportation, and in habitats. The early habitats are envisioned as a series of small units hugging the inside walls of a crater. But Selenopolis—the city-state of lunar civilization and the lunar biosphere—will be a network of enclosures gradually expanding to cover many square miles of the lunar surface, and some parts of the subsurface.

The enclosures comprise sections that are several miles long, with interior dimensions of 3200 feet across at floor level, and 1600 feet high to the center of a curved ceiling. The sections are joined at nodal points that serve as power, supply, and climatic control centers.

Selenopolis embodies urban, rural, agricultural, industrial, and resort areas. Each section is separated from the other by a solid but transparent "curtain," because each has a different Earth-like climate and season. Normal atmospheric conditions for Earth are maintained. In the beginning, simulated Earth climates will include continental, dry subtropical, and semi-arid, with climatic cycles where applicable. Other sections will have climates that are adjusted to their special agricultural functions in order to maximize plant growth (measured as yield per unit area and number of crops per annum). This will be accomplished primarily through CO_2 enrichment of the atmosphere, and by temperature, humidity, and suitable irradiation cycles, all coordinated to achieve the optimum combination. Figure 2 shows an agricultural zone.

Resort areas can include a winter section with snow, a subsurface lake for boating, a "sunbelt" with "lunar desert" views from a clubhouse that also overlooks the Alan B. Shepard low-gravity golf course, *etc.* Selenians may also enjoy the Krafft A. Ehricke rotating swimming pool.

Selenopolis interiors are illuminated by sunlight reflected through the ceiling by a mirror system. Since a lunar day is 14 Earth-days long, some of the mirrors are colored to provide the same time changes and sky colors experienced on Earth from morning to night and from season to season.

Selenians will not be bound to their biological environmental niches. In comfortable space vehicles, or in transport vehicles with interior shirtsleeve environments, they can tour the coasts of mare, the mountains, the cliffs of the southern highlands, the province of large craters stretching from the eastern coast of the Mare Nubium to the South Pole, and more.

With establishment of Selenopolis, the development of lunar habitation reaches its conclusion in the sense that a new environmental niche—a lunar biosphere honeycombed with ecological niches—has been created. But Selenopolis is open-ended, growing with

Figure 2. A large agricultural zone in Selenopolis is accented by occasional high-rises whose unconventional shapes are permitted by lunar gravity, maximizing the use of surface space.

its population and advancing technologies. In principle, the overall complex could eventually house many hundred million people. Such a large complex is never completed, just as development of a continent is never completed. Like the giant cathedrals of the Middle Ages, Selenopolis will be the work of many generations.

ENERGY

Selenopolis cannot be built with yesterday's technology. For economy, and because of the long lunar night, fusion energy is as fundamental and as indispensible to the Selenosphere as the Sun's energy is for the terrestrial biosphere.

The lunar environment is exposed to the nuclear influences of the Sun and the cosmos, and there are no liquid or gaseous means for circulation. The absence of any circulations makes the Moon suitable for the installation of nuclear power plants and for the underground use of nuclear energy, uniform or explosive. Nuclear technology will destroy no natural balance. It is most effective because it has the lowest entropy and mass. The open, peaceful use of nuclear energy, including release of nuclear detonation energy underground, is permitted by the International Treaty on Outer Space (1967).

Large deuterium-tritium (D-T) fusion power plants are the logical lunar base power source for many reasons. Their operation is easier on the Moon than on Earth because of high lunar vacuum. For example, the lunar vacuum facilitates maintenance of a high vacuum in the plasma chamber without size restrictions, reducing neutron flux density per unit of wall area, thereby reducing associated wall erosion and embrittlement, and impurities. It also simplifies reactor construction and maintenance and facilitates the use of superconducting magnets.

The D-T reaction is the comparatively "easiest" realizable fusion reaction. Most importantly, an excess of tritium can be bred and stored in the D-T reactor, decaying into Helium-3 (He-3) and an electron. This extremely rare helium isotope is practically non-existent on Earth. It is an important fuel for a D-He-3 fusion reaction.

After the D-T reaction, the D-He-3 is the next more difficult fusion process to attain. But as soon as the D-T reaction is operative, it becomes important for terrestrial fusion technology to advance to the D-He-3 reaction, because no radioactive tritium is employed and the reaction is almost completely "clean." Only about 7% of the released energy is carried away in neutrons, so virtually no radioactive isotopes are generated. The reaction essentially produces protons and alpha particles, that is, charged particles that can be confined magnetically. Wall-loading problems practically disappear, resulting in greatly improved power plant economy. Moreover, the proton-helium reaction plasma is a highly valuable resource for processing heat for material extraction and waste recycling, as well as for generating further electric power. Therefore, it will be highly desirable to develop the techniques for generating the higher temperatures needed to ignite a D-He-3 plasma.

But the He-3 must be produced somewhere, and the Moon is the best candidate. Therefore, large D-T fusion power plants become not only the chief power source for lunar industry and civilization, but valuable fuel factories supplying He-3 for terrestrial use and for use by interplanetary and possibly cislunar spacecraft. D-T plants should begin operation in the latter part of DS-4, D-He-3 plants in DS-5. D-He-3 power plants are also of interest because the proton output, combined with the electron output from tritium decay and with lunar electrons, forms hydrogen.

If Earth depends on the Moon for its He-3, the Moon depends on Earth for its deuterium and lithium—a case of mutual self-interest, which historically has formed more enduring and reliable arrangements among peoples than ideology or idealistic cooperation.

A 1,000 gigawatt-year (8,760 billion kilowatt-hours, almost four times the electricity consumed in the United States in 1980), generated by a D-T fusion reaction at a thermoelectric conversion of 0.33 and a triton yield of 2, produces 168.9 metric tons of excess tritium. Of that about 84 tons are converted to He-3 12.3 years later (and more each year thereafter, if the D-T fusion process continues at that level or increases). Eighty-four tons of He-3 suffice to generate more than a 500 gigawatt-year in D-He-3 fusion reactors on Earth at a profit of $44 billion annually per $0.01 of profit on the kilowatt-hour.

In terms of mass, the supply requirements are no problem at all, because of the extremely low entropy level of the fusion plasma. For the generation of a 1,000 gigawatt-year at 100% efficiency, 112.6 tons of deuterium and 777 tons of natural lithium are consumed annually, assuming a yield of 2 tritons for each triton burned. If the fusion reaction operates at 50% efficiency and if in this case all the unused deuterium is lost (an unlikely possibility), 225 tons of deuterium must be supplied to the Moon annually.

In the first case, the transportation requirement is a modest 13 ascents annually with a derivative of the Shuttle at 68-ton payload. In the second case, it is 15 ascents at a cost of about $1 billion plus procurement of the deuterium and lithium. Selling He-3 can therefore be quite profitable for the Selenian economy and is an example that demonstrates the continuing growth from DS-2 through DS-5.

At a conversion efficiency of 0.33, which is very conservative considering this state of technology, the waste heat per 1,000 gigawatt-year of a D-T power plant complex amounts to some 4.1×10^{13} kilocalories per 24 hours. This is enough energy to warm the atmosphere in almost 155 km^2 of 1,000 m $\times$ 500 m Selenopolis half-cylinders, from 0°–25° C or to heat 2 billion tons of water by 20°C. Because of insulation and thermal flux control, one such power complex could climatize a 30,000 km^2 sector of Selenopolis, capable of accommodating a population of one to ten million (at U.S. to European population density).

NUCLEAR POWER FOR RESOURCE EXTRACTION AND PROCESSING

Extraction of the Moon's raw materials will involve separation and refining. Processes using electrolysis, chemical, or thermal methods reduce, and can also be used to separate, different semi-metallic and metallic minerals from each other. Electrolysis poses high energy requirements. The thermal-mechanical method involves centrifugal separation after elements or compounds are partially or completely melted. Chemical reduction is complex and requires carbon and hydrogen. These chemicals would have to be imported from Earth because they are not initially available in relevant quantities. But abundant low-cost nuclear energy can be. The issue is adequate processing heat at the lowest possible cost.

The sources of heat for reduction are, in order of increasing temperature: the high-temperature reactor (HTR), with reactor cooling gas outlet temperatures of 900°–950° C; nuclear-electric arcs (these are less suitable for quantity production); solar concentrator-heater ovens, for temperatures up to several thousand degrees and low mass requirement (these are inoperative, of course, during the 354-hour-long lunar night); underground atomic ovens (UAO) stoked by small fission or fusion detonations; and the plasma from a fusion reactor, once this technology is developed. For early application and high-mass flow, day and night, the HTR and UAO are the most effective methods, besides solar heating during the day.

Figure 3 shows a typical arrangement for a strip mining-beneficiation-refining facility. It uses electrolytic and thermal-centrifugal, as well as other reduction and separation furnaces. Strip-mined lunar crude is fed into the furnaces by conveyor belts that also serve to transport tailings and slag back to the strip mine zone. The entire system is powered by a pebble bed thorium HTR that breeds U-233 fuel from thorium-232, thus

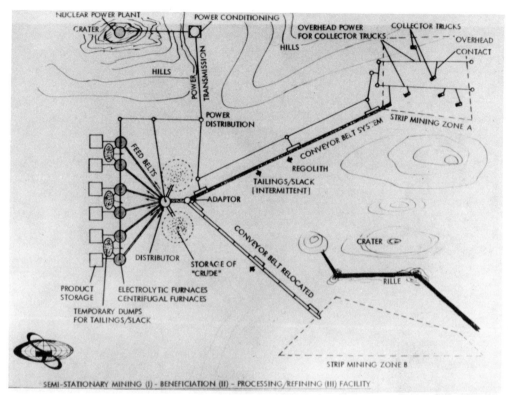

Figure 3. Lunar Development Stage Three includes establishment of the first strip mining-beneficiation-processing/ refining facility as part of the Central Lunar Processing Complex. Surface mined material is sorted and fed into furnaces. Metals, silicon, and other outputs are removed from the furnaces to storage as industrial feedstock for semi-finished and finished products, for transport to circumlunar orbit and further processing at 0 g, or for export to customers in circumterrestrial orbits and on Earth.

avoiding production of the more dangerous plutonium–239. Later, a combination of a thorium breeder HTR and molten–salt reactor (MSR) will be used; the former using the excess to generate more electricity. The MSR is in some respects even more convenient to maintain than the pebble bed HTR.

The nuclear power facility will be located at some short distance in a suitable crater or canyon, whose walls provide safe shielding.

NUCLEAR PULSE TECHNOLOGY FOR UNDERGROUND DETONATIONS

Since my first publication on lunar detonation mining in 1972, I have investigated a broad variety of operational modes for this technique. Only a brief overview can be presented here.

Figure 4.   Three operational techniques for lunar underground mining made possible by small nuclear detonations.

Fortunately, the Moon's crust is much poorer than the Earth's in elements that could be turned into undesirable radioisotopes by neutrons released during an underground nuclear detonation. The detonation process itself releases radioisotopes if it is fission. Fusion detonation is preferable.

Figure 4 depicts three operational techniques for nuclear detonation mining in underground caverns formed by small nuclear detonations. The caverns are blasted in bedrock (solid mare basalt, a pre-Imbrian rock). Because of the rock's total dryness, there is no steam generation to crack the cavern walls (as would occur on Earth, leading to collapse of the ceiling). Detonations in this environment produce highly compressed, glassy walls. They may be so water-impermeable that the cavern could be used as a water reservoir without the danger of losses.

After blasting, the initial nuclear detonation heats up the interior lunar rock. This produces intense thermal radiation from the walls, since their heat conductivity is poor by terrestrial standards, and the heat remains concentrated in them. This means that the detonation energy is essentially dissipated productively in the processed material. Radioactivity declines rapidly after a detonation and is at a minimum when mining begins. Moreover, the mine is cleared out by robots. The mined, deoxidized material itself is not radioactive and can readily be handled after removal.

Figure 4 shows, at left, single-detonation oxygen extraction; in the center, a technique for absorbing detonation-produced oxygen with other elements to produce life-supporting and industrial materials; and at far right, an atomic oven.

Lunar crude is fed into the atomic oven from the overlying loose breccia layer, either as fines and sand or as crushed breccia, to provide a maximum surface-to-mass ratio for maximum outgassing (reduction) efficiency. Oxygen evaporates out, and the deoxidized elements form an enriched ore, which may subsequently be separated into its constituents.

A more advanced UAO arrangement can involve a series of ovens, sequentially alternating between hot state for gas extraction, and cold state for mining the reduced material. There is practically no limit to the amount of lunar materials that can be processed annually by a combination of UAOs and nuclear-electric power plants.

TRANSPORTATION

While production of low-entropy fusion energy becomes a major consideration during later DS, the first prerequisite for lunar development is low-cost Earth-Moon transportation and low-cost transportation between the Moon's surface and circumlunar orbit. This must involve maximum use of existing Earth-Moon transport capabilities, along with development of a novel class of vehicles, facilities, and techniques that use the unique lunar and cislunar environments to advantage.

The Diana Fleet

A fleet of ships is needed for carrying out all aspects of lunar and cislunar development. In DS-1 and 2, the payloads required for lunar prospecting, establishment of the CLSS, and pilot surface facilities can be delivered to the Moon and CLO by combinations of the Shuttle, a Shuttle derivative (the Heavy-Load Launch Vehicle, HLLV), and Centaur. The latter will be modified to fit the Shuttle orbiter payload bay and is referred to here, unofficially, as Centaur II. The Centaur design allows sizing for various mission requirements. With a small maneuver module for lunar capture, it can deliver between 5.6 and 12 tons into circumlunar orbit, providing adequate capability for one-way transport of automated systems.

In DS-2, Shuttle capability is assumed to be increased to that of the HLLV and Centaur II to a cluster of rockets as drive for a large, first-generation Geolunar Transport-I to establish the CLSS. DS-2 would establish two new transportation systems—Moon Ferry and a highly cost-effective nuclear-electric geolunar freighter.

In DS-3, the advanced HLLV is still adequate. But an enlarged Geolunar Transport-II is needed, along with a LULOX filling station in CLO. The chemical transport should be designed for the Hohmann Braking Maneuver (HBM), single-module version. It should have variable thrust and modular design to permit as many geolunar transport functions as possible.

My key conclusion after extensive investigations concerning chemical transports operating between NEO and CLO is that the flight modality is the key to the relation between launch mass in NEO and cargo capacity as well as vehicle propellant load. A flight modality without HBM and without LULOX would shift an economic development of the Moon far into the future.

The decisive factor is not the limited reduction in propellant and launch cost possible with Earth launch vehicles. It is the cost-effective geolunar flight modalities with correspondingly designed interorbital transport vehicles (IOT). A flight modality involving HBM alone already yields a decisive improvement in cargo capacity. At first, the HBM advantage is economical for delivering to the Moon the freight needed to build a LULOX production capability. But the heavier deliveries required for Earth-Moon transportation involve the burden of carrying all the needed oxygen from Earth; therefore HBM cannot be used. The solution will be LULOX filling stations in NEO as well as CLO. The economic superiority of a LULOX modality lies not in an improved freight-carrying capacity, but in a decisive reduction of the Earth-NEO supply requirements, hence transport costs.

The most economical modality uses both LULOX and HBM. It yields the lowest propellant factor for a geolunar round-trip with chemical propulsion, because only hydrogen needs to be supplied. This corresponds to a supply I_{sp} of 12,400–15,200 seconds, which could otherwise only be attained by a very advanced nuclear propulsion system.

For cost-effective supply of a LULOX depot in NEO, a specialized LULOX freighter capable of HBM is required. It will bring LULOX as slush that is allowed to thaw slowly during transit, reducing evaporation losses.

For cislunar space operations involving missions between CLO, GSO, and 48-hour orbits, solar-powered electric propulsion systems will provide superior payoffs. They are ideally suited to this region because it has negligible Earth shadow effects, no radiation belt effects on photovoltaic cells, and very weak gravitational forces. At a given service life of the electric propulsion system, more round-trips can be made between CLO and GSO or 48-h orbit than between CLO and NEO, and more cargo can be transported. Cislunar freighters are also suitable for transporting service personnel to maintenance jobs in GSO and back to their lunar homes.

In comparing ion and magnetoplasmadynamic (MPD) drives, the latter show certain advantages, such as three orders of magnitude higher thrust density and potentially longer thruster operating life.

The Lunar Slide Lander

If lunar transportation is based on oxygen/hydrogen, and if LULOX is used, the price for this advantage is the cost of importing H_2 from Earth. It is therefore of vital importance to minimize this import. Fortunately, this can be done.

Figure 5. Nuclear-powered sweeper, designed for 1/6 g, prepares and maintains an 80-km-long runway for the Lunar Slide Lander.

Conventional lunar descent and ascent are prime consumers of H_2. But for ascent, there is a way to capture and reuse the H_2 (see next section). And for descent, the amount of H_2 required can be reduced to a fraction of that used by conventional systems.

A new concept, the Lunar Slide Lander (LSL), is proposed. It lands on a long, flat area and transfers its momentum to the lunar surface material, thus requiring very little propellant to land. This concept is not only decisively cost-saving, it avoids release of increasing exhaust gases as lunar traffic grows, preserving the Moon's valuable high vacuum for cost-effective ascent and for industrial uses.

In the DS-3 discussion, it was suggested that Cynthia, the Central Lunar Processing Complex, be established in the far western part of Oceanus Procellarum because, among other reasons, a flat area for slide landing exists. Covered with dust and sand, it can be turned into a suitable landing strip with little preparation. I envision a landing strip sweeper (Fig. 5) that can easily cleanse the area of larger stones down to do a depth of 20–50 cm. My calculations show that the LSL braking surfaces go down only a few centimeters.

Touchdown at velocities of up to 5500 km/h, bringing the LSL to a halt along an 80-km-long landing strip through interaction with glassy-sandy material, introduces a new branch of spaceflight dynamics, for which I propose the term *harenodynamics* (from harenosus, Latin for sandy). Harenodynamics encompasses the dynamics of flow, boundary layer formation, and pressure, temperature, and gas (O_2) release conditions in the boundary layer, at high speed flow of sand along harenodynamic brakes. The LSL's most critical component is its braking assembly, particularly the linings. It is desirable to manufacture the linings on the Moon from lunar materials. A harenomechanical and harenothermo-dynamic data base must be established to define target characteristics.

The lunar environment provides many advantages for slide landing. Vacuum permits high-speed LSL approach without temporary communication blackout due to ionized boundary layer formation. The absence of atmospheric effects, and superb sky and ground visibility (including optical signals at night) permit high predictability and automation of approach navigation. The LSL body as a whole is not subject to aerodynamic heating, which probably simplifies the design and results in lower structural mass. Possibilities for aborting the landing exist practically to touchdown.

The LSL descends from low CLO along an elliptic path. For elliptic descent from 10, 20, and 40 km, the supercircular velocity excess at perilune is about 5, 10, and 20 m/s. Therefore, a retromaneuver of only a few meters per second reduces the speed to subcircular and causes the LSL to approach the surface at a shallow downward path angle.

The approach phase is followed by a supporting vertical thrust phase, whose purpose is threefold: control of the touchdown point; fine adustment of the vertical velocity component for smooth touchdown; and initial support of most of the lander weight, to control deceleration and stability during the high-velocity phase.

The vertical thrust, which replaces the aerodynamic lift of a landing aircraft, is eventually terminated in the third and final phase. The slide time is about two minutes, of which supporting thrust is needed for at most 90 seconds. Average LSL weight during this period is about 0.8 of its full weight due to the centrifugal effect during the high-speed phase. The supporting thrust of the LSL at 16 tons full lunar weight requires, under these conditions, a propellant consumption equivalent to a velocity change of less than 120 m/s—much less than the almost 1700 m/s that must be eliminated at conventional landing. LSL propellant consumption is therefore reduced to less than 10% of that required for conventional landing, corresponding to a hydrogen expenditure per unit mass of payload of $P_H \sim 0.01$.

Figure 6 shows one of my LSL concepts. A very controlled touchdown begins with the aft edge which, together with the supporting thrust, stabilizes the vehicle for careful ground contact of the main drag vanes. These are spring supported, inclined, and in adjustable yaw position, in order to hurl the lunar material away from the vehicle.

I have examined a multitude of other aspects for the slide landing process—navigational, harenodynamic, emergency options, *etc.* Suffice it to say that none appears to pose serious problems.

The Partially Enclosed Launch Track

For takeoff and transporting cargo from the Moon, I have considered alternatives ranging from Apollo I conventional takeoff to electromagnetic takeoff. The latter requires

Figure 6. Touching down at 5500 km/h after descending from circumlunar orbit to the Moon, the O_2/H_2 Lunar Slide Lander transfers its momentum to the surface materials and requires very little propellant to land. This concept will involve a small fraction of the H_2 needed for landing by conventional methods, rendering lunar access an affordable matter of routine.

a very extensive launch structure, very high g loads, and an enormous power input of the order of millions of kilowatt seconds during the acceleration phase. I propose instead a Partially Enclosed Launch Track (PELT), employing a chemical catapult concept that accelerates the LSL and freight to ascent speed.

Growth of exports is the bottom line of the lunar industrial economy. Ascent loads must eventually exceed descent loads by a widening margin. But also, for lunar self-sufficiency, resources such as hydrogen and water must be maximized at every opportunity. Therefore, the point of the PELT concept is not only to gain an economical lunar ascent capability, but to recycle the water and hydrogen from exhaust materials generated during lunar ascent. The concept also helps preserve the lunar vacuum.

The 12.4-mile-long facility comprises the launch tracks, payload platform, O_2/H_2 booster (catapult) and external data platform. The latter, which is linked to the booster, is located outside to provide more space for odd-shaped payloads inside the PELT.

The LSL and cargo mount on the payload platform. The platform runs on frictionless magnetic cushions to prevent material contact. The tracks are mounted on a small tubular enclosure that contains the chemical booster having four or less engines. The booster is connected to the payload platform by a dorsal fin that fits through a slot in the top of the tubular enclosure. As the booster accelerates the payload to release speed, the slot is closed magnetically behind the fin, preventing escape of exhaust gases.

To catapult a payload toward a slightly eliptical orbit, which is then circularized for payload rendezvous with the CLSS, the PELT's booster reaches a release speed of about 1750 m/s (about 6800 ft/s). The exhaust velocity is over 4000 m/s. (*De facto* specific impulse is of the order of 30,000 to 40,000 seconds, which could be matched only by a very advanced nuclear pulse or mirror fusion system.)

After payload release, the booster is slowed down and retrieved. Meanwhile its exhaust water (with the hydrogen excess, if any, burned by oxygen injection behind the nozzles) has been automatically collecting at the back of the tube. Here it can be easily condensed by cold oxygen. If a 40-ton payload had been launched, about 7 tons of water would have been produced. It can either be used as water, in which case the propellant performs double duty, or it can be electrolytically separated into O_2 and H_2 for reuse as propellant.

One may assume an initial phase where the LSL returns to orbit as cargo carrier with high launch acceleration from a relatively short and simple Enclosed Launch Track (ELT). Figure 7 shows the ELT interior as an LSL is launched. The still light personnel traffic is handled by conventional vertical landing and takeoff. When the PELT is built,

Figure 7.   The Enclosed Launch Track accelerates a cargo-carrying Lunar Slide Lander to ascent speed, catapulting the vehicle into the trajectory for rendezvous with the Circumlunar Space Station.

it can grow in length, permitting progressively lower acceleration levels down to values suitable for acceleration-sensitive cargo and personnel. The PELT will eventually become a track for electromagnetic propulsion, once sufficiently high power levels are available to release about 60 million kWh in three minutes.

ELT and PELT construction will be greatly facilitated by the use of local materials and extensive automation.

THE EXTRATERRESTRIAL IMPERATIVE AND THE OPEN WORLD

The extraterrestrial imperative facing the human species was outlined in a paper that I presented in 1972: "Overshadowed by the limitations of the terrestrial environment, the scenario of world development will undergo fundamental changes in the next 30 years. But the need for resources will continue to grow. Therefore, the emphasis on opening new environments to industrial operations will grow. Since environments that are removed from our biosphere answer both the need for continued industrial growth and for reducing the industrial burden on the biosphere, the opening of extraterrestrial environments will become increasingly attractive, commensurate with economic viability.

"Thus, one of the fundamental changes in the world development scenario will be the transition from the classic closed-world to an open-world development model. The open world adds open-space and lunar-type environments to the terrestrial environment. The nearest of the second group is our Moon." The conclusions of the 1972 paper remain unchanged. They were adopted after the "limits to growth" precepts gradually became recognized as falsehoods.

In the past, human growth could unfold in a monoglobal framework. In the future, human civilization needs to be polyglobal. The Moon is the first step. It is a seventh continent, almost as large as the Americas. It is large enough to support a civilization. It alone offers the opportunity to create a strong exo-industrial economy based on highly advanced nuclear, cybernetic, and material processing technologies, ultimately turning large parts of the once-barren lunar surface into a lush oasis of life, capable eventually of exporting even foodstuffs to orbiting installations, if not to Earth (Fig. 8).

In an open world, there are no limits to growth—only to mindless multiplication.

THREE-DIMENSIONAL CIVILIZATION

For Selenians, creation will be the major trade. The human being and its technosphere arrive first and then create a biosphere. This kind of creation will require an approach that is rational and effective without precedent. It will require an intensive learning process, particularly for a life form that also possesses extraordinarily destructive tendencies.

In their modularized, adaptive, growing bio-niches, Selenian societies can develop a plurality of life-styles and new social structures. Whatever the details of the human relationships may be, Selenians will be space "amphibians" traveling with ease between surface and orbital gravities and later, in the gravities of Mars, as well as on the moons of Jupiter or Saturn.

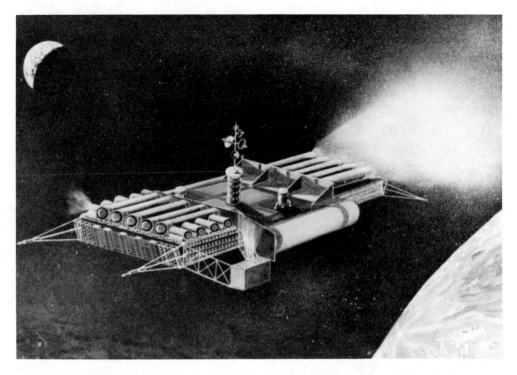

Figure 8. *A cislunar superfreighter, powered by lunar oxygen and aluminum powder, carries lunar bounty produced by Selenians for their terrestrial customers.*

For those coming generations who are born on the Moon and who are not transfixed by the beauties of Earth, and for those who remain on the Moon except for vacation or business trips to Earth, there will be no place in geolunar space more beautiful, vital, or abundant with the future than the Moon and the selenosphere created on it.

The Moon is a cause and a consequence. It offers even more than important contributions to overcoming our critical problems and achieving the essential technological advances. Selenopolis, symbol of a civilized Moon, is a new beginning of such magnitude that it can be compared only with man's emergence from the shady shelter of forests into the light of the open savannas.

The Moon is the touchstone of the human future. Instead of searching for and speculating about life elsewhere, we will put it there. Forthwith, civilization will be three-dimensional, and life will be polyglobal. Living at the ethereal shores of heliocentric space, the Selenians will be the Cosmopolynesians of the solar system, navigating between worlds. They will build the bridge between a dim past under terrestrial skies, where the great legends of human emergence tower, and a deathless civilization in a stellar future whose shadows beckon and long to be given substance.

REFERENCES

Ehricke K. A. (1957) The anthropology of astronautics. *Astronautics II*, Nov., 26–27. American Rocket Society.

Ehricke K. A. (1971) The extraterrestrial imperative. *Bull. Atomic Sci.*, Nov.,18–26.

Ehricke K. A. (1972) Lunar industries and their value for the human environment on Earth. *Acta Astronaut. 1*, 585.

Ehricke K. A. (1979) The extraterrestrial imperative, part 1: Evolutionary logic. *J. Br. Interplanet. Soc., 32*, 311–317.

Ehricke K. A. (1986) *The Seventh Continent: Industrialization and Settlement of the Moon.* The Krafft A. Ehricke Institute, La Jolla, CA. In press.

Epilogue:
Address Given at a Tricentennial Celebration, 4 July 2076, by Leonard Vincennes, Official Historian of Luna City

Ladies, Gentlemen, and Robots:

*I*t is a great honor to be asked to give Luna City's official Tricentennial Address, celebrating the three hundredth anniversary of the independence of the nation of my forefathers, the United States of America, Earth.

It is also a weighty responsibility, for this date marks not only the three hundredth year of American independence, but also the seventy-fifth year since the permanent settlement of our own great nation of Luna by pioneers from the United States.

I am extremely gratified, therefore, to see such a large and enthusiastic turnout for this anniversary dinner. Not only have we filled the Assembly Hall tonight, but this speech is being beamed to every nation on Earth. Believe me, the warmth of your applause and support is just about all that's holding me up right now. Even in one-sixth g, my knees are knocking here behind the podium.

As a historian, I have always been fascinated by the words of Astronaut Sheila Davidson on that historic occasion, seventy-five years ago, when she became the first American to set foot on the Moon in twenty-nine years: "We're back, and this time we're going to stay!" I hasten to assure you, I'm not going to stay up here at the podium for very long!

The reason for my fascination, as a historian, with Davidson's words is that they bring into clear focus the parallels between the first landings and eventual settlement on the Moon and the earlier European discovery and eventual settlement of the New World of the Americas.

*T*he Vikings, the Polynesians, perhaps even the Carthegenians, all "discovered" America centuries before Columbus. And, of course, every human being who had ever looked into the night sky had "discovered" the Moon. However, it was Columbus' landings in America that actually awakened the Europeans to the fact that a whole new world existed on the other side of the Atlantic Ocean. More importantly, Columbus' pioneering voyages proved that the Atlantic could be crossed, and that the ocean was not a barrier but a *highway—a highway that could lead to riches beyond Europe's wildest imagination. It took nearly a century, however, for the Europeans to realize this. At first they were disappointed that Columbus had not reached Asia. At one point, Columbus was returned forcibly to Spain, under arrest, in chains, humiliated and disgraced.*

Eventually, Europe realized that this New World had fantastic new wealth to offer. They plundered it for gold and silver at first, the kind of hard metallic wealth that they could easily understand. Later, when they had established permanent colonies in the Americas, new kinds of wealth crossed the Atlantic to enrich Europe: tobacco, maize (Indian corn), and the lowly potato made more fortunes for more people than the gold of the Incas or the silver of the Aztecs.

Later still, a treasure of infinitely greater worth was discovered in North America. We celebrate that treasure tonight: the development of large-scale democracy, the realization that government should be based not on the whims of kings, but on the will of the people. This great leap forward in individual human freedom culminated in the Declaration of Independence, which established the United States of America as a separate and free nation.

Note the parallels with the history of our own nation of Luna.

The first men to land on the Moon had no idea of what they had accomplished. They had come, basically, for the adventure of it, for the thrill of setting foot where no human being had stepped before, for the honor, in the words of a popular dramatic entertainment of that era, of "boldly going where no man had gone before."

Those brave astronauts had been sent to the Moon mainly as a result of the competition between the two great superpowers of the mid 20th Century: the United States of America and the Union of Soviet Socialist Republics. The two nations were engaged in a "space race" at the time. Once the United States reached the Moon, the race was declared finished, and no one set foot on the Moon again for nearly three decades.

Although the Moon receded from the attention of the politicians and the general public, that handful of Apollo astronauts had accomplished something far more important than most people realized. They had proved, much like Columbus nearly five centuries earlier, that space is not a barrier. It is a highway, a road to riches.

The three decades between Neil Armstrong and Sheila Davidson saw the development of a powerful space technology that began to transform the Earth. Not only did Earth-orbiting satellites take on global tasks such as communications, weather observation, and resource monitoring; but planetary probes began the investigation of the other worlds of our solar system; orbiting telescopes searched the distant stars; and the first tentative experiments in space manufacturing were begun. All the industrialized nations of the Earth began to work in space, and soon enough the less-developed nations began to see space technology as a means to enrich their own economies.

Then, seventy-five years ago, astronauts returned to the Moon. The permanent habitation of this world began.

Like the Conquistadores of old, the first wave of settlers came to the Moon in search of mineral wealth. Not gold or silver, of course. They sought aluminum and silicon, oxygen and titanium for the factories in orbit near the Earth, and water. In those pioneering days, water was far more precious than gold in space, and also more expensive.

Soon they found other forms of wealth, which today are the basis of Luna's economy. We export energy to Earth, and our mining expeditions to the asteroids have been so successful that last year they supplied more than fifty percent of the raw materials used by factories on Earth itself. The energy and natural resources we supply to Earth have raised the standard of living across the planet, and have helped in no small way to usher in the era of peace and international cooperation that has marked the twenty-first century and made it so different from the twentieth.

Most important, however, have been the social, moral, scientific, and spiritual riches that we have gained since permanently settling the Moon.

As I look out on this distinguished audience, I see men and women from every part of Earth. I see people who were born here, whose children have migrated even farther from their ancestral home and now live in the great habitation complexes that ply the trade routes out to the asteroid belt. I see robots, whose intelligence is different from that of human but no less valued and no less revered. Without your untiring contributions, the human habitation of the Moon could never have been accomplished.

We, here on the Moon, have achieved a new level of society, integrating men and women of all races, all religions, and all political backgrounds into a harmonious, productive, prosperous community. We serve as an example to the nations and peoples of Earth, many of whom are still striving to reach the equality and freedom that we take for granted as citizens of the nation of Luna.

And, as you all know, it was our own Farside Observatory that first picked up the faint microwave signals from the region of the Veil Nebula, in the constellation Cygnus. While as yet the astronomers have not been able to determine exactly which star these signals originate from, they have conclusively demonstrated that the signals are purposeful, that intelligent life does exist elsewhere in our galaxy.

The parallels with the American experience are manifest, and just as America reached out beyond the limits of planet Earth to establish this human settlement here and help it grow into an independent community, we of Luna have already started to send our children deeper into the wilderness of space. Someday we may send them beyond the reaches of our solar system altogether.

Where will those children of ours be a quarter century from now, on the first centennial of Luna's original settlement? They will go as far as our own vision and faith *allow them to go. And they will send back to us new wealths of knowledge, new vistas of hope, new visions of the universe.*

Thank you.

This speech was "ghost written," somewhat in advance of the date on which it is to be delivered, by Ben Bova, president of the National Space Institute and author.

INDEX

Page citations refer to the first page of each article that deals with the indexed term.

ISBN 0-942862-02-3